A handbook of the Yemen Flora

For Paul and Claire

A handbook of the Yemen Flora

J R I Wood

With colour illustrations
by Hugo Haig-Thomas

Royal Botanic Gardens, Kew

First published 1997

General Editor: J. M. Lock

Cover illustration: *Atractylis kentrophylloides* by Hugo Haig-Thomas

Typeset by The Comp-Room, Aylesbury, UK;
Cover design and page make-up by Media Resources,
Information Services Department, Royal Botanic Gardens, Kew

ISBN 1 900347 31 8

Printed in The European Union
by
Whitstable Litho Printers Ltd, UK

CONTENTS

The author and Hugo Haig-Thomas are very grateful to the following three artists who prepared the line drawings:

Ela Bullon

Jane Tippett

Sidney Wood

They would also like to express their sincere gratitude to the following donors who have helped with the costs of the production of the colour illustrations for this book:

Miss Masako Fujii

Mr Yasuro Ishikawa

Miss Naoko Kitazawa

Miss Mariko Miura

Mr Masahito Nara

Mr Masatsugu Ohtake

Mr Joji Sugimoto

Mr Y Terada

Mr N Tomita

Mr Hifumi Uematsu

Mr Masato T Yoshida

and, especially, Mr Kei Kawaguchi

PREFACE

This is the first full account of the plants occurring in North Yemen, the former Yemen Arab Republic before its union with the south, the country with the highest rainfall and most varied scenery in the Arabian peninsula. It aims to state what is known about the vegetation of this fascinating country and covers all 1750 species of fern and flowering plant known to occur naturally, including trees, shrubs and grasses. Additionally, descriptions of all commonly cultivated species are provided and particular attention is given to species of economic importance. Every effort has been made to keep this book up-to-date since it was completed in 1982 despite unforeseen delays in publication.

It is hoped that the book will be of interest not only to botanists but to all with an interest in the plants of Yemen whether it is only because they want to know what grows around them or because their work as foresters, range ecologists or agriculturalists requires knowledge of Yemen plants. Since the book is aimed both at the professional and at the amateur the use of scientific language has been kept to a minimum.

The publication of a flora such as this often acts as a brake on botanical research rather than a spur to further discoveries. It is sincerely hoped that this book will encourage people to find out more about the plants of Yemen. Many gaps remain to be filled and huge areas of Yemen are poorly known botanically, particularly in the southeast close to the former border with South Yemen. Many groups of plants are incompletely known and further collecting of these plants is much needed. There are many interesting taxonomic and ecological research projects which an enterprising amateur or, hopefully, a botany student at Sana'a University, might undertake. Many areas requiring further research will be apparent in the text.

It must be obvious that a book like this cannot be written without a great deal of help from others. So many people have helped in various ways and at various times that it is inevitable that some will be omitted.

Undoubtedly my greatest debt is to my wife. Not only has she typed a substantial part of the manuscript and prepared a large proportion of the line drawings but she has always given me support, accompanying me on a number of memorable journeys, tolerating my absence from home on numerous occasions and giving me the time both to do the field work and to write the book. Quite simply this book would never have been produced without her.

The colour pictures which add so much to the attractiveness of the book are the work of Hugo Haig-Thomas. All are original watercolours painted with a single exception from plants growing in Yemen. Most of them have never been illustrated before and the majority are unknown outside southwest Arabia. They required an enormous amount of time and much dedicated work, often at very inconvenient times. I am also indebted to Hugo Haig-Thomas for drawing my attention to Wadi Sukhnah which has proved one of the most interesting plant localities in Yemen.

The remaining line drawings are the work of Ela Bullon and Jane Tippett. I owe a particular debt to Ela Bullon who has devoted a great deal of time and care to her drawings despite having no connections with Yemen.

I have had a lot of help from botanical institutions and individual botanists. Ever since I first became interested in Yemen plants in 1972, Miss D. Hillcoat of the British Museum (Natural History) has generously shared her unique knowledge of Arabian plants and has helped with the identification of literally thousands of specimens. All of us who have worked on Arabian plants over the past forty years owe her a considerable debt. I am also grateful to the keeper of Botany and the staff of the Natural History Museum who have freely allowed me to use their facilities and particularly to A. O. Chater, Dr N. Robson and Dr D. Sutton who have all given valuable help and advice.

The staff of the Royal Botanic Gardens at Kew have always been generous with their time and their knowledge. I am particularly grateful to P. S. Green, the Keeper of the Herbarium, who has not only allowed me to use the facilities at Kew but has also organised loans of specimens from botanical institutions in Copenhagen, Florence, Geneva and Hamburg and arranged for me to visit the Natural History Museum in Paris and the Botanical Institute in Florence. It seems invidious to name individuals at Kew but all the following have read and criticized parts of the text: D. Bridson (*Rubiaceae*), Dr R. K. Brummit (*Acanthaceae*), S. Carter (*Euphorbia*), Dr T. Cope (*Gramineae*), Dr P. Cribb (*Orchidaceae*), J. Dransfield (*Palmae*), D. Field (*Asclepiadaceae*), R. Harley (*Labiatae*), F. N. Hepper (*Lemnaceae*, *Scrophulariaceae*), S. S. Hooper (*Cyperaceae*), C. Jeffrey (*Compositae*, *Cucurbitaceae*), G. Lewis (*Leguminosae*), B. Matthew (*Iridaceae*), R. D. Meikle (*Nyctaginaceae*), R. M. Polhill (*Loranthaceae*), A. Radcliffe-Smith (*Euphorbiaceae*), C. C. Townsend (*Amaranthaceae*, *Umbelliferae*), Dr B. Verdcourt (*Convolvulaceae*) and G. E. Wickens (*Crassulaceae*). Obviously the final result is my responsibility but their co-operation has led to a great many improvements. I am also grateful to Dr W. D. Clayton and Dr F. Davies who allowed me to make use of unpublished manuscripts which have enabled me to update certain sections of this book.

I have also had help from the Royal Botanic Garden, Edinburgh, in particular from I. Hedge and A. Miller. The former not only arranged for me to visit Edinburgh and organised the loan of specimens but has given much useful advice particularly on Labiatae. Tony Miller has helped in numerous ways on a great variety of matters, not least in letting me use the results of his own expeditions to Yemen and in passing on the records of other collectors which have come his way.

The late Dr Steinberg generously allowed me access to the results of his explorations in the Ṣa'dah region and I am also grateful to Professor Moggi and Dr P. Cuccuini of the Instituto Botanico in Florence who helped me to go through Dr Steinberg's specimens after his death. Others who have helped include Professor L. Boulos, J. B. Gillett (Commiphora), P. Jaeger (*Solanum*), J. J. Lavranos (particularly with succulent *Asclepiadaceae*) and Dr Mats Thulin (*Rhynchosia*, *Acacia*). I owe a particular debt to Dr R. B. Faden of the Smithsonian Institute who has been a model correspondent over many years and has generously shared his wide knowledge of *Pteridophyta* and *Commelinaceae*.

Others have helped in indirect but nonetheless important ways. Hugh Leach first aroused my interest in Yemen with his accounts of the excitement of travelling there while John Baldry arranged my first trip into the Yemen countryside. I owe a considerable debt to my employers, the British Council and the Ministry of Education in Sana'a (1974–80) as well as to German Technical Assistance and the Ministry of Agriculture (1981), all of whom tolerated and in some cases actively supported my botanical activities; particularly was this the case with the British Council representatives Clive Smith and Bill Charlton and I am grateful to the former for drawing my attention to Wadi Aḥjar and to the latter for being such a phlegmatic travelling companion on some of my longer and more exciting journeys. I am also very grateful to Ekkehard Nolte for arranging so interesting a programme when I was working for his project in 1981. He is perhaps the first to have taken seriously the need for nature conservation in Yemen. Others who have helped are Jim Clissold who has searched after a number of problematic plants in the southeast border region and Professor R. B. Sergeant who has provided several interesting sources for Arabic names. Finally, like so many others, I have benefited from Etienne Renaud's unrivalled knowledge of all aspects of Yemen and particularly from his insistence on the potential botanical interest of the Markah region of J. Milḥān.

One glaring omission from this list of acknowledgements must be obvious. Over the ten years or so I have travelled and worked in Yemen, I have received assistance from hundreds and probably thousands of Yemenis. They have provided all sorts of information, have acted as guides, provided transport and given hospitality with the generosity for which the Arabs are famous. Wherever I have travelled I have been welcomed into all types of home even though a stranger with an abstruse and perhaps sometimes scarcely comprehensible purpose in my travels. Yemen may not be the most comfortable country in the world but I have travelled alone with a security which can be found in few other places. It was both a fascinating and often an exhilarating experience. Yemen is changing fast. My early travels were made almost entirely on foot or by donkey or camel but in later times mostly by car. Ideas are changing too and it is my sincere hope that the people of Yemen will come to enjoy the plants of their country just as I have done.

THE DISCOVERY OF THE YEMEN FLORA

Interest in the plants of Yemen goes back thousands of years. To the ancient world Yemen was the land of myrrh and incense with a reputation for a much richer vegetation than it really had. With Arab writers the chief interest lay in the value its plants had for agriculture and medicine. Catalogues containing plants found in Yemen were drawn up by Dinawari in the Kitab al Nabat (*c* 895 A.D.) and by Zabidi in the Taj al 'Arus (*c* 1780 A.D.), both of which give the uses to which the plants could be put. Neither of these works, however, were specifically about Yemen but the most interesting pre-Linnean work, the Bughyat al Fallahin, was. This was composed in about A.D. 1370 A.D. by the Rasulid Sultan al-Malik al Afdal al 'Abbas ibn 'Ali and gives a fuller account of agricultural practice than anything published before or since; in particular there is extensive information on a great variety of cereals, pulses, fruit trees, vegetables and miscellaneous plants of economic importance. Another Rasulid Sultan al-Malik al Muzaffar Yusuf ibn 'Umar also wrote a work of some interest, a treatise on medicaments entitled 'Al-Mut'amad'.

Unlike Arab writers, Europeans have only recently become interested in the economic aspects of Yemen plants as a result of their involvement in the development projects of the 1970s and '80s. Hitherto their interest has largely been in the botanical exploration of Yemen and the classification of its plants, a task which is now nearing completion. Although there were precursors, it is reasonably accurate to state that modern botany began with Linnaeus whose work gave an enormous boost to the study of plants worldwide. It is appropriate therefore that the first European botanist to visit Yemen was the Swede, Pehr Forsskal, who was one of Linnaeus' own pupils.

Forsskal came to Yemen as one of a six-man expedition sent by the Danish government. The expedition was fraught with troubles from the beginning but it eventually reached Luḥayyah on the Yemen coast at the end of 1762 after spending a period of about a year in Egypt. The happiest and most successful part of the expedition now began. In January 1763 Forsskal explored the Wadi Mawr from Luḥayyah making extensive collections, particularly around Kudmīyah (Kudmie). In February the expedition moved south through the Tihama to the village of Mighlāf near the Wadi Surdūd. From here Forsskal visited the foothills of J. Milḥān after which he named the genus *Melhania*. The expedition then moved on through Ḍaḥī to Bayt al Faqīh.

Bayt al Faqīh was to be the expedition's base for almost the whole of March and April but Forsskal was rarely there. Most of March he spent in the southern part of J. Raymah around the village of Ḥadīyah, possibly the richest part of Arabia floristically. Ḥadīyah was then much better known than it is now since it was frequently visited by coffee merchants but it is still an excellent base for botanical exploration and is spectacularly situated with a magnificent backdrop of mountains. It is not possible to identify all the villages Forsskal visited but it is clear that he must have climbed up the main valley through the pass at Jirjir and thence via Bughah (Boka) to Kusmah (Kurmah) and the adjacent peak, J. Bārad (Barah). He descended to 'Alūjah (Aludje) and thence returned to Ḥadīyah by way of J. Aswad. In late March he went with Carsten Niebuhr on a long donkey journey through Waṣāb As Sāfal (Uafad) via Suq Ar Rubū' (Roboa) to 'Udayn whose botanical riches Forsskal does not seem to have appreciated. They then climbed to Jiblah and went on to Taiz through the pass now known as Najd al Aḥmar which he referred to as Maḥarras. They returned to Bayt al Faqīh through Ḥays and the Wadi Zabīd.

The expedition's troubles now began in earnest. Several members were suffering from debilitating fevers (probably malaria) but they made their way slowly to Mukhā by way of Zabīd and Mawshij. They were delayed by the governor of Mukhā for about six weeks and here one of their number died. In June they were at last allowed to begin their journey to Sana'a and travelled through Mawza' and Dorebat to Taiz. There were further delays here and when they were at last allowed to leave, Forsskal himself had caught the fever. His interest in plants remained despite his deteriorating health and some plants, notably *Plectranthus ovatus*, must have been collected on this last journey. By the Sumārah pass, however, his condition was so bad that he had to be tied to his camel and early in July he died at Yarīm.

The remaining four members of the expedition continued to Sana'a and then returned to Mukhā in time to catch a ship to India. By the time they arrived in Bombay, Niebuhr was the only survivor but he eventually succeeded in returning safely to Europe through Iran.

Botanical science owes him a considerable debt. Not only did he ensure that the majority of Forsskal's specimens returned to Europe where they are still housed at Copenhagen, but he also published Forsskal's notes in the *Flora Aegyptiaco-Arabica* (1775). He did this at his own expense as the Danish government had lost all interest in the expedition and its results.

Forsskal's work is the main source book for all studies on the Yemen flora and it contains a wealth of information. It is remarkable how much Forsskal was able to see and record over a short time amidst considerable difficulties. It is a fitting tribute that many plants occurring in the main text of this book are named after him.

No further botanical exploration took place until the 1820s and 1830s when a number of botanists and travellers made collections. These include Ehrenberg, Bové and Botta but not the famous Ethiopian collector, W. Schimper who never in fact visited Yemen. Ehrenberg was the first and came in the company of Hemprich in 1825. He was mostly interested in marine plants and visited many of the islands in the Red Sea. The only places in Yemen that he went to were Kamaran Island and the Wadi Mawr which he explored from Luḥayyah as Forsskal had done sixty years earlier. His collections were sent to Berlin but were never fully written up. As a result his contribution to our knowledge of the Yemen flora is less than it might have been though he found a few plants which had eluded Forsskal.

N. Bové came in the winter of 1830–31. He worked as

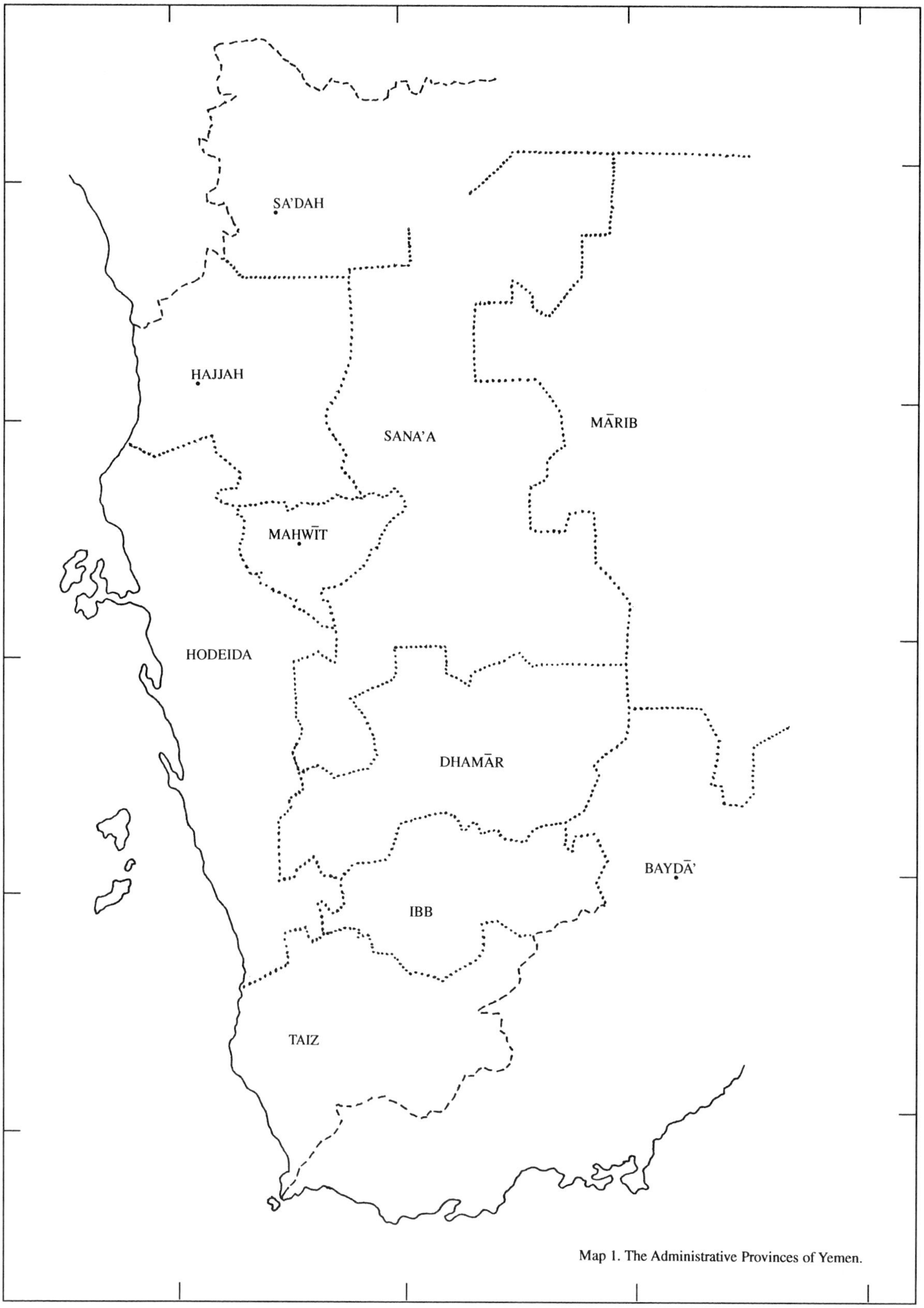

Map 1. The Administrative Provinces of Yemen.

director of gardens for Ibrahim Pasha in Cairo and, like Ehrenberg, collected only on the coast. He went to Kamaran Island and Hodeida where he found *Rhizophora mucronata*. It is interesting that many casual collections of plants were made on the Red Sea islands and along the coast in the days of sea travel in the nineteenth and early twentieth century while today these islands are the least accessible parts of Yemen. Besides Ehrenberg and Bové, at least two other collectors, Faurot and Mordtmann are known to have visited Kamaran Island before the first World War and similarly our knowledge of Ḥanish Island is based solely on the collections by a British naval officer, Lieutenant E. Slade, around 1880. No botanist has visited any of these islands for at least fifty years.

Much more important than any of these other early 19th-century collectors was P. E. Botta who visited Yemen on behalf of the Paris Natural History Museum in 1836. Botta made valuable collections but he was far from methodical and most of his specimens are unlocalised and those localities that are given are often hard to trace. He arrived at Hodeida and travelled from there to Ḥays. He explored J. Ra's and then made his way through Shar'ab and Mikhlāf to Taiz. Apart from Kāhim which is in Shar'ab I have been unable to trace any of his localities in this area including J. Māmarah (Maamara) where he found many plants. From Taiz he explored Janad and climbed to the summit of J. Ṣabir, visiting the villages of Hadnān and Hajaf (Haguef). He appears to have returned to Ḥays by the same way he had come and then went through Mawshij to Mukhā where he became seriously ill.

Botta was not a botanist and did not write up his own collections. Consequently, many of his specimens have languished at Paris ever since though others have been described as new by Decaisne, Deflers, Jaubert and Spach and other botanists. Many plant names commemorate Botta's name including *Arisaema bottae* and *Vernonia bottae* but it is not for his botanical collections in Yemen that Botta is well-known. Some years later he was appointed French consul in Mesopotamia and was the first man to appreciate the significance of certain mounds at Nineveh. It is for his contribution to archaeology rather than botany that Botta is famous.

No further botanical work was done till Deflers' visit in 1887. Alexandre Deflers lived in Egypt for many years and made it his task to explore the plants of southwest Arabia. His itinerary is described in detail in his book *Voyage au Yemen* and only the main details are given here.

He arrived in Hodeida at the end of March and after a short excursion to Manẓar, set off for the interior. He followed the line of the present road as far as Bājil and then the old Turkish road to Sana'a. This goes by way of 'Ubal, Ḥajjaylah, 'Usil, Attārah and Ḥajarah to Manākhah before descending the east side of the Ḥarāz and again following the line of the present road. He stayed in Sana'a for about a month and was able to visit Shibām, Kawkabān, Thilā and 'Amrān as well as most of the mountains in the vicinity of Sana'a. He left Sana'a on July 29th and reached Hodeida in less than a month, travelling by way of Taiz. He mostly followed the lines of the present main roads except that he went from Taiz directly to Ḥays by way of Ḥajdah and Madrūbah.

Deflers' book was published two years later and contains a complete list of the plants he found together with descriptions of new species. Although the book gives the first description of Yemen's vegetation it is rather disappointing. Deflers is remarkable not so much for what he saw but for what he did not see. He was the first botanist to visit the Manākhah area, the interior of Yemen and the Sumārah-Ibb region if we exclude Forsskal's last journey. Although the last part of his journey was done in haste much of his journey was quite leisurely. He spent a week at Ḥajjaylah and ten days at Manākhah but failed to find many of the plants which are frequent in these regions. His book also contains many errors of identification.

In the 1890s Deflers made a series of expeditions into the hinterland of Aden. These are of much greater value and it seems that he developed a knowledge of the flora of what was the People's Democratic Republic of Yemen which has not been equalled before or since. He also developed an interest in succulent *Asclepiadaceae* in which he was something of an expert. During the course of these expeditions he entered into the foothills of the Ḥujarīyah and was the first person to appreciate the interest of this region. Amongst the species he described from here were *Blepharispermum yemense* and *Wendlandia arabica*.

Deflers was followed in 1889 by G. A. Schweinfurth who is Forsskal's only equal among botanists to visit Yemen. Schweinfurth deserves to be more widely known than he appears to be since he was one of the foremost explorers of central Africa and had a generosity of character unusual among that group of men. He had already travelled extensively through what is now the Sudan, Uganda, Zaire and Ethiopia and had been one of the few people to visit the island of Socotra. He was a careful and systematic botanist, almost unique amongst his contemporaries in his knowledge of plants both in the field and the herbarium.

Schweinfurth's purpose in coming to Yemen was to refind as many of the plants discovered by Forsskal as possible. He was in Yemen for less than three months and so thorough was he that in retracing his footsteps I have rarely been able to find anything that he missed.

He travelled from Hodeida to Khalīfah and thence to Ḥillah on J. Bura'. Here he only climbed to the coffee country at about 1000m before returning and making his way to J. Milḥān. On J. Milḥān he collected widely between Walajah and 'Ukaybir and then went across the Tihama to Ḥajjaylah. Next he climbed the Ḥarāz mountains following the route used by Deflers. In the Ḥarāz he explored the villages and side valleys along the route extremely thoroughly and found many plants that Deflers had missed. Finally he returned by way of Bājil to Hodeida.

Schweinfurth published many of his results in a series of articles printed as a supplement to the Bulletin of Boissier's Herbarium. These were unfortunately never completed but those published contain a wealth of information which is invaluable to anyone who seriously wishes to study the Yemen flora.

Apart from minor collections by Botez (1911) and de Benedictis (1926) both of whom penetrated to the interior of Yemen, no further advance in our knowledge of the Yemen flora took place till the journeys of C. Rathjens and H. von Wissmann between 1927 and 1938. Rathjens and von Wissmann were basically geographers but interested themselves in a great variety of subjects, one of which was botany. Both collected at Sana'a and along the old motor road which ascended to Sana'a through Madīnat al 'Abīd and Ḥammām 'Alī. Rathjens also collected around Ṭawīlah, Ḥajjah and Suq al Amān. Wissmann travelled along the Turkish road to Sana'a which Deflers had used and also made an interesting journey through Sukhnah, Bilad Aṭ Ṭa'am and Salafīyah to Madīnat al 'Abīd before going to Sana'a. Their plant records were written up by O. Schwartz in the *Flora von tropischen Arabien* (1939) and Wissmann has also published a number of articles of a phytogeographical and ecological nature based on his work in Yemen.

The results of one important expedition in the 1930s were not included in Schwartz's *Flora von Tropischen Arabien*. This was the journey in 1937–8 by Hugh Scott and E. Britton who travelled from Aden by way of Taiz and Ibb to Sana'a. Hugh Scott subsequently wrote what is perhaps the best travel book on Yemen, *In the High Yemen*, but the scientific results were never published even though some of the interesting plants of the Ibb region were found for the first time by this expedition.

Few other botanists came to Yemen until the 1970s but three need to be mentioned. In 1945 Ahmed Khattab made an extensive journey through the Tihama to Taiz, and Sana'a returning by the Madīnat al 'Abīd road. A checklist was published much later (1971) and his collections are now housed at Cairo. Particularly interesting were the journeys of G. Popov and W. Zeller in 1962 since they were the first collectors to penetrate the eastern regions of Yemen, reaching the Jawf, Mārib and Bayḍā. Their collections are now at the Natural History Museum in London but their results have never been published.

In recent years a great deal of botanical research has taken place in Yemen. Particularly important have been the extensive collections made on behalf of botanical institutes. F. N. Hepper (1975), A. Radcliffe-Smith and S. Henchie (1977) have collected for Kew Gardens. A. J. Miller alone in 1978 and with D. Long (1981) collected for the Edinburgh Botanic Garden. C. Steinberg (1979, 1980) and P. Cuccuini (1981) have added considerably to the Florence herbarium by their collections in the Majz area which have been assisted by B. Lanza and N. Steinberg. D. Podlech has collected for Munich University (1981). K. Müller-Hohenstein and U. Diel (1980–82) have travelled widely in Yemen on behalf of Bayreuth University making plant collections from different areas.

Amongst private individuals, J. Lavranos who came in search of succulents sometimes alone and sometimes with L. E. Newton (1974, 1976, 1977) has made many valuable records and described a number of new species from Yemen. Other collectors include: S. Chaudhary (1976–78) who worked as a lecturer in botany in Sana'a University; B. Acres, M. Brunt, P. Henry and D. Pratt all of whom worked with the Montane Plains and Wadi Rima' project between 1974 and 1977; D. Wood (1978–79) who collected seeds for German Technical Assistance; E. Westinga, a range ecologist with the Rada' project; L. Gillespie (1979) a traveller and botany student. Smaller but useful contributions have been made by D. Carden (1974–76), B. Clissold (1982), P. Christmann (1982), H. Ellenberg (1981), J. Firebrace (1980), J. Fleurentin (1976–81), B. Goodman (1980), H. Haig-Thomas (1977–80), P. Heath (1977), Kamal Ibrahim (1964), L. Kassasian (1982), J. King (1982), T. Larsen (1980–81), H. Leach (1971), P. S. Madge (1980), J. M. Ritchie (1974–76), F. Weber (1980).

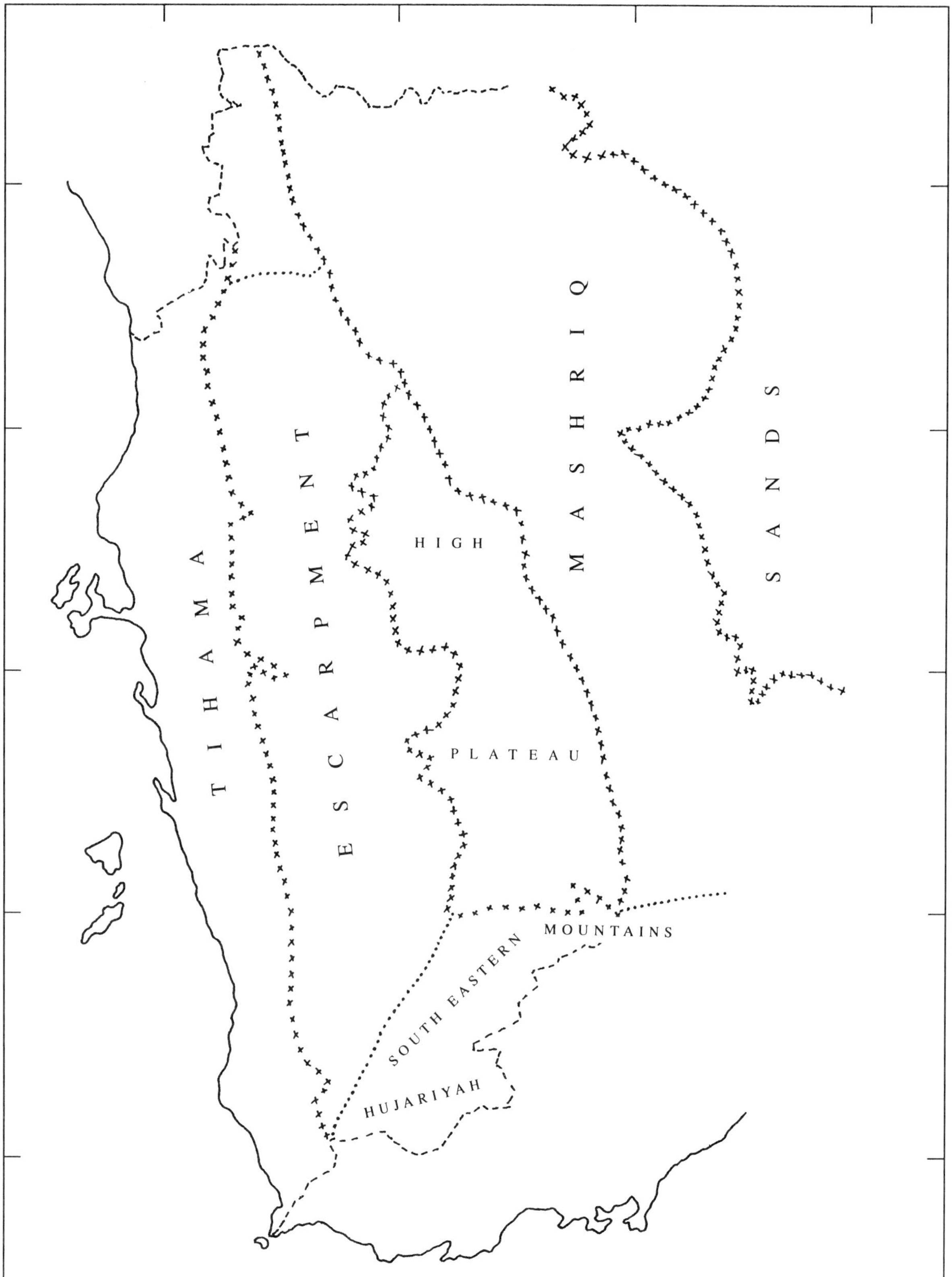

Map 2. The Topographical Regions of Yemen.

TOPOGRAPHICAL REGIONS

Five basic regions are recognised in this book: the Tihama, the Escarpment, the High Plateau, the Mashriq and the Sands. They differ from each other in topography, geology, climate and the pattern of human settlement. Each of the five regions has its own distinctive vegetation types and many plants are restricted to one region only and may be abundant in that region though absent from all others.

THE TIHAMA

This is the traditional name for the coastal plain of Yemen. It extends from north to south along the whole length of the country and from the Red Sea coast inland for about fifty kilometres though it is slightly narrower in the south. It is almost entirely flat rising imperceptibly from sea level to 400–500m at the foot of the escarpment and is broken only by the occasional small volcanic or limestone outcrop. With the exception of these outcrops the soils are alluvial. Most of the Tihama consists of sands of recent origin but nearer the escarpment there are extensive gravel and silt deposits washed down from the mountains.

The weather is uniformly hot and humid, temperatures rarely falling below 15°C at night and rising to about 35°C during the day and with little variation between summer and winter. Rainfall is low but increases steadily towards the escarpment. Agriculture and hence human settlement is mostly on the rich silt plains in the east where there is sufficient rainfall or along the various wadis that cut through the Tihama. Underground water is usually present along the wadi lines and the periodic floods which come down from the mountains can be diverted for irrigation. Consequently standing crops can be found at all seasons in the Tihama and the country has a characteristic pattern of dense settlement with extensive agriculture contrasting with great tracks of barren, unexploited land.

THE ESCARPMENT

The escarpment region consists of all the western mountains of Yemen together with the valleys and occasional plains enclosed within them. The western limits of the escarpment can be clearly seen on any physical map of Yemen and are equally readily appreciated on the ground. In some cases such as J. Bura' the mountains rise sheer out

Map 3. Quaternary alluvial deposits including silts, loess, sands and gravels.

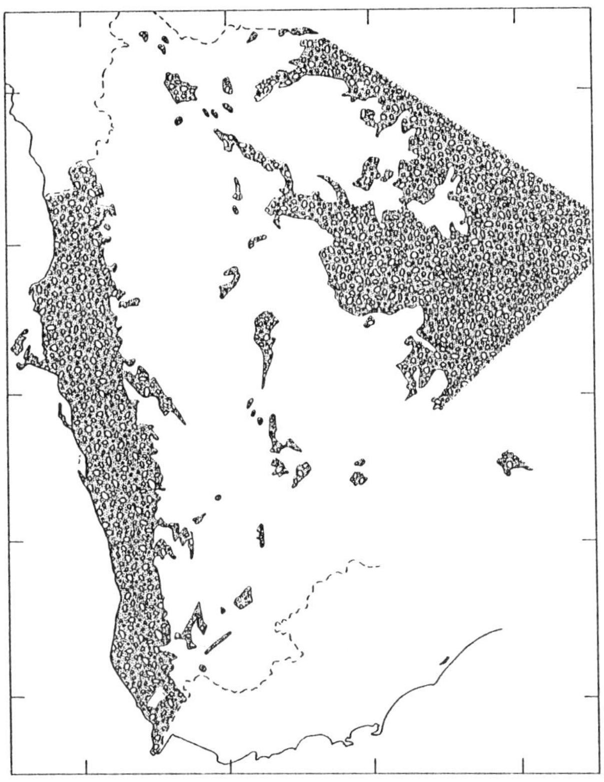

Map 4. Distribution of *Lasiurus scindicus* (●), a species entirely restricted to the Tihama, and *Calligonum crinitum* (▼), a species entirely restricted to the Sands.

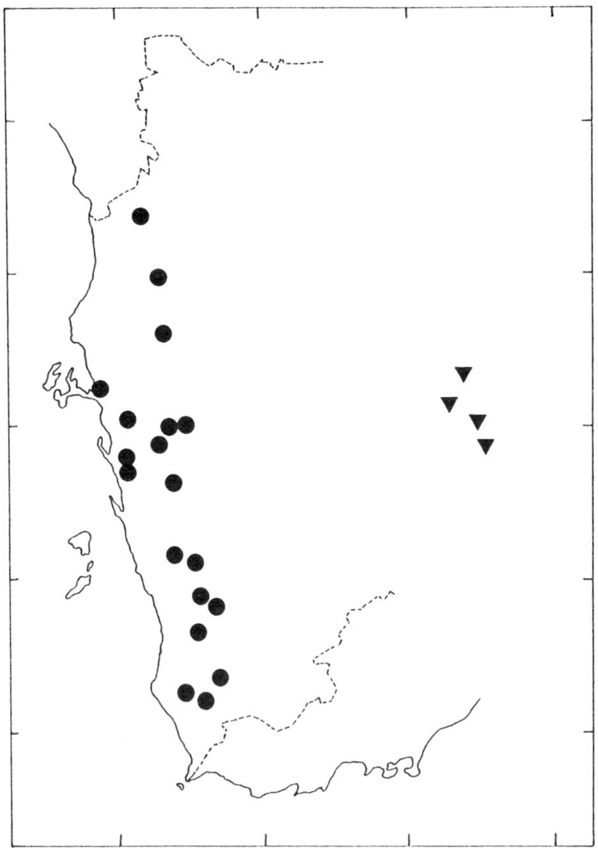

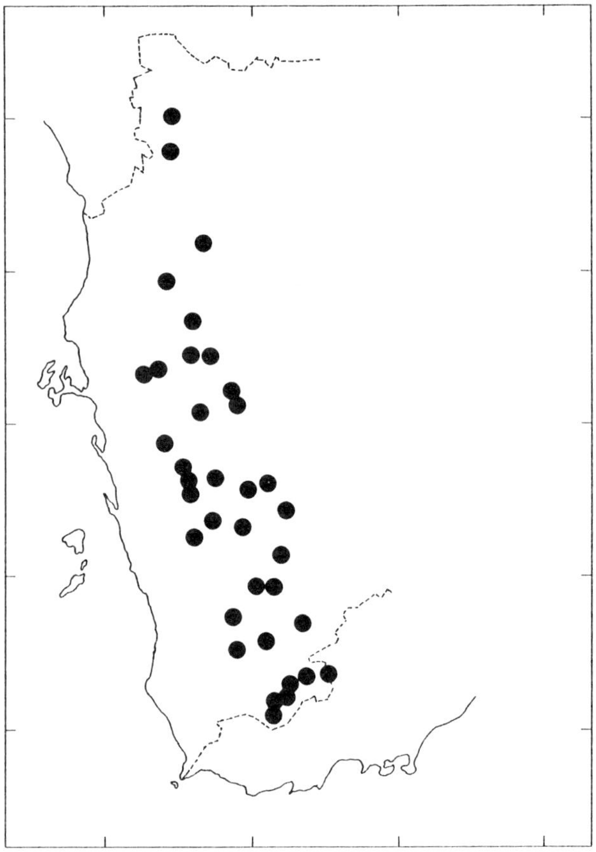

Map 5. Distribution of *Barleria bispinosa*, a species abundant on the escarpment but unknown elsewhere.

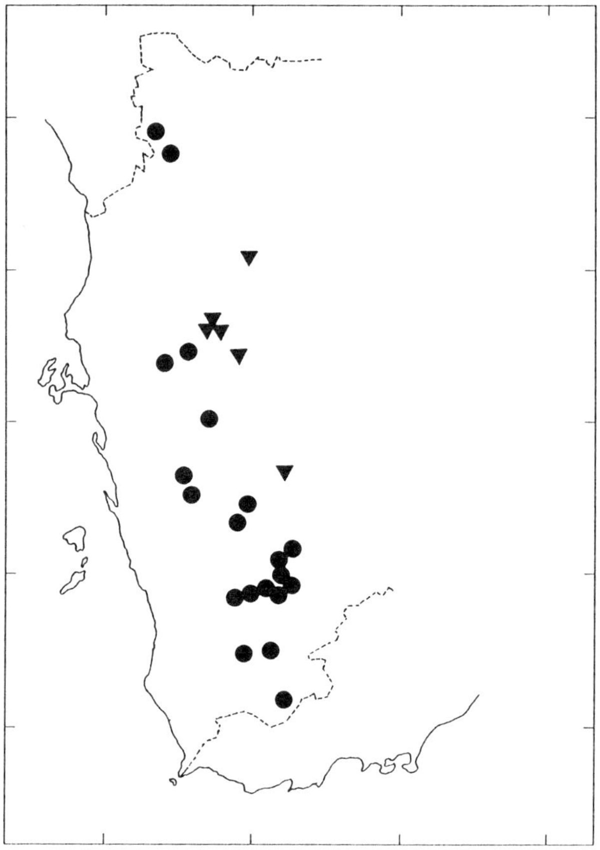

Map 6. Distributions of *Argyrolobium confertum* (▼), a species restricted to the inner escarpment and *Argyrolobium rupestre* (●), a species restricted to the outer escarpment.

of the Tihama plain for 2000m but even in the gaps which the main roads follow the boundary is quite clear. On the main road to Sana'a, for example, the escarpment region is entered through a small defile known as Bāb Al Ibl. West of this defile the Tihama stretches out, east is the escarpment.

The escarpment region is not a single steep incline but a series of mountain ranges cut by deep valleys. In very general terms there are a number of distinct mountain ranges in the west (the outer escarpment) and an almost continuous line of mountains lying to the east of this (the inner escarpment), whose watershed marks the eastern limits of the escarpment region. This picture though is greatly simplified. In some places there is no outer escarpment and south of Jabal Manār there is no inner escarpment at all. Thus in some respects this region is a disordered series of mountains whose average height at its western limits is around 500m and in the east around 2400m though many of the peaks are well over 3000m high and J. Nabī Shu'ayb the highest point in Arabia, is 3666m. The escarpment region is about one hundred kilometres wide and stretches the whole length of Yemen from north to south.

Geologically the escarpment is almost entirely formed of the Yemen volcanic series, a complex of basalts and metamorphosed rocks which give rise to very fertile soils. Some of the most spectacular mountains though are at least partially formed of granite and huge granite blocks are a conspicuous feature of the Ḥarāz and J. Bura'. There are extensive areas of limestone in the north particularly around Maḥwīt and J. Sharafayn. Sandstone is much less common but there are large deposits around Ṭawīlah and in the Ḥujarīyah. The Ḥujarīyah, though, like its northern counterpart, Khawlan Ash Sham, is not really typical of the escarpment and is dealt with separately below.

Climatically the escarpment is very varied. At low altitudes temperatures are similar to those of the Tihama; the higher altitudes are cooler but temperatures rarely, if ever, fall below freezing. Humidity is very variable but generally high. Rainfall is closely related to topography. Steep, high, west-facing slopes have the highest rainfall and in some areas notably around Ibb, the annual rainfall averages more than 1000mm and this is the highest in Arabia. East-facing slopes are often in rain shadow and great variations can be observed over quite short distances.

Below about 1800m agriculture is mostly limited to the wadis where flood waters can be trapped and used to irrigate the richer alluvial plains along which the wadis flow. The mountain sides are steep, rocky and covered with a thin *Acacia-Commiphora* bushland. The main exception to this picture is the coffee country of the central escarpment (J. Raymah, Bura', Milḥān, J. Sāfān in the Ḥarāz, Maḥwīt)

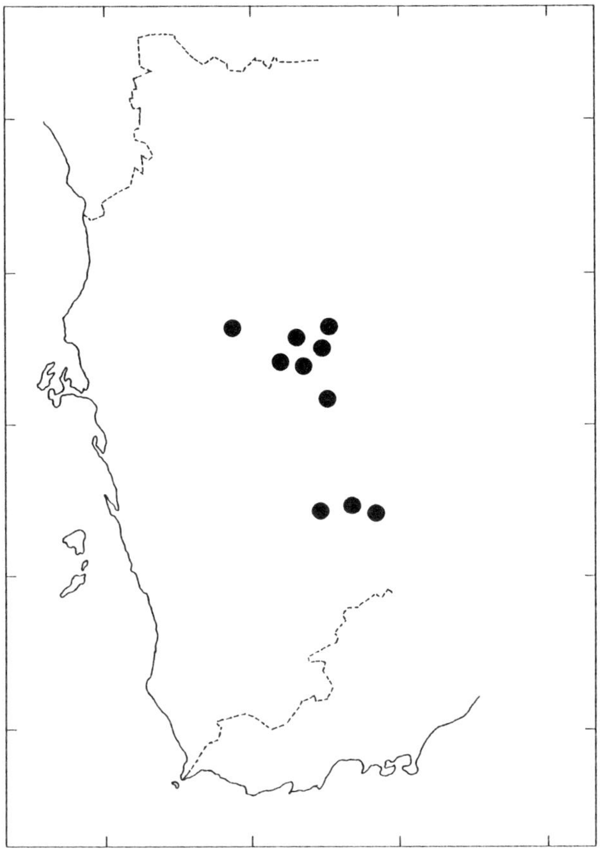

Map 7. Distribution of *Cleome amblyocarpa*, a species entirely restricted to the high plateau.

where terrace cultivation is found at low altitudes. Above 1800m rainfall is generally sufficient to support agriculture particularly the cultivation of sorghum and qat. The hills are covered in a network of terraces which conserve the rich volcanic soil and every conceivable corner of land is cultivated. Population density is high and natural vegetation limited to cliffs and other inaccessible places.

THE HIGH PLATEAU

The high plateau or high Yemen as it is sometimes called, consists of a series of plains at an elevation of between 2000 and 3000m separated from each other by mountain ranges many of which rise to over 3000m. The most southerly of the intermontane plains is the Qa' al Ḥaql just north of the Sumārah pass and the most northerly the Qa' Ash Shams south of the great limestone massif on which Khamir is situated. The high plateau is about 50 kilometres wide and in general the plains on the west are higher than those in the east, the most westerly, the Qa' Saḥmān in Bani Maṭar, lies at about 3000m whereas the most easterly, the Qa' Rada', is at about 2000m.

The high plateau region is not so apparent on a map as it is on the ground. As has been noted the watershed on the inner escarpment marks the western boundary. The traveller coming from the west can immediately appreciate the change of country whether he comes over the watershed above Kuḥlān or Suq al Aman or across the Sumārah pass. The latter is the most dramatic. Standing at the top of the pass the traveller can look west, and if the cloud permits, see a vast panorama of valleys and mountain ranges but east after a short descent is laid out the patchwork of the densely cultivated Qa' al Ḥaql. The eastern boundary of the high plateau is marked by small passes such as the Naqil Ghaylān or the 'Azum pass east of Jiḥānah. The boundary often coincides with a geological change. The town of Rada' lies exactly on the boundary of the two regions. West of Rada' the cultivated plains stretch out whereas east of the town there is a jumble of sandstone hills, almost entirely uncultivated and of quite different appearance. Unlike the other regions of Yemen the high plateau does not extend the whole length of the country. The Sumārah-Manār range marks its southern limit and the limestone ridge around Khamir the northern limit, the distance between the two being about two hundred and fifty kilometres.

With the exception of this northern limestone massif and a few sandstone outcrops in the Bani Ḥushaysh and around the Wadi Ḍahr, the mountains surrounding the high plateau are almost entirely volcanic. Indeed a feature of the high plateau is the black lava plain of Ḥamdān-Arḥab and the area around J. Isbil. This is the result of quite recent volcanic activity which has changed a formerly fertile plain into a black wilderness. The intermontane plains themselves are composed of loess and various alluvial silts and are generally very fertile.

The climate of the high plateau is fairly uniform. Rainfall is moderate though decreasing slightly from south to north. The temperature is cool, seldom rising above 30°C by day. Frosts are common at night particularly in November and December. Humidity is low except during the rainy season and most areas have twelve hours of sunshine almost every day.

The high plateau is densely populated and many of the largest towns including Sana'a are situated in this region. Agriculture is largely limited to the plains. Rain is sufficient to grow cereals except in the north but underground water is plentiful and irrigation is used in many areas for the cultivation of qat, fruit trees and alfalfa. The surrounding hills and surprisingly large areas of plain are uncultivated but heavily grazed by sheep and goats and this is the likely explanation for the treeless aspect of most of the high plateau.

Although many plants are restricted to the high plateau, many of the species common in this region are also found on the escarpment while others extend into the western parts of the Mashriq.

THE MASHRIQ

The Mashriq is an Arabic term meaning the east and I have adopted this name for the eastern dipslope mountains of Yemen. This is a vast and little-known region famous for its warring tribes and ancient ruins. It extends for about 100 kilometres from the eastern edge of the high plateau to the edge of the Great Arabian Desert and from South Yemen north to the Saudi Arabian border. Its eastern boundary is

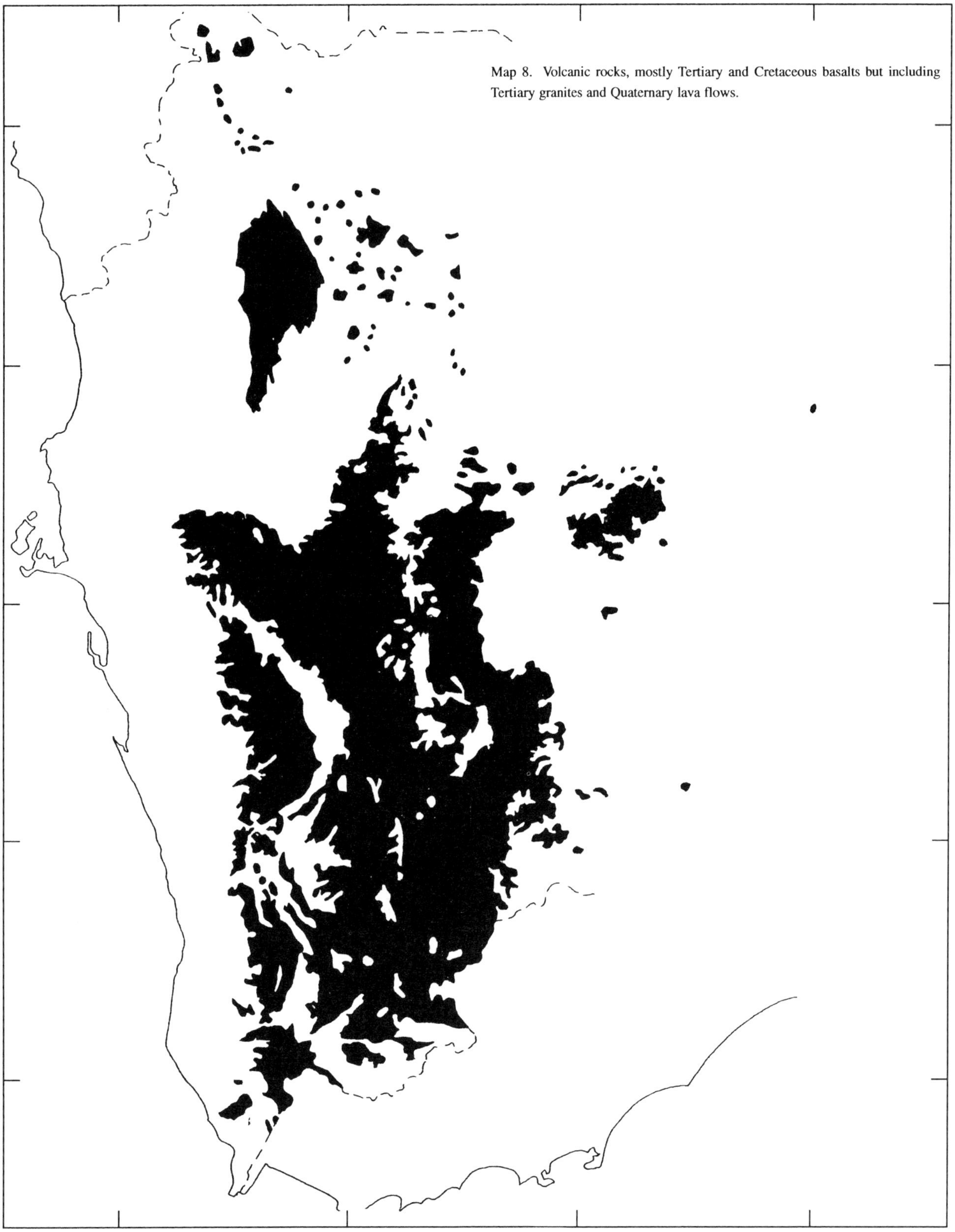

Map 8. Volcanic rocks, mostly Tertiary and Cretaceous basalts but including Tertiary granites and Quaternary lava flows.

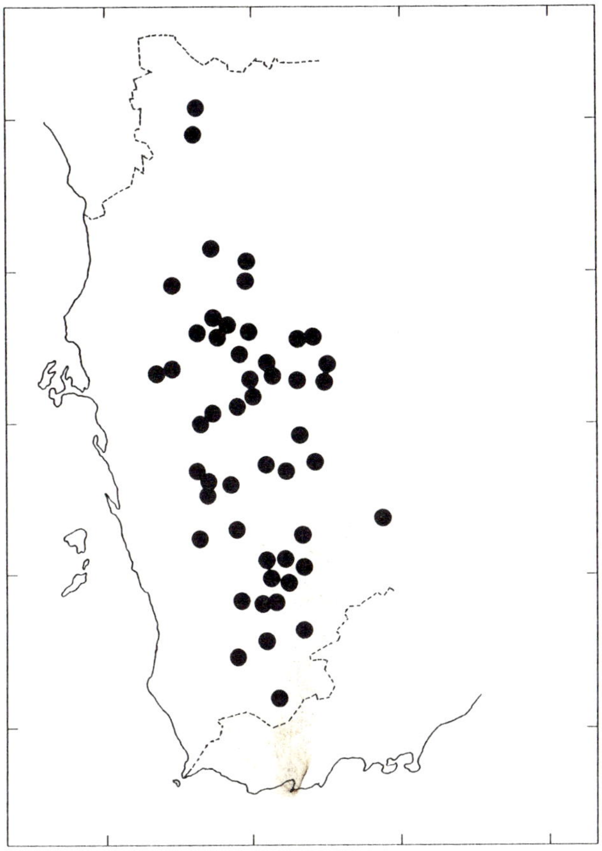

Map 9. Distribution of *Osteospermum vaillantii*, a species common both on the escarpment and the high plateau but unknown elsewhere.

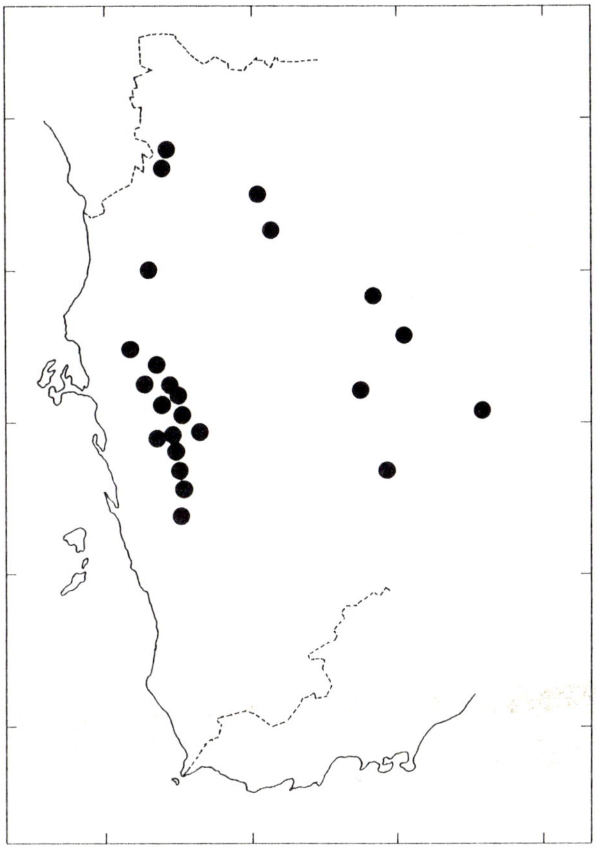

Map 10. Distribution of *Acacia oerfota*, a locally common species in the eastern Tihama and the Mashriq.

drawn along a line connecting the last settlements in the east of Yemen, Khabb, Mārib and Ḥarīb. For much of its length its western boundary is marked by the passes at the east of the high plateau but north of Ḥūth it follows the watershed of the streams flowing east and west until Ṣa'dah. From Ṣa'dah to the Saudi Arabian border its boundary lies along the foot of the escarpment running parallel to and just west of the new Ṣa'dah-Majz road.

The predominant impression of the Mashriq is space. An area like Ḥarf Sufyān can be taken as typical. Here are vast uneven plains stretching away to distant ridges on the far horizon, broken only by isolated mountains standing out of the plain. In some areas such as the Jawf, the mountains recede entirely while in others such as the Rada'-Mārib area, the mountains are numerous though mostly low and rounded. Not all mountains in the Mashriq are insignificant, some reach 2500m but the average height falls from about 2000m in the west to 1000m at the edge of the Sands.

The more open scenery is a reflection of the geology. The great plains of Ṣa'dah and the Jawf are, of course, composed of recent alluvial deposits but apart from some spectacular volcanic plugs to the south and east of Ṣa'dah and the lava flows between Sirwāḥ and Mārib, the rocks of the Mashriq are much older than those of the escarpment and high plateau. Ancient sandstones predominate in Ṣa'dah province whereas limestone is prevalent over huge areas west and south of the Jawf. The rounded hills of Bayḍā' province are the result of the long weathering of the ancient pre-cambrian granites, schists, slate and marble from which they are formed.

The Mashriq is very arid. Rainfall can be expected at some time every year but is very unpredictable in quantity. Humidity is low. Days are hot and cloudless but temperatures fall rapidly after sunset and nights can be surprisingly cold especially in winter. Frosts occur in some areas but no meteorological data are available.

Agriculture and human settlement are limited to valleys where underground water and the occasional flood can be used to irrigate the land. Population is low and there are considerable areas of scrub remaining despite extensive goat-herding.

Although many plants are restricted to the Mashriq, a large part of the flora is composed of species found in the Tihama or in drier parts of the escarpment which reappear in the Mashriq after being absent on the high plateau. Many species found on the high plateau also occur sporadically in the high, western parts of the Mashriq.

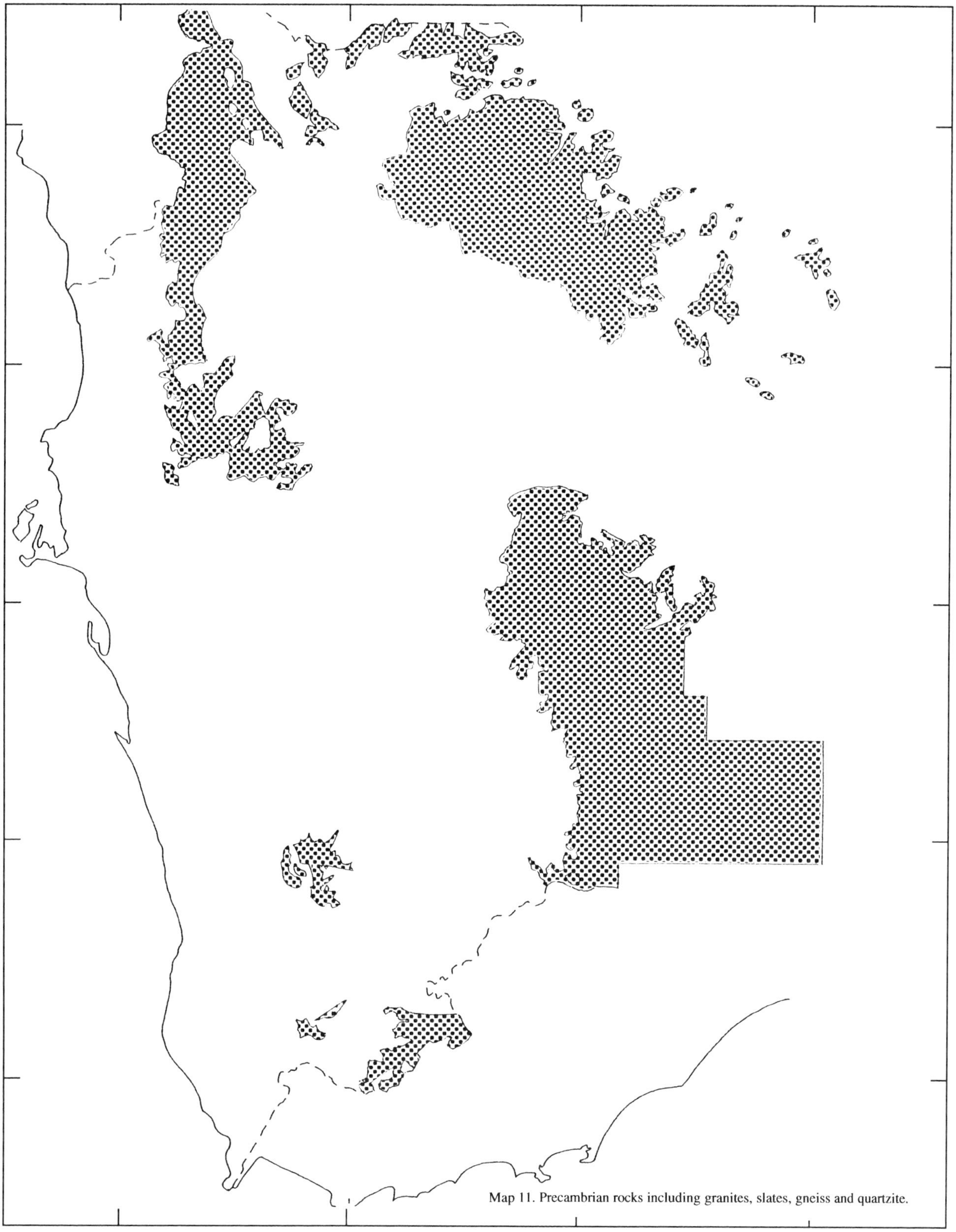

Map 11. Precambrian rocks including granites, slates, gneiss and quartzite.

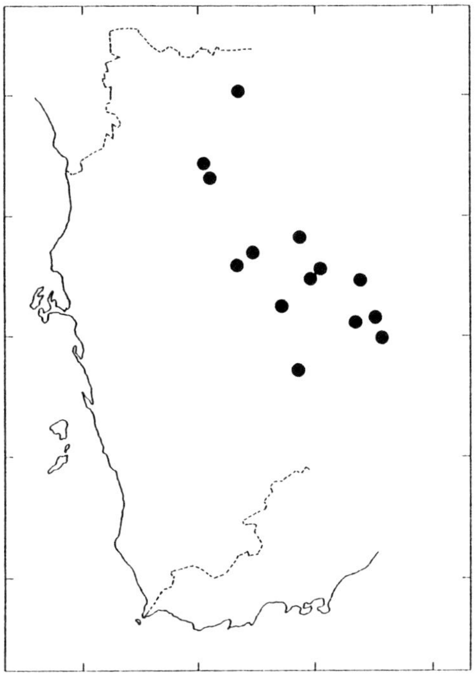

Map 12. Distribution of *Rhazya stricta*, a species restricted to the Sands and the Mashriq.

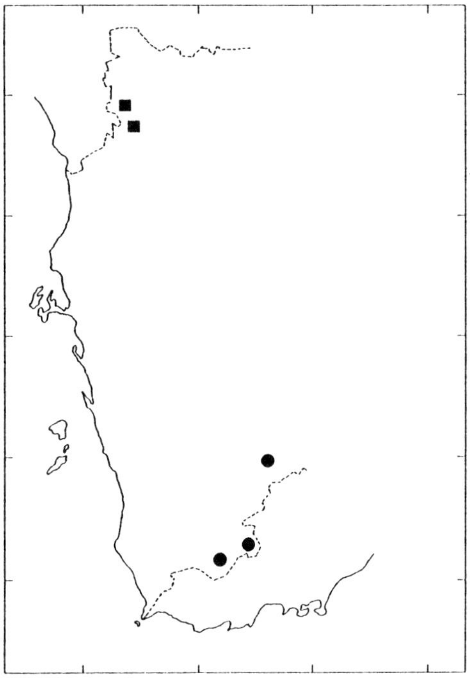

Map 13. Distribution of *Senecio asirensis* (■), a species restricted to Khawlan Ash Sham, and *Plectranthus hyemalis* (●), a species restricted to the southeastern mountains.

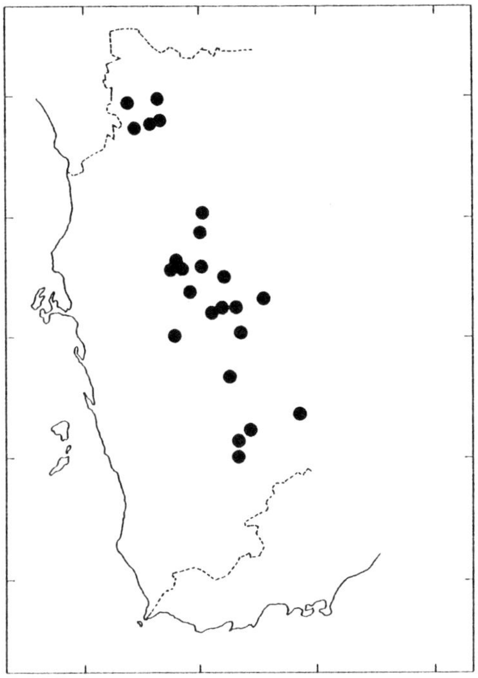

Map 14. Distribution of *Alkanna orientalis*, a species common both on the high plateau and in Khawlan Ash Sham.

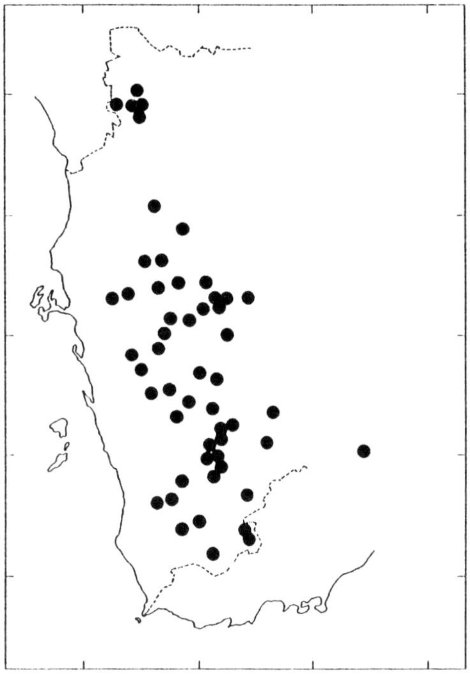

Map 15. Distribution of *Hypoestes forskalei*, a species widespead on the escarpment, high plateau and southeastern mountains but absent elsewhere.

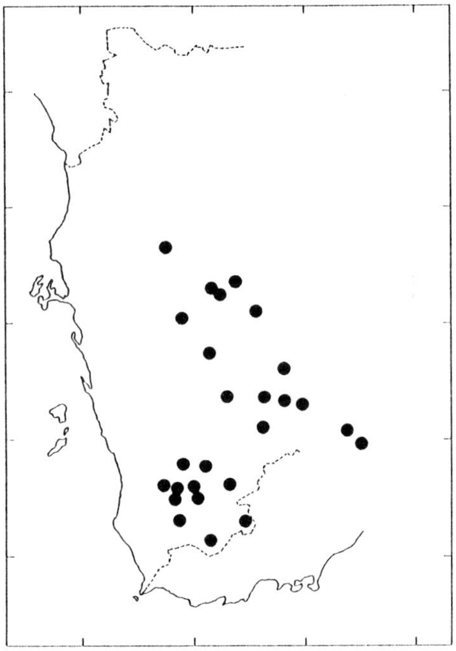

Map 16. Distribution of *Heliotropium bottae*, a species virtually restricted to the southeastern mountains and the high plateau.

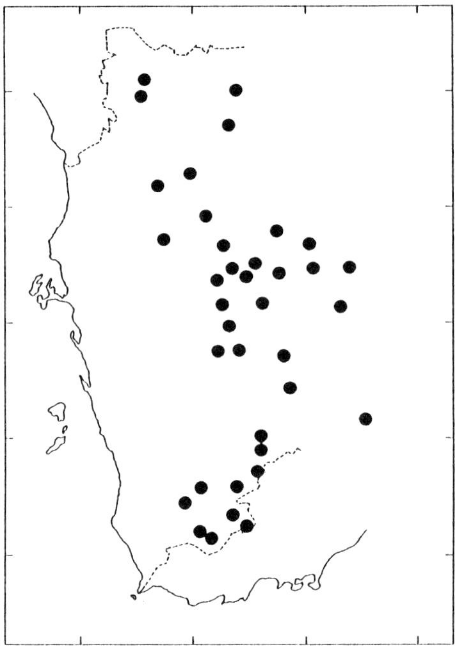

Map 17. Distribution of *Lycium shawei*, a species restricted to the high plateau, the Mashriq and the southeastern mountains.

THE SANDS

This is the traditional bedouin name for the Great Arabian Desert which lies at the east of Yemen and stretches across Arabia to Oman. It is a vast sea of sand broken by the occasional rock outcrop or stony ridge. Temperatures are high, humidity low and rainfall is limited to the occasional local storm that breaks droughts lasting literally for years. There is no settled population but a few bedouin are not afraid to travel the sands in search of whatever fodder they can find for their camels.

THE NORTH AND THE SOUTHEAST

The division of Yemen into five basic regions requires some qualification in the north and southeast of the country.

It has already been noted that the high plateau region reaches its northern limit between Khamr and Ḥūth. Between Ḥūth and Ṣa'dah the line of the inner escarpment is not clear and the mountains are comparatively low. The division between the Mashriq and the escarpment is marked formally by the line of the watershed, but in fact plants from both regions intermingle and the division is not very satisfactory.

North of about 16° 30′ the mountains rise up again in the district of Khawlan Ash Sham and the Mashriq and the escarpment are clearly separated. Khawlan Ash Sham does have a number of species unique to it but it is also interesting as a microcosm of the Yemen flora. In a mere sixty kilometres from Malāḥiṭ to Ṣa'dah one passes from the Tihama through an escarpment flora into a region with close resemblances to the high plateau and thence into the Mashriq around Ṣa'dah. Indeed one of the most striking features of plant distribution in Yemen is the reappearance of many characteristic high plateau plants in the central parts of Khawlan Ash Sham around Sāqayn, Jifān and Madaq.

The high plateau does not extend to the extreme south of Yemen either. At J. Mānar, the inner escarpment turns abruptly to the east thus marking the southern limits of the high plateau region. The area south and east of lines drawn from this point to Al Bayḍā' in the east and Bāb al Mandab in the south enclose the region referred to in this book as the southeastern mountains.

Quite a lot of plants are restricted to this region particularly to the most distinctive part, the Ḥujarīyah, where there is a concentration of endemic species. The region is also interesting because species which are geographically separated elsewhere in Yemen occur here promiscuously mixed together. Thus not only do species widespread on both the high plateau and the escarpment such as *Hypoestes forskalei* occur but also species otherwise strictly limited to the escarpment or to the high plateau or to the high plateau and Mashriq. The explanation for these patterns of plant distribution lies in the strange combination of features that this region enjoys. The geology is most similar to that of the Mashriq with extensive areas of sandstone and pre-cambrian rocks as well as volcanic outcrops. The climate consists of a relatively low rainfall, perhaps similar to that of the high plateau, combined with the frequent mists which are an escarpment feature. Indeed the region bears a close resemblance to Khawlan Ash Sham in the far north which is reflected in the abundance of juniper in both areas.

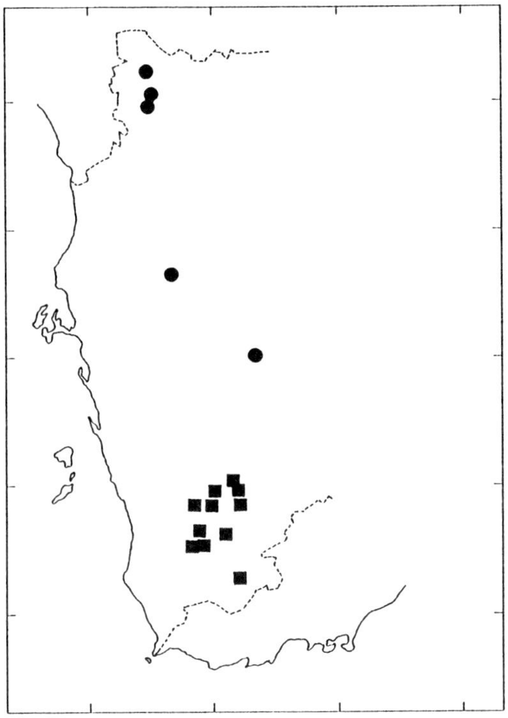

Map 20. Distribution of *Lavandula dentata* (●), a species restricted to the north of Yemen for phytogeographical reasons, and *Asystasia guttata* (■), a species restricted to the south for similar reasons.

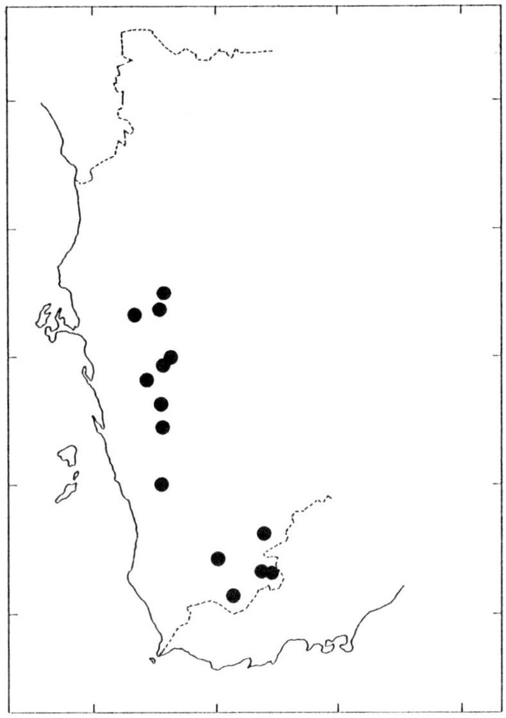

Map 21. Distribution of *Plectranthus hadiensis*, a species restricted to areas in which mists are prevalent.

seasons. This means that moisture is rapidly evaporated and the effects of all but the heaviest rainstorms are negligible. Rain needs to fall at regular intervals over several weeks in order that sufficient moisture be sustained for plants to grow to maturity. Thus occasional showers outside the main rainy seasons have almost no effect on the vegetation.

In some areas of Yemen, however, prolonged periods of mist or low cloud are a common phenomenon. Along the inner escarpment mist is largely limited to the high peaks and is unusual except in late afternoon. In the Ḥujarīyah, the Ḥushā'-Sawraq range, and most of the outer escarpment from J. Raʼs northwards mists are very important. Particularly in winter mists can persist for days without a break even at altitudes as low as 500m. The effects are quite striking. The ground is covered with moisture, evaporation is much reduced or ceases altogether and any rainfall that does occur can soak through to the roots of plants. The area that most benefits from mist is the coffee country on the western slopes of the central escarpment. Here at the driest time of year (February) crops can be grown, the hills are covered with areas of green grassland and numerous plants can be found in flower.

Mist is crucial in explaining a number of plant distribution patterns in Yemen. It is certainly the explanation for the concentration of ferns in the central escarpment since the warm, damp, sunless atmosphere created by the mists is ideal for the growth of ferns. It is the only probable explanation for the curious distribution of *Plectranthus hadiensis*, *Priva adhaerens* and *Scrophularia arguta*, species whose distribution is clearly centred on the central escarpment and the Ḥujariyah, two regions which have little in common besides the prevalence of mist. It also plays an important part in explaining why certain species such as *Agrocharis melananthera*, *Erica arborea*, *Helichrysum forskahlii*, *Parietaria debilis* and *Carex brunnea* occur at altitudes as low as 2000m on the outer escarpment but are restricted to high mountains above 2600m on the inner escarpment. This phenomenon is so common that it cannot be merely coincidence and it seems that the plants are following the mist though whether in search of moisture or shade or both is not clear. The *Erica* is in fact peculiarly adapted to benefit from mist since the moisture condenses on its fine needle-like leaves and adds significantly to the amount of water available to the plant.

ROCKS, SOILS AND MINERALS

The geology of Yemen is poorly known. The only geological map is the Geologic Map of the Yemen Arab Republic (Sana'a) by M. J. Grolier and W. C. Overstreet prepared for the Yemen government by the U.S. Geological Survey (1978). This map is based on satellite photographs which have been interpreted with the aid of the few ground surveys available. Many parts are acknowledged to be uncertain.

The most important formations with respect to plant distribution are shown on Maps 3, 8, 11, 22 and 23, and are

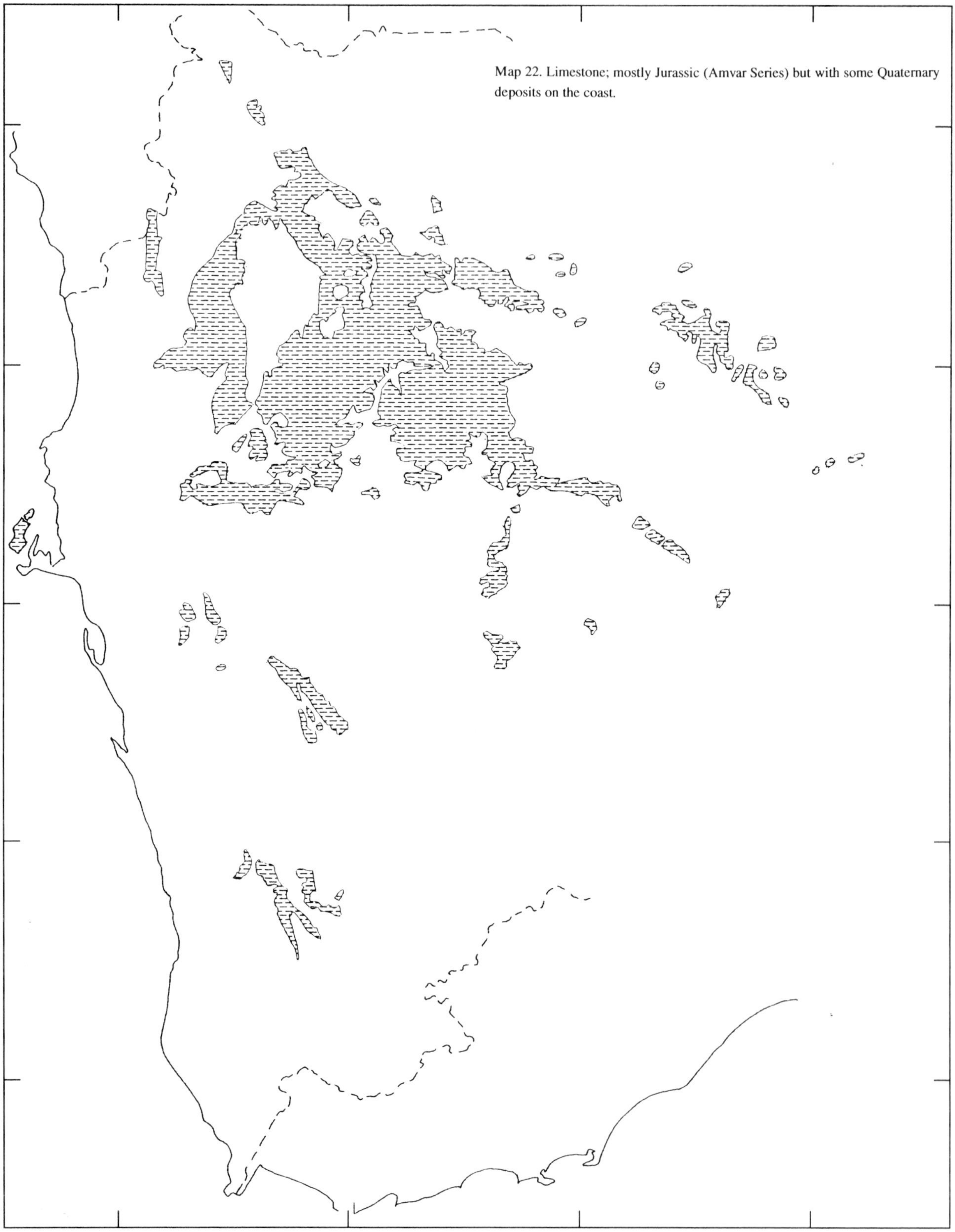

Map 22. Limestone; mostly Jurassic (Amvar Series) but with some Quaternary deposits on the coast.

Map 23. Sandstone, mostly Jurassic (Kohlan Series) but including extensive Ordovician deposits in the Sa'dah area.

discussed below, but no systematic attempt has been made to analyse the flora of each.

Volcanic rocks

Although much of the escarpment and high plateau consists of volcanic rocks and they are by no means rare elsewhere, comparatively few species can be singled out as characteristic of such formations. Amongst the species showing a marked preference for volcanic soils are *Andrachne aspera*, *Cleome schweinfurthii*, *Cometes abyssinica*, *Euphorbia uzmuk*, *Fagonia bruguieri*, *Forsskalea tenacissima*, *Lavandula pubescens*, *L. coronopifolia*, *Scrophularia deserti*, all *Trichodesma* spp. and *Tricholaena teneriffae*, all of which are frequent on lavas and other distinctively volcanic formations. A few other species such as *Alkanna orientalis*, *Anarrhinum forskahlii* and *Salvia merjamie* may favour volcanic soils but it is difficult to be certain of this.

Limestone

Although huge areas of limestone exist no plants are restricted to this rock except for a few very local species such as *Cysistemon heliocharis* and *Diplotaxis sp. A*. Several species, however, do show some preference for limestone though their calcicole tendencies may be obscured by their occurrence on base-rich volcanic rocks. Amongst these species are *Aloe fleurentinorum*, *Commicarpus mistus*, *Euphorbia balsamifera*, and *Heliotropium sessilistigma*. A few other species are noticeably more common on limestone though by no means restricted to this rock. This is true of *Commiphora gileadensis* and *C. myrrha* which are much more frequent on the limestone hills around Bājil than on the volcanic hills.

Sandstone

Sandstone occurs in scattered places around Yemen notably around Ṣa'dah, between Shibām and Maḥwīt, around Sana'a in the Wadi Ḍahr and Bani Ḥushaysh, at Rada' and in the Turbah area. Several species show a marked preference for sandstone including *Aristida congesta*, *Centaurea schimperi*, *Chascanum sessilifolium*, *Cyperus wissmannii*, *Notoceras bicorne*, *Sporobolus pellucidus*, *S. tourneuxii* and perhaps *Pterocephalus frutescens*. Most of these species occur on sandstone plain or in small gullies rather than on steep hillsides. An interesting group of species restricted to sandstone are those found on wet seepages on cliffs and include both *Fuirenia* spp, *Epipactis veratrifolia* and *Schoenus nigricans*.

Sand

Many species are restricted to sand dunes in the Tihama and the Sands such as *Dipterygium glaucum* and *Leptadenia pyrotechnica* and these communities are described fully in the next chapter. Of particular interest are *Mollugo cerviana* and *Gisekia pharnaceoides* both of which are widespread on sand dunes in the Tihama but also occur sporadically in run offs in sandstone country on the escarpment and high plateau.

Salt

High levels of salt are most obviously associated with the Red Sea coast but they are not restricted to it. Scattered over the interior of Yemen are salt springs (near Suq al Jammah and Ḥawbān near Taiz) and various hot mineral springs of which the most famous are Sukhnah, Ḥammām 'Alī and Ḥammām Damt. Salt also accumulates along wadis in the Mashriq where floods are insufficient to wash away the salt deposited as the streams evaporate.

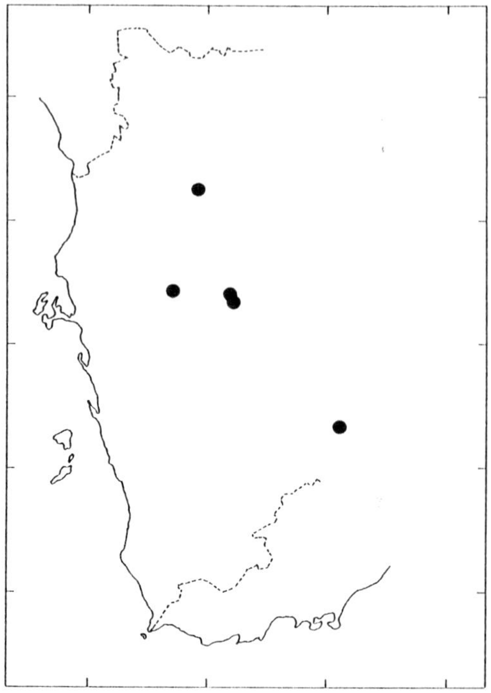

Map 24. Distribution of *Sporobolus pellucidus*, a species restricted to sandstone.

Many halophytic species are entirely coastal such as *Limonium cylindrifolium* and *Halopeplis perfoliata*, but many others are found both at the coast and inland where their presence can almost always be taken to indicate the presence of high salt levels. Species in this category are *Aeluropus lagopoides*, *Cyperus effusus*, *Fimbristylis cymosa*, *Paspalum vaginatum*, *Salsola imbricata*, *Sporobolus spicatus* and *Suaeda monoica*. A few halophytes are found inland only and these include *Atriplex leucoclada*, *Juncus socotranus*, *Scirpoides holoschoenus*, *Sporobolus consimilis* and *Suaeda aegyptiaca*. A further rather larger group consists of those plants that though not halophytes are surprisingly tolerant of salt and are often found in halophytic communities. Examples are *Cyperus laevigatus*, *Phoenix caespitosa*, *Salvadora persica*, *Senra incana*, *Tamarix* spp., *Trianthema crystallina* and *Zygophyllum simplex*.

Nitrogen

High levels of nitrogen in the soil are usually associated with villages, sewage outlets and manure heaps. Certain species are particularly associated with such habitats (see section 2:29 in the next chapter) and include *Chenopodium murale*, *Gynandropsis gynandra*, *Urtica urens* and *Withania somnifera*.

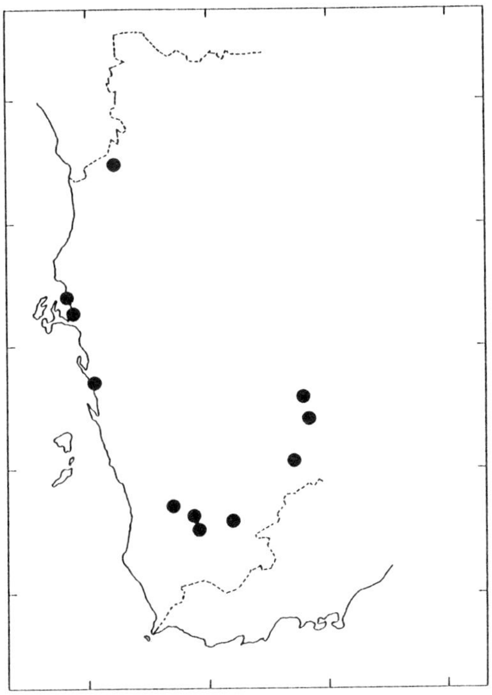

Map 25. Distribution of *Sporobolus spicatus*, a species restricted to damp places with a high salt content.

Alluvium
Fine alluvial silts and loess deposits are characteristic of many areas of Yemen including the eastern Tihama, the plains around Taiz airport, the high plateau, the Ṣa'dah area and the Jawf. Probably the only species found in all of these areas are *Cynodon dactylon* and *Zizyphus spina-christi* but other species are commonly associated with such deposits including *Dactyloctenium scindicum* (Tihama), *Launea nudicaulis*, *Acacia gerrardii* and *Peganum harmala*. It is noteworthy that all soils examined in Yemen have a constantly high pH value, usually above 8.

MAN AND ANIMALS

Man and his domestic animals have played a large part in modifying the vegetation of Yemen and its present-day appearance is largely the result of their activities. None of the revolutionary changes they have brought about have been recorded but it is not difficult to deduce what they have been.

Clearance
It is almost certain that much of Yemen was covered in open woodland in prehistoric times. Although parts of this forest were probably swept by periodic fires, no wholesale destruction is likely to have taken place till about 1000 B.C. when the first major human settlements were made. The forests were then cleared to make way for fields and to provide wood for house construction and fuel. This process is still continuing and large numbers of trees are felled each year, particularly in the Mashriq and the foothills of the escarpment, as there is a continuing demand for firewood in the large towns.

The clearance of the forests has undoubtedly played a large part in reducing the number of tree species found in Yemen. Many are now extremely rare. *Bersama abyssinica* has not been seen since the nineteenth century. Only two surviving trees are known of *Adansonia digitata*, *Bridelia scleroneura* and *Brucea antidysenterica*. Small numbers of *Lannea fruticosa*, *Piliostigma thonningii* and *Stereospermum kunthianum* are known from one or two places but it seems likely that they were much more common in earlier times. Many similar examples will be apparent later in this book.

Grazing
If human interference had been limited to tree felling, the damage would have been fairly quickly rectified by regeneration. However man brought grazing animals in his train and these ensured that this did not take place. The most destructive animal is the goat which will eat almost any vegetation including seedlings and shoots from tree stumps thus making regeneration of the forests impossible. Goats are largely responsible for the treeless scenery of much of Yemen today.

If the animal population increases above a certain level, the effect is even more deleterious as all vegetation except the most resistant plants disappears. In this situation erosion can take place rapidly and once fertile areas can turn into desert. It is impossible to say how much this has happened in Yemen but the ruins of ancient civilizations in the Jawf and at Mārib suggest that these desert areas were once a good deal more fertile than they are today. The plants that survive are often inedible spiny shrubs like *Indigofera spinosa* or bitter poisonous plants like *Rhazya stricta* and *Anisotes trisulcus* resistant even to the most hungry goat.

Terraced Cultivation
Cultivation would not have been possible on the mountains without the development of a remarkable system of soil conservation. Over the centuries a network of terraces has been built up over the hillsides to prevent the soil being washed away. An elaborate system of water channels ensures that flood waters descending the mountains pass from terrace to terrace depositing their silt as they go. Other practices have helped preserve the soil: trees have been planted on field borders; the roots of harvested crops are allowed to remain in the soil over the dry winter period and ploughing does not take place till the following spring so soil is not easily eroded by the wind.

The terraced fields have not only provided a haven for many introduced weeds but have enabled many native species to extend their range. Native plants like *Cyperus rotundus*, *Commelina forskalaei*, *Eragrostis ciliaris* and *Oxalis corniculata* have undoubtedly become more frequent through the agency of man. Ironically two endemic species *Phyllanthus hodjelensis* and *P. tenellus* var. *arabicus* are virtually restricted to terraces and might well become extinct if the terraces are destroyed.

Unfortunately this is now a possibility. Agriculture has been in rapid decline since the 1960s and the most obvious

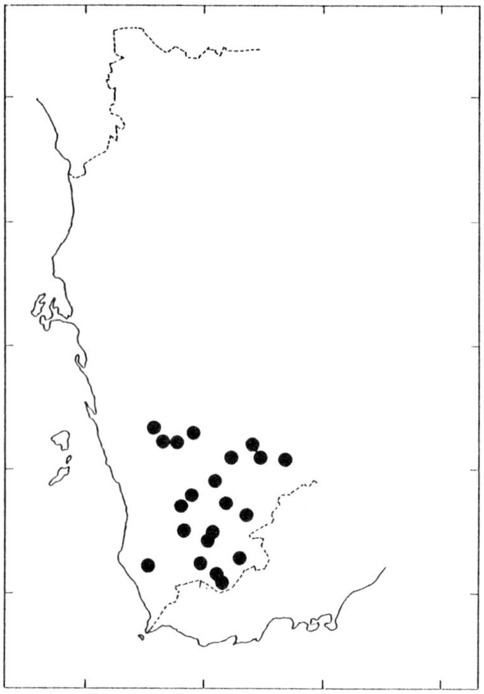

Map 26. Distribution of *Argemone mexicana*, an introduced species spreading from the south.

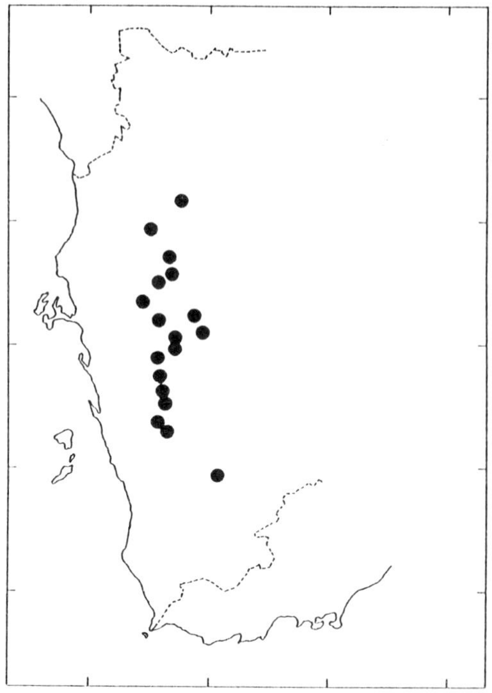

Map 27. Distribution of *Tridax procumbens*, an introduced species spreading north and south from the hinterland of Hodeida.

sign of this is neglected, uncultivated terraces. This is now a common sight in marginal areas such as the lower slopes of the central escarpment and can be expected to become more frequent. If the terraces are not maintained, the walls may collapse and the soils carefully built up over centuries may be washed away in a year or two. The hillsides will become worthless except for the coarsest grazing.

Introductions

Since prehistoric times man has been introducing plants. Cereals were probably the first to be brought to Yemen closely followed by the fruit trees of the Mediterranean region. In medieval times qat, coffee and mangoes were introduced. After the discovery of America potatoes, tomatoes, tobacco and maize made their appearance. This century has seen the introduction of *Eucalyptus* and many ornamental trees and this process is unlikely to slow down.

Most of the plants deliberately introduced by man have remained under his control but others have escaped. *Opuntia dillenii* has become a pest in the foothills of the central escarpment. *Jatropha curcas* is now one of the most common plants in low altitude wadis while *Annona squamosa* is locally an important constituent of scrub. Some plants such as *Iris albicans* or *Ruta chalepensis* have become so well established that their status is uncertain.

Not all plants have been introduced deliberately. Most of the really successful introductions have arrived accidentally, some from the Mediterranean region but most from America. *Amaranthus spinosus*, *Argemone mexicana*, *Datura stramonium*, *Flaveria trinerva*, *Tridax procumbens* and *Xanthium spinosum* are amongst the most common and conspicuous plants in Yemen today. A curious and rather interesting feature of many of these species is that their distribution often gives a clue to the place of their introduction. Thus *Argemone mexicana* is only found in the south of the country suggesting it has been introduced from Aden whereas *Tridax procumbens* is absent from the south suggesting that it may have been introduced with traffic along the Hodeida – Sana'a road.

THE VEGETATION OF YEMEN

This chapter outlines the main plant communities found in Yemen. While every effort has been made to give as full and as accurate an account as possible, many small, but distinctive communities have been omitted while others remain to be discovered and described.

Each plant community is treated as a separate entity but the reality on the ground is rarely as simple as the outline on paper. One community merges imperceptibly into another and the boundaries are far from clear. Disparate elements may be introduced by a change of soil or agricultural practice. Slight changes of climate or gradient can effect striking changes on the ground. It would be both impossible and confusing to attempt to detail all such modifications in the basic vegetation types but it is hoped that anyone examining the vegetation of Yemen will find most of the major types he comes across described in the following pages.

1. THE TIHAMA

Although this area is well known botanically no account of its vegetation has been published apart from the description by Hepper (1977). A comprehensive study of the plant communities would be an interesting research project since the region is easily divided into distinct zones each with their characteristic vegetation.

Regrettably nothing is known of the vegetation of the Red Sea islands but it seems reasonable to postulate that it is similar to that of the coastal regions of the Tihama.

(1:1) *Submarine sea grass communities*
The first communities dominated by flowering plants are in fact at 1–2m depth in the Red Sea where various of the 'sea grasses' (*Cymodoceaceae*, *Hydrocharitaceae*) are abundant but nothing is known of their relative frequency or their precise ecology.

(1:2) *Coastal mangrove swamp*
Along the beach there is an almost continuous fringe of the White Mangrove *Avicennia marina* which extends from the Saudi Arabian border south to the Wadi Sihām. In some areas, notably around Mīdī and Luḥayyah, the *Avicennia* has colonised large areas of mud flat around the margins of creeks and lagoons. No other species of flowering plant is associated with *Avicennia* though the Red Mangrove, *Rhizophora mucronata*, can be expected to occur.

I have no explanation for the almost complete absence of *Avicennia* south of the Wadi Sihām since there are many apparently suitable places for it to colonise.

(1:3) *Coastal sabkhah*
Behind the Mangroves, the coast is usually quite flat but since tidal variation is minimal in the Red Sea, the coast is not liable to inundation. However strong on-shore winds carry salt water for up to at least a kilometre inland and this has brought about a distinct habitat known locally as sabkhah. The sand becomes saturated with salt water and takes on a dark silty texture and develops a hard surface crust which is often coated with salt crystals.

Much of the sabkhah is completely barren but along its margins and on scattered low hummocks and ridges, a distinct vegetation develops, dominated by *Suaeda vermiculata*. Often associated with it is *Aeluropus lagopoides* and less commonly *Cressa cretica*. The only other species characteristic of this habitat are *Limonium axillare* and *L. cylindrifolium* both of which are local. *Dichanthium foveolatum*, *Salsola spinescens* and *Suaeda monoica* occur occasionally as adventives from 1:6.

(1:4) *Shingle beaches*
In a few places the beach rises relatively steeply to a shingle bank just above the high tide mark and this is the habitat of *Atriplex farinosa*, *Halopyrum mucronatum*, *Zygophyllum album* and *Z. coccineum* which occur in abundance around Khawkhah and Mīdī.

(1:5) *Brackish lagoons at the wadi mouths*
The only other coastal habitat of note is found at the outflow of the major wadis. These cannot usually be identified by the presence of a distinct flood bed but are marked by extensive plantations of date palms which are grown where fresh water lies close to the surface. In some places fresh water actually emerges on the beach or in small lagoons or creeks. At Al Fāzah at the mouth of the Wadi Zabīd there is a large lagoon famous for its bird life where there are large colonies of *Cyperus laevigatus* and *Schoenoplectus littoralis* which are typical of this habitat. Other species found at the outflow of wadis are *Fimbristylus cymosa*, *Senra incana* and *Sporobolus spicatus*.

(1:6) *Coastal sand dunes*
Behind the sabkhah lies an area of sand dunes which extends inland for about twenty kilometres and is found almost the whole length of Yemen. These dunes are generally treeless except where underground water is reasonably close to the surface along the lines of the main wadis. Here the native vegetation would seem to consist of very open woodland of *Hyphaene sp. A* but these wadi lines are often heavily populated and the *Hyphaene* has been replaced by date palm plantations and irrigated fields. Apart from the *Hyphaene* and cultivated palms the only trees occurring are *Acacia ehrenbergiana* and *A. tortilis*.

A characteristic feature of the coastal sand dunes is the local dominance of particular species notably *Suaeda monoica* and *Salsola spinescens* near to the sea and *Jatropha pelargoniifolia*, *Leptadenia pyrotechnica* and *Odyssea mucronata* further inland. The dominance of the first two species near the sea is probably explained by their tolerance of salt but it is not clear why one should be dominant to the exclusion of the other nor what it is that determines the dominance of *Jatropha*, *Leptadenia* or *Odyssea* in a particular place.

Besides the species which are locally dominant several others are widespread and often common in this region, in effect characterising the area as a whole. They are *Aerva javanica*, *Dichanthium foveolatum*, *Dipterygium glaucum*,

Fagonia indica, *Heliotropium pterocarpum*, *Panicum turgidum* and *Tephrosia purpurea*. Other species are uncommon and are either intrusives from other vegetation types or one of the following species which though uncommon are characteristic of coastal dunes: *Indigofera semitrijuga*, *Kohautia caespitosa*, *Limeum humile*, *Polygala irregularis*, *Tribulus pentandrus* and *Zygophyllum simplex*.

(1:7) *Salvadora, Cadaba and Capparis scrub on silt deposits (0–200m)*
Towards the east of the coastal sand dune region are extensive beds of fine silt which sometimes extend into broad plains. They are rather local in occurrence, often small and usually in the vicinity of the larger wadis. They were probably formed at the point where wadis dried up in prehistoric times and deposited their silt load. As nowadays water is mostly diverted from the wadis at their point of entry into the Tihama, these silt beds are not being added to and are in fact subject to considerable wind erosion along their western side. In some places all that is left are hummocks of silt each capped with bushes of *Salvadora* or *Cadaba rotundifolia* and surrounded by mobile sand dunes. The silt beds often show some signs of salinity and this probably explains why they are not extensively cultivated. Today their appearance is of a baked, brown plain whose surface though hard, when disturbed breaks up easily into a very fine dust which any traveller going east from Husaynīyah or from Qutay to Khalīfah will be only too familiar with.

These silt pans are commonly dominated by *Salvadora persica* (Qutay-Khalīfah; Mukhā-Mafraq), *Cadaba rotundifolia* (Mawza'; between Hodeida and Manṣūrīyah) or *Capparis decidua* (Ḥaraḍ; Husaynīyah), the latter sometimes mixed with *Balanites aegyptiaca*. The three species rarely grow together and it is possible they have slightly different ecological requirements. All form dense scrub about 2m high with periodic open areas of bare ground. The most characteristic associates are climbing plants growing in the protection of the dominant bush and include *Cissus quadrangularis*, *Corallocarpus schimperi*, *Cyphostemma ternatum*, *Sarcostemma viminale* and *Pentatropis nivalis*. Very few species colonise the bare ground but those that do include *Cassia senna*, *Jatropha glauca*, *Lasiurus scindicus* as well as *Trianthema triquetra* and *Zygophyllum simplex* which may indicate the presence of salt.

About twenty-five kilometres east of the coast there is a distinct change in the nature of vegetation but the eastern half of the Tihama is not as uniform as the western part and in particular there is a dramatic division at about the line of the Wadi Zabīd.

(1:8) *Acacia-Commiphora scrub on gravel steppe in the southern Tihama (100–500m)*
South of the Wadi Zabīd are extensive stony and gravelly steppes whose vegetation bears some resemblance to the *Acacia-Commiphora* orchard described by Vesey-Fitzgerald from the Saudi Arabian Tihama. In Yemen this steppe country lies between 200 and 500m and is best developed around Al Mafraq, Nukaybah and Mawza'. The whole area appears to have a rather low rainfall, probably because the escarpment lying to its east is comparatively low and has gentler gradients than further north.

The vegetation is of an open bushland type with a large variety of small trees 2–4m high. They include *Acacia edgeworthii*, *A. ehrenbergiana*, *A. hamulosa*, *A. tortilis*, *Capparis decidua*, *Commiphora gileadensis*, *C. myrrha*, *Euphorbia cuneata*, *Grewia tenax* (rather untypical), *Maerua crassifolia*, *Maytenus senegalensis*, *Rhigazom somaliense* and *Saltia papposa*. Many of these species provide support for climbing plants which include *Cissus quadrangularis*, *Kedrostis gijef* and *Ochradenus baccatus*. The ground flora consists of a number of tough, often spiny undershrubs which include *Aloe niebuhriana*, *Blepharis ciliaris*, *Blyttia spiralis*, *Campylanthus yemenensis*, *Chascanum marrubifolium*, *Euphorbia triaculeata*, *Fagonia indica*, *Farsetia longisiliqua*, *Indigofera spinosa*, *I. spinulifera*, *Schweinfurthia pedicellata* and *Seddera latifolia*. Characteristic grasses are the tough perennials *Cymbopogon schoenanthus* and *Lasiurus scindicus*. Some annual species appear after rain, notably *Aristida adscensionis* and *A. mutabilis* but annuals are not characteristic of this habitat though *Anticharis arabica* and *Cleome paradoxa* are virtually restricted to it.

Several of the characteristic species of this community occur in the northern Tihama around Ḥaraḍ, an area with a similarly low rainfall and it seems that that area is transitional to some of the communities described by Vesey-Fitzgerald.

A somewhat similar vegetation is also found in the Mashriq (4:3).

Map 28. Distribution of *Aloe vera* (●), a species found mainly on gravel ridges in the east of the Tihama north of the Wadi Zabid (1:15) and *Acacia hamulosa* (▼), a species characteristic of gravel steppe in the southern Tihama (1:8) and at low altitudes in the Mashriq (4:3).

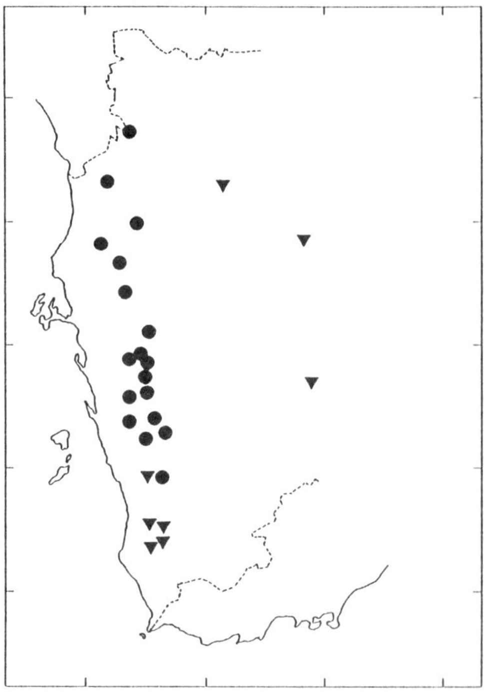

(1:9) *Acacia ehrenbergiana woodland on the inland dunes of the central Tihama. (50–200m)*
East of the twenty-five kilometre line north of the Wadi Zabīd, the sand dunes continue for a further ten kilometres or so to the east and are found around many of the larger towns of the central Tihama such as Bayt al Faqīh, Zaydīyah and Qanāwis. These inland dunes receive a higher rainfall than the coastal dunes and seem also to be richer in plant nutrients. They were obviously able to support woodland and some of this remains, notably around Zaydīyah. The dominant species is *Acacia ehrenbergiana*, often associated with but rarely replaced by *Acacia tortilis*. Almost no other tree species has been found in this woodland but *Commiphora quadricincta* may have been an associated species. As with other bushland communities in the Tihama the dominant Acacias support a number of climbers including *Cissus quadrangularis*, *Cocculus pendulus*, *Maerua oblongifolia* and *Pentatropis nivalis*. The ground flora is very species poor and usually consists only of *Tephrosia purpurea* which is often dominant.

(1:10) *Cultivated sand dunes inland in the central Tihama (50–200m)*
In most places the woodland has been completely cleared and the dunes are planted with bulrush millet which flourishes if the rainfall is sufficient. In late autumn following the rains the dunes are covered in plants and a good example of their vegetation can be seen around Qutay. The only trees are an occasional *Acacia ehrenbergiana* and *Zizyphus spina-christi* but the ground flora includes *Aerva javanica*, *Aristida mutabilis*, *Blepharis ciliaris*, *Boerhavia diffusa*, *Cassia holosericea*, *C. italica*, *Cenchrus biflorus*, *Citrullus colocynthis*, *Convolvulus prostratus*, *Crotalaria microphylla*, *Dipterygium glaucum*, *Euphorbia scordifolia*, *Gisekia pharnaceoides*, *Heliotropium pterocarpum*, *Indigofera argentea*, *I. sessiliflora* (rare), *Jatropha pelargoniifolia*, *Leptodenia pyrotechnica*, *Leptothrium senegalense*, *Mollugo cerviana*, *Octhochloa compressa*, *Panicum turgidum*, *Rhynchosia schimperi* and *Tephrosia purpurea*.

This is clearly a much enriched version of the sand dune vegetation outlined in 1:6 and differs not only in the absence of certain plants but additionally in the lack of any clear dominating species. It is probable that the vegetation of the coastal dunes is essentially natural while that of the interior dunes has been grossly altered by man when the woodland was felled.

(1:11) *Dobera parkland in the silt plains of the eastern Tihama (200–500m)*
At the extreme east of the Tihama lie heavily cultivated silt plains. They occupy a comparatively narrow belt of country and are best developed in the plains immediately below the central escarpment particularly around Sa'īd Miṣr, east of Khalīfah, between Bājil and Bab al Ibl, around al Kadān and east of Dayr Khushm; they are virtually absent north of 'Abs.

The silt plains were undoubtedly wooded in prehistoric times. The dominant tree was probably *Dobera glabra* which still occurs in abundance often giving a park-like effect to these plains and enabling them to be spotted from a distance. It was probably associated with *Acacia ehrenbergiana* and *A. oerfota* both of which occur in abundance in this region. Other trees which survive and may have been important are *Acacia abyssinica*, *A. tortilis*, *Delonix elata*, *Ficus sycomorus* and *Zizyphus spina-christi*.

There is very little ground flora under *Dobera* itself but in the numerous open areas between the trees, along the margins of tracks and between cultivated land, dense secondary scrub has developed dominated by *Indigofera oblongifolia* and more locally *Abutilon pannosum*. In well-trodden or more heavily grazed areas the dominant species are usually *Dactyloctenium scindicum* and *Corchorus depressus*. Associated species are few but include *Cassia italica*, *C. senna*, and *Ocimum forskolei*.

(1:12) *Vegetation around villages in the silt plains of the eastern Tihama (200–500m)*
The silt plains are densely populated and numerous villages are found. Trees are usually planted and as well as ornamentals such as *Moringa oleifera* and *Parkinsonia aculeata*, *Balanites aegytiaca*, *Hyphaene thebaica* and *Zizyphus spina-christi* are commonly grown for shade. Villages are well-trodden places and there flourishes a group of plants with long prostrate stems which are well adapted to withstand trampling, including *Boerhavia diffusa*, *B. repens*, *Corchorus depressus*, *Trianthema portulacastrum* and *Zaleya pentandra*. The villages are usually surrounded by thorn fences which provide a refuge for inland plants like *Acalypha fruticosa*, *Peristrophe paniculata* and *Ruellia patula*, succulents like *Caralluma subulata* and *Opuntia dillenii* and climbers such as *Luffa acutangula*, *Momordica balsamina* and *Pentatropis nivalis*.

(1:13) *Wadi margins in the eastern Tihama (200–500m)*
The silt plains are cut by a series of wadis descending from the escarpment. None of the wadis has permanent flowing water much beyond their point of entry into the Tihama since all available water is diverted into irrigation channels. Floods occasionally break through the artificial barriers during the rainy season but the characteristic streamside vegetation developed on the escarpment (2:13) is not found.

The banks of the wadis are marked by a fringe of *Tamarix arabica* or *T. aphylla*, an abundance of *Calatropis procera* and *Withania somnifera* but very few other species. *Cassia senna* is often the only plant to occur on the sandy bed itself. In a few areas, mainly in the northern Tihama, *Desmostachya bipinnata* is abundant along the margins.

(1:14) *Weed flora on the silt plains and on irrigated land in the eastern Tihama (200–500m)*
Most of the silt plains are cultivated with sorghum and they have developed a rich weed flora. This is almost identical to that found on the rich irrigated land near the major wadis and the weed flora of these two slightly differing areas are treated as one below. The list of species found as field weeds is very long but the following are typical of this area: *Abutilon pannosum*, *Alysicarpus glumaceus* (usually slightly stony fields), *Amaranthus graecizans*, *Boerhavia diffusa*, *Brachiaria reptans*, *Cassis italica*, *C. nigricans*, *Cleome viscosa* (mainly irrigated land), *Commelina forskalaei*, *Convolvulus glomeratus*, *Corchorus depressus*,

C. trilocularis, *Crotalaria retusa*, *Cucumis melo*, *Cymbopogon caesius*, *Cynodon dactylon*, *Cyperus rotundus*, *Dactyloctenium scindicum*, *Dichanthium annulatum*, *Digera arvensis*, *Digitaria ciliaris*, *Echinochloa colona*, *Eragrostis ciliaris*, *Euphorbia granulata*, *E. indica*, *Gynandropsis gynandra* (irrigated land), *Heliotropium longiflorum*, *H. pallens*, *H. strigosum* var. *bicolor*, *Indigofera oblongifolia*, *Ipomoea eriocarpa*, *Launea intybacea*, *Ocimum forskolei*, *Paspalidium desertorum*, *Polygala erioptera*, *Priva tenax*, *Pulicaria jaubertii* (near villages) *Rhynchosia memnonia* (becoming dominant in the dry season), *Schoenefeldia gracilis*, *Schouwia purpurea*, *Sesbainia leptocarpa*, *Solanum coagulans*, *Striga hermonthica* (parasitic on sorghum), *Tribulus bimucronatus*, *Trichodesma zeylanica* and *Urochloa trichopus*.

(1:15) *Gravel ridges in the eastern Tihama (300–500m)*
Rising out of the silt plains are small gravel ridges which sometimes extend into broad plains. These areas of alluvial gravel may be entirely treeless or may have any of the trees found on the silt plains but *Dobera* and *Acacia oerfota* are the most likely to occur. The ground flora is quite poor usually consisting of large quantities of *Blepharis ciliaris* with smaller amounts of *Aerva javanica*, *Cissus quadrangularis*, *Indigofera spinosa*, *Jatropha glauca* and *Lasiurus scindicus*. A particularly characteristic species is *Aloe vera* which sometimes occurs in great abundance.

The poverty in species stands in marked contrast to the richness of the flora on the surrounding cultivated plains.

(1:16) *The escarpment foothills (300–500m)*
The gravel ridges become more extensive towards the east of the silt plains and merge with the foothills of the escarpment. These carry a vegetation similar to that found on isolated hills in the east of the Tihama, notably those around Bājil and Khalīfah. In many respects this is transitional between the vegetation of the gravel ridges (1:15) and the Acacia-Commiphora bushland of the escarpment (2:1) since it has many of the species of both these vegetation types.

The foothill vegetation, however, is far from uniform and would repay detailed investigation. In particular there are marked differences between the flatter ridges on which villages are often situated and the steeper slopes. The dominant species on the flatter ground are usually *Euphorbia inarticulata* and *E. cactus* and this vegetation bears a strong superficial resemblance to the Euphorbia scrub around Taiz (2:9). Associated with the Euphorbias are often *Acacia ehrenbergiana*, *A. oerfota*, *Aloe vera*, *Balanites aegyptiaca*, *Caralluma subulata*, scattered *Dobera glabra*, *Pavonia flavo-ferruginea* and *Sansevieria ehrenbergii*. On steeper slopes these species are uncommon or absent and the dominant plants are *Acacia mellifera*, *Commiphora gileadensis* and *C. myrrha*. These are often associated with *Cadaba glandulosa* and *C. rotundifolia*, and more rarely with *Barleria hochstetteri*, *Cadaba longifolia*, *Euphorbia arabica*, *Seddera virgata* and *Tephrosia nubica*, none of which spread onto the flatter ground. Another variation is found on the steep sides of ravines cut by wadis through stony ridges. Here *Adenium obesum* is the most abundant species and can transform the appearance of arid ravines with its brilliant pink flowers.

Despite the variations outlined in the previous paragraph, there is in fact a unity about the foothill vegetation epitomised by the shrub *Grewia tenax*, abundant here but almost unknown elsewhere in Yemen. Throughout the foothills it is associated with the two small shrubs *Acalypha fruticosa* and *Anisotes trisulcus*, both of which occur in great abundance. Other common shrubs are *Cadaba farinosa*, *Cordia sinensis*, *Euphorbia cuneata*, *Maytenus senegalensis*, *Premna resinosa* and *Salvadora persica*, while scattered trees of *Boscia angustifolia*, *Delonix elata* and *Maerua crassifolia* occur. Climbing plants are much in evidence and include *Ceropegia variegata*, *Cissus quadrangularis*, *C. rotundifolius*, *Kedrostis gijef* (dry country), *Leptadenia arborea*, *Pentatropis nivalis* and *Sarcostemma viminale*. Common herbaceous plants include *Barleria acanthoides*, *Blepharis ciliaris*, *Brachiaria deflexa*, *B. leersioides*, *B. ramosa*, *Caralluma penicillata*, *C. quadrangula*, *Chrysopogon plumulosus*, *Commelina albescens*, *Commicarpus helenae*, *Corbichonia decumbens*, *Dactyloctenium scindicum*, *Dalechampia scandens*, *Ecbolium viride*, *Endostemon tenuiflorus*, *Indigofera oblongifolia*, *I. spinosa*, *I. tinctoria*, *Jatropha glauca*, *Orthosiphon pallidus*, *Pavonia arabica*, *Pupalia lappacea*, *Ruellia patula*, *Solanum forskalaei*, *Talinum portulacifolium*, *Tetrapogon cenchriformis* and *T. tenellus*.

Wetland habitats are very rare in the Tihama and I only know of three all lying quite close to the escarpment foothills.

(1:17) *Saline swamp*
North of Suq al Jammah is a large area of saline swamp which contains many halophytic species including *Cyperus effusus*, *Fimbristylis cymosa*, *F. sieberiana*, *Cyperus laevigatus*, *Phoenix caespitosa*, *Sporobolus consimilis* and *S. spicatus*. A remarkable feature of halophytic communities growing around salt springs such as this is how little they are affected by altitude. All the above species, with the exception of the first two, could be expected at Damt, 'Ujlī or Rada' at 2000m. No other comparable group of plants shows anything like this tolerance of altitude variation and it is possibly explained by the fact that salt springs are almost always to some degree thermal so allowing plants intolerant of cold to grow at higher altitudes than might otherwise be expected.

(1:18) *Freshwater pools*
Between Sa'īd Miṣr and 'Alūjah is an area of marsh which is unique on account of some large permanent pools which support the only truly aquatic flora found in the Tihama. *Najas graminea* is common on the bottom of the ponds and the leaves of *Potamogeton nodosus* cover the surface.

(1:19) *Freshwater marsh*
At Dair 'Uraj below Hillah on J. Bura' there is a spring surrounded by an area of marsh, in turn flanked by seasonally inundated rough scrub. The marsh contains *Cyperus alopecuroides*, *C. articulatus*, *C. laevigatus*, *Fimbristylis cymosa*, *F. sieberiana*, *Ludwigia octovalvis* and *Phyla nodiflora*. *Indigofera articulata* is dominant in the surrounding rough scrub but in more open places *Boerhavia*

diffusa, *Erichloa fatmense*, *Ocimum forskolei*, *O. tenuiflorum* and *Paspalidum desertorum* and *Solanum glabratum* are all common.

Although this spring clearly produces fresh water and is the source of drinking water for the neighbouring villages, the flora cited above suggests some degree of salinity which is probably caused by rapid evaporation and salt deposition in the heat of the Tihama sun.

2. THE ESCARPMENT

The escarpment region is far the most complicated region in its vegetation. The great variations in altitude and rainfall are reflected in a great variety of plant communities. It is impossible to do justice to all of these and the following account has concentrated on the most important and the most distinctive.

(2:1) *Acacia – Commiphora bushland (500–1600m)*
Almost everywhere between 500 and 1600m the escarpment consists of steep stony hillsides almost entirely devoid of good soil which is washed away in heavy rainstorms. Cultivation is limited to the valleys where some soil has accumulated and there are a few villages near the permanent streams in the larger wadis where irrigation is possible.

The basic vegetation of the hillsides is an open *Acacia – Commiphora* bushland about 2–4m high. In remote areas and also in areas of high rainfall the bushes are about 5m apart but near settlements and roads a good deal of clearance has taken place, and in some areas the hills are almost devoid of bushes. The most characteristic tree is undoubtedly *Acacia mellifera*. It is often associated with *A. asak*, which locally becomes the dominant species, *Commiphora sp. A* and less frequently with *C. kataf*.

Underneath the dominant *Acacia – Commiphora* trees are a number of associated shrubs which often occur in great abundance, notably *Acalypha fruticosa*, *Adenia venenata*, *Adenium obesum*, *Anisotes trisulcus*, *Cadaba farinosa*, *Grewia tembensis*, *Maytenus forsskaoliana* and *Ormocarpum yemenense*. Succulents are also very prominent and include the climbers *Cissus quadrangularis* and *C. rotundifolius* as well as many undershrubs ½–2m high, notably *Aloe sp. B*, *Caralluma penicillata*, *C. quadrangula*, *Eulophia petersii*, *Euphorbia cactus*, *E. inarticulata*, *Kalanchoe glaucescens*, *Kleinia odora* and *Sansevieria ehrenbergii*. The spiny shrubs, particularly the Euphorbias, provide protection for a number of species in this vegetation type including *Digitaria pennata*, *Hibiscus deflersii*, *Hyparrhenia hirta* and *Pupalia lappacea*.

During the dry season the ground flora seems to be very sparse and largely limited to a few spiny perennials, particularly *Blepharis ciliaris* and *Indigofera spinosa*, and the almost ubiquitous *Aerva javanica* and *Cassia italica*. After the rains a less impoverished ground flora is in evidence and includes *Aristida adscensionis*, *Becium filamentosum*, *Boerhavia diffusa*, *Brachiaria leersioides*, *Commelina forskalaei*, *Commicarpus helenae*, *Eragrostis papposa*, *Ocimum forskolii*, *Pegolettia senegalensis*, *Ruellia grandiflora*, *R. patula*, *Sida ovata*, *Tetrapogon cenchriformis* and *T. tenellus*.

All the species mentioned in this section can be found in both low and high rainfall areas. High rainfall areas are discussed separately in 2:6 below but it is worth mentioning that low rainfall areas are somewhat distinctive too. They are characterised by the relative frequency of *Commiphora myrrha*, *Euphorbia cuneata* and *Salvadora persica* while the ground flora is marked by an abundance of *Aerva javanica* and the presence of *Fagonia indica*, *Farsetia longisiliqua* and *Seddera latifolia*. *Cadaba heterotricha*, though rare, is a good indicator of drier than average conditions.

(2:2) *Dry valleys in Acacia-Commiphora bushland (500–2000m)*
Along drainage lines, in valley bottoms where there has been some soil accumulation and on rough ground near cultivation, a different vegetation has developed. Trees 10m or more in height are quite frequent; *Combretum molle*, *Ficus populifolia* (mainly in the south), *F. salicifolia*, *F. sycomorus*, *Trichelia emetica* and especially *Zizyphus spina-christi* are characteristic but usually occur only as isolated individuals. Shrubs are not particularly common but *Acalypha fruticosa* usually occurs in great abundance and the introduced *Jatropha curcas* is sometimes frequent. Other typical plants are *Abutilon bidentatum*, *Acalypha indica*, *Achyranthes aspera*, *Amaranthus spinosus*, *Commicarpus plumbagineus*, *Ecbolium viride*, *Hibiscus vitifolius*, *Solanum incanum* and *Withania somnifera*. Grasses are significantly rare or absent in this habitat.

(2:3) *Weeds of cultivated land in the Acacia-Commiphora bushland zone (500–1600m)*
Zizyphus spina-christi and sometimes *Cordia abyssinica* are found round cultivated land but the weed flora itself is usually rather poor consisting only of *Amaranthus graecizans*, *Commelina forskalaei*, *Cyperus rotundus*, *Eragrostis cilianensis* and *Gynandropsis gynandra*. The grassy field borders characteristically support *Bothriochloa insculpta*, *Cenchrus ciliaris*, *Cynodon dactylon*, *Euphorbia hirta*, *Heliotropium longiflorum* and *Solanum glabratum*.

(2:4) *Impoverished Acacia-Commiphora bushland between 1600 and 2000m*
An impoverished, open version of Acacia-Commiphora bushland often extends from 1600 to 2000m and good examples can be seen on the Sana'a-Hodeida road around Maghrabah and in Ḥaymat Al Kharijīyah but it is by no means limited to this area. *Acacia asak* and *A. mellifera* are much less frequent, often disappearing altogether but are replaced by *Acacia etbaica* though it is rarely as abundant as the other two Acacias are. *Commiphora* may be completely absent or represented by scattered bushes of *C. schimperi*. All the shrubs noted at lower altitudes become rare except for *Adenium obesum*, *Grewia tembensis* and *Maytenus forsskaoliana*, all of which may appear to be more abundant through the absence of many of their associates. Not all is lost however as *Barleria bispinosa*, *Cadia purpurea*, *Pavetta longiflora*, *Psiadia punctulata*, *Rhamnus staddo* and *Segeretia thea* become frequent at this altitude and provide much of the low scrub cover. In some areas, notably the Ḥujarīyah, *Ficus salicifolia* is important extending from the valleys onto the open hillsides, though never forming pure stands. *Capparis cartiliginea* is another characteristic shrub in this vegetation type though it is virtually restricted to

perpendicular cliffs. Of the succulents occurring at lower altitudes only *Euphorbia cactus* and *Kleinia odora* remain frequent, the *Euphorbia* sometimes growing in pure stands. *Aloe vacillans* replaces *A. sp. B.*

The ground flora is usually poor even after heavy rain. *Leucas alba*, *Seddera arabica* and *Tragia pungens* are the most typical plants but *Aerva javanica*, *Aristida adscensionis*, *Boerhavia diffusa*, *Cassia italica*, *Hyparrhenia hirta* and *Sida ovata* are often abundant.

(2:5) *Abandoned fields 1500–2000m*

Agriculture is in rapid decline in most areas associated with the impoverished Acacia bushland of 2:4. Abandoned fields and terraces are common. These are invaded by *Laggera decurrens* which frequently becomes dominant and may be associated with *Heliotropium ramosissimum* and *Pulicaria jaubertii*.

(2:6) *Acacia-Commiphora bushland in rainshadow areas of the central escarpment between 600 and 1500m.* A richer type of *Acacia-Commiphora* bushland than that described in 2:1 occurs in the rainshadow areas of the central escarpment. These areas enjoy a higher rainfall than much of the *Acacia-Commiphora* bushland found in Yemen but they dry out very quickly after rain as the prolonged mists so characteristic of the western slopes of the central escarpment are not found here. Good examples of this type of bushland are found around 'Usil and Mawsanah in the Ḥarāz, east of Maḥābishah, around Madīnat al 'Abīd and south of Mishrāfah in Waṣab As Sāfal.

All species noted in 2:1 are also abundant in this type of bushland which is distinguished principally by the greater number of species present. Most characteristic is the abundance and diversity of the shrub genus *Grewia*. In addition to *G. tembensis*, which is found everywhere, *G. schweinfurthii*, *G. tomentosa* and *G. velutina* are all locally common and various other Grewias, though rather rare, are virtually restricted to these areas and include *G. arborea*, *G. erythraea*, *G. flavescens* and *G. mollis*. Other important scrub constituents are the shrubs *Abrus bottae*, *Cordia ovalis*, *Ehretia obtusifolia*, *Ochna inermis* and *Tarenna graveolens*, all of which are usually 2–3m high. Two succulent shrubs are sometimes important; the native *Aloe sabaea* occurs in great abundance in parts of the Ḥarāz and the introduced *Opuntia dillenii* has become locally dominant on cliffs to which it has been introduced by baboons.

The ground flora is surprisingly rich after heavy rain and contains a number of interesting elements besides those common to other types of *Acacia-Commiphora* bushland on the escarpment. Several species including *Barleria acanthoides*, *Chrysopogon plumulosus*, *Cleome scaposa*, *Corbichonia decumbens*, *Dalechampia scandens*, *Endostemmon tenuiflorus*, *Enneapogon cenchroides* and *Talinum portulacastrum* are equally characteristic of the foothill vegetation (1:18). Another group consisting of *Abutilon fruticosum*, *Barleria parviflora*, *Chrysopogon plumulosus*, *Commelina albescens*, *Heliotropium rariflorum*, *Leptothrium senegalense*, *Phyllanthus maderaspatensis*, *Psilotrichum gnaphalobryum*, *Solanum cordatum*, *Tetrapogon villosus* and *Vernonia cinerascens* can also occur in the open bushland of the Mashriq (4:1), sometimes in abundance. A further group is characteristic of the valley scrub communities on the west of the central escarpment (2:17) and includes *Crossandra johanninae*, *Cyperus rubicundus*, *Justicia flava* and *Pancratium maximum*. Finally there are the plants which are largely restricted to this type of bushland and include *Barleria argentea*, *Caralluma deflersiana*, *Chascanum laetum*, *Crotalaria deflersii*, *Echidnopsis scutellata*, *Euphorbia arabica*, *Hibiscus sidiformis*, *Mariscus amauropus*, *Megalochlamys violacea*, *Melhania ovata*, *Ocimum spicatum*, *Pavonia hildebrandtii*, *Polygala senensis*, *Ruellia praetermissa*, *Tephrosia schweinfurthii* and *Vernonia spatulata*.

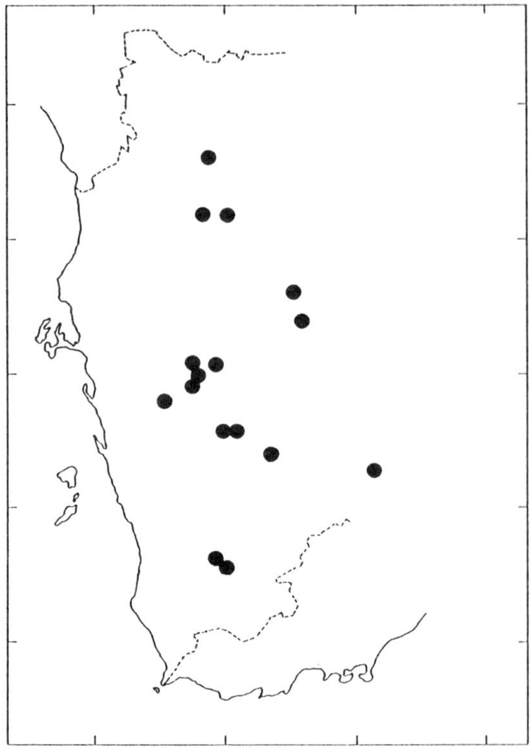

Map 29. Distribution of ***Heliotropium rariflorum***, a species widespread in rain shadow areas of the escarpment and also in the Mashriq.

(2:7) *Acacia-Commiphora bushland in the southwest escarpment (500–1300m)*

South of the Wadi Zabīd and particularly in Taiz province a rather different type of *Acacia-Commiphora* bushland develops below about 1300m. All the species noted in 2:1 are still abundant but certain additional species are found and these serve to distinguish this type of vegetation which is best developed in the western Ḥujarīyah around Barakānī, Bathayjah and particularly Bani 'Umar where there is a remarkable remnant of mature Acacia-Commiphora 'parkland' on J. Jirdad.

The species that best distinguishes this type of bushland is the flat-topped *Acacia tortilis* which is abundant below 1300m in Taiz province. Other trees which appear locally are *Sterculia africana* and *Cadaba heterotricha* (on cliffs). Of shrubs about 2m high *Commiphora spp.* are extremely abundant, perhaps more so than elsewhere on the escarpment. *Grewia schweinfurthii* and *G. gillettii* are both locally frequent but *G. tembensis* is perhaps less common than further north.

Other shrubs that are restricted to this part of the escarpment and are sometimes important are *Hildebrandtia africana*, *Jatropha spinosa*, *J. variegata* (in the Ḥujarīyah), *Rhus flexicaulis* (abundant on J. Jirdad), *Turraea parvifolia* and the succulents *Euphorbia schimperi* and *E. uzmuk* (both rather local). *Flueggia virosa* is much more abundant here than elsewhere on the escarpment.

The ground flora is marked by the local abundance of *Cymbopogon schoenanthus* and *Pulicaria somaliensis* and the replacement of *Aloe sp. B* by *A. inermis* and *A. niebuhriana*. Other species found in this area and absent further north are *Barleria orbicularis*, *B. proxima*, *Brachiaria ovalis* (rare), *Echidnopsis squamulata*, *Pergularia tomentosa* and *Tetrachaete elionuroides* (rare).

(2.8) *Weeds of stony fields in Taiz province (1000–1400m)*
Associated with the bushland of the southwestern escarpment (2:7) is an interesting weed flora developed on stony fields in Taiz province. On the field margins can be found *Heliotropium zeylanicum*, *Indigofera articulata*, *I. oblongifolia*, *I. spinosa*, *Melhania ovata* and *Rhynchelytrum repens*. The most common crop is sorghum which is underplanted with various legumes and amongst these grow *Bidens biternata*, *Cleome angustifolia*, *Crotalaria pycnostachya*, *Cucumis prophetarum*, *Eragrostis cilianensis*, *Heliotropium ramosissimum*, *H. strigosum*, *Kohautia aspera*, *Physalis micrantha*, *Sida alba* and *Striga asiatica*.

(2:9) *Succulent Euphorbia scrub in the southeast mountains between 1200 and 2000m.*
In parts of the southeast mountains the *Acacia-Commiphora* bushland is entirely replaced by one of the most remarkable plant communities in Yemen – the succulent *Euphorbia* scrub. This is best developed between 1200 and 2000m around Taiz and Qā'idah but it is not restricted to this area and examples can be found in parts of the Ḥujarīyah, in the Wadi Banā and probably elsewhere in the little-known, southeastern border region.

The *Euphorbia* scrub is far from uniform, varying in particular with altitude, so that the type around Taiz at *c* 1400m differs from that found around Qā'idah at *c* 1800m. The similarities though are greater than the differences and it is best to treat the two types together.

The characteristic feature of this type of scrub is the scarcity of all species of *Acacia*, *Commiphora* and *Grewia* and their replacement by various species of succulent *Euphorbia*. At Taiz the dominant *Euphorbias* are *E. cactus*, *E. inarticulata* and *E. schimperi* all standing 1–2m high. At Qā'idah they are joined by the tree species *E. ammak* and *E. parciramulosa*. No other species above 1m are found in these communities except *Zizyphus spina-christi* in gullies, the occasional *Acacia mellifera*, *A. nilotica* subspecies **kraussiana** and *Commiphora sp. A* on the hills round Taiz and the occasional *A. gerrardii* at Qā'idah.

There are many features among the undershrubs and the ground flora which are common to both the Taiz and the Qā'idah types. Succulent shrubs are particularly common and include *Adenia venenata*, *Adenium obesum*, *Eulophia petersii*, *Kleinia odora*, *Sansevieria ehrenbergii*, *S. forskaliana* and *Sarcostemma viminale*. Common non-succulent shrubs are *Grewia tembensis*, *Maerua sp. D*, *Pavetta longiflora*, *Psiadia punctulata* and *Segeretia thea*. The ground between the Euphorbias is mostly bare except for hardy undershrubs like *Indigofera spinosa*, *Ruellia patula* and *Seddera arabica*. It is in the protection of the Euphorbias themselves that most herbs grow and here can be found *Digitaria nodosa*, *Hibiscus deflersii*, *H. vitifolius*, *Hyparrhenia hirta*, *Themeda triandra* and *Tragia pungens*. Two parasites, *Plicosepalus acaciae* and *Striga gesneroides* are common, while *Cistanche phelypaea* is frequent in disturbed places near fields.

In other respects the Taiz and Qā'idah types of Euphorbia scrub are rather different. At Taiz the most prominent species after the three Euphorbias is *Cissus rotundifolius* but this is absent from Qā'idah where it is partially replaced by *Cyphostemma digitatum*. Both *Caralluma penicillata* and *C. quadrangula* are abundant at Taiz whereas at Qā'idah the only prominent *Caralluma* is *C. cicatricosa*. Similarly *Aloe inermis* and *A. niebuhriana* at Taiz are replaced by *A. vacillans* and *A. rivierei* at Qā'idah. The other species frequent at Taiz are mostly characteristic of lower, drier situations and include *Acalypha fruticosa*, *Megalochlamys violacea*, *Pulicaria jaubertii* and *Salvadora persica*.

At Qā'idah several elements are exceptionally well represented. First among these are succulents. In addition to the many species already mentioned *Ceropegia arabica*, *C. rupicola*, *Kalanchoe deficiens*, *K. glaucescens*, *K. yemensis*, *Plectranthus cylindraceus* and *Senecio hadiensis* are common. Plants with bulbs and other storage systems are also characteristic. Many such as *Chlorophytum tordense*, *Crinum album* and *Ceropegia bulbosa* are rare but others appear in great abundance following the rains. These include *Cyanotis nyctitropa*, *Dorstenia barnimiana*, *Mariscus schimperi* and *Scilla hyacinthina*. The final element is a ragbag of species found in rough, open scrub including *Acacia hockii*, *Barleria prionitis*, *Becium filamentosum*, *Gomphocarpus fruticosus*, *Heliotropium bottae*, *Eragrostis papposa*, *Hypoestes forskalei*, *Justicia flava* and *Selaginella yemensis*.

(2:10) *Marsh and its surrounding bushland in high rainfall areas of Ibb province (1600–1800m)*
Associated with the Euphorbia scrub at Qā'idah is an interesting area of marsh and associated bushland in the Wadi Sūdān. Although this area is probably unique it bears strong resemblances to areas of marsh between Ibb and Suq As Sabt at the same altitude, to the vegetation of the rich middle altitude wadis (2:15) and to that found at 2500m in the Wadi Banā (2:16).

The Wadi Sūdān flows through a broad, shallow valley between the Euphorbia scrub. There is no large wadi bed, but the permanent water is diverted into small scrub-lined ditches which serve to irrigate the heavily cultivated valley.

Along the ditches are trees and large shrubs 2–10m high amongst which are *Arundo donax*, *Carissa edulis*, *Cordia africana*, *Debregeasia saenab*, *Dombeya schimperiana*,

Ehretia cymosa, *Erythrococca abyssinica*, *Maesa lanceolata*, *Myrica salicifolia*, *Pluchea discoridis* and *Vernonia amygdalina*.

In open places along the ditch sides there has developed quite a rich marsh vegetation which includes *Agrostis viridis*, *Apium graveolens*, *A. nudiflorum*, *Bacopa monnieri*, *Blumea lacera*, *Cirsium vulgare*, *Commelina diffusa*, *Cyperus laevigatus*, *Epilobium hirsutum*, *Eriochloa meyeriana*, *Equisetum ramosissimum*, *Fimbristylis sieberiana*, *Kanahia laniflora*, *Laggera crispata*, *Lotus corniculatus*, *Lythrum hyssopifolium*, *Mentha longifolia*, *Paspalidium geminatum*, *Plantago major*, *Persicaria maculosa*, *P. salicifolia*, *P. senegalensis*, *Phyla nodiflora*, *Potentilla reptans*, *Pteris dentata*, *P. vittata*, *Ranunculus multifidus*, *Samolus valerandi* and *Sonchus oleraceus*.

Along many of the ditches there are thorn fences to prevent animals going amongst the growing crops. Here is developed an interesting association which is common in such situations particularly in Ibb province. Several of the tall shrubs mentioned above grow here but most of the typical plants are herbs, undershrubs or climbers, particularly *Arthraxon prionodes*, *Asystasia gangetica*, *Clematis simensis*, *Commelina imberbis*, *Dicanthium aristatum*, *Digitaria ternata*, *Dyschoriste perrottetii*, *Eleusine indica*, *Hypoestes triflora*, *Phaulopsis imbricata*, *Pterolobium stellatum*, *Rubus sp. E*, *Senecio bojeri*, *Sida urens* and *Teramnus labialis*.

On grassy banks above the marsh *Bothriochloa insculpta* and *Indigofera spicata* are characteristic plants.

(2:11) *Bulb-rich communities on stony hills around Ibb (1800–2000m)*

North and west of Ibb are a number of rounded stony hills which bear a distinctive vegetation obviously related to the Euphorbia scrub hills of the Qā'idah region. These hills are remarkable for their spring bulbs.

Some of the hills are virtually devoid of scrub while others have a rather dense growth of *Acacia hockii* which sometimes forms impenetrable masses 1–2m high. On steeper slopes *Euphorbia ammak* is abundant everywhere but the only other shrub species present are *Carissa edulis*, *Euphorbia parciramulosa* and *Pavetta longiflora* apart from the climbers *Cyphostemma digitatum* and *Jasminum grandiflorum*.

For much of the year the hills appear to be quite barren with scattered rosettes of *Aloe vacillans* and the dried-up remains of *Hyparrhenia hirta* in the protection of the few shrubs. A remarkable change takes place around April at the onset of the summer rains. The first thing to appear is the leafless orchid *Holothrix aphylla* and this is quickly followed by *Microchloa kunthii* and *Craterostigma pumilum* whose violet flowers carpet small hollows a few days after the rains begin. In late May the hills are probably at their best when they are covered with *Chlorophytum tetraphyllum*, *Dorstenia barnimiana* and *Scilla hyacinthina*. These die down in June and are replaced by *Bonatea steudneri*, *Cyanotis nyctitropa*, *Habenaria macarantha*, *Plectranthus barbatus* and *Striga asiatica*.

Later in the season the hills are rather dull and give no evidence of their rich spring flora.

(2:12) *Weeds of arable land and field borders around Ibb (1600–2300m)*

The Ibb area is intensively cultivated and the rolling hills north, south and west of the town are covered in fields that develop a rich weed flora under the sorghum which is almost the only crop grown in this area.

The field borders are mostly grassy and *Andropogon greenwayi*, *Hyparrhenia hirta*, *Pennisetum villosum* and *Themeda triandra* are all common. Particularly interesting is *Hyparrhenia dregeana* which is restricted to this habitat and comes into flower at the end of the rainy season. Often associated with the grasses are *Crotalaria spinosa* and *Guizotia scabra* while *Euphorbia inaequilatera* is frequent in bare patches. Long-established field borders are sometimes invaded by *Acacia hockii*.

Numerous weeds are found under the Sorghum and many of them are restricted to the Ibb region or virtually so. Common weeds include *Ammi majus*, *Anagallis arvensis*, *Aspilia helianthoides*, *Brachiaria comata*, *B. eruciformis*, *Commelina forskalaei*, *Convolvulus arvensis*, *Eragrostis barrelieri*, *E. cilianensis*, *Erucastrum arabicum*, *Geranium ocellatum*, *Guizotia scabra*, *Ipomoea eriocarpa*, *Kohautia coccinea*, *Oldenlandia corymbosa*, *Solanum villosum* subsp. *puniceum*, and *Spermacoce sphaerostigma*. *Striga hermonthica* is a common parasite on sorghum roots.

(2:13) *Wadis in the lower regions of the escarpment (500–1000m)*

The escarpment is deeply cut by a series of wadis flowing westwards. The more important have permanent flowing water as they pass through the slopes of the escarpment and all can be transformed into raging torrents during the rainy season. The steep slopes above the wadis carry one of the forms of *Acacia-Commiphora* bushland described in 2:1,6,7 but distinct vegetation types develop in and beside the wadis.

The permanently flowing part of the wadi usually covers only part of the flood bed. Along its margins can be found *Bacopa monnieri*, *Eclipta prostrata*, *Heliotropium ovalifolium* and *Phyla nodiflora*. *Typha domingensis* grows in pools and deeper water. In drier parts of the flood bed *Amaranthus spinosus*, *Cassia senna*, *Ricinus communis* and *Tridax procumbens* are often established.

The edges of the flood bed are lined with scattered large trees. The most common are *Ficus sycomorus*, *F. vasta*, *Trichelia emetica* and *Breonadia salicina*. The introduced *Ceiba pentandra* sometimes occurs. The only shrubs that occur are likewise probably introduced. The giant grass, *Arundo donax*, *Jatropha curcas* and *Pandanus odoriferus* are all frequent but their distribution is probably directly influenced by man. Two undershrubs are particularly characteristic – *Datura metel* and *Kanahia laniflora*, the latter often growing in abundance along banks washed by the permanent watercourse. No other species are particularly associated with this habitat but *Cassia occidentalis* and *Commicarpus plumbagineus* often occur.

(2:14) *Weeds of irrigated fields along the major wadis in lower regions of the escarpment (500–1500m)*

Wherever the valleys are broad enough fields are constructed and water is diverted out of the wadi to irrigate them. Many of the larger wadis are lined with fields of bananas, maize or papayas.

The field borders are almost invariably grassy, most of the sward being formed from *Cenchrus ciliaris* and *Cynodon dactylon*. Associated with these grasses are *Boerhavia diffusa*, *Euphorbia hirta*, *Ocimum forskolei*, *Sida alba* and *Solanum glabratum*.

The irrigated fields themselves have a rich weed flora which bears a strong resemblance to that found on the silt plains in the east of the Tihama (1:14). Many species are common to both habitats including *Amaranthus graecizans*, *Cleome viscosa*, *Commelina forskalaei*, *Corchorus trilochularis*, *Cyperus rotundus*, *Echinochloa colona*, *Gynandropsis gynandra*, *Ipomoea eriocarpa* and *Solanum coagulans*. Other species however are found principally in the irrigated wadi fields notably *Amaranthus spinosus*, *Croton lobatus*, *Dactyloctenium aegyptium*, *Euphorbia indica*, *Micrococca mercurialis*, *Phyllanthus fraternus*, *Physalis micrantha*, *Rottboellia cochinchinensis*, *Ruellia patula*, *Setaria verticillata* and *Solanum villosum* subsp. *puniceum*.

(2:15) *The lush middle-altitude wadis (1200–1500m)*
Permanent flowing water can be found in quite a few wadis up to about 1500m but very rarely higher than this (see below 2:16). The wadi vegetation is generally at its richest towards the upper limits of permanent flowing water around 1400–1500m and particularly in narrow, fertile valleys, flanked by steep mountains which attract a high rainfall. The best examples are the 'Udayn wadis and Wadi Sukhnah but other good and more accessible examples are Wadi Ḍabāb near Taiz and Wadi Sharas between Ḥajjah and Kuḥlān.

The steep banks above the wadis are generally covered with one of the types of *Acacia-Commiphora* bushland described in 2:1, 6, 7 but there are a few species particularly characteristic of bushland in the wadis notably *Lannea fruticosa* and *Ozoroa insignis*, the grass *Hyparrhenia coleotricha*, *Kalanchoe lanceolata* and *Plectranthus tenuiflorus*.

All the trees and shrubs found growing beside wadis below 1000m also occur at the higher altitudes discussed here and are at least equally common. Indeed one of the distinguishing features of these middle-altitude wadis is the abundance of trees. At lower altitudes (2:13) trees occur sporadically along the banks but here they often line the wadi margins and sometimes their branches meet overhead transforming the wadi bed into a leafy tunnel. The most abundant tree is *Breonadia salicina* which often grows in pure stands. Figs are very common; *F. vasta* is the most prominent especially at 'Udayn but *F. sycomorus*, *F. exasperata* and *F. salicifolia* are sometimes frequent while *F. ingens* is common on river cliffs with *Aloe yemenica*. Other characteristic but less common trees are *Cordia africana*, *Nuxia oppositifolia*, *Syzygium guineense*, *Trichelia emetica* and *Zizyphus mucronata*. *Arundo donax* often forms impenetrable stands but shrubs are not particularly prominent and those that do occur such as *Acanthus arboreus*, *Jatropha curcas*, *Maesa lanceolata*, *Oncoba spinosa*, *Pluchea discoridis* are of weedy origin or more characteristic of other areas. Woody climbers are sometimes frequent and include *Caesalpinia bonduc* in the south, *Pterolobium stellatum* and *Rhoicissus revoilii*.

Most middle-altitude wadis are cultivated wherever possible. Water is diverted and there are large irrigated plantations of bananas, coffee and other crops. The weed flora is very similar to that occurring at lower altitudes outlined in 2:14 above but *Laportea aestuans* is common in shade under coffee and *Acalypha ciliata* grows in irrigated sorghum fields. A particularly characteristic vegetation is found where the irrigation channels break down and water pours down the face of the retaining wall. Here grow the ferns *Adiantum capillus-veneris* and *Pteris vittata*, almost always associated with *Bidens pilosa*, *Equisetum ramosissimum* and *Veronica anagallis-aquatica*.

The fields are usually surrounded by thorn fences which provide a refuge for many plants particularly where they adjoin the wadi bed itself. The fences are often festooned with *Commicarpus plumbagineus* or one of the twiners such as *Coccinea grandis*, *Ipomoea obscura*, *Neonotonea wightii* or *Rhynchosia malacophylla*. The most common plant living in the protection of the fence is probably *Acalypha fruticosa* which may be associated with less common species such as *Blainvillea gayana*, *Brachiaria comata*, *Digitaria velutina*, *Panicum maximum*, *Sigesbeckia orientalis* and *Urochloa panicoides*. Where the ground is wet *Asystasia gangetica*, *Commelina benghalensis*, *C. imberbis*, *Ocimum gratissimum*, *O. suave* and *Sida urens* are likely to be found. On rough ground near the fences grow *Abutilon bidentatum*, *Amaranthus spinosus*, *Bidens biternata*, *Cassia obtusifolia*, *C. occidentalis*, *C. sophera*, *Conyza bonariensis*, *Cynodon dactylon*, *Triumfetta rhomboidea* and the ubiqutous *Withania somnifera*.

Kanahia laniflora is the most characteristic species growing by the margins of the stream. *Bacopa*, *Eclipta*, *Phyla* and *Typha* are as common as they were at lower altitudes (2:13) but are often joined by *Chenopodium ambrosioides*, *Colocaisia antiquorum*, *Mentha longifolia*, *Paspalidium geminatum* and *Persicaria senegalensis*, all of which seem to prefer places where some silt has been deposited.

(2:16) *Permanent flowing wadis at high altitude (2400–2500m)*
The only permanent flowing wadi at high altitudes that I know of is the Wadi Banā where there is permanent water at Saddah at 2500m.

Here the natural vegetation is greatly modified and many introduced trees such as *Schinus molle*, *Eucalyptus spp* and *Prunus armenaica* are common. Of native trees *Acacia origena* is probably the most common on the banks. Others that occur include *Debregeasia saenab*, *Ficus sur*, *Maesa lanceolata*, *Myrica salicifolia*, *Salix subserrata*, *Tamarix aphylla* and *T. arabica*.

Where the slopes are steep, the banks above the watercourse are entirely devoid of any special vegetation while in other areas where the slopes are gentle extensive marsh has developed. The marsh bears a strong resemblance to that found on the high plateau (3:9), nearly all species being common in temperate countries and including *Apium graveolens*, *A. nodiflorum*, *Berula erecta*, *Carex distans*, *Cirsium vulgare*, *Dicrocephala chrysanthemifolia*, *Epilobium hirsutum*, *Fimbristylis sieberiana*, *Lotus corniculatus*, *Lythrum hyssopifolium*, *Myosotis laxa*, *Nasturtium micro-*

phyllum, *Pennisetum macruorum*, *Persicaria maculosa*, *Potentilla reptans*, *Rorippa islandica*, *Rumex steudelii*, *Sporobolus africanus*, *Trifolium fragiferum* and *Veronica anagallis-aquatica*.

In the wadi bed itself *Kanahia laniflora* grows beside the water while true aquatics such as *Marsilea aegyptiaca*, *Potamogeton pusillus* and *Ranunculus rionii* grow in pools where the water is deeper. As at lower altitudes most of the flood bed consists of gravel and rocks and this is colonised by introduced weeds of temperate origin including *Argemone mexicana*, *Capsella bursa-pastoris*, *Conium maculatum*, *Datura stramonium*, *Galinsoga parviflora*, *Lactuca serriola* and *Poa annua* which are all doubtless at home here amongst the cool and damp of this valley.

(2:17) *Valley forest in the central escarpment (500–1000m)*
The last wadi type of vegetation to be considered is the most remarkable. This is the valley forest (or more appropriately valley scrub) found between 500 and 1000m principally in the central escarpment. This is best developed in the valley above Suq As Sabt, J. Bura', but good examples can be found above 'Alūjah, below Ḥadīyah and around Suq Ar Ribāṭ in J. Raymah, above Ḥillah on J. Bura', in the Wadi Yūr and Wadi Ḥillah on J. Milḥān, at Rajamah and in the Wadi Ḥār in the Ḥarāz and in the Wadi Līyah in Khawlan Ash Sham. All these valleys have a lot in common. They all face west or south, they all receive a relatively high rainfall but reduced hours of sunshine at certain seasons because of mist. The sides rise up very steeply from the wadi bed. They are entirely uncultivated and almost completely devoid of human settlement. Although there is usually flowing water throughout the rainy season, all the wadis dry up completely for about half the year, though occasional rock pools persist.

At the entrance to the valleys at about 500m the vegetation is essentially that of the foothills (1:16) but it soon changes into the distinctive valley forest vegetation. This is developed in the bottom of the valleys, along and immediately above the flood bed but does not usually extend far up the sides of the valleys whose upper slopes carry the widespread *Acacia-Commiphora* bushland described in 2:1. Indeed all the species characteristic of that bushland also occur in the valley forests with the exception of *Commiphora myrrha*, *Euphorbia cuneata* and *Digitaria pennata* which do not tolerate the higher rainfall of these valleys.

The most abundant tree in the valley forests is probably *Combretum molle* which may well have been dominant in many areas as it still is today above Suq As Sabt. However it is not by any means the largest tree present and huge scattered specimens of *Ficus vasta*, *F. sycomorus* (both riparian), *Tamarindus indica*, *Celtis toka* and *Mimusops laurifolia* can often be found. The last two trees are particularly characteristic of the valley forest and are sometimes abundant though they are not known in other vegetation types in Yemen. Other common trees are *Acacia abyssinica*, *A. asak*, *Berchemia discolor*, *Breonadia salicina* (riparian), *Ficus glumosa*, *Trichelia emetica*, *Zizyphus mucronata* and *Z. spina-christi*.

The shrub layer is commonly dense and impenetrable. In addition to the common shrubs of *Acacia-Commiphora* bushland, *Bauhinia tomentosa*, *Carissa edulis*, *Cordia ovalis*, *Diospyrus mespeliformis*, *Grewia schweinfurthii*, *G. tembensis* (large-leaved form), *G. trichocarpa*, *G. villosa*, *Oncoba spinosa*, *Premna resinosa* and *Tarenna graveolens* are all characteristic and often frequent. The palm *Phoenix caespitosa* sometimes forms large communities in which almost no other species is found.

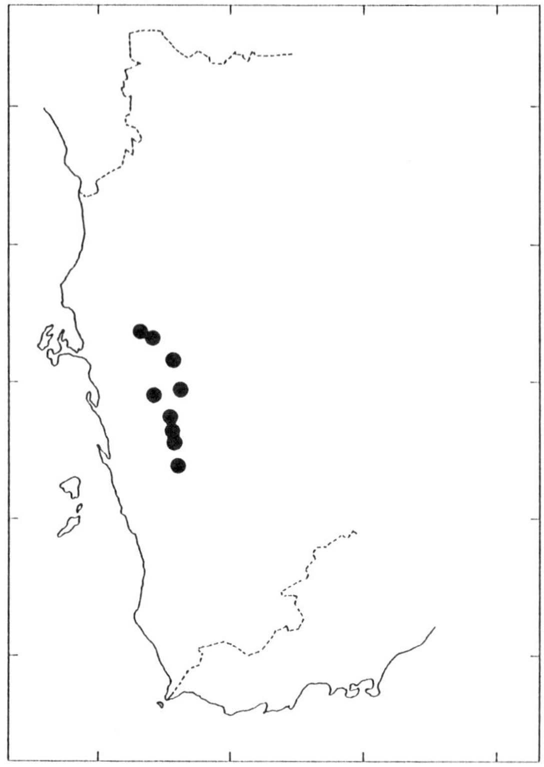

Map 30. Distribution of *Celtis toka*, a species restricted to the valley forest communities of the central escarpment.

Many shrubs are covered with the parasite *Cassytha filiformis* or one of the frequent climbers which include *Abrus bottae*, *A. precatorius*, *Caucanthus edulis*, *Cissus rotundifolia*, *Clitoria ternatea*, *Coccinea grandis*, *Cocculus hirsutus*, *Ipomoea obscura*, *Jasminum fluminense* and *Rhoicissus revoilii*. The scrambler *Commicarpus plumbagineus* is also abundant.

The ground flora is not strikingly rich and is mostly composed of widespread species such as *Indigofera spinosa*, *Pupalia lappacea* and *Ruellia patula*, all of which grow in open places. There are, however, a number of species frequent in the valley forests which are mostly uncommon elsewhere. They include *Actinopteris radiata*, *Aneilema forskalii*, *Blepharis maderaspatensis*, *Commelina erecta*, *Crossandra johanninae*, *Desmodium gangeticum*, *Enteropogon macrostachyus*, *Evolvulus alsinoides*, *Indigofera tinctoria*, *Lepidagathis scariosa*, *Melhania velutina*, *Plectranthus arabicus*, *Selaginella imbricata*, *Tephrosia pentaphylla*, *T. pumila* and the introduced *Tridax procumbens* which is often abundant in the wadi bed.

There are very few species found in the flood bed itself but around the occasional pool *Adiantum capillus-veneris*, *Cyperus alterniflorus*, *Kanahia laniflora* and perhaps *Pentodon pentandrus* may be found.

One of the most remarkable features of the valley forests is the large number of rare relic species restricted to them. The list is quite long and includes *Antiaris toxicaria*, *Bridelia scleroneura*, *Endostemon gracilis*, *Mucuna pruriens*, *Piliostigma thonningii* as well as others which appear in the main text.

(2:18) *Degenerate scrub on eroded slopes of the central escarpment between 1000 and 2000m*

At about 1000 m the valley forest communities come to an end. The valleys usually open out, the *Acacia-Commiphora* bushland disappears and terraced fields and villages can be seen. Natural vegetation is restricted to steep slopes or long-abandoned terraces. In such situations in the central escarpment there develops an open bushland 1–2m high which extends from about 1000 to 2000m. Characteristic features of this bushland are the absence of trees including almost all species of *Acacia* and *Commiphora* and a high degree of erosion since almost all topsoil is washed away by heavy rain.

Low shrubs are the predominant plants but no one species establishes more than a local dominance and that is unusual. Amongst the main species that can be found are *Acokanthera schimperi*, *Barbeya oleoides*, *Barleria bispinosa*, *B. prionitis*, *Cadia purpurea*, *Canthium schimperianum*, *Carissa edulis*, *Dichrostachys cinerea*, *Dodonaea viscosa*, *Ehretia obtusifolia*, *Euclea racemosa*, *Grewia tembensis*, *Kleinia odora*, *Lantana viburnoides*, *Maytenus parviflora*, *Pavetta longiflora*, *Rhus natalensis*, *Rumex nervosus*, *Segeretia thea*, *Tarconanthus camphoratus* and *Teclea nobilis*. There is not a very clear difference between the shrubs and the ground flora since the latter are mostly wiry perennials or tough succulents and include *Adiantum incisum*, *Aerva lanata*, *Arthraxon prionodes*, *Ceropegia arabica*, *Clutia richardiana*, *Conyza incana*, *C. pyrrhopappa*, *Cynoglossum lanceolatum*, *Ecbolium gymnostachyum*, *Eustachys paspaloides*, *Gomphocarpus fruticosus*, *Hyparrhenia hirta*, *Indigofera trita*, *Kalanchoe glaucescens*, *Kickxia scalarum*, *Launea massauensis*, *Leptochloa rupestris*, *Plectranthus cylindraceus*, *P. hadiensis*, *Polygala tinctoria*, *Rhynchosia totta*, *Ruellia patula* and *Selaginella yemensis*. Climbers are not particularly common but *Asparagus africanus*, *Clematis hirsuta*, *Cyphostemma digitatum* and *Jasminum fluminense* all occur.

(2:19) *Terraced cultivation on exposed hillsides in the central escarpment (1000–1800m)*

The terrace system begins at about 1000m and where this has survived a distinct vegetation has grown up on the surrounding land, in the fields themselves and on the retaining walls.

Scattered tall trees are often present, usually *Breonadia salicina*, *Cordia africana* or *Terminalia brownii*. Shrubs are infrequent and mostly found on terrace walls and by tracks. Any of those listed in 2:18 may occur but the most common are *Annona squamosa*, *Dichrostachys cinerea*, *Ehretia obtusifolia*, *Flueggia virosa*, *Segeretia thea* and *Vernonia abyssinica*. Open grassy areas are also quite frequent on steep slopes around the terraces and these are probably artificially maintained as a source of animal fodder. Common grasses are *Arthraxon prionodes*, *Bothriochloa insculpta*, *Cymbopogon caesius*, *Heteropogon contortus*, *Hyparrhenia hirta* and *Sehima nervosa*. Few associated species are found in this type of grassland but legumes such as *Indigofera oblongifolia* and *Rhynchosia minima* may occur.

The retaining walls of the terraces have quite a rich flora even at these low altitudes. Common plants are *Adiantum incisum*, *Arthraxon prionodes*, *Crepis rueppellii*, *Micromeria imbricata*, *Oxalis corniculata* and *Selaginella yemensis*.

Most terraces are poor and stony, bearing some resemblance to those found in Taiz province (2:8). Many are being abandoned and are in the process of reversion into the type of scrub described in 2:18. In this situation the division between the weeds found amongst crops and those found on the margins is blurred and they are not fully distinguished here. Much the most characteristic species is *Rhynchelytrum repens* which is often dominant on the margins and colonises the looser soil amongst the crops. Other typical species are *Abutilon bidentatum*, *Bidens biternata*, *Boerhavia diffusa*, *Cassia fallacina* (mostly on abandoned terraces), *Chrysanthellum americanum*, *Crotalaria pycnostachya*, *Heliotropium zeylanicum*, *Melhania incana*, *Portulaca quadrifida*, *Reichardia tingitana*, *Seddera arabica* (margins), *Sida ovata* and *Tridax procumbens*. The introduced species *Malvastrum coromandelium*, *Sida cordifolia* and *Waltheria indica*, though local, are very characteristic of field borders and disturbed ground in this area.

(2.20) *The coffee terraces of the central escarpment* (1000–1800m)

On many of the west-facing slopes of the central escarpment a different type of terraced cultivation is found to that described in 2:19. On J. Bura', J. Sāfān, in the Ja'farīyah region of J. Raymah and in narrow, shaded valleys elsewhere in the central escarpment between 1000 and 1800m, coffee is extensively grown. The coffee country is usually partially shaded from its aspect and enjoys prolonged periods of mist. Although shade trees are commonly planted around coffee in less favoured areas this is not a very common practice in the central escarpment though *Breonadia salicina* and *Cordia abyssinica* are fairly frequent and *Michelia champaca* is only known here.

The coffee bushes themselves provide additional shade and the weed flora that grows on the coffee terraces is comprised of species that are at least partially tolerant of shade. Common species are *Acalypha indica*, *A. lanceolata*, *A. racemosa*, *Amaranthus spinosus*, *Cassia occidentalis*, *C. sophera*, *Celosia digyna*, *Commelina benghalensis*, *Digitaria velutina*, *Echinochloa colona*, *Eragrostis macilenta*, *Geranium biuncinatum*, *G. trilophum*, *Laportea aestuans*, *Oxalis corniculata*, *Phyllanthus tenellus*, *Portulaca oleracea*, *P. quadrifida*, *Sida alba* and *Sisymbrium erysimoides*. A few species are only known in this habitat but all of them are rare. They are *Cardamine africana*, *Girardinia diversifolia* and the endemic *Phyllanthus hodjelensis*.

The retaining walls of the terraces are also rich in shade-loving species. Ferns are particularly frequent. *Actinopteris semi-flabellata*, *Adiantum incisum* and *Hypodematium crenatum* are common everywhere. *Ceterach dalhousiae* and *Arthropteris* orientalis are locally frequent while *Christella dentata* is often found by ditches where it is associated with *Setaria plicatilis*. Other characteristic species found on terrace walls are *Cyanotis nyctitropa*, *Dorstenia foetida*, *Mariscus schimperi*, *Parietaria debilis*, *Plectranthus hadiensis*, *Pouzolzia parasitica* and *Scrophularia arguta*.

(2.21) *Shaded gullies at 1600–2000m in the central escarpment*
The upper limit for coffee in the central escarpment is about 1800m and the end of the coffee country seems often to be marked by ridges at about 2000m. The ridges are broken by narrow defiles and gullies or lined with cliffs. In places where these face north or west, sunlight is reduced to an hour or two each day and frequent mists keep the vegetation damp throughout the year. A very specialised vegetation is found in such gullies and is best developed below Jirjir in the Ḥadīyah region and around Ṣa'bah on J. Milḥān but very similar conditions can be observed around Mirzin on J. Bura', near Bayt Bashārī and elsewhere around Maḥwīt and on the escarpment around Maḥābishah.

The cliffs lining the gullies are sometimes almost devoid of vegetation but in other places are quite rich. All the characteristic species are at least partially succulent and include *Aeonium sp. A*, *Aloe pendens*, *A. yemenica*, *Euphorbia ammak*, *E. fruticosa*, *Rumex nervosus* and *Saponaria umbricola*.

In the gullies themselves shrubs are not particularly numerous but odd plants of *Acanthus arboreus*, *Canthium schimperianum*, *Carissa edulis*, *Celtis africana*, *Erythrococca abyssinica*, *Myrsine africana*, *Otostegia fruticosa*, *Pavetta villosa*, *Phyllanthus ovalifolius*, *Rosa abyssinica*, *Spiniluma oxyacantha* and *Turraea holstii* occur. In the gully north of Ṣa'bah, the *Celtis* and *Erythroccoa* are particularly abundant and are the main species forming a small area of woodland at the bottom of the gully.

Scrambling and climbing plants are frequent. Various species of *Rubus* are sometimes dominant. In the Ṣa'bah gully *Pupalia grandiflora* is in great abundance. Other typical scramblers and twiners are *Clematis hirsuta*, *Cyphostemma digitatum*, *Ipomoea ficifolia*, *Rhoicissus tridentata*, *Senecio bojeri*, *S. hadiensis*, *Solanum terminale*, *Stephania abyssinica*, *Vigna vexillata* and *Zehneria scabra*.

Ferns are the other characteristic group found in the gullies. Several fairly frequent species occur like *Asplenium aethiopicum*, *A. trichomanes*, *Pteris cretica* and *P. serrulata*, but a much larger group is entirely restricted to this habitat and includes *Asplenium protensum*, *A. rutifolium*, *A. varians*, *Loxogramme lanceolata*, *Pteris catoptera* and *Tectaria gemmifera*.

Many other species favour the damp humus-rich floors of the gullies and the following are frequent in this habitat: *Achyranthes aspera*, *Bidens pilosa*, *Campanula edulis*, *Crepis rueppellii*, *Desmodium repandum*, *Dichrocephala chrysanthemifolia*, *D. integrifolia*, *Droguetia iners*, *Galinsoga parviflora*, *Galium aparinoides*, *Geranium arabicum*, *Hybanthus enneaspermus*, *Hypoestes triflora*, *Justicia caerulea*, *Oplismenus campositus*, *Oxalis corniculata*, *Pelargonium alchemilloides*, *Plectranthus hadiensis*, *P. lanuginosus*, *Poa schimperiana*, *Pouzolzia parasitica*, *Scadoxus multiflorus* and *Torilis arvensis*.

(2:22) *Montane grassland on J. Raymah (1500–2500m)*
A traveller going from Ḥadīyah to Kusmah in J. Raymah would pass through several zones. He would be in valley forest below Ḥadīyah (2:17) and thence would pass up the valley with secondary scrub (2:18) and exposed terraced cultivation (2:19) on either side before entering the coffee region (2:20) at the head of the valley. Next he would climb out of the coffee region through the gully below Jirjir (2:21) and then enter another remarkable region – the montane grasslands. These grasslands extend along the northwest slopes of the mountains between Ḥadīyah and Kusmah from about 1500m at Bani Sa'īd to 2950m on J. Bārad and are virtually restricted to this part of J. Raymah.

The grasslands are almost certainly not the natural climax but are preserved as fodder for domestic animals. They do not seem to be much grazed but the vegetation is plucked by the women who store it in their houses for later use. The hillsides are mostly very steep and the top soil is shallow. The grassland is thus extremely vulnerable to any change of agricultural practice and would deteriorate fast, for example, through overgrazing.

The main areas of grassland lie between 2000 and 2500m and those around Jirjir can be taken as reasonably typical. The sward is made up of many grass species of which the following are the most common: *Andropogon distachyos*, *Aristida ferrilateris*, *Arthraxon prionodes*, *Brachypodium retusum*, *Eragrostis papposa*, *Hyparrhenia hirta*, *Sporobolus pyramidalis* and *Themeda triandra*. Other grasses though less important in forming the sward are characteristic of this community and include *Digitaria diagonalis*, *Eragrostis schweinfurthii*, *Eustachys paspaloides*, *Gastridium phleoides*, *Michrochloa kunthii*, *Poa schimperiana* and *Tripogon leptophyllus* whilst *Eragrostis longifolia* and *Hyparrhenia formosa* are unknown elsewhere. The sedges *Fimbristylis ovata* and *Carex brunnea* are also frequent while several other sedges are restricted to this habitat.

The grasslands are very rich in herbs and many species are present including *Agrocharis melananthera*, *Aloe vacillans*, *Anthospermum herbaceum*, *Argyrolobium rupestre*, *Asplenium adiantum-nigrum*, *Astragalus atropilosulus*, *Bellardia trixago*, *Bonatea steudneri*, *Campanula edulis*, *Cheilanthes farinosa*, *Commelina africana*, *Crassula alba*, *Crepis rueppellii*, *Cycniopsis humifusa*, *Dichondra repens*, *Eulophia streptopetala*, *Eriosema longipedunculatum*, *Euphorbia schimperianum*, *Geranium arabicum*, *Helichrysum forskahlii*, *H. nudifolium*, *Hybanthus enneaspermus*, *Hypoxis angustifolia*, *Kalanchoe alternans*, *Lactuca inermis*, *Linum volkensii*, *Macrotyloma axillare*, *Micromeria abyssinica*, *M. imbricata*, *Pelargonium alchemilloides*, *Pellaea quadripinnata*, *Pentas lanceolata*, *Phaulopsis imbricata*, *Picris scabra*, *Piloselloides hirsuta*, *Pimpinella menachensis*, *Plectranthus barbatus*, *Polygala abyssinica*, *Scabiosa columbaria*, *Scadoxus multiflorus*, *Scutellaria arabica*, *Selaginella yemensis*, *Senecio lyratus*, *Silene burchellii*, *Sonchus saudensis*, *Swertia polynectaria*, *Trifolium semipilosum* and *Vernonia leopoldii*.

Small rock outcrops are fairly frequent in the grasslands and *Adiantum poirettii*, *Aloe pendens* and *Plectranthus hadiensis* are largely restricted to their vicinity.

Trees and shrubs are not likely to be frequent in grassland but several species occur as scattered individuals, though never in quantity. These include *Acacia origena*, *Berberis holstii*, *Clematis simensis*, *Erica arborea*, *Heteromorpha arborescens*, *Jasminum grandiflorum*, *Maytenus parviflora*, *Myrica salicifolia*, *Rhamnus prinoides*, *Rhoicissus tridentata*, *Rosa abyssinica*, *Rubus spp.* and *Spiniluma oxyacantha*.

This unique area of grassland has no exact equivalent elsewhere in Yemen. Probably the most similar area is on the western slopes of J. 'Ayfū in the Mudhaykhirah region where several of the distinctive species such as *Eriosema* and *Spiniluma* also occur. It would not be reasonable though to expect such a huge area of grassland to be uniform and variations do occur and some of these have equivalents elsewhere in Yemen.

In shallow stony grassland a distinctive flora comprising *Dyschoriste radicans*, *Cyperus obtusiflorus*, *C. rubicundus* and *Tephrosia elata* can be found and this has equivalents elsewhere in the central escarpment and in the Ibb region.

At lower altitudes in the Bani Sa'īd area the grassland is thinner and obviously merging into the type found on abandoned terraces (2:19). Characteristic species at 1500–1800m are *Argyrolobium arabicum*, *Buchnera hispida*, *Chlorophytum tordense*, *Crotalaria natalitia*, *Heteropogon contortus*, *Justicia flava*, *Sporobolus angustifolius*, *S. festivus* and *Triumfetta flavescens*. Communities with these species are found on J. Bura' and J. Milḥān and rather similar ones are found in the Ibb area on the lower slopes of J. Sumārah and J. Ba'dān.

(2:23) *High-altitude grassland on the southern escarpment mountains (2500–3200m)*

The higher grasslands above 2500m are exemplified by those found around the high peaks of the Ibb region such as J. Ta'kar. Many species which are common in the grasslands of J. Raymah as low as 2000m are restricted in the Ibb region to peaks above about 2700m. The probable explanation for this discrepancy is the influence of mist which is extremely unusual below 2700m in the Ibb area but abundant at quite low altitudes in J. Raymah. Among the species affected in this way are the ferns *Adiantum poirettii* and *Pteris cretica*, the grass *Eragrostis schweinfurthii*, the shrubs *Erica arborea* and *Rhamnus prinoides* and numerous herbs notably *Helichrysum forskahlii*, *H. nudifolium* and *Anthospermum herbaceum*.

Several species are present at Kusmah and on the Ibb mountains which serve to distinguish this higher grassland. The shrub *Hypericum revolutum* is the most conspicuous. Orchids are also very characteristic and some like *Habenaria lefebureana* and *Satyrium brachypetalum* can occur in local abundance while others like *Habenaria schimperiana* occur as scattered individuals. *Cheilanthes marantae*, *Cyanotis vaga*, *Dianthus uniflorus*, *Merendera abyssinica*, *Oreoschimperella arabiae-felicis*, *Romulea fischeri*, *Sporobolus africanus*, *Striga yemenica* and *Thesium humifusum* are examples of grassland species occurring both in the Ibb region and on J. Raymah.

(2:24) *Shaded gullies at high altitudes on the escarpment mountains (2800–3100m)*

Another feature which is common to both the Ibb mountains and Kusmah is the existence of damp shaded gullies near the mountain summits around 2800–3100m. These gullies are a highland version of those described in 2:21 and like them experience prolonged periods of mist and few hours of sunshine each day. They are found on J. Ṣabir, on almost all the Ibb mountains, at Kusmah and at Shirnamah on J. Maswar.

The most characteristic highland gully species is *Arisaema bottae* which sometimes fills the gully bottoms excluding almost all other species. Often associated with it are the lilies *Crinum album*, *Kniphofia sumarae* and *Scadoxus multiflorus*. Several other characteristic species are temperate, notably *Anagallis foemina*, *Geranium robertianum* and *Anthriscus sylvestris*, the latter sometimes as abundant as the *Arisaema*. Ferns are common and *Cystopteris fragilis*, *Polystichum fuscopaleacum* and *Pleopeltis macrocarpa* are virtually restricted to these gullies. Other plants commonly found are *Achyranthes aspera*, *Alectra sessiliflora*, *Arisaema flava*, *Bromus leptoclados* and *Geranium arabicum*, while the steep rocky slopes above often carry *Albuca abyssinica*, *Aloe rubroviolacea*, *Arabis alpina*, *Minuartia filifolia*, *Rosa abyssinica* and *Umbilicus intermedius*.

(2:25) *Marshland in high-altitude grassland (2500–3000m)*

Small springs are fairly frequent in the high-altitude grassland described in 2:23 and have a distinctive flora. This shows little variation and the following ample from near Kusmah is quite typical: *Agrostis viridis*, *Alchemilla cryptantha*, *Dolichos trilobus*, *Epilobium hirsutum*, *Pennisetum macruorum*, *P. thunbergii*, *Pycreus elegantulus*, *Ranunculus multifidus*, *Sanicula elata* and *Veronica anagallis-aquatica*.

(2:26) *Terraced cultivation in the highlands (2000–3000m)*

Most of the escarpment between 2000 and 3000m is not put down to grassland but is heavily cultivated. The terrace system is very extensive and covers whole hillsides in many areas. At this altitude rainfall is almost always sufficient to grow qat, Sorghum or one of the other cereals or pulses. Beneath the crops a large weed flora is found. The commonest species are probably *Amaranthus graecizans*, *Anagallis arvensis*, *Chenopodium murale*, *C. schraderianum*, *Convolvulus arvensis*, *Conyza stricta*, *Datura stramonium*, *Eragrostis cilianensis*, *Erucastrum arabicum*, *Misopates orontium* and *Solanum villosum*. At very high altitudes (2800–3100m) the species are more obviously temperate in origin and *Bromus pectinatus*, *Calendula arvensis*, *Euphorbia helioscopia*, *Urtica urens*, *Veronica persica* and *V. polita* are quite common.

Field margins are usually grassy but bare patches may be colonised by any of the plants above. Species which are characteristic of field margins but do not usually occur as weeds include *Andropogon greenwayi*, *Convolvulus siculus*, *Cynoglossum lanceolatum*, *C. yemenense*, *Digitaria abyssinica*, *Eleusine floccifolia*, *Euphorbia inaequilatera*, *E. schimperiana*, *Medicago lupulina*, *Pennisetum villosum* and *Reichardia tingitana*.

The terrace walls have a very distinctive flora and provide a refuge for many of the species found in the grassland areas (2:22). Many plants occur here casually but the following are probably the most frequent: *Arthraxon prionodes*, *Asplenium aethiopicum*, *Campanula edulis*, *Ceterach officinarum*, *Cheilanthes pteridioides*, *Commelina africana*, *Conyza hochstetteri*, *Crassula schimperi*, *Crepis rueppellii*, *Hypodematium crenatum*, *Hypoestes forskalei*, *Micromeria imbricata*, *Minuartia filifolia*, *Nepeta deflersiana*, *Osteospermum vaillantii*, *Oxalis corniculata*, *Pentas lanceolata*, *Phagnalon harazianum*, *P. scalarum*, *Reichardia tingitana*, *Rumex nervosus*, *Silene burchellii*, *Solanum villosum*, *Sonchus saudensis* and *Verbascum bottae*.

(2:27) *Rough scrub on steep eroded slopes in the highlands (2000–2900m)*
On steep slopes cultivation is not possible and without the protection of the terraces, the topsoil is washed away. Heavily eroded slopes of this type are common around 2500m and can be seen on J. Sumārah, Najd al Aḥmar and at Manākhah as well as in more remote areas.

Few trees can get established on these slopes but odd plants of *Acacia origena*, *Buddleja polystachya*, *Ficus palmata*, *Heteromorpha arborescens* and *Rhus retinorrhaea* occur. Much more typical are coarse shrubby herbs about 1m in height which often densely cover the slopes. Typical colonists are *Acanthus arboreus*, *Achyranthes aspera*, *Achyrocline schimperi*, *Annarhinum forskahlii*, *Bidens biternata*, *Cineraria abyssinica*, *Cleome schweinfurthii*, *Clutia spp.*, *Commicarpus arabicus*, *Conyza incana*, *C. pyrrhopappa*, *Echinops spinosissimus*, *Euphorbia schimperiana*, *Helichrysum foetidum*, *Hyparrhenia hirta*, *Hypoestes forskalei*, *Nepeta deflersiana*, *Osteospermum vaillantii*, *Plectranthus barbatus*, *Pennisetum setaceum*, *Pulicaria petiolaris*, *Rumex nervosus*, *Senecio hadiensis*, *S. schimperi*, *Solanum incanum*, *S. platacanthum*, *Themeda triandra*, *Vernonia leopoldi* and *Withania somnifera*. Small herbs are mostly uncommon but *Micromeria imbricata*, *Minuartia filifolia* and *Vermifrux abyssinica* are quite frequent.

(2:28) *Damp flushes on cliffs at high altitudes (2000–3000m)*
On cliffs and on the steep slopes described in 2:27 small flushes are quite frequent. *Rhus retinorrhaea* and *Centaurothamnus maximus* are often common in the vicinity but the distinctive plants are mostly found growing on the permanently wet rock slopes. *Primula verticillata* is the most conspicuous and characteristic species and it is often associated with *Adiantum capillus-veneris*, *Agrostis viridis*, *Conyza sp. A*, *Pteris serrulata* and *P. vittata*. On sandstone at Turbah and Aḥjar *Epipactis veratrifolia* and *Schoenus nigricans* are common while in some places *Argyrolobium confertum*, *Blumea bovei* and *Samolus valerandi* occur.

(2:29) *Disturbed habitats around highland villages (2300–2900m)*
A rather distinctive vegetation develops in mountain villages, particularly those situated around 2500m along the inner escarpment such as Ḍawrān, Manākhah, Shibām, Ṭawīlah and Kuḥlān. The exact reason for this distinctive vegetation is not clear but it is probably connected to the

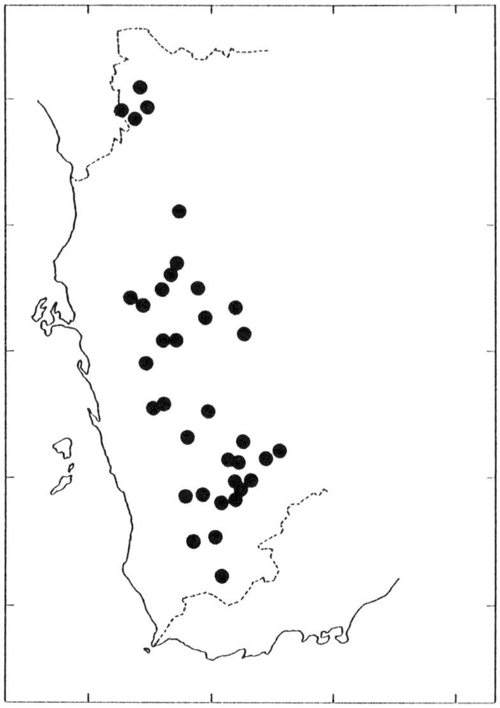

Map 31. Distribution of *Acacia origena*, showing the probable extent of prehistoric highland forest.

frequent disturbance of the habitat by human activity and the high nitrogen level related to the presence of raw sewage.

Most of the vegetation is in fact composed of species more characteristic of eroded hill slopes (such as *Commicarpus arabicus* or *Senecio schimperi*), of terrace walls (*Parietaria debilis*) or of cultivated land (*Solanum villosum*). Many of the plants, however, are special to villages. Some have become established as the direct result of human acitivity, notably *Opuntia ficus-indica*. Others are weeds which though occurring in fields are more characteristic of waste ground in villages. Amongst these are *Amaranthus hybridus*, *Erodium cicutarium*, *E. malacoides*, *Malva parviflora*, *M. verticillata*, *Polycarpon tetraphyllum*, *Sisymbrium irio*, *Urtica urens* and *Xanthium spinosum*. The most interesting group, however, consists of those species mostly and in a few cases entirely restricted to waste ground and walls in villages. These species are *Dichrocephala chrysanthemifolia*, *Hyoscyamus albus*, *Kedrostis foetidissima*, *Laggera crispata*, *Lepidium armoracea*, *Marrubium vulgare*, *Pavonia kraussiana*, *Pollichia campestris*, *Senecio harazianus*, *Silene yemensis* and *Sisymbrium orientale*.

(2:30) *Acacia origena highland forest (2000–3000m)*
All the best land between 2000 and 3000m is cultivated and the vegetation types hitherto described in this altitude range are nearly all at least partially artificial. It is interesting to speculate, however, about the nature of the natural vegetation before the arrival of man. It seems probable that

the annual rainfall of about 400–800mm would have allowed the development of extensive areas of woodland, though none of this survives today except for some small fragments.

A few relics found between 2300 and 2500m in the Ḥarāz near Bayt Mughlad on J. Shibām and Bayt Shamrān on J. Masār may provide some clues as to the nature of the prehistoric woodland. Much the most common tree is *Acacia origena* which forms a rather open woodland about 5–6m high. Other trees of a similar height associated with it are *Cordia africana* (rare, perhaps not typical), *Cussonia holstii*, *Dombeya schimperiana*, *Ehretia cymosa* (rare, perhaps not typical), *Euphorbia ammak*, *Grewia mollis*, *Olea europaea* (frequent) and *Pittosporum viridifolium*. The only common shrubs are *Carissa edulis* (about 2m high), *Indigofera arrecta* and *Myrsine africana* (about 50cm high) but wherever there are rock outcrops a much more varied shrub layer is evident, which includes *Acanthus arboreus*, *Heteromorpha arborescens*, *Maytenus parviflora*, *Pavetta longiflora*, *Rhoicissus tridentata*, *Rhus retinorrhaea*, *Senecio hadiensis* and *Vernonia leopoldii*. The ground flora is very species poor, probably because of shade, the ground being covered with thin grassland composed of *Andropogon greenwayi*, *Hyparrhenia hirta*, *Pennisetum villosum* and *Themeda triandra*. Amongst these can be found the occasional flowering plant, notably *Justicia flava* and *Plectranthus barbatus*.

No comparable relics are known elsewhere but it is possible that *Buddleja polystachya* (on rocky slopes), *Celtis africana* and *Rosa abyssinica* were also important in this sort of woodland as they survive in many areas at this altitude, notably in Wadi Aḥjar. In the south the woodland may well have been richer in species. Certainly considerable quantities of *Debregeasia saenab*, *Ficus sur*, *Hypericum revolutum*, *Myrica salicifolia* and *Nuxia congesta* can be found on J. Ba'dān at the same sort of altitude and as these occur sporadically on all the main mountains from J. Ṣabir north to Kusmah it seems probable that they were important constituents in prehistoric *Acacia origena* woodland in this area.

(2:31) *Juniperus procera highland forest (2000–2800m)*

The *Acacia origena* woodland was not the only highland forest in prehistoric times. In the north much of Khawlan Ash Sham and in the southeast, the Ḥujarīyah and the Sawraq-Ḥushā range, were probably covered in *Juniperus procera* forest above about 2000m. Even though both these areas receive frequent mist, they appear to have a lower annual rainfall than is usual in other areas of the escarpment at similar altitudes and so the terrace system is not well developed.

In neither area are there well-developed Juniper woodlands as still exist in Saudi Arabia (Wissmann 1972). Instead the Juniper occurs as scattered individuals less than 5m high, though locally growing in groups particularly in remote areas. Associated trees are rather few. At high altitudes *Olea europaea* and *Acacia origena* are quite frequent while below about 2300m they are joined by *Acacia etbaica* and *Ficus salicifolia*. *Dracaena serrulata* is locally common in Khawlan Ash Sham, notably on J. Minmār.

Shrubs and large woody herbs are frequent and are in fact the most characteristic species in this formation. The shrub layer is mainly ½–1m in height and the only common shrub taller than this is *Dodonaea viscosa* which sometimes forms large areas of bushland. Common low shrubs or woody herbs are *Aerva lanata*, *A. javanica*, *Cadia purpurea*, *Clutia lanceolata*, *Euphorbia sp. aff. fruticosa*, *Euryops arabicus*, *Kleinia odora*, *Lantana viburnoides*, *Lavandula pubescens*, *Otostegia fruticosa*, *Psiadia punctulata*, *Rhamnus staddo*, *Segeretia thea*, *Solanum schimperianum* and *Tarconanthus camphoratus*.

The ground flora usually consists of tough perennials interspersed with tufts of grass. The number of grass species found in this type of vegetation is considerable but the following are especially common: *Aristida adscensionis*, *Chrysopogon plumulosus*, *Enneapogon schimperianum*, *Eragrostis papposa*, *Heteropogon contortus*, *Hyparrhenia hirta*, *Tetrapogon villosus* and *Themeda triandra*. Typical low perennials are *Achryocline glumaceum*, *Becium serpyllifolium*, *Commicarpus grandiflorus*, *C. sinuatus*, *Cucumis prophetarum*, *Hypoestes forskalei*, *Justicia flava*, *Osteospermum vaillantii*, *Periploca visciformis* (on rocks), *Pulicaria petiolaris* and *Teucrium yemense*.

There are some differences between the Juniper communities in Khawlan Ash Sham and those in the southeast. A number of species more typical of the high plateau such as *Stipagrostis obtusa* are quite common in Khawlan Ash Sham and the scrambling shrubs *Asparagus africanus* and *Ochradenus baccatus* are noticeably more frequent there. In addition species with a northern distribution such as *Ephedra pachyclada*, *Euphorbia sp. aff. schimperi*, *Lavandula dentata*, *L. citriodora* and *Plectranthus asirensis* are entirely absent from the southeast though frequent in Khawlan Ash Sham. In contrast some species such as *Pavetta longiflora* are more frequent in the southeast while others such as *Crotalaria squamigera* and *Wendlandia arabica* (in gullies) are entirely restricted to this region.

The Juniper communities are clearly closely related to those found on J. Lawz (3:1) which are really only an isolated high altitude version. They bear an even stronger resemblance however to the Juniper woodlands found in the Asir region of Saudi Arabia. There are also relationships with the Juniper woodlands of Ethiopia and more surprisingly with those in Oman and Afghanistan (Mandeville 1977). One of the characteristic species of such communities in Oman is *Monotheca buxifolia*. This is very rare in Yemen but perhaps significantly its two known localities lie in Khawlan Ash Sham and in the Ḥujarīyah close to the Juniper communities.

(2:32) *The mountain summits (3000–3666m)*

The mountain summits above 3000m are cold, exposed and windswept and trees and shrubs cannot establish themselves. A few erect plants eke out an existence in the protection of boulders and cliffs but nearly all the characteristic species are prostrate or cushion-like plants such as *Polygala rupicola* and *Pterocephalus frutescens*.

The highest point in Arabia is J. Nabī Shu'ayb which reaches 3666m. The complete list of species occurring above 3500m is of some interest. If we ignore taxonomic difficulties, the plants fall into four distinct phytogeographical groups. The largest consists of species restricted to southwest Arabia, the mountains of Ethiopia and neighbouring regions of Kenya and comprises *Anthemis tigrensis*, *Artemisia abyssinica*, *Cineraria abyssinica*, *Conyza variegata*, *Lepidium armoracea*, *Minuartia filifolia*, *Salvia merjamie*, *Stipa tigrensis*, *Trifolium semipilosum* and *Vermifrux abyssinica*. The second largest group consists of species endemic to southwest Arabia: *Cichorium bottae*, ***Crepis sp. A, Delosperma harazianium, Dianthus uniflorus, Festuca cryptantha, Nepeta deflersiana, Phagnolon harazianum, Potentilla dentata, Scutellaria arabica*** and ***Thymus laevigatus***. The third group consists of ruderal species common in temperate areas: ***Anagallis foemina***, *Conium maculatum*, *Erodium cicutarium*, *Marrubium vulgare*, *Malva parviflora*, *Medicago lupulina*. The smallest group consists of miscellaneous widespread species: *Agrocharis melananthera*, *Dichrocephala chrysanthemifolia*, *Echinops spinosissimus*, *Eragrostis papposa*, *Helichrysum forskahlii* and *Poa leptoclada*.

A large proportion of the recorded species are dwarf prostrate or cushion plants. Particularly common at high altitudes is the endemic thyme, *Thymus laevigatus*, which is probably the most frequent species on bare stony hillsides above 3200m everywhere in Yemen. Another interesting group consists of *Cichorium bottae* and *Dianthus uniflorus* which together with the endemic *Convolvulus thymoides* often form extensive mats on earthy banks at high altitudes. Wherever the soil is broken, they are likely to be associated with *Filago prolifera*.

The number of ruderal species at high altitudes is also interesting, though their presence is partly explained by the existence of a shrine at the top of J. Nabi Shu'ayb. Species of temperate origins or those in genera centred on temperate regions such as *Thymus* and *Dianthus* are noticeably more common at high altitudes.

(2:33) *Aquatic habitats on the summit ridges (3000m)*
The high mountain ridges are often quite flat and pools are frequent in small hollows. They are most common on the Sumārah-Manār ridge at about 3000m but can also be found around Kawkabān and in the Nabī Shu'ayb area.

Most of the pools are quite small, rarely exceeding 4m in width but it seems that many are permanent or nearly so as they often support true aquatics like *Myriophyllum spicatum*, *Persicaria amphibia*, *Potamogeton spp.*, *Ranunculus rionii* and *Zannichellia palustris*, though none of these species are particularly common.

Several species are virtually restricted to the margins of these pools. The most characteristic is *Odontelytrum abyssinicum* which sometimes occurs in great abundance with the stems floating out over the water surface. Other species are *Anagallis serpens*, *Carex stenophylla*, *Cyperus rigidifolius*, *Eleocharis palustris*, *Mimulus gracilis* and *Schoenoplectus inclinatus*.

The most interesting plants, however, are those that appear on the dried-up mud of the pools as the waters recede at the end of the rainy season. *Crassula hedbergii*, *Gnaphalium unionis*, *Limosella africana*, *L. macrantha* and *Marsilea aegyptiaca* are restricted to this habitat in Yemen.

3. THE HIGH PLATEAU

This region is easily accessible and possibly the best-known botanically since it has a good road system and Sana'a and many of the larger towns are situated in it. Accounts of its vegetation are given by Hepper (1977) and Westinga (1980). The latter only covers the Rada' area but his vegetation types coincide more closely with mine and I have freely used his information to supplement my own in the following account and that of the Mashriq.

The high plateau is the smallest region and is generally quite uniform though the southern parts around Yarīm are wetter than elsewhere and the northeast region is mostly limestone with few wetland habitats. Some of the valleys southeast of Yarīm have a similar vegetation to that of the escarpment and are not considered further here. Parts of Khawlan Ash Sham, particularly the areas around Sāqayn, Jifān and Madaq are very similar to the high plateau though geographically isolated from it and their high mountain vegetation in particular is virtually identical to that outlined in 3:1 below.

One of the most salient characteristics of the high plateau is the absence of trees. Apart from the Juniper in 3:1 and odd patches of *Acacia gerrardii*, *A. origena*, *Tamarix* and *Zizyphus spina-christi*, there are almost no trees. It is very difficult therefore to reconstruct what the natural vegetation was before the arrival of man. It may well have been treeless except in protected valleys where the few relic trees observed probably flourished in greater numbers.

(3:1) *High montane vegetation (2500–3500m)*
Mountains above about 2500m carry a vegetation distinct from those below this altitude. The best-developed high montane vegetation is on J. Madhbūr in the J. Lawz range and this is described below but similar, though often depauperate versions can be found on many of the mountains surrounding the high plateau including the mountains around the Yislah pass, J. Nuqum and J. Aybān near Sana'a, J. Kanin, the mountains around the Qa' Bakīl including the eastern slopes of J. Ḍawrān and even the mountains in the immediate vicinity of Yarīm.

J. Madhbūr lies between about 2800 and 3300m and is remarkable for the large colony of *Juniperus procera* which is common here and forms open woodland in remote, inaccessible parts of the mountain. It is the only tree present and it seems reasonable to postulate that it was a good deal more frequent in prehistoric times on this and similar mountains. Associated with it are a few shrubs including *Berberis holstii*, *Dodonaea viscosa* (very rare), *Rhamnus staddo*, *Rosa abyssinica* and *Segeretia thea*. A surprising absence is *Acacia origena* but this is only found on the mountains on the west of the high plateau. The ground flora consists of scattered undershrubs and tough perennials with a good deal of bare rocky ground between them; huge rock slabs are characteristic of J. Lawz. Among the plants recorded from J. Madhbūr which I would regard as typical of this habitat are *Aloe tomentosa*, *Anagyris foetida* (rare), *Andropogon distachyos*, *Annarhinum forskahlii*, *Asparagus africanus*, *Astragalus atropilosulus* (rare), *Campylanthus pungens*, *Clutia lanceolata*, *Conyza incana*, *Delosperma harazianum* (usually on summit ridges),

Dianthus uniflorus, Echinops spinosissimus, Ephedra pachyclada, Euryops arabicus, Felicia abyssinica, F. dentata, Gomphocarpus sinaicus, Hyparrhenia hirta, Hypoestes forskalei, Kleinia semperviva, Lavandula pubescens, L. citriodora, Macowania ericifolia, Micromeria imbricata, Minuartia filifolia, Nepeta deflersiana, Osteospermum vaillantii, Psiadia punctulata, Pterocephalus frutescens, Pulicaria petiolaris, Scutellaria arabica, Silene schimperiana, Stachys yemenensis, Stipa tigrensis, Teucrium yemense, Thymus laevigatus, Vermifrux abyssinica. There are in fact a few species which are probably typical of these high mountains not recorded here because of the exceptionally high altitude; they include *Euphorbia sp. aff. schimperi, Meriandra benghalensis* and *Solanum schimperianum.*

This type of vegetation is obviously very similar to that outlined in 2:31 and is apparently only a high altitude version of it. Rather surprisingly there is no exact equivalent below 2500m on the high plateau.

(3:2) *High-altitude cliffs (2500–3500m)*

Cliffs above 2500m have a rather specialised flora and provide a refuge for many plants otherwise restricted to the escarpment. These include *Centaurothamnus maximus, Ceterach officinarum, Cheilanthes coriacea, Clematis simensis, Crassula schimperi, Phagnalon harazianum, Rumex nervosus* and many others. Two species, *Tylophoropsis heterophylla* and *Zehneria scabra* are particularly characteristic, being more common in this habitat than anywhere else in Yemen.

(3:3) *Hills, rock outcrops and stony plain between 2000 and 2500m*

Hills lying between 2000 and 2500m carry a very different flora to that at higher altitudes. The difference is not easy to account for fully. Part of the explanation lies in altitude as many of the species found below 2500m do not flourish at higher altitudes. Rainfall is also a factor since many of the lower mountains lie towards the east of the high plateau and their flora is obviously related to that outlined in the Mashriq under 4:1 below. Another factor is probably the gentler contours of these smaller hills so that run-off is less pronounced and erosion consequently less severe. Whether these explanations fully account for the differences in the vegetation is, however, doubtful.

The following list of species was composed from an old volcanic cone near Shamlān and from sandstone hills south of the Wadi Rijām in the Bani Ḥushaysh. It was expected that the vegetation would be very different but with the exception of a few species specified below, it was remarkably similar in the two localities.

Both localities were completely without trees or large shrubs and it is very difficult to speculate as to what the original vegetation was, though it was presumably richer than that found today, which is greatly modified by heavy grazing and consists of a low, rather thin ground cover. The following species were seen: *Achyrocline schimperi* (volcanic rocks), *Andropogon crossotas, Aerva javanica, Alkanna orientalis* (volcanic rocks), *Andrachne aspera* (volcanic rocks), *Annarhinum forskahlii, Argyrolobium arabicum, Blepharis ciliaris, serpyllifolium* (volcanic rocks), *Campylanthus pungens, Caralluma plicatiloba, C. quadrangula, Cenchrus ciliaris, Chascanum sessilifolium* (sandstone), *Cheilanthes catenensis, Chrysopogon plumulosus, Cleome schweinfurthii* (volcanic rocks), *Cometes abyssinica* (volcanic rocks), *Commelina albescens, Crotalaria emarginella, Cyperus obtusiflorus, Dipcadi viride, Echinops spinosissimus, Enneapogon desvauxii* (sandstone), *Eragrostis papposa, Fagonia spp., Farsetia longisiliqua, Felicia abyssinica, F. dentata, Fingerhuthia africana* (sandstone), *Forsskalea tenacissima* (volcanic rock), *Gnidia somaliensis, Helichrysum pumilum, Heliotropium stendneri, Indigofera arabica, I. phillipsae, Kleinia odora, Lavandula pubescens, L. stricta* (volcanic rock), *Leucas inflata, Lotus quinatus, Melhania stipulosa, Micromeria imbricata, Notoceras bicorne* (sandstone), *Osteospermum vaillantii, Polygala erioptera* (sandstone), *Portulaca quadrifida, Pulicaria jaubertii* (volcanic rock), *Salvia aegyptiaca, S. schimperi, Seddera secundiflora* (sandstone), *Silene schimperiana, Stipagrostis ciliata, S. hirtigluma, S. obtusa, Tetrapogon villosum, Teucrium yemense, Trichodesma africana* (volcanic rocks), *Volutaria abyssinica.*

This type of vegetation can be found in many parts of the high plateau including all the low hills around Sana'a and the Wadi Ḍahr, in the western Bani Ḥushaysh, much of Ḥamdān, on the hills around 'Amrān and to the south of Sana'a in Sanḥān, parts of Al Ḥadā', around Rada' and near Riṣābah but not as far as I know in the Yarīm area.

(3:4) *Cliffs between 2000 and 2500m*

As noted in 3:2 above, cliffs provide a refuge for plants from other, more favoured regions. The cliff slopes below 2500m also carry a distinct vegetation even though all the plants found on open hills or rocky plain (3:3) may occur there sometimes, even in abundance. The following species, however, are particularly characteristic of cliff slopes and it is interesting that a large proportion are shrubs, which clearly find a refuge in such inaccessible places from the heavy grazing of the surrounding areas: *Acanthus arboreus, Aloe vacillans, Asparagus africanus, Commicarpus sinuatus, Digitaria nodosa, Hyparrhenia hirta, Meriandra benghalensis, Ochradenus baccatus, Otostegia fruticosa, Pavetta longiflora, Pennisetum setaceum, P. stramineum, Polygala abyssinica, Rosa abyssinica* and *Sarcostemma vanlessenii.*

(3:5) *The montane plains (2200–2500m)*

The lower hills and rock outcrops merge into the flat intermontane plains with their alluvial or loess soils. Most of the plains lie at around 2300m and have a distinct though rather poor flora. The plains are heavily grazed and largely treeless, though occasional groups of *Acacia gerrardii* can be found and it is possible they were covered in an open bushland composed of this tree and the odd *Tamarix* and *Zizyphus spina-christi* in prehistoric times.

The plains are heavily cultivated but considerable areas are left for rough grazing and the main species to be found are *Aristida adscensionis* (often dominant over considerable areas after the rains), *Boerhavia diffusa, Cassia italica, Caylusea hexagyna, Cenchrus ciliaris, Centaurea pseudosinaica, Commicarpus pedunculosus, C. sinuatus, Cynodon dactylon, Echinops spinosissimus, Euphorbia schimperiana* (very abundant), *Fagonia indica*, *F. schweinfurthii* (both abundant), *Gomphocarpus sinaicus*, *Gnidia somaliensis*, *Launea nudicaulis*, *Leucus inflata*, *Lycium shawei*, *Peganum harmala* (abundant), *Picris sp.A.*, *Pulicaria crispa*, *Robbairea delileana*, *Solanum glabratum*, *S. incanum* and *Withania somnifera*.

The plains around Yarīm, notably the Qa' al Ḥaql, differ from the rest of the high plateau in the absence of *Peganum harmala*, probably as a result of the greater rainfall in that area. In contrast the plains near Raydah from the Qa' al Bawn eastward have a lower rainfall and their vegetation is poorer, resembling that described in 4:4 below.

(3:6) *Sandy plain in the Bani Ḥushaysh*

One area of the intermontane plains is sufficiently different to warrant closer attention. This is the area at the west of the Bani Ḥushaysh and in the Bani al Ḥārith. The surrounding mountains are composed of sandstone and the plains are formed of light sandy soil which supports a vegetation unique in Yemen containing a number of plants absent or very rare elsewhere.

The plain is not uniform. Much of it is treeless but in other areas scattered *Acacia gerrardii* and *Zizyphus spina-christi* occur, giving it a parkland appearance. Some parts are extremely barren, notably the plain north of Ghirās where the flat, sandy surface is devoid of almost all vegetation except the grass, *Aristida congesta*. In other areas there is much underground water and a lot of grapes and qat are grown under irrigation. Some areas have a remarkably rich ground flora. In one small, apparently barren area of about 20 square metres south of the Wadi Rijām I was able to find all the following species: *Anastatica hierochuntica*, *Aristida adscensionis*, *A. congesta*, *A. pennei*, *Astragalus fatmensis*, *A. sieberi*, *A. tribuloides*, *Blepharis ciliaris*, *Boerhavia sp. A.*, *Calotropis procera*, *Caylusea hexagyna*, *Centaurea pseudosinaica*, *Chryzophora oblongifolia*, *Citrullus colocynthis*, *Cleome amblyocarpa*, *C. brachycarpa*, *Commicarpus sinuatus*, *C. helenae*, *Cynodon dactylon*, *Echium sericeum*, *Enneapogon desvauxii*, *Eragrastis barrelieri*, *E. papposa*, *Euphorbia inaequilatera*, *Fagonia brughieri*, *F. schweinfurthii*, *Farsetia longisiliqua*, *Indigofera phillipsae*, *Laggera decurrens*, *Launea capitata*, *L. nudicaulis*, *Lotononis platycarpos*, *Lotus sp. A.*, *Morettia parviflora*, *Peganum harmala*, *Picris sp. A.*, *Polycarpaea repens*, *Polygala erioptera*, *Pulicaria crispa*, *P. jaubertii*, *Salsola bottae*, *Salvia aegyptiaca*, *Sporobolus ioclados*, *S. tourneuxii*, *Stipagrostis obtusa*, *Telephium sphaerospermum*, *Tribulus terrestris*.

(3:7) *Weeds of cultivated land (2000–3000m)*

Large parts of the high plateau are cultivated and a distinct weed flora has developed. The richness of the flora seems to depend mainly on the amount of water available and many weed species are largely restricted to the vicinity of ditches. Others, notably *Chenopodium spp.*, are much more abundant in areas irrigated with sewage water.

Typical of many fields in the high plateau is the weed flora recorded by Westinga from Rada' at 2100m: *Anchusa orientalis*, *Calendula arvensis*, *Cynodon dactylon*, *Datura stramonium*, *Erucastrum arabicum*, *Euphorbia peplus*, *Malva parviflora*, *Reichardia tingitana* and *Solanum villosum*.

At high altitudes toward the west, the number of species is likely to be greater. Near Shibām at 2500m, the following species were growing together in a field: *Anagallis arvensis*, *Cardaria draba*, *Chenopodium procerum*, *C. schraderianum*, *Convolvulus arvensis*, *Diplotaxis erucoides*, *Erodium cicutarium*, *E. malacoides*, *Euphorbia helioscopia*, *Geranium biuncinatum*, *Launea nudicaulis*, *Malva verticillata*, *Pseudognaphalium luteo-album*, *Senecio schimperi*, *Sisymbrium irio*, *Sonchus oleraceus*, *Tragus racemosus*, *Verbena officinalis* and *Xanthium spinosum*.

These two lists do not exhaust the weed flora known from the high plateau and many other species occur, including *Amaranthus hybridus*, *Avena abyssinica*, *Galium tricornutum*, *Heliotropium europaeum*, *Hibiscus trionum*, *Kickxia elatine*, *Medicago lupulina*, *Veronica persica* and *V. polita*.

(3:8) *Weeds of abandoned fields*

A distinctive vegetation develops in abandoned fields on the montane plains, which is to some extent intermediate between 3:5 and 3:7 but consists mainly of species which are not particularly characteristic of either. These species include *Artemisia abyssinica*, *Centaurea pseudosinaica*, *Citrullus colocynthis*, *Cleome amblyocarpa*, *Convolvulus arvensis*, *Conyza stricta*, *Cynodon dactylon*, *Euphorbia schimperiana*, *Heliotropium ramossissimum*, *Laggera decurrens*, *Launea nudicaulis*, *Onorpordum arabicum*, *Peganum harmala*, *Pulicaria crispa* and *Xanthium spinosum*. *Acacia gerrardii* is almost always the first tree to colonise.

(3:9) *Wetland habitats*

The high plateau is very rich in wetland habitats particularly south and west of Sana'a.

Most submerged and floating aquatic plants are restricted to man-made cisterns where typically one species, such as *Potamogeton pusillus* or *Myriophyllum spicatum*, is dominant to the exclusion of all others. Some of the ponds in the Yarīm area and in Bani Maṭar seem to be more or less permanent and several interesting species have been found here, notably *Utricularia australis*.

There are no large wadis at all on the high plateau but there are a few permanent streams, particularly around Baw'ān and near Ribāṭ in the Qa' al Ḥaql. The vegetation of these stream sides is not particularly interesting but typical species include *Myosotis laxa*, *Pennisetum macruorum* and *Ranunculus multifidus*.

Much more typical of the high plateau are ditches flowing into flat marshy grassland in which there are often small seasonal ponds. Examples of this type of habitat can be found all over the Yarīm, Dhamār and Al Ḥadā' regions and also in Bani Maṭar and near Shibām in the north. The vegetation is surprisingly uniform, though each marsh usually has one or two species which are rather unusual. Two samples are given below. The first is the marshy valley north of Wa'lān familiar to many picnickers from Sana'a and the second is a marshy area between Dhamār and Shirqabī. Both lie at about 2300m and both have the following species in common: *Apium nodiflorum*, *Carex stenophylla*, *Cirsium vulgare*, *Cyperus rigidifolius*, *Eleocharis palustris*, *Equisetum ramosissimum*, *Juncus inflexus*, *J. punctorius*, *J. socotranus*, *Lotus corniculatus*, *Plantago lanceolata* and *Potentilla reptans*. Peculiar to the Wa'lān marsh were *Agrostis viridis*, *Berula erecta* (open water), *Cyperus laevigatus*, *Cynodon dactylon*, *Eleusine floccifolia*, *Isolepis setacea*, *Juncus bufonius*, *Lythrum hyssopifolia*, *Mentha longifolia*, *Persicaria amphibia*, *Plantago major*, *Polygonum aviculare*, *Ranunculus multifidus*, *Rumex dentatus*, *R. steudelii*, *Salix mucronata* (planted), *Schoenoplectus lacustris* (open water), *Spergularia marina*, *Taraxacum sp. A.* and *Verbena officinalis*. Several of the recorded species occur in Wa'lān because of the existence of open water while others, notably *Cyperus laevigatus* and the *Spergularia*, suggest a fairly high salt level. The species peculiar to the Dhamār marsh were *Carex distans*, *Centella asiatica*, *Crepis rueppellii*, *Falkia oblonga*, *Helictotrichon elongatum*, *Pulicaria inuloides*, *Pycreus elegantulus*, *Schoenoplectus inclinatus* and *Trifolium fragiferum*. This group is not particularly remarkable, though the *Helictotrichon* and *Pycreus* are at least as characteristic of marsh on the escarpment as on the high plateau.

The last type of wetland habitat on the high plateau is the seasonally flooded mud flat. This is quite uncommon and as far as I know there are only three, one in Al Hada' (Sad Kumayn), one at Ribāṭ and one at Rada'. This last is the largest and most interesting. The typical mudflat species are found as the mud dries out at the end of the rainy season and include *Althaea ludwigii*, *Anthemis tigrensis*, *Crypsis schoenoides*, *Launea nudicaulis*, *Marsilea aegytiaca*, *Phyla nudiflora*, *Polygonum corrigioloides* and *Pseudognaphalium luteo-album*. The mudflats at Rada' are clearly saline and also have a remarkable assemblage of halophytic plants including *Aeluropus lagopoides*, *Atriplex leucoclada*, *Diplachne fusca*, *Spergularia marina*, *Suaeda aegyptiaca* and *Trianthema crystallina*. Around the edge of the mudflats are grassy banks where there is again an interesting group which includes *Cichorium bottae*, *Cirsium vulgare*, *Crepis rueppellii*, *Falkia oblonga* and *Taraxacum sp. A*. Significantly, the only other place where these are all associated is the flooded grassland at Ribāṭ.

4. THE MASHRIQ

This large area is comparatively poorly known botanically partly because of difficulties of access and partly because the flora can only be easily identified after significant rainfall, which is not a frequent occurrence. It is fairly clear though that the flora is nothing like as diverse as that of the escarpment and the number of distinct plant communities is small.

(4:1) *Sparse bushland on exposed mountains above 1800m*
Mountains above about 1800m predominate in the southern and western parts of the Mashriq. There is little difference in the flora whether the hills are formed of limestone, sandstone or one of the older pre-cambrian rocks. As on the high plateau the tree cover has been removed and the hills are almost completely devoid of soil. Consequently run-off is very rapid when there is rain, and erosion is severe.

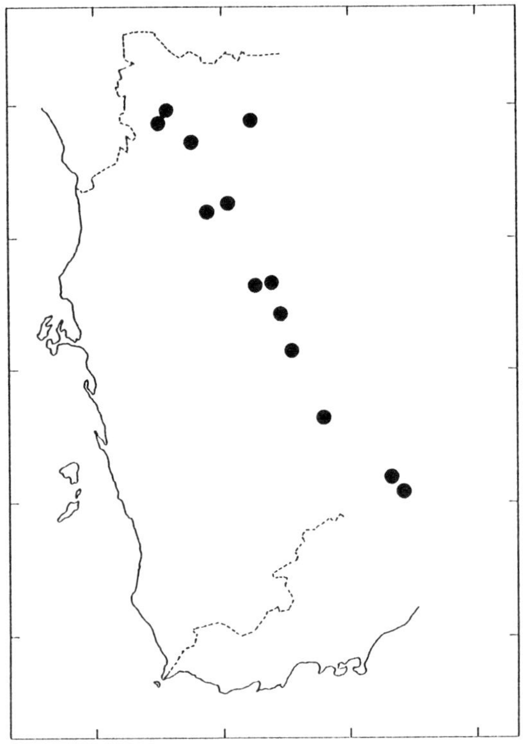

Map 32. Distribution of *Euphorbia balsamifera*, a species occurring on exposed hillsides above 1500 m in the Mashriq.

It seems likely that in prehistoric times these hills were covered in an open bushland dominated by various species of *Acacia*, particularly *A. etbaica* but including *A. gerrardii*, *A. mellifera* and *A. nilotica* subspecies *kraussiana* (in the Rada' area) and isolated bushes can be found as relics. Today the flora consists mainly of tough perennials, many of them spiny, and shows obvious similarities with that of some of the hills around the high plateau (3:3). Characteristic species include *Abutilon fruticosum*, *Aloe doei*, *A. fleurentinorum*, *Aristida adscensionis*, *Barleria hillcoatiae*, *B. parviflora*, *B. proxima*, *Caralluma hexagona*, *C. quadrangula*, *Chrysopogon plumulosus*, *Commicarpus ambiguus*, *C. mistus*, *Crotalaria emarginella*, *Dicliptera suffruticosa*, *Digitaria nodosa*, *Dipcadi viride*, *Eragrostis papposa*, *Euphorbia balsamifera*, *Fagonia brughieri*, *F. indica*, *F. schweinfurthii*, *Farsetia longisiliqua*, *Helichrysum pumilum*, *Heliotropium bottae*, *H. rariflorum*, *Hibiscus somaliensis*, *Leptothrium senegalense*, *Melhania stipulosa*, *Morettia parviflora*, *Polygala erioptera*, *Sarcostemma viminale*, *Seddera arabica*, *S. latifolia*, *Solanum cordatum*, *Stipagrostis ciliata*, *S. obtusa*, *Tetrapogon villosus*, *Vernonia cinerascens*.

Good examples of this type of vegetation can be seen in many places in the Mashriq including the Majz area, the Ḥuth area, between the Ghaylān and Fardah passes in Nihm and between Rada' and Dhī Nā'im.

(4:2) *Cliffs above 1800m*

A rather specialised vegetation is found on cliffs where there is some protection both from grazing and the full effects of the sun. It is in such habitats that typical escarpment plants find a precarious refuge and there are scattered records of such plants from all over the higher parts of the Mashriq. Probably the best developed of these specialised floras is that on the escarpment west of Dhī Nā'im which includes a large number of typical escarpment species but that area is not really typical of the Mashriq and is best considered part of the southeast mountains. A better example is that of the Fardah pass where the following escarpment species can all be found: *Abutilon bidentatum*, *Capparis cartiliginea*, *Ficus salicifolia*, *Grewia tembensis*, *Launea massauensis*, *Megalochlamys violacea*, *Pegolettia senegalensis*, *Periploca visciformis*, *Pulicaria petiolaris*, *Ruellia grandiflora*, *R. patula* and *Xerophyta arabica*. Other species not present here but which I would regard as characteristic of this habitat are *Hibiscus micranthus*, *Lantana viburnoides*, *Otostegia fruticosa*, *Tragia pungens* and *Tripogon multiflorus*.

(4:3) *Acacia bushland between 1000 and 1800m*

Below about 1800m the Mashriq consists mostly of stony plains and small mountains. No matter what the underlying rock there seems to be a great uniformity about the vegetation of such areas which contrasts strongly with that described in 4:1 above. The ground flora is much poorer but in contrast the number of shrub species is much larger. Open bushland prevails throughout this region and though much has been cleared, much still remains. Important shrub species include *Acacia campoptila*, *A. hamulosa*, *A. mellifera*, *A. oerfota*, *Cadaba farinosa*, *Commiphora myrrha*, *Euphorbia cuneata*, *Jatropha spinosa*, *Maerua crassifolia*, *Ochradenus baccatus*, *Periploca aphylla*, *Salvadora persica*. The ground flora consists of few species but *Aerva javanica*, *Blepharis ciliaris* and *Cymbopogon schoenanthus* are locally abundant. Several species characteristic of 4:1 also occur, such as *Crotalaria emarginella* and *Solanum cordatum*, and it is likely that others can be found in greater frequency when rain does fall.

Certain analogies with the flora of the escarpment foothills and the stony steppe of the southern Tihama are apparent and this is confirmed by the presence of certain rarer species in this area notably *Cadaba heterotricha*, *Caralluma acutangula*, *Chascanum marrubifolium*, *Delonix elata*, *Pergularia tomentosa*, *Schweinfurthia pedicellata* and *Tephrosia nubica*.

(4:4) *Alluvial plains (1200–1900m)*

Around Ṣa'daḥ and in the Jawf there are extensive loess and alluvial plains analogous to those on the high plateau (3:5). Although the soil is presumably quite fertile, such plains are surprisingly barren except where irrigated. Typical trees are *Acacia gerrardii*, *Tamarix aphylla*, *T. arabica*, *Zizyphus spina-christi* and more rarely *Acacia ehrenbergiana*. The ground flora consists of *Aristida adscensionis*, *Cynodon dactylon*, *Fagonia indica*, *F. schweinfurthii*, *Lycium shawii*, *Peganum harmala*, *Pulicaria crispa* and *Withania somnifera*.

(4:5) *Riparian vegetation along dry wadis (1100–1900m)*

The Mashriq is dissected by numerous wadis which are dry for nearly the whole year but experience occasional spates during the rainy season. The actual wadi bed is sandy or gravelly and has little vegetation but the margins are rich in species. Nearly all the shrub and tree species noted in 4:1-4 can be found along the margins of the wadis or on the cliffy slopes above them but some are particularly characteristic of this habitat. These include *Acacia campoptila*, *A. ehrenbergiana* (below 1700m), *A. gerrardii* (above 1700m), *Salvadora persica*, *Tamarix aphylla*, *T. arabica* and *Zizyphus spina-christi*. These riparion trees often support the climbers *Coccinea grandis*, *Coculus pendulus*, *Ephedra aphylla*, *Ochradenus baccatus*, *Pentatropis nivalis* and *Zehneria anomala*.

The sandy margins of the wadis usually have *Calotropis procera* and *Rhazya stricta* in abundance and at lower altitudes *Cassia senna*, *Desmostachya bipinnata* and *Pennisetum divisum* may also be frequent. Where richer soils have accumulated, a rather different flora occurs, including *Lycium shawii*, *Pluchea dioscorides*, *Solanum glabratum*, *S. incanum* and *Withania somnifera*. These species are reminiscent of the high plateau and this impression is reinforced by the occasional appearance of *Flaveria trinerva*, *Laggera decurrens* and *Xanthium spinosum* in broader valleys where there is a greater accumulation of alluvial soils.

(4:6) *Wetland habitats*

Although the Mashriq is extremely dry, there are occasional springs and a few, short, semi-permanent streams. Distinct plant communities have developed in these habitats and though quite diverse in many respects, they all show strong halophytic features for one of the most striking features of wetland habitats in the Mashriq is the high salt level.

The extremes are represented by the valleys north of Rada' and some stream sides in Nihm just west of the Fardah pass. The Rada' valleys are possibly the most saline localities in Yemen apart from the Red Sea coast. Much of the damp ground is entirely barren and covered with deposits of salt crystals. Vegetation is restricted to *Juncus socotranus*, *Suaeda aegyptiaca* and *Tamarix arabica* with the occasional *Cynodon dactylon* on hummocks. In the Nihm locality, in contrast, the salt level is quite low and few of the following species are notably salt-loving: *Agrostis viridis*, *Blumea bovei*, *Cynoglossum lanceolatum*, *Kanahia laniflora*, *Mentha longiflora*, *Plantago major*, *Pseudognaphalium luteo-album*, *Pulicaria inuloides*, *Scirpoides holoschoenus*, *Tamarix arabica* and *Typha domingensis*.

These aquatic communities are not only diverse but quite rich in species. One of the most notable is at 1700m at 'Ujlī between Ṣa'daḥ and Al Ḥarf. Here can be found *Blumea bovei*, *Centella asiatica*, *Fimbristylus sieberiana*, *Juncus punctorius*, *Panicum coloratum*, *Saccharum ravennae*, *S. spontaneum*, *Sporobolus consimilis*, *S. ioclados* and *S. kentrophyllus*. Another interesting one is in Sirwah district at 1200m, where the species include *Blumea bovei*, *Cyperus laevigatus*, *Juncus rigidus*, *Saccharum ravennae*, *Scirpoides holoschoenus*, *Senra incana*, *Suaeda aegyptiaca* and *Tamarix aphylla*.

5. THE SANDS

Very little is known of this vast area of desert and nothing at all about the vegetation of the occasional rock outcrop found in it. Most of the region, however, consists of mobile sand dunes, some of great size. Photographs taken from the centre of the Empty Quarter suggest that the more eastern sands are very barren but those close to the inhabited part of Yemen at about 1000m altitude support a surprising amount of vegetation.

The most prominent species on the sand dunes are the tough perennial shrubs *Calligonum crinitum*, *Rhazya stricta*, *Salvadora persica* and, much less importantly, *Leptodenia pyrotechnica.* All four are specially adapted to their environment; the first has succulent leaves; the last is leafless and the other two have thick leathery leaves; all have extensive root systems which bind the sand on which they grow so they are usually found on low hummocks of stabilised sand. Other perennial species found include *Aerva javanica*, *Centropodia forsskalii*, *Cyperus conglomeratus*, *Dipterygium glaucum*, *Heliotropium digynum* and *Pennisetum divisum*.

Ephemeral annual species which are so characteristic of desert regions in many parts of the world are not important here. *Tribulus arabicus* and *T. omanense* are characteristic examples of this sort of desert plant and the latter appears in very great abundance after heavy rain. No other annual species have definitely been found on the Sands but *Eremobium aegytiacum*, *Fagonia ovalifolia* and *Triraphis pumilio* occur at the edge of this region and will probably be found on the Sands when they are thoroughly explored.

THE COMPOSITION OF THE YEMEN FLORA

Yemen has always been of interest to plant geographers because it lies on the border between Asia and Africa and its high mountains have provided an avenue by which plants from the temperate regions of the world could penetrate into the tropics. The present chapter will attempt to analyse the composition of the Yemen flora by examining the various geographical groupings into which the species found in Yemen have been divided by plant geographers and considering their relative importance. It is premature to attempt to do this on any mathematical basis since the world distribution of many species is uncertain and even when it is known many species do not fall into a clear-cut geographical category.

Despite the difficulties in giving exact statements, certain broad trends are readily apparent in the composition of the Yemen flora. It is clear that the flora is tropical rather than temperate and African rather than Asian. Temperate species are comparatively few in number and are rare below about 2000m. Species whose distribution is primarily in tropical Africa are much more numerous but many important African genera are unrepresented in Yemen and it seems that the Red Sea has provided an effective barrier against the immigration of many African plants. Endemic species are fairly numerous and are found in genera of both tropical and temperate origin, though a large proportion are in succulent genera such as *Aloe*, *Caralluma* and *Euphorbia*.

1. THE TEMPERATE ELEMENT

Plants whose distribution is centred on the temperate regions of the world belong to many different families. The largest families in Yemen, the *Compositae*, *Leguminosae* and *Gramineae* all have important temperate elements while several medium-sized families, such as the *Cruciferae*, *Caryophyllaceae*, *Rosaceae* and *Umbelliferae*, are predominantly temperate.

In Yemen, species of temperate distribution are mostly found at high altitudes. They can be divided into several distinct groupings on the basis of their geographical distribution and this is discussed below.

(1:1) *Cosmopolitan Plants*

Most plant geographers include a ragbag of species within this category. In fact very few species are truly cosmopolitan and perhaps the only genuine example in Yemen is *Pteridium aquilinum*, a species almost cosmopolitan in places with good rainfall, being abundant for example both in the jungles of Southeast Asia and the moors of western Ireland.

Most so-called cosmopolitan species, however, are either widespread tropical species such as *Cynodon dactylon* or *Adiantum capillus-veneris* which penetrate the colder regions of the world in mild, usually coastal regions or else temperate species which penetrate the tropics on mountains.

It is this last category that concerns us here and probably thirty or so species in the Yemen flora fall into this group. Many of these are weeds such as *Anagallis arvensis*, *Capsella bursa-pastoris*, *Poa annua*, *Polygonum aviculare* and *Stellaria media* and their status as native species is open to some question since they are virtually restricted to man-made habitats. Most others are plants of wetland habitats such as *Ceratophyllum demersum*, *Juncus bufonius*, *Lemna minor*, *Utricularia australis*, *Verbena officinalis*, *Veronica anagallis-aquatica* and *Zannichellia palustris* and some of these, like the *Utricularia*, may have been casual introductions by birds. The few remaining cosmopolitan species of temperate origin are rather interesting. They include the ferns *Cystopteris fragilis* and *Asplenium adiantum-nigrum* and the parasite *Orabanche ramosa*, the former dispersed by spores and the latter with minute seeds, all alike capable of distribution over great distances enabling them to colonise any suitable ground.

(1:2) *Boreal plants*

Included here are plants whose distribution is centred on the cooler regions of the northern hemisphere. A few of these are found both in America and Eurasia including *Arabis alpina* and *Poa pratensis* but the majority are found in Eurasia only. Like the cosmopolitan species they are mostly weeds (*Arenaria leptoclados*, *Lolium perenne*, *Medicago lupulina*, *Misopates orontium*) or plants of wetland habitats (*Carex distans*, *Epilobium hirsutum*, *Myosotis laxa*, *Nasturtium microphyllum*) and it is difficult to know whether the former are recent introductions or not. A few boreal species show a rather more interesting pattern of distribution in Yemen. *Anthriscus sylvestris* and *Geranium robertianum* are hedgerow plants in temperate regions but in Yemen are restricted to high mountain gullies. A small group comprising *Lathyrus pratensis*, *Lotus corniculatus*, *Potentilla reptans* and *Rubus caesius*, are plants of rough grassland in temperate regions but are restricted to aquatic habitats in Yemen. *Scabiosa columbaria* is the only temperate grassland species which occurs in abundance on the rocky hills of Yemen.

(1:3) *Mediterranean plants*

Plants whose main area of distribution is around the Mediterranean Sea are fairly well represented and probably amount to around thirty in number. Although all are restricted to higher altitudes in Yemen, they are in other respects rather heterogenous and can be broken down into a number of small groups.

Several are weeds of waste ground and fields principally on the high plateau and may be ancient introductions. This group includes *Diplotaxis erucoides*, *Hyoscyamus albus*, *Marrubium vulgare*, *Sisymbrium irio* and *Urtica pilulifera*. A second group comprising *Ceratonia siliqua*, *Iris albicans*, *Myrtus communis* and *Ruta chalepensis* may also be introductions but appear in apparently native surroundings in Yemen. *Alkanna orientalis*, *Lavandula dentata* and *L. pubescens* comprise a third group; they are common on dry rocky ground in the East Mediterranean and extend in similar habitats through the Hejaz to their southernmost limits at various points along the mountains in Yemen. Not all Mediterranean plants are found in dry areas. An interesting group which includes *Bellardia trixado*, *Erica arborea*, *Ononis reclinata* and *Scrophularia arguta* are restricted to areas of prolonged mist particularly in the central escarpment. Possibly the largest group, however, consists of those species restricted to high mountains around 3000m principally in the Ibb area. It includes *Anagyris foetida*, *Biserrula palecinus*, *Ferula communis*, *Lamarkia aurea*, *Linum strictum*, *Sedum hispanicum*, *Umbilicus intermedius* and *Vulpia bromoides*, many of which can be found growing together in Yemen. All, for example, except the first two grow on the summit ridge of J. Ta'kar.

(1:4) *Irano-Turanian plants*
This group of plants has its distribution centred on the Caspian Sea region and is poorly represented in Yemen. Characteristic Irano-Turanian genera like *Astragalus*, *Centaurea* and *Salsola* have few representatives in southwest Arabia and very few species with a typical Irano-Turanian distribution have been found. One of the few examples is *Oryzopsis holciformis* but there is some doubt about the taxonomic status of the Yemen plant which has been clearly isolated from the main population for a long time. Other examples are less satisfactory but include *Sterigmostemum sulphureum* (casual in Yemen), *Polygonum corrigioloides* (virtually restricted to Iraq in the Irano-Turanian region) and *Pterocephalus pulverulentus* (restricted to the mountains in the interior of Palestine and Syria). *Cotoneaster racemiflorus* and *Epipactis veratrifolia* are often cited as examples of Irano-Turanian plants but both occur in the Mediterranean islands and extend east into the Himalayas. *Anarrhinum forskahlii* is widespread in the mountains from Syria to Iran but is at least equally common in the mountains of Yemen and Ethiopia and is as much part of the Afromontane flora as of the Irano-Turanian flora.

2. THE SAHARA-SINDIAN ELEMENT

Plants with a Sahara-Sindian distribution are those which are distributed across the desert regions of the Old World from the west coast of North Africa across Arabia and southern Iran to Sind in India. No families are restricted to or even centred on this region but certain genera are. Examples include *Farsetia*, *Fagonia*, *Stipagrostis* and *Tribulus*.

Many Sahara-Sindian species are distributed over the whole of this vast region and the complete list of those occurring in Yemen is considerable. They include widespread species of rocky ground such as *Aerva javanica*, *Cassia italica*, *Forsskalea tenacissima* and *Ochradenus baccatus*, plants of the Tihama and inland sands such as *Calatropis procera*, *Capparis decidua*, *Convolvulus prostratus*, *Leptadenia pyrotechnica*, *Panicum turgidum* and *Pennisetum divisum*, several species of the dry foothills including *Heliotropium rariflorum*, *Seddera latifolia* and *Solanum forskalii* and many of those found on the high plateau and rocky plains of the Mashriq including *Althaea ludwigii*, *Anastatica hierochuntica*, *Cleome amblyocarpa*, *Dicanthium annulatum*, *Oligomeris linifolia*, *Pulicaria crispa* and *Stipagrostis ciliata*.

Many Sahara-Sindian species are not found throughout this region. In particular there appears to be a division corresponding to the deserts in the west of Egypt and Sudan which separates the eastern Sahara-Sindian species from those found further west. There are hardly any species which are restricted to the western Sahara-Sindian region and extend east as far as Yemen. *Cadaba glandulosa*, *Euphorbia scordifolia* and *Tribulus mollis* are among the few examples. On the other hand, species from the eastern Sahara-Sindian region (Zohary's Nubo-Sindian province) are numerous and important in Yemen. They include such varied species as *Blepharis ciliaris*, *Cymbopogon jwarancusa*, *Dipterygium glaucum*, *Fagonia indica*, *Farsetia stylosa*, *Haplophyllum tuberculatum*, *Indigofera argentea*, *Periploca aphylla*, *Rhazya stricta*, *Solanum cordatum* and *Tamarix aphylla*. Several Saharo-Sindian species are restricted to the Arabian part of this region but are not particularly important numerically or in the vegetation of Yemen. They include *Fagonia ovalifolia* and *Tribulus arabicus*.

Some dispute has taken place over whether the Sahara-Sindian region is really part of the tropical or temperate plant domains. Clearly both temperate and tropical elements are present but a number of the most important species belong to tropical families such as the *Apocynaceae*, *Asclepiadaceae* and *Capparaceae*. The impression that the Sahara-Sindian flora is derived from a tropical one is reinforced by the existence of a number of species which combine a Sahara-Sindian distribution with a Sudano-Zambesian distribution. Examples include such common and characteristic plants as *Balanites aegyptiaca*, *Cadaba farinosa*, *Capparis cartiliginea*, *Salvadora persica*, *Sporobolus ioclados*, *Withania somnifera* and *Zizyphus spina-christi*.

Whatever the affinities of the vegetation of the Sahara-Sindian region, it is clear that most of the vegetation of the drier parts of Yemen is broadly Sahara-Sindian. Species such as *Capparis decidua*, *Rhazya stricta* and *Salvadora persica* are locally dominant and nearly all the most widespread and abundant species in Yemen (*Aerva javanica*, *Blepharis ciliaris*, *Withania somnifera*, *Zizyphus spina-christi*) belong to this group.

3. THE TROPICAL ELEMENT

Plants whose distribution is centred on the tropics are very numerous in Yemen, particularly in the Tihama and on the escarpment. Tropical families such as *Acanthaceae*, *Asclepiadaceae* and *Euphorbiaceae* are well represented while significant numbers of *Commelinaceae* and *Cucurbitaceae* can be found. The tropical flora contains many disparate elements and these are examined below.

(3:1) *Cosmopolitan plants*
As noted in 1:1 above, some cosmopolitan species are essentially tropical but in cooler areas occur in protected habitats. Examples found in Yemen include *Adiantum capillus-veneris*, *Equisetum ramosissimum* and *Cynodon dactylon* but this category is not important in Yemen.

(3:2) *Pantropical plants*
There are many species found in Yemen which are distributed throughout the tropical regions of the world. As with other widely distributed species, their status as native plants is often doubtful and they are mainly found in waste places or wetland habitats. Examples of pantropical weeds include such abundant species as *Aristida adscensionis*, *Cassia occidentalis*, *Euphorbia hirta*, *Gynandropsis gynandra* and *Sida alba*. Plants of wet places include *Commelina diffusa*, *Cyperus laevigatus*, *C. rotundus*, *Echinochloa colona*, *Eclipta prostrata* and *Typha domingensis*. The remaining pantropical species do not fall into clear-cut categories. They include the common shrub *Dodonaea viscosa*, a number of climbers such as *Cardiospermum halicacabum*, *Momordica balsamina* and *Vigna vexillata*, plants of disturbed bushy places such as *Achyranthes aspera* and *Triumfetta rhomboidea* and a number of species of open stony ground such as *Evolvulus alsinoides and Heteropogon contortus*. The *Cyperaceae* seem to be particularly rich in pantropical species perhaps because of their preference for wetland habitats.

(3:3) *Palaeotropical plants*
These are plants restricted to the tropical regions of the old world and like the last category they are well represented in Yemen. Although most representatives of this group are restricted to the lower and warmer parts of the escarpment they are otherwise very heterogenous. Palaeotropical species include trees and shrubs like *Carissa edulis*, *Flueggia virosa* and *Tamarindus indicus*, aquatic plants like *Centella asiatica* and *Najas graminea*, weeds like *Dactyloctenium aegyptium*, *Ipomoea eriocarpa* and *Micrococca mercurialis*, climbers like *Teramnus labialis*, plants of damp shaded places like *Asystasia gangetica*, *Desmodium repandum* and *Hewittia scandens* and common species of bushland such as *Commelina benghalensis*, *Pupalia lappacea* and *Themeda triandra*.

(3:4) *Sudano-Deccanian plants*
These are plants distributed across the lowland savannah regions of Africa which extend eastwards to similar situations on the plains of India. This category is well represented in Yemen and consists almost entirely of species found in Acacia-Commiphora bushland, particularly in the rich escarpment wadis like Wadi Sukhnah. Typical plants in this category include *Acalypha ciliata*, *A. fruticosa*, *Actinopteris semiflabellata*, *Bothriochloa insculpta*, *Brachiaria deflexa*, *Cordia sinensis*, *Dicoma tomentosa*, *Eragrostis aspera*, *Hibiscus micranthus*, *Indigofera hochstetteri*, *Lepidagathis scariosa*, *Pegolettia senegalensis*, *Pergularia daemia*, *Plectranthus barbatus*, *Ruellia patula*, *Sclerocarpus africanus* and *Stylosanthes fruticosa*.

(3:5) *Deccanian plants*
Deccanian plants are those whose distribution is centred on the plains of India. This group is extremely poorly represented in Yemen, thus showing clearly how little connection there is between the floras of India and Yemen. Among the few examples of Deccanian plants found are *Crotalaria hirta*, *Dicanthium aristatum*, *Mariscus cyperinus* and *Orthosiphon thymiflorus*, but most are not really typical since all except the *Crotalaria* extend east of India and might, in fact, be better considered as examples of palaeotropical species with an extremely restricted (or non-existent) African range.

(3:6) *Sudano-Zambezian plants*
Plants whose distribution is centred on the great savannah plain regions of Africa which stretch from Senegal east to the Sudan and thence southwards through East Africa to South Africa are known as Sudano-Zambezian plants. The popular impression of this region is of extensive grassy plains with scattered trees and the characteristic plants, apart from trees and grasses, are species which readily coexist with grass or climbers which thrive in the protection of the trees. The most typical tree genus is undoubtedly *Acacia*, which gives the region its well-known appearance, and other important genera include *Aloe*, *Crotalaria* and *Indigofera*, all of which tolerate the grazing to which the grasslands are subjected.

It is not surprising therefore that a large proportion of the tree species found in Yemen are widespread throughout the Sudano-Zambezian region. Examples include *Acacia gerrardii*, *A. hockii*, and *A. seyal*, *Adansonia digitata*, *Breonadia salicina*, *Bridelia scleroneura*, *Combretum molle*, *Cordia abyssinica*, *Ficus sycomorus*, *Piliostigma thonningii*, *Teclea nobilis*, *Trichelia emetica* and many others. In the same way, many of the climbers found throughout this region extend into Yemen and examples include *Cayratia gracilis*, *Cissus quadrangularis*, *Clematis hirsuta*, *Dregea abyssinica*, *Kedrostis hirtella*, *Sarcostemma viminale* and *Stephania abyssinica*. Sudano-Zambezian grasses are likewise well represented in the Yemen flora and include *Digitaria horizontalis*, *Panicum maximum*, *Pennisetum unisetum* and *Rhynchelytrum repens*. Almost all other widespread Sudano-Zambezian species which occur in Yemen are grassland plants such as *Crotalaria senegalensis*, *C. spinosa*, *Cyanotis lanata*, *Hypoestes forskalei*, *Hypoxis angustifolia*, *Indigofera costata*, *Sida ovata* and *Tephrosia pumila*. It is interesting that, with a few exceptions, Sudano-Zambezian plants are unrepresentated among field weeds, aquatics and plants of damp shaded habitats.

Like the Sahara-Sindian region, the Sudano-Zambezian region can be divided into subregions since many plants do not extend over this whole vast area. Many species, for example, are found in the Sudanian savannah regions south of the Sahara desert but do not extend southwards through East Africa. Some plants with this type of distribution have the eastern limits of their range in Yemen. Curiously, apart from *Cocculus pendulus* and the parasite *Tapinanthus globiferus*, all these species are trees or shrubs and include *Celtis toka*, *Lannea fruticosa* and *Ozoroa insignis*. *Adenium obesum* should also be included here although it does extend southwards as far as Tanzania.

In contrast, with the notable exception of *Acacia mellifera* and *A. tortillis*, those species restricted to the eastern part of the Sudano-Zambezian region from South Africa north to Sudan and extending into Yemen are almost entirely herbaceous. They include *Commelina imberbis*, *Crotalaria natalitia*, *Dyschoriste radicans*, *Indigofera schimperi*, *Plectranthus hadiensis*, *Rhynchosia totta* and many grasses such as *Aristida congesta*, *Hyparrhenia dregeana*, *H. variabilis*, *Setaria plicatilis* and *Sporobolus panicoides*.

Of most importance, however, for the Yemen flora are the *Sudano-Zambezian* species restricted to northeast Africa (the Afro-oriental domain) and separated from Arabia only by the Red Sea. Possibly the most characteristic genus of this area is *Commiphora*, which is very poorly represented further west but co-dominant with *Acacia* in large areas of bushland here and in Yemen. Many of the characteristic Acacias of Yemen are restricted to the Afro-oriental domain including *Acacia asak*, *A. etbaica*, *A. hamulosa* and *A. oerfota*. Other important species are *Barleria acanthoides*, *Caralluma penicillata*, *Cissus rotundifolius*, *Dobera glabra*, *Euphorbia cactus*, *Ficus vasta*, *Grewia schweinfurthii*, *G. velutina*, *Hibiscus deflersii*, *Indigofera oblongifolia*, *I. spinosa*, *Mumusops laurifolia*, *Seddera arabica* and *Terminalia brownii*.

The South Arabian domain is the last subregion that needs to be considered. A separate domain is clearly justified by the existence of a number of important species endemic to this area but of obvious Sudano-Zambezian affinity. Apart from *Anisotes trisulcus* which is only known from a single locality in Africa, the species include one *Acacia* (*A. campoptila*), at least one *Commiphora* (*C. kataf*), almost all Aloes including *A. sabaea*, *A. vacillans* and *A. sp. B*, *Abrus bottae*, *Barleria bispinosa*, *Caralluma quadrangula*, *Cyanotis nyctitropa*, *Euphorbia ammak*, *E. inarticulata*, *Ormocarpum yemenense* and *Pavetta longiflora*. Many other endemic species belonging to common Sudano-Zambezian genera occur though they are of lesser ecological importance. Among the many examples are *Boscia arabica*, *Ceropegia arabica*, *Crotalaria squamigera*, *Indigofera arabica*, *Phyllanthus hodjelensis*, *Rhus flexicaulis* and *Vernonia bottae*.

It is clear from this discussion that the eastern plains of the Tihama, all the escarpment below 2000m, and much of the Mashriq are clothed with a vegetation, albeit somewhat depauperate, which is characteristic of the Sudano-Zambezian region to which almost all the characteristic dominant species are restricted.

4. THE MONTANE ELEMENT

It is convenient to compare the mountains of Yemen and those of Africa to islands. Thus, although the surrounding lowlands support a vegetation characteristically Saharo-Sindian or Sudano-Zambezian, the mountains provide a refuge for many species of a different type. Some of these, as already noted, are plants of temperate regions but the vast majority are species entirely restricted to the mountains of the tropical regions of the world. Three distinct groups are discernible in Yemen.

(4:1) *The Himalayan group*

Rather few montane species found in Yemen are essentially Himalayan and most in fact extend into Ethiopia or some other part of Africa. This small group comprises *Ajuga bracteosa*, *Arisaema flava*, *Asplenium varians*, *Cyanotis vaga*, *Cynoglossum lanceolatum*, *Debregeasia saenab*, *Pycnocycla glauca* and *Tripogon purpurescens*.

(4:2) *The Afromontane group*

Although several of the best-known species of the Afromontane flora such as the giant Lobelias and Senecios do not extend to Yemen, the affinities of the Yemen montane flora are clearly with that of the African mountains and particularly with those in Ethiopia.

One of the most obvious features of the Afromontane flora is the fact that the majority of its characteristic species belong to genera which are essentially temperate in their distribution. This applies both to species which are widespread in the African mountains such as *Agrostis lachnantha*, *Alchemilla cryptantha*, *Astragalus atropilosulus*, *Bromus leptoclados*, *Crassula alba*, *Helictotrichon elongatum*, *Koeleria capensis*, *Lactuca inermis*, *Poa leptoclada*, *Ranunculus multifidus* and *Silene burchellii*, and to those species which are of very restricted distribution, being found only in the mountains of Ethiopia and southwest Arabia. Within this latter group are *Dianthus uniglumis*, *Lotus quinatus*, *Merendera abyssinica*, *Primula verticillata*, *Rosa abyssinica*, *Rumex nervosus* and *Thesium radicans*. Trees are poorly represented but *Myrica salicifolia* is a good example. All the species hitherto mentioned belong to genera widespread in the northern hemisphere but many Afromontane species belong to genera which are essentially South African. This group is poorly represented in Yemen but includes *Helichrysum forskahlii*, *Macowania ericifolia*, *Pentaschistis pictigluma*, *Sebaea microphylla* and *Schoenoxiphium sparteum*.

Not all Afromontane species belong to obviously temperate genera. In some cases such as *Acacia origena*, *Crotalaria quartiniana*, *Erythrococca abyssinica*, *Isoglossa punctata* and *Vernonia leopoldii*, they belong to characteristic lowland genera. In the case of orchids, the Ethiopian mountains seem to have been a centre of speciation and *Disa pulchella*, *Habenaria lefebureana*, *Holothrix arachnoidea* and *Satyrium brachypetalum* all belong to genera which are most abundant in Africa. In other cases the Afromontane species belong to large cosmopolitan genera and individual species have evolved on the African mountains. Curiously, all the following examples are of species essentially restricted to the mountains of Ethiopia and Yemen: *Arisaema bottae*, *Conyza incana*, *Crepis rueppellii*, *Cyperus fissus*, *Eragrostis longifolia*, *Kyllinga chlorotropis*, *Senecio hadiensis* and *Solanum schimperianum*.

A few genera have evolved in the Afromontane region and occur in both Ethiopia and Yemen. The most interesting is undoubtedly *Barbeya oleoides* which is not only in a genus of its own but in a distinct family often thought to be allied to the *Urticaceae*. The only other examples are the legume *Vermifrux abyssinica* and the grasses *Odontelytrum abyssinicum* and *Phaenanthoecium kostlinii*.

(4:3) *The Yemen endemic group*

All species hitherto considered are found elsewhere but a not inconsiderable proportion of the Yemen montane flora consists of endemic species. Although a few belong to mainly South African genera (*Delosperma harazianum* and *Helichrysum arwae*) and several are in cosmopolitan genera (*Convolvulus thymoides*, *Senecio harazianus*, *Solanum platacanthum*) the vast majority belong to genera centred on the north temperate regions of the world, particularly the Mediterranean region. Examples include *Cichorium bottae*, *Cynoglossum bottae*, *Dianthus uniflorus*, *Festuca cryptantha*, *Galium kahelianum*, *Nepeta deflersiana*, *Picris scabra*, *Pimpinella menachensis*, *Rubus arabicus*, *Saponaria umbricola*, *Swertia polynectaria*, *Thymus laevigatus* and *Verbascum melhanense*. Of particular interest is *Centaurothamnus maximus*, the only example of a genus restricted to the montane regions of southwest Arabia.

5. SOME MINOR PATTERNS OF DISTRIBUTION

(5:1) *Red Sea species*
A significant number of species are restricted to the coastal regions of the Red Sea and the Gulf of Aden. Although some submaritime species are included in this number, such as *Halopeplis perfoliata* and *Limonium axillaris*, the majority are species of the coastal plains and are mostly of distinct Sudano-Zambezian affinity. They include *Barleria lanceata*, *Blyttia nivalis*, *Commiphora gileadensis*, *C. quadricincta*, *Crotalaria microphylla*, *Cyphostemma ternatum*, *Euphorbia triaculeata*, *Heliotropium pterocarpum*, *Indigofera semitrijuga*, *I. tritoides*, *Odyssea mucronata*, *Pavonia kotschyi*, *Rhigazom somaliense*, *Schweinfurthia pedicellata*, *Seddera virgata*, *Taverniera lappacea* and *Tephrosia quartiniana*. It is interesting that this list contains most of the species found in the drier regions of the Tihama which are not of obvious Sahara-Sindian distribution.

(5:2) *The Sahara connection*
A small group of species comprising *Argyrolobium arabicum*, *Phagnolon scalarum* and *Tripogon multiflorus* have their distribution centred on the dry mountains of the central Sahara such as the Tibesti and extend eastwards as far as Arabia. There seem to be few other connections between these two mountain regions.

(5:3) *The Canary Islands connection*
A connection between the flora of the Canary Islands and that of the southern Red Sea area has long been noted. Most of the species having a disjunct distribution connecting our area with the Canary Islands are found in Somalia rather than Arabia and *Euphorbia balsamifera* is the only example occurring in Yemen. However the genera *Aeonium* and *Campylanthus* show the same disjunct distribution, though they are represented by different species in the two areas. Another curious connection is the superficial similarity of several of the succulent Euphorbias found in the Canary Islands with those found in Yemen.

(5:4) *The Yemen-Somalia connection*
The connection between the montane flora of Yemen and Ethiopia is wellknown and has been discussed in 4:2 above. There is, however, an important connection between the vegetation of Yemen and that of Somalia. South Arabia and the Horn of Africa have a distinctive Sudano-Zambezian type of vegetation which is characterised by the existence of several genera restricted to this area, notably *Caucanthus*, *Hildebrandtia*, *Rhytidocaulon* and *Tetrachaete* and it is clearly also a centre for speciation in a number of larger genera including *Campylanthus*, *Caralluma*, *Commicarpus*, *Cystostemon* and *Seddera*. It is not surprising, therefore, that these genera are rich in undescribed and endemic species such as *Campylanthus yemenensis*, *Commicarpus arabicus*, *Rhytidocaulon macrolobum* and many species of *Caralluma*.

Although plants restricted to Somalia and Yemen are found in scattered localities throughout Yemen, including such unlikely regions as the higher, wetter escarpment mountains where *Hibiscus purpureus*, *Pavetta villosa*, *Plectranthus cicatricosus* occur, two centres are immediately apparent. Plants characteristic of stony plain on the high plateau reappear on the Somali highlands and many of these species are largely or entirely restricted to these two areas. Thus both on cliffs south of Wadi Ḍahr and on stony plains near Sheikh in Somalia I have personally seen *Barleria parviflora*, *Chascanum sessilifolium*, *Dicliptera suffruticosa*, *Gnidia somaliensis*, *Helichrysum pumilum*, *Indigofera phillipsae* and *Tephrosia heterophylla*. Other species with a similar distribution are *Cystostemon heliocharis*, *Melhania stipulosa* and *Salsola bottae*.

The other group of plants occurring elsewhere only in Somalia are found in the Acacia bushland regions of the escarpment. Even the more widespread of these species like *Caucanthus edulis*, *Hildebrandtia africana*, *Ocimum spicatum* and *Pulicaria somaliensis* have a strong southern tendency in their distribution but the others are all restricted to the south of Yemen, particularly Taiz province. They include *Asystasia guttata*, *Barleria mucronifolia*, *Blepharispermum yemense*, *Plectranthus caerulescens*, *Tetrachaete elionuroides*, *Turraea parvifolia*, *Vernonia spatulata* and *Wendlandia arabica*.

(5:5) *The Yemen-Namibia connection*
A connection has long been noted between plants occurring in Somalia and Namibia. The connection between Yemen and Namibia is extremely tenuous and supported only by *Fingerhuthia africana* and *Osteospermum muricatum*, species unknown in the territory between Somalia and Namibia.

(5:6) *The 'Udayn-Dhofar connection*
There is an unexpected connection between species occurring at 'Udayn and in the Dhofar mountains. It is most dramatically illustrated by *Ormocarpum dhofarense*, an endemic species known only from these two places, but *Eulophia guineensis* and *Peperomia pellucida* are found in Arabia only at these two localities while *Desmodium ospriostreblum* and the genus *Lannea* are known elsewhere only from Wadi Sukhnah, a wadi with a very strikingly similar vegetation to that found at 'Udayn.

THE PLAN OF THE FLORA

The flora contains descriptions of all native and naturalised species of flowering plants, ferns and fern allies recorded from North Yemen. In addition descriptions are given of all commonly cultivated plants, including all those of economic importance but no attempt has been made to include species introduced casually as garden flowers or those trialled in agricultural or forestry experiments.

The boundaries are those of the Yemen Arab Republic as conventionally accepted in 1982 though nearly all the eastern boundaries are undefined. These boundaries are not completely accepted and do not necessarily correspond to those used by earlier botanists. In particular the concept of Arabia Felix used by many earlier travellers does not coincide with that of North Yemen but is a much wider area extending to Aden and Jeddah. I have included all old records from Qa'ṭabah and the Ḥujarīyah even though it is not always clear which side of the old border the plants were collected on and I have excluded records from Jizan and the neighbouring islands, mainly those of Ehrenberg, since this is now part of Saudi Arabia.

The order of families adopted is that of Arthur Cronquist in *An Integrated System of Classification of Flowering Plants* (1981) and I have in general followed his concept of the limits of the families though I have maintained the unity of the Leguminosae and there are a few other differences. In addition I have not been able to include some of the new family concepts established later in the 1980s, in particular the splitting of the *Liliaceae*, as these would involve excessive editorial costs.

Short descriptions of the salient features of families and genera are given. It needs to be stressed that these descriptions only refer to plants found in Yemen and may not be accurate for plants found in other countries. The purpose of these descriptions is partly to provide information so that the reader can be certain that his plant belongs to that family or genus but also to save repetition of common features in the descriptions of the species that follow. It is obvious therefore that information contained in the generic and family descriptions will not necessarily be repeated under each species and that generic descriptions will not be given where there is only one species in that genus.

Dichotomous keys to genera and species are provided. It is expected that readers will rely mainly on these to identify their plants and that having arrived at a particular species in the key they will then check their plant against its description. No key to families is provided. It was felt that such a key would be cumbersome and would not be used by many readers. Anyone wanting a key to families should refer to *Thonner's Analytical Key to the Families of Flowering Plants* by R. Geesink, A. Leewenberg, C. Ridsdale and J. Veldkamp (1981).

Every effort has been made to bring the nomenclature up to date and some unfamiliar names will be apparent. The citation of synonyms has been kept to a minimum but all names used in O. Schwartz *Flora von Tropischen Arabien* (1939) are cited and anyone wishing to trace names used by earlier botanists should refer to that work. Names which are misapplied in Schwartz are indicated by the words 'sensu Schwartz' following the species name. This is a convenient device but slightly unfair to Schwartz who in many cases had merely taken over the names used by his contemporaries or cited plants which he had not seen and were in fact mis-identified by others, particularly Deflers. Apart from the names used in Schwartz, the only other synonyms cited are those widely used in contemporary floras or in the few recently published articles on the Yemen flora. Very occasionally synonyms have been given to clarify some specific taxonomic or nomenclatural problem.

The species without an accepted name are designated species A, B, C, etc. This device is used in two situations. In some cases, for example *Picris*, the species are being actively investigated by various specialists who can be expected to publish the names of these plants within the next few years. In most cases however (*Aloe*, *Grewia*, *Hibiscus*, *Hyphaene*, *Pavetta*, etc.) it would be premature to describe new species since the groups are either extremely complex or the collected material is inadequate or both. Few contemporary botanists are prepared to describe new species from a single collection as it is impossible to know the range of variation of a species from one specimen and many mistakes have been made in the past. Further material of all such plants would be welcome.

Specimens are only cited for undescribed species or for species where there is known to be taxonomic uncertainty. The research worker will be able to find out my concept of these species by referring to the cited specimens which can mostly be found in the herbaria at Kew and the British Museum (Natural History).

Arabic names are given where there is firm evidence that a particular name is used for a particular species in Yemen. In general, names are not given when they have been recorded by a single observer or from a single locality. It is necessary to stress that the names given are those used in Yemen and they may not have their classical Arab meaning, nor may they apply elsewhere in the Arab world. One of the best examples is 'Zaytūn' which is used thoughout the Arab world for the olive but in Yemen refers to the guava. Arabic names are transliterated according to the system below but there are many local variants and there is no certainty that all have been correctly spelt.

The distribution and ecology of each plant within Yemen is given after each description. Every effort has been made to get the information as accurate as possible but there are bound to be errors and omissions. The altitude range is not an estimate but is based on the highest and lowest localities from which the plant has been recorded. The cited localities are based on field records as well as herbarium specimens. Where no observer is cited it means that I have seen the plant at that particular locality.

Information on the distribution of plants outside Yemen is only given for some species. Thus endemic species and those with a very restricted or unusual distribution are usually indicated. Accurate information on the world-wide distribution of many species is simply not available. The taxonomic information in many floras has not been updated and it is very difficult to know how reliable much of the

System of transliteration.

Alif	ا	a	Ḍad	ض	ḍ
Ba	ب	b	Ṭa	ط	ṭ
Ta	ت	t	Ẓa	ظ	ẓ
Tha	ث	th	'Ayn	ع	'
Jīm	ج	j	Ghayn	غ	gh
Ḥa	ح	ḥ	Fa	ف	f
Kha	خ	kh	Qaf	ق	q
Da	د	d	Kaf	ك	k
Dha	ذ	dh	Lam	ل	l
Rā	ر	r	Mim	م	m
Zā	ز	z	Nun	ن	n
Sīn	س	s	Ha	ه	h
Shīn	ش	sh	Waw	و	w
Ṣad	ص	ṣ	Ya	ى	y

The vowels are indicated by a - i - u and a length mark is added where necessary.

information given really is without extensive research.

Some effort has been made to restrict the technical language used but it is inevitable that a considerable number of words will be unfamiliar to the beginner. Although no glossary is provided in this book, many words used to describe characteristic features in certain large families, notably grasses, legumes and composites, are explained in the family descriptions. For a good glossary, the reader is referred to *Flowering Plants of the World* edited by V. H. Heywood, a book which contains a lot of useful information on the plant families represented in the Yemen flora.

PTERIDOPHYTES

These are the ferns and fern allies, the only non-flowering plants covered in this book.

The fern allies (*Selaginella, Equisetum*) have elongate stems without scales. The leaves are numerous, simple, 1-nerved and small. The sporangia are borne in the axils of the leaves (*Selaginella*) or in a terminal cone-like head (*Equisetum*).

In the true ferns the stem (rhizome) is usually creeping and densely covered in scales or hairs. The leaves (fronds) are usually petiolate with a simple or compound lamina (blade). In compound laminae the primary divisions or leaflets are called pinnae which may in turn be divided into pinnules. In a few species the pinnules are also divided and their divisions are called segments. The mid-vein of the lamina (a continuation of the petiole) is called the rhachis and that of the pinnae a costa. The sporangia are borne on the lower surface of the frond or at the margins. They are usually in distinctively shaped clusters, called sori, which may be covered with a special structure called an indusium.

Pteridophytes favour moist, shaded habitats. In Yemen they are particularly common in the mist zones of the high mountains and the central escarpment but are almost completely absent from the Tihama and the Mashriq. Many species are rare and restricted to certain specialised microclimates. The two most remarkable are both narrow, almost sunless gullies which act as natural funnels taking mist from one valley to another. One of these lies north of Aṣ Ṣa'bah, J. Milḥān, and the other below Jirjir in the Ḥadīyah region of J. Raymah. Jirjir is also remarkable as the centre for the only extensive areas of mountain grassland in Yemen and these too have a number of ferns restricted to them. Clearly these areas and the half-a-dozen or so other localities which recur in the distribution notes are rather vulnerable but fortunately there is no immediate threat to them. It is to be hoped that the rare ferns that have survived in isolation for so long will remain undisturbed in the future.

None of our species is endemic. Many, though rare in Yemen, have a wide distribution elsewhere, some like *Pteridium aquilinum* being almost cosmopolitan, others like *Christella dentata* being nearly pantropical while still others such as *Asplenium adiantum-nigrum* are widespread in temperate regions. Only *Cheilanthes coriacea*, *Negripteris scioana*, *Selaginella imbricata* and *S. yemensis* are restricted to southwestern Arabia and northeastern tropical Africa.

All ferns are called 'Mushat' or 'Mushayt' in Yemen.

1. Stems jointed. Leaves simple, whorled, fused into a tube-like sheath. Sporangia in terminal cone-like heads — 3. *Equisetum*
Stems not jointed. Leaves usually compound, never whorled or joined into a sheath. Sporangia not in terminal cone-like heads — **2**

2. Fronds shamrock-like, with 4 obdeltoid-obovate pinnae at the apex of the petiole — 25. *Marsilea*
Fronds not as above — **3**

3. Plants with tiny 1-nerved leaves, never exceeding 0.5cm long — **4**
Plants with small or large fronds, at least 3cm long and usually much more — **5**

4. Leaves few, scale-like 1. *Psilotum*
Leaves numerous, moss-like 2. *Selaginella*
5. Lamina wedge- or fan-shaped, dichotomously divided into ± linear segments 11. *Actinopteris*
Lamina not wedge- or fan-shaped, not dichtomously branched **6**
6. Fronds simple, lobed or unlobed **7**
Fronds compound, pinnate to quadripinnate **10**
7. Fronds unlobed **8**
Fronds deeply lobed, ± pinnatifid 21. *Ceterach*
8. Rhizome erect. Fronds dimorphic, the fertile ones without a lamina, bearing two rows of large sporangia near the tip 4. *Ophioglossum*
Rhizome creeping. Fronds uniform. Sporangia in sori **9**
9. Sori elongate. Undersurface of frond without scales 24. *Loxogramme*
Sori round. Undersurface of frond with scattered scales 23. *Pleopeltis*
10. Fronds simply pinnate **11**
Fronds bi-quadripinnate at least in the lower pinnae **13**
11. Petiole black or dark brown. Pinnae less than 2cm long **12**
Petiole pale. Pinnae more than 4cm long 12. *Pteris*
12. Sori medial, often seeming to cover the whole undersurface of the pinnae. Fronds not rooting at the tips 20. *Asplenium*
Sori distinctly marginal, never covering the undersurface of the pinnae. Fronds often rooting at the tips 8. *Adiantum*
13. Fronds dimorphic. Plant growing in low-altitude swamps 9. *Ceratopteris*
Fronds uniform. Plants not growing in swamp, though sometimes by ditches and streams **14**
14. Underside of frond covered with white powder, brown hairs or scales **15**
Underside of frond naked **16**
15. Sori of minute separate black dots. Undersurface of frond white, the rhachis scaly 6. *Negripteris*
Sori not as above. Undersurface of frond white or covered in hairs or scales but if white, the rhachis not scaly 5. *Cheilanthes*
16. Upper part of frond clearly bi-pinnatifid. Gemmae present 15. *Tectaria*
Upper part of frond bi-quadripinnate. Gemmae absent except in *Asplenium protensum* **17**
17. Petiole and rhachis densely scaly, the scales ± continuous to the top of the rhachis **18**
Scales mostly at the base of the petiole, never continuous along the whole length of the petiole and rhachis **21**
18. Some of the pinnae more than 6cm long. Petiole pale. Sori rounded **19**
All pinnae less than 4cm long. Petiole black or dark purple-brown. Sori various, but not rounded **20**
19. Frond strongly tripinnate. Pinnules without aristate teeth 16. *Dryopteris*
Frond ± bipinnate. Pinnules with aristate teeth 17. *Polystichum*
20. Pinnules ± orbicular. Sori marginal 5. *Cheilanthes*
Pinnae elongate, not rounded. Sori medial 20. *Asplenium*
21. Small delicate parsley-like fern with pinnules less than 2mm broad. Sori marginal, ± covered by the reflexed margins of the pinnules 10. *Onychium*
Not as above, the pinnules at least 2mm broad. Sori various **22**
22. Sori marginal or apparently so **23**
Sori clearly medial **27**
23. Sori ± continuous along the margins of the pinnules **24**
Sori not continuous along the margins of the pinnules **26**
24. Lamina pubescent 13. *Pteridium*
Lamina glabrous **25**
25. Lamina bipinnate 12. *Pteris*
Lamina tri-quadripinnate 7. *Pellaea*
26. Petiole black. Sori borne on reflexed marginal lobes 8. *Adiantum*
Petiole green. Sori not borne on reflexed marginal lobes 20. *Asplenium*
27. Petiole hairy **28**
Petiole glabrous **30**
28. Fronds ± glabrous. Pinnae with white dots on the upper surface. Indusium glabrous 22. *Arthropteris*
Fronds densely hairy. Pinnae without white dots. Indusium hairy **29**
29. Lamina narrowly elliptic, tapering at the base 14. *Christella*
Lamina ovate-deltoid, not tapering at the base 18. *Hypodematium*
30. Sori rounded 19. *Cystopteris*
Sori elongate 20. *Asplenium*

1. PSILOTUM *Sw.*

1. Psilotum nudum (*L.*) *Beauv.*
Twiggy, repeatedly forked, glabrous, almost leafless plant up to c 20cm high. Leaves widely spaced, scale-like, up to 1.5mm long. Sporangia borne near the tips of the upper branches.

Only known from rocks in the wadi above 'Adūf near Turbah at 1900m. (K. Gordon)

2. SELAGINELLA *P. Beauv.*

Clubmosses. Herbs with erect or prostrate stems and numerous tiny 1-veined, ligulate leaves, which are dimorphic in our species.

1. Erect, tufted plant with leafless shoots 10–20cm long *S. imbricata*
Prostrate plants with leafy stems **2**
2. Leaf margin white, ciliate. Branch tips recurved when dry. (Abundant plant of dryish habitats) *S. yemensis*
Leaf margin green, ± glabrous. Branch tips not recurved when dry. (Rare plant of wet places) *S. goudotana*

1. Selaginella goudotana *Spring*
A small prostrate moss-like plant, either mat-forming or

more usually consisting of individual fronds, the branches not becoming recurved when dry. Leaves pale green with a glabrous green margin.

The Yemen plant is variety **abyssinica** (*Spring*) *Bizarri*

Only known from a wet rock face at 2800m on J. Ba'dān.

2. Selaginella imbricata (*Forssk.*) *Spring*

An erect, tufted plant with several frond-like shoots to 30cm. Shoots leafless below but highly branched and leafy above. Leaves dark green above, very pale below, with an entire glabrous margin. One leaf form has both margins green; the other has one green margin and one with a broad hyaline margin.

This plant like *Anastatica hierochuntica* is strongly hygroscopic. The fronds are normally curled up into a loose ball but expand rapidly after rain. They quickly return to their curled-up condition if the rains do not continue.

On cliffs and rocky slopes from 500 to 1500m in areas of rainfall above about 400mm; widespread on the escarpment in the south but becoming rarer north of Bājil and not recorded north of the Wadi Mawr. It is abundant only in areas of good rainfall such as 'Udayn.

3. Selaginella yemensis (*Swartz*) *Spring*

Ḥuqa', Ḥuqil

A much branched prostrate plant, the branch tips becoming recurved when the plant is dry. Leaves dark green with a white, ciliate margin.

Frequent on the escarpment between 1100 and 2700m growing on terrace walls, cliffs, broken rocky slopes and in rough grassland. Its distribution is closely related to the presence of mist since it is only abundant in areas with good rainfall or those enjoying prolonged winter mists such as the Ḥujarīyah. It is very frost sensitive, only climbing to 2700m in protected valleys such as Wadi Aḥjar.

3. EQUISETUM *L.*

1. Equisetum ramosissimum *Desf.*

Horsetail. Stems arising at intervals from a black underground rhizome, green, jointed, hollow, finely ridged longitudinally, often with whorls of branches from the nodes. Leaves whorled, united at the base into a tube-like sheath ending in small black teeth. Sporangia borne in a terminal cone-like head.

Fertile plants are uncommon in Yemen.

Widespread on the escarpment and high plateau from 700 to 2900m, usually in permanently wet muddy places – marsh, ditches and slow-flowing streams.

4. OPHIOGLOSSUM *L.*

1. Ophioglossum polyphyllum *A. Br.*

O. reticulatum sensu Hepper

A small fern to c 5cm producing annual dimorphic fronds. Sterile lamina c 2cm long, broadly lanceolate with a cuneate base. Fertile frond spike-like, attached to the lower part of the sterile lamina. Sporangia in 2 rows on the upper part of the fertile frond.

Although no other *Ophioglossum* has been found in Yemen, other species might occur. *O. polyphyllum* can be distinguished from other species by the presence of the previous year's dead petiole bases.

An uncommon but perhaps overlooked species found near Ẓulmah (J. Ḥubaysh), J. Shibām (Ḥarāz) and a number of localities on the high plateau between 2000 and 2900m extending east to Rada' and Nihm (Miller). It is characteristic of small sandy depressions where water accumulates after rain and, like other plants typical of this specialised habitat such as *Craterostigma* spp. and *Michrochloa kunthii*, it appears very soon after the first rains.

5. CHEILANTHES *Swartz*

Small or medium-sized tufted ferns. Petiole polished dark red-brown in colour. Lamina bi-tripinnate usually densely scaly or with white powder beneath. Sori marginal, ± covered by the reflexed margin of the pinnules; indusium absent.

1. Lamina covered in white powder on the undersurface — *C. farinosa*
   Lamina not covered in white powder beneath — 2
2. Undersurface of lamina ± glabrous — *C. pteridioides*
   Undersurface of lamina densely covered in scales or hairs — 3
3. Lamina broadly triangular, c 5cm long — *C. coriacea*
   Lamina narrowly lanceolate or oblong, at least 10cm long — 4
4. Undersurface of lamina woolly. Pinnules orbicular — *C. catenensis*
   Undersurface of lamina scaly. Pinnules oblong — *C. marantae*

1. Cheilanthes catenensis (*Cosenti*) *H.P. Fuchs*

Notholaena vellae (Ait.) Desv.

Fern to 30cm, the whole plant densely covered with white or reddish-brown woolly hairs, scales absent. Petiole very short or absent; lamina narrowly lanceolate, dark green above, bipinnate; pinnules orbicular with a truncate base.

A local fern of rock crevices between 1200 and 3000m on the inner escarpment, the high plateau and in Khawlan Ash Sham. It is northern in distribution being absent south of J. Ḍawrān except for an isolated record near Ibb. Although usually in moderate shade it does not tolerate high rainfall and is perhaps the most drought-resistant of Yemen ferns.

2. Cheilanthes coriacea *Decne.*

Small fern to 15cm. Petiole and rhachis covered in thin chaff-like scales. Lamina ± deltoid, bipinnate, dark green above, covered in brown scales beneath; pinnules ± linear in outline with a rounded tip.

A common and drought-resistant fern of terrace walls and rock crevices on mountains from c 1400 to 2900m. It occurs all along the escarpment, on the mountain ranges round the high plateau and in the southeast mountains as far east as Dhī Nā'im. It has a very similar distribution to *Actinopteris semiflabellata* but occurs more commonly at higher altitudes than that species.

3. Cheilanthes farinosa (*Forssk.*) *Kaulf.*

Our largest *Cheilanthes* reaching 50cm. Petiole glabrous except for a few scales below; lamina broadly ovate, bi-tripinnate, green above, covered in white powder below; pinnules narrowly triangular or oblong in outline.

A rare fern of a few localities between 2000 and 2900m in the southern mountains (J. Ṣabir; J. Sumārah (Miller); J. Ba'dān; between Ribāt and Saddah) and the central escarpment (J. Milḥān (Shaḥir and Ṣa'bah) and J. Raymah (J. Kabūrah and Ẓallamlam above Ḥadīyah)). Except in the Ḥadīyah region where it grows in grassland, it is a plant of broken, scrubby cliffs.

4. Cheilanthes marantae (*L.*) *Domin*

Notholaena marantae (L.) Desv.

Fronds up to 30cm; petiole and rhachis covered in dense reddish-brown hair-like scales; lamina narrowly lanceolate or oblong, bipinnate, green above, densely covered in brown scales beneath.

A rare fern of most high mountains from the Ibb region, through Waṣāb and 'Utmah to the high peaks of the central escarpment. It occurs between 2000 and 3000m and is completely absent from the inner escarpments. It grows on open grassy slopes.

5. Cheilanthes pteridioides (*Reich.*) *C. Chr.*

C. fragrans (L.f.) Swartz

Fronds up to 15cm; petiole and rhachis covered in numerous chaff-like scales; lamina narrowly lanceolate, bi- to tripinnate, green above, the underside of the small, ± orbicular pinnules almost completely covered by the sori.

A characteristic fern of rock crevices and terrace walls on the higher mountains growing as low as 1800m near Ḥadadah (J. Raymah) but much more typical of the 2500 to 3200m range.

6. NEGRIPTERIS *Pic.-Ser.*

1. Negripteris scioana (*Chiov.*) *Pic.-Ser.*

A fragile tufted fern to c 40cm, but commonly much less. Petiole and rhachis dark shiny brown with narrow red-brown scales; lamina narrowly ovate, bipinnate, green above, covered in white powder beneath. Sori black, dot-like on the margins of the pinnules which are recurved at maturity.

This fern has been confused with *Cheilanthes farinosa* from which it differs in having scales all along the rhachis and in the black dot-like sori. In *C. farinosa* the brown sori are ± continuous on the margin.

An uncommon and perhaps overlooked fern of rock crevices in dry country, preferring damp rock overhangs and drainage lines where bryophytes also flourish; perhaps calcicole. It is found mainly in the southern half of Yemen between 900 and 1800m but reaches the exceptional height of 2600m above Ḥiliyal on J. Sumārah (Miller).

Outside Yemen it is only known from Somalia and Ethiopia.

7. PELLAEA *Link*

Medium-sized tufted ferns. Petioles shiny, dark brown; lamina tri-quadripinnate, coriaceous and glabrous in Yemen species. Sori marginal, partially covered by the reflexed margins; indusium absent.

Lamina tripinnate, narrowly ovate	*P. viridis*
Lamina quadripinnate, deltoid	*P. quadripinnata*

1. Pellaea quadripinnata (*Forssk.*) *Prantl*

Fronds 30–50cm long; lamina deltoid, quadripinnate, green; pinnules c 1–2cm long.

Only known from the grassland country at c 1800 to 2200m on the northwest slopes of J. Ẓallamlam and J. Kabūrah above Ḥadīyah in J. Raymah. It was first collected here by Forsskal in 1762.

2. Pellaea viridis (*Forssk.*) *Prantl*

Similar to the last species, but the lamina narrowly ovate, tripinnate, dark green; pinnules 1.5–4cm long.

A very rare fern of the central escarpment. It was found by Forsskal in the middle-altitude mountains above Ḥadīyah, but has not been refound there. I only know it from a wall in a densely shaded coffee grove between Qimah and Mirzin on J. Bura' where it grows with *Pteris catoptera*.

8. ADIANTUM *L.*

Maidenhair ferns. Medium-sized tufted ferns to 80cm. Petioles dark reddish-brown or black, shiny; lamina pinnate to quadripinnate; pinnae and pinnules stalked. Sori borne on the marginal lobes of the pinnae; indusium absent.

1.	Fronds pinnate, hairy especially on the rhachis, often rooting at the tips	*A. incisum*
	Fronds bi quadripinnate, glabrous, never rooting at the tips	2
2.	Pinnules mostly fan-shaped	*A. poirettii*
	Pinnules cuneate	*A. capillus-veneris*

1. Adiantum capillus-veneris *L.*

Erect fern to 60cm, glabrous except for the base of the petiole which is scaly. Petiole c 15cm long; lamina bi-tripinnate, ovate, glabrous; pinnules cuneate on delicate capillary stalks, not deciduous when old.

One of the commonest ferns in Yemen occurring between 600 and 3000m in all mountain areas except the Mashriq. It is always found on rocks in the vicinity of ± permanent water: on waterfalls, by springs, along streams or water channels and even in wells. It tolerates almost complete shade.

2. Adiantum incisum *Forssk.*

A. caudatum sensu Schwartz

Sprawling tufted fern to 30cm, hairy in all its parts. Petiole very short, usually about 2cm long; lamina pinnate, narrowly lanceolate.

This species is unique among Yemen ferns in its arching fronds which root at the tips.

A common fern of the entire escarpment region from Aḥkūm to J. Rāziḥ. It is found in scrub and on terrace walls from 600 to 2000m.

3. Adiantum poirettii *Wikstr.*

A. thalictroides Schlechtend

Very similar to *A. adiantum-nigrum*, but differing in its fan-shaped pinnules which fall with age, leaving the capillary stalks naked.

A local and perhaps overlooked fern of areas with good rainfall and plentiful mist. It is thus frequent at 2000m in the meadow country above Ḥadīyah and also occurs frequently at c 2500 to 3000m on the high escarpment mountains, particularly in the Ibb region.

9. CERATOPTERIS *Brongn.*

1. Ceratopteris cornuta (*P. Beauv.*) *Le Priéur*

C. thalictroides sensu Schwartz

Glabrous, somewhat succulent, tufted fern to 40cm, almost without scales. Petiole 5–15cm long; lamina narrowly ovate, strongly dimorphic; sterile fronds ± bipinnatifid with deltoid segments up to 15 × 10mm in size; fertile fronds usually larger than the sterile, ± bipinnate, segments linear, c 15 × 2–3mm in size. Sori marginal, ± covered by the recurved margins; indusium absent.

Only collected twice in Yemen and on both occasions in low-altitude swamp: 'Ajārah near Ḥajjaylah (Schweinfurth), near Suq Al Rūbū', Waṣāb As Sāfal (600 to 900m).

10. ONYCHIUM *Kaulf.*

1. Onychium divaricatum (*Poir.*) *Austin*

O. melanolepis (Decne.) Kunze

Small tufted fern to 20cm. Petiole yellow-green, glabrous except for a very few scales below; lamina glabrous, deltoid, tripinnate, green; pinnules less than 2mm wide. Sori covering the underside of the pinnules, ± hidden by the reflexed margins; indusium absent.

Sterile fronds have a strong resemblance to parsley leaves.

Locally common from 2500 to 3000m in rock crevices and terrace walls on a few mountains in the Ibb region – from the head of the Wadi Banā south to J. Ṣabir, Taiz.

11. ACTINIOPTERIS *Link*

Small ferns which are green when fresh but silvery when dry. Petiole deeply furrowed with a brown central ridge, scales few; lamina much shorter than the petiole, dichotomously divided into ± linear segments. Sori ± marginal (but often appearing to cover the whole undersurface), covered by the recurved margins. Indusium absent.

Schwartz's records under *A. australis* (L.f.) Link are an amalgam of the two following species.

Lamina cuneate, each segment ending in a single point — *A. semiflabellata*
Lamina fan-shaped, the segments of fertile fronds ending in several teeth — *A. radiata*

1. Actiniopteris radiata (*Swartz*) *Link*

Tufted fern with fronds to 40cm. Lamina fan-shaped, the segments terminating in 2 ± equal teeth.

A local fern of the outer escarpment usually found in partial shade in scrub and on rocky slopes from 500 to 1400m, only really common, though, in low-altitude valley scrub in the central escarpment.

2. Actiniopteris semiflabellata *Pic.-Ser.*

Tufted fern with fronds to 20cm. Lamina ± wedge-shaped, the segments ending in a single point.

A very drought-resistant fern which occurs between 700 and 2600m along the whole escarpment and extends through the southeastern hills to Bayḍā' and through Khawlan Ash Sham to Majz. Although occurring on cliffs and other natural habitats it is most typical of terrace walls where it may be the only fern present.

12. PTERIS *L.*

Large tufted ferns, 50–100cm high. Petiole yellow-brown, scaly only below; rhachis ± glabrous except for a few scales at the intersections with the pinnae; lamina in Yemen species pinnate or bipinnate. Sori elongate, marginal, not obscured by the weakly reflexed margins; indusium absent.

1. Pinnae simple — 2
   Pinnae pinnatisect — 3
2. Lamina with 10–40 pairs of pinnae, the lowest reduced in size — *P. vittata*
   Lamina with 2–7 pairs of pinnae the lowest not reduced in size — *P. cretica*
3. Pinnules entire — *P. catoptera*
   Pinnules toothed — *P. serrulata*

1. Pteris catoptera *Kunze*

Vigorous fern to 1½m. Petiole shiny, up to c 50cm long; lamina broadly lanceolate to subdeltoid, bipinnate, the basal pair of pinnae forked to the base, the segments untoothed, not exceeding 15mm long.

A rare fern found between 1500 and 1800m in densely shaded gullies in the central escarpment; below Jirjir, Ḥadīyah; between Qimah and Mirzin, J. Bura'; in several places around Ṣa'bah, J. Milḥān.

2. Pteris cretica *L.*

Fern to 60cm. Petiole shiny, 10–30cm long; lamina ovate, pinnate with 2–7 pairs of pinnae, the pinnae linear with a cuneate base, the basal pair sometimes forked, the sterile pinnae serrulate, slightly wider than the fertile pinnae.

An uncommon but locally frequent fern of mist regions in the southern mountains and the central escarpment, preferring grassy places: below Dimarayn, J. Ṣabir; above Jabin, J. Ba'dān; J. Kabūrah, Ḥadīyah; Shaḥir, J. Milḥañ; Bayt Bashārī, Maḥwīt.

3. Pteris serrulata *Forssk.*

P. dentata Forssk.

Vigorous fern to 1½m. Petiole shiny, up to 50cm long; lamina ovate or lanceolate, bipinnate, the basal pair often forked but not from near the base, the segments serrate, 10–30mm long.

Common on the escarpment between 1500 and 2900m growing in rocky places by permanent water, often in partial shade, particularly by irrigation channels, seepages on cliffs and streamsides.

4. Pteris vittata *L.*
Fern to c 70cm. Petiole dull brown, less than 10cm long; rhachis thinly scaly; lamina lanceolate, bipinnate with 10–30 pairs of pinnae, the pinnae ± linear with a truncate base, always simple, reduced in size towards the base of the lamina.

A common fern of the escarpment between 1500 and 2700m, often near water and usually favouring the banks of small streams.

13. PTERIDIUM *Scop.*

1. Pteridium aquilinum (*L.*) *Kuhn*
Bracken. Fronds arising at intervals from a subterranean rhizome, very vigorous, reaching 2m. Petiole glabrous except below; scales absent; lamina deltoid, tripinnate, coriaceous, glabrous above but tomentose below. Sori continuous along the revolute margin.

A cosmopolitan fern known in Yemen only from the grasslands at 2000 to 2300m on the northwest slope of J. Ẓallamlam above Ḥadīyah.

14. CHRISTELLA *Léveillé*

1. Christella dentata (*Forssk.*) *Holtt.*
Dryopteris dentata (Forssk.) C. Chr.
D. mauritiana sensu Schweinfurth
Cyclosurus dentatus (Forssk.) Ching
Vigorous tufted fern to 70cm high. Petiole with a few light brown scales, hairy especially above; lamina pinnate, pubescent, characteristically light green in colour with the pinnae regularly decreasing in length towards the base so giving the lamina a narrowly elliptic outline. Sori orbicular on the veins; indusium present, hairy.

Quite frequent between 800 and 1900m in the central escarpment, where it grows mostly by shaded ditches in the coffee groves; elsewhere known only from Wadi Ḥadabah, 'Udayn.

15. TECTARIA *Cav.*

1. Tectaria gemmifera (*Fée*) *Alston*
Tufted fern to 50cm. Petiole yellow-brown with a few scales below; lamina ovate, bipinnate below and, in marked contrast, bipinnatifid above, the costae hairy. Sori round; indusium present.

This fern can be recognised by the presence of gemmae (asexual buds which produce a new plant) on the upper part of the rhachis at junctions with the costae. The only other Yemen fern with this feature is *Asplenium protensum*.

A very rare plant of damp shaded gullies at c 1700m in the central escarpment: below Jirjir, Ḥadīyah; north of Ṣa'bah, J. Milḥān.

16. DRYOPTERIS *Adans.*

1. Dryopteris inaequalis (*Schlechtend.*) *O. Kuntze*
D. rigida sensu Schwartz
Vigorous tufted fern to 40cm. Petiole densely scaly below; lamina lanceolate, tripinnate, the segments with toothed margins; rhachis and costae with light brown scales. Sori on the veins, rounded; indusium reniform.

One of a number of ferns found between 2000 and 3000m on grassy slopes, terrace walls and cliffs in areas of plentiful mist. It is rare at high altitudes on J. Ṣabir and J. Manār, but much more frequent in the central escarpment being found on J. Raymah, J. Milḥān and Maḥwīt.

17. POLYSTICHUM *Roth*

Fairly large ferns with scaly rhizomes and petioles. Fronds tufted, the lamina lanceolate, bipinnate, the pinnules with aristate teeth on the margins. Sori on the veins, round, peltate.

Pinnules serrate. Marginal bristles less than 1mm long ... *P. sp. A*
Pinnules, at least the basal ones, pinnatisect. Marginal bristles, or some of them, more than 1mm long ... *P. fuscopaleacum*

1. Polystichum fuscopaleacum *Alston*
Rather similar to both *Dryopteris inaequalis* and the next species and best distinguished from both by the densely scaly petiole, rhachis and costae and the long aristate bristles which terminate the teeth on the pinnules. The basal pinnules are rather distinct, being pinnatisect with one much enlarged lobe on the upper side.

Only known at 2900m in a shaded limestone gully separating Shirnamah from the main part of J. Maswār (Wood 3318).

2. Polystichum sp. A
Similar to the last species but only moderately scaly and the pinnules serrate, never pinnatisect, the teeth terminating in a very short bristle.

Only known at 1700m from a deep shaded gully leading north from Aṣ Ṣa'bah, J. Milḥān (Wood 3462).

18. HYPODEMATIUM *Kunze*

1. Hypodematium crenatum (*Forssk.*) *Kuhn*
Dryopteris crenata (Forssk.) O. Kuntze
Tufted fern to 50cm but commonly less. Petiole scaly at base, hairy above; lamina ± deltoid, tripinnate, grey-green, hairy on both surfaces. Sori rounded; indusium grey, hairy.

Frequent on rock crevices and terrace walls along the whole escarpment from 900 to 2600m, but usually absent from low altitudes except in the coffee country of the central escarpment.

19. CYSTOPTERIS *Bernh.*

1. Cystopteris fragilis (*L.*) *Bernh.*
Small, tufted fern to 30cm. Fronds delicate; petiole pale brown with a few scales near the base; lamina thin-textured, light green, glabrous, narrowly lanceolate, bipinnate with toothed pinnae. Sori ovate, on the veins beneath; indusium present.

A rare fern of shaded rocks above 2800m present on nearly all the suitable mountains in the south (J. Ṣabir, J. Ta'kar; Al Manār on J. Ḥubaysh, J. Ba'dān, J. Manār) but elsewhere only known from J. Maswār.

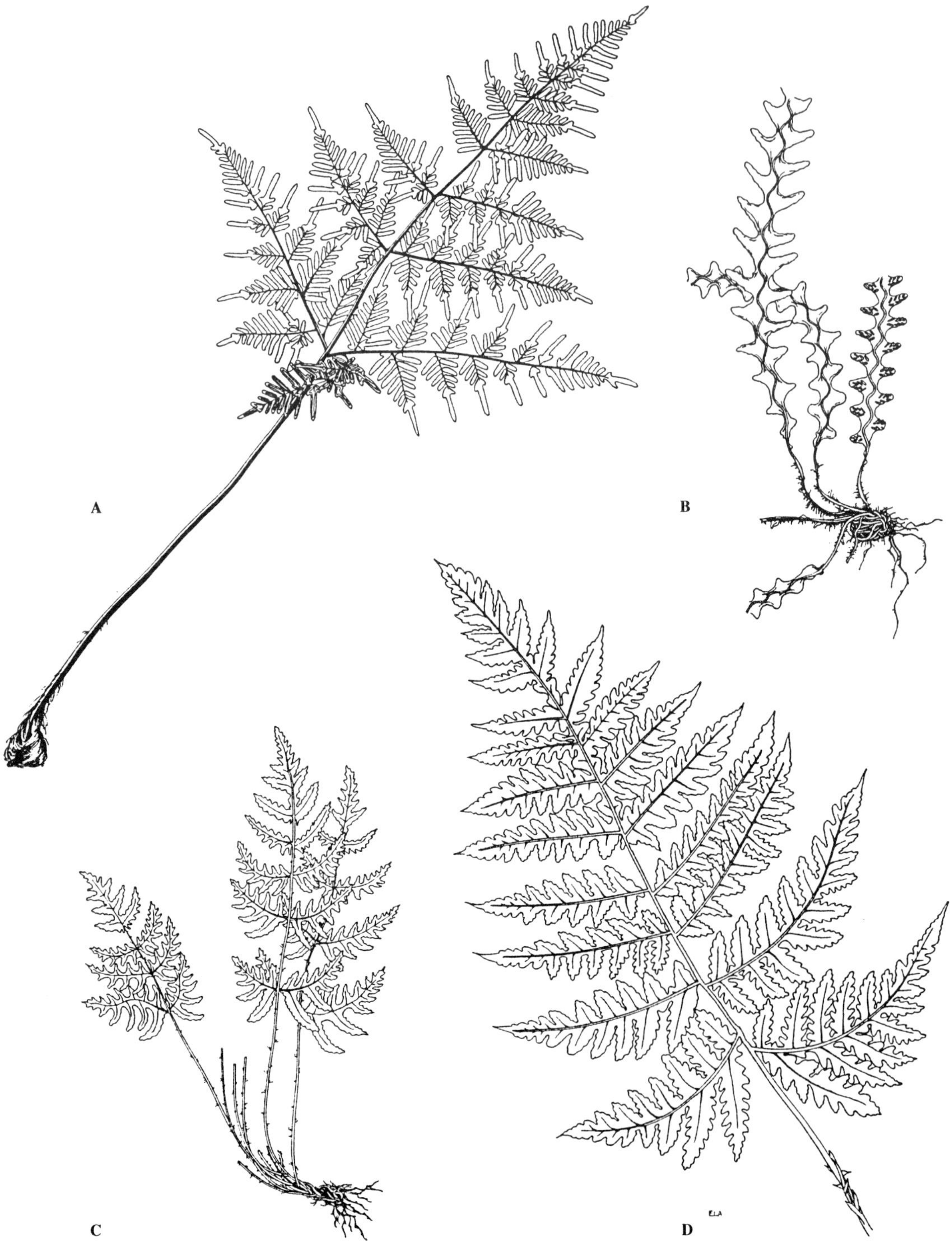

Fig. 1. **A**. *Pellaea quadripinnata*; **B** *Ceterach dalhousiae*; **C** *Negripteris scioana*; **D** *Tectaria gemmifera*.

20. ASPLENIUM *L.*

In Yemen mostly small ferns with tufted fronds to 30cm. Lamina pinnate to tripinnate. Sori linear or elliptic on one side of a vein, but often seeming to cover the whole undersurface at maturity; indusium attached to the vein so often exposing one side of the sori.

1. Fronds pinnate, the pinnae toothed but never lobed — *A. trichomanes*
   Fronds bi- to tripinnate or if simply pinnate, the pinnae deeply lobed — 2
2. Rhachis of the lamina densely scaly or hairy — 3
   Rhachis of the lamina ± glabrous — 4
3. Petiole base covered in numerous red-brown hair-like scales. Pinnae terminating in long narrow points. Gemmae absent — *A. aethiopicum*
   Petiole base not densely covered in red-brown hair-like scales. Pinnae not ending in long narrow points. Gemmae present on the rhachis — *A. protensum*
4. Petiole black — *A. adiantum-nigrum*
   Petiole green — 5
5. Segments acute. Sori clearly medial — *A. varians*
   Segments rounded. Sori apparently marginal — *A. rutifolium*

1. Asplenium adiantum-nigrum *L.*

Tufted fern to 25cm. Petiole black, swollen at the base, covered in thin chaffy scales; rhachis green, nearly glabrous; lamina deltoid, bi- to tripinnate, coriaceous, dark shiny-green.

An uncommon fern of rock crevices, terrace walls and cliffs. It is most typical of the highest peaks of the escarpment reaching 3500m on J. Nabī Shu'ayb but descends to 2000m in the mist regions of the central escarpment.

2. Asplenium aethiopicum (*Burm. f.*) *Becherer*

A. filare (Forssk.) Alston
A. praemorsum Swartz

Densely tufted fern to 30cm but usually much less, the whole plant covered in numerous hair-like scales. Petiole black below becoming greener along the rhachis above; lamina lanceolate, dark green, bipinnate, the pinnae extending into long narrow points, pinnatifid, the segments terminating in several unequal teeth.

Common along the whole escarpment from 1800 to 3000m. It is the characteristic fern of terrace walls in areas of good rainfall where it is an important constituent of the terrace wall flora with *Campanula edulis*, *Pentas lanceolata* etc.

3. Asplenium protensum *Schrad.*

Tufted fern to 50cm. Petiole and rhachis dark purple-brown, covered in numerous white crisped hairs and scattered scales; gemmae often present on the rhachis above; lamina oblong, dull green, pinnate, the pinnae deeply pinnatifid with the lobes toothed at the tips.

A very rare fern known only at 1700m in the gully leading north of Aṣ Ṣa'bah, J. Milḥān.

4. Asplenium rutifolium (*Berg.*) *Kunze*

A. achilleifolium sensu Schwartz

Tufted fern to 20cm. Petiole and rhachis green, scales numerous below but becoming few above; lamina obovate to oblong, rather fleshy, green, bipinnate (recalling an *Achillea* leaf) with linear pinnules. Sori appearing to be marginal.

Yemen plants are assigned to variety **bipinnatum** (*Forssk.*) *Schelpe.*

A very rare fern originally collected by Forsskal in the 'mountain woods of Yemen' near Bulgose (Buljawz). It has only been found once subsequently at 1800m in the gully below Jirjir, above Ḥadīyah (Wood 2241) which must be close to Forsskal's original locality

5. Asplenium trichomanes *L.*

A. viride sensu Hepper

Tufted fern to 15cm. Petiole and rhachis dark brown, the former up to 2cm long, scales absent or very few; lamina narrowly oblong, pinnate, with small rounded toothed pinnae scarcely longer than broad.

An uncommon fern of rock crevices and cliffs from 2000 to 3000m in the mist regions of the central escarpment and on a few other high mountains especially in the Ibb region.

6. Asplenium varians *Hook. & Grev.*

Tufted fern to 10cm. Petiole and rhachis green, scales very few; lamina bipinnate, lanceolate, the pinnules with sharp teeth.

Only known from deeply shaded rocks in the gully below Jirjir, above Ḥadīyah, J. Raymah (c 1800m).

This Yemen station is of some interest as it forms a geographical link between the isolated localities for this species in Kenya and its main area of distribution in East Asia.

21. CETERACH *DC.*

Small tufted ferns. Fronds simple, pinnatifid, glabrous on the upper surface but scaly below. Sori linear; indusium absent.

Undersurface of lamina densely covered with shiny brown scales which hide the sori — *C. officinarum*
Undersurface of lamina sparsely covered in scales so the sori are clearly visible — *C. dalhousiae*

1. Ceterach dalhousiae (*Hook.*) C.Chr.

C. cordatum auct.

Small tufted fern. Petiole dark brown, c 1cm long; lamina up to 15cm long, scales thinly covering the undersurface so the sori are clearly visible.

A rare fern found between 1500 and 2000m in a few densely shaded localities mainly in the central escarpment: in the gully below Jirjir, above Ḥadīyah; between Mirzin and Aṭ Ṭuba', J. Bura'; in gullies around Ṣa'bah, J. Milḥān; on cliffs, Raḥbān, Maḥwīt; in the valley above Marātibah, J. Ḥabashī.

Fig. 2. **A** *Asplenium trichomanes*; **B** *Pleopeltis macrocarpa*; **C** *Asplenium varians*; **D** *Adiantum incisum*.

2. Ceterach officinarum *DC.*
C. cordatum auct.
Small tufted fern. Petiole black, c 2cm long; lamina up to 10cm long, scales numerous, brown, shiny, completely covering the undersurface so the sori are hidden.

A frequent fern of terrace walls and rock crevices from 2500 to 3300m. It is found on all the high mountains of the escarpment and high plateau regions.

22. ARTHROPTERIS *J. Sm.*

1. Arthropteris orientalis *(J.F. Gmel.) Posth.*
Fronds arising individually from a rhizome, up to 40cm high. Petiole and rhachis brown, hairy, with a very few scales below; lamina lanceolate, bipinnatifid, almost glabrous, with pale white dots above. Sori round, covering most of the undersurface; indusium present, glabrous.

A very local species of the coffee regions of four mountains: J. Ṣabir (between Misrākh and Ḥasabān), J. Milḥān (between Ḥabāsh and Ṣa'bah), J. Bura' (near Mirzin), J. Raymah (Bani Dawān and the Ja'farīyah area, where it is abundant). It grows between 1000 and 2000m and is frequently found on rocky banks and in gullies in the shade of the coffee bushes.

23. PLEOPELTIS *Willd.*

1. Pleopeltis macrocarpa *(Willd.) Kaulf*
Small fern to 10cm with fronds arising at intervals from a rhizome. Lamina simple, unlobed, lanceolate, covered with peltate scales beneath. Sori in two rows on either side of the midrib, ± rounded; indusium absent.

A rare fern of shaded high-altitude cliffs between 2700 and 2900m in the southern mountains: above Ḥiliyal, J. Sumārah; on the ridge between J. Sumārah and J. Manār; above Jabīn, J. Ba'dān near Ḥiṣn 'Arūs, J. Ṣabir.

24. LOXOGRAMME *(Blume) Presl*

1. Loxogramme lanceolata *(Swartz) C. Presl*
Small fern to 15cm, with fronds arising at intervals from a rhizome. Lamina simple, unlobed, somewhat fleshy, lanceolate. Sori arranged in two rows on either side of the midrib, elongate; indusium absent.

A rare plant of damp shaded gullies between 1700 and 1900m in the central escarpment: below Jirjir, above Ḥadīyah; north of Ṣa'bah, J. Milḥān.

25. MARSILEA *L.*

Aquatic ferns which can only be identified reliably when sporocarps are present. The following is the only species positively identified from Yemen, but others might be expected to occur.

1. Marsilea aegyptiaca *Willd.*
Rhizome creeping. Lamina consisting of 4 digitate obovate-obdeltoid shamrock-like pinnae at the apex of the erect petiole. Sporocarps arising from the base of the petiole, pedicellate, cuboid, grooved and appearing saddle-shaped.

Very variable vegetatively. Plants in standing water are glabrous with petioles up to 30cm long and entire pinnae. Plants in dry places are pubescent with petioles often less than 5cm long and the pinnae with crenate margins. Sporocarps are usually only found on plants on pool margins where the water has receded.

Locally abundant from c 1800 to 3100m on and around the high plateau. It is found in various aquatic habitats: saline mudflats near Rada'; in pools at the edge of the Wada Banā as far east as Nādirah; in a cistern in limestone country at Ḥūth but is particularly a feature of exposed summit pools around Kawkabān and on the Sumārah – Manār range. In many suitable places it is inexplicably absent.

CUPRESSACEAE

JUNIPERUS *L.*

1. Juniperus procera *Hochst. ex Endl.* Firūwish, 'Ar'ar
J. excelsa M. Bieb.
J. macropuda sensu Wissmann
Unmistakable. Yemen's only native conifer. Resinous evergreen tree of untidy appearance rarely exceeding 10m in Yemen. Bark brown, fibrous, peeling easily. Juvenile trees have needle-like leaves c 2cm long but mature trees have scale-like leaves. Cones small (4mm long), solitary, blue-grey, berry-like.

Found between 1800 and 3300m mostly on mountains with medium rainfall and locally abundant in Khawlan Ash Sham, Jabal Lawz and the higher mountains of Taiz province (J. Sawraq, J. Ṣabir, J. Ḥabashī and the Ḥujarīyah mountains). Although Juniper woodland is probably the natural climax in the areas mentioned, the woodland which is such a feature of the Asir region in Saudi Arabia is nowhere found in Yemen and this is presumably the result of tree felling and goat grazing over many centuries.

On the main west-facing escarpment mountains the Juniper is rare and mostly found as scattered individuals. Various observers have suggested this scarcity is the result of human interference since the Juniper is common in similar situations in Africa (see for example von Wissmann: *Die Juniperus – Gebirgswälder in Arabien*). I find this explanation very hard to accept. The Juniper is abundant in the Ḥujarīyah with its dense human population but completely absent from J. Milḥān with a minute population and huge tracts of inaccessible scrub.

A possible explanation for the Juniper's distribution in Yemen may lie in the relationship of mist and rainfall. It has long been recognised that the water supply of plants with needle-like leaves can be increased by the capture of mist particles on the leaves. It is therefore not surprising that the Juniper is only abundant in areas of prolonged winter mist. Its rarity in the Ibb area and the central escarpment where mists are frequent would seem to be due to an intolerance of high summer rainfall. While accepting that this does not explain its African distribution, I can see no other explanation for its distribution in Yemen. It is interesting that *Erica arborea*, a species with similar needle-like leaves has an almost complementary distribution, being abundant only in mist-covered areas of the central escarpment.

The Cypress, **Cupressus sempervirens** *L.*, is commonly cultivated, particularly in the Sana'a region. Several cultivars are grown but forma **sempervirens** with a narrow crown is the most common. There are fine mature trees around some of the old Imamic palaces.

EPHEDRACEAE

EPHEDRA *L.*

Dioecious shrubs with long stems and minute opposite leaves. Flowers in small spikes, the female developing into a cone.

Blue-green climbing shrub with long fragile branches.
 Inflorescence of pedunculate heads — *E. aphylla*
Green shrub with short stiff erect branches. Inflorescence
 sessile, axillary — *E. pachyclada*

1. Ephedra aphylla *Forssk.*
E. alte C.A. Mey.
Shrub, climbing high over bushes. Stem woody; branches blue-green, scabrous, fragile. Leaves opposite, c 2mm long. Inflorescence dull green, the flowers in terminal pedunculate heads. Cone berry-like.

Flowers in spring.

Collected three times growing over *Acacia* by wadis in the Mashriq between 1300 and 2000m: Wadi Nuwāsh, north of Rada' (Westinga); Watadah; Sirwāḥ. It is probably more frequent than these few records suggest. (Wood 3151, 3155)

2. Ephedra pachyclada *Boiss.*
E. alte sensu Schwartz
E. fragilis sensu Deflers
Undershrub to ½m. Branches green, minutely scabrid, stiff, erect. Leaves opposite, c 2mm long. Inflorescence axillary, yellow. Cone dry, largely enclosed by the white-margined bracts.

The Yemen plant is subspecies **sinaica** *Freitag & Maier-Stolte.*

Flowers in spring.

In rock crevices, often on cliffs between 2000 and 3300m. It is common on all the mountains round the high plateau and also occurs in the rain-shadow area of Khawlan Ash Sham.

MAGNOLIACEAE

MICHELIA *L.*

1. Michelia champaca *L.*
Small tree to 8m. Leaves narrowly elliptic, entire with an acuminate tip, up to 20cm long. Flowers orange, fragrant, made up of whorls of various sized perianth segments not clearly differentiated into petals or sepals. Fruits spirally arranged on a thick peduncle, dry, tuberculate opening on one side to release the single seed.

Flowers in May and June.

In coffee terraces between 700 and 1000m in the central escarpment: in the main valley above Ḥadīyah, J. Raymah; Maghārib, J. Sāfān, Ḥarāz. Probably introduced from India.

ANNONACEAE

ANNONA *L.*

1. Annona squamosa *L.* Safarjil, Khirmūsh
Custard Apple. Shrub 2–4m high. Leaves without stipules, alternate, petiolate, usually elliptic, obtuse, slightly blue-green, entire, c 10cm long. Flowers arising opposite the leaves, solitary, borne on a recurved pedicel; perianth segments in 3s, narrowly triangular, yellow brown. Fruit globose, dull green (blackish when ripe), c 5cm in diameter, consisting of numerous aggregated berries. The brown seeds are surrounded by sweet, edible pith.

An introduction which is cultivated but also well naturalised particularly along wadis in the central escarpment where it thrives between 400 and 1000m though reaching as high as 1500m.

LAURACEAE

CASSYTHA *L.*

1. Cassytha filiformis *L.* Tushām
Yellow-green twining parasite with few reduced scale-like leaves and a small spicate inflorescence, thinly pilose except for the new growth which is tomentose. Flowers very small, up to 2mm long, without petals but consisting of 6 pale yellow petal-like sepals. Fruit a black drupe ± enclosed by the sepals.

Locally abundant between 300 and 1000m, mostly in bushland on the lower slopes of the central escarpment and in similar areas north to J. Marran but sometimes found in *Acacia* scrub in the Tihama, particularly around 'Abs; absent south of the Wadi Rima'. It is usually found festooning low shrubs and rarely grows on trees above 3m high.

PIPERACEAE

PEPEROMIA *Ruiz & Pav.*

Herbs, commonly epiphytic in other countries but growing on rocks in Yemen. Leaves entire, rather fleshy, dark, often bronze-green. Inflorescence a dense spike of minute crowded flowers without a perianth. Fruit a small sessile berry.

1. Plant pubescent. Leaves opposite. — *P. blanda*
 Plant glabrous. Leaves alternate — 2
2. Leaves oblanceolate to elliptic, always exceeding 2cm in length — *P. abyssinica*
 Leaves broadly ovate to obliquely elliptic, less than 2cm in length — *P. pellucida*

1. Peperomia abyssinica *Miq.*
Trailing glabrous perennial herb with erect fleshy stems to 20cm. Leaves alternate, variable in shape but usually ellip-

tic or oblanceolate, c 2–4cm long, base cuneate, often bronze in colour. Flower spikes c 4–8cm long.

Only known from cliff ledges above Ḥiliyal, J. Sumārah at c 2600m.

2. Peperomia blanda (*L.*) *Kunth*
P. arabica Miq.
Very similar to the previous species in general appearance but pubescent and with opposite, always elliptic leaves.

The Yemen plant is subspecies **leptostachya** (*Hook. & Arn.*) *Düll.*

Uncommon on damp shaded cliffs, walls and gullies between 1100 and 1800m, usually growing with rare fern species: Marātibah, J. Ḥabashī; Wadi Dūr, 'Udayn; in the great wadi above Ḥadīyah, quite frequent; around Aṣ Ṣa'bah, J. Milḥān; on the escarpment below Maḥābishah.

3. Peperomia pellucida (*L.*) *Kunth*
Delicate prostrate glabrous fleshy annual herb with slender stems up to 10cm long. Leaves alternate, broadly ovate to obliquely elliptic, c 1–2cm long, base shallowly cordate, usually thin in texture and green in colour. Flower spikes c 1–2cm long.

Only known from around 'Udayn market at 1500m where it is abundant on rocks in open scrub.

ARISTOLOCHIACEAE

ARISTOLOCHIA *L.*

1. Aristolochia bracteata *Retz* Līyah
Prostrate trailing glabrous, glaucous perennial herb. Leaves alternate, ovate, c 4–5cm long, base hastate to subcordate, margin irregularly crenate. Flowers solitary, axillary consisting of an enlarged tubular calyx with a prominent purple-brown petaloid lip. Fruit a shortly cylindrical capsule, c 2cm long.

Widespread and frequent on the alluvial plains in the east of the Tihama and on the escarpment from 200 to 1500m, usually growing along wadi beds and in stony seasonally flooded fields.

NYMPHAEACEAE

NYMPHAEA *L.*

1. Nymphaea coerulea *Sav.*
Waterlily. Rhizomatous plant producing floating leaves on a long petiole. Leaves large, rounded, cordate, entire. Flowers solitary on the surface of the water; petals numerous, white or bluish; stamens numerous.

Known from a single pool at c 1900m at the head of the wadi above 'Adūf near Turbah.

Although this is the only modern record of the occurrence of waterlilies in Yemen, they are mentioned by the fourteenth-century writer Al Afḍal al Abbās ibn 'Alī ibn Raṣul in the *Bughyat al Fallāḥin*.

CERATOPHYLLACEAE

CERATOPHYLLUM *L.*

1. Ceratophyllum demersum *L.*
Entirely submerged, dark green, rather stiff perennial aquatic herb. Leaves whorled, often crowded into a dense mass at the top of the stem, once or twice forked, the segments capillary with finely dentate margins. Flowers solitary, sessile in the axils of the leaf whorls.

Easily confused with sterile specimens of *Myriophyllum spicatum* with which it sometimes grows, but distinguished by the leaves which are pinnate in *Myriophyllum*, but forked in *Ceratophyllum*.

A rare plant growing in ponds in scattered localities between 700 and 2400m: between Taiz and Ḥays (Deflers); Ḥūth; near Suq al Jum'ah, J. Ash Shirq; Bayt Bādī, 'Amrān; Suq Ar Rūbū', Waṣāb As Sāfal.

RANUNCULACEAE

Herbs or climbing plants. Leaves sheathing, without stipules and often palmate. Inflorescence terminal; flowers bisexual, regular; petals present or absent, free; stamens numerous. Fruit of dry achenes.

1.	Climbing plants. Leaves opposite	1. *Clematis*
	Erect or aquatic herbs. Leaves alternate	**2**
2.	Petals present	3. *Ranunculus*
	Petals absent	2. *Thalictrum*

1. CLEMATIS *L.*

Woody climbers. Leaves opposite, pinnate; leaflets 1–7. Inflorescence a many-flowered panicle with prominent, cream petaloid sepals. Fruit of silky, white achenes.

Both the following species flower mainly in November–December.

Leaflets pubescent, some lobed	*C. hirsuta*
Leaflets glabrous at least above, never lobed	*C. simensis*

1. Clematis hirsuta *Guill. & Perr.* Ḥalal, Sharijah
C. wightiana Wall
Vigorous climber with deltoid to orbicular leaflets, which are dull green, pubescent, irregularly toothed and frequently lobed.

Frequent along watercourses, in cliff gullies, in scrub and on terrace walls in areas of good rainfall along the whole escarpment between 700 and 2900m, but most typical of areas of high rainfall at middle altitudes.

2. Clematis simensis *Fres.* Ḥalal, Sharijah
Similar to the previous species but leaflets ovate with a long acuminate tip, dark shiny green above, ± glabrous, regularly toothed but never lobed.

In similar conditions to the previous species but not recorded below 1700m and more typical of the higher mountain regions of the escarpment.

2. THALICTRUM *L.*

1. Thalictrum minus *L.*

Slightly glandular perennial to 1m. Leaves 2–3 pinnate, up to 25cm long; leaflets about as broad as long, deeply dentate. Inflorescence a large open panicle; flowers without petals, small; stamens long, yellow, pendent. Fruit a head of achenes.

T. minus is a polymorphic species. Our plants agree well with Ethiopian examples, which have been called *T. schimperianum* A. Rich. but this is doubtfully distinct from *T. minus*.

Frequent from 1500–3000m on all the wetter southern mountains from Turbah north to Waṣāb and Iriyān; completely absent further north except for an isolated record near the waterfall at Baw'ān. It is found on cliffs and rocky gullies where it is protected from grazing. (Wood 72/145, 75/501.)

3. RANUNCULUS *L.*

Annual or perennial herbs usually with palmate leaves. Flowers regular; petals 5, white or yellow; stamens and carpels numerous. Fruit a head of achenes.

Terrestial plant with yellow flowers	*R. multifidus*
Aquatic plant with white flowers	*R. rionii*

1. Ranunculus multifidus *Forssk.*

R. forskoehlii DC.

Buttercup. Erect perennial to 50cm. Leaves pinnate with pinnatifid segments. Flowers in lax panicles; petals yellow.

Frequent on the escarpment and high plateau from 1500 to 3000m growing in marsh and by streams.

2. Ranunculus rionii *Lagg.*

R. aquatilis sensu Schwartz

Water Crowfoot. Aquatic herb lacking floating leaves and usually totally submerged except for the flowers and fruit. Leaves capillary, dark green. Flowers solitary; petals white, up to 5mm long. Achenes c 1mm long.

A local plant found in ponds between 2500 and 3000m on the high plateau, particularly around Yarīm. Like several other aquatics it is usually abundant in one pond, but absent from others. (Wood 1306, Hepper 5636.)

Ranunculus rionii is often confused with related species but is best distinguished by its small achene. Its closest ally, *R. trichophyllus* Chaix., has achenes at least 2mm long while *R. aquatilis L.* has petals 10cm long.

Nigella sativa *L.* is cultivated on the high plateau and has been found in the Qa' Bakīl (Hepper) and Arḥab. It is an annual with pale blue petaloid sepals. The black seeds are sold in the markets as 'Habbah Sawda' and are used in cooking.

BERBERIDACEAE

BERBERIS *L.*

1. Berberis holstii *Engl.* — Ḥadhḥadh

B. forskaliana C.K. Schneider

B. petitiana C.K. Schneider

Spiny shrub to 3m, though commonly much less, the stems pale yellow-grey. Leaves borne in alternate clusters in the axils of the 3 – fid spines, oblanceolate, 2–4cm long, reticulate on both surfaces, entire on flowering shoots except for the cuspidate tips but often toothed on sterile shoots. Flowers in few-flowered axillary cymes, yellow, c 5mm long, rarely seen in Yemen. Fruit a purple drupe.

The type of *B. forskaliana* is a sterile shoot with very short internodes. All fertile plants I have seen in Yemen have internodes 10–15mm long and agree well with *B. holstii*.

A local plant of montane grassland and cliff gullies found on the escarpment from J. Ṣabir north to J. Rāziḥ and on mountains surrounding the high plateau; like many other species it occurs much lower in the grasslands of the central escarpment (1400m) than on the more arid inner mountains where it reaches 3200m on J. Lawz.

MENISPERMACEAE

Dioecious climbing plants. Leaves alternate, without stipules. Flowers inconspicuous, greenish; petals variable in number, up to 6. Fruit drupaceous.

1.	Leaves peltate	3. *Stephania*
	Leaves not peltate	**2**
2.	Leaves cordate	1. *Tinospora*
	Leaves not cordate	2. *Cocculus*

1. TINOSPORA *Miers*

1. Tinospora sp. A

Glabrous climber with distinctive peeling bark. Leaves ovate, shortly acuminate, c 6cm long, the base cordate. Inflorescence not seen.

Only known from a sterile specimen collected in *Salvadora* scrub at 300m between Bayt al Faqīh and Manṣūrīyah. It may be an African or Indian species or may be distinct. (Wood 2668.)

2. COCCULUS *DC.*

Dioecious climbing shrubs. Leaves simple or lobed. Male and female inflorescences similar, consisting of small axillary cymes.

Plant glabrous or appressed hairy. Leaves 1–3-veined	*C. pendulus*
Plant spreadingly hairy. Leaves 3–5-veined	*C. hirsutus*

1. Cocculus hirsutus (*L.*) *Diels*

Densely pubescent woody climber with thin, ovate – elliptic, often 3-lobed, mucronate leaves, up to 8 × 3cm, which are 3–5-veined from the base and diminish regularly in size towards the ends of the branches. Inflorescence of shortly pedunculate cymes.

Locally frequent from 400 to 1500m in the scrubby foothills of the central escarpment, but not found south of the Wadi Zabīd or north of the Wadi Mawr; inland much rarer and restricted to the lusher wadis such as Wadi Sukhnah, Wadi 'Annah, and Wadi 'Ayn 'Alī, Ḥajjah.

2. Cocculus pendulus *(J. R. & G. Forst.) Diels*
Adpressed pubescent or glabrous climber with coriaceous, oblong, usually entire, 1–3-veined leaves up to 3 × 1.5cm but diminishing in size towards the tip of shoots. Inflorescence similar to *C. hirsutus*.

Locally frequent from 100 to 500m in *Acacia ehrenbergiana* scrub in the Tihama; more scattered along sandy wadis up to 2000m in the Mashriq, where it climbs over bushes lining the wadis.

3. STEPHANIA *Lour.*

1. Stephania abyssinica *(Dillon & A. Rich.) Walp.*
Glabrous herbaceous climber. Leaves peltate, orbicular, acuminate, 5–12cm long, dark green above, but much paler below. Flowers in small axillary cymes.

A very local plant of damp gullies, terraces and broken rocky ground found between 1600 and 2700m at Mudhaykhirah (J. 'Ayfū, J. Qarad) and in the central escarpment (J. Milḥān; J. Ḥufāsh; above Ḥadīyah, J. Raymah).

PAPAVERACEAE

Herbs with latex. Leaves alternate, deeply lobed. Flowers solitary, mostly terminal; sepals 2, petals 4–6, showy, quickly deciduous. Fruit usually a capsule as in our species.

Leaves not spiny. Flowers red or purple	1. *Papaver*
Leaves spiny. Flowers yellow	2. *Argemone*

1. PAPAVER *L.*

1. Papaver dubium *L.*
Erect, hispid annual to 50cm, with white latex. Leaves pinnatifid, usually 3–5 lobed. Petals 4, crumpled in bud, orange-red, 2–4cm long. Capsule obovoid-oblong, 1.5–2.5cm long, the stigmas persistent as a disc on the top.

A rare field weed found from 2300 to 3000m on the ridge between J. Sumārah and J. Manār, at Kawkabān and in Wadi Ḍahr (Ahmed Khattab).

Papaver somniferum *L.*, the Opium Poppy, known locally as 'Khishkhāsh' is cultivated in the Ibb region. Cultivators are usually evasive as to its precise localities but J. Fleurentin has photographed it above Dhī Sufāl. It is a distinctive plant with glaucous leaves and pale purple-brown petals. The dried capsules are sold in the Taiz and Ibb markets. I have no knowledge of its use as a narcotic in Yemen, but the seeds are used in cooking and traditional medicine.

2. ARGEMONE *L.*

1. Argemone mexicana *L.*
Short-lived perennial herb to 1m, latex yellow. Leaves sessile, pinnatifid, spiny, glaucous with white veins. Flowers large, solitary; petals 6, yellow, 2–4cm long. Capsule ellipsoid, spiny, glaucous, 3–4cm long.

Two distinct colour forms occur; the bright yellow form occurs below 2000m and the pale yellow form above this height.

A weed introduced from America found on roadsides, dry wadi beds and less frequently in fields between 100 and 2800m. It is probably a relatively recent introduction through Aden since it appears to be spreading northwards. At present it is widespread in the south from Mawza' east to Damt but has not been recorded north of the Wadi Rima' apart from an isolated occurrence at Sana'a.

FUMARIACEAE

Very similar to *Papaveraceae* in which it is often included, but the flowers irregular.

FUMARIA *L.*

Annual herbs with weak sprawling stems. Leaves grey, 2–4 pinnatisect. Inflorescence in racemes with small white or pale pink flowers. Fruit an indehiscent nutlet.

Sepals 1.5–2mm long	*F. abyssinica*
Sepals 0.5–1.5mm long	*F. parviflora*

1. Fumaria abyssinica *Hamm.*
Decumbent herb with pink flowers 5-7mm long in shortly pedunculate racemes; sepals 1.5–2mm long, wings of upper petal bent upwards.

Recorded by Schweinfurth above Ḥillah. J. Bura' and around Manākhah; recently found only in Khawlan Ash Sham (D. Wood 1031).

2. Fumaria parviflora *Lam.*
Very similar to the previous species but flowers often white, up to 6mm long only; sepals 0.5–1.5mm long; wings of upper petal erect, spreading or deflexed.

An uncommon field weed found above 1500m on the escarpment and high plateau but rare below 2400m except in the central escarpment.

BARBEYACEAE

BARBEYA *Schweinf.*

1. Barbeya oleoides *Schweinf.* Kimb
Small dioecious tree to 5m with red-brown fissured bark and delicate pendulous branches. Branchlets white-tomentose when young, becoming brown and glabrous with age. Leaves opposite, entire, narrowly lanceolate, up to 6 × 3cm, dark olive green above, white tomentose below. Male and female flowers in 3-flowered subsessile axillary cymes;

perianth with 3 (−4) prominent, accrescent, papery, ovate-oblong segments c 15mm long; petals absent. Fruit an ovoid, 1-seeded drupe, c 9mm long.

Flowers in winter.

A local plant of steep scrubby slopes between 700 and 1800m on the western mist-covered slopes of the central escarpment. Abundant on J. Bura', Milḥān and the Ḥarāz but infrequent on J. Raymah and Maḥwīt. It is also known from J.'Alab in the north of Khawlan Ash Sham (Steinberg) and this is presumably the southernmost outpost of the Asir colony of this interesting species which is the sole member of the Barbeyaceae. Outside Arabia it is only known from Ethiopia.

ULMACEAE

Trees with simple alternate, stipulate leaves. Flowers without petals, commonly male and bisexual.

Leaves scabrid above, white tomentose below — *Trema*
Leaves scabrid on both surfaces, not tomentose below — *Celtis*

1. CELTIS *L.*

Trees with monoecious flowers

Leaves entire — *C. toka*
Leaves serrate — *C. africana*

1. Celtis africana *Burm. f.* — Sharjab

C. kraussiana Bernh.

Deciduous tree to about 15m, bark smooth, grey. Branchlets rusty pubescent when young, becoming darker and ± glabrous with age. Leaves ovate, up to 11 × 7cm, but commonly less, prominently 3-veined from the rounded slightly unequal base, apex acuminate, margin serrate, scabrid and thinly pubescent on both surfaces. Male flowers pale yellow, pedicellate in clusters in the lower leafless branchlet nodes; female flowers 1–2 in the axils of the young leaves. Fruit a pedicellate, globose, brown drupe about 4mm long.

Flowers in spring.

Widespread but rare and scattered on the escarpment between 1000 and 2700m from the Ḥujarīyah to Khawlan Ash Sham; also a single plant near Kumaym in Al Ḥadā' on the high plateau. It is often found as an immature bush on cliffs and other places protected from grazing and very rarely grows into a mature tree. I can only remember seeing mature trees on J. Ṣabir, in Wadi Aḥjar, at Ḥaydān and on J. Rāziḥ where it is tolerably frequent. It was probably common in prehistoric highland forest.

2. Celtis toka (*Forssk.*) *F. N. Hepper & J. R. I. Wood* — Sulām, Tuqāḥ

C. integrifolia Lam.

Tree to c 20m, bark smooth, grey. Branchlets glabrous, green, scabrid. Leaves dark green, ovate to elliptic, up to c 9 × 5cm, prominently 3-veined from the rounded, unequal base, apex acuminate, margin entire, scabrid on both surfaces. Flowers yellow-green in small axillary cymes, the male below and the female mostly above. Fruit a pedicellate, dark brown, ellipsoid drupe about 1cm long.

Widespread and locally abundant as a riverine tree in the foothills of the central escarpment between 400 and 800m, just entering Waṣāb As Sāfal south of Mishrāfah. It is particularly abundant in the plain between Ḥajjaylah and Rajamah in the Ḥarāz.

2. TREMA *Lour.*

1. Trema orientalis (*L.*) *Bl.*

T. guineense (Schumach.) Ficalho.

Small tree to c 10m with a slender trunk, spreading crown and smooth grey bark. Branchlets white tomentose. Leaves ovate to lanceolate, up to c 13 × 6 cm, with a slightly unequal, shallowly cordate to rounded base, gradually tapering to an acute apex, margin serrate, upper surface dark green, scabrid, under surface white tomentose. Flowers bisexual, yellow green, in subsessile or shortly pedicellate axillary clusters. Fruit a subglobose black drupe, c 5mm long.

Recorded by Schweinfurth above Ḥillah, J. Bura' (900m) and at Mawad near Ḥajjaylah, Ḥarāz (800m) but not seen at all in the central escarpment subsequently and it may be extinct in that area as it is said to be a very short-lived tree. I only know it from 1400m in Wadi Ḍabāb near Taiz where there are a number of scattered trees and where it has shown signs of increasing in recent years.

CANNABACEAE

Indian Hemp or Marihuana, **Cannabis sativa** *L.*, known in Yemen as 'Ḥashish', is an annual herb about 2m high with alternate, deeply palmatisect leaves and an erect, glandular, branched inflorescence. Though officially prohibited, both the Ṣa'dah and Ibb regions have a long-standing reputation for its cultivation. It also occurs sporadically, though not spontaneously, in the gardens of younger North Americans and Europeans in Sana'a and Taiz.

MORACEAE

A family of trees, shrubs and herbs with latex. Leaves alternate, with stipules. Flowers unisexual, very small, in spikes, heads or fleshy receptacles. Fruits various but commonly multiple, formed by the union of the fruit with the fleshy receptacle as in figs and mulberries.

1. Herbs — 1. *Dorstenia*
   Shrubs or trees — **2**
2. Anthers and/or stigma exserted through the mouth of the receptacle. Leaves obovate-elliptic, the base very unequal, strongly reticulate below — 3. *Antiaris*
   Anthers and/or stigma not exserted through the mouth of the receptacle. Leaves various but always broadest around the middle or base which is ± equal, not strikingly reticulate below — 2. *Ficus*

1. DORSTENIA *L.*

Fleshy herbs. Receptacle ± concave, open, with male and

female flowers mixed.

Leaves ovate – cordate. Receptacle elongate, narrowly oblong — *D. barnimiana*
Leaves lanceolate, elliptic or oblanceolate. Receptacle orbicular — *D. foetida*

1. Dorstenia barnimiana *Schweinf.*

Low stemless herb. Leaves and inflorescence arising from a globose underground tuber at the onset of the spring rains. Leaves numerous, ovate-cordate, up to c 4 × 4 cm, dentate. Receptacles elongate, narrowly oblong, overtopped by the numerous bract arms.

Abundant on stony uncultivated ground between 1200 and 2500m in the area bounded by Shar'ab, 'Udayn, Riḥāb and Qā'idah where it is commonly associated with *Scilla hyacintha*. Like that species it is scarce and very sporadic elsewhere on the escarpment but extends south to Turbah and north to the Wadi Lā'ah.

2. Dorstenia foetida (*Forssk.*) *Schweinf.* Qusār

A glabrous herb with a thick branched fleshy tuberous stem with numerous leaf scars, usually very low but occasionally up to 20cm high. Leaves dark green, lanceolate, elliptic or oblanceolate, up to 7 × 2cm with a cuneate base, borne in a terminal rosette at the end of the branches, deciduous in the dry season. Receptacles orbicular, overtopped by the bract arms.

Widespread in rock crevices and on terrace walls between 200m (on the first foothills at Khalīfah) and 2500m along the whole escarpment, but particularly abundant in areas of good rainfall. Probably also in the southeastern mountains and known certainly from Qa'bayṭah and Damt.

2. FICUS *L.*

Trees or shrubs with white latex. Flowers enclosed within a fleshy receptacle, 'the fig', which has a small pore-like opening at the apex known as the ostiole. Receptacle many-seeded.

Figs are abundant throughout much of Yemen but are of relatively little ecological or economic importance. They usually occur as scattered individuals and are rarely dominant in any plant community. The wood is of little value even as firewood. *Ficus carica* provides the well-known fig, but their only other economic use is as shade, particularly for coffee and banana plantations. They are especially valuable for this purpose as they have the ability to conserve soil moisture.

1. Leaves, petioles, figs and young branchlets scabrid — *F. exasperata*
   Petioles, figs and young branches not scabrid, though the leaves sometimes so — **2**
2. Leaves serrate or dentate, sometimes 3-lobed. Figs pear-shaped — **3**
   Leaves entire or obscurely repand, never 3-lobed. Figs ± globose — **4**
3. Figs up to c 1.5cm long — *F. palmata*
   Figs exceeding 1.5cm long, in Yemen commonly 2–4cm — *F. carica*
4. Leaves very large, commonly exceeding 15cm broad — *F. vasta*
   Leaves small to medium-sized, up to 10cm broad — **5**
5. Leaves smooth, ovate – cordate, with an acuminate apex, about as broad as long — *F. populifolia*
   Leaves various but not as above and always longer than broad — **6**
6. Leaves ± lanceolate, gradually tapering to an acuminate apex, usually glabrous, never scabrid or repand — **7**
   Leaves ovate to elliptic, usually abruptly contracted towards the apex, always hairy at least on the petiole; sometimes scabrid or repand — **9**
7. Petioles and figs tomentose — *F. ingentoides*
   Petioles and figs glabrous or glabrescent — **8**
8. Leaf base cordate. Leaves not reticulate above but with veins raised below. Figs wrinkled in fruit — *F. ingens*
   Leaf base usually cuneate, rarely cordate. Leaves reticulate above. Figs not wrinkled when dry — *F. salicifolia*
9. Figs sessile on the branchlets. Leaves smooth, not scabrid or repand — *F. glumosa*
   Figs pedunculate on old branches. Leaves scabrid or repand — **10**
10. Leaves entire, usually scabrid above — *F. sycomorus*
    Leaves obscurely repand, smooth — *F. sur*

1. Ficus carica *L.* Balas

Cultivated fig. Spreading deciduous shrub to c 4m, bark grey. Branchlets pubescent or glabrous, not scabrid. Leaves exceedingly variable: palmately lobed or unlobed, scabrid on both surfaces or not, glabrous or pubescent, but in Yemen usually unlobed, scabrid on both surfaces, ovate, acute with a cordate base, up to c 20 × 15cm, sometimes pubescent below, but always dentate. Fig peduculate on the branchlets, pear shaped, purple-black and glabrous when ripe, usually 2–4cm long.

Figs ripe in June and July.

Commonly cultivated in the valleys around Sana'a and Ṣa'dah and eastwards in irrigated gardens to Mārib. Its distribution on the escarpment is uncertain but it does not appear to be common and the only certain localities I know of it in are Manākhah (Schweinfurth) and J. Ṣabir.

2. Ficus exasperata *Vahl* Hishrif

Tree to c 20m, bark smooth, grey. Branchlets scabrid, red-brown. Leaves unlobed (3-lobed on coppice shoots), elliptic, up to 12 × 7cm, apex shortly acuminate, base rounded, margins obscurely dentate, scabrid on both surfaces. Figs pedunculate on the branchlets, subglobose, red-brown, scabrid, about 12mm long.

A tree of wadi sides usually by permanent water in areas of good rainfall between 600 and 1900m. Widely distributed in the central escarpment but elsewhere only known from the Wadi Lā'ah above Ḥammām Shamsān, Wadi Sukhnah and the 'Udayn region (Jabalayn and Wadi Ḥadabah); almost always in small quantity.

3. Ficus glumosa *Del.* Mudaḥ, Dhalīf

In Yemen a small tree or shrub to 4m. Branchlets longi-

Fig. 3. Leaves of *Ficus* species. **A** *F. exasperata*; **B** *F. sycomorus*; **C** *F. carica*; **D** *F. glumosa*; **E** *F. populifolia*; **F** *F. palmata*; **G** *F. sur*; **H** *F. ingens*; **J** *F. vasta*; **K** *Antiaris toxicaria*; **L** *F. salicifolia*; **M** *F. ingentoides*.

tudinally wrinkled, pubescent with tufts of pilose hair at the leaf nodes. Leaves smooth, dark green, ovate or ovate-oblong, up to 11 × 7cm but commonly much less, base cordate, apex rounded to very shortly acuminate, glabrous above, glabrous or pubescent below. Figs sessile on the branchlets, globose, red, glabrous or pubescent, about 6mm long.

Both the typical plant with pubescent undersurfaces to the leaves (e.g. Wood 75/188 from Bani 'Umar, Hujariyah) and var **glaberrima** *Martelli* with the undersurface glabrous (e.g. Wood 2038 from J. Bura') occur in Yemen.

Quite common in valley scrub from 600 to 1500m in the central escarpment extending to Maḥwīt and J. Sharafayn. Elsewhere rare in similar habitats and known from Bani 'Umar, Wadi Sukhnah, Wadi Sharas and J. Marran. Its distribution has a distinctly northern tendency.

4. **Ficus ingens** (*Miq.*) *Miq.* — Bidaḥ, Ḍarīf

F. lutea Vahl

Small tree to 5m with smooth grey bark. Branchlets yellow-brown, finely pubescent. Leaves smooth, dull green, entire, ovate to lanceolate, up to c 20 × 7cm, base cordate, apex acute, glabrous on both surfaces. Figs shortly pedunculate on the branchlets, globose, dull red, glabrous, c 8mm long, strikingly wrinkled when dry.

Very easily confused with *F. salicifolia* var. *latifolia* from which it can best be distinguished by its pubescent branchlets, leaves with veins raised on the lower surface and figs wrinkled when dry.

Widespread on the escarpment between 400 and 1900m in areas of average to good rainfall, but almost always growing on wadi cliffs in partial shade.

5. **Ficus ingentoides** *Hutch.*

Identical to *F. ingens* in general appearance but differing from it in having tomentose branchlets, petioles and figs. In addition, the subsessile figs are not wrinkled when dry.

Known only from the central valley of Jabalayn, 'Udayn at c 1900m. (Wood 75/299), but Steinberg 5999 from Majz may be this species.

Outside Yemen this appears to be a rare species restricted to Ethiopia and Sudan.

6. **Ficus palmata** *Forssk.* — Balas

F. pseudosycomorus Decne.

The wild species from which *F. carica* is probably derived and, like it, very variable. Deciduous shrub to 4m. Branchlets glabrous, smooth, often red-brown. Leaves scabrid or more rarely, smooth, dull green, 3-lobed or unlobed, serrate, ovate to lanceolate, up to c 9 × 6cm, base cordate to cuneate, apex acute, pubescent or glabrous on the undersurface. Figs pedunculate on the branchlets, pear-shaped, purple-black and glabrous when ripe, up to 1.5cm long.

Widespread and abundant on the escarpment and high plateau between 500 and 3000m, but less common in the lower part of its altitude range. It grows on cliffs, rocky slopes and terrace walls, rarely in great quantity and is distinctly more frequent around villages and cultivation, along tracks and generally near human activity than in remote scrub.

7. **Ficus populifolia** *Vahl* — Mudaḥ

Small tree to 5m with smooth grey bark. Branches often pendulous. Branchlets yellow-green, glabrous. Leaves smooth, yellow-green, entire, heart-shaped, up to 10 × 10cm, base widely cordate, apex acuminate, glabrous on both surfaces. Figs with peduncles c 15mm long, borne on the branchlets, globose, brown, glabrous, c 8mm long.

Abundant between 400 and 1400m on cliffs and rocks by wadis in Taiz province, but less common and restricted to the escarpment foothills further north and completely absent north of J. Milḥān.

8. **Ficus salicifolia** *Vahl* * — Ithāb, Thāb

Small tree to 6m with smooth grey bark. Branchlets purple-brown, glabrous. Leaves smooth, dark shiny green, entire, lanceolate to elliptic, up to 12 × 3cm, base cuneate to rounded, apex acute, glabrous on both surfaces, reticulate above especially when dry. Figs sessile or shortly pedunculate on the branchlets, globose, dull red, glabrous, c 6mm long.

In addition to the type, variety **latifolia** *Hutch.* occurs. This has leaves 3–6cm broad and the base ± cordate. It is common around Ḥayfān (Wood 74/209).

Widespread and common between 500 and 2300m on the escarpment and southeastern mountains; widespread but very scattered on the high plateau and in the Mashriq. It probably tolerates a rainfall of between 100 and 1000mm per annum and is thus absent from the Ibb area and present in drier conditions than any other fig. In dry country it usually grows on cliffs but elsewhere it grows on open hillsides and particularly along wadis. It is especially abundant in the east Ḥujarīyah where it is the commonest tree and gives the scenery of that area a very distinctive appearance especially in winter when it is the only green object in an otherwise brown landscape.

9. **Ficus sur** *Forssk.* — Sūr, Yasār

F. capensis Thunb.

Tree to 10m high, bark smooth, brown. Branchlets tomentose when young, becoming glabrous, red-brown. Leaves smooth, dull green, obscurely repand-dentate, ovate or more commonly elliptic, up to 13 × 7cm, base ± cordate, apex shortly acuminate, usually slightly pubescent below. Figs in panicles in the old wood, globose to obovoid, dull orange, finely puberulous, c 1cm long, 1.5cm broad.

Common between 2200 and 2900m on the southern mountains centred on Ibb extending south to J. Ḥabashī and J. Ṣabir, north to Kusmah, J. Raymah and east to the Wadi Banā area. Elsewhere very rare and known only from Ḥaddah and Wadi Ḍahr by irrigation ditches and by springs in the sandstone country at Aḥjar and Ṭawīlah. On some of the Ibb mountains, notably J. Ba'dān, it is in great abundance and was clearly an important constituent of prehistoric highland forest. Schweinfurth's record from above Ḥillah, J. Bura' at 1000m is much lower than I have ever seen it.

10. **Ficus sycomorus** *L.* — Suqām, Bidaḥ, Khanas

Tall spreading tree to c 20m, bark grey. Branchlets red-brown, scaly, ± glabrous except for a ring of hairs at the

*Recently treated as *Ficus cordata* subsp. *salicifolia* (Vahl) C.C.Berg (Kew Bull. 43: 82 (1988)).

leaf nodes. Leaves slightly scabrid above, dull green, entire, broadly ovate, up to c 10 × 7cm, base cordate, apex obtuse to rounded, pubescent or glabrous below. Figs in panicles on the old wood, globose to obovoid, dull orange, tomentellous c 2cm long, 2.5cm broad.

Common on the escarpment and southeastern mountains, generally growing by wadis and less commonly around fields on alluvial plains or in villages; also in the Tihama as isolated trees usually in villages; found between 200 and 2000m.

11. Ficus vasta *Forssk.* Tālūq, Tawlāq, Khisāl
Tall and often massive spreading tree to 30m with grey bark, sometimes buttressed and sometimes sending down aerial roots, the largest tree in Yemen except for the Baobab. Branchlets red-brown, pubescent. Leaves smooth, bright green, entire, broadly ovate to suborbicular, up to 30 × 25cm, base cordate, apex obtuse to rounded; glabrescent on the upper surface, pubescent below. Figs shortly pedunculate on the branchlets, globose, brown, pubescent, c 1.5cm long.

As a wild species probably restricted to the lusher valleys between 500 and 1500m in the high rainfall areas principally around 'Udayn and in the central escarpment. In this area it is abundant as a riverine tree and occasionally extends up the valley sides. It is often planted as a shade tree for coffee. Elsewhere on the escarpment and more rarely on the high plateau it is widely planted up to c 2200m, particularly by wells and in markets where it provides shade for the crowds that gather there.

3. ANTIARIS *Lesch.*

1. Antiaris toxicaria *Lesch.* Khalah
A. challa (Schweinf.) Engl.
Large deciduous tree to 20m in Yemen with a smooth grey trunk. Branchlets longitudinally wrinkled, rusty tomentose. Leaves entire (sometimes dentate on coppice shoots), obovate – elliptic, up to 15 × 8cm, apex obtuse, base ± cordate, very unequal, upper surface scabrid, undersurface strongly reticulate, slightly scabrid, pubescent at least on the veins. Flowers usually dioecious. Fruit drupe-like, single-seeded, scarlet and velvety when ripe, c 1.5cm long.

A very rare plant of valley scrub in the central escarpment. Recorded by Schweinfurth at Rajamah, Ḥarāz (900m) and above Ḥillah, J. Bura' (800m). At the former locality there is now a single mature tree, while I have failed to find it at the latter locality despite repeated searches. There are, however, a number of bushes and small trees between 600 and 800m in the valley forest above Suq As Sabt, J. Bura'.

The Mulberry, **Morus nigra** *L.*, known in Yemen as 'Tūt', is planted in and around Sana'a and in a few other places such as J. Ṣabir. It is a tree with large ovate-cordate leaves and dark purple 'blackberry'-like fruits which are ripe in June and July.

URTICACEAE

Herbs or shrubs often with stinging hairs. Leaves simple, alternate or opposite, with or without stipules. Flowers small, usually unisexual, without petals. Fruit usually a dry achene.

1.	Leaves opposite	**2**
	Leaves alternate	**3**
2.	Plant with stinging hairs	1. *Urtica*
	Plant without stinging hairs	3. *Droguetia*
3.	Inflorescence pedunculate	**4**
	Inflorescence sessile	**5**
4.	Leaves palmately lobed. Plant with stinging hairs	3. *Girardenia*
	Leaves unlobed. Plant without stinging hairs	2. *Laportea*
5.	Leaves entire	**6**
	Leaves serrate	**7**
6.	Herb	6. *Parietaria*
	Undershrub	4. *Pouzolzia*
7.	Shrub	5. *Debregeasia*
	Herb	**8**
8.	Leaves white woolly below, clinging to passing seeds etc.	7. *Forsskalea*
	Leaves pubescent and green below, not clinging	4. *Pouzolzia*

1. URTICA *L.*

Herbs with stinging hairs (Nettles). Leaves opposite, incised-serrate in our species. Flowers monoecious in axillary inflorescences.

Female flowers in long pedunculate globose heads	*U. pilulifera*
All flowers in spike-like racemes	*U. urens*

1. Urtica pilulifera *L.*
Tall annual to c 80cm. Leaves dark green, ovate, 2–6cm long. Male flowers spicate, female flowers in long pedunculate globose heads, the perianth inflated.

Only known from waste places in Manākhah village at 2400m where it has persisted for a number of years.

2. Urtica urens *L.*
Erect or ascending annual, usually less than 50cm. Leaves green, ovate, 1–4cm long. Male and female racemes similar, spike-like, up to 2cm long.

A common plant of waste ground in villages, sewage outlets, village gardens and other nitrogen-rich places between 2000 and 3000m on the escarpment and high plateau.

2. LAPORTEA *Gaud.*

1. Laportea aestuans (*L.*) *Miq.*
Fleurya aestuans (*L.*) Miq.
Nettle-like erect annual to 50cm without stinging hairs; stem pilose. Leaves alternate, ovate, 6–9cm long, coarsely serrate. Flowers bisexual in long pedunculate, axillary panicles exceeding the leaves.

Fig. 4. **A** *Pouzolzia parasitica*; **B** *Debregeasia saenab*; **C** *Urtica pilulifera*; **D** *Laportea aestuans*.

Found on the escarpment between 700 and 1600m, usually as a weed under coffee and more rarely in damp shaded places around springs and irrigation channels. Frequent in suitable habitats in the central escarpment but rare elsewhere: Wadi Ḍabāb and Ḥasabah Asfal near Taiz; in the coffee groves by the 'Udayn wadis; Wadi Lā'ah near Suq Ath Thulūth.

3. GIRARDINIA *Gaud.*

1. Girardinia diversifolia (*Link.*) *Friis* Muḥaraq

G. condensata (Hochst.) Wedd.

Erect monoecious or dioecious herb to 1½m with stinging hairs. Leaves alternate, deeply lobed up to 15cm long, the margins very coarsely serrate. Inflorescence axillary, spicate; male spikes catkin-like up to 4cm long; female spikes rather shorter, densely covered in stinging hairs.

Found by Forsskal above Ḥadīyah and still present there in some abundance as a weed of the coffee terraces between 900 and 1800m, but restricted to the main valley and unknown elsewhere in Arabia.

4. POUZOLZIA *Gaud.*

Herbs or shrubs without stinging hairs. Leaves alternate with distinctive red-brown filiform stipules. Flowers monoecious.

Herb. Leaves serrate — *P. parasitica*
Shrub. Leaves entire — *P. mixta*

1. Pouzolzia mixta *Solms*

P. arabica Deflers
P. hypoleuca Wedd.

Shrub to 1½m with brown, somewhat succulent stems. Branchlets pubescent. Leaves entire, ovate, acuminate, c 2–5cm long, dark green above, white tomentose below. Flowers sessile, crowded in the leaf axils, the male and female mixed.

A rare plant of cliffs, rocks and terrace walls, usually near semi-permanent flowing water between 600 and 1500m in areas of good rainfall on the escarpment: Wadi Dur, 'Udayn; Wadi Rijān, 'Alujah, J. Raymah; Wadi Ḥillah, J. Milḥān; Wadi Yūr; Wadi Ḥijān, near 'Usil and below Maghārib, J. Sāfān, Ḥarāz; Wadi Sukhnah, Ḥaymat ad Dakhilīyah.

2. Pouzolzia parasitica (*Forssk.*) *Schweinf.*

Herb with pubescent stems. Leaves serrate, ovate to elliptic, acuminate, c 4–6cm long, pubescent and green on both surfaces. Flowers sessile, crowded in the leaf axils, the male and female mixed.

An uncommon plant of terrace walls and rocks in scattered localities in high rainfall areas of the escarpment between 1000 and 2000m: between Misrākh and Ḥasabān, J. Ṣabir; just north of Ibb; abundant in the main wadi above Ḥadīyah; below Bani Dawan, Al Jabin, J. Raymah; between Aṣ Ṣa'bah and Ḥabāsh, J. Milḥān; Wadi Ḥijān and J. Sāfān, Ḥarāz; by a waterfall below Manākhah to the north-west.

5. DEBREGEASIA *Gaud.*

1. Debregeasia saenab (*Forssk.*) *F.N. Hepper & J.R.I. Wood* Sinab

D. bicolor (Roxb.) Wedd.
D. salicifolia (D. Don) Rendle

Monoecious shrub to 3m, with dark red-brown pubescent stems; stinging hairs absent. Leaves clustered towards the ends of the branchlets, alternate, serrate, lanceolate, c 6–10cm long, dark green on the upper surface, white tomentose below. Flowers sessile, crowded into berry-like clusters in the leaf nodes.

Widespread in areas of good rainfall on the escarpment from 600 to 3000m from the Ḥujarīyah north to Khawlan Ash Sham but rare below 1800m. It usually grows by small permanently flowing wadis but at higher altitudes it occurs on cliffs and rocky hillsides. It is commonly associated with *Myrica* and *Maesa*.

6. PARIETARIA *L.*

Herbs with alternate, entire leaves. Stinging hairs absent. Flowers bracteate, bisexual, in sessile clusters in the leaf axils.

1. Perennial. Leaves elliptic, base cuneate — *P. judaica*
   Annual. Leaves ovate, base rounded — 2
2. Bracts inconspicuous, linear lanceolate — *P. debilis*
   Bracts conspicuous in fruit, membranous, ovate-cordate — *P. alsinifolia*

1. Parietaria alsinifolia *Del.*

Very similar to the following species from which it can only be reliably separated by the bracts which are ovate-cordate, membranous and particularly conspicuous in fruit.

Found by Schweinfurth in the coffee region of J. Bura' above Ḥillah at 900m, but not recorded again.

2. Parietaria debilis *Forst. f.*

Slender, ± decumbent, leafy annual herb with sparsely pubescent, green stems. Leaves thin, ovate, c 1–3cm long, base rounded, apex shortly acuminate. Flowers minute, the perianth less than 1mm long; bracts very small, linear-lanceolate, not membranous.

Widespread but rather local on the escarpment from about 1000 to 3300m on J. Nabī Shu'ayb; not recorded in Khawlan Ash Sham. Its distribution is closely related to the occurrence of winter mists since it is generally restricted to high mountains but descends to low altitudes in areas such as the Ḥujarīyah and the central escarpment where low-altitude winter mists are common. It tolerates a wide variation in rainfall as long as mists are frequent. It is almost always found on terrace walls but also occurs rarely on cliffs.

3. Parietaria judaica *L.*

Spreading perennial herb with pubescent, often red-brown stems, somewhat woody below. Leaves elliptic, c 2–3cm long, base cuneate, apex shortly acuminate. Flowers with a perianth 2–3mm long; bracts small, elliptic, not membranous.

Common on walls and rocks in Manākhah and several

nearby villages (2000–2400m); probably a relatively recent introduction as it was not recorded by the early botanists.

7. FORSSKALEA *L.*

1. Forsskalea tenacissima *L.* Ḥamshad
Much branched, perennial, hispid 'clinging' herb to c 60cm, stem purplish, woody below. Leaves alternate, crenate, obovate, c 2cm long, base long cuneate, apex rounded, the upper surface grey-green, the undersurface white-tomentose. Flowers monoecious in dense axillary bracteate clusters, the bracts long ciliate.

Widely distributed between 200 and 2500m but absent from sand, alluvial deposits and the eastern part of the Mashriq. It is usually an indicator of recent volcanic activity and is common on lava flows, old volcanic cones etc. Thus in the Tihama it is restricted to the Mawza area but is particularly common on the high plateau in Hamdān-Arḥab and around J. Isbīl on the quarternary lavas.

8. DROGUETIA *Gaud.*

1. Droguetia iners (*Forssk.*) *Schweinf.*
Perennial trailing herb without stinging hairs. Leaves opposite, serrate, dark green, ovate, acuminate, about 1cm long; stipules prominent, reddish. Flowers monoecious in sessile clusters in the leaf axils.

A small easily overlooked plant of shaded walls and rocks between 1400 and 2500m in high rainfall areas on the escarpment: near Dimarayn, J. Ṣabir; in the valley north of Ibb; J. Ḥisn Madnān, Waṣāb Al 'Ali; in the gully below Jirjir above Ḥadīyah and near Ḥadādah, J. Raymah; 'Ukaybir and near Aṣ Ṣa'bah, J. Milḥān.

JUGLANDACEAE

The Walnut, **Juglans regia** *L.*, known in Yemen as 'Jawz' is frequently cultivated on the high plateau and in a few places on the escarpment above 2000m such as Manākhah. It is particularly common in the irrigated valleys around Sana'a such as Ḥaddah and Wadi Ḍahr. It is a large spreading tree with aromatic pinnate leaves. The nuts are sold in the markets in Sana'a and elsewhere.

MYRICACEAE

MYRICA *L.*

1. Myrica salicifolia *Hochst. ex A. Rich.* Thū'am
Slightly aromatic dioecious shrub or small tree to c 8m. Branchlets grey-brown, puberulous. Leaves alternate, lanceolate, up to c 9 × 2.5cm, apex acute, base cuneate, slightly unequal, margin entire or obscurely dentate, both surfaces dotted with yellow glands. Flowers in catkins: male catkins, yellow, fragrant, c 2cm long; female catkins greenish. Fruit a globose, warty drupe often with a white waxy covering, 3–4mm long.

Flowers January to March.

Locally common between 1700 and 2900m in Ibb province, sometimes by streams where it is associated with *Debregeasia* and *Maesa* (e.g. Wadi Banā; Wadi Sūdān) and sometimes on open hillsides where it is doubtless a relic of highland forest (e.g. J. Ba'dān; Mudhaykhirah region). It also occurs commonly in the highland grasslands on the mountains near Kusmah, notably J. Bārad and J. Ḥuzār. Elsewhere it is extremely rare, occurring as isolated bushes: J. Ḥabashī; on the Naqil Jūbirān, 'Akāhilah, Ḥujarīyah; near Al 'Urr, Ḥaymat al Kharijīyah; J. Miswar; Markah, J. Milḥān.

CASUARINACEAE

Species of the mainly Australian genus **Casuarina** *L.* have been introduced to Yemen in recent years and are commonly planted in Sana'a and Taiz. Three Casuarinas are known to occur in Yemen, ***C. cunninghamiana*** *Miq.*, **C. equisetifolia** *L.* and **C. glauca** *Sieb. ex Spreng.*, but their exact frequency and distribution is unknown. They are all rapid growing monoecious trees with drooping grey-green branches, their appearance somewhat similar to the indigenous tamarisks. The branches are jointed and have minute scale-like leaves arranged in whorls around the joints. The male flowers are borne in terminal spikes towards the top of the tree while the female flowers are borne in cone-like heads lower down the tree. The three species can be distinguished by the following key:

1. Branches relatively stout, 0.7–0.9mm in diameter, with 14–16 ribs. Leaves transversely banded — *C. glauca*
Branches relatively slender, 0.5–0.7mm in diameter, with 7–9 ribs. Leaves banded or not — **2**
2. Branches prominently ribbed. Leaves pale, not banded — *C. equisetifolia*
Branches obscurely ribbed. Leaves transversely banded — *C. cunninghamiana*

NYCTAGINACEAE

Herbs, shrubs or climbers. Leaves usually opposite, without stipules. Flowers in cymes or panicles; petals absent, the perianth formed by the fused sepals which are entirely petaloid in *Mirabilis* but in our other genera the upper part is petaloid and deciduous while the lower part forms a sheath round the ovary and enlarges to enclose the indehiscent fruit technically known as the anthocarp.

1. Woody climber with axillary falcate spines — 4. *Pisonia*
Herbs, rarely shrubby below, unarmed — **2**
2. Flowers more than 3cm long, very conspicuous. Fruit a globose eglandular nut enclosed by the bracts — 1. *Mirabilis*
Flowers less than 2cm long. Fruit elongate, often sticky-glandular, not enclosed by the bracts — **3**
3. Fruit with large raised glands especially near the apex, not conspicuously longitudinally ridged — 3. *Commicarpus*
Fruit glabrous or with minute glands, conspicuously longitudinally ridged — 2. *Boerhavia*

1. MIRABILIS *L.*

1. Mirabilis jalapa *L.* Warad Al Layl

Marvel of Peru. Erect, tuberous, nearly glabrous perennial herb to 1m. Leaves ovate, acuminate. Flowers in crowded terminal cymes, opening in the evening; perianth funnel-shaped, with a long tube and 5-lobed, magenta (rarely yellow or white) petaloid limb: bracts fused below, calyx-like, enclosing the black nut-like fruit.

Originally from tropical America, this species was established as long ago as the eighteenth century and is now widely naturalised between 1000 and 2500m near villages and on roadsides but has only become abundant where there is permanent water or very high rainfall. In consequence it is particularly common around Ibb and 'Udayn where it is ± dominant in irrigated coffee plantations by the wadis.

2. BOERHAVIA *L.*

Herbs with opposite leaves which are usually strikingly pale below. Flowers usually in panicled cymes; perianth pink, ± campanulate, with or without a tube; stamens and style not exserted. Fruit with prominent longitudinal ridges, glabrous or with minute sessile sticky glands, usually ellipsoid.

A very complex genus of which there is no satisfactory treatment. The following account should work in Yemen but can only be regarded as a stop-gap. I recognise four species, but a fifth, *Boerhavia rubicunda* Steud. (*B. elegans* Choisy) should be looked out for as it is known from close to the Yemen border in both Saudi Arabia and South Yemen. It can be easily recognised by the very lax panicle with the flowers solitary on capillary peduncles.

1. Inflorescence a leafy panicle, the flowers in small, axillary cymes with the peduncle up to 1cm long only — *B. repens*
   Inflorescence an almost leafless panicle but if somewhat leafy (because of immaturity), some peduncles at least 1.5cm long — 2
2. Stem and leaves densely sticky glandular-pilose — *B. sp. A*
   Stem and leaves pubescent; stalked glands absent or rarely a few at the top of the panicle — 3
3. Stems pink, up to c 50cm long. Leaves pubescent. Inflorescence usually sticky — *B. sp. B*
   Stems dark maroon often exceeding 1m in length. Leaves nearly glabrous. Inflorescence not sticky — *B. diffusa*

1. Boerhavia diffusa *L.* Ruqmah, Quṣīf

B. repens var. diffusa (L.) Boiss.
B. adscendens sensu Schwartz

Probably perennial herb with numerous prostrate to ascending sparsely pubescent, dark maroon stems often exceeding 1m in length. Leaves broadly ovate, up to 3.5cm long, entire, glabrous or sparsely pubescent, dark green above. Inflorescence a large diffuse ± leafless panicle; perianth pink. Fruit ellipsoid to obconical, c 3–4mm long, usually covered in numerous yellowish glands but sometimes almost glabrous.

Distinguished from other species by its very vigorous habit and eglandular stem and leaves.

One of the most widespread and common species in Yemen, being found from sea level to 2400m in all regions except the coastal fringe and the Sands though it extends as far east as Jūbah. It is uncommon though in very arid country. It grows in a correspondingly varied group of habitats including roadsides, field borders, sand dunes, rocky flood beds, open Acacia scrubland and cliffs. (Wood 74/200, 2999, 3149, Hepper 5675.)

2. Boerhavia repens *L.* Ruqmah

Annual herb with prostrate, sparsely pubescent, pale pink stems rarely exceeding 50cm in length. Leaves lanceolate, ovate or elliptic, up to 3cm long, entire or often sinuate margined, sparsely puberulent, dull green above, grey below. Inflorescence a leafy panicle, sometimes much reduced so that the inflorescence is of few flowered axillary cymes only, the peduncles up to c 1cm long; perianth pale pink or white. Fruit ellipsoid, c 4mm long, sparsely covered in glands or sub-glabrous.

Although very variable this is usually easily recognised by the prostrate habit, leafy inflorescence and very pale pink flowers.

Frequent between 100 and 1000m in the silt plains in the east of the Tihama and in the alluvial basins of the larger wadis on the escarpment growing on banks around fields, on roadsides, on mud roofs and on tracks. Much rarer at higher altitudes but reaching 1700m near Qā'idah. (Wood 539, 2975, 3277, Hepper & Wood 5838.)

3. Boerhavia sp. A

B. repens var. viscosa Choisy
B. glutinosa sensu Deflers

Annual herb with prostrate or ascending, densely glandular-hairy green stems up to about 80cm in length. Leaves ovate, up to 3.5cm long, usually entire but sometimes sinuate margined, densely glandular hairy, yellow green above, whitish below. Inflorescence a compact nearly leafless panicle; perianth pink. Fruit narrowly obconical, c 4mm long, densely covered in sticky glands.

Readily recognised by its sticky-glandular hairy indumentum which usually results in the plant being covered with a layer of fine sand.

Found on the escarpment, high plateau and the Mashriq from 500 to 2400m, but its exact distribution is not known. It shows a marked preference for gullies in sandstone country (the Ṣa'dah area, Bani Ḥushaysh, Wadi Ḍahr, Rada' etc.) and on the escarpment is usually found in gullies filled with fine silt though it does extend on to stony hillsides in open bushland at Mafḥaq, Khamīs Bani Sa'd and elsewhere. (Wood 75/320, 3204, 3231, Steinberg 939, 1372.)

Wood 3229 from Bani Manṣūr differs from the typical plant in being densely hairy, but not glandular or sticky, but is probably only an aberrant form as it was growing with the typical plant.

4. Boerhavia sp. B

Delicate, probably annual herb, with numerous prostrate pubescent pink stems which rarely exceed 40cm in length. Leaves ovate, up to 2.5cm long, usually sinuate margined, thinly pubescent, green above, white below. Inflorescence a small, usually leafless panicle (rarely rather leafy and then the cymes with peduncles at least 1.5cm long); perianth pink. Fruit obconical, c 4mm long, usually densely sticky-glandular, but sometimes very sparsely so.

Distinguished by its delicate habit, pink stems and nearly leafless inflorescence which is commonly rather sticky.

Apparently a local plant of stony plain in rather arid country. In suitable habitats between 200 and 1500m in the Tihama (e.g. Barḥ, 'Abs) and inland to the Wadi Badr, Al 'Ashshah, Mafḥaq and doubtless elsewhere. (Wood 2082, 2923, 3062, 3230.)

3. COMMICARPUS *Standl.*

Herbs, often woody below and scrambling over other bushes. Leaves opposite, rather thick. Perianth typically funnel-shaped; stamens and style long exserted. Fruit obscurely longitudinally ridged, ± club-shaped and covered in sticky raised glands especially towards the apex.

This genus has been the subject of considerable misunderstanding in contemporary as well as older floras and little reliance can be placed on the species concepts within either; Schwartz appears to have separated the species out correctly though mostly recognising them as varieties. In fact the species are clearcut both taxonomically and ecologically and the confusion has to be put down to the poor condition of most herbarium material and the mistaken belief that *Commicarpus* is as complex as the related genus, *Boerhavia*.

1. Flowers crowded into a dense head at the apex of a long, stout peduncle — *C. pedunculosus*
   Flowers in panicles composed of umbels or whorls — 2
2. Flowers whorled. Perianth broadly funnel-shaped with a short, inconspicuous basal tube — 3
   Flowers in umbels (rarely with a single whorl). Perianth narrowly funnel-shaped with a well-developed basal tube — 4
3. Flowers pale pink, sessile or shortly pedicellate. Stamens usually 2 — *C. helenae*
   Flowers rich mauve, distinctly pedicellate, the pedicels mostly 5–15mm long. Stamens usually 3 — *C. ambiguus*
4. Perianth white. Stems entirely glabrous except for the finely pubescent young growth — *C. plumbagineus*
   Perianth pink. Stems pubescent, glandular or finely scabrid at least around the leaf nodes — 5
5. Plant very sticky. Stems densely glandular-pilose — *C. grandiflorus*
   Plant not sticky (except the fruit). Stems finely pubescent or scabrid, never glandular
6. Leaves sinuate lobed. Flowers subsessile, the pedicels less than 3mm long — *C. sinuatus*
   Leaves entire or weakly lobed. Pedicels 5–10mm long — 7
7. Stem, uniformly pubescent. Glands at the apex of the fruit distinctly stalked — *C. mistus*
   Stem sparsely scabrid, mostly near the leaf nodes. Glands all sessile — *C. arabicus*

1. Commicarpus ambiguus *Meikle*

C. boissieri auct.

Perennial herb, woody below, with numerous, ± erect, glabrous stems to 50cm. Leaves ovate, up to 7cm long, glabrous, entire. Inflorescence a leafy panicle; the flowers in whorls; pedicels c 6–10mm long; perianth rich mauve, broadly funnel-shaped with a very short tube, c 3mm long; stamens usually 3. Glands maroon in colour, in distinct whorls on the fruit, those at the apex distinctly raised.

Readily recognised by its whorled, pedicellate, mauve flowers.

Widespread between 1900 and 2400m in the Mashriq extending from Bayḍā' to Ṣa'dah; apparently uncommon, though perhaps overlooked. It grows on arid rocky hills where it is often associated with *C. mistus*, *Barleria proxima* and *Campylanthus pungens*.

2. Commicarpus arabicus *Meikle*

Boerhavia plumbaginea var. forskalei Schweinf.

Vigorous perennial herb with numerous, long, somewhat glaucous, finely scabrid stems, often climbing high over bushes. Leaves ovate, up to 5cm long, ± glabrous, entire or weakly lobed. Inflorescence a leafy panicle, the flowers arranged in umbels; pedicels c 8mm long; perianth pink or (rarely) red, narrowly funnel-shaped with a distinct tube, c 4–6mm long; stamens usually 3. Glands at the apex of the fruit only, ± sessile.

A rather nondescript species usually recognisable by its vigorous habit, finely scabrid stems and large pink flowers.

Widely distributed between 1800 and 2900m on the escarpment and high plateau from Turbah north to Sāqayn but rather scattered in occurrence, growing mostly in disturbed ruderal habitats such as roadsides and thorn fences round fields. It is apparently endemic to Yemen but will probably be found in the neighbouring parts of South Arabia.

3. Commicarpus grandiflorus (*A. Rich.*) *Standl.*

Boerhavia plumbaginea var. grandiflora (A. Rich.) Schweinf.

Perennial herb with numerous, decumbent or ascending, densely pilose-glandular stems. Leaves ovate (almost triangular), up to 6cm long, glandular-pilose especially below, entire. Inflorescence a leafy panicle, the flowers in umbels; pedicels 5–8mm long; perianth pink, narrowly funnel-shaped with a distinct tube, c 7mm long; stamens 3. Glands, numerous, sessile, covering the entire fruit.

Distinguished from all other species by its sticky glandular-pilose indumentum.

Widely distributed but rather local between 1100 and 2400m on the escarpment and high plateau. It grows on roadsides and in scrub but usually near human activity; it appears to avoid areas of high rainfall.

Plants similar to *C. grandiflorus* but with white flowers and glaucous stems have been found twice. They may be the hybrid *C. grandiflorus* x *plumbagineus*, but this has not been reported elsewhere. The records are from Bani Manṣūr, Ḥaymat al Kharijīyah and from between Raḍmah and Damt. (Wood 2204, 2415.)

4. Commicarpus helenae (*J.A. Schultes*) *Meikle*
Ruqmah

Boerhavia verticillata sensu Schwartz
Commicarpus stellatus (Wight) Berhaut
Perennial herb with numerous, mostly decumbent, glabrous stems. Leaves ovate, up to 3cm long, glabrous, sinuate-lobed. Inflorescence an almost leafless panicle; flowers in whorls; pedicels very short (less than 2mm long) or absent; perianth pale pink, broadly funnel-shaped with a very short tube, c 2mm long; stamens 2. Glands mostly in whorls on the fruit, those at the apex distinctly raised.

Readily recognised by the small, almost sessile, pale pink, whorled flowers.

Widespread and abundant on stony ground in open *Acacia-Commiphora* scrubland between 400 and 1500m on the escarpment; also in the Mashriq up to 2300m from Rada' north to Ṣa'dah but apparently local.

5. Commicarpus mistus *Thulin.*
C. squarrosus auct.
Boerhavia squarrosa Heimerl
Perennial herb, woody below and with numerous erect densely pubescent stems to 50cm. Leaves ovate to rhomboid, up to 3cm long, thinly pubescent, entire. Inflorescence a nearly leafless panicle, the flowers in umbels; pedicels c 6–9mm long, usually pink in colour; perianth magenta, narrowly funnel-shaped with a distinct tube, c 5mm long; stamens 2–3. Fruit with prominent long-stalked glands at the apex.

Similar in habit to *C. ambiguus* with which it often grows, but distinguished by its umbellate flowers and pubescent stem.

Widely distributed from 1100 to 2700m on arid rocky hills in the Mashriq and eastern part of the high plateau from Bayḍā' to Majz; also occurring very locally on the southeastern mountains at Damt, Qa'ṭabah (Wissmann) and J. Jirdad, Bani 'Umar, Ḥujarīyah. It appears to favour limestone hills.

6. Commicarpus pedunculosus (*A. Rich.*) *Cuf.*
Boerhavia pedunculosa A. Rich.
Perennial herb with numerous, long, prostrate, finely pubescent stems. Leaves ovate, up to 4cm long, thinly pubescent, ± entire. Inflorescence a long-pedunculate dense head of almost sessile flowers; perianth magenta, narrowly funnel-shaped with a distinct tube, c 6mm long; stamens 5–6. Fruit with a few raised glands at the apex only.

Found between 2100 and 2500m on stony volcanic plains and roadsides on the high plateau: frequent within a 15 km radius of Sana'a, but elsewhere very rare and only known from Shibām and the Qa' Bakīl in 'Anis.

7. Commicarpus plumbagineus (*Cav.*) *Standl.*
Ghurmil, Khardal

Boerhavia plumbaginea var. *dichotoma* (Vahl) Aschers. & Schweinf.
Perennial herb with numerous, long, glabrous, sometimes glaucous stems, often climbing high over bushes. Leaves ovate, up to 5cm long, glabrous, entire. Inflorescence a leafy, thinly pubescent panicle, the flowers arranged in umbels; pedicels c 4–7mm long; perianth white, narrowly funnel-shaped with a distinct tube, c 6–8mm long; stamens 3. Glands sessile, mostly at the apex of the fruit.

Easily recognised by its white flowers and glabrous stems.

Abundant between 400 and 2400m on the escarpment especially in areas of high rainfall where it is a conspicuous species scrambling over low scrub as in the Ibb area. In dry country and at higher altitudes it is only found in sheltered valleys; elsewhere it is restricted to a few places such as Wadi Rakwah near Majz (Steinberg), Ḥaddah (Rathjens) and Wadi Ḍahr.

8. Commicarpus sinuatus *Meikle*
Boerhavia plumbaginea var. *viscosa* Boiss.
Perennial herb with numerous, long, shortly pubescent, procumbent or climbing stems. Leaves ovate, up to 5cm long, glabrous or thinly pubescent, sinuate-lobed. Inflorescence a leafy panicle, the flowers arranged in umbels; pedicels absent or very short (up to 2mm); perianth pale magenta, narrowly funnel-shaped with a distinct tube, c 6–10mm long; stamens 3. Glands sessile, scattered all over the fruit.

Readily recognised, especially in the field by the strongly sinuate-lobed leaves and pale magenta, umbelled flowers.

Essentially a plant of the high plateau and similar areas in Khawlan Ash Sham where it is frequent in stony gullies, on cliffs, on rocky plain and roadsides. It extends westwards on the escarpment in dry country to Rujum, Mafḥaq and Madīnat Al 'Abid. Although absent from the Ibb region it is widespread in Taiz province and often abundant particularly in the Ḥujarīyah. Found between 1000 and 2600m.

4. PISONIA *L.*

1. Pisonia aculeata *L.*
Dioecious climbing shrub with pale brown, tomentose branchlets and solitary, weakly falcate axillary spines. Leaves opposite or alternate, obovate to elliptic, up to 9cm long, acute or obtuse, entire, softly tomentose below. Flowers in head-like corymbs, greenish, the male with long exserted stamens and the female with an elongated ovary. Fruit narrowly ellipsoid, c 22mm long, ridged, the ridges with a line of stalked glands.

A local species of riverine scrub between 500 and 1400m, mostly in the central escarpment: Wadi Annah, 'Udayn; 'Alūjah and Suq Ar Ribāṭ, J. Raymah; in the Ḥarāz around Ḥajjaylah in Wadi Madfar (Schweinfurth), Wadi Ḥār and at Rajamah.

Although known from a number of localities in Yemen this pantropical climber is extremely rare in northeast tropical Africa and unknown elsewhere in Arabia.

Bougainvillea spectabilis *Willd.*, an armed woody climber with alternate leaves is cultivated in Sana'a and Taiz for its spectacular mauve or red flowers which are in fact the brightly coloured bracts. It is usually (always?) sterile and is unlikely to become naturalised.

AIZOACEAE

An ill-defined family of herbs, often succulent. Leaves alter-

nate or opposite, with or without stipules, usually entire. Flowers bisexual, regular, solitary or in cymes; petals absent, sometimes replaced by petal-like staminodes. Sepals usually 5, free or united. Fruit usually a capsule. Most species are characteristic of dry sandy or salty conditions.

1. Flowers with numerous prominent mauve petal-like staminodes — 2
   Flowers without staminodes or, if present, staminodes neither conspicuous nor mauve — 3
2. Leaves subcylindrical. Stem absent or very short — 10. *Delosperma*
   Leaves obovate, flattened. Stems up to 40cm long — 5. *Corbichonia*
3. Flowers in diachasal umbellate cymes. Leaves whorled, either all in a basal rosette or, if stem leaves, linear — 6. *Mollugo*
   Flowers all axillary. Leaves not in a basal rosette, if whorled, never linear — 4
4. Leaves all alternate — 1. *Aizoon*
   Leaves opposite — 5
5. Stems densely stellate-pubescent — 7. *Glinus*
   Stems glabrous or scaly, never stellate-pubescent — 6
6. Flowers in shortly pedunculate umbels. Calyx and undersurface of leaves with white 'streaks' — 8. *Gisekia*
   Flowers sessile in the leaf axils. Calyx and young leaves without white streaks — 7
7. Leaves ± orbicular, the petiole without scarious wings. Fruit of 2 hemispherical mericarps — 9. *Limeum*
   Leaves always longer than broad, the petiole with scarious wings at the base. Fruit a capsule — 8
8. Style 1 (best seen on young fruit) — 3. *Trianthema*
   Styles 2 — 9
9. Flowers solitary — 2. *Sesuvium*
   Flowers in clusters — 4. *Zaleya*

AIZOON *L.*

1. Aizoon canariense *L.*

A prostrate grey annual or perennial herb with thick rather woody often zigzag stems, which sometimes persist for years after the plant has died, indumentum variable but commonly both scaly and pubescent. Leaves alternate, oblanceolate or obovate, usually c 2mm long. Flowers numerous, solitary in the axils of the leaves or the branches, styles 5; calyx with triangular lobes, yellow inside, persistent round the fruiting capsule, the whole bearing a striking superficial resemblance to the fruit of *Malva parviflora*.

An uncommon plant occurring in widely scattered localities between 1800 and 2400m in the Mashriq and on the high plateau, usually growing on exposed gravel ridges and in similar stony places: Damt; Rada' region; Bani al Hārith; Wadi Ḍahr; Arḥab; Shahārah; Majz (Steinberg).

2. SESUVIUM *L.*

1. Sesuvium sesuvioides (*Fenzl*) *Verdc.*

Trianthema polysperma Hochst. ex Oliv.

Prostrate, straggling succulent annual herb, the stems warted but otherwise glabrous, up to 30cm long. Leaves mostly opposite, elliptic 1–2cm long, the petiole with scarious wings at the base. Flowers solitary, sessile in the leaf axils; calyx pink inside; styles 2. Fruit a capsule, opening by a lid.

A very rare plant collected by Deflers on sand at Manẓar and not recorded since.

3. TRIANTHEMA *L.*

Succulent annual herbs. Leaves opposite, the petiole with scarious wings at the base. Flowers axillary, sessile; sepals joined; style 1. Fruit a capsule.

1. Flowers solitary. Stems ± glabrous, never scaly — *T. portulacastrum*
   Flowers in clusters. Stem scaly on young parts — 2
2. Leaves linear, glabrous or with a few scales — *T. triquetra*
   Leaves obovate or orbicular, very scaly — *T. crystallina*

1. Trianthema crystallina (*Forssk.*) *Vahl*

Prostrate annual herb, at first rather slender but eventually much-branched and mat-like, stems scaly when young, becoming glabrous and white with age. Leaves obovate or orbicular, 5–10mm long, grey and densely scaly. Flowers in small axillary clusters; calyx 1–1.5mm long.

Common on sand dunes near the sea (also on the Red Sea Islands), becoming less common inland in the Tihama but penetrating the mountains in the south as far as Taiz where it is found as a field weed near the salt springs at Ḥawbān at 1500m. It also grows at Mārib and might occur elsewhere at the edge of the sands. It is clearly very drought resistant and salt tolerant.

2. Trianthema portulacastrum *L.*

T. monogyna L.

Prostrate or ascending, ± glabrous herb, usually much branched. Leaves broadly obovate, 1–4cm long. Flowers solitary in the leaf axils, partially hidden by the sheathing petiole bases, usually pink; calyx 4–5mm long.

Widespread in the Tihama and the foothills up to 800m, but unknown south of Ḥays and generally infrequent. It is usually found on waste ground in villages and more rarely as a field weed on rich soils.

3. Trianthema triquetra *Willd.*

T. sedifolia Vis.

Prostrate, much branched, mat-forming annual herb, the stems scaly and often reddish when young. Leaves very shortly petiolate, linear (or at most narrowly oblanceolate), 5–15mm long, glabrous or with a few scales. Flowers in small axillary clusters; calyx 2–3mm long.

A rare (or perhaps overlooked) plant of bare alluvial plain at the beginning of the foothills around 500–600m: around Ḥajjaylah (Schweinfurth and Deflers); near Bāb al Ibl east of Bājil.

4. ZALEYA *Burm. f.*

1. Zaleya pentandra (*L.*) *Jeffrey*

Trianthema pentandra L.

Much branched, prostrate, slightly succulent annual herb, the stems up to 40cm long, pubescent when young but becoming glabrous and woody with age. Leaves opposite, oblanceolate or elliptic, 1–3cm long, grey on both surfaces but paler and sometimes pubescent below, the petiole with scarious wings at the base. Flowers in sessile axillary clusters; calyx pink or green inside, the lobes with scarious margins; styles 2. Fruit a capsule.

Widely distributed in the Tihama, on the escarpment and high plateau up to 2400m, growing as a weed of roadsides, tracks, waste places and disturbed stony ground, but rather scattered in occurrence and less common than might be expected.

5. CORBICHONIA *Scop.*

1. Corbichonia decumbens (*Forssk.*) *Exell.*

Orygia decumbens Forssk.

Branched, decumbent, glabrous annual or perennial herb. Leaves alternate, fleshy, obovate-spathulate, apiculate, 1–4cm long. Inflorescence of lax leaf-opposed racemes; flowers with numerous mauve petal-like staminodes; sepals free, green with scarious margins, c 4mm long. Fruit a capsule.

A local plant of dry stony hills in Acacia bushland found between 200 and 1500m in the southern Tihama, the escarpment foothills, the rainshadow areas of the central escarpment and the dry hills of Maqbanah and Waṣāb; absent from the Ḥujarīyah and everywhere north of the Wadi Surdūd, apart from an isolated record from between 'Abs and Suq al Khamīs.

6. MOLLUGO *L.*

Annual herbs. Leaves whorled, the basal leaves forming a rosette, the stem leaves sometimes absent. Inflorescence of lax, diachasial, umbellate cymes; flowers inconspicuous, greenish; sepals free. Fruit a capsule.

All leaves in a basal rosette, persistent; stem leaves absent — *M. nudicaulis*

Both stem and rosette leaves present, the rosette leaves withering by flowering time — *M. cerviana*

1. Mollugo cerviana (*L.*) *Ser.*

Small glabrous herb with several delicate, erect pale brownish stems to 15cm. Rosette leaves sessile, oblanceolate or linear, up to 2.5cm long, withering by flowering time; stem leaves sessile, linear, up to 1.5cm long, whorled. Flowers in lax, leafy, umbellate, diachasial cymes with fine capillary pedicels.

A common ephemeral of sand dunes in the Tihama which occurs erratically elsewhere most usually in run-offs in sandstone country both on the escarpment in the Shibām-Ṭawīlah-Maḥwīt region and in the Mashriq at Rada' and Majz (Steinberg); it also grows eccentrically in the coffee terraces of J. Bura' and reaches as high as 2500m on a roadside on J. Masār in the Ḥarāz.

2. Mollugo nudicaulis *Lam.*

Glabrous herb with a rosette of petiolate, oblanceolate leaves up to 6cm long and several erect leafless inflorescences composed of lax, umbellate diachasial cymes, the pedicels delicate but not finely capillary as in the previous species.

A rare plant found between 400 and 1100m in the foothills and rainshadow areas of the escarpment: in stony wadi beds at Walajah, J. Milḥān and below Suq As Sabt and Ḥillah, J. Bura'; in stony fields, Ḥajdah; on open stony hillsides between Malāḥiṭ and Suq al Jammah, between Khamīs Banī Sa'd and Al Qidam and between Riḥāb and Waṣāb.

7. GLINUS *L.*

Prostrate annual herbs. Leaves opposite or whorled, entire or obscurely serrate. Flowers in axillary clusters; staminodes usually present; sepals free. Fruit a capsule.

Fruiting calyx 5–8mm long, 2.5–4.5mm broad; staminodes simply bifid or absent — *G. lotoides*

Fruiting calyx 7–13mm long, 6.5–8.5mm broad; staminodes multifid — *G. setiflorus*

1. Glinus lotoides *L.*

Densely grey-stellate-pubescent herb with long prostrate stems. Leaves opposite or in whorls of 3–4, obovate, up to 2cm long, the margin entire or obscurely serrate. Flowers shortly pedicellate in axillary clusters of up to 10, greenish-white; staminodes absent or up to 9, bifid, whitish.

A rare plant growing in seasonally inundated sandy places in the Tihama and the Mashriq reaching 1800m at Ḥūth where it grows by a large cistern: Wadi Mawr (Forsskal); J. Milḥān, presumably in the foothills (Forsskal); between Sa'īd Miṣr and 'Alūjah; Zabīd; Mārib.

2. Glinus setiflorus *Forssk.*

Similar to the last species but the flowers yellowish, in axillary clusters of up to 4 only, the staminodes 9–18, yellow, multifid and the fruiting calyx larger, with the dimensions given in the key.

A very rare plant collected at about 100m in a seasonally flooded shaded place at Ḍaḥī by Forsskal. This species has been generally neglected so the fact that it has not been refound is probably without significance.

8. GISEKIA *L.*

1. Gisekia pharnaceoides *L.* — Ruqmah

Procumbent or ascending, glabrous, often wine-red, annual herb with short stems to c 30cm. Leaves opposite, oblanceolate or oblong, up to 2cm long. Flowers in shortly pedunculate or subsessile, axillary umbels, inconspicuous, green; sepals free, the margin hyaline. Fruit of 5 free carpels.

A distinctive feature of the plant is the presence of white crystalline 'streaks' on the undersurface of the leaves and the calyx.

A very widespread plant with a similar distribution to *Mollugo cerviana*. It is a common ephemeral on sand dunes in the Tihama south to Ḥays and occurs in similar situations at Mārib. It has been found on the escarpment in the sandstone areas round Ṭawīlah and Aḥjar where it reaches 2600m, on the high plateau near Sana'a and in the Mashriq at Majz, near Ṣa'dah and in the Damt region.

9. LIMEUM *L.*

1. Limeum humile *Forssk.*

L. indicum Stocks ex T. Anders.

Small, much branched, prostrate, annual herb, glabrous or glandular-puberulent especially on the calyx. Leaves opposite, shortly petiolate, broadly elliptic or orbicular, mucronate, up to 8mm long. Flowers sessile in the leaf axils, solitary or in small clusters, minute, greenish; sepals free, the margin hyaline. Fruit of 2 hemispherical mericarps.

A rare plant found mostly on coastal sands in the Tihama but reaching 200m: Luḥayyah (Forsskal); between Luḥayyah and Khawbah; Mukhā; Mafraq.

10. DELOSPERMA *N.E.Br.*

1. Delosperma harazianum (*Deflers*) *Poppendieck & Ihlenfeldt*

Mesembryanthemum harazianum Deflers

Dwarf prostrate succulent herb, stems very short or absent. Leaves opposite, sessile, subcylindrical to obscurely triangular in cross-section, 1–2.5cm long, densely covered in minute scales. Flowers solitary, sessile, terminal, with numerous prominent mauve petal-like staminodes and a yellow centre formed by the numerous stamens; calyx 5-lobed, fleshy, cup-like round the fruiting capsule.

The flowers are only open in full sunshine.

A local plant of rock crevices and small gaps between stones on exposed mountain ridges from 2500 to 3500m mostly on the inner escarpment: on the Sumārah-Manār ridge extending north to the hills around Yarīm; J. Ḍawrān; J. Nabī Shu'ayb; Kawkabān; J. Maḥdad; J. Lawz range; J. Shibām, Ḥarāz. This species is endemic to southwest Arabia and only known elsewhere from the Asir region of Saudi Arabia.

CACTACEAE

OPUNTIA *Miller*

Prickly Pear. Succulent shrubs with flattened jointed stems. Leaves very small, subulate, quickly deciduous. Branches, flowers, spines and barbed bristles all arising from the cushion-like 'areoles'. Flowers solitary with many perianth segments, the green sepals merging into the yellow petals. Fruit a spiny, fleshy, many-seeded pear-shaped berry. Both the following species are introduced from America.

Areoles lanate. Fruit purple when ripe. *O. dillenii*
Areoles not lanate. Fruit orange-yellow when ripe *O. ficus-indica*

1. Opuntia dillenii (*Ker-Gawl.*) *Haw.*

Tīn, Tīn Tūrkī, Tīn Shawkī

Erect shrub to 2m. Joints somewhat glaucous, glabrous except for the tuft of hairs which arise from the areoles together with the pale yellow spines and barbed bristles. Flowers bright yellow. Fruit c 5cm long, purple when ripe.

Thoroughly naturalised in the escarpment foothills between 200 and 1200m from Ḥadīyah north to the Wadi Mawr apparently centred on the Bājil-Wadi Surdūd line. It is reported to have been deliberately introduced at Ḥiṣn Az Zahir by the Imam who wished to make ink from the fruit. It has been spread partly by people, though the fruit is not particularly palatable, but mostly by baboons. Unlike *O. ficus-indica* therefore this species is as common on remote rocks as by villages and is widely regarded as a pest. It is possible, though, that it will allow forest regeneration as neither grazing animals nor people can penetrate its thickets.

2. Opuntia ficus-indica (*L.*) *Miller*

Birmūsh, Tīn, Tīn Turkī, Tīn Shawkī

Erect shrub to 3m, similar to the previous species. Joints dull green, glabrous; areoles not lanate; spines absent or, if present few; barbed bristles present. Flowers yellow. Fruit 5–8cm long, orange-yellow when ripe.

Widespread between 1500 and 3000m (rarely descending to 600m) on the escarpment and high plateau and often the most common plant around villages. Its spread seems to be almost entirely the result of human activity and it is encouraged both as a hedge plant and for its fruit which are ripe in June and July. The unarmed variety though, has not yet been used as a fodder crop in Yemen.

CHENOPODIACEAE

Herbs or shrubs, often succulent. Leaves mostly alternate, without stipules, commonly fleshy and/or mealy, sometimes absent. Flowers small, greenish, usually bisexual, solitary or in cymose clusters, forming a spicate or paniculate inflorescence; petals absent; perianth formed of 1–5 sepals, rarely absent. Fruit a nut.

Most species are very salt-tolerant and consequently *Chenopodiaceae* are common along the Red Sea coast, some growing in salt marsh. They are also found by salt springs inland (such as Ḥawbān near Taiz) and in desert depressions where salt has accumulated after the evaporation of water. Several species are common field weeds.

1. Leaves absent. Stem jointed 5. *Arthrocnemum*
   Leaves present, though sometimes much reduced. Stem not jointed (though appearing so in *Halopeplis*) 2
2. Fruiting perianth with spreading spiny teeth. Plant grey-villous 3. *Bassia*
   Fruiting perianth unarmed. Plant not villous 3
3. Leaves broad, flat, at least 1cm long 4
   Leaves narrow, terete, or, if broad, minute and fleshy 5
4. Flowers unisexual, the fruit enclosed by 2 small bracteoles. Leaves white-mealy on both surfaces, eglandular 2. *Atriplex*
   Flowers bisexual, the fruit enclosed by the persistent perianth segments. Leaves mostly green, sometimes grey and mealy, often with yellowish gland dots 1. *Chenopodium*
5. Stem succulent. Leaves orbicular, fleshy, enclosing the stem, resembling swollen nodes 4. *Halopeplis*
   Stem not succulent. Leaves not as above 5
6. Leaves linear 0.5–2cm long. Fruiting perianth not winged 6. *Suaeda*
   Leaves ovate, ± 1mm long. Fruiting perianth winged 7. *Salsola*

1. CHENOPODIUM *L.*

Annual or perennial herbs. Leaves flat, toothed, lobed or pinnatifid, rarely ± entire; indumentum also variable, mealy, glandular-pubescent or gland-dotted, rarely glabrous. Flowers bisexual, usually in cymes; bracteoles absent; perianth unchanged in fruit, ± enclosing the nutlets.

1. Plant pubescent, aromatic. Leaves dotted with yellow glands, often pinnatifid **2**
Plant ± mealy but otherwise glabrous, rarely aromatic. Leaves neither gland-dotted nor pinnatifid **4**
2. Inflorescence composed of flowers in diachasial cymes in the axils of leaves and bracts. Lower leaves pinnatifid **3**
Inflorescence composed of small sessile clusters of flowers in the axils of the leaves and the bracts. Leaves coarsely dentate, but not pinnatifid *C. ambrosioides*
3. Some glands shortly but distinctly stalked. Leaf lobes mostly acute *C. procerum*
All glands sessile. Leaf lobes mostly rounded *C. schraderianum*
4. Leaves entire. Plant stinking of rotten fish *C. vulvaria*
Leaves almost always toothed. Plant not foetid **5**
5. Leaf bases mostly truncate (a few rounded or sub-cordate) *C. fasciculosum*
Leaf bases cuneate **6**
6. Seeds with a sharp, prominent keel. Young growth sparsely mealy *C. murale*
Seeds with obtuse keel. Young growth densely grey-mealy **7**
7. Leaves almost as broad as long *C. opulifolium*
Leaves at least 1½ times as long as broad *C. album*

1. Chenopodium album *L.*
Very variable, much-branched, annual herb, usually erect, grey-green and mealy, mostly 30–50cm high but sometimes reaching 1½m. Leaves ovate-rhomboid or lanceolate, narrowed to a cuneate base, up to c 6cm long, at least 1½ times as long as broad, the margin shallowly but acutely dentate, sometimes entire. Inflorescence a leafy panicle made up of crowded cymes or spikes. Seeds with an obtuse keel, nearly smooth.

Widespread on the escarpment and high plateau growing as a field weed between 1000 and 2800m, but apparently infrequent and scattered in occurrence. It is most common in irrigated vegetable fields around Sana'a. (Wood 2168.)

2. Chenopodium ambrosioides *L.*
Pubescent annual or perennial with yellow, sessile glands, an unpleasant oily-aromatic smell and erect branched woody stems to 1½m. Leaves lanceolate with a cuneate base, up to 5cm long, the margin coarsely and acutely dentate, rarely entire. Inflorescence paniculate composed of short spikes made up of small sessile clusters of flowers in the axils of leaves and bracts.

A comparatively recent introduction from America now thoroughly naturalised by streams and in waste ground in areas of good rainfall (Ibb, for example) between 600 and 2700m on the escarpment north to the Wadi Lā'ah; much more common in Ibb and Taiz provinces than elsewhere.

3. Chenopodium fasciculosum *Aellen*
Variable erect or spreading, much-branched, mealy annual herb to 40cm. Leaves ovate with an abruptly rounded or truncate base (rarely subcordate), up to 5cm long, the margin with numerous spreading acute teeth of varying size. Inflorescence a short leafy panicle made up of crowded, divaricately-branched cymes.

An uncommon plant of disturbed rocky slopes and high-altitude fields found from 2600 to 3200m on a few high escarpment mountains: J. Ta'kar; J. Sumārah; J. Ḥiṣn Madnān, Waṣāb al'Ālī; J. Shibām, Ḥarāz; J. Maḥdad; J. Maswar. (Wood 1660.)

4. Chenopodium murale *L.*
Erect or spreading, much-branched, green, sparsely mealy annual herb to 50 (−80)cm. Leaves ovate-rhomboid or lanceolate, narrowed to a cuneate base, up to 5cm long, the margin with numerous acute, forward-pointing teeth. Inflorescence an irregular, sometimes small, panicle formed of leafy divaricately branched cymes. Seeds with a prominent acute keel, minutely pitted.

A widespread and often common weed of roadsides, fields, coffee terraces and waste ground around villages, growing wherever there are suitable habitats on the escarpment, high plateau and in the Mashriq from 900 to 2500m.

5. Chenopodium opulifolium *Schrad.*
Erect, much-branched, grey-green, mealy annual herb to 1½m, similar in appearance to *C. album* but often more vigorous. Leaves ovate-rhomboid with a cuneate base, 2–4cm long, about as long as broad or slightly longer, the margin entire, obscurely lobed or with a few small teeth. Inflorescence a very grey mealy panicle of crowded spikes and cymes. Seeds with an obtuse keel, obscurely furrowed.

An uncommon but probably overlooked plant of cultivated and waste land and disturbed wadi margins from 1000 to 2400m on the escarpment: Ṣayfarah, Taiz; Suq As Sabt, Ibb; Manākhah (Schweinfurth) and Wadi Ḥijān, Ḥarāz; above Ḥillah, J. Bura' (Schweinfurth); Suq al Badr, Khawlan Ash Sham. (Wood 1789.)

6. Chenopodium procerum *Hochst. ex Moq.*
Erect, much-branched, yellow-green, pubescent annual herb to 2m with yellow, sessile and shortly stalked glands and a strong aromatic smell. Leaves elliptic with an abruptly cuneate base, up to 10cm long, pinnatifid with acute lobes. Inflorescence a narrow sparingly leafy terminal panicle composed of diachasial cymes in the axils of the leaves and the bracts.

A local plant growing in fields and disturbed rocky ground around villages from 2000 to 2700m on the escarpment from J. Ṣabir north to Shahārah.

7. Chenopodium schraderianum *Schult.*
C. foetidum Schrad.
C. botrys sensu Schwartz
Erect, branched, yellow-green, pubescent annual herb to 1m, with sessile, yellow, dot-like glands and an aromatic

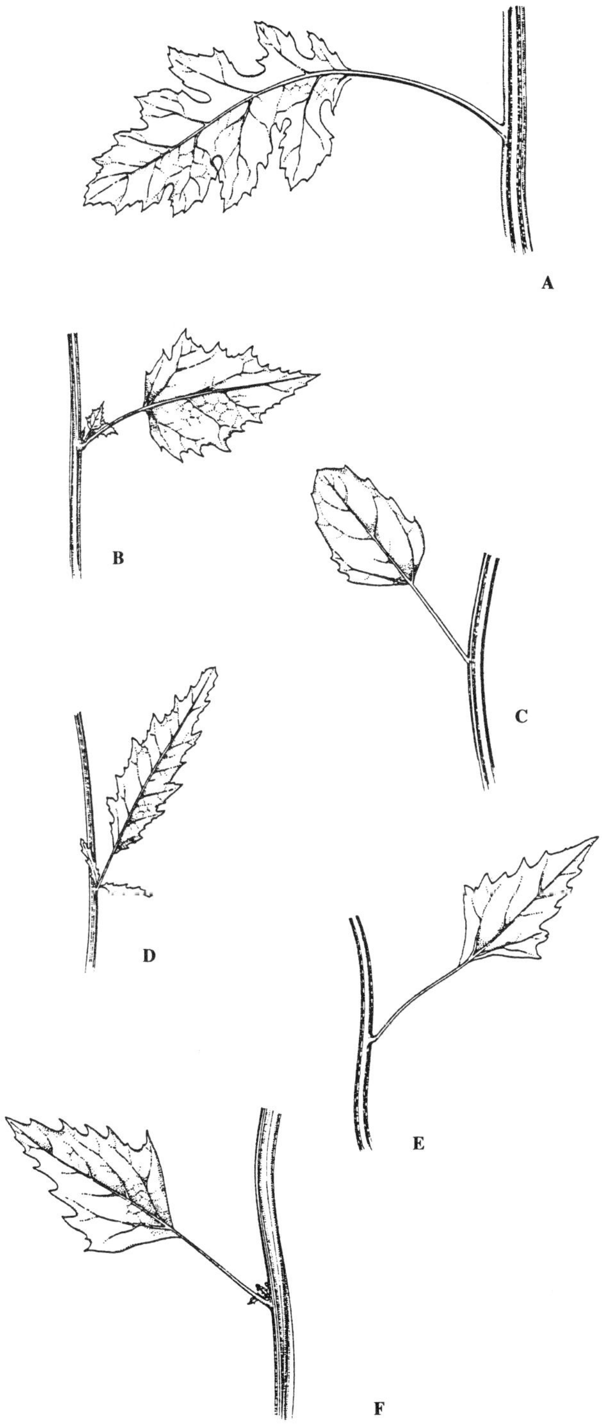

Fig. 5. Leaves of *Chenopodium* species. **A** *C. procerum*; **B** *C. fasciculosum*; **C** *C. opulifolium*; **D** *C. ambrosioides*; **E** *C. album*; **F** *C. murale*.

smell. Leaves oblong with a cuneate base, up to 6cm long, deeply pinnatifid with obtuse lobes. Inflorescence a narrow terminal panicle composed of small diachasial cymes in the axils of the leaves and the bracts.

C. schraderianum has the appearance of a small version of *C. procerum*.

The commonest *Chenopodium* in Yemen. It is widespread and abundant as a weed of cultivation and waste places (less commonly on disturbed, eroded, hill slopes) growing from 1200 to 3000m wherever there are suitable habitats on the escarpment, high plateau and in the Mashriq.

8. Chenopodium vulvaria *L.*

Decumbent, much-branched, grey-mealy annual, stinking of rotten fish. Leaves ovate-rhomboid or obovate with a cuneate base, up to 2.5cm long, the margin entire or with an obscure angle on the rhomboid-shaped leaves. Inflorescence of small terminal and axillary cymes forming a loose panicle.

A rare plant only known from irrigated vegetable fields at Sana'a (2200m).

Chenopodium bryoniifolium Bunge (*C. atripliciforme* Murr.) is reported by Schwartz from Yemen on the basis of a record by Wissmann from Ghirās. This is a distinctive species with 3-lobed, entire leaves but its distribution is centred on Siberia and its occurrence in Yemen seems highly improbable. It is very difficult to guess what the plant might have been, though an odd form of *C. album* seems most likely.

2. ATRIPLEX *L.*

Annuals or perennial undershrubs. Leaves flat, densely white-mealy, toothed or entire. Flowers unisexual, the male with 5 perianth segments, the female without a perianth but with 2 persistent bracteoles which ± enclose the fruit.

Leaves entire with a truncate or cordate base	*A. farinosa*
Leaves dentate with a cuneate base	*A. leucoclada*

1. Atriplex farinosa *Forssk.* 'Antikār

Erect, white-mealy, branched, monoecious undershrub to 1½m. Leaves ovate, entire, 2–4cm long, the base truncate or cordate. Inflorescence a terminal panicle made up of short axillary spikes; male flowers with 5 yellow stamens; female flowers enclosed by 2 suborbicular entire bracteoles.

A very local plant of coastal sand hills just above the high tide mark but abundant where it occurs: north and south of Mawshij; Mīdī.

2. Atriplex leucoclada *Boiss.*

Obione leucoclada (Boiss.) Ulbr.

Decumbent or ascending much-branched, grey-mealy monoecious undershrub with pale stems to c 50cm. Leaves triangular, ovate or ± oblong, up to 4cm long, abruptly narrowed into a cuneate base, the margin irregularly dentate. Inflorescence a lax leafy terminal panicle made up of long spikes of rather distant flower clusters; female flowers enclosed by 2 triangular, dentate bracteoles.

A locally abundant plant of salty alluvial plain found

from 1500 to 2300m almost always in association with *Suaeda aegyptiaca*: Ḥawbān near Taiz; in a belt from Rada' north to Jiḥānah, thence to Sana'a (Rathjens) Arḥab and Khamr.

3. BASSIA *All.*

1. Bassia muricata (*L.*) *Murr.*

Densely villous, much-branched, spreading annual with pink woody stems. Leaves linear or linear-triangular, acute, grey, villous, 4–5mm long. Flowers in small sessile clusters in the leaf axils forming a large leafy panicle; perianth with 5 segments, becoming enlarged in fruit and developing 5 radiating spines, each about 2mm long.

An uncommon plant of roadsides and field borders from 2000 to 2300m in the Mashriq restricted to the area bounded by Bani al Ḥārith, the Wadi Khārid and the Fardah pass. It is a widespread plant of the North African deserts.

4. HALOPEPLIS *Bunge ex Ung.-Sternb.*

1. Halopeplis perfoliata (*Forssk.*) *Bunge ex Aschers. & Schweinf.*

Stout, branched, glabrous, succulent perennial to 80cm, the stem succulent but not jointed. Leaves rudimentary, almost globose, surrounding the stem and looking like curious swollen fleshy nodes. Inflorescence a panicle formed of short dense spikes, the flowers sunk into the rhachis of the spike in groups of 3; perianth segments 3.

Abundant on sand banks by the sea at Mīdī but unknown elsewhere in Yemen.

5. ARTHROCNEMUM *Moq.*

1. Arthrocnemum fruticosum (*L.*) *Moq.*

Salicornia fruticosa L.

Glasswort. Erect, branched, leafless, glabrous, glaucous undershrub with jointed succulent cylindrical stems to 1m. Inflorescence of terminal cylindrical or club-shaped spikes similar to the stem; flowers in groups of 3, shorter than the joint into which they are inserted, falling to leave a tripartite hollow; perianth segments 4; seeds grey, hairy.

A salt marsh plant known only from Mukhā where it was collected by Forsskal but it has not been seen since.

Although Forsskal's plant is undoubtedly named correctly, the common *Arthrocnemum* of the Red Sea is in fact *A. macrostachym* (Moric.) Moris & Delponte (*A. glaucum* (Del.) Ung.-Sternb.) which differs from *A. fruticosum* in the flowers which are as long as the joint in which they are inserted and fall to reveal a single hollow. The seeds are black, tuberculate and glabrous.

6. SUAEDA *Forssk. ex Scop.*

Herbs or shrubs. Leaves linear, fleshy. Flowers bisexual or monoecious; perianth segments 5, fleshy.

1. Leaves fleshy, rigid, ± cylindrical, not mealy. Perianth not enlarging in fruit — 2
   Leaves fleshy but neither rigid nor cylindrical, mealy. Perianth often enlarging and becoming spongy in fruit — *S. aegyptiaca*
2. Shrub 1–3m high. Leaves 1–2.5cm long — *S. monoica*
   Undershrub less than 1m high. Leaves 0.5–1cm long — *S. vermiculata*

1. Suaeda aegyptiaca (*Hasselq.*) *Zohary* Milaḥ, Harm

Schanginia aegyptiaca (Hasselq.) Aellen
S. hortensis (Forssk.) Moq.
Suaeda baccata Forssk.
S. hortensis Forssk.

Much-branched, shrubby annual or short-lived perennial to 1m. Leaves linear, soft and fleshy, flattened above, slightly mealy, 1–2cm long. Flowers bisexual, in sessile, axillary clusters; perianth fleshy, c 1mm long in flower, sometimes enlarging and becoming spongy in fruit.

A locally abundant species of roadsides, waste ground and alluvial basins but always where there is a high salt level. It has been found between 100 and 2300m at Mawza', Ḥawbān near Taiz and in a belt of country on the borders of the high plateau and the Mashriq from Rada' northwest to Jiḥānah, Sana'a, Arḥab and Nihm. It often grows under *Tamarix* in association with *Atriplex leucoclada*. The forms with an enlarged perianth are mostly found in the drier localities around the high plateau.

2. Suaeda monoica *Forssk. ex J.F. Gmel.* 'Asal

Glabrous, much-branched shrub to 3m. Leaves linear, cylindrical, firm and fleshy, always green, 10–25mm long. Flowers often monoecious, in sessile axillary clusters; perianth fleshy, c 1mm long, not enlarging in fruit.

Locally abundant and sometimes dominant on sand in salty areas but not usually occurring by the Red Sea itself. It is found from sea level to 1100m in scattered colonies in the Mashriq near Barāqish in the Jawf and various places in the Tihama (Yakhtul-Khawkah; Tuḥaytah near Zabīd; Hodeida; Luḥayyah and along the Wadi Mawr as far as Zuḥrah; 'Abs-Ḥaraḍ).

3. Suaeda vermiculata *Forssk* Dāluq, 'Asal

S. fruticosa Forssk. ex J.F. Gmel.

Glabrous, much-branched undershrub to c 70cm. Leaves linear, cylindrical, firm and fleshy, green or purplish, 5–10mm long. Flowers bisexual, in sessile axillary clusters; perianth fleshy, c 1mm long, not enlarging in fruit.

An abundant and locally dominant plant of the salt marsh and subkhah of the Red Sea littoral where the land is in range of salt spray. It occurs in all suitable localities and is often associated with *Limonium spp.*, and *Aeluropus lagopoides*.

7. SALSOLA *L.*

Our species are glabrous or pubescent undershrubs with very small, alternate, sessile, ovate leaves. Flowers bisexual; perianth segments 5 developing a conspicuous membranous wing in fruit.

1. Young branchlets green, papillose, but otherwise glabrous. Flowering perianth 2–3mm long — *S. bottae*
   Young branchlets dirty-white, pubescent. Flowering perianth less than 1mm long — 2
2. Branchlets with spinescent tips. Leaves c 0.5mm

broad. Plant not foetid *S. spinescens*
Branchlets without spinescent tips. Leaves c 1.5mm broad. Plant foetid *S. imbricata*

1. Salsola bottae *(Jaub. & Spach) Boiss.* Jithjāth
Much-branched, nearly glabrous undershrub with many prostrate or ascending branches. Branchlets green, papillose, not spine-tipped. Leaves scale-like, ovate, acute, green, glabrous, c 1mm broad. Flowers sessile, solitary in the leaf axils forming lax spicate inflorescences; flowering perianth 2–3mm long; fruiting perianth including the wings, 5–6mm in diameter.

A locally abundant Mashriq plant growing on alluvial and stony plain between 1000 and 2400m from Ḥarīb northwestwards in an arc through Mārib and Arḥab to Raydah, but only known definitely further north at Majz. It also occurs on the high plateau near Wa'lān and in the Bani Ḥushaysh and may also be in the Ḥujarīyah though it is not clear whether Deflers' record from west of J. Munīf was within North or South Yemen.

S. bottae is only known from tropical Arabia and Somalia.

2. Salsola imbricata *Forssk.*
S. foetida sensu Schwartz
Much-branched, prostrate, pubescent undershrub stinking of rotten fish. Branchlets dirty-white, pubescent, not spine-tipped. Leaves broadly-ovate, rounded, grey-green, pubescent, c 1.5mm broad. Flowers sessile in the leaf axils forming dense spicate inflorescences; flowering perianth minute, less than 1mm long; fruiting perianth including the wings, 2–3mm broad.

A rare plant of a few isolated localities near the sea in the Tihama and between 1100 and 2000m in the Mashriq: Luḥayyah (Forsskal); Hodeida; Barāqish in the Jawf; 20 kms north of Rada' (Wood 1184, 3094).

I have followed Botschantzev and Boulos in using the name *Salsola imbricata* Forssk. for the Arabian plant previously known as *S. baryosma* (Schult.) Dandy though I have considerable doubts as to whether they are correct to do so. The problems were clearly set out by Schweinfurth and it seems that in the absence of an authentic specimen collected by Forsskal it is impossible to be certain what *S. imbricata* really is. The description could fit either this species or the next but *S. spinescens* is much more common in Yemen and as far as I know, the only species growing at the type locality (Luḥayyah) today. Unfortunately the rather fanciful illustration in Forsskal's *Flora Aegyptiaco-Arabica* could not be *S. spinescens* but the illustration should probably not be relied on.

3. Salsola spinescens *Moq.*
Ḥarm (Tihama) Ḥamḍ (Mashriq)
S. forskalii Schweinf.
Similar to the previous species but a rounded bush up to 70cm high with many short, erect or spreading, entwined spine-tipped branches and very small leaves c 0.5mm wide. It is not foetid.

Locally abundant in three areas of the Tihama and the Mashriq, growing from sea level to 1900m: frequent on sand dunes within about 20km of the Red Sea from Mīdī south to the Wadi Sihām, but like *Suaeda monoica* absent from the actual Red Sea littoral; abundant on arid stony hills from Rada' north east to Mārib and Ḥarīb; in the Majz area (Steinberg).

Salsola longifolia Forssk. is recorded by Schwartz on the basis of a collection by Wissmann from near Hodeida. The plant has not been recollected and the occurrence of this North African species in Yemen seems very unlikely. All other records of *S. longifolia* in Schwartz were taken wholesale from Blatter and are (fide Botschantzev) *Sevada schimperi* Moq., a species that might well occur in Yemen though listed separately by Schwartz.

Beta vulgaris *L.* and **Spinacia oleracea** *L.*, both called 'Sabānakh', are grown for their leaves which are eaten as spinach but cultivation is limited to market gardens round the larger towns, particularly Sana'a.

AMARANTHACEAE

Herbs or shrubs. Leaves opposite or alternate, entire, without stipules. Flowers usually bisexual in bracteate, often spictate inflorescences, the fertile flowers sometimes subtended on either side by a modified sterile flower; perianth segments usually 5, free, commonly dry and scarious. Fruit a capsule, dehiscing transversely or irregularly.

Most species in this family are either annual weeds or perennial species of dry bushland.

1. Inflorescence and stems densely white woolly 7. *Aerva*
Inflorescence and stems various, but never white woolly **2**
2. Leaves alternate **3**
Leaves opposite **6**
3. Shrub. Lateral sterile flowers transformed into tufts of plumose hair in fruit 4. *Saltia*
Herbs. Lateral sterile flowers absent or, if present, not transformed into tufts of plumose hair **4**
4. Flowers in spikes, the spikes all arising from the leaf axils, the individual flowers apparently solitary but in fact in 3s, the lateral pair bract-like 2. *Digera*
Inflorescence various but, if spike-like, some or all spikes terminal. Flowers in reduced cymes of more than 3 flowers **5**
5. Bracts scarious. Fruit with many seeds 1. *Celosia*
Bracts not scarious. Fruit with a single seed 3. *Amaranthus*
6. Sterile flowers modified into hooked spines so the fruit is 'clinging' 5. *Pupalia*
Sterile flowers not modified into hooked spines **7**
7. Inflorescence axillary, sessile 9. *Alternanthera*
Inflorescence terminal, pedunculate **8**
8. Inflorescence spicate 8. *Achyranthes*
Inflorescence a lax panicle 6. *Psilotrichum*

1. CELOSIA *L.*

Herbs with alternate leaves. Perianth segments scarious. Fruit with numerous seeds.

1. Inflorescence an uninterrupted spike. Perianth segments c 10mm long — *C. argentea*
   Inflorescence interrupted. Perianth segments less than 5mm long — **2**
2. Inflorescence sparingly branched, often spicate. Perianth segments delicate, white — *C. trigyna*
   Inflorescence much branched, paniculate. Perianth segments dull brown with a dark central line — *C. polystachia*

1. Celosia argentea *L.*

Erect, branched, glabrous annual to 1m. Leaves lanceolate below, linear above, up to 10cm long. Inflorescence a dense, continuous, terminal, cylindrical spike up to 15cm long; perianth segments scarious, shining, purple, c 10mm long.

In Yemen only variety **cristata** (*L.*) *O. Kuntze* with sessile inflorescences has been found.

Recorded by Schweinfurth as a cultivated plant at Bājil and J. Bura', but only seen recently as a roadside casual at Suq al Amān near Ḥajjah.

2. Celosia polystachia (*Forssk.*) *C.C. Townsend*

C. populifolia Moq.
Achyranthes polystachia Forssk.
Sparsely pubescent branched untidy scrambler attaining 1½m with the support of bushes. Leaves dark green, ovate-lanceolate, up to 8cm long. Inflorescence of spike-like racemes massed together into a terminal panicle; perianth segments slightly scarious only, dull brown with a dark central line, c 3mm long.

A local plant of dry bushland found at Qa'ṭabah in the southeastern mountains (Deflers) and along the foothills of the central escarpment where it grows mostly in valley scrub between 200 and 1000m: between Suq Al Aḥad and Suq ar Rubū, Waṣāb As Sāfal; Mishrāfah, Wadi Rima'; below Ḥadīyah; Khalīfah (Schweinfurth); Wadi Surdūd (Forsskal); Walajah, J. Milḥān (Schweinfurth); Wadi Yūr; Wadi Sara'.

3. Celosia trigyna *L.*

Erect to decumbent glabrous straggling annual to c 30cm. Leaves ovate, up to 8cm long. Inflorescence of simple or sparingly branched, interrupted, terminal, spike-like racemes; perianth segments silvery-white, often flushed pink, very delicate, c 5mm long.

A common field weed found between 400 and 1900m along all the escarpment but with a marked preference for high rainfall areas and particularly characteristic as a weed of coffee terraces in the central escarpment.

2. DIGERA *Forssk.*

1. Digera muricata (*L.*) *Mart.* — Sanafī, Dijr

D. arvensis Forssk.
An erect, little-branched, almost glabrous annual to c 50cm. Leaves alternate, lanceolate or ovate, up to c 6cm long. Inflorescence of somewhat interrupted axillary spikes; flowers in 3s, the central one pink, fertile, c 4mm long, the lateral pair sterile, green, bract-like. Fruit l-seeded.

An abundant field weed in the rich alluvial plains in the east of the Tihama, which grows more rarely along wadis in the escarpment up to 1000m.

3. AMARANTHUS *L.*

Annual monoecious herbs with alternate, prominently nerved leaves. Flowers in panicled spikes or axillary clusters. Bracts often spinescent. Perianth segments linear or lanceolate. Fruit l-seeded, usually opening transversely. All species are weeds and many are comparatively recent introductions.

1. All spikes axillary — *A. graecizans*
   Terminal as well as axillary spikes present — **2**
2. Plant armed with spikes at the leaf nodes — *A. spinosus*
   Plant without spines at the leaf nodes — **3**
3. Perianth segments usually 3 — *A. viridis*
   Perianth segments usually 5 — *A. hybridus*

1. Amaranthus graecizans *L.* — Da'da'

A. angustifolius Lam.
A. sylvestris Vill.
A. aschersonianus Chiov.
Erect or decumbent, usually glabrous, annual herb. Leaves variable in shape, rhomboid, ovate or oblanceolate, up to 4cm long. Inflorescence of small, pale green axillary clusters; bracts shorter than the perianth, spinescent; perianth segments 3. Capsule wrinkled.

Widely distributed as a field weed, particularly of irrigated land, from sea level to 2700m in the Tihama, escarpment and high plateau, but more characteristic of low altitudes. This species is sometimes used as a 'spinach' vegetable, but apparently less commonly in Yemen than in some other Arab countries.

2. Amaranthus hybridus *L.*

A. hypochondriacus L.
A. chlorostachys Willd.
A. caudatus sensu Deflers
Erect or decumbent annual herb to 1m, stems usually reddish and ± glabrous below, but pubescent above. Leaves rhomboid, elliptic or oblanceolate, up to 12cm long. Inflorescence of dense axillary and terminal, green or reddish spikes, sometimes elongating and becoming pendulous with age; bracts spinescent, exceeding the perianth; perianth segments 5. Capsule smooth.

Two subspecies are found in Yemen. Subspecies **hybridus** is characterised by the inflated style base which is best observed on young fruit. In other respects it is very variable but the common plant in Yemen (Wood 692) is a decumbent plant with rather short, broad inflorescences. Other forms occur though, including a very robust erect plant with long, drooping inflorescences (Wood 74/300). In Yemen subspecies *hybridus* usually has green inflorescences but reddish ones also occur. Subspecies **cruentus** (*L.*) *Thell.* (*A. paniculatus* L.), however, always has red inflorescences. The style base is narrow and uninflated.

Subspecies *hybridus* is widely distributed on tracksides, waste ground and as a garden weed on the escarpment and high plateau between 1300 and 2900m. It is the only common *Amaranthus* at higher altitudes. Subspecies *cruentus* is a rare garden escape known only from Sana'a.

3. Amaranthus spinosus *L.*

Erect branched almost glabrous annual herb to 1m, usually armed with axillary spines. Leaves lanceolate to ovate, up to 7cm long. Inflorescence of dense axillary and terminal greenish spikes up to 7.5mm broad; bracts about equalling the perianth, not spinescent; perianth segments 5. Capsule smooth.

First recorded in the 1930s and now abundant in the Tihama and on the escarpment up to 2000m. It is found on waste ground, as a field weed and particularly in wadis. It favours high rainfall areas, irrigated land and permanent streams in hot country.

4. Amaranthus viridis *L.*

A. gracilis Desf.

Erect glabrous annual herb to 50cm. Leaves ovate to rhomboid, up to 6cm long. Inflorescence of rather few dense axillary and terminal green spikes mostly c 5mm broad; bracts shorter than the perianth, not spinescent; perianth segments 3. Capsule wrinkled.

A local weed of irrigated fields mostly in the Tihama and foothills up to 1000m, but also recorded in Sana'a.

The two following species of *Amaranthus* have been cultivated in Yemen:

Amaranthus caudatus *L.* is similar to *A. hybridus* subspecies *cruentus* but usually has red flowers in long drooping spikes. It may further be distinguished by its broad overlapping female perianth segments and inflated style bases. It is cultivated for ornament in Sana'a and has also been found in Bājil (Schweinfurth).

Amaranthus lividus *L.* was recorded by Forsskal as a cultivated vegetable in the Wadi Mawr near Kudmīyah. Like *A. viridis* it has only 3 perianth segments but can be distinguished from all our other species by the emarginate apex of its leaves and the indehiscent fruit.

4. SALTIA *R. Br. ex Moq.*

1. Saltia papposa (*Forssk.*) *Moq.* Say'am, San'am

Spinescent shrub to 3m, stem grey-brown, glabrous. Leaves alternate, narrowly linear-oblanceolate, 1–1.5cm long. Inflorescence of pedunculate axillary spikes up to 4cm long; flowers in 3s, the central one fertile, pale brown, villous, c 4mm long, the lateral pair sterile, bract-like, becoming transformed in fruit into conspicuous feathery pappus.

Abundant in the arid foothills and stony steppe of the southern Tihama from Waṣāb As Sāfal southwards; also in the southeastern mountains around Rāḥidah, at Qa'ṭabah and on the Naqīl al Ḥajlah between Jūbah and Ḥarib. It grows from 100 to 1200m and is commonly associated with *Commiphora myrrha*.

5. PUPALIA *Juss.*

Perennial herbs with opposite leaves. Inflorescence of interrupted terminal bracteate spikes, each bract subtending a small cluster of flowers; flowers in 3s, the central one fertile, the lateral pair sterile, modified into hooked spines which lengthen rapidly in fruit.

Large scrambling undershrub. Perianth segments
7–9mm long ... *P. grandiflora*
Erect or decumbent herb. Perianth segments
± 5mm long ... *P. lappacea*

1. Pupalia grandiflora *Peter*

A very vigorous scrambling undershrub. Leaves narrowly ovate, up to 10cm long, sparsely pubescent. Burrs green, 1.5–2cm broad; perianth segments pubescent, 7–9mm long; spines c 9mm long.

Known only from a damp shaded gully descending northwards from Aṣ Ṣa'bah, J. Milḥān at c 1700m where it is in great abundance scrambling over *Erythrococca abyssinica* and other shrubs.

2. Pupalia lappacea (*L.*) *Juss.* Ḥamshad

Decumbent to erect perennial herb to c 50cm. Leaves ovate up to c 4cm long, variable in indumentum, but usually densely pubescent, often silky when young. Burrs whitish, c 1cm broad; perianth segments villous, ± 5mm long; spines c 6mm long.

The Yemen plant is variety **velutina** (*Moq.*) *Hook. f.*

Widely distributed on rocky ground everywhere except for the driest parts of the Mashriq and mountains above 2500m; completely absent from sand. It is particularly common in Acacia scrub in the escarpment foothills and a large collection of burrs in one's clothing is one of the penalties of walking in such areas.

6. PSILOTRICHUM *Blume*

1. Psilotrichum gnaphalobryum (*Hochst.*) *Schinz*

P. cordatum Moq.

Much-branched, pubescent perennial to 1m, the stems often rather woody. Leaves opposite, ovate, up to 3cm long. Inflorescence large, lax, terminal, consisting of panicled spikes, the spikes few-flowered; flowers solitary; perianth segments green, villous, c 3.5mm long.

A local plant growing between 500 and 2000m in rocky Acacia scrub on the escarpment from Taiz north to Al 'Ashshah. It is characteristic of rain shadow areas in the central escarpment and also occurs in the southeastern mountains at Qa'ṭabah (Deflers) and east of Rada'. It is often associated with *Vernonia cinerascens*.

7. AERVA *Forssk.*

Woolly undershrubs with alternate leaves. Flowers in axillary spikes or terminal panicled spikes; bracts and perianth segments scarious, densely covered in long white woolly hairs.

Inflorescence terminal *A. javanica*
Inflorescence axillary *A. lanata*

1. Aerva javanica (*Burm. f.*) *Juss. ex Schult.*

Ra', Ra'in, Ruā

A. persica (Burm f.) Merrill

An erect much-branched undershrub to 1m, stems densely white stellate-hairy. Leaves oblanceolate, elliptic or linear, up to 6cm long, grey-green above, whitish below. Inflorescence a terminal panicle of white, leafless cylindrical spikes c 2–3cm long.

Very probably the most widespread plant in Yemen being found from the Red Sea coast east to the Sands at Mārib. It is only absent from mountains above 2500m and the wetter parts of the escarpment. It grows on sand dunes, rocky plain, stony hillsides and as a weed of dry stony fields. The inflorescences were much used traditionally to stuff pillows and mattresses.

In addition to the typical plant described above, variety **bovei** *Webb* also occurs. This is generally smaller in all parts than the type with leaves up to c 2.5cm long and the flower spikes less than 3mm broad. (Wood 74/256.) Its exact distribution is not known but it is found in more arid areas.

2. Aerva lanata (*L.*) *Juss. ex Schult.*

Erect or decumbent undershrub to 40cm, stems densely white hairy, but stellate hairs absent. Leaves obovate or spathulate, up to 2cm long, dull green above, grey below. Inflorescence of white axillary spikes 1–2cm long, decreasing in size upwards, the uppermost not subtended by leaves.

Widespread and frequent on the escarpment and high plateau from 1200 to 2500m. It is found around boulders on open stony plain, on rocky hillsides and on cliffs.

8. ACHYRANTHES *L.*

1. Achyranthes aspera *L.*

Ḥamshad, Muḥaṭ.

A very variable erect or straggling decumbent herb to c 1½m, stems pubescent. Leaves opposite, broadly ovate to rhomboid, up to c 10cm long, usually pubescent and green below. Inflorescence of long lax terminal spikes; flowers solitary, deflexed, reddish-tinged, shiny; bracts and perianth segments spinescent, glabrous, the segments c 6mm long.

Abundant between 500 and 2900m on the escarpment, high plateau and southeastern mountains, growing on roadsides, in waste ground, on field borders and eroded mountain slopes; more common above 1500m. In autumn the fruit is a great nuisance, readily becoming attached to socks and trousers.

The common Yemen plant described above is variety **pubescens** (*Moq.*) (*C. Townsend*) but variety **sicula** *L.* also occurs. It has adpressed pilose stems, leaves silvery below and perianth segments 4–5mm long. Its exact distribution is not known, but it is probably quite common.

9. ALTERNANTHERA *Forssk.*

Herbs with opposite leaves. Inflorescence of short, almost capitate, axillary spikes.

Stem glabrous. Perianth white, soft in fruit *A. sessilis*
Stem pubescent. Perianth yellow-brown, spiny in fruit *A. pungens*

1. Alternanthera pungens *Kunth*

Prostrate annual herb with hairy stems. Leaves obovate or elliptic, up to 2.5cm long. Flower spikes 2–3 together in the leaf axils, sessile, shortly cylindrical, up to 1.5cm long; perianth segments unequal in length, two longer than the other 3, becoming spiny and yellow-brown in fruit.

An introduction from America, this species is an unpleasant weed of paths and waste ground found between 600 and 2000m on the escarpment. Although now widespread, it is noticeably more common in Taiz province suggesting that it may have been introduced through Aden.

2. Alternanthera sessilis (*L.*) *R.Br. ex Roth.*

Decumbent herb with glabrous stems. Leaves oblanceolate or elliptic, up to 5cm long. Flower spikes usually solitary in the leaf axils, globose, c 5–8mm long; perianth segments, equal, white remaining soft.

A rare plant of grassy stream banks between 300 and 1300m: Wadi Ḍabāb, Taiz; between Sa'īd Miṣr and 'Alūjah.

Gomphrena globosa *L.*, a small herb with attractive pink terminal flower heads is commonly cultivated at Taiz. Elsewhere I have only seen it grown in the villages of the coffee region of J. Bura' above Ḥillah where it was found by Schweinfurth but it was also seen at Bayt al Faqīh by Forsskal.

PORTULACACEAE

Herbs, often with fleshy leaves. Flowers regular; sepals 2; petals 5. Fruit a capsule.

Leaves with stipules modified into hairs. Flowers yellow 1. *Portulaca*
Leaves without stipules. Flowers purple 2. *Talinum*

1. PORTULACA *L.*

Herbs, glabrous except for the stipules which are modified into tufts of hair. Flowers yellow, sessile, solitary or in small clusters at the ends of the branches and ± enclosed by leaves. Flowers open in the morning.

1. Stipular hairs very few, falling quickly so the plant appears completely glabrous. Leaves rounded at the apex. *P. oleracea*
   Stipular hairs numerous and persistent. Leaves acute at the apex **2**

2. Erect herb with linear, terete leaves *P. foliosa*
Prostrate herb with lanceolate, flattened leaves *P. quadrifida*

1. Portulaca foliosa *Ker-Gawl.*
Erect annual herb to 20cm in Yemen. Leaves opposite to ± alternate, linear, terete, acute, c 6–12mm long; stipular hairs persistent. Flowers yellow, solitary or in small clusters at the top of the branches.

Still known only from Ḥillah, J. Bura' where it was first found by Schweinfurth. It persists in scrubby grassland above the suq at c 600m.

2. Portulaca oleracea *L.* Rijil
Annual herb with numerous procumbent or spreading branches. Leaves alternate, obovate or oblanceolate, flattened, rather fleshy, 1–3cm long, the apex rounded; stipular hairs inconspicuous, falling quickly. Flowers yellow, in small terminal clusters.

Widespread but rather scattered in occurrence between 100 and 2400m in the east of the Tihama, on the escarpment and rarely on the high plateau. It is usually found as a field weed, often in rich irrigated fields along wadis where it may be a relic of cultivation. It is occasionally cultivated in small patches and sold in the markets particularly in Sana'a where it is popular as a vegetable among foreign Arabs.

3. Portulaca quadrifida *L.* 'Uraymbah, Burar
Prostrate annual herb with a creeping stem, often rooting at the nodes. Leaves opposite, usually lanceolate (rarely ovate or elliptic), flattened, fleshy, acute, 5–10mm long; stipular hairs numerous, persistent, conspicuous. Flowers yellow or orange, solitary at the ends of the branches.

Generally common but very inconspicuous except when in flower, occurring throughout the Tihama, escarpment and high plateau from 100 to 3000m wherever open gravelly or stony conditions are found. It also grows on infrequently disturbed cultivated land such as coffee terraces and in derelict fields.

2. TALINUM *Adans.*

1. Talinum portulacifolium (*Forssk.*) *Aschers. ex Schweinf.*
T. cuneifolium (Vahl) Willd.
Glabrous shrubby perennial with erect rod-like branches terminating in long racemes, up to 1½m high. Leaves alternate, rather fleshy, obovate to oblanceolate, 3–5cm long. Flowers purple, conspicuous when open in the evening. Capsule globose with a smooth shell-like casing; pedicels recurved in fruit.

Widely distributed between 500 and 1700m along all the escarpment, but uncommon except in the central escarpment. It is found in scrub, overgrown thorn fences and on bushy banks around fields and by tracks.

BASELLACEAE

ANREDIA Juss.

1. Anredia cordifolia (*Ten.*) *Van Steenis*
Boussingaultia cordifolia Ten.
Glabrous perennial fleshy twining herb. Rhizomes with potato-like tubers. Leaves ovate, entire, c 2–4cm long. Flowers small, white, in dense axillary spikes, but very rarely seen in Yemen.

Thoroughly naturalised at 2400m in Manākhah village where it grows in abundance on walls, steep slopes and in waste places. Also cultivated in Sana'a.

CARYOPHYLLACEAE

Herbs or (rarely) undershrubs, the stems swollen at the leaf nodes. Leaves opposite (rarely alternate), simple, entire, with or without stipules. Inflorescence cymose; flowers bisexual, regular; sepals 4–5, free or joined; petals 5, free, sometimes absent, stamens 5 or 10. Fruit a capsule or achene.

A mainly temperate family. Most species are either weeds or found in arid desert habitats and the family is poorly represented on the escarpment. *Dianthus uniflorus* and *Saponaria umbricola* are endemic.

1. Leaves with scarious stipules. Flowers small and insignificant, the petals always less than 3mm long **2**
Leaves without stipules. Flowers insignificant or prominent **11**
2. Leaves alternate 11. *Telephium*
Leaves opposite **3**
3. Flowers sessile in the leaf axils. Fruit red, fleshy 17. *Pollichia*
Flowers and fruit not as above **4**
4. Fruit surrounded by a prickly ball of modified and enlarged bracts. Woody undershrub 16. *Cometes*
Fruit not as above. Herb **5**
5. Fruit a 1-seeded achene. Bracts numerous, conspicuous, scarious 15. *Paronychia*
Fruit a capsule. Bracts inconspicuous, scarious or not **6**
6. Leaves subulate. Plant glandular, pubescent or glabrous **7**
Leaves elliptic to obovate. Plant glabrous **9**
7. Calyx minute, less than 1.5mm long. Plant pubescent 12. *Polycarpaea*
Calyx 2.5–5mm long. Plant glandular or glabrous **8**
8. Stipules joined. Leaves in opposite pairs. Petals usually pink 10. *Spergularia*
Stipules entirely free. Leaves whorled. Petals always white 9. *Spergula*
9. Flowers subsessile in terminal clusters. Sepals pinkish-white 12. *Polycarpaea*
Flowers pedicellate in branched terminal cymes. Sepals green **10**
10. Bracts and stipules entirely scarious 14. *Polycarpon*
Bracts and stipules with a green herbaceous midrib 13. *Robbairea*

11. Sepals free or joined only at the base **12**
Sepals joined to form a calyx tube **15**
12. Leaves subulate. Perennial 7. *Minuartia*
Leaves narrowly oblong to ovate. Annual **13**
13. Capsule much longer than the calyx. Styles 4 5. *Cerastium*
Capsule about equalling the calyx. Styles 3 **14**
14. Petals bifid or absent 6. *Stellaria*
Petals entire 8. *Arenaria*
15. Calyx with an epicalyx of bracts at its base 2. *Dianthus*
Epicalyx of bracts absent **16**
16. Calyx with 10 veins 1. *Silene*
Calyx obscurely 5-veined **17**
17. Calyx with 5 green wings. Glabrous annual 3. *Vaccaria*
Calyx unwinged. Glandular-pubescent perennial 4. *Saponaria*

1. SILENE *L.*

Herbs. Inflorescence of cymes, sometimes reduced to racemes; calyx tubular, 10-veined; petals with a narrow claw; styles usually 3. Fruit a capsule opening with 6 teeth.

1. Leaves narrowly linear, glabrous. Calyx more than 3cm long *S. schimperiana*
Leaves elliptic or obovate, rarely linear, always hairy. Calyx less than 2cm long 2
2. Calyx 10–15mm long, the capsule borne on a short stalk within the calyx 3
Calyx 7–10mm long, the capsule sessile within the calyx *S. gallica*
3. Plants with a distinct basal rosette, the rosette leaves spathulate, mucronate *S. yemensis*
Plants with or without a basal rosette, rosette-leaves, if present, oblanceolate or elliptic, never spathulate *S. burchellii*

1. Silene burchellii *Otth. ex DC.*

S. chirensis A. Rich.
S. schweinfurthiana Rohrb.
S. flammulifolia Steud. ex A. Rich.
Very variable perennial herb with a decumbent or erect stem to 30cm, the basal rosette usually withered at flowering time, indumentum varying from crisped pubescent to almost hispid. Lower leaves linear, oblanceolate or elliptic, 2–5cm long. Inflorescence cymose or reduced to a raceme; calyx tubular to clavate, prominently 10-veined, 11–25mm long; petals bifid, usually white with purplish veins, but sometimes dark red, pink or purple. Capsule borne on a stalk 5–6mm long. The flowers open at night.

S. burchellii is extremely variable particularly in leaf shape, flower colour and in the seed which is always flattened with 2 wings but is variously grooved between the wings which may or may not be undulated. Some forms are indistinguishable except by their perennial habit from the Mediterranean *S. colorata* Poir. which was recorded from Ghirās by Wissmann and is best excluded for the time being from the Yemen flora on geographical grounds. In most specimens the capsule is longer than its stalk but in some cases (eg. Wood 3332) the stalk is longer than the capsule. Some forms have glandular pedicels including the type of *S. flammulifolia*.

Widespread and frequent from 1800 to 3300m on the escarpment and high plateau growing on rocky hillsides, in scrubby grassland (particularly in the central escarpment), on terrace walls and in disturbed ground around cultivation.

2. Silene gallica *L.*

Erect, branched, annual herb to 40cm, pubescent below, viscid above. Leaves oblanceolate, 1–4cm long. Inflorescence of branched racemes; calyx ovoid, green-nerved, 7–10mm long; petals white or pink, entire or emarginate but not bifid; capsule sessile within the calyx.

A casual found in a ditch at Yāzil at 2800m.

3. Silene schimperiana *Boiss.*

S. macrosolen Steud. ex A. Rich.
Woody based perennial with many erect stems to 50cm, glabrous below but sticky on the stem and inflorescence above. Leaves narrowly linear, glaucous, glabrous, 2–5cm long. Inflorescence lax, paniculate; calyx narrowly cylindrical, pinkish-veined, 3–4cm long; petals pale pink, shortly bifid, much longer than the calyx; capsule shorter than its 15–25mm long stalk.

Widespread from 2300 to 3200m on rocky hills on and around the high plateau but infrequent, often overlooked and frequently chewed down by goats; rare on the outer escarpment but known from J. Ṣabir, J. Ḥiṣn Madnān in Waṣāb al 'Ālī and J. Shibām in the Ḥarāz.

4. Silene yemensis *Deflers.*

Similar to the last species but softly pubescent with a persistent basal rosette whose leaves are spathulate, mucronate, c 2cm long.

Quite frequent on cliffs, terrace walls and rocky banks between 2000 and 3000m on the escarpment from Kusmah and Ḍawrān north to Shahārah; particularly frequent around Manākhah and Shibām.

2. DIANTHUS *L.*

Perennial herbs with linear leaves. Flowers solitary with an epicalyx of bracts appressed to the lower part of the calyx; calyx tubular; petals pink, long-clawed, often dentate or laciniate but not bifid; styles 2. Capsule with 4 teeth.

1. Prostrate mat-forming plant. Flowers subsessile *D. uniflorus*
Plant neither prostrate nor mat-forming. Flowers borne on long pedicels 2
2. Calyx 3–5cm long. Leaves more than 3cm long *D. longiglumis*
Calyx 1.5–1.8cm long. Leaves 1–3cm long *D. deserti*

1. Dianthus deserti *Kotschy.*

Glabrous perennial with weak leggy stems which attain 50cm in the protection of bushes. Basal leaves withering early; stem leaves few, widely spaced and about equalling the internodes, linear, 1–3cm long, 1–3mm wide. Inflorescence much branched, the flowers solitary or few together at the ends of the branches; calyx 1.5–1.8cm long; petals pink, dentate, longer than the calyx.

Known only from J. Um Laylah and J. 'Alab (Steinberg) in Khawlan Ash Sham.

2. **Dianthus longiglumis** *Del.*

Glabrous, pendulous perennial, the stems stout, branched, leafless below, but densely leafy at the branch tips. Leaves slightly pungent, linear, scarious-margined near the base, 4–12cm long, 3–6mm wide. Flowers solitary or paired on almost leafless stalks arising from the leaf axils and reaching 20cm in length; calyx 3–5cm long; petals pink, fimbriate, much exceeding the calyx.

A spectacular species restricted to basaltic cliffs at c 2700m on J. Ḍawrān and known outside Yemen only from one or two localities in Ethiopia.

3. **Dianthus uniflorus** *Forssk.*

Glabrous, prostrate, mat-like perennial. Leaves subulate, scabrid margined, 8–12mm long. Flowers solitary, subsessile; calyx 1–1.2cm long; petals pink, dentate, longer than the calyx.

Very attractive, especially when extensive mats are covered in pink flowers following the rains, and endemic to southwest Arabia. It is frequent from 2200 to 3400m on and around the high plateau where it grows with *Cichorium bottae* on grassy banks around fields and on mountain tops. It is much rarer on the escarpment but occurs in small quantity on nearly all the high mountains; in Khawlan Ash Sham only at Madaq (Steinberg).

The Carnation, **Dianthus caryophyllus** *L.*, 'Qaranful', is occasionally cultivated particularly in small gardens on the escarpment mountains.

3. VACCARIA *Medic.*

1. **Vaccaria pyramidata** *Medic*

Erect glabrous annual to 50cm, repeatedly branched above. Leaves sessile, ovate or lanceolate, glaucous, 4–9cm long. Flowers in much-branched diachasial cymes; pedicels long; calyx tube inflated below, whitish, green-winged on the angles, 12–17mm long; petals long-clawed, pale purple. Capsule, 4-toothed.

A rare plant found once only growing in a field of Fenugreek at 2300m between Sana'a airport and Arḥab.

4. SAPONARIA *L.*

1. **Saponaria umbricola** *J.R.I. Wood*

Woody based perennial, much branched at the base with short decumbent or ascending, brittle branches, glabrous except on the inflorescence. Leaves oblong-oblanceolate, 1.5–3cm long. Inflorescence of terminal cymes, glandular-hairy; calyx tubular, obscurely veined, 3–4mm long; petals cuneate, white; styles 2. Capsule 4-toothed.

A rare endemic plant of shaded cliffs found between 1600 and 2900m on the northern escarpment: in the gully below Jirjir, above Ḥadīyah; between Markah and Ṣa'bah, J. Milḥān; near Bayt Bashārī, Maḥwīt; Shirnamah, J. Maswar; Shahārah; Kuḥlān (Miller).

5. CERASTIUM *L.*

1. **Cerastium octandrum** *Hochst. ex A. Rich.*

Erect, softly villous, simple or branched annual herb to 10cm. Leaves lanceolate or oblong, 10–22mm long. Flowers in small diachasial cymes; sepals free, 4–6mm long; petals absent or present, white, shortly bifid, shorter than the sepals; styles 4. Capsule much longer than the sepals, opening with 8 teeth.

A rare plant of short, disturbed montane grassland between 3000 and 3200m: J. Ta'kar; J. Manār.

6. STELLARIA *L.*

Annual herbs. Flowers in diachasial cymes though sometimes reduced to a single flower; sepals free; petals 5 or absent, white, deeply bifid; styles 3. Fruit a capsule dehiscing with 6 teeth.

1. Seeds 0.8–1.3mm broad, dark reddish-brown ... *S. media*
   Seeds less than 0.8mm broad, light yellowish-brown ... *S. pallida*

1. **Stellaria media** (*L.*) *Vill.*

Chickweed. Decumbent annual with a glabrous or hairy stem. Leaves ovate, 1–2.5cm long. Flowers often solitary, sometimes in cymes; petals almost always present, small; sepals 3–7mm long; stamens 3–10. Seeds dark reddish-brown, 0.8–1.3mm broad.

Much confused with the next species but apparently a rare plant of disturbed, damp, shady places found in a few places between 2200 and 3000m, principally in the central escarpment. Confirmed records are from J. Ṣabir and Ḥaddah (Wood 74/310, 1973).

2. **Stellaria pallida** (*Dumort.*) *Piré*

Similar in habit to the last species but petals almost always absent, the sepals less than 3mm long, the stamens 1–3 and the seeds light yellowish-brown, less than 0.8mm broad.

Probably in similar situations to *S. media* but only confirmed from near Mirzin, J. Bura' (Wood 2122).

7. MINUARTIA *L.*

1. **Minuartia filifolia** (*Forssk.*) *Mattf.*

Small glabrous or pubescent perennial with short, decumbent, moss-like stems from a woody rootstock. Leaves subulate, 7–10mm long. Flowers in a terminal cyme; calyx c 4mm long; petals entire, white, slightly exceeding the calyx; styles 3. Fruit a 3-toothed capsule.

Widespread and frequent on cliffs, terrace walls and rocky slopes throughout the escarpment and high plateau from 1500 to 3300m.

Known only from southwest Arabia and Ethiopia.

8. ARENARIA *L.*

1. Arenaria leptoclados (*Reichb.*) *Guss.*

A. serpyllifolia sensu Schwartz

Decumbent, grey-green, scabrid-puberulent annual, much branched at the base. Leaves ovate, 2–6mm long. Flowers in diffuse diachasial cymes; sepals lanceolate, equalling the capsule, 3–4mm long; petals white, entire. Capsule narrowly conical, opening with 6 teeth; seeds less than 0.5mm wide.

Widespread and quite frequent on track sides and terrace walls and as a weed of cultivation in the wetter parts of the escarpment from 1700 to 3200m.

A. leptoclados has been much confused with *A. serpyllifolia* which differs in its ovate sepals, broader capsule which breaks with a pop when squeezed and larger seeds. There is no firm evidence that it occurs in Yemen.

9. SPERGULA *L.*

1. Spergula fallax (*Lowe*) *E.H.L. Krause*

S. flaccida (Roxb.) Aschers.

Glabrous annual herb, branched at the base and with ascending stems to 10cm. Leaves apparently whorled, subulate, 8–15mm long, the stipules scarious. Flowers in compound cymes; sepals 4–5mm long; petals white. Capsule with 5 teeth; seeds with a broad hyaline wing.

A rare weed of fields and waste places found between 1500 and 2500m: Ḥajjah; Manākhah; Raydah (Hepper).

10. SPERGULARIA (*Pers.*) *J. & C. Presl.*

Herbs with decumbent stems. Leaves opposite, rarely whorled, linear; stipules scarious, joined at the base to surround the node. Inflorescence glandular, cymose; sepals free, green with scarious margins; styles 3. Capsule with 3 teeth.

Seeds light brown, usually winged. Stipules forming a sheath on young shoots — *S. marina*
Seeds dark brown, never winged. Stipules joined at the base only, not forming a sheath even on young shoots — *S. rubra*

1. Spergularia marina (*L.*) *Griseb.*

Much-branched annual or short-lived perennial herb with numerous decumbent branches. Leaves opposite, fleshy, up to 3cm long, the stipules joined for about half their length to form sheaths on young shoots. Sepals 2.5–4mm long; petals pale pink. Capsule exceeding the sepals; seeds light brown usually mixed winged and unwinged but sometimes all unwinged.

An uncommon plant found between 1900 and 2400m mostly on the high plateau growing on bare ground which is seasonally flooded with salty water: on the escarpment west of Dhī Nā'im; on the salt flats north of Rada'; northeast of Dhamār (Acres); Al Zīl, Ḥadā'; in the wadi north of Wa'lān.

2. Spergularia rubra (*L.*) *J. & C. Presl*

Similar to the last species but usually procumbent and slightly smaller in all parts. Leaves less than 2.5cm long, not fleshy; stipules joined at the base only, not forming sheaths even on young shoots. Petals pink. Seeds dark brown, unwinged.

A rare plant of paths and trodden ground around villages between 2600 and 3000m: in a village on the summit ridge of J. Sumārah; Shibām.

11. TELEPHIUM *L.*

1. Telephium sphaerospermum *Boiss.*

Glabrous annual herb, much branched at the base with many decumbent branches. Leaves alternate, narrowly elliptic, glaucous, 5–12mm long; stipules scarious. Flowers subsessile in crowded, branched, terminal racemes; sepals glaucous with scarious margins, 3–3.5mm long; petals white. Capsule opening with 3 teeth.

An uncommon plant of sandy plain and gravelly hillslopes between 1900 and 3000m mostly on the high plateau: between Milāḥ and Sanabān; Sana'a (Wissmann); Shamlān; Bani Ḥushaysh; Qa'al Bawn; Majz (Steinberg).

12. POLYCARPAEA *Lam.*

Annual or perennial herbs. Leaves opposite or whorled, with scarious stipules. Flowers in terminal cymes or clusters; sepals scarious, free, not keeled; petals minute, whitish; stigmas 1–3. Capsule 3-toothed.

Glabrous annual. Flowers in dense terminal clusters — *P. spicata*
Pubescent perennial. Flowers in terminal cymes — *P. repens*

1. Polycarpaea repens (*Forssk.*) *Aschers. & Schweinf.*

Pubescent perennial with many short, spreading branches from a woody rootstock. Leaves subulate with revolute margins, rather stiff, 3–8mm long. Flowers in terminal cymes, pedicellate, minute; sepals 1–1.5mm long, green in the centre, but with broad scarious margins.

Apparently an uncommon but almost certainly overlooked plant found from 1900 to 2400m on sandy plain between Ṣa'dah and Majz and on the high plateau around Sana'a and in the Bani Ḥushaysh.

2. Polycarpaea spicata *Wight ex Arn.*

Glabrous annual herb with red-brown stems 5–12cm high arising from a small basal rosette. Leaves spathulate, 5–15mm long. Flowers in dense, terminal, umbellate clusters; sepals 2.5–3mm long, white-scarious apart from a pinkish midrib.

Only known from sandy plain by the sea, Ṣalīf.

13. ROBBAIREA *Boiss.*

1. Robbairea delileana *Milne-Redhead*

Polycarpaea prostrata Decne. p.p.

Prostrate, glabrous, glaucous annual herb with a persistent basal rosette and many radiating branches. Leaves obovate below, linear above, 5–10mm long; stipules scarious with a green herbaceous midrib. Flowers in lax cymes; bracts similar to the stipules; sepals with a broad, green keel and scarious margins 1.5–2.5mm long; petals pale pink, clawed, equalling the sepals. Capsule 3-toothed.

Frequent on sandy and alluvial plain between 1900 and 3000m on the high plateau and in similar areas around Ṣa'dah and Majz; on the escarpment only near Ḥajarah in the Ḥarāz.

14. POLYCARPON *Loefl. ex L.*

1. Polycarpon tetraphyllum (*L.*) *L.*
Glabrous annual herb, much branched at the base with numerous spreading or ascending branches c 5cm long. Leaves 2–4 together, oblanceolate or elliptic, 5–8mm long; stipules entirely scarious. Inflorescence dense, cymose, much branched; bracts scarious; sepals with green keels and scarious margins, longer than the petals, 1.5–2mm long. Capsule 3-toothed.

A weed of fields, tracksides and terrace walls found on most of the larger escarpment mountains between 1800 and 3100m, but only descending below 2500m in the mist regions of the central escarpment.

15. PARONYCHIA *Miller*

1. Paronychia chlorothyrsa *Murb.*
P. sinaica sensu Schwartz
Small perennial herb, much branched at the base with many decumbent, pubescent branches less than 5cm long. Leaves grey-green, pubescent, subulate, 3–6mm long; stipules lanceolate, scarious. Flowers in the axils of the leaves near the tips of the branches, ± concealed by the conspicuous, scarious, silvery, suborbicular bracts; calyx deeply 5-lobed, the lobes subulate, c 4mm long; petals absent. Fruit an achene.

Frequent on exposed stony plain and mountain ridges between 2000 and 3200m on and around the high plateau showing a preference for sandstone country; also at Majz (Steinberg).

16. COMETES *L.*

1. Cometes abyssinica *R. Br.*
Small undershrub, much branched from the woody base with many erect or ascending, pubescent branches to 30cm. Leaves opposite or whorled, sessile, linear or lanceolate, 1–3cm long; stipules scarious. Flowers in dense, pedunculate clusters arising from the axils of the leaves towards the tips of the branches; sepals green with a spinescent tip; petals minute, yellow-green. Fruit an achene surrounded by prickly ball of enlarged bristly bracts.

A plant of rocky, often volcanic, hills between 1100 and 2800m; frequent on the high plateau but elsewhere only known from Shahārah and the Ḥujarīyah where it is unexpectedly widespread.

17. POLLICHIA *Ait.*

1. Pollichia campestris *Ait.*
Small undershrub, much branched at the base with many short, procumbent or ascending, grey-pubescent branches. Leaves opposite or whorled, sessile, linear or oblanceolate, 8–15mm long; stipules scarious. Flowers in dense, sessile, axillary clusters; petals minute. Fruit an achene surrounded by the enlarged fleshy red calyx, c 1.5mm long.

A local plant of scattered occurrence on terrace walls and rocky hillsides, usually near villages, which is found from 1500 to 2500m on the escarpment and high plateau from 'Utmah and Raḍmah northwards; most common in the Kawkabān, Maḥwīt, Ḥajjah region and often associated with *Galium kahelianum* and *Pavonia burchellii*.

POLYGONACEAE

Herbs or shrubs. Leaves alternate, usually entire in our species, the stipules modified into a scarious tubular sheath (ochreae). Flowers small, without petals; sepals free, imbricate, 3–6, forming a persistent sometimes petaloid perianth which encloses the usually trigonous or lens-shaped, nut-like fruit.

A mostly temperate family represented in Yemen mainly by species found at high altitudes or near water.

1.	Fruiting perianth winged, armed or toothed	3
	Fruiting perianth neither winged, armed nor toothed	2
2.	Flowers in axillary clusters	3. *Polygonum*
	Flowers in terminal heads or spikes	4. *Persicaria*
3.	Nearly leafless desert shrub	1. *Calligonum*
	Herbs or shrubs, leafy	4
4.	Fruiting perianth winged, the wings sometimes with bristle-like teeth, but never hard and spiny	3. *Rumex*
	Fruiting perianth with 3 hard spines, never winged	5
5.	Fruiting perianths axillary, each with 3 spines at the apex	6. *Emex*
	Fruiting perianths in terminal spike-like racemes, each with 3 spines in the middle	2. *Oxygonum*

1. CALLIGONUM *L.*

1. Calligonum crinitum *Boiss.* Abal
Shrub to 3m with many jointed brittle white branches; branchlets green, slender, glabrous, often arising in clusters. Leaves minute. Flowers pedicellate, in axillary clusters; perianth segments 5, white or pink, c 3mm long. Fruit narrowly conical, c 12mm long, covered in yellowish bristles which are up to 1.5cm long at maturity.

The Yemen plant is subspecies **arabicum** (*Sosk.*) *Sosk.*

Abundant in the sands from Mārib and Ḥarīb eastwards. It is one of the few species found in this region and plays an important role in stabilising sand dunes.

2. OXYGONUM *Burch.*

1. Oxygonum sinuatum (*Meisn.*) *Dammer*

O. atriplicifolium (Meisn.) Martel var. *sinuatum*.

Usually decumbent, glabrous or hairy weedy annual. Leaves ± elliptic in outline, deeply sinuate-lobed, up to 4cm long. Inflorescence a leafless spike-like raceme up to 25cm long; flowers rather distant, pink, slightly exceeding the scarious fringed bracts; perianth segments 5. Fruit with 3 hard spreading spines from near the middle.

A common weed of roadsides, field borders and open stony hills in areas of good rainfall between 600 and 2200m on the escarpment, but apparently absent from Taiz province.

3. POLYGONUM *L.*

Prostrate herbs. Leaves small, narrow; ochreae silvery, lacerate. Flowers in small axillary clusters; perianth segments 5, petaloid, ± enclosing the nut-like fruit.

Leaves linear-spathulate, up to 2mm wide. Pedicels usually longer than the perianth *P. corrigioloides*
Leaves elliptic-oblong, more than 2mm wide. Pedicels absent or very short, never longer than the perianth *P. aviculare*

1. Polygonum aviculare *L.*

A usually prostrate, much branched glabrous annual. Leaves elliptic-oblong, c 1–4 cm long, the stem leaves larger than those on the inflorescence. Flowers pink or green, solitary or in clusters of 2–5, ± sessile in the leaf axils. Nut trigonous, shiny, equalling or larger than the perianth.

The Yemen plant (Wood 983) is quite uniform though various names have been applied to it in different institutions. It does differ from European examples of *P. aviculare* is not having a dull, punctate nut but in other respects seems identical.

Widespread but infrequent as a field weed between 2300 and 2900m on the high plateau, usually growing on irrigated land or near ditches.

2. Polygonum corrigioloides *Jaub. & Spach*

A much branched prostrate glabrous annual. Leaves linear-spathulate, up to 1cm long, the stem leaves the same size as those on the inflorescence. Flowers pink, in clusters of 4–6 (very rarely less), in the leaf axils, distinctly pedicellate, the pedicels up to 3 times the perianth. Nut trigonous, shiny, shorter than the perianth.

Widely distributed, but not very common, between 1800 and 2800m on the high plateau north to Ḥūth, usually growing on dired up mud by ponds or in depressions where rain water collects.

4. PERSICARIA *L.*

Herbs. Leaves entire. Inflorescence of dense terminal pedunculate heads or spikes. Perianth segments 5, petaloid, ± enclosing the nut-like fruit.

1. Flowers in a pedunculate head. Leaves triangular or rhomboid *P. nepalensis*
   Flowers in a spike. Leaves lanceolate 2
2. Leaf base rounded *P. amphibia*
   Leaf base cuneate 3
3. Peduncles and usually the perianth covered in numerous sessile orange or yellow glands 4
   Peduncles and perianth eglandular 5
4. Annual. Ochreae truncate or fringed (especially above). Nut lens-shaped, the faces never dimpled *P. lapathifolia*
   Perennial. Ochreae never fringed. Nut lens-shaped, the faces dimpled *P. senegalensis*
5. Ochreae clearly fringed 6
   Ochreae truncate, not fringed *P. glabra*
6. Bracts ciliate with conspicuous bristles *P. barbata*
   Bracts not, or very shortly, ciliate, the bristles less than 1mm long 7
7. Spikes lax, slender, usually more than 5cm long *P. salicifolia*
   Spikes dense, stout up to 3cm long *P. maculosa*

1. Persicaria amphibia (*L.*) *S. F. Gray*

Polygonum amphibium L.

A perennial with two distinct forms. The aquatic plant is glabrous with a trailing stem, which roots at the nodes, and floating petiolate lanceolate leaves with a rounded cordate base up to c 17cm long. The terrestial form has an erect stem and rather smaller subsessile pubescent leaves. The ochreae are brown, truncate, never fringed. Flowers pink in a dense usually solitary eglandular spike c 3–4cm long. Nut lens-shaped, shiny.

An infrequent but characteristic plant of ponds between 2300 and 3000m on the high plateau from J. Sumārah north to Bani Maṭar.

2. Persicaria barbata (*L.*) *Hara*

Polygonum barbatum L.

Erect, branched pubescent perennial to 50cm. Ochreae brown, fringed. Leaves lanceolate, up to 15cm long. Flowers pink or greeny-white in a dense eglandular spike c 3–5cm long. Nut trigonous, shiny.

Superficially similar to several other species but easily recognised by its hairy peduncles and conspicuously long-ciliate bracts with bristles up to 5mm long.

A rare plant of stream sides between 600 and 1800m in high rainfall areas: below Ḥadīyah; between 'Ukaybir and Shaḥir, J. Milḥān; Wadi Sukhnah (Wood 1050, 1215).

3. Persicaria glabra (*Willd.*) *Gomez de la Maza*

Polygonum glabrum Willd.

Erect branched glabrous perennial to 1m. Ochreae large, brown, truncate, not fringed. Leaves lanceolate (oblong, oblanceolate) up to 20cm long. Flowers pink in a dense eglandular spike 3–7cm long. Nut lens-shaped, shiny.

This is our only species with truncate ochreae and a

completely eglandular inflorescence except for *P. amphibia*.

Only known with certainty at 800m beside the Wadi Līyah in Khawlan Ash Sham (Wood 3041), but Schweinfurth's records from Ḥajjaylah and 'Attārah in the Ḥarāz may well be correct.

4. Persicaria lapathifolia (*L.*) *S. F. Gray*

Polygonum lapathifolium L.

A very variable erect or ascending, usually branched annual to 70cm. Ochreae brown, entire or often fringed especially above. Leaves lanceolate, up to 20cm long, glabrous or pubescent, sometimes with a large blackish spot. Flowers white or dull pink in a dense glandular spike c 3–5cm long. Nut usually lens-shaped, shiny.

Only known from a streamside at Sanafān (Hepper 6168) and by a lake at Yarīm (Acres 17k) both at about 2400m.

5. Persicaria maculosa *S. F. Gray*

Polygonum persicaria L.

Erect, usually branched, nearly glabrous annual to 70cm. Ochreae brown, fringed. Leaves lanceolate, up to 15cm long, often with a large blackish spot. Flowers pink in a dense eglandular spike. Nut lens-shaped, shiny.

Easily confused with several other species but distinguished by its dense eglandular flower spike and fringed ochreae.

A casual by steams and ponds and more rarely on waste ground between 1800 and 2600m; quite frequent in Ibb province but elsewhere rare and known only from near Ḥajarah in the Ḥarāz, Ṭawīlah, Wadi Aḥjar and Wadi Ḍahr (Wood 819, 2556; Smith & Henchie 5041).

6. Persicaria nepalensis (*Meisn.*) H. Gross

Polygonum nepalense Meisn.

Branched, straggling annual to 20cm, stems glabrous except for a few stalked glands above. Ochreae brown, truncate, not fringed. Leaves triangular or rhomboid, up to 4cm long, usually sparsely hispid. Flowers white in a pedunculate head subtended by a sessile bract. Nut lens-shaped, scarcely shiny.

A rare field weed found between 2700 and 3100m on a few southern mountains: J. Barad, Kusmah, J. Raymah; J. Ḥiṣn Madnān, Waṣāb al 'Ali; J. Ba'dān; J. Ta'kar.

7. Persicaria salicifolia (*Willd.*) *S. F. Gray*

Polygonum salicifolium Willd.

P. serrulatum Lag.

Decumbent to ascending, branched, nearly glabrous perennial herb to 70cm. Ochreae brown, fringed. Leaves lanceolate, up to 15cm long. Flowers pink in a long slender lax eglandular spike up to 9cm long, often strikingly zig-zag when immature. Nut trigonous or lens-shaped, shiny.

Readily distinguished by the long lax spike in which the flowers are distant from each other.

A rare plant of marsh and stream sides at comparatively low altitudes being found between 500 and 1700m on the escarpment: Wadi Hiḍrān, Taiz; Wadi Sūdan, Qā'idah; near Suq As Sabt, Ibb; Walajah, J. Milḥān (Schweinfurth); around Ḥajjaylah (Deflers and Schweinfurth); Attārah in the Ḥarāz (Wissmann).

8. Persicaria senegalensis (*Meisn.*) *Sojak*

Polygonum senegalense Meisn.

Vigorous erect branched perennial to 2½m, usually glabrous in Yemen but elsewhere varying from glabrous to tomentose. Ochreae brown, truncate, not fringed. Leaves lanceolate, strikingly acuminate, up to 25cm long, covered in sessile yellow glands below. Flowers pink, orange or white in dense glandular spikes up to 10cm long. Nuts lens-shaped with dimpled faces, shiny.

Usually easily identified by its vigorous habit and glandular inflorescence and leaves; this can be confirmed by examination of the nuts.

Much the most common *Persicaria* in Yemen, growing on the escarpment between 500 and 2300m but with a strong southern tendency, not being found north of Maḥwīt. It is usually found by large permanent or semi-permanent streams and forms large communities on the stony flood bed. It is thus one of the few species characteristic of the great wadis such as Wadi Zabīd.

5. RUMEX *L.*

Herbs or shrubs. Leaves entire. Flowers in whorls in branched paniculate inflorescences. Perianth segments in 2 whorls of 3, the outer small and inconspicuous, the inner enlarging in fruit, so making the fruit appear to be trigonous and winged, the wings often toothed. The fruit may also have reddish tubercles which are of taxonomic importance.

1. Shrub. Leaves succulent, all lanceolate with a cuneate base — *R. nervosus*
   Herbs. Leaves not succulent, the basal with an abruptly narrowed, truncate or cordate base — **2**
2. Fruiting perianth toothed. Leaves not triangular — **3**
   Fruiting perianth winged but without teeth. Leaves triangular-hastate — *R. vesicarius*
3. Teeth on the fruiting perianth hooked at the apex — *R. steudelii*
   Teeth on the fruiting perianth ± straight — **4**
4. Fruiting perianth c 6mm long, the teeth mostly very short, less than 2mm long — *R. obtusifolius*
   Fruiting perianth c 4mm long, the teeth exceeding 2mm in length — *R. dentatus*

1. Rumex dentatus *L.*

Erect annual to 60cm, much branched and bushy in habit. Basal leaves ovate with a truncate base, up to 18cm long, but usually less. Inflorescence yellow-green in colour. Fruiting perianth c 4mm long, each segment with a prominent tubercle, the margins armed with straight teeth c 2.5–4mm long.

A local but characteristic plant of the high plateau between 2200 and 2600m. It grows in waste ground (Ḍawrān; Khamr) and more commonly by ditches (Shibām; Tin'am; between Yarīm and Raḍmah etc.).

2. Rumex nervosus *Vahl* 'Athrub

Much-branched ascending to erect dioecious shrub to 2m. Leaves lanceolate to elliptic, rather fleshy, up to c 8cm long, mostly crowded on small lateral branchlets. Inflorescence a lax reddish leafless panicle. Fruiting perianth c 7mm long, without tubercles, the margins with broad scarious red wings.

One of the most abundant and conspicuous plants of the escarpment between 1500 and 2900m which penetrates the western parts of the high plateau to Yarīm, the Yislah pass and Wa'lān. It is common on cliffs and eroded mountain slopes where it is associated with *Acanthus arboreus* but is particularly characteristic of terrace walls which are often festooned with its prominent red fruiting inflorescences. The leaves are valued for wrapping bunches of qat whose freshness is preserved by their succulence.

3. Rumex obtusifolius *L.*

Erect, branched perennial to 1½m. Basal leaves ovate-oblong with a cordate base, very large, up to 25cm long. Inflorescence red-brown in colour. Fruiting perianth c 6mm long, each segment (in Yemen) with an obscure tubercle, the margins armed with straight stout teeth mostly 1mm or less long.

A local species found between 2400 and 3000m on the high plateau extending west to J. Masār (Ḥarāz), Ṭawīlah and J. Miswar. It grows as a field weed (e.g. near Dhamār), as a weed of waste ground (at Kawkabān) but most commonly by ditches or ponds.

In Europe *R. obtusifolius* hybridises freely with other species and, though no hybrids have been recorded from Yemen, it can be expected to hybridise with *R. dentatus* and *R. steudelii* where it grows with them.

4. Rumex steudelii *Hochst. ex A. Rich.* Ṭibil

R. nepalensis sensu Schwartz

Very similar to the previous species with which it can be easily confused but usually less robust in habit with smaller leaves and fruiting perianths 4–5mm long only. It is best distinguished by the marginal teeth on the fruit which are about 3mm long and strongly hooked.

Widespread and frequent on the escarpment from 1800 to 3200m, growing by ditches and ponds and more rarely in waste places in areas of good rainfall. On the high plateau it is rare and largely replaced by *R. dentatus* and *R. obtusifolius*.

5. Rumex vesicarius *L.*

Decumbent to erect glaucous annual to c 30cm. Leaves triangular – hastate, fleshy, up to c 4cm long. Inflorescence a small reddish panicle; the flowers paired on a short delicate pedicel, the primary flower much more vigorous and concealing the secondary. Fruiting perianth c 15mm long, the margins with broad, scarious, veined, reddish wings which are bent back to conceal the tubercles.

Widespread between 2000 and 2400m on the high plateau growing on roadsides and loose open volcanic hillslopes; eleswhere only known from Shahārah.

6. EMEX *Campd.*

1. Emex spinosa (*L.*) *Campd.*

Erect, glabrous, monoecious annual herb to 30cm, superficially resembling *Rumex*. Leaves ovate, truncate at the base, 3–7cm long, Male flowers in terminal and axillary pedunculate clusters; female flowers sessile in the leaf axils; perianth segments 6, the outer 3 joined, becoming woody in fruit and developing hardened spines at the tips.

An uncommon field weed found between 2300 and 2900m, mostly along the inner escarpment: J. Sumārah; Yarīm (Hepper); Wadi Ḍahr; Shibām; Manākhah (Schweinfurth); Bayt 'Adhāqah.

PLUMBAGINACEAE

Herbs or shrubs with alternate, entire leaves, without stipules. Flowers bisexual, regular; calyx tubular; petals free or joined. Fruit a 1-seeded capsule.

Calyx much shorter than the corolla, densely covered in stalked glands, lacking petaloid lobes 2. *Plumbago*
Calyx about as long as the corolla, eglandular, with petaloid lobes 1. *Limonium*

1. LIMONIUM *Miller*

Sea lavenders. Perennial undershrubs growing in salty conditions usually near the sea. Inflorescence of terminal leafless panicles, made up of small spikes which in turn are composed of small 1–5 flowered spikelets, each spikelet with 3 scale-like bracts; calyx funnel-shaped with 5 petaloid scarious lobes, enclosing and largely concealing the short-tubed corolla.

Leaves oblanceolate, flattened *L. axillare*
Leaves cylindrical *L. cylindrifolia*

1. Limonium axillare (*Forssk.*) *Kuntze*

Statice axillaris Forssk.

Erect undershrub to 40cm. Leaves oblanceolate, up to 5cm long, gradually narrowed into the petiole, fleshy, grey-green, glabrous. Inflorescences erect; bracts dull red; flowers purple or violet, c 4mm long.

Locally common in sabkhah along the coast from Ṣalīf north to Ibn Abbās, Khawbah and Luḥayyah.

2. Limonium cylindrifolium (*Forssk.*) *Kuntze*

Statice cylindrifolia Forssk.

Virtually identical to the previous species except for the leaves which are sessile and narrowly cylindrical.

Locally common in sabkhah along the coast from Mukhā north to Yakhtul and Mawshij.

2. PLUMBAGO *L.*

1. Plumbago zeylanica *L.*

Perennial, much-branched undershrub with long decumbent stems, often scrambling over other vegetation. Leaves ovate, up to c 6mm long, with a cuneate base narrowed into a winged petiole. Inflorescence paniculate, consisting of many branched terminal spikes; calyx tubular, c 1cm long, densely covered in stalked glands; corolla white, with a long tube c 2.5cm long. Capsule enclosed in the persistent calyx.

Locally common in rocky scrub particularly along small wadis, growing between 600 and 2600m in areas of good rainfall along the escarpment; most common in the south, thinning out northwards but reaching as far as Shahārah.

Plumbago auriculata *Lam.* (*P. capensis* Thunb.) is a similar undershrub with larger pale blue flowers. It is sometimes cultivated in Taiz and Sana'a.

OCHNACEAE

OCHNA *Schreb.*

1. Ochna inermis (*Forssk.*) *Schweinf.* Binat

Small unarmed shrub to 2m. Leaves alternate, oblong-elliptic, obtuse, glabrous, serrulate, the upper surface shiny, up to 5cm long. Flowers solitary, pedicellate, pleasantly scented, 10–15mm in diameter; petals yellow; stamens numerous. Fruiting sepals rigid, c 1cm long, bright red, spreading as the fruit ripens to reveal up to 5 black drupe-like mericarps.

This shrub is deciduous and the new leaves, which appear in March and April at about the same time as the flowers, are a rich bronze colour. The conspicuous yellow flowers last only about a week and the leaves gradually turn to a dark shiny green by May by which time the distinctive fruit can be seen. After being one of the most distinct shrubs in spring, this plant is undistinguished and inconspicuous throughout the rest of the year.

Locally common on the escarpment between 1000 and 2000m from 'Arūq in the Ḥujarīyah north to the Wadi Lā'ah growing in scrub on ungrazed rocky slopes particularly, but not exclusively, in areas of high rainfall, notably the Ibb-'Udayn region and the central escarpment. (Wood 1036)

As well as the common form described above, a second form differing in its lighter green leaves which are elliptic to rotund and only about 3cm in length has occasionally been found (Wood 1877, 3280). It appears to favour drier Acacia scrub and grows at Bir Ghubāḥ in Shar'ab, 'Udayn, Madīnat al 'Abīd, 'Usil in the Ḥarāz, Wadi Sukhnah, between Mafḥaq and Banī Manṣūr and between Ḥajdah and Ḥays. No floral characters are known to separate the two forms but no plants intermediate in leaf character have been seen in Yemen and the form is probably worthy of some recognition.

GUTTIFERAE

HYPERICUM *L.*

The Yemen species are undershrubs with opposite, entire leaves, lacking stipules, and prominent yellow flowers.

Flowers solitary. Leaves less than 1cm broad — *H. revolutum*
Flowers in cymes. Leaves more than 1cm broad — *H. quartinianum*

1. Hypericum quartinianum *A. Rich.*

Undershrub reaching only about 1m in Yemen. Leaves up to 10 × 3.5cm but usually rather smaller, grey-green, ovate-lanceolate, acute, cordate and clasping at the base. Inflorescence a few-flowered terminal cyme.

Flowers in March and April.

A rare plant of northwest-facing cliffs at about 2000m in J. Raymah: Na'mah above Ḥadīyah; in a gully north of Al Jabīn.

2. Hypericum revolutum *Vahl* 'Abīya', Ibiz

H. lanceolatum auct.

Usually a small undershrub, but sometimes growing into a small tree 3m high especially near Kusmah, J. Raymah. Leaves about 10 × 3mm, dark green, linear-lanceolate, acute with a shortly attenuate base. Flowers solitary.

Flowers in early summer, but a shy flowerer.

Locally abundant below terrace walls, in high-altitude grassland and in scrub on the higher altitude escarpment mountains from J. Ṣabir north to Kuḥlān. It grows above 2400m and reaches 3200m on J. Ta'kar.

TILIACEAE

A family of trees, shrubs and herbs closely related to the *Malvaceae* but with the stamens free or connate at the base only and the anthers 2-celled.

1.	Fruit elongate, pod-like	3. *Corchorus*
	Fruit not elongate and pod-like	**2**
2.	Undershrubs with softly spiny fruit	2. *Triumfetta*
	Trees and shrubs with unarmed fruit	1. *Grewia*

1. GREWIA *L.*

Shrubs with simple, alternate, toothed often stellate-hairy leaves. Flowers solitary or in few-flowered cymes. Petals clawed, white or yellow. Fruit a drupe.

An important genus in Yemen since it is one of the main constituents of scrub in the foothills below about 1500m. All species in Yemen come into flower after the spring rains and continue flowering sporadically until the autumn.

It is also one of the most taxonomically complex genera in the Yemen flora. Unfortunately there is no modern treatment of the genus. The only attempt this century is by Burrett in Engler's *Botanischer Jahrbucher* in 1911. This is in fact little more than an annotated list of African species

with a number of specimens cited under each species. I have therefore attempted an account of the Yemen species which is avowedly 'splitting' but which nevertheless goes some way to doing justice to the diversity of forms found in the country. I have relied almost entirely on material I have collected myself simply because there is very little else. The number of East African specimens at Kew and the British Museum is not very large and very little of it has fruit. It is difficult to see how a more satisfactory description of the genus can be made until more material is collected.

Not all species are controversial. *G. flavescens*, *G. schweinfurthii* and *G. villosa* are as far as I know distinctive and constant. The remaining species fall into three groups: *G. tembensis*, the *G. tenax* group (species 2, 8, 15) and the yellow-flowered species with white or grey felted leaves. I have regarded the three forms of *G. tembensis* as varieties of that species largely because they appear to differ only in leaf characters and have never been separated as distinct species. *G. tenax* itself is a well-defined species though it certainly could be regarded as an extreme example of one of the assortment of forms generally included under *G. erythraea*. That 'species' though is merely a name applied to a whole aggregate of forms and this is discussed in more detail below. *G. sp. C* differs from both *G. tenax* and *G. erythraea* in its large flowers but has only been collected once. In the yellow flowered complex leaf shape is always distinctive, but this is closely correlated with features of the inflorescence and fruit. Three species (*G. gillettii*, *G. sp. A* and *G. trichocarpa*) and perhaps also *G. mollis* represent a continuum of characters and are not very well defined though they should be recognised easily enough in most cases. The term 'species', in fact, has been used as a convenient label for forms whose exact status is unclear. Intermediates exist though they are uncommon and it is far from clear whether they are hybrids or simply evidence of the polymorphic nature of the species. It is possible that further collecting and detailed taxonomic study using modern methods may eventually elucidate the exact status of the plants.

In view of the comments made above, the following key needs to be used with caution. It should be effective for well over 90% of specimens collected but some forms will show characters intermediate between two species. It should be noted that the number of lobes on the fruit may be reduced by abortion.

1. Leaves ± orbicular, cordate, up to 8cm broad — *G. villosa*
   Leaves various but never with the above combination and not exceeding 5cm broad — 2
2. Petals white. Fruit 4-lobed (often less by abortion) — 3
   Petals yellow. Fruit 1–2 lobed, rarely obscurely 7-lobed — 6
3. Inflorescence a 1–3 flowered cyme. Filaments pink — *G. tembensis*
   Inflorescence a solitary axillary flower. Filaments white — 4
4. Leaves orbicular, coarsely crenate. Petioles exceeding 6mm long — *G. tenax*
   Leaves obovate to elliptic, serrate. Petioles less than 6mm long — 5
5. Petals and sepals c 2cm long — *G. sp. C*
   Petals and sepals less than 1cm long — *G. erythraea*
6. Leaves green below — 7
   Leaves white or grey felted below — 8
7. Leaves oblong-lanceolate; bracteoles green, much shorter than the unopened flowers; fruit obscurely 7-lobed, dull yellow — *G. flavescens*
   Leaves rhomboid; bracteoles red, equalling the unopened flowers; fruit unlobed, black — *G. schweinfurthii*
8. Petals orbicular. Sepals c 15mm long — *G. arborea*
   Petals linear to elliptic. Sepals 10mm or less in length — 9
9. Leaves elliptic, rounded or obtuse, petioles less than 5mm long — *G. tristis*
   Leaves acute to acuminate, but if obtuse then petioles longer than 5 mm or leaf oblanceolate to obovate — 10
10. Fruit mostly 2-lobed — 11
    Fruit unlobed (very rarely a 2-lobed fruit is present) — 12
11. All cymes solitary in the leaf axils. Petioles c 5mm long. Leaves up to 3.5cm long — *G. gillettii*
    Some cymes paired in the leaf axils. Petioles c 10mm long. Leaves up to 7cm long — *G. sp. B*
12. Some leaves rhomboid, softly grey tomentose on both surfaces — *G. velutina*
    Leaves not strikingly rhomboid, not tomentose on both surfaces — 13
13. Leaves up to 5cm long. Cymes usually solitary in the leaf axils — *G. sp. A*
    Leaves up to 10cm long. Cymes 1–3 in the leaf axils — 14
14. Leaves ovate, usually more than 3.5cm broad — *G. trichocarpa*
    Leaves oblong to oblanceolate, less than 3.5cm broad — *G. mollis*

1. Grewia arborea (*Forssk.*) *Lam.* — Ṣirak

G. fallax K. Schum.

Shrub or small tree to 5m. Leaves oblong to obovate, up to about 8 × 4cm, apex obtuse to sub-acute, margins serrate, becoming coarsely and irregularly dentate towards the apex; base unequal, rounded; upper surface dark shiny green and glabrous; undersurface strongly white-felted. Inflorescence a solitary 1–3-flowered axillary cyme; sepals yellow-brown, up to 15mm long; petals yellow, orbicular, up to 8mm long. Fruit 1–2-lobed.

This species is larger in all its parts than other yellow-flowered Grewias in Yemen and is difficult to mistake.

A very rare plant of bushland between 800 and 1000m: between Mishrāfah and Al Masbaḥ in Waṣāb as Sāfal, probably the location of Forsskal's type collection (Wood 3174); Wadi Yūr (Wood 3473); J. Masanah, Ḥujarīyah (Deflers).

2. Grewia erythraea *Schweinf.*

Shrub to 3m. Leaves subsessile or with short petioles up to 5mm long, obovate to elliptic, up to 3 × 1.5cm, apex rounded to obtuse, margins dentate, base cuneate to rounded, green on both surfaces, glabrous to densely stellate-hairy. Flowers solitary in the leaf axils; sepals grey-green, c 8–9mm long; petals and filaments pure white. Fruit 4-lobed, glabrous or stellate hairy.

The most extreme form of this species is represented by Wood 3213 collected on sandstone hills north of Rada'. This has almost sessile obovate leaves with a cuneate base; the fruit and both surfaces of the leaves are densely stellate-hairy. Similar plants are found elsewhere in Arabia, particularly in Oman, and this obviously represents one end of a continuum of forms with *G. tenax* at the opposite end. Following convention, I have included all intermediate forms as well as the extreme form under the name *G. erythraea*. None of the specimens I have seen from Yemen are identical and each can be placed at some point along the continuum between the extreme forms. Indumentum, petiole length, leaf dentation and shape, are closely correlated. Thus, forms with elliptic leaves tend to be subglabrous, long petiolate and rather coarsely dentate (Wood 2089).

G. erythraea is a rare plant of scattered localities between 5000 and 2100m mainly on the escarpment and in the Mashriq: Al Barḥ; J. Milḥān; between Mishrāfah and Al Masbaḥ, Waṣāb As Safāl; near Mawsanah in the Ḥarāz; between Bani Manṣūr and Maghrabah, Ḥarāz; Sanabān; Rada'; J. Balaq, Mārib (Miller); Majz (Steinberg). It is usually found on scrubby hills with other species of *Grewia* and *Acacia*.

3. **Grewia flavescens** *Juss.*

Shrub to 4m with distinctive fluted bark. Twigs scabrid-stellate-hairy. Leaves oblanceolate, up to 8 × 3cm, apex acute to somewhat acuminate, margin serrate, base rounded to subcordate, green stellate-hairy on both surfaces and characteristically rough to the touch. Inflorescence a solitary 2–3 flowered axillary cyme; sepals green with whitish hairs, up to 12mm long; petals yellow, linear, c 7mm long. Fruit obscurely 3–4 lobed, yellow-brown, shiny with bristly stellate hairs.

An uncommon shrub found on open scrubby hills between 1200 and 1500m in areas of good rainfall: Wadi Dūr, 'Udayn; between 'Alūjah and Kusmah, J. Raymah; 'Usil, Ḥarāz; between Mafḥaq and Bani Manṣūr; Wadi Sukhnah.

4. **Grewia gillettii** *Sebsebe* — Nasham

G. bicolor Schwartz p.p.

Shrub or small tree to 3m. Leaves with petioles less than 4mm long, obovate to oblanceolate, up to 3.5 × 1.5cm, apex acute to obtuse, margins serrate, base equal, cuneate (rarely rounded); upper surface green and minutely stellate-hairy; undersurface grey-felted. Inflorescence a solitary 1–3-flowered axillary cyme on a short flowering branch; sepals grey, c 10mm long; petals yellow, linear, c 5mm long. Fruit 2-lobed.

Scattered along the western escarpment between 500 and 1400m from the Ḥujarīyah north to at least Al 'Ashshah. It is characteristic of very dry country and especially the first foothills rising from the Tihama.

5. **Grewia mollis** *Juss.* — Nasham

Shrub or small tree to 3m. Leaves with petioles c 6mm long, oblong to oblanceolate, always broadest at or above the middle, up to 9 × 3.5cm, apex acute to acuminate, margins serrate, base somewhat unequal, cuneate to rounded; upper surface green, glabrous to minutely puberulent; undersurface grey-felted. Inflorescence on long ± flattened flowering branches consisting of cymes growing 1–3 together in each leaf axil and each cyme containing 1–3 flowers; sepals grey, c 7mm long; petals yellow, elliptic, c 5mm long. Fruit unlobed.

My concept of this plant is represented by Wood 1232 and 2030; some forms (such as Wood 2995 from near Madīnat Al 'Abīd) seem to be intermediate between this species and *Grewia trichocarpa*.

Scattered along the escarpment between 1000 and 2000m usually on scrubby hills, but the exact distribution is not certainly known.

6. **Grewia schweinfurthii** *Burrett* — Burtūm

Shrub to 4m. Leaves elliptic to rhomboid, up to 6 × 3.5cm, apex acute, margin serrate, base rounded to broadly cuneate; dark green and thinly scabrid-stellate-hairy on both surfaces. Inflorescence a solitary, 2–3-flowered axillary cyme; sepals grey, up to 10mm long equalled in bud by the distinctive reddish subulate bracteoles; petals yellow, linear, c 9mm long. Fruit unlobed, globose, black, thinly pilose.

Locally common on open scrubby hills along the western escarpment from 500 to 1500m; sometimes one of the most important constituents of scrub as in the valley forest on J. Bura' and on the limestone hills above Bani 'Umar in the Ḥujarīyah.

7. **Grewia tembensis** *Fresen.* — Khadar, Shawḥat, Niba'

Shrub to about 4m. Leaves vary variable but always serrate with a rounded to subcordate base, the upper surface green, glabrous or minutely stellate-hairy. Inflorescence a solitary 2–3-flowered axillary cyme; sepals grey outside, dull pink inside, up to 7mm long; petals white, linear, c 5mm long. Fruit 4-lobed, glabrous, red when mature.

This species is divisible into at least three forms in Yemen:

The first form was collected by Schweinfurth under the name *Grewia membranacea* A. Rich. and is characterised by its long petioles (c 10mm long) and its large, ovate, acute leaves up to 6 × 4cm. It is an uncommon plant, restricted to the lush valley forest communities between 500 and 1000m in the central escarpment: J. Raymah, J. Bura', J. Milḥān and Al Sharafayn (Wood 1893, 3285).

The second form is characterised by its short petioles (c 3mm long) and its small, elliptic to rotund, obtuse leaves reaching 14 × 10mm. This is a very widespread plant in Yemen occurring in all areas between 1000 and 2500m on exposed stony hills. It is very drought resistant, but is only really abundant in well-drained stony scrub on the escarpment. In some areas, such as Ḥaymat al Kharijīyah, it is an important constituent of scrub (Wood 1064, 2926, 3227).

The third form is probably referrable to variety **ellenbeckii** *Burrett* and is similar to the last form but has leaves which are densely white-felted below. I only know this definitely from a denuded hillside at about 1400m near Ḍabī, 'Abūs, Ḥujarīyah (Wood 75/89), but it may have been overlooked.

Intermediates between the forms appear to be very rare. The only example I have come across in Yemen is Wood 1475, collected at Misrākh near Taiz which is intermediate between the first two forms, but other intermediates are known in Saudi Arabia and elsewhere.

8. Grewia tenax *(Forssk.) Fiori* Khadar
G. populifolia Vahl
Low shrub rarely exceeding 1m. Leaves on rather delicate petioles, c 9mm long, orbicular, up to 15mm across, margin coarsely crenate, dark green on both surfaces, completely glabrous. Flowers solitary in the leaf axils; sepals dull green, c 8mm long; petals and filaments pure white. Fruit 4-lobed, completely glabrous.

With its orbicular, crenate, glabrous, long-petiolate leaves, this is the best demarcated white-flowered species; its characters appear to be quite constant throughout its range. Examples include Wood 1046 and 3257.

Abundant and very characteristic of scrub on stony plain in the Tihama and in the foothills of the escarpment from c 200 to 1100m, though rarely found above about 600m.

9. Grewia trichocarpa *Hochst. ex A. Rich.* Nasham
Grewia mollis var. *trichocarpa* (Hochst. ex A. Rich.) Burrett
Grewia mollis var. *petitiana* (A. Rich.) Burrett
Shrub or small tree to 3m. Leaves with petioles about 8mm long, ovate, up to 10 × 5cm, usually of characteristic thin texture, apex acuminate, margin coarsely dentate to serrate, base rounded; upper surface green, ± glabrous; undersurface grey-felted. Inflorescence a solitary or paired 1–3-flowered axillary cyme on a long flowering branch; sepals grey, c 8mm long; petals yellow, narrowly elliptic, c 5mm long. Fruit unlobed.

Apparently a local plant of shaded areas between 500 and 1600m in the central escarpment: Walajah, J. Milḥān and Ḥajjaylah (Schweinfurth); near 'Usil in the Ḥarāz (Wood 1472); above Suq As Sabt, J. Bura' (Wood 3292); below Shamsān, Maḥābishah (Wood 3262).

Grewia mollis var. *petitiana* is probably the same as *G. trichocarpa* but I am very uncertain whether Schwartz's records of variety *petitiana* should be included within my concept of *G. trichocarpa* outlined above.

10. Grewia ? tristis *K. Schum.*
Shrub to 2m. Leaves with petioles less than 4mm long, elliptic, up to 4.5 × 2.5cm, apex obtuse to rounded, margins coarsely serrate, base equal, rounded to subcordate; upper surface green and very shortly stellate-hairy; undersurface grey tomentose. Inflorescence a solitary or paired 1–3-flowered axillary cyme; sepals grey but with numerous tufts of short orange hair, c 10mm long; petals yellow, elliptic, c 5mm long. Fruit not known.

The sole Yemen collection (Wood 3258) agrees very well with collections made in Somalia (e.g. Glover & Gilliland 1246) and Kenya (e.g. Faden & Faden 74/1282) which have been called *G. tristis*. The Yemen plant differs only in having some cymes paired whereas in East African plants the cymes are solitary. In African plants the fruit is 2-lobed but unfortunately my Yemen specimens have no fruit.

Only known from a single bush at 400m on an exposed scrubby hillside in the gorge of the Wadi Surdūd about 4km west of Khamīs Bani Sa'd.

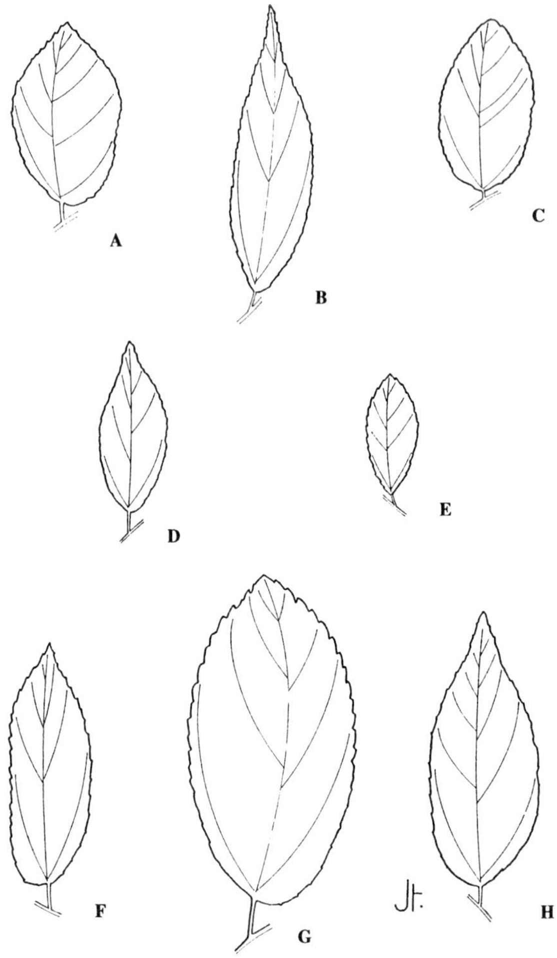

Fig. 6. Leaves of *Grewia* species. **A** *G. velutina*; **B** *G. mollis*; **C** *G. tristis*; **D** *G. sp. A*; **E** *G.* aff. *bicolor*; **F** *G. sp. B*; **G** *G. arborea*; **H** *G. trichocarpa*.

11. Grewia velutina *(Forssk.) Vahl* Nasham
G. bicolor variety *canescens* (A. Rich.) Burrett p.p.
Shrub reaching about 4m. Leaves with petioles 6–8mm long, obovate to rhomboid, up to 5 × 3cm, apex acute, margin coarsely serrate, base somewhat unequal, rounded; upper surface grey-green, undersurface grey, both surfaces shortly but distinctly tomentose. Inflorescence on both long and short flowering branches, consisting of single or paired 1–3-flowered axillary cymes; sepals grey, c 7–8mm long; petals yellow, elliptic, c 5mm long. Fruit unlobed.

On scrubby hills from 600 to 1600m along the escarpment from Shar'ab north to Khawlan Ash Sham. Although rare in the southern part of its range it is generally common, especially in areas of good rainfall in the central escarpment (Wood 1625, 2029, 2573, 3050).

12. Grewia villosa *Willd.*
Shrub to 3m. Leaves with petioles 15–20mm long, ± orbicular, up to 8cm long, apex rounded, margin serrate, base

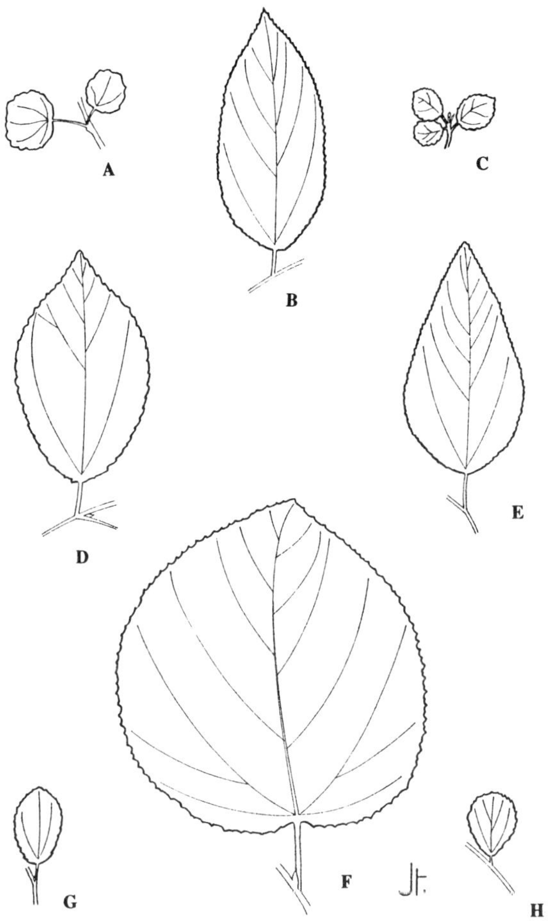

Fig. 7. Leaves of *Grewia* species. **A** *G. tenax*; **B** *G. flavescens*; **C** *G. tembensis*; **D** *G. schweinfurthii*; **E** *G. tembensis*; **F** *G. villosa*; **G-H** *G. erythraea*.

cordate; upper surface green, pubescent with impressed veins; undersurface grey-felted and villous. Flowers in subsessile axillary clusters; sepals brown; petals reddish-brown and much shorter than the sepals. Fruit unlobed, dull yellow, wrinkled.

A common constituent of *Acacia* scrub between 600 and 1500m along the escarpment, but rarer in areas of poor rainfall.

13. Grewia sp. A Nasham

Shrub or small tree to 4m. Leaves with petioles 3–5mm long, obovate to oblanceolate, up to 5 × 3cm, apex acute to acuminate, margins serrate, base cuneate to rounded; upper surface shiny green, ± glabrous; undersurface grey-felted. Inflorescence a solitary (very rarely paired) 1–3-flowered axillary cyme on a short flowering branch; sepals grey, c 8mm long, petals yellow, narrowly elliptic, c 6mm long. Fruit 1 (–2)-lobed.

This plant is represented by Wood 75/195, 1012 and 3018. It is intermediate in character between *G. gillettii* and *G. trichocarpa*.

Found between 1500 and 2100m in Taiz province: Dhubhān plateau, Ḥujarīyah; Misrākh; Shar'ab and As Salam; Taiz (Lavranos & Newton). It is usually found on field borders and in rocky gullies protected from grazing.

14. Grewia sp. B.

Shrub to 3m. Leaves with petioles c 10mm long, oblong (strikingly parallel sided) up to 7 × 2.5cm, apex acute, margin serrate, base unequal, rounded to subcordate; upper surface dark green, ± glabrous; undersurface grey-felted. Inflorescence on long and short flowering branches, consisting of solitary and paired, 1–3-flowered axillary cymes; sepals yellow-grey, c 10mm long; petals yellow, narrowly elliptic, c 6mm long. Fruit (1–) 2-lobed.

The affinities of this plant are very hard to guess at. Its leaves recall *G. mollis* but its fruit are similar only to *G. gillettii*. The petioles resemble neither species.

Known from a single collection (Wood 1928) made in valley scrub at about 600m just above Suq Ar Ribāṭ, J. Raymah.

15. Grewia sp. C

Low undershrub. Leaves with petioles c 4mm long, obovate, c 2.5 × 1.5cm, apex rounded, base cuneate, margin dentate. Flowers solitary in the leaf axils; sepals grey-green, c 2cm long; petals and filaments pure white, the petals c 2cm long. Fruit glabrous.

Readily distinguished from other related species by the very large flowers.

Known only from a cultivated silty wadi near Mārib (Miller).

2. TRIUMFETTA *L.*

Our three species are closely related and share the following common features. They are vigorous bushy plants with ovate to orbicular, serrate, shortly acuminate or 3-lobed hairy leaves. The inflorescence consists of almost naked spike-like racemes of small yellow flowers. The fruit is a small hairy capsule about 2–3mm in diameter covered in hooked prickles.

1.	Fruit globose; prickles glabrous	*T. rhomboidea*
	Fruit ovoid; prickles hairy	2
2.	Leaves tomentose	*T. flavescens*
	Leaves very sparsely stellate-hairy	*T. pentandra*

1. Triumfetta flavescens *Hochst. ex A. Rich.*

A low perennial undershrub to about 50cm, usually of yellow-green colour; stem often with red blotches, black dots and more spreadingly hairy than in the following species. Leaves softly tomentose on both surfaces. Capsule ovoid with hairy prickles.

Characteristic of poor shallow grassland between 1800 and 2100m in the central escarpment, but apparently very local or overlooked: between Bani Dawān and Al Jabin, J. Raymah; near Maghrabah, J. Bura'; frequent near 'Ukaybir, J. Milḥān; 'Attārah in the Ḥarāz; between Bayt 'Adhāqah and Suq ath Thalūth in the Wadi Lā'ah.

2. Triumfetta pentandra *A. Rich.*
Erect shrubby annual to 1½m. Leaves more commonly 3 lobed than in the other species. Capsule ovoid with hairy prickles.

This species lacks the distinctive stem markings and the tomentose leaf indumentum of the previous species. It can be distinguished from the next, which it superficially resembles, by its hairy capsule prickles.

A rare plant of valley scrub between 500 and 1200m in the central escarpment: above 'Alūjah in J. Raymah; above Ḥillah and Suq As Sabt, J. Bura'; Walajah, J. Milḥān; 'Usil in the Ḥarāz. The last two localities are based on Schweinfurth's records. The plant has not been seen there recently.

3. Triumfetta rhomboidea *Jacq.*
Perennial undershrub to 1m. Leaves suborbicular to ovate, usually shortly acuminate, but often obscurely 3 lobed, somewhat tomentose below but never so on the upper surface. Capsule globose with glabrous prickles.

The commonest *Triumfetta* in Yemen. It has been found in areas of good rainfall along much of the escarpment from Misrākh north to Jebel Marran. Although recorded from 500 to 1900m it is usually found in shaded wadis and is particularly characteristic of the lush middle-altitude wadis such as Wadi Ḍabāb, Wadi Ḥadabah and Wadi Sukhnah.

3. CORCHORUS *L.*

Herbs with small, yellow, solitary or paired, axillary flowers and usually crisped simple hairs. Fruit a subsessile, pod-like capsule opening from the apex by valves.

1. Prostrate perennial. Leaves not exceeding 2cm long — *C. depressus*
   Erect annuals. Leaves exceeding 2cm long — 2
2. Capsule terminating in 3 spreading teeth — *C. tridens*
   Capsule with a simple beak — 3
3. Stem glabrous. Capsule 5-valved — *C. olitorius*
   Stem sparsely pubescent. Capsule 3-valved — *C. trilocularis*

1. Corchorus depressus (*L.*) *Christens.* Waykī
C. antichorus Raensch.
Mat-forming perennial with taproot and numerous thick stiff procumbent branches. Leaves dark green, appressed hairy, elliptic, up to c 2cm long, equal to or shorter than the petioles, margin crenate, base lacking the 2 aristate teeth found in the following three species. Capsule 10–15mm long, beaked, straight or slightly curved, opening by 4 valves.

An abundant weed of fields and alluvial plain in the Tihama and the broader valleys entering the foothills thus reaching places such as Aṭ Ṭūr, Ḥajjaylah, Madrūbah. Only once found inland – near Al 'Ashshah. Absent from both mobile sand and all rocky hills.

2. Corchurus olitorius *L.* Mulukhīyah
Erect annual to about 1m, glabrous except for the petioles. Leaves narrowly ovate to lanceolate, up to 9 × 4cm, much longer than the petioles, margin serrate, base with 2 aristate teeth up to 15mm long. Capsule 3–6cm long, beaked, straight, opening by 5 valves, much thicker (c 4mm) than in the following two species.

At present a rare and probably recently introduced weed of irrigated fields in lush valleys between 1000 and 1500m: Misrākh and Wadi Ḍabāb near Taiz; Wadi Dūr, 'Udayn; Wadi Sukhnah; between Shahārah and Al 'Ashshah.

This plant is widely cultivated as a soup vegetable in Egypt and is becoming popular in Yemen where it can be regularly bought in the major markets. It is being cultivated around the main towns and will probably become a well-established weed in the near future.

3. Corchorus tridens *L.* Mulukhīyah
Erect or sprawling annual, usually less than 30cm, ± glabrous except on the petioles. Leaves linear-lanceolate, up to 10 × 1.5cm, much longer than the petioles, margin serrate, base with 2 aristate teeth up to 8mm long. Capsule 3–4cm long, straight, with 3 conspicuous spreading teeth at the apex.

Sterile specimens can usually be safely identified by the distinctive linear lanceolate leaves.

Grows in scattered localities in the Tihama and the escarpment up to 1500m. It is characteristic of fields on light soils and grows in the sand dune country around Bayt al Faqīh and Manṣūrīyah where Pennisetum millet is cultivated.

4. Corchorus trilocularis *L.* Mulukhīyah
Erect or sprawling annual up to about 50cm, the stem distinctly pubescent. Leaves lanceolate to oblong, up to 6 × 3cm, much longer than the petioles, margin serrate, base with 2 aristate teeth up to about 5mm long. Capsule c 4–6cm long, straight, beaked, opening by 3 valves.

Sterile specimens can usually be identified by the pubescent stem which is absent in the two previous species.

A common but not very important weed of fields and waste places in the Tihama and on the escarpment up to about 1600m.

STERCULIACEAE

A family of trees, shrubs and herbs similar to the *Malvaceae* but with the filaments united below into a delicate tube and the anthers 2-celled.

1. Trees or large shrubs — 2
   Herbs sometimes woody at the base — 3
2. Petals present, very showy — 2. *Dombeya*
   Petals absent — 1. *Sterculia*
3. Glandular annual — 4. *Hermannia*
   Eglandular perennial — 4
4. Flowers crowded into dense head like panicles — 5. *Waltheria*
   Flowers in lax few-flowered cymes or solitary — 3. *Melhania*

1. STERCULIA *L.*

1. Sterculia africana (*Lour.*) *Fiori* Difrī, Takhūb

S. arabica T. Anders.
S. platanifolia sensu Schwartz
Culhamia hadiensis J. F. Gmel.

A spreading tree to about 12m high which is deciduous in the dry season. Bark smooth and flaky leaving greyish patches so the trunk resembles that of a plane tree. Leaves cordate, orbicular or 3–5 lobed with an acuminate apex, up to 8cm long, thinly pubescent to nearly glabrous. Flowers in small axillary panicles with the calyx reddish inside. Fruiting carpels 3–5, spreading, ovoid, acuminate up to 8cm long, densely grey-green tomentose.

Flowers in spring, when the tree is leafless.

An infrequent tree found in the dry western foothills from 300 to 1000m from Musalla in Aḥkūm (Ḥujarīyah) north to the Wadi Sihām near Al Khalīfah. It grows in small numbers on exposed scrubby hillsides.

2. DOMBEYA *Cav.*

1. Dombeya schimperianum *A. Rich.* Tanqah

Small tree to 5m with densely pilose branchlets. Leaves with pilose petioles, ovate, cordate with a long acuminate apex, up to 8cm long, scabrous on both surfaces. Flowers in axillary umbels, very showy, recalling apple blossom; petals about 1cm long, obovate, pale pink. Fruit a small capsule rarely seen in Yemen.

Flowers erratically throughout the year.

Very widely but thinly distributed along the western escarpment between 1700 and 2900m from J. Ḥabashī north to Ḥaydān in Khawlan Ash Sham. It is usually found as isolated trees in small scrub remnants, but sometimes occurs in some abundance as in the *Acacia origena* woodland at Bayt Mughlad near Ḥajarah (Ḥarāz).

3. MELHANIA *Forssk.*

Perennial woody based plants often densely stellate hairy. Leaves grey-felted beneath at least when young. Flowers solitary or in lax axillary cymes, usually open in the evening; epicalyx of 3 free bracts which in some species enlarge in fruit becoming membranous; petals yellow. Fruit a capsule.

The genus was named by Forsskal after Jebel Milḥān where he found *M. velutina*. It appears to be centred on the southern Red Sea area.

1. Epicalyx bracts not enlarging in fruit — 2
   Epicalyx bracts enlarging in fruit — 4
2. Leaves usually exceeding 5cm broad, tomentose, not grey-felted below when mature — *M. velutina*
   Leaves less than 5cm broad, ± appressed hairy, strongly grey-felted below — 3
3. Leaves ovate. Epicalyx bracts linear-lanceolate — *M. ovata*
   Leaves oblong. Epicalyx bracts broadly ovate — *M. incana*
4. Leaves very large exceeding 5 × 3cm — *M. phillipsae*
   Leaves less than 5 × 3cm — 5
5. Seeds muricate — *M. muricata*
   Seeds smooth or sparsely tuberculate — 6
6. Leaves ovate-elliptic, serrate, up to 2cm long — *M. denhamii*
   Leaves linear-oblong, ± entire, up to 4cm long — *M. stipulosa*

1. Melhania denhamii *R. Br.*

Erect undershrub to about ½m. Leaves ovate – elliptic, c 3 × 1.5cm, serrate, softly tomentose on both surfaces. Flowers in 1–2 flowered cymes; epicalyx bracts reddish, cordate, enlarging in fruit. Seeds smooth.

This species differs from *M. muricata* in its tomentose indumentum, shorter leaves and smooth seeds.

Known only from a single locality: the Wadi Mawr region where it was found by Ehrenberg.

2. Melhania incana *Heyne*

M. steudneri Schweinf.

Prostrate or ascending woody perennial to about 25cm. Leaves oblong with a rounded base, up to c 5.5 × 1.5cm, serrate, green and appressed hairy above but grey-felted below. Flowers in 1–3 flowered cymes. Epicalyx bracts broadly ovate, not enlarging in fruit.

Frequent between 600 and 2000m on the outer escarpment from the Wadi Rima' near Madīnat Al 'Abīd north to Madaq in Khawlan Ash Sham (Steinberg). It is most common in the central escarpment growing in derelict fields and in short rather sparse grassland in areas of average to good rainfall.

3. Melhania muricata *Balf. fil.*

Prostrate or ascending undershrub to about 20cm. Leaves usually ovate to lanceolate with a truncate rounded apex, up to 5 × 2cm but commonly less, crenate, appressed hairy above and slightly grey-felted below especially when young. Flowers in 1–2 flowered cymes; epicalyx bracts cordate, enlarging in fruit. Seeds strongly muricate.

This is known with certainty only from about 1300m on the exposed limestone hill of J. Jirdad, Bani 'Umar, Ḥujarīyah. However it seems that *M. muricata* is a polymorphic and ill-defined species except for its seed character and I would also place Wood 1634 here. This was collected in the rocky foothills between Bājil and Khamīs Bani Sa'd. Its muricate seeds place it immediately in *M. muricata* but its serrate, rather tomentose leaves recall *M. denhamii*.

4. Melhania ovata (*Cav.*) *Spreng.*

Prostrate or ascending undershrub to about 25cm. Leaves ovate with a sub-cordate base, up to c 4 × 3cm, serrate, very shortly tomentose above, grey-felted beneath. Flowers in 1–3 flowered cymes; epicalyx bracts linear, not enlarging in fruit.

A local plant of open scrub often associated with *Barleria orbicularis* and *Indigofera spinosa*, found between 600 and 2000m. Widespread in the south (though absent from the Ḥujarīyah), apparently centred on the Shar'ab and Maqbanah areas, becoming rarer further north and there restricted to the rainshadow communities of the central escarpment and not found north of the Wadi Surdūd basin.

The records in Schwartz are probably mainly errors for *M. incana*, but it is impossible to be completely certain of this.

5. Melhania phillipsae *Bak. f.*

Vigorous, erect, tomentose undershrub to 1m. Leaves ovate, cordate based, up to 9 × 5cm, weakly serrate, densely tomentose on both surfaces and slightly grey-felted below. Flowers in 3–5 flowered cymes; epicalyx bracts cordate, enlarging in fruit and becoming reddish.

Only known on volcanic rocks at c 1300m on the lower slopes of J. Shahārah.

6. Melhania stipulosa *J.R.I. Wood*

M. muricata sensu Schwartz

Prostrate undershrub. Leaves oblong, up to c 4 × 5cm but commonly much smaller, ± entire, the base cuneate, the apex truncate, apiculate, adpressed hairy above and grey-felted below. Flowers (always?) solitary; epicalyx bracts cordate, enlarging in fruit. Seeds smooth or very sparsely tuberculate.

This plant has 10mm long, persistent, filiform stipules which are more conspicuous than in other species.

An inconspicuous plant of exposed stony plain and hill tops scattered in the Mashriq and the high plateau from the Shirqabī pass in Maghrib 'Ans and Dhī Nā'im north to Majz (Steinberg). It grows between 1800 and 2500m and is commonly associated with two other small neglected species *Chascanum sessilifolium* and *Dicliptera suffruticosa* similarly restricted to southwest Arabia and the Somali highlands.

7. Melhania velutina *Forssk.*

M. ferruginea A. Rich.

Erect undershrub to 1m, red-brown tomentose especially on young growth. Leaves ovate, subcordate, up to 9 × 6cm, crenate, softly tomentose and green on both surfaces, only the youngest leaves being grey-felted below. Flowers in 1–3 flowered cymes. Epicalyx bracts broadly ovate, not enlarging in fruit.

An uncommon plant found mostly in dense valley scrub between 500 and 1200m and restricted to the central escarpment mountains including Maḥwīt.

4. HERMANNIA *L.*

1. Hermannia modesta (*Ehrenb.*) *Mast.*

Slender glandular hairy annual herb reaching about 20cm in height. Leaves linear, entire or somewhat dentate below, about 2cm long. Flowers solitary, red, on long capillary peduncles from the leaf axils. Fruit a truncate capsule.

Only known from loose rocky country in open scrub at 1500m near Mafḥaq.

5. WALTHERIA *L.*

1. Waltheria indica *L.*

An erect woody perennial to about 1½m usually dying back during the dry season. Leaves lanceolate to oblong, c 7 × 2cm, serrate, stellate-hairy especially below. Inflorescence consisting of dense axillary head-like panicles on peduncles c 1.5cm long; petals yellow, persistent round the fruit.

An introduction from America now locally common between 500 and 1200m in the central escarpment from J. Milḥān south to 'Alūjah in J. Raymah and also in the Ḥarāz near 'Usil. It has not appeared yet at Ḥadīyah.

The ornamental **Brachychiton diversifolius** (*G. Don*) *Terr.* is cultivated in Sana'a and Taiz. It has a distinctive dark green trunk and ovate-acuminate or 3-lobed leaves.

BOMBACACEAE

Large trees with digitate stipulate leaves. Flowers subtended by an epicalyx; sepals and petals 5, free.

Tree with massive trunk. Leaves with 5 digitate leaflets — 1. *Adansonia*
Tree with slender trunk. Leaves with nine or more leaflets — 2. *Ceiba*

1. ADANSONIA *L.*

1. Adansonia digitata *L.*

Baobab. A tree of massive girth reaching about 15m high, deciduous in winter. Leaves long petiolate, digitate with 5 oblong, acuminate entire leaflets up to 12cm long. Flowers appearing before the leaves, long-pedunculate, solitary, pendulous, large, white, said to be pollinated by bats. Fruit a woody, felted, indehiscent, oblong-ellipsoid pod up to 20cm long.

Known with certainty from two trees at about 1200m – one at Samsarah in the Ḥujarīyah and the other near Bani Dawān, J. Raymah (J. Firebrace). It seems impossible to be certain whether these are relics of a larger native population or introductions.

2. CEIBA *Miller*

1. Ceiba pentandra (*L.*) *Gaertn.* Ḥarīr, Quṭīn

Kapok. An erect tree to 25m with a characteristic straight, smooth, grey, buttressed trunk, slender, spiny branches arising at right angles to it and a very sparse crown, the whole recalling an overgrown sprouting telegraph pole. Leaves long petiolate, digitate, with 9 or more elliptic to lanceolate, entire, acute leaflets to 15cm long. Flowers white, clustered. Fruit an ellipsoid capsule 10–30cm long, containing woolly seeds.

An introduced tree which is found between 500 and 1200m around irrigated fields in the wadis passing through the foothills to the Tihama; usually near villages and often solitary.

MALVACEAE

A family of herbs and undershrubs usually with stellate hairs. Leaves alternate, stipulate, palmately nerved and often palmately lobed. Flowers solitary or in racemes, bisexual, regular; calyx of 5 sepals, often subtended by an epicalyx; petals 5, free. A distinctive feature of the family is the staminal column, a hollow tube of joined filaments which is particularly conspicuous in the large-flowered genera such as *Hibiscus*; the anthers are 1-celled. Fruit of free mericarps or a capsule.

1. Epicalyx present **2**
Epicalyx absent **9**
2. Epicalyx bracts 3 **3**
Epicalyx bracts 5–12 **6**
3. Shrubs. Epicalyx much larger than the calyx **4**
Herbs. Epicalyx inconspicuous, shorter than the calyx **5**
4. Seeds pilose. Grey tomentose plant with predominantly maroon flowers 9. *Senra*
Seeds woolly. Villous to glabrous plants with predominantly yellow flowers 11. *Gossypium*
5. Flowers mauve. Leaves palmately lobed 2. *Malva*
Flowers yellow. Leaves unlobed 3. *Malvastrum*
6. Fruit a capsule **7**
Fruit of free mericarps **8**
7. Flowers purple. Capsule with 1 seed per loculus 8. *Kosteletzkya*
Flowers yellow, red or white, sometimes with a dark centre. Capsule with more than 1 seed per loculus 10. *Hibiscus*
8. Epicalyx bracts shorter than the calyx 1. *Althaea*
Epicalyx bracts longer than the calyx 7. *Pavonia*
9. Fruit a capsule 10. *Hibiscus*
Fruit of free mericarps **10**
10. Mericarps always 5, much longer than the calyx 5. *Wissadula*
Mericarps usually more than 5 but, if 5, equalling or shorter than the calyx which ± encloses them **11**
11. Mericarps 8–40, if less then with more than 1 seed in each mericarp 6. *Abutilon*
Mericarps 5–10, always 1-seeded 4. *Sida*

1. ALTHAEA *L.*

1. Althaea ludwigii *L.*

Prostrate, sometimes mat-forming, perennial. Leaves deeply palmately lobed. Flowers sessile or with pedicels up to 15mm long, solitary or in few-flowered clusters in the leaf axils; epicalyx of 6–9 bracts; calyx villous; petals small, mauve.

Superficially similar to *Malva* but readily distinguished in the field by the deeply lobed leaves.

A local but probably overlooked plant of the intermontane plains of the high plateau found c 2000–2400m from Rada' and Dhamār north to Raydah; also in the Majz area (Steinberg). It is found on uncultivated alluvial plain and derelict fields.

The related Hollyhock, **Alcea rosea** *L.* (*Althaea rosea* (*L.*) Cav.), known as 'Khubayz franji' is commonly cultivated in gardens particularly on the high plateau but also in the large middle-altitude towns such as Taiz. It is a tall perennial up to 3m high with long racemes of pink or white flowers.

2. MALVA *L.*

Prostrate or ascending weedy annual herbs with shallowly palmately lobed leaves which are ± orbicular in outline. Flowers in subsessile, axillary clusters; epicalyx of 3 bracts; petals pale mauve, very small.

The following two species are superficially identical. They are both called 'khubayz', meaning small bread, on account of the resemblance of the fruit to the flat loaf of the Middle East.

Epicalyx bracts linear. Calyx not accrescent in fruit, the mericarps thus being clearly visible *M. parviflora*
Epicalyx bracts oblong. Calyx accrescent in fruit so that the mericarps are ± hidden *M. verticillata*

1. Malva parviflora *L.*

Differs from the following species in the key characters above, in its shorter petals which ± equal the calyx and its rugose mericarps. It is best identified in the field by the mericarps which are not enclosed by the fruiting calyx.

A widespread weed of waste places, gardens, field borders and tracks from 1000m to the very summit of J. Nabī Shu'ayb at 3666m. It is found on the escarpment, the high plateau and in irrigated fields east to Ḥarīb.

2. Malva verticillata *L.*

Similar to the last species except for the key characters, the petals that slightly exceed the calyx and the smooth mericarps, which are enclosed by the fruiting calyx.

The distribution of this species is very similar to the last with which it is sometimes intermixed. It is generally rather more common but has not been found above 3200m nor in the Mashriq.

3. MALVASTRUM *A. Gray*

1. Malvastrum coromandelianum (*L.*) *Garcke*

Branched ascending appressed hairy perennial up to about 20cm in Yemen. Leaves ovate, unlobed, strongly dentate. Flowers in axillary and terminal clusters, subsessile, epicalyx bracts 3; petals small, yellow. Mericarps pilose.

Probably a recent introduction to Yemen, but now locally common between 600 and 1600m by tracksides on the western slopes of J. Milḥān and J. Bura' – mostly in the coffee country.

4. SIDA *L.*

Mostly low annual or perennial plants with woody stems. Leaves simple. Flowers small, often solitary in our species. Epicalyx absent. Mericarps 1-seeded.

1. Plant pilose. Leaves cordate at base — 2
   Plant pubescent, but never pilose. Leaves rounded or cuneate at base — 3
2. Mericarps 5, not awned — *S. urens*
   Mericarps 10, awned — *S. cordifolia*
3. Mericarps 5. Petioles usually with a spine-like tubercle at their base — *S. alba*
   Mericarps 7–10. Petioles without a spine-like tubercle at their base — 4
4. Leaves ovate. Mericarps without awns. — *S. ovata*
   Leaves narrowly elliptic to rhomboid. Mericarps awned. — *S. rhombifolia*

1. Sida alba *L.*

Erect woody-stemmed pubescent annual to about 50cm. Leaves ovate or elliptic, to c 3.5cm long, base rounded. Flowers solitary or in few-flowered axillary racemes, white or cream. Mericarps 5, each with 2 awns.

A common weed on the escarpment between 500 and 1500m. It is usually found in derelict fields, tracksides, coffee groves and very stony agricultural land where there is not too much shade from the growing crops.

2. Sida cordifolia *L.*

Erect woody-stemmed pilose annual to about 30cm. Leaves ovate or orbicular, to about 4cm long, base cordate. Flowers axillary, solitary or less commonly in clusters, mainly at the top of the stem, pale yellow. Mericarps 10, each with 2 awns, the hairs on which are deflexed.

An introduced weed found in waste places between 500 and 1000m along the outer escarpment from the Wadi Rima' at Mishrāfah north to the Wadi Līyah. It is only found in high rainfall areas and is now locally common in parts of the central escarpment as around Suq Ar Ribāṭ, J. Raymah.

3. Sida ovata *Forssk.*

Prostrate or ascending woody-stemmed pubescent perennial. Leaves ovate, up to 2.5cm long, base rounded, undersurface grey. Flowers solitary, orange. Mericarps 7–8, unawned.

A common plant of the escarpment and the silt plain at the east of the Tihama from c 200–2400m. It is also recorded from Rada', Raydar (Hepper) and Majz and may occur elsewhere in the southeast and the Mashriq. It grows on field borders, open stony slopes and roadsides.

4. Sida rhombifolia *L.*

Decumbent or ascending woody pubescent perennial. Leaves more shortly petiolate than in our other species (c 5mm), narrowly elliptic to rhomboid, up to c 6cm long, base cuneate, undersurface light green-felted and contrasting with the dark green upper surface. Flowers solitary or in axillary racemes, pale yellow. Mericarps 9–10, awned.

A rare plant found by small ditches in areas of high rainfall between 1600 and 2000m: between Ibb and Suq As Sabt; J. Ẓallamlam above Ḥadīyah, J. Raymah; above Suq Ath Thalūth, Wadi La'ah.

5. Sida urens *L.*

A pilose usually prostrate annual herb. Leaves ovate, up to 6cm long, base cordate. Flowers in axillary racemes, small, pale orange. Mericarps 5, not awned.

A plant of scrubby vegetation around ditches and watercourses in areas of good rainfall between 800 and 2600m along the escarpment from Misrākh north to Shahārah. It is most common in the Ibb – 'Udayn region and the central escarpment. It is commonly associated with plants like *Erythrococca abyssinica* and *Hypoestes triflora*.

5. WISSADULA *Medic.*

1. Wissadula rostrata (*Schumach.*) *Hook. f.*

W. hernandioides (L'Herit.) Garcke

Erect pubescent undershrub to 1½m. Leaves ovate, up to 12cm long, base cordate, apex acuminate, margin entire, upper surface green, glabrous, undersurface grey-felted. Flowers small, orange, in leafless terminal panicles. Mericarps 5, beaked, much longer than the calyx; seeds usually 3.

Recorded by Forsskal from the Wadi Mawr and Waṣāb As Sāfal, but only subsequently found twice – on both occasions in the wettest areas of the central escarpment: 'Ukaybir, J. Milḥān (Schweinfurth); in riverine forest at 600m above 'Alūjah, J. Raymah.

6. ABUTILON *Mill.*

Tall erect perennials usually exceeding 1m in height. Leaves usually simple, ovate, with a cordate base. Flowers usually axillary, solitary or in panicles, opening in the evening; epicalyx absent. Fruit of 5 or more mericarps each with 2–3 seeds except for *A. longicuspe*. Most species occur in disturbed habitats.

1. Flowers pale purple. Mericarps 1-seeded — *A. longicuspe*
   Flowers yellow or orange. Mericarps 2–3 seeded — 2
2. Mericarps rounded on the outside — *A. pannosum*
   Mericarps with an acute point on the outside — 3
3. Flowers yellow with a dark centre. Stem with numerous brown glandular hairs — *A. sp. A*
   Flowers pure yellow. Brown glandular hairs absent — 4
4. Mericarps usually 8, with awns more than 2mm long — *A. ramosum*
   Mericarps 10 or more, unawned — 5
5. Grey-canescent plant. Mericarps 10–12 — *A. fruticosum*
   Plant not canescent, spreadingly hairy. Mericarps 15 or more — 6
6. Mericarps about 15, c 4mm wide. Stipules filiform c 5mm long — *A. bidentatum*
   Mericarps about 20, c 8mm wide. Stipules linear c 10mm long — *A. hirtum*

1. **Abutilon bidentatum** *Hochst. ex. A. Rich.*

A. indicum sensu Schwartz

Erect rather scruffy softly pilose, glandular perennial to about 1m. Leaves ovate, acute with a cordate base, up to about 6cm long, coarsely dentate, green above, grey below. Flowers orange-yellow, solitary or paired in the leaf axils or in small leafy axillary panicles. Mericarps about 15, acute.

A widespread plant found from sea level to about 2500m. It is abundant on the escarpment and occurs sporadically in the alluvial plains in the east of the Tihama, on the high plateau and in the Mashriq. It is usually found in disturbed habitats, typically roadsides, field borders and thorn fences round fields and houses.

2. **Abutilon fruticosum** *Guill. & Perr.*

Erect grey-canescent woody undershrub to about 50cm. Leaves ovate, cordate based, up to c 4mm long, apex obtuse to rounded, margin denticulate, grey on both surfaces. Flowers yellow, usually solitary in the leaf axils. Mericarps about 10, acute.

A widespread plant characteristic of dry stony hills between 300 and 2500m. It is thus found in the Tihama foothills, in rainshadow areas of the central escarpment, in the southeastern mountains, the high plateau and the Mashriq.

3. **Abutilon hirtum** (*Lam.*) *Sweet*

A. figarianum Webb.

A. graveolens sensu Schwartz

This species resembles *A. bidentatum* but differs in its larger, more orbicular leaves which may reach 10cm long; its more numerous axillary panicles which have more flowers; its longer linear stipules and its mericarps which are 20 or more in number and about 11 × 8mm in size.

Apparently a rare species mostly of roadsides in the rain-shadow areas of the central escarpment: Jabal Ash Shirq near Madīnat Al 'Abīd; between Mawsanah and Bāb Bahal in the Ḥarāz; Rajamah near Ḥajjaylah; Wadi Yūr; between Muḥarraq and Maḥābishah; Shahārah (Wood 1957, 2434, 3443).

4. **Abutilon longicuspe** *Hochst. ex A. Rich.*

Erect perennial to about 2½m high. Stems tomentose and with long hairs present. Leaves broadly ovate, up to 18cm long, base cordate, apex acuminate, margin dentate, dark green above but grey-felted below. Flowers pale purple in terminal and axillary, nearly leafless panicles. Mericarps 20 or more, rounded.

A very distinctive species on account of its size, purple flowers and 1-seeded mericarps.

Found between 2600 and 2900m on disturbed rocky slopes and gullies on high mountains in the Ibb area: J. Ba'dān; J. Sumārah near Ḥiliyal; at the head of the Wadi Banā between Ribāṭ and Saddah.

5. **Abutilon pannosum** (*Forst. f.*) *Schlecht.* Ra'in

A. muticum (Del.) Webb

Erect perennial to 1½m, tomentose but with a few long hairs. Leaves ovate to orbicular, up to 8cm long, base cordate, apex acute to rounded, dentate, dull green on both surfaces. Flowers orange-yellow, sometimes with a dark centre, axillary, sometimes solitary, more usually in small leafy panicles. Mericarps usually more than 20, rounded.

Throughout the Tihama extending up some of the broad valleys of the foothills to about 700m. Also known from Ḥarīb (1100m) at the edge of the Sands. Although found by the coast, it is most characteristic of the alluvial plain in the east of the Tihama where it occurs in enormous abundance and is often dominant in derelict fields and by irrigation canals (Wood 74/447, 758).

6. **Abutilon ramosum** (*Cav.*) *Guill. & Perr.*

Erect woody perennial to about 1m, stem pubescent with a few longer hairs. Leaves broadly ovate, or very slightly trilobed, up to about 10cm long, base truncate to slightly cordate, apex shortly acuminate, margin coarsely crenate, dull green and glabrescent on both surfaces. Flowers pale yellow in 2–4 flowered lateral cymes. Mericarps about 8, acute, with an awn about 2mm long.

Scattered along the escarpment between 300 and 1500m, from Misrākh north to Shahārah with an isolated record by Schweinfurth from the Tihama at Khalīfah. It is usually found on field borders and tracksides in the lusher wadis.

7. **Abutilon sp. A**

Erect branched perennial to about 1m, the stems, petioles and peduncles covered in a dense indumentum of brown glandular hairs intermixed with a few very long slender white hairs. Leaves broadly ovate, very large up to 15cm long, base cordate, apex shortly acuminate, upper surface green, the undersurface softly grey-canescent. Flowers orange with a purple centre in small axillary panicles; petals c 15mm long. Mericarps c 15–20, acute.

This very distinctive species is only known from a single collection (Wood 2526) made just above Suq Ath Thalūth in the Wadi Lā'ah at c 1200m. It was growing in low scrub by a village.

7. **PAVONIA** *Cav.*

Small undershrubs or herbs. Flowers solitary, long pedicelled in the leaf axils. Epicalyx present consisting of 5–6 bracts. Fruit of 5 1-seeded mericarps. Most species are characteristic of dry country in the escarpment foothills.

1.	Epicalyx bracts filiform, ciliated, forming a basket round the fruit	2
	Epicalyx bracts lanceolate to elliptic, not ciliated and not forming a basket round the fruit	5
2.	Leaves 3-lobed	*P. triloba*
	Leaves unlobed	3
3.	Flowers pink. Leaves entire or with a few teeth near the apex	*P. arabica*
	Flowers yellow. Leaves toothed	4
4.	Leaves crenate. Mericarps broadly winged	*P. kotschyi*
	Leaves serrate. Mericarps scarcely winged	*P. hildebrandtii*
5.	Flowers yellow, c 20mm long. Mericarps muricate	*P. flavo-ferruginea*
	Flowers orange, c 15mm long. Mericarps reticulate	*P. burchellii*

1. Pavonia arabica *Hochst. & Steud. ex Boiss.*
Erect, glandular-pubescent and pilose, branched perennial. Leaves very variable but always unlobed and densely pubescent on both surfaces, ovate to elliptic, base cordate, apex truncate, emarginate or obtuse, the margin entire or, more usually, with (1–) 3 (–5) obscure teeth near the apex. Petals pink. Mericarps villous, unwinged.

An uncommon plant of arid stony hills and rocky plain in the escarpment foothills found mainly between 300 and 600m from Dayr Khushm south to Bani 'Umar (Ḥujarīyah) where it reaches an exceptional 1200m.

2. Pavonia burchellii (*DC.*) *R.A.Dyer*
P. kraussiana Hochst.
Procumbent or ascending undershrub, stem persistently pubescent or pilose, some long hairs always present. Leaves usually weakly 3-lobed, ovate, crenate, base cordate, the undersurface strongly pubescent and markedly paler than the upper surface. Epicalyx bracts broadly elliptic, obtuse. Petals orange, about 1.5cm long. Mericarps woody, pubescent, reticulate.

A locally frequent plant of scrubby cliffs and rocky slopes, usually near villages, between 1500 and 2800m on the high plateau and the escarpment, but becoming very rare southwards and absent south of the Sumārah pass except for J.'Ayfū and J. Ba'dān. Particularly characteristic of cliff-side villages like Ghaymān, Kuhlān, Shahārah, Rāziḥ etc. (Wood 815, 2938).

3. Pavonia flavo-ferruginea (*Forssk.*) *F.N. Hepper & J.R.I. Wood*
P. glechomifolia (A. Rich.) Garcke.
P. patens (Andr.) Chiov.
Low glandular-pubescent undershrub, the older branches chestnut-brown, glabrescent. Leaves unlobed, ovate to orbicular, crenate, base cordate, ± green and glabrescent on both surfaces. Epicalyx bracts lanceolate, acuminate. Petals pale lemon-yellow with a dark purple base, about 2cm long. Mericarps woody, pubescent, muricate.

A local, but characteristic plant of scrubby slopes in the first foothills below the escarpment often growing in the protection of fences and other bushes such as *Acalypha fruticosa*, *Anisotes* etc. Inland it is rarer but found at a few places such as Al 'Ashshah; Wadi Ḥidrān, Taiz; Bani 'Umar, Ḥujarīyah. Known from 200 to 1400m from the Wadi Mawr southwards (Wood 777, 1364, 2924).

This species and *P. burchellii* are commonly regarded as conspecific and treated under the name *P. patens* (Andr.) Chiov. The occasional intermediate (Wood 724 from Mafḥaq) does occur, but the status of such intermediates is unclear and it seems best to treat such widely differing plants as separate species until they can be proved to be conspecific.

4. Pavonia hildebrandtii *Gurke ex Ulbr.*
Hibiscus flavus sensu Forsskal
Erect, stellate-hispid, branched annual. Leaves ovate, serrate, base cordate, apex obtuse, undersurface greyer and more strongly pubescent than the dull green upper surface. Petals yellow. Mericarps villous, scarcely winged, about half the length of the filiform epicalyx bracts.

Found by Forsskal in the Wadi Mawr area, but known to me only from four localities between 1000 and 1700m in open *Acacia-Commiphora* scrub in the rainshadow areas of the central escarpment: southwest of Madīnat Al 'Abīd; Mafḥaq; Mawsanah in the Ḥarāz; on the dip slope east of Maḥābishah.

5. Pavonia kotschyi *Hochst. ex Webb*
Erect, stellate-hispid, woody annual. Leaves elliptic to orbicular, crenate, base and apex rounded, pubescent on both surfaces. Petals yellow. Mericarps pubescent, broadly winged, scarcely overtopped by the filiform epicalyx bracts.

Found once only, on sand at Bayt al Faqīh (200m).

6. Pavonia triloba Guill. & Perr.
Erect annual, thinly pubescent with a few scattered long hairs. Leaves 3-lobed, base cordate to truncate, margin with a few teeth, pubescent especially below. Petals pink. Mericarps pubescent, unwinged.

A rare plant of dry stony country in the foothills and central escarpment from 300–1200m: west of Al Barḥ; between Attārah and 'Usil; between Bāb Baḥal and Mawsanah; between Malaḥiṭ and Suq Al Jammah; Wadi Līyah.

8. KOSTELETZKYA *C. Presl*

1. Kosteletzkya adoensis (A. Rich.) Mast.
Perennial herb with numerous pilose, irritant, stellate hairs and long decumbent stems. Leaves ovate, usually 3-lobed below, c 3–6cm long, cordate based. Flowers solitary or in few-flowered axillary clusters, pale purple. Capsule winged; seeds glabrous, one per loculus.

A curious feature of this plant is the narrow line of pubescence extending from one internode to the next.

Only known from damp hollows in dense scrub near a pool at c 1600m between Ibb and Suq As Sabt.

9. SENRA Cav.

1. Senra incana Cav.
Erect, branched, softly grey-tomentose undershrub usually about 1m high. Leaves palmately 3 (–5) lobed, dentate, base cordate. Flowers solitary, axillary, mostly towards the ends of the branchlets thus forming leafy racemes; petals yellow with a maroon base, almost enclosed by the large epicalyx bracts. Capsule ovoid completely enclosed by the bracts; seeds pilose.

Apparently a rare species of hot saline bushy habitats. Found by Ehrenberg in the Wadi Mawr area. Known by me from the Red Sea coast (Yakhtul – Mawshij; Manẓar – Ṭa'if), inland in the Tihama at Ḥays, and in the Mashriq from a spring at 1200m due north of Sirwāḥ on the road from Mārib to the Fardah pass.

10. HIBISCUS *L.*

Annual or perennial plants, often shrubby. The indumentum is variable but, in addition to stellate hairs, several species have prickles or irritating spinules. Leaves simple or palmately lobed. The flowers are usually solitary, axillary, red or yellow and often very showy, but short lived and withering during the day. Epicalyx usually present. The fruit is a dry capsule whose seeds are very variable in appearance. There is always more than 1 seed per loculus.

1. Epicalyx absent — **2**
   Epicalyx present — **3**
2. Flowers yellow. Leaves orbicular below, usually with three separate leaflets above — *H. sidiformis*
   Flowers white or very pale pink. Leaves ovate or 3-lobed but not with 3 leaflets — *H. lobatus*
3. Inflorescence consisting of 2–7-flowered axillary and terminal cymes — *H. rhabdotospermus*
   Flowers solitary in the axils of the upper leaves, sometimes forming a lax terminal raceme — **4**
4. Petals red, pink or white. Seeds woolly — **5**
   Petals yellow. Seeds not woolly — **10**
5. Leaves deeply pinnatisect — *H. somaliensis*
   Leaves not pinnatisect — **6**
6. Calyx lobes 8–10mm long — *H. aponeurus*
   Calyx lobes less than 5mm long — **7**
7. Leaves densely stellate-tomentose — **8**
   Leaves sparsely stellate-hairy — **9**
8. Flowers pale pink, c 6mm long — *H. micranthus*
   Flowers bright red, c 10mm long — *H. purpureus*
9. Petals white, c 8mm long — *H. sp. A*
   Petals bright crimson, c 15mm long — *H. deflersii*
10. Annuals. Leaves deeply lobed or dissected to at least half their length — **11**
    Perennials. Leaves entire or shallowly lobed, never cut to half their length — **13**
11. Stem with falcate spines. Calyx lobes each with a large gland — *H. cannabinus*
    Stem without falcate spines. Calyx lobes without glands — **12**
12. Capsule apiculate. Calyx not enlarging much in fruit, veins green — *H. palmatus*
    Capsule not apiculate. Calyx enlarging in fruit, veins deep purple — *H. trionum*
13. Petals pure yellow. Petioles less than 1cm long — *H. aethiopicus*
    Petals yellow with a dark base. Petioles exceeding 1cm long — **14**
14. Epicalyx bracts spathulate. Leaves grey tomentose below — *H. panduriformis*
    Epicalyx bracts not spathulate. Leaves not grey tomentose below — **15**
15. Leaves lobed, hirsute — **16**
    Leaves unlobed, sparsely hairy — **17**
16. Epicalyx bracts narrowly lanceolate. Capsule unwinged, glabrous — *H. ovalifolius*
    Epicalyx bracts subulate. Capsule winged, hairy — *H. vitifolius*
17. Leaves serrate. Seeds glabrous — *H. dongolensis*
    Leaves crenate. Seeds pubescent — *H. owariensis*

1. Hibiscus aethiopicus *L.*

Decumbent, perennial hispid herb often covered in red-brown hairs. Leaves very shortly petiolate, simple or weakly 3-lobed, ovate or elliptic, up to 4cm long. Flowers large, yellow, on pedicels up to 5cm long.

This is easily recognised by its pure yellow flowers.

A very rare plant of shallow rocky grassland between 1600 and 2500m in the central escarpment: Shaḥir, J. Milḥān; Banī Sa'id and J. Kabūrah above Ḥadīyah, J. Raymah.

2. Hibiscus aponeurus *Sprague & Hutch.*

Robust, erect, branched perennial to 1m densely covered in grey hairs. Leaves very variable but never lobed, oblong, ovate or orbicular, up to 6cm long and thus larger than in other red-flowered species. Flowers red, the petals c 15mm long; pedicels 3cm long; calyx 8–10mm long in flower, about twice as long as the epicalyx. Seeds woolly.

Very similar to *H. purpureus* from which it is best distinguished by the calyx lobes which are 1cm long.

A rare plant found between 1200 and 1600m on rain-shadow hills in the northwest: on limestone at Al 'Ashshah (Wood 2491) and Maḥābishah (Wood 3141); on volcanic rock, Shahārah.

3. Hibiscus cannabinus *L.*

Erect annual to 1½m, thinly pubescent and with scattered spines, some bulbous based. Leaves 5-lobed almost to the base, the lobes narrowly lanceolate, up to 10cm long. Flowers subsessile, large, yellow with a dark centre. Seeds thinly pubescent.

A widespread but uncommon plant of irrigated fields along wadis on the escarpment between 500 and 1800m, e.g. Wadi Ḍabāb, Taiz; Wadi Rima', Mishrāfah; Wadi Līyah, Khawlan Ash Sham.

4. Hibiscus deflersii *Schweinf. ex Cufod.*

Erect, sparsely pubescent undershrub with long straggly branches reaching 2m in the protection of bushes. Leaves orbicular to ovate, always simple, more coarsely dentate than in other related species, rather small and not usually exceeding 3cm long. Flowers crimson, the petals c 15mm long, pedicels c 4cm long; epicalyx about equalling the calyx. Seeds woolly.

Abundant on scrubby hills on the escarpment and the southeastern mountains from 500 to 1900m.

The striking red flowers of this species can usually be found throughout the dry season when the hills are otherwise barren. Outside southwest Arabia it appears to be rare and is only known from Somalia and Eritrea.

5. Hibiscus dongolensis *Del.*

Vigorous, sparsely pubescent perennial with long decumbent or ascending branches up to 1m long. Leaves unlobed, orbicular to ovate, up to 10cm long, coarsely serrate, base truncate, apex acuminate. Flowers large, yellow with a maroon centre, crowded into a short raceme at the apex of the stem, shortly pedicellate; epicalyx about equalling the calyx, the bracts subulate. Capsule villous; seeds glabrous.

Scattered between 600 and 1800m along the whole escarpment, but never abundant and often growing as individual plants. It appears to be very intolerant of grazing and is usually found on cliffs or in thorn hedges. It may require a relatively humus-rich soil.

6. **Hibiscus lobatus** (*Murr.*) *Kuntze*

Weakly ascending, villous annual with straggly branches up to 30cm long. Leaves ovate, unlobed to deeply 3–5 lobed, up to 12cm long. Flowers shortly pedicellate, small, white, clustered at the apex of the stem; calyx lobes green. Seeds tuberculate.

This and *H. sidiformis* are the only two *Hibiscus* species in Yemen which are without an epicalyx.

Known only from valley scrub in the Wadi Qabayn, just west of 'Udayn Suq.

7. **Hibiscus micranthus** *L.f.*

Much-branched perennial to 2m high with numerous rather stiff ascending or erect branches, all covered in dense colourless stellate hairs. Leaves variable in shape and size, ovate-deltoid to lanceolate, usually simple but sometimes shallowly 3-lobed, c 1–4cm long. Flowers pale pink, the petals c 5mm long, often recurved; pedicels 3–5cm long; epicalyx about as long as the calyx. Seeds woolly.

This is one of the first plants to bloom after the spring rains when it is covered in flowers. Later in the season it flowers more sporadically and is easily overlooked.

Scattered all over the escarpment and high plateau in scrub and on rocky slopes from 1400 to 2700m, but apparently absent from high rainfall areas. It is readily eaten by goats and is commonly found in chewed down condition. (Wood 2660, 2741)

8. **Hibiscus ovalifolius** (*Forssk.*) *Vahl*

H. calyphyllus Cav.

Much-branched, decumbent perennial with short rather stiff branches. Leaves hispid, ovate to orbicular, usually weakly 3-lobed, up to 6cm long, base truncate, apex acute. Flowers solitary on pedicels up to 30mm long, large, yellow with a maroon centre; epicalyx equalling or slightly shorter than the calyx, the bracts lanceolate. Capsule glabrous, seeds pubescent.

This species is usually easily recognised by the distinctive dark purple leaf margin not found in other Yemen *Hibiscus* species.

A local plant of terrace walls and rocky banks along the whole western escarpment, found between 1500 and 2200m in areas of good rainfall (Wood 1729, 2566).

9. **Hibiscus ? owariensis** *P. Beauv.*

Shrub to 2m with long, usually rather lax branches. Leaves unlobed, sparsely hispid, broadly ovate, up to 14cm long, base cordate, apex acuminate. Flowers solitary or in pairs on very short pedicels up to 15mm long, large, lemon-yellow with a maroon centre; epicalyx equalling the calyx, the bracts subulate. Capsule shortly stellate-hairy; seeds pubescent.

An uncommon plant found between 600 and 1500m restricted to dense scrub on the margins of permanently flowing wadis in areas of good rainfall in the escarpment: above 'Alūjah, J. Raymah; above Ḥillah, J. Bura'; Wadi Sukhnah; Wadi Ḥabāb between Madīnat Al 'Abīd and Hammām 'Ali.

Yemen plants (Wood 1376, 2956) agree quite well with West African examples of this species but *Hibiscus* is in great need of revision and some doubt must remain as to their correct identity.

10. **Hibiscus palmatus** *Forssk.*

H. intermedius A. Rich.

Annual with procumbent or ascending branches not usually exceeding 60cm long. Leaves up to 7cm long, the lower ones deeply palmately lobed for about ⅔ of their length with narrowly triangular lobes, becoming merely coarsely inciso-dentate on the inflorescence. Flowers very pale yellow with a dark centre; the petals and pedicels rather short, both c 1cm long. Seeds shiny black with a few appressed hairs.

Easily recognised by its apiculate capsule.

A local plant of field borders, tracksides and open, often disturbed ground in scrub in the foothills of the escarpment from 500 to 1400m. It appears to have a strong southern tendency and is very widespread in Taiz province but has not been found north of 'Abs and Al 'Ashshah. In the central escarpment it is a plant of the first foothills but attains higher altitudes in drier areas.

11. **Hibiscus panduriformis** *Burm. f.*

Erect, branched undershrub to c 1½m, tomentose with the hairs mixed with spinules. Leaves ovate, rarely shallowly lobed, c 4–10cm long, grey-felted below. Flowers borne on short pedicels c 10mm long, yellow with a dark centre. Seeds densely pubescent.

This plant can be easily recognised by the spathulate epicalyx bracts.

In some abundance as the weed of a rather marshy field at 300m at Dayr 'Urāj in the Tihama below Ḥillah, J. Bura'. It has also occurred as a casual at Taiz.

12. **Hibiscus purpureus** *Forssk.*

H. meidiensis Ulbr.

Much-branched perennial with numerous, usually short, erect or ascending branches rarely exceeding 50cm high. Leaves ovate-deltoid, mostly simple but some weakly 3-lobed, up to 3cm long. Flowers brilliant carmine with petals c 10mm long; pedicels 3–4cm long; epicalyx about equalling the calyx. Seeds woolly.

The whole plant is covered in a dense brownish stellate-hairy indumentum which is most prominent on the inflorescence.

This is the common red-flowered *Hibiscus* found between 1800 and 2800m along the whole escarpment. It is rarely in abundance and is usually found around bushes on stony slopes and cliffs. (Wood 1775, 3236).

Outside Yemen this species is only known from Somalia where it appears to be very rare.

13. **Hibiscus rhabdotospermus** *Garcke*

Perennial, with several erect or ascending stems, to about 50cm, densely pubescent and with spinules. Leaves ovate, up to 10cm long, never lobed, base truncate or somewhat rounded. Flowers very pale yellow tinged with pink;

epicalyx bracts subulate. Seeds glabrous, dark brown with distinctive pale transverse ridges.

This plant is easily recognised by the inflorescence which is a leafy, axillary or terminal 1–7 flowered cyme.

Only known from waste ground and field borders at about 1000m in Maqbanah: Ḥajdah; between Ḥajdah and Madrūbah; by the Wadi Rasyān on the road to Ash Shūb.

14. Hibiscus sidiformis *Baill.*

Erect or spreading, branched, pubescent annual to about 25cm high. Leaves extremely variable; lower leaves orbicular, c 2cm diameter, base cordate, margin crenate; upper leaves usually trifoliolate, the leaflets very variable in shape. Flowers lemon yellow without a dark centre, on pedicels 1–2cm long, calyx lobes scarious with a green midrib and margins. Seeds glabrous, tuberculate.

This species is easily recognised by its yellow flowers which lack an epicalyx.

A widespread but very local plant of stony slopes and abandoned cultivation on the escarpment between 600 and 1500m: Ḥajdah; Bani Miftāḥ between Riḥāb and Waṣāb al 'Ali; Rajamah near Ḥajjaylah; in various places in Wadi Qaṣab, Ḥarāz; Wadi Sukhnah; between Ṭūr and Suq Al Amān; Malāḥiṭ. It is usually found in the rainshadow areas associated with *Melhania ovata*, the small Barlerias, *Endostemon tenuiflorum* etc.

15. Hibiscus somaliensis *Franch.*

A prostrate, hispid, woody undershrub with numerous short stems up to 15cm long. Leaves ± ovate, up to 2cm long, deeply pinnatisect, often by reduction 3-lobed. Flowers red (similar in colour to *H. deflersii*), the petals c 8mm long; pedicels 1–3cm long; epicalyx much shorter than the calyx. Seeds woolly.

The prostrate habit and pinnatisect leaves make this species distinct.

A rare plant of exposed rocky hills between 1700 and 2000m in the Mashriq: on limestone, between Ḥūth and Al 'Ashshah; on the hills east of Rada'; Dhī Nā'im.

16. Hibiscus trionum *L.*

Prostrate annual herb with short stems to about 15cm long covered with distinctive stellate-hispid tubercles. Leaves palmately lobed, small, up to 2cm diameter. Flowers yellow with a dark centre, borne on pedicels c 2cm long. Seeds muricate.

Easily identified by the calyx which enlarges in fruit becoming scarious with purple veins.

A local field weed found between 1000 and 2600m on the high plateau and the richer cultivated areas of Taiz and Ibb provinces.

17. Hibiscus vitifolius *L.*

Erect or sprawling, much-branched, pilose, glandular undershrub to about 1m with spinules on the stem. Leaves 3–5 lobed, c 3–8cm long, base cordate, under surface with stellate hairs. Flowers large, yellow with a maroon centre, on pedicels c 3cm long; epicalyx about half as long as the calyx, the bracts subulate. Capsule villous, winged; seeds tuberculate.

This is the commonest large yellow-flowered *Hibiscus* in Yemen. Its winged capsule is distinctive.

Widespread in Yemen between 1000 and 2500m on the escarpment, southeastern mountains and in isolated places on the high plateau and in the Mashriq. It is usually found on roadsides, around villages, in thorn fences around fields and in other partially disturbed habitats.

18. Hibiscus sp. A

Erect, sparsely hairy undershrub to 1m, similar to *H. micranthus* but the leaves sparsely hairy and less than 3cm long; the petals almost pure white with a faint red flush, up to 8mm in length; epicalyx much shorter than the calyx.

Only known from scrub in the valley just above 'Alūjah, J. Raymah (Wood 1061).

In addition to the native species, various forms of **Hibiscus rosa-sinensis** *L.* are cultivated in gardens in Sana'a and Taiz.

The Okra, **Abelmoschus esculentus** (*L.*) *Moench.*, known as 'Bāmiyah' is widely cultivated and sold in the markets of the major towns. The fruit is eaten as a vegetable but it is not as popular as it is around the Mediterranean and is mainly eaten by the more sophisticated.

11. GOSSYPIUM *L.*

Erect hirsute undershrubs to about 2m high. Leaves palmately 3–7 lobed, commonly black dotted. Flowers solitary in the leaf axils, lemon yellow, often with a dark purple centre. Fruit a woody capsule ± enclosed by the large ovate epicalyx bracts. Seeds with woolly hairs.

This genus comprises the various cotton plants, known in Arabic as 'Quṭn'. Cotton is an important crop in the Tihama particularly in the Zabid area where great efforts have been made to develop its cultivation using the flood waters of the Wadi Zabid. At the present time it is probably Yemen's most important agricultural export. Cotton is also casually cultivated in gardens and around villages as high as Sana'a and east to Ḥarīb, but it is of no economic importance outside the Tihama.

***Gossypium* is a notoriously complex genus taxonomically and I have made no attempt to study the cultivated forms in detail. The following key serves to distinguish the four species known to occur in Yemen, though *G. arboreum* is the only species definitely naturalised.**

1. Leaves lobed to about a quarter of their length, each lobe being an equal-sided triangle in outline — *G. hirsutum*
Leaves lobed at least half way, the lobes of varying shape but not forming an equal-sided triangle — **2**

2. Leaf lobes acuminate. Flowers (including bracts) 4–6cm in diameter — *G. barbadense*
Leaf lobes acute or obtuse. Flowers less than 4cm in diameter — **3**

3. Leaf lobes ± elliptic, the base of the lobe about half its width at the widest point. Epicalyx bracts with 6 or more teeth — *G. herbaceum*

Leaf lobes lanceolate or oblong-lanceolate, the base of the lobe at most slightly narrower than the widest point. Epicalyx bracts with 0–8 teeth *G. arboreum*

1. Gossypium arboreum *L.*

G. rubrum Forssk.

A vigorous shrub to 3m, distinguished from other species by its leaves which are divided to more than half their length with lanceolate or oblong-lanceolate lobes, scarcely narrowed below their mid-point. The flowers are c 3cm in diameter, the epicalyx bracts with 0–8 teeth, but showing variation in number on the same plant. The capsule (best seen on immature fruit) is nearly conical in shape and lacks the distinct shoulders of *G. herbaceum*.

Well naturalised in scrub between 500 and 700m in the central escarpment at Ḥadīyah (Forsskal), 'Alūjah (Wood 1366) and in the Wadi Ḥār in the Ḥarāz (Wood 3368). It also grows at 1200m in Bani 'Umar, Ḥujarīyah (Wood 1437) and an unusually hairy form grows amongst *Cadaba rotundifolia* at about 20km between Hodeida and Manṣūrīyah (Wood 3173).

Gossypium herbaceum *L.* (*G. arboreum* sensu Forssk.) is easily confused with *G. arboreum* but the leaf lobes are elliptic with the base about half the width of the broadest point. The epicalyx bracts usually have numerous teeth (more than 6) and the capsule has distinct shoulders, being ± subglobose in shape with an apical beak. It was recorded by Forsskal in the Wadi Mawr and by Ahmed Khattab from Bayt al Faqīh.

Gossypium barbadense *L.* is distinguished by its narrowly triangular leaf lobes which have long acuminate tips and by its very large flowers which are 5–6m in diameter. This is the most commonly cultivated cotton and occurs not only as a field crop but also in old village gardens in Taiz (Wood 3417), on J. Sāfān (Wood 3361) and doubtless elsewhere.

Gossypium hirsutum *L. variety* **punctatum** (*Schum.*) *Hutch.* (*G. punctatum* Schum.) has shallowly lobed leaves, the lobes being the shape of an equilateral triangle. The flowers are intermediate in size (3–5cm) and the plant is usually very hairy. This was collected by Wissmann in Sana'a in the 1930s.

Thespesia populnea (*L.*) *Soland. ex Corr.* has been much planted in recent years particularly in Taiz and Hodeida. It is especially valuable because it is very tolerant of salt and can be planted close to the sea. There are some fine old specimens at Khawkhah. It is a shrub or small tree with ovate, cordate, usually unlobed leaves and large yellow flowers. The capsule is woody.

FLACOURTIACEAE

ONCOBA *Forssk.*

1. Oncoba spinosa *Forssk.* Dhajayj, 'Unqub

Shrub or small tree to 4m, armed on the branchlets with straight spines up to 5cm long. Leaves alternate, elliptic, acuminate, serrate, up to 8 × 5cm. Flowers solitary, terminal on short branchlets, c 5cm in diameter, resembling a small rose; petals large, white; stamens numerous. Fruit globose, green, woody, edible, 4–5cm in diameter.

Flowers in summer.

Common in the wetter parts of the escarpment from c 600–1800m, but absent from the Ḥujarīyah and not recorded from Khawlan Ash Sham. It is particularly abundant in 'Udayn and the valley scrub of the central escarpment but is less restricted to valley communities than most other shrub species.

VIOLACEAE

1. HYBANTHUS *Jacq.*

1. Hybanthus enneaspermus (*L.*) *F. Muell.*

Variable annual or perennial herb with numerous spreading branches. Leaves alternate, up to 7cm long, lanceolate, entire or with a few obscure teeth. Flowers spurred with 5 violet petals, the lower one of which is much longer than the others and may reach 15mm. Fruit a capsule.

Two varieties occur in Yemen: Variety **enneaspermus** is definitely annual, ± ascending in habit, the pedicels only 5–8mm long. It is a rare plant only known from the valleys above Suq As Sabt and Ḥillah at c 600m on J. Bura' (Schweinfurth 356, Wood 2171). Variety **latifolius** (*De Wild.*) *Engl.* is a woody perennial, usually prostrate in habit, the pedicels 10–20mm long. This is a characteristic plant of the outer escarpment being found between 1200 and 2500m from Turbah to J. Rāziḥ; it is particularly common on steep, mist-covered, west-facing cliffs. Apart from a single record from the Kuḥlān cliffs it appears to be totally absent from the inner escarpment (Wood 74/228).

Viola odorata *L.*, the Sweet Violet, was cultivated in Yemen from the Middle Ages at least until the nineteenth century. It has not been seen recently.

TAMARICACEAE

TAMARIX *L.*

Trees or shrubs with slender pendulous rather brittle branches. Leaves minute, alternate, scale-like or apparently absent, often covered in salt crystals. Flowers white or pink in slender spike-like racemes. Fruit a capsule, seeds with a tuft of hairs.

The following three species of Tamarisk are all known as 'Athl' or 'Ithl'. They are superficially identical and can only be distinguished by careful examination of the flowers and

leaves. Indeed some botanists regard the last two species (*T. arabica* and *T. mascatensis*) as forms of the very widespread *T. nilotica* (Ehrenb.) Bunge.

The natural distribution of all three species is obscured by widespread planting, particularly on the high plateau – they are all tolerant of saline conditions where no other trees will grow. They are of some importance in the traditional economy partly for their timber but more especially for their branchlets which are laid across the roof joist and from the framework onto which mud and plaster is fixed. Many plants are therefore extensively pollarded and their natural rounded shape is not obvious.

1. Leaves sheath-like, without a blade, superficially appearing to be absent — *T. aphylla*
   Leaves with a small, scale-like blade — **2**
2. Lobes of nectar-bearing disc gradually narrowed into the filaments — *T. mascatensis*
   Lobes of nectar-bearing disc clearly distinct from the filaments — *T. arabica*

1. Tamarix aphylla (*L.*) *Karst.*
T. articulata Vahl
Tree or shrub with brown fissured bark to 8m. Leaves reduced to sheaths, glaucous. Flowers pink or white, in racemes up to 7cm long.

Widely distributed in the Tihama, the drier parts of the escarpment, the high plateau and the Mashriq from sea level to 2600m. It is usually found on alluvial plain, along wadis and drainage lines, particularly where flooding is infrequent and the salt content is high. It is thus common by the wadis draining into the Jawf and the Sands and along the wadis petering out into the Tihama.

2. Tamarix arabica *Bunge.*
T. nilotica sensu Schwartz
Similar to the other two species but distinguished on close examination by the presence of small, clasping glaucous leaves, 1–1.5mm long and by the filaments which are clearly distinct from the lobes of the nectar-bearing disc.

Widespread in Yemen with the same distribution as the previous species with which it often grows.

3. Tamarix mascatensis *Bunge.*
T. nilotica sensu Schwartz
Identical to *T. arabica* except for the filaments which are wide at the base and merge into the lobes of the nectar-bearing disc.

The exact distribution of this species is not known but it is probably widely distributed in the same habitats as the other species.

FRANKENIACEAE

FRANKENIA *L.*

1. Frankenia pulverulenta *L.*
Procumbent, setulose-hairy, much-branched annual herb. Leaves opposite, obovate, up to 8mm long. Flowers with a tubular calyx and small pale pink petals, sessile, solitary or in clusters.

Known from a single collection by D. Pratt from the edge of a seasonal swamp in the Qa' Būsān.

PASSIFLORACEAE

ADENIA *Forssk.*

1. Adenia venenata *Forssk.* 'Adan
Dioecious climbing shrub with axillary tendrils. Branches long, slender, whip-like, growing from the apex of the 1m high, narrowly conical, tapering, succulent trunk which bears a superficial resemblance to the low-altitude form of *Adenium obesum*. Leaves petiolate with two prominent dark glands at the bottom of the leaf-blade, dark green, glabrous, c 5–8cm long, 3–5 lobed, the lobes rounded at the apex. Inflorescence subsessile on short lateral shoots. Flowers with yellow petals and a corona, the male longer than the female. Fruit large, ellipsoid, c 3–4cm long, pale green with prominent dark veins.

Generally distributed between 500 and 2000m on the escarpment and southeastern mountains but, unlike *Adenium* with which it often grows, never occurring in large numbers. It is found as scattered individuals in Acacia and Euphorbia scrub and has a particular preference for steep rocky slopes and cliffs.

CARICACEAE

The Pawpaw, **Carica papaya** *L.*, which is known in Yemen as ''Amb' is commonly cultivated below about 2000m, but it only really flourishes in the alluvial plains of the lusher escarpment wadis. It is a dioecious tree with large palmately lobed leaves clustered in a crown at the apex of the trunk, which reaches about 4m in height. The fruit ripens throughout the year, but does not travel well and is mainly sold from the roadsides near where it is grown. Although it is only eaten as a fruit in Yemen, it has the lesser-known quality of being a meat tenderizer.

CUCURBITACEAE

A distinctive family of monoecious or dioecious climbing herbs with tendrils arising beside the base of the alternate, often palmately lobed, leaves. Stipules absent. Flowers mostly regular, usually solitary or in racemes. Petals free or joined, usually yellow-green in colour. Fruit various but commonly a fleshy berry.

This tropical family (virtually restricted to the Tihama and the escarpment in Yemen) is rather difficult. Many species are dioecious and it is not always possible to find the fruit which is distinctive in most of our species. I have therefore attempted to draw up a key which depends largely on vegetative characters. The leaves, though rather variable, are distinctive and with a little practice can be used to distinguish the small number of species in Yemen.

1. Plant with a long ciliate, stipule-like bract at the petiole base — 9. *Ctenolepis*
   Plant without such a bract — **2**

2. Tendrils (2–) 3–6 fid. Very robust plants with large flowers, 4cm or more in diameter 1. *Luffa*
Tendrils simple or bifid. Plants less robust, the flowers much smaller **3**
3. Stem scabrid or hispid and very rough to the touch. Leaves scabrid or hispid **4**
Stem glabrous or hairy but not obviously rough to the touch. Leaves smooth or not **6**
4. Leaf lobes pinnatisect. Fruit orbicular 4. *Citrullus*
Leaves unlobed or palmatilobed. Fruit various but not orbicular **5**
5. Stems woody below. Fruit beaked 7. *Kedrostis*
Stems not woody below. Fruit not beaked 3. *Cucumis*
6. Stem distinctly woody even on young growth 6. *Zehneria*
Stem not woody **7**
7. Leaves palmately lobed, the lobes sharply sinuate-dentate. Male and female flowers solitary, subtended by a prominent bracteole 2. *Momordica*
Leaves unlobed or, if lobed, the lobes not sinuate-dentate. Flowers not as above, bracteole absent **8**
8. Leaves punctate above. Fruit not beaked **10**
Leaves not punctate above, or only very slightly so. Fruit beaked **9**
9. Fruit indehiscent or opening longitudinally, more than 3cm long or, if less, the leaves strongly scabrid 7. *Kedrostis*
Fruit opening transversely with a lid, less than 3cm long. Leaves glabrous or hairy, at most slightly scabrid 8. *Corollocarpus*
10. Leaves glabrous below. Corolla long, exceeding 1cm 5. *Coccinia*
Leaves glabrous to tomentose below. Corolla c 5mm long 6. *Zehneria*

1. LUFFA *Mill.*

Monoecious climbing or trailing herbs with large palmately lobed leaves. Tendrils usually 3–6 fid. Male flowers long pedunculate, racemose; female flowers shortly pedunculate, solitary; petals free, large, yellow. Fruit elongate, ± cylindrical or ellipsoid, dry.

Fruit smooth without prominent ribs. Leaves green, very sparsely hairy, deeply lobed *L. cylindrica*
Fruit with 10 prominent raised ridges. Leaves grey-green, densely hairy, very shallowly lobed *L. acutangula*

1. Luffa acutangula (*L.*) *Roxb.*

L. forskalii Schweinf.
Vigorous trailing herb with a pubescent stem and ovate – cordate shallowly lobed, grey-green, densely scabrid-hairy leaves up to 20cm long. Petals pale yellow, c 3–4cm long. Fruit with 10 prominent raised ribs, ellipsoid, c 8–10cm long, glabrescent; seeds without wing-like margins.

Quite common in villages, by irrigation ditches and particularly on stony flood beds between 300 and 600m where the Tihama meets the foothills from Mishrāfah in the Wadi Rima' north to Al Kadan on the Wadi Surdūd. Elsewhere very rare and only known near Suq Ar Rubū', Waṣāb As Sāfal and at Ḥajdah where it reaches 1000m.

2. Luffa cylindrica (*L.*) *M.J. Roem.* Lūfah

Vigorous climbing herb, similar to the previous species. Stem sparsely pubescent. Leaves deeply lobed to about half their length, green, scabrid but very sparsely hairy or glabrous. Petals deep yellow, c 4cm long. Fruit cylindrical to ellipsoid, unribbed, c 6–30cm long, glabrous; seeds with membranous wing-like margins.

An uncommon cultivated plant in the Tihama towns like Bājil and Zabīd and more rarely on the escarpment up to 1500m at Taiz. It is usually found in gardens and always close to houses. The plant is the source of loofahs.

2. MOMORDICA *L.*

1. Momordica balsamina *L.*

Monoecious climbing perennial (?). Stem glabrous or thinly pubescent. Leaves glabrous or thinly pubescent, slightly punctate, ± reniform in outline, c 4–6cm long, always deeply palmately lobed, the lobes sinuate-dentate, the main teeth apiculate. Tendrils unbranched. Male and female flowers solitary, long pedunculate, the male peduncle with a prominent veined bract just below the flower; petals free, yellow with green veins, c 10–12mm long. Fruit c 3cm long, ovoid-ellipsoid, orange-red, soft and fleshy, slightly ridged and warted; seeds red.

Widespread and generally common in the Tihama and on the escarpment from 100 to 1200m. It is a plant of the larger wadis and the alluvial plain at the east of the Tihama and grows on fences and in bushland near streams and irrigation ditches. It is rare away from cultivated areas.

3. CUCUMIS *L.*

Climbing or more commonly trailing, monoecious herbs. Stem and leaves usually scabrid-hairy. Leaves simple or palmately lobed. Tendrils unbranched. Male flowers sessile or shortly pedicellate, solitary or in small clusters; female flowers solitary; petals pale yellow, united below, variable in size, but generally small, 5–10mm long. Fruit fleshy but firm walled.

1. Fruit (and ovary) smooth, hairy *C. melo*
Fruit (and ovary) covered in bristles or fleshy spines or pustules **2**
2. Fruit covered in low bristle-tipped pustules *C. figarei*
Fruit covered with elongate spines at least 5mm long **3**
3. Leaves and stem spreadingly hairy. Fruit red when ripe *C. metuliferus*
Leaves and stem mostly adpressed hairy, sometimes ± spreadingly hairy. Fruit yellow when ripe *C. prophetarum*

1. Cucumis figarei *Naud.*

C. pustulatus sensu Schwartz
C. ficifolius A. Rich, var. *ficifolius* Naud.
Perennial trailing herb. Stems scabrid with falcate spine-

like hairs mixed with a few soft spreading hairs. Leaves scabrid, ovate, c 4–7cm long, usually shallowly and obtusely 3–5 lobed, base shallowly cordate, margin obscurely sinuate. Male flowers solitary, shortly pedicellate. Fruit on a 3–7cm long peduncle, ellipsoid, 5–6cm long, covered in low bristle-tipped pustules, bright yellow when ripe.

Apparently a local plant of open stony and gravelly plain, earth banks by fields, open scrub etc. in high rainfall areas between 500 and 2200m. It has been found from Misrākh south of Taiz north to Shahārah but nearly all records are from the Ibb-'Udayn region and the central escarpment, in both of which it is tolerably frequent.

2. **Cucumis melo** *L.*

C. dipsaceus sensu Schwartz.

Annual trailing herb. Stems scabrid with stiff spreading hairs of varying length. Leaves scabrid, ovate-cordate, mostly 3–4cm long, unlobed or shallowly but acutely 3–5 lobed, margin dentate. Male flowers in clusters of 1–4, shortly pedicellate. Fruit on a 1–4cm long peduncle, oblong-ellipsoid, 2–10cm long, pubescent but without pustules or spines, yellow when ripe.

Widespread but uncommon, growing on earth banks by fields and as a field weed on the alluvial plains in the east of the Tihama; rare in similar habitats on the escarpment reaching 1300m near Taiz; possibly more common than the few records suggest.

This species is also cultivated for melons which are known as 'Battikh' in Arabic. Cultivated forms have larger leaves and fruit than the wild form and the ripe fruit may be brown, green or yellow in colour and globose or ellipsoid in shape. The most common form in Yemen is the musk melon, called 'Shammām'. Melons are mainly cultivated on irrigated land and have a similar distribution to the wild form of the species.

3. **Cucumis metuliferus** *E. Mey. ex Naud.*

Climbing or trailing herb. Stem hispid with long spreading hairs. Leaves hispid, ovate-cordate, c 3–8cm long, slightly broader than long, unlobed to deeply and acutely 3 (–5) lobed, margin sinuate. Male flowers in clusters of 1–4, sessile or shortly pedicellate. Fruit on a 2–6cm long peduncle, ellipsoid, 6–10cm long, covered in scattered fleshy bristle-tipped spines, orange-red and rather fleshy when ripe.

Found in a number of scattered localities on the escarpment between 500 and 1500m from Wadi Ḍabāb near Taiz, north to J. Bura'. Its ecology is uncertain but it appears to be characteristic of lush valleys in the central escarpment and the Ibb-'Udayn region, but is by no means frequent.

4. **Cucumis prophetarum** *L.*

A very variable trailing or climbing perennial herb. Stem scabrid or hispid with adpressed or spreading hairs. Leaves scabrid to weakly hispid, ovate, c 2–6cm long, unlobed or often deeply but obtusely 3–5 lobed, base shallowly cordate, margin sinuate or serrate. Male flowers shortly pedicellate, solitary or in small clusters. Fruit on a 1.5–3cm long peduncle, ellipsoid, c 2–5cm long, pubescent or glabrous but always with bristle-tipped fleshy spines, yellow when ripe.

This species is divisible into two subspecies in Yemen: Subspecies **prophetarum** is a grey-green plant with adpressed scabrid hairs. The fruit spines are rather short and do not exceed 5mm long. Its distribution is not accurately known but it favours drier localities. (Wood 74/282) Subspecies **dissectus** (*Naud.*) *Jeffrey* (*C. ficifolius* A. Rich. var. *echinophorus* Naud.) is a green hispid plant with green hispid hairs. The fruit spines reach 12mm in length. In addition the leaves are commonly deeply and narrowly lobed and the fruit is always glabrous. It seems to favour high rainfall areas. (Wood 2560)

Both subspecies are widespread in Yemen and the species as a whole is found from sea level to 2800m in all areas except the Sands. Its occurrence though is scattered and it never occurs in large quantity. It grows in open habitats on stony plain, on roadsides, in dried-up wadi beds, on field borders and in derelict fields.

The Cucumber, **Cucumis sativus** *L.*, 'Khiyāh', is widely cultivated in Yemen for its edible fruit. The fruits are extremely variable in shape and various cultivars are grown. The short ellipsoid fruited variety is the most common.

4. CITRULLUS *Eckl. & Zeyh.*

1. **Citrullus colocynthis** (*L.*) *Schrad.* Hanẓil

Colocynthis vulgaris Schrad.

Monoecious trailing perennial herb. Stems shortly retrorse scabrid. Leaves scabrid, narrowly ovate in outline, up to 10cm long, palmately lobed with the lobes pinnatisect. Tendrils simple or bifid. Male and female flowers solitary on short pedicels; petals united below, yellow-green c 1cm long. Fruit shortly pedunculate, globose, 5–10cm diameter, glabrous, mottled green when young becoming yellow with maturity.

The fruit, known as the desert gall, dries out slowly and is said to break off from the parent plant and roll over the ground in strong winds, eventually breaking open and scattering its seed.

Widespread in Yemen from sea level to 2400m but absent from almost all the escarpment except for a few arid places in Taiz province. It grows in the Tihama and at Mārib on sand dunes and in sandy fields, locally in abundance. It is more scattered in occurrence on the high plateau and in the Mashriq but grows on alluvial plain and in gravelly wadi beds.

The water melon, **Citrullus lanatus** (*Thunb.*) *Matsum. & Nakai* (*C. vulgaris* Schrad.) called 'Ḥabḥab' in Yemen is very similar except when in fruit but can be distinguished by the soft, spreading hairs on the stem. The fruit is either ellipsoid and pale green or globose and dark green. It is widely cultivated on irrigated land in the Tihama and in parts of the high plateau, particularly at Rada' and north-east of Sana'a.

5. COCCINIA *Wight & Arn.*

1. Coccinia grandis (*L.*) *Voigt*

C. moghadd (J.F. Gmel.) Schweinf.
Cucumis inedulis Forssk.
Dioecious climbing perennial. Stems glabrous. Leaves glabrous, strongly punctate, ovate – cordate in outline, c 5–8cm long, deeply or shallowly palmately 3–5 lobed but never unlobed, margin entire or slightly sinuate. Tendrils simple. Male and female flowers solitary, pedicellate; petals united into a bell-shaped corolla, pale yellow with green veins, c 1.5–2.5cm long. Fruit on a 1–4cm long peduncle, ellipsoid, c 3–6cm long, glabrous, red, very fleshy.

A widespread and common plant found between 200 and 2000m on the alluvial plains in the east of the Tihama and on the escarpment. There is a solitary record from beside a wadi between Rada' and Dhī Nā'im. It grows on fences and on bushes along irrigation ditches and wadis where there is either permanent water, frequent floods or good rainfall.

6. ZEHNERIA *Endl.*

Perennial climbers. Tendrils simple. Flowers very variable, but petals small, united at the base. Fruit small, globose.

Glabrous plant with woody stems. Leaves entire, deeply lobed — *Z. anomala*
Usually hairy plant with herbaceous stems. Leaves ± dentate, unlobed or shallowly lobed — *Z. scabra*

1. Zehneria anomala *C. Jeffrey*

Monoecious or dioecious woody climbing perennial. Stems slender, glabrous, woody with grey bark. Leaves ± glabrous, slightly punctate, fleshy, entire, small, c 1–2cm long, deeply 3–5 lobed, the lobes obtuse and very variable in shape. Male flowers 2–7, in subsessile clusters; female flowers solitary, subsessile; petals c 2mm long. Fruit on a short peduncle up to 4mm long, globose, c 8mm long, red.

Although rarely found in flower or fruit, this species is readily recognised by its small fleshy leaves and slender grey-barked stems.

Widely distributed between 250 and 2000m throughout Yemen in scrub on rocky hills in low rainfall areas but rare and very easily overlooked since it normally grows as single plants in the protection of spiny shrubs particularly *Acacia* species. It has been found in the escarpment foothills (Khalīfah; Ḥajjaylah), in rainshadow areas on the escarpment (Riḥāb-Waṣāb; Mafhaq; east of J. Sharafayn etc.), in the southeastern mountains (Barakānī; Dhī Nā'im) and in the Mashriq (Khaywan, Wadi Nuwāsh, north of Rada' etc.).

2. Zehneria scabra (*L.f.*) *Sond.*

Melothria tomentosa Cogn.
M. punctata (Thunb.) Cogn.
M. maderaspatana sensu Schwartz
Dioecious climbing or trailing perennial, very variable in stem and leaf indumentum (see below). Leaves dentate, ovate or ± triangular in outline, base cordate, unlobed or very shallowly 3–5 lobed, 3–4cm long. Male and female flowers sessile or pedicellate, but usually in clusters; petals 3–4mm long, cream. Fruits usually in clusters, globose, c 6–8mm diameter, orange when ripe, rather shiny.

This species is divisible into 2 subspecies in Yemen: Subspecies **scabra**. The stems are usually pubescent and the leaves dark green and densely scabrid – punctate above, but grey – tomentose below. Both male and female flowers are in sessile or very shortly pedicellate clusters and the seeds are relatively large, being 3–5mm long at their widest point. It is characteristic of stream sides and rocky gullies on high-altitude escarpment mountains and more rarely on the high plateau and the surrounding mountains. It is found from J. Ṣabir north to Khawlan Ash Sham and east to J. Lawz from 2300 to 3100m.

Subspecies **argyrea** (*A. Zimm.*) *C. Jeffrey*. This has glabrous or sparsely hairy stems and leaves. The male flowers are ± umbellate on a peduncle c 2–3cm long; the female flowers are in few-flowered pedicellate clusters. The seeds are rather small, not exceeding 3mm at their widest point. It is apparently a rare plant of scrubby banks near water in high rainfall areas between 300 and 1600m; between Sa'īd Miṣr and 'Alūjah; between 'Ubāl and Ḥajjaylah; between Ibb and Suq As Sabt (Wood 807, 921, 1985).

7. KEDROSTIS *Medik.*

Dioecious or monoecious climbing herbs or shrubs. Tendrils simple or bifid. Male flowers in sessile or pedunculate, rather lax racemes; female flowers solitary or in small clusters, sessile or almost so; petals united below, yellow or greenish. Fruit sessile or nearly so, fleshy, beaked, indehiscent or dehiscing longitudinally.

1. Stems scabrid, woody — *K. gijef*
   Stems never woody and, at most, slightly scabrous — 2
2. Leaves lobed. Plant not foetid — 3
   Leaves unlobed. Plant foetid — *K. foetidissima*
3. Fruit tapering at both ends. Stem glabrescent — *K. leloja*
   Fruit sharply truncate at the base. Stem ± persistently hairy — *K. hirtella*

1. Kedrostis foetidissima (*Jacq.*) *Cogn.* Khuliyah

Stinking monoecious climbing perennial. Stem glabrous to pubescent. Leaves scabrid, hairy especially below, sagittate or hastate, unlobed, c 5–8cm long, margin ± entire. Tendrils simple. Male flowers in shortly pedunculate umbels; female flowers 1–2, shortly pedicellate; petals c 3–4mm long, yellow. Fruit sessile or on peduncles up to 15mm long, pear-shaped, 1–2.5cm long including the long beak, hairy, red, very fleshy.

Widespread and common on the escarpment from 600 to 2300m. It grows on waste ground, in villages, on walls, in between boulders and in scrub by ditches and streams. It is usually found close to human activity and may favour organically rich habitats.

2. Kedrostis gijef (*Forssk.*) *Jeffrey*

Corallocarpus gijef (Forssk.) Hook. f.
Dioecious woody climbing perennial. Stems scabrid, woody with pinkish bark. Leaves scabrid, ± punctate, c 4–6cm long, lobed, the lobes usually narrowed below, base ± cuneate, margin sinuate. Tendrils simple. Male flowers

in sessile or shortly pedunculate racemose clusters; female flowers solitary or in small clusters; petals yellow, c 2mm long. Fruit shortly pedunculate, ± ovoid, shortly beaked 12–15mm long, glabrous, dull red, fleshy.

A characteristic plant of scrub in rather arid conditions in the foothills between 200 and 1100m. It is particularly common in the southwestern foothills but turns up elsewhere wherever there is a gap in the escarpment as at Mishrāfah, Khamīs Banī Sa'd, Ṭūr etc.

3. Kedrostis hirtella (*Naud.*) *Cogn.*

Monoecious climbing perennial. Stems sub-glabrous to thinly spreadingly hairy. Leaves softly hairy, c 4–6cm long, 3–5 lobed, margin sinuate. Tendrils simple or bifid. Male flowers in lax racemes on peduncles c 5cm long; female flowers solitary, shortly pedicellate; petals greenish, 4–6mm long. Fruit shortly pedunculate, conical with a long tapering beak, c 4–9cm long, very sparsely hairy, orange-red. The base of the fruit is very distinctive, when 'pressed' it appears to be truncate but in fact the apex of the peduncle is inserted into and ± concealed by the campanulate base.

Known only from between Bayt al Faqīh and Manṣūrīyah where it grows in *Salvadora* scrub at 200m.

4. Kedrostis leloja (*Forssk.*) *Jeffrey*

Monoecious climbing perennial. Stems glabrescent. Leaves ± scabrid-hairy, c 4–6cm long, 3–5 lobed, the lobes obovate or ovate, margin sinuate. Tendrils simple or bifid. Male flowers in racemes on peduncles 2–7cm long; female flowers solitary, shortly pedicellate; petals greenish, 2–4mm long. Fruit shortly pedunculate, narrowly ellipsoid, gradually narrowed into the apex and the base, 5–8cm long, glabrous, orange.

Collected by Forsskal at Luḥayyah and not seen since in Arabia.

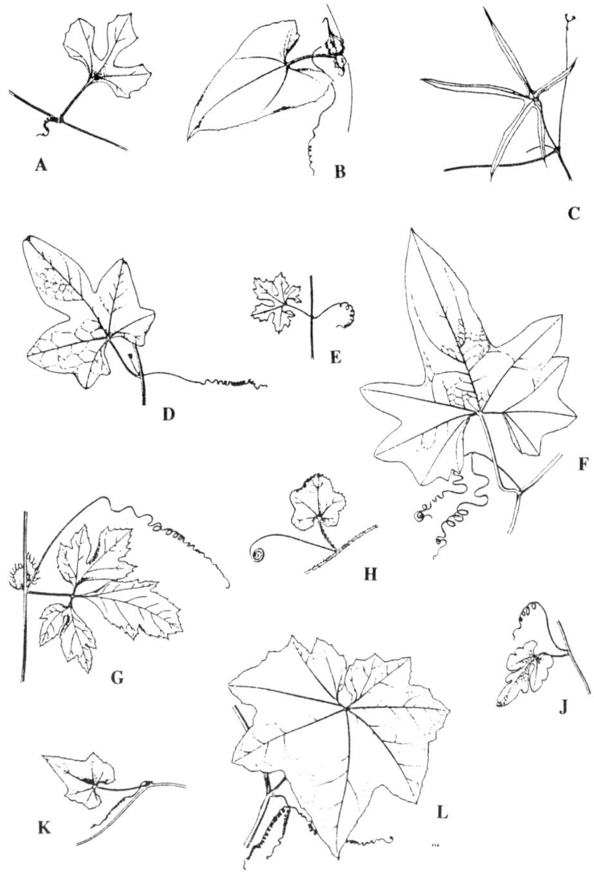

Fig. 8. Leaves of Cucurbitaceae. **A** *Kedrostis gijef*; **B** *Kedrostis foetidissima*; **C** *Corallocarpus sp. A*; **D** *Coccinia grandis*; **E** *Momordica balsaminea*; **F** *Luffa cylindrica*; **G** *Ctenolepis cerasiformis*; **H** *Cucumis melo*; **J** *Cucumis prophetarum* subsp. *dissectum*; **K** *Zehneria scabra*; **L** *Luffa acutangula*.

8. CORALLOCARPUS *Hook. f.*

Very similar to *Kedrostis* but the fruit opening transversely with a lid to expose the red seed mass.

Stem glabrous. Fruit pedunculate, bottle-shaped, c 3cm long ... *C. sp. A*
Stem sparsely hairy. Fruit ± sessile, ovoid, 1–1.5cm long ... *C. schimperi*

1. Corallocarpus schimperi (*Naud.*) *Hook. f.*

C. erostris (Schweinf.) Hook. f.

Monoecious perennial climbing herb. Stem slightly rough, pubescent, the hairs of different lengths. Leaves densely hairy, slightly tomentose to weakly scabrid, ovate-cordate, c 3–6cm long, unlobed or more commonly 3–5 lobed, margin sinuate, tendrils simple. Male flowers in long-pedunculate, rather dense, elongate racemes; female flowers in sessile clusters; petals yellow, c 2mm long. Fruit ± sessile, ovoid, shortly beaked, 1–1.5cm long, glabrous or minutely hairy, red.

Apparently a very local but possibly overlooked species of bushland by wadis in the foothills and the Tihama around Bājil: Marāwi'ah (Deflers); Wadi Sāra'; Wadi Surdūd below Khamīs Banī Sa'd; Wadi Sihām near Khalīfah; above Suq As Sabt, J. Bura'.

2. Corallocarpus sp. A.

Monoecious climbing perennial herb. Stem glabrous. Leaves glabrous, triangular – ovate, ± hastate, c 3–5cm long, unlobed or more commonly 3–5 lobed with narrowly oblong lobes, apex apiculate, margin entire. Tendrils simple. Male flowers in 10–15mm long dense racemes on peduncles 5–6cm long; female flowers solitary, pedicellate; petals c 2mm long, yellow-green. Fruit on a 1cm long peduncle, ±

bottle-shaped, long-beaked, c 3cm long, fleshy, glabrous, orange-red.

A rare plant known from a few places between 1000 and 1800m on the escarpment: in *Euphorbia* scrub near Qā'idah (Wood 1760), north of the Suq, 'Udayn (Wood 2009), between Walajah and 'Ukaybir, J. Milḥān (Wood 1203); Shahārah.

9. CTENOLEPIS *Hook. f.*

1. Ctenolepis cerasiformis (*Stocks*) *Hook. f.*

Blastania fimbristipula (Fenzl.) Kotschy & Peyr.

Monoecious climbing perennial. Stem ± glabrous. Leaves dark green, punctate, scabrid-hairy on the veins, c 4–9cm long, palmately 3–5 lobed almost to the base, the lobes distinctly narrowed below, base cordate, margins sinuate. Tendrils unbranched. Male flowers in small racemes on slender peduncles c 1–3cm long; female flowers solitary, shortly pedicellate; petals very small c 1.5mm long, yellow. Fruit on a short peduncle c 3mm long, globose, c 1cm diameter, glabrous, red when ripe.

Readily distinguished from all other Yemen cucurbits by the stipule-like bract at the base of the petiole which is ± orbicular and long-ciliate margined.

Quite common in scrub by ditches and watercourses in the foothills of the central escarpment extending west in the Tihama to Sa'īd Miṣr, Bājil and Al Kadan. Elsewhere rare and recorded in the Wadi Mawr by Ehrenberg and at Ramādah and Nashimah in Taiz province where it reaches 1300m.

In addition to the cultivated species mentioned under the appropriate genera, several other cucurbits are grown in Yemen.

The bottle gourd, **Lagenaria siceraria** (*Molina*) *Standl.* (*Lagenaria vulgaris* Ser.) known in Yemen as 'Dubbah' is commonly cultivated around villages in the Tihama and the escarpment up to c 1600m. It is a large, densely hairy plant with white flowers and the distinctive bottle-shaped fruit. The fruit is very hard-shelled and is used to store liquids, particularly milk and oil. With the advent of plastic containers, its use and cultivation are likely to become infrequent.

Several species of **Cucurbita** are cultivated as marrows ('Kūsah') or pumpkins ('Kash') probably including **C. maxima** *Duchesne*, **C. moschata** (*Duchesne*) *Duchesne* and **C. pepo** *L.* They are found in most large markets and are used to bulk out and flavour soups and stews.

SALICACEAE

SALIX *L.*

Dioecious trees and shrubs with alternate, simple, deciduous, stipulate leaves. Flowers in catkins. Fruit a capsule.

All our species are introductions, but appear wild.

1.	Leaves dark green, glabrous when mature	2
	Leaves silky adpressed hairy on both surfaces when mature, rarely glabrescent and then blue-green	*S. alba*
2.	Stamens 2. Leaves lanceolate; neither sticky nor fragrant	*S. mucronata*
	Stamens usually 5. Leaves elliptic; sticky and fragrant when young	*S. pentandra*

1. Salix alba *L.*

White Willow. A very variable tree to 25m, branchlets brown, grey or yellowish, not brittle. Leaves lanceolate, acuminate, c 5–10cm long, about 5–8 times as long as broad, serrate, densely silky, adpressed – hairy on both surfaces when mature, rarely glabrescent and then blue-green on both surfaces. Catkins on short leafy shoots, the male c 2.5 – 5cm long, the female rather smaller but elongating in fruit; stamens 2.

Recorded by Manzoni at Taiz (1500m), doubtless an introduction, but the tree does not appear to be there today.

2. Salix pentandra *L.*

Shrub or small tree to 7m. Branchlets brown, shiny, not brittle. Leaves elliptic, acuminate, 5–12cm long, about 2–4 times as long as wide, serrate, glabrous, dark green, sticky and fragrant when young. Catkins similar to *S. alba* but the stamens usually 5.

Flowers in February.

Known only from Raydah (D. Wood).

3. Salix mucronata *Thunb.*

S. subserrata Willd.

Shrub or small tree to 10m. Branchlets red-brown, brittle. Leaves lanceolate, acuminate, 4–8cm long, 3–5 times as long as broad, serrate, glabrous, dark green. Catkins similar to *S. alba*; stamens 2.

Flowers in March as the new leaves appear.

Locally common by permanent streams between 2100 and 3000m around the high plateau; around the head of the Wadi Banā and the Yarīm area; Wadi Maswar, Jiḥānah; above Wa'lān; Wadi Ḍahr; by the streams southwest of J. Nabī Shu'ayb particularly around Baw'ān.

Easily propagated by cuttings and so likely to become more frequent.

CAPPARACEAE

A family of trees, shrubs and herbs closely related to the Cruciferae. The flowers are usually in bracteate racemes and may be regular or irregular with 4 or no petals. A characteristic feature is the presence of a gynophore, an elongated stalk connecting the ovary and the receptacle, which is found in most species.

In Yemen the family includes a number of trees and shrubs which are remarkable for their drought resistance.

1. Plant spiny — 7. *Capparis*
   Plant unarmed — **2**
2. Herbs, sometimes shrubby below but not exceeding 1½m — **3**
   Trees or shrubs — **5**
3. Fruit a podlike capsule — **4**
   Fruit 1-seeded, nut-like, winged — 1. *Diptergium*
4. Stamens free — 2. *Cleome*
   Stamens attached to an elongated androgynophore — 3. *Gynandropsis*
5. Calyx tube present — 4. *Maerua*
   Calyx tube absent — **6**
6. Gynophore with a long ligulate appendage. Fruit cylindrical, torulose — 5. *Cadaba*
   Long ligulate appendage absent. Fruit globose 6. *Boscia*

1. DIPTERYGIUM *Decne.*

1. Dipterygium glaucum *Decne.* Halka, 'Alka

Glabrous, bushy perennial with numerous rigid straight branches from a shrubby base, reaching c 50cm. Leaves shortly petiolate, elliptic, c 2–3mm long, but quickly deciduous. Flowers in long terminal racemes, distant, yellow. Fruit elliptic with a broad membranous wing.

Abundant on sand dunes and in sandy fields throughout the Tihama; also in the Sands east of the Jawf, Mārib and Ḥarīb. It has been found from sea level only to 1300m, but suitable habitats are rare at higher altitudes in Yemen.

2. CLEOME *DC.*

Annual or perennial branched herbs, often foetid. Leaves simple or digitately 3–9-foliolate. Flowers in terminal racemes, usually irregular; petals often long-clawed, stamens free. Fruit a long, pod-like capsule.

The species in Yemen are remarkable for their clear-cut ecological requirements.

1. Leaves simple — **2**
   Leaves digitately 3–9-foliolate — **3**
2. Leaves broadly ovate — *C. scaposa*
   Leaves narrowly lanceolate — *C. monophylla*
3. Plant glabrous — **4**
   Plant hairy — **5**
4. Leaves filiform, c 1mm wide. Petals c 5mm long — *C. angustifolia*
   Leaves linear, at least 3mm wide. Flowers showy with petals exceeding 10mm long — *C. paradoxa*
5. Petals pure yellow, at least 5mm long — **6**
   Petals purple or yellow with purple lines, very small, c 3mm long — **7**
6. Fruit c 1cm long — *C. brachycarpa*
   Fruit c 6–8cm long — *C. viscosa*
7. Fruit ± glabrous when mature — *C. schweinfurthii*
   Fruit strongly glandular — *C. amblyocarpa*

1. Cleome amblyocarpa *Bair. & Murb.*

C. arabica sensu L. (1756)

Erect, glandular, foetid annual to 50cm. Leaves 3-foliolate with narrowly obovate leaflets. Petals c 3mm long, purple with a paler limb. Capsule c 50–60 × 10mm, glandular, not stipitate, straight or falcate, very pale green; seeds lanate.

Locally frequent as a weed of abandoned fields and less commonly of roadsides in the high plateau region; always between 2000 and 2400m.

2. Cleome angustifolia *Forssk.*

C. diandra Burch.

Erect, glabrous, glaucous, woody annual to 1m. Leaves 3–7-foliolate with filiform leaflets. Sepals c 3mm long, soon deciduous; petals c 5mm long, orange-yellow with mauve veins. Capsule c 50–60 × 3–4mm, veined, stipitate, straight or slightly curved, green; seeds pilose.

Locally abundant as a weed of stony fields between 1000 and 1500m around Taiz and in the Ḥujarīyah. Elsewhere a very local plant of well-drained stony hills reaching 2400m in the Banī Ḥushaysh.

3. Cleome brachycarpa *Vahl.*

Much-branched, bushy perennial with a woody base, to 30cm, glandular and scabrid. Leaves 3 (–4)-foliolate with elliptic leaflets about twice as long as broad. Petals c 5mm long, yellow. Capsule c 10 × 3mm (excluding the persistent 4mm long style), very shortly stipitate, straight, green; seeds glabrous.

A local plant of sandy places from sea level to 2300m; rare on dunes in the Tihama and usually found close to the sea: Hodeida – Manẓar; Ṣalīf; below Ḥillah (Schweinfurth); Luḥayyah (Forsskal); much more frequent along sandy drainage lines in the Mashriq but absent from Baydā' province.

4. Cleome monophylla *L.*

Erect or sprawling, glandular and pubescent annual herb to 30cm. Leaves simple, lanceolate. Petals c 3mm long, mauve. Capsule c 60 × 2mm, long-beaked, stipitate, ± straight, densely glandular with parallel green veins; seeds glabrous.

A very rare plant of grassy field borders and roadsides in areas of good rainfall; Ibb (c 1800m) and Suq al Jum'ah, J. Ash Shirq (c 2000m).

5. Cleome paradoxa *R. Br.*

Erect glabrous annual to 1½m. Leaves 5-foliolate with linear leaflets. Flowers very showy and attractive; sepals c 6–8mm long, ± persistent; petals yellow with a purplish limb, c 25mm long. Capsule c 150 × 3.5mm, beaked, long stipitate (c 20mm), slightly curved, green; seeds pubescent.

A very rare plant of arid rocky country, only recorded

after the heavy autumnal rains of 1977; between Jūbah and Ḥarīb (H. Haig-Thomas and P. Heath); Barḥ to Mafraq.

6. Cleome scaposa *D.C.*

Erect, scabrid, sparsely glandular annual to 30cm, usually simple but becoming much branched after heavy rain. Leaves simple, broadly ovate, about as broad as long. Petals c 3mm long, dull yellow with a mauve midrib. Capsule c 25 × 1mm, unbeaked, not stipitate, straight or somewhat curved, glabrous, green; seeds glabrous.

Locally common in sand in the Tihama. Also a frequent constituent of the flora of the foothills and rainshadow areas of the central escarpment where it occurs on open stony hillsides up to 1500m.

7. Cleome schweinfurthii *Gilg.*

C. arabica sensu Deflers

Erect, glandular, foetid perennial from a woody base to about 80cm. Leaves 3-foliolate with oblong leaflets about 9mm broad. Petals c 3mm long, yellow with strong purple veins. Capsule c 80 × 4mm, beaked, stipitate, ± straight, glabrous, light green with raised parallel veins; seeds pilose.

Widespread but scattered on the escarpment and high plateau from c 1500 to 2800m, almost always on exposed volcanic rock and frequently on cindery lavas of recent origin.

Outside southwest Arabia this plant is only known from the Red Sea Hills of Ethiopia and Sudan.

8. Cleome viscosa *L.*

Erect, glandular annual to about 60cm. Leaves 5-foliolate with oblong leaflets up to 15mm broad. Petals c 11mm long, yellow. Capsule c 70–80 × 3mm, beaked but not stipitate, ± straight, strongly erect, green with raised parallel veins; seeds glabrous.

A common weed of fields with rich alluvial soil between 200 and 1500m, mainly found on the silt plain in the east of the Tihama and in irrigated fields along the wadis in the escarpment foothills.

3. GYNANDROPSIS *DC.*

1. Gynandropsis gynandra (*L.*) *Briq.* 'Unīs

Pedicellaria pentaphylla (L.) Schrank.

Erect, branched, glandular annual to 70cm. Leaves 5-foliolate. Petals pale purple with a long filiform claw; stamens attached to an elongated androgynophore. Capsule c 70 × 2mm, very long stipitate (c 20mm), very shortly beaked, straight or curved.

A weed, apparently preferring nitrogen-rich soils as it is frequently found in villages around dung heaps and sewage outlets; frequent c 100–1500m on the plain at the east of the Tihama and in the wadis of the escarpment foothills; also in irrigated fields at Mārib.

4. MAERUA *Forssk.*

The genus was named by Forsskal from the Arabic name, 'Miru', which is used for *Maerua crassifolia* in the Tihama. It consists of unarmed trees, shrubs and climbers with distinctive dark green, coriaceous, simple or 3-foliolate leaves. Flowers axillary or terminal, solitary, in clusters or racemes but always with the sepals united into a persistent tube. Petals 0 or 4, pale yellow-green. Stamens numerous. Fruit globose or cylindrical, torulose.

The genus is in much need of revision. I have found it almost impossible to relate our species to described species in Africa. I have therefore resorted to labelling four species A, B, C, D. The problem is compounded by the fact that most of these species are very local and have rarely been collected in both flower and fruit.

1. Leaves all l-foliolate — 2
   Leaves all or mostly 3-foliolate — 4
2. Leaves very small, not exceeding 10mm long and commonly much less — *M. crassifolia*
   Leaves exceeding 10mm long — 3
3. Scrambler. Leaves oblong with a rounded base — *M. oblongifolia*
   Erect shrub. Leaves elliptic with a weakly cordate base — *M. sp. A*
4. Whole plant pubescent. Leaf apex hooded — *M. sp. B*
   Plant with upper leaf surface glabrous. Leaf apex not hooded — 5
5. Undersurface of leaves strongly reticulate, glabrous — *M. sp. C*
   Undersurface of leaves not reticulate, minutely puberulent — *M. sp. D*

1. Maerua crassifolia *Forssk.* Mirū, Siraḥ

Tree to 5m with smooth grey bark, typically of gnarled aspect with short chewed-off branches, but in protected places producing long, thin, somewhat pendulous branches. Leaves simple, 5–10mm long, obovate with a cuneate base and usually a rounded apex. Flowers without petals. Fruit up to 5cm long, strongly torulose.

Plants from the Tihama (Wood 75/226, 1182) seem always to have glabrous leaves whereas plants from the Mashriq (Wood 3153) have pubescent leaves and fruit.

An important constituent of scrub in the Tihama especially on stony plain south of the Wadi Zabīd and along the foothills further north; virtually absent from sand. It is also one of the most frequent trees in the Mashriq reaching about 1800m. It is one of the most drought-resistant of all Yemen trees.

2. Maerua oblongifolia (*Forssk.*) *A. Rich.*

Scrambling shrub, climbing to 3m on other bushes. Leaves simple, up to 5 × 3cm, oblong with a rounded base and an obtuse, apiculate apex. Flowers with petals. Fruit up to c 2cm long, torulose.

Frequent in the protection of *Acacia* and other bushes in dry rocky country in the foothills of the western escarpment, the southeastern mountains and the Mashriq reaching about 1800m northeast of Rada'.

3. Maerua sp. A

Bush to 2m. Leaves simple, c 4 × 3cm, elliptic with a weakly cordate base and a retuse, apiculate apex. Flowers without petals. Fruit up to 10cm long, torulose.

This is represented by Wood 2425 from Al Ḥaraḍah, Damt, c 1900m, Wood 2425 from Rada', c 2100m, and an unnumbered collection by Westinga from about 30km north of Rada' at c 1800m. In the first two localities it was found on barren volcanic rock slopes.

M. sp. A appears to be identical to plants from Somalia which have been called *M. thomsonii* T. Anders. but they do not agree well with the type material of that plant.

4. Maerua sp. B

Small tree to 3m, pubescent in all its parts. Leaves 3-foliolate; leaflets up to 9 × 2cm, narrowly elliptic with a cuneate base and an apiculate, hooded apex, strongly reticulate on both sides. Flowers with petals. Fruit not seen.

Found once only, near a village at c 1500m between 'Alūjah and Sha'af, J. Raymah (Wood 2885).

5. Maerua sp. C

Small glabrous tree to 5m. Leaves 3-foliolate; leaflets up to 8 × 3cm, oblong with a cuneate base and an obtuse or retuse apex, strongly reticulate below. Flowers with petals. Fruit globose.

This is represented by two collections: Hepper & Wood 5992 from the wadi above 'Alūjah, J. Raymah at 600m and Wood 946 from the Turbah cliffs at 1900m. However, it is almost certainly more frequent and field records from the cliffs at Bāb Baḥal, Ḥarāz and above Suq Ar Ribāṭ, J. Raymah, as well as Schweinfurth's record of *M. variifolia* Gilg. & Bened. from above Ḥillah, J. Bura', probably belong here.

6. Maerua sp. D

M. cylindricarpa sensu Schwartz

Small bush to 2m. Leaves frequently both 1 and 3-foliolate; the leaflets up to 6 × 2.5cm but commonly less, oblong-lanceolate with a cuneate base and an obtuse apiculate apex, minutely but densely pubescent below. Flowers with petals. Fruit club-shaped.

Abundant between 1300 and 1800m in the *Euphorbia* scrub from Taiz north to Mikhlāf and Qā'idah. There are many collections from Taiz including Scott and Britton 293, Wood 74/329, 1519 etc. A field record from Najd Dhamrān, Qabayṭah probably belongs here as does Wood 3082 from Rada' which, however, differs from the Taiz plant in a number of small details.

5. CADABA *Forssk.*

This genus is called after the Yemen plant name 'Qaḍab' which is the local name for *Cadaba rotundifolia*. It consists of trees and shrubs with entire leaves. The flowers are in short racemes or corymbs and petals may be present or absent. The fruit is a cylindrical torulose pod. The genus is characterised by the long ligulate appendage to the gynophore, known as the nectary.

All species found in Yemen are strongly drought-resistant.

1.	Leaves glandular hairy	*C. glandulosa*
	Leaves eglandular	2
2.	Leaves grey-canescent	3
	Leaves green, glabrous when mature	4
3.	Leaves longer than broad, up to 10mm wide	*C. farinosa*
	Leaves about as broad as long, c 25mm wide	*C. heterotricha*
4.	Leaves orbicular	*C. rotundifolia*
	Leaves lanceolate	*C. longifolia*

1. Cadaba farinosa *Forssk.*

Small shrub to 2m, frequently growing in the protection of other bushes. Leaves c 15 × 8mm but very variable in size, oblanceolate, obtuse or sometimes mucronate, grey-canescent. Flowers in few-flowered racemes with pale greenish yellow petals, much visited by insects; stamens inserted about half-way from the base of the gynophore. Fruit c 2–2.5cm long, strikingly orange inside.

Common throughout the escarpment foothills up to about 1800m, though perhaps more typical of *Acacia-Commiphora* than of *Euphorbia* scrub. It also occurs rarely in the Mashriq where it is perhaps overlooked as it occurs in virtual desert conditions at Barāqish in the Jawf.

2. Cadaba glandulosa *Forssk.*

Low glandular shrub to 1½m. Leaves c 15 × 13mm, orbicular to broadly elliptic, apiculate, grey-green. Flowers solitary or in few-flowered racemes, lacking petals; nectary prominent, orange, c 15mm long. Fruit c 1.5cm long.

A drought-resistant shrub of a few areas of sand in the Tihama, mostly east of Hodeida, but more frequent on the more arid foothills of the escarpment, reaching to about 1400m.

3. Cadaba heterotricha *Stocks* 'Uknī

A small tree to 5m, often with a gnarled twisted trunk and grey canescent branches. Leaves c 2.5 × 3cm, orbicular-obovate with an obtuse, often emarginate apex. Flowers in terminal corymbs, with petals; nectary 2-lipped, brown. Fruit c 2cm long.

A local tree of cliff faces in dry country between 500 and 1400m. Frequent in Taiz province but much rarer elsewhere along the escarpment north to the Wadi Mawr region. In the Mashriq known from Al 'Ashshah and the Naqīl al Ḥajlah between Jūbah and Ḥarīb.

4. Cadaba longifolia (*R. Br.*) *DC.*

Small glabrous shrub with long slender branches to 2m. Leaves up to 6 × 1cm, narrowly lanceolate, acute or mucronate, bright green. Flowers in terminal corymbs; petals reddish; stamens inserted about half-way from the base of the gynophore. Fruit c 1cm long, glandular.

A rare shrub of arid foothills found between 500 and 800m: Wadi Rusyān, Maqbanah; on hills east of Bājil; Dayr Khushm.

5. Cadaba rotundifolia *Forssk.* Qaḍab

Low shrub to 2m, glabrous except for the young shoots which are pubescent. Leaves up to 3cm long, orbicular, dark green. Flowers in terminal racemes, without petals; nectary yellow. Fruit up to 6cm long.

Frequent on sand in the Tihama and locally forming communities on dried-up silt laid down by ancient wadis, particularly where there are indications of salinity. Less common in the drier foothills but reaching 1000m.

6. BOSCIA *Lam.*

Small trees with simple coriaceous leaves. Flowers in small racemes, yellow-green, scented; petals absent. Fruit globose, hard.

Leaves 2–5cm long, reticulate below	*B. angustifolia*
Leaves 1–2cm long, not reticulate	*B. arabica*

1. Boscia angustifolia *A. Rich.* Thū', 'Āmas

Small tree to 4m, usually with a stout, characteristically grey-barked trunk. Leaves oblanceolate, usually with a mucronate apex, up to 5cm long, yellow-green, the under surface reticulate and often minutely pubescent. Flowers in small axillary racemes, usually seen in the autumn.

Frequent from 300 to 1800m on exposed hillsides in the foothills and on the lower slopes of the western escarpment from 'Udayn north to Khawlan Ash Sham. Largely absent from the valley forest communities but tolerant of considerable drought.

2. Boscia arabica *Pestalozzi*

Similar to the previous species but leaves cuneate with an emarginate apex, glabrous, grey-green and not exceeding 2cm long.

Only known from a single collection (Wood 1026) from the Wadi Rusyān, Maqbanah, at c 700m.

This species is endemic to southern Arabia being known from Maqbanah east to Dhofar.

7. CAPPARIS *L.*

Shrubs with simple leaves and stipular spines.

1.	Petals red. Leaves soon deciduous	*C. decidua*
	Petals white. Leaves persistent	**2**
2.	Leaves glabrous, suborbicular	**3**
	Leaves tomentose below, elliptic	*C. tomentosa*
3.	One sepal much larger than the others	*C. cartiliginea*
	Sepals ± equal	*C. spinosa*

1. Capparis cartiligenea *Decne.* Laṣaf, Niṣaf

C. galeata Fres.

Low, spreading, somewhat glaucous shrub to 2m. Leaves orbicular-broadly oblong, mucronate, up to 3 × 4cm. Flowers solitary; petals white, becoming pink with age; one sepal much larger than the others. Fruit club-shaped.

Widespread on the escarpment, the southeastern mountains and in the Mashriq from 500 to 2000m, but always on cliffs or at least steep rocky slopes.

2. Capparis decidua (*Forssk.*) *Edgw.* Sidad

Yellow green shrub to 4m with long leafless spiny branches. Flowers in axillary clusters of 3–5, red. Fruit cherry-like, edible.

Common on silt beds in the Tihama where it often forms impenetrable thickets. It is rare both on sand dunes and the stony steppe in the south of the Tihama. Not seen above 500m.

3. Capparis spinosa *L.*

Trailing shrub with branches 2m long. Leaves orbicular-elliptic c 3 × 2.5cm. Flowers solitary; petals white; sepals ± equal in size. Fruit club-shaped.

Plants from Yemen have mucronate leaves and somewhat stiff pubescent branches and may be referable to var. **canescens** *Boiss.*

A rare plant of sandy plain, often near irrigated cultivation between 1000 and 2300m in the Mashriq; Mārib; Barāqish (Jawf); Ḥūth; Banī al Ḥārith (Miller).

4. Capparis ? tomentosa *Lam.*

Woody scrambling shrub, forming dense thickets. Leaves elliptic, obtuse, 3–6cm long, the undersurface usually softly tomentose. Flowers solitary in the leaf axils or, more commonly, in terminal racemes, petals white with red veins, stamens pink. Fruit round, orange.

There is some doubt as to the identity of this plant as it has not been seen in flower in Yemen.

Only known from Wadi Sukhnah in Ḥaymat ad Dakhilīyah where it grows on river cliffs at 1500m.

CRUCIFERAE

Herbs with alternate leaves which lack stipules. The flowers are usually in racemes. The family is usually easily recognised by the 4 free sepals and petals, 6 stamens and dehiscent pod-like capsules.

The family is centred on the Mediterranean and southwest Asia. In Yemen there are few African species but several characteristic of the Sahara-Sindian desert region. The majority however are essentially Mediterranean and occur as weeds and casuals of varying frequency.

1.	Fruit less than 3 times as long as broad. Shape various, but never parallel-sided	**2**
	Fruit more than 3 times as long as broad. Commonly parallel-sided though sometimes constricted between the seeds	**7**
2.	Flowers yellow. Flowering racemes up to 50cm long	7. *Rapistrum*
	Flowers mauve or white. Flowering racemes less than 30cm and commonly very short	**3**
3.	Fruit ± sessile	13. *Anastatica*

Fruit pedicellate **4**
4. Fruit triangular-obcordate 12. *Capsella*
Fruit ovate-orbicular **5**
5. Flowers mauve. Pod up to 20mm in diameter 8. *Schouwia*
Flowers white. Pod less than 5mm in diameter **6**
6. Leaves amplexicaul. Fruit ± cordate 11. *Cardaria*
Leaves petiolate. Fruit not cordate 10. *Lepidium*
7. Leaves trifoliolate 15. *Cardamine*
Leaves simple, pinnatisect or pinnate, never trifoliolate **8**
8. Leaves amplexicaul **9**
Leaves petiolate, rarely sessile above and then never amplexicaul **11**
9. Plant stellate hairy. Flowers white. 16. *Arabis*
Plant ± glabrous, except below, never stellate hairy. Flowers yellow or mauve. **10**
10. Flowers yellow. Pod c 3–5cm long 1. *Brassica*
Flowers mauve. Pod c 5–8cm long 9. *Moricandia*
11. Leaves entire **12**
Leaves serrate, sinuate lobed, pinnatisect or pinnate **15**
12. Plant stellate hairy **13**
Plant with simple or branched hairs, but not stellate hairy **14**
13. Leaves oblanceolate. Fruit borne on a slender pedicel much narrower than the fruit itself 21. *Eremobium*
Leaves elliptic. Fruit borne on a stout pedicel about as thick as the fruit itself 20. *Morettia*
14. Flowers orange or purplish. Fruit entire at the apex 14. *Farsetia*
Flowers white. Fruit bifid at the tip 19. *Notoceras*
15. Flowers white, mauve or dark-veined **16**
Flowers pure yellow **19**
16. Glabrous or nearly glabrous aquatic plant with white flowers. Leaves pinnate 17. *Nasturtium*
Hairy plants, often with dark-veined flowers, not usually in aquatic habitats. Leaves not pinnate **17**
17. Fruit constricted between the seeds, breaking transversely 6. *Raphanus*
Fruit not constricted between the seeds, opening longitudinally **18**
18. Pod terete, long-beaked; pedicel up to 4mm long 5. *Eruca*
Pod flattened, short-beaked; pedicel more than 4mm long 4. *Diplotaxis*
19. Seeds in two rows in the pod **20**
Seeds in one row in the pod **21**
20. Pod short, less than 1cm long, not stipitate 18. *Rorippa*
Pod more than 2cm long, clearly stipitate above the calyx 4. *Diplotaxis*
21. Plant grey stellate-tomentose, fruit breaking transversely 22. *Sterigmostemum*
Plant glabrous or hairy, but not grey or stellate-tomentose. Fruit opening longitudinally **22**
22. Pod 3–6 seeded, appressed against the axis 3. *Hirschfeldia*
Pod more than 6 seeded, not appressed to the axis **23**
23. Pod with a beak 5–10mm long 1. *Brassica*
Pod beakless or with a short beak less than 5mm long **24**
24. Fruit beaked. Fruit valves 1-veined 2. *Erucastrum*
Fruit not beaked. Fruit valves 3-veined 23. *Sisymbrium*

1. BRASSICA *L.*

Annual herbs with a long tap root. Flowers in terminal racemes, yellow. Fruit a torulose pod with a long beak.

All species found in Yemen are of cultivated origin and only two of the following, *B. juncea* and *B. napus*, have been found as casuals.

Upper leaves petiolate *B. juncea*
Upper leaves sessile *B. napus*

1. Brassica juncea (*L.*) *Czern.*

Green annual herb to 1m distinguished from all other *Brassica* species in Yemen by the petiolate stem leaves. The lower leaves are lyrate – pinnatifid, the upper leaves ± entire. Petals yellow, 6–9mm long.

A rare casual known only from Sana'a (Wood 2284).

2. Brassica napus *L.* Khardal

Glaucous annual herb to about 1m. The leaves are ± glabrous, the lower pinnatifid and petiolate but the upper entire and clasping. Petals yellow, 10–18mm long, the open flowers not overtopping the buds.

Commonly cultivated between 2200 and 3000m on the escarpment and high plateau for the oil that can be obtained from its seeds (Rape seed oil). It occasionally persists as a weed of cultivated and waste land.

Several other species of *Brassica* are cultivated but have not been observed as escapes. **Brassica oleracea** *L.*, 'Kubaj', is grown as a vegetable (cabbage, cauliflower) particularly around Sana'a. **Brassica rapa** *L.* (*B. campestris L.*), the turnip, was recorded around Sana'a by earlier botanists but has only been seen recently at Jiblah (Hepper). It is easily confused with *B. napus* but the leaves are usually bristly and the open flowers overtop the buds on the inflorescence. The Chinese cabbage, **Brassica chinensis** *L.* has been recently introduced and is grown in the Sana'a area.

2. ERUCASTRUM *C. Presl*

Annuals or perennials. Flowers in terminal racemes, yellow. Pod linear, torulose, shortly beaked.

Racemes bare or with a few leaves below, 8–25 flowered *E. arabicum*
Racemes bracteate to the apex, 20–35 flowered *E. abyssinicum*

1. Erucastrum abyssinicum (*A. Rich.*) *O.E. Schulz*

Very similar to *E. arabicum* in which it has often been included but differing in the bracteate racemes which are leafy to the very top and bear 20–35 bright yellow flowers.

Only known from a single collection made between Sana'a and Thilā (Steinberg 800).

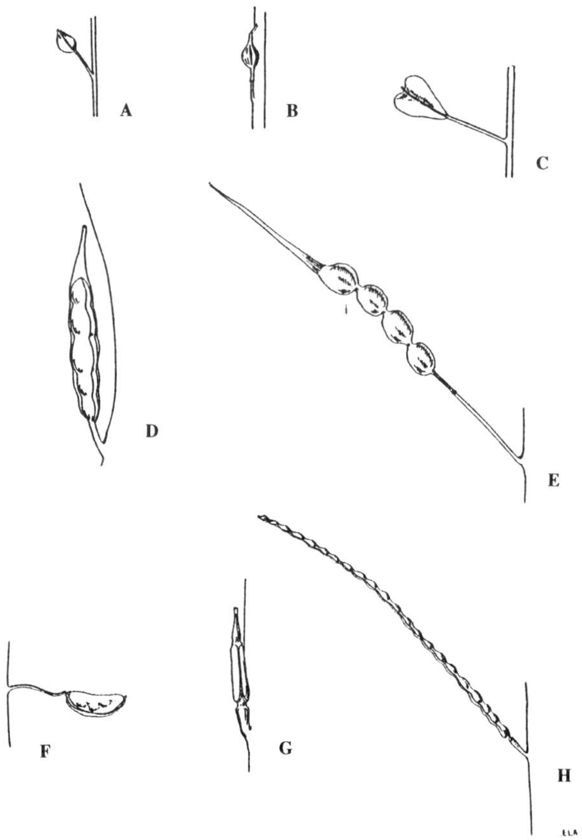

Fig. 9. Pods of Cruciferae. **A** *Lepidium armoracea*; **B** *Rapistrum rugosum*; **C** *Capsella bursa-pastoris*; **D** *Eruca sativa*; **E** *Raphanus raphanistrum*; **F** *Rorippa palustris*; **G** *Hirschfeldia incana*; **H** *Sisymbrium irio.*

2. **Erucastrum arabicum** *Fisch. & Mey.*

A very variable, erect or decumbent annual or perennial herb which may be glabrous or hispid. Lower leaves pinnatifid, upper leaves usually simple. Racemes with up to 25 pale yellow flowers, usually completely leafless but sometimes with a few bracts near the base. Pod up to 30mm long, the seeds in 1 row.

Four varieties can be distinguished in Yemen: *

Variety (a). A decumbent or ascending annual, almost glabrous except for a few stiff hairs. Almost all leaves pinnatifid, the terminal lobe oblong-lanceolate. The majority of pods straight, c 30mm long, over-topping the flowers. (Wood 1121, 2060)

The commonest weedy crucifer in Yemen, found from 1000–2500m in arable land east to Ḥarīb but less common in villages and small domestic gardens than in open fields.

Variety (b). Glabrous, dark green perennial with numerous, ascending, usually entwined stems. Almost all leaves pinnatifid, the terminal lobe lanceolate. Flowers numerous. Pods very narrow, strongly torulose, curving upwards, c 15mm long, never overtopping the flowers. Plant not very fertile, the pods often not forming properly. (Wood 75/68, 2646).

On cliffs from 2100 to 2800m: Akāḥilah near Turbah; Dimarayn, J. Ṣabir; J. Ḍawrān.

Variety (c). Hispid, green perennial with erect stems. Basal leaves coarsely pinnatifid, the terminal lobe rounded, forming the widest part of the leaf. Upper stem leaves simple. Pods strongly torulose, curved upwards. (Wood 1665, 2252, 3069)

On cliffs 2700–3100m: J. Sumārah; Kuḥlān; J. Mada'.

Variety (d). Erect, sparsely hispid green perennial. Basal leaves pinnatifid; terminal lobe narrowly ovate. Upper stem leaves simple. Flowers large. Pod torulose, strongly curved downwards. (Wood 1674)

Known from sandstone cliffs at Kawkabān, c 2900m.

3. HIRSCHFELDIA *Moench*

1. **Hirschfeldia incana** (*L.*) *Lagrèze-Fossat*

Branched annual to 1m, the stem covered with short white hairs below. Leaves pinnatifid with a large elliptic terminal lobe. Flowers yellow. Pod 10–15mm long, beaked, erect and ± appressed to the stem. Seeds 3–6 in a single row.

Casual on a roadside at Ma'bar, c 2400m.

4. DIPLOTAXIS *DC.*

Annual or perennial herbs with leafless racemes. Stigma 2 lobed. Fruit a flattened pod with seeds in two rows, beak very short or absent.

1.	Flowers white, rarely mauve, or white with mauve veins	*D. erucoides*
	Flowers yellow	2
2.	Plant hairy. Pod pendent	*D. harra*
	Plant ± glabrous. Pod slightly ascending	3
3.	Leaves pinnatifid	*D. tenuifolia*
	Leaves obovate, coarsely serrate but not pinnatifid	*D. sp. A*

1. **Diplotaxis erucoides** (*L.*) *DC.* Ḥalqab

Sparsely hispid annual, usually branched at the base with several erect stems 30–50cm high. Leaves mainly basal, coarsely pinnatifid. Flowers usually white, but often pale mauve or with mauve veins. Pods 20–40mm long with a short conical beak, spreading to slightly ascending.

A common field weed of the high plateau, especially round Sana'a; rare elsewhere between c 1700 and 2900m and known only from Ḥayfān (Deflers), the Turbah plateau and Manākhah.

2. **Diplotaxis harra** (*Forssk.*) *Boiss.*

Hispid perennial, branched at the base with several erect stems 30–70cm high. Leaves obovate, coarsely toothed. Flowers yellow. Pod cylindrical, 30–40mm long, pendulous, stipitate, not beaked.

Recorded in four widely scattered localities between 2000 and 2300m in the high plateau and the Mashriq: Ṣa'dah; Bayt Baws near Sana'a; Sanḥān; Sanabān. It was growing on open stony slopes along drainage lines.

3. **Diplotaxis tenuifolia** (*L.*) *DC.*

Glabrous, somewhat glaucous perennial with an erect,

*Recent research has clarified these varieties: Variety (a) is *Erucastrum arabicum*; Variety (b) is *Erucastrum woodiorum* Jonsell; Varieties (c) and (d) are *Erucastrum meruensis* Jonsell subsp. *yemenense* Jonsell.

branched stem to c 80cm. Lower stem leaves deeply and narrowly pinnatifid with ± linear lobes; upper stem leaves linear-lanceolate. Flowers yellow. Pod cylindrical, 20–50mm long, ascending, stipitate, shortly beaked.

A casual, found at 2600m at Ḥuṣnah, Aḥjar.

4. **Diplotaxis sp. A** *

Glabrous perennial from a woody rootstock, with many branched stems to c 50mm. Leaves obovate, coarsely toothed. Flowers yellow. Pod cylindrical, 25–35mm long, ascending, stipitate, with a persistent beak c 2mm long.

A very rare endemic plant of limestone cliff ledges at c 2700m above Kuḥlān and on J. Maḥdad. (Wood 2250, 2518, 2691; Miller 236)

5. ERUCA *Miller*

1. **Eruca sativa** *Miller* Jirjīr

E. vesicaria (L.) Cav. ssp *sativa* (Miller) Thell.

Erect, sparsely hispid annual to 50cm. Leaves lyrate-pinnatifid with a large terminal lobe, coarsely dentate. Petals pale yellow or white with dark veins. Pod 15–25mm long, erect, with a long beak c 5mm long; seeds in 2 rows.

A recent introduction from Egypt which is cultivated by Egyptian teachers wherever there is a school. It is now becoming popular as a salad vegetable among Yemenis and occurs as a casual, especially around Taiz and Sana'a.

6. RAPHANUS *L.*

1. **Raphanus raphanistrum** *L.*

Wild Radish. Erect, branched annual herb to 60cm with a slender taproot, bristly especially below. Lower leaves pinnatifid; upper leaves coarsely toothed. Flowers yellow or white with dark veins. Pod up to 50mm long including the 20mm long conical beak, ribbed, strongly constricted and jointed between the seeds, breaking up transversely at the joints when mature.

A rare casual of wet places in the high plateau: roadside ditch, Yāzil, at 2800m (yellow-flowered form); in the gorge above Tin'am at 2500m (white-flowered form).

The cultivated radish, **Raphanus sativus** *L.* ('Fijil', 'Baql';), is grown in nearly every Yemeni village, except at the coast. The common cultivar has lilac flowers and a long white taproot. Both the leaves and the root are eaten as a salad.

7. RAPISTRUM *Crantz*

1. **Rapistrum rugosum** (*L.*) *All.*

Annual with an erect, usually much-branched stem to 60cm, hispid below but glabrous above. Lower leaves pinnate; upper leaves simple, coarsely dentate. Flowers yellow. Fruit 4–10mm long consisting of two segments, the lower cylindrical, the upper globose, beaked. Plants are not usually very fertile in Yemen and most flowers abort.

All plants from Yemen seem to belong to ssp **orientale** (*L.*) *Arcangeli*.

A Mediterranean plant found as a casual in four places between 2000 and 2600m: Turbah; Saddah in the Wadi Banā; Al Rawḍ in Wadi Ḍahr; near Ad Dan, Waṣāb al 'Ālī. In all localities it was found near markets.

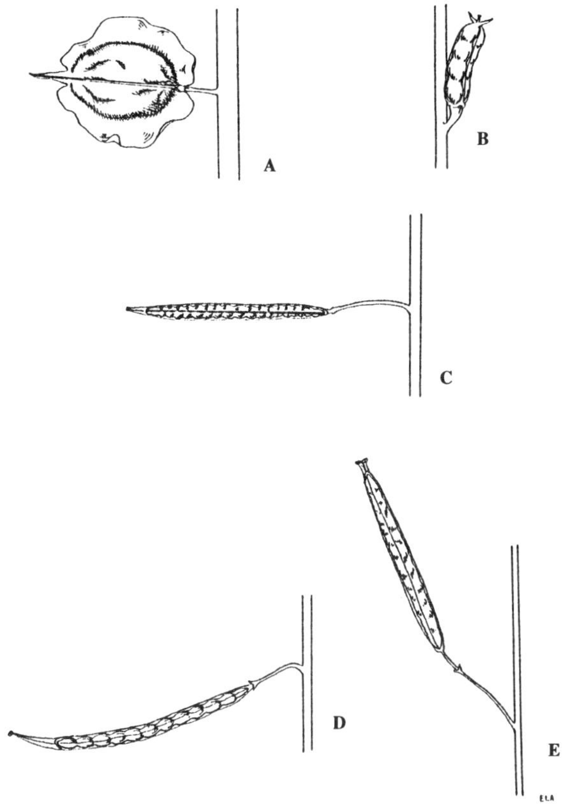

Fig. 10. Pods of Cruciferae. **A** *Schouwia purpurea*; **B** *Notoceras bicorne*; **C** *Diplotaxis erucoides*; **D** *Erucastrum arabicum*; **E** *Diplotaxis sp. A.*

8. SCHOUWIA *DC.*

1. **Schouwia purpurea** (*Forssk.*) *Schweinf.* Fayjil

Glabrous, branched, woody annual to 1m. Leaves amplexicaul, fleshy, ovate, simple, obscurely sinuate, up to 10cm long and 8cm wide. Flowers mauve. Fruit orbicular, compressed, winged, 10–20mm wide, the apex cuspidate with a beak c 5mm long.

An abundant weed of sandy fields in the Tihama extending infrequently into the escarpment foothills up to 1400m.

9. MORICANDIA *DC.*

1. **Moricandia sinaica** (*Boiss.*) *Boiss.*

Glabrous branched perennial to 1m. Leaves clasping, ovate, entire, glaucous, fleshy, up to 8cm long. Flowers mauve. Pod cylindrical, c 8cm long with a short beak.

Persistent in three roadside localities between 1800 and 2500m: Ḍawrān; between Khamr and Ḥūth; Ḥarf Sufyān.

* Recently described as *Diplotaxis kohlaanensis* A.G. Miller & J.A. Nyberg

10. LEPIDIUM *L.*

Annual or perennial herbs, usually smelling of cress. Flowers small, white, in dense terminal racemes. Fruit orbicular – elliptic, usually 2-seeded.

1.	Leaves ovate, at least 10mm wide, simple	*L. latifolium*
	Leaves linear, c 2–3mm wide, pinnatisect below	2
2.	Perennial. Fruit acute at the apex	*L. armoracia*
	Annual. Fruit retuse at the apex	3
3.	Plant glabrous	*L. sativum*
	Plant hairy	*L. bonarense*

1. Lepidium armoracia *Fisch. & Mey.*

L. schweinfurthii Thell.

Sparsely pubescent, prostrate or ascending perennial to 15cm, usually much branched. Basal leaves pinnatisect; stem leaves linear to lanceolate, simple, entire or dentate, c 20mm long and 3mm wide. Fruit c 3mm long, unwinged, acute.

Schweinfurth's collection from the Turkish Hospital at Manākhah was divided by Thellung into 3 taxa but I agree with recent studies by Jonsell that consider these all to be conspecific. The prostrate form with short stems and entire leaves is found in exposed places at high altitudes on J. Sumārah and J. Nabī Shu'ayb (Wood 72/75). More bushy branched plants are found in more protected places at lower altitudes (Wood 684, 2253).

A plant of waste ground and tracks, in and around villages between 2400m and 3600m. It occurs in scattered localities along the inner escarpment and the western part of the high plateau but is known elsewhere only from Shahārah and Manākhah.

Outside Yemen it is restricted to Ethiopia and Tanzania.

2. Lepidium bonarense *L.*

Erect, unbranched, pubescent annual to 30cm, resembling a robust form of the last species. All leaves pinnatisect. Fruit c 3.5mm long, winged, retuse at the apex.

Originally from America, this introduced species is only known from a trackside at 2100m at Akāhilah near Turbah.

3. Lepidium latifolium *L.*

Much-branched perennial with glabrous stems to 1m. Lower leaves broadly ovate to lanceolate, serrate, simple, up to 30cm long. Upper leaves sessile, entire and ± bract-like. Fruit c 2mm long, unwinged, obtuse, hairy.

Recorded by Deflers from a number of places around Sana'a. Probably cultivated and not collected since.

4. Lepidium sativum *L.* Ḥalf, Ḥarf

Cress. Erect, unbranched, glabrous annual to 40cm. Leaves pinnate below, pinnatisect above. Fruit 5–6mm long, broadly winged, notched.

Cultivated at high altitudes on the escarpment and in scattered places on the high plateau. Occasionally persisting as a weed for a year or two. It is much valued in traditional medicine.

11. CARDARIA *Desv.*

1. Cardaria draba (*L.*) *Desv.* Qumbār

Erect pubescent perennial to 40cm. Leaves obovate, coarsely toothed, clasping. Flowers white. Fruit borne on a pedicel c 10mm long, cordate, 3–4mm long, the style persistent.

Probably introduced fairly recently and occurring as a scattered weed of roadsides and fields between 2200 and 2900m on the high plateau and at Manākhah.

12. CAPSELLA *Medic.*

1. Capsella bursa-pastoris (*L.*) *Medic.*

Shepherd's Purse. Erect annual to 30cm, glabrous or with simple or stellate hairs. Leaves oblanceolate, entire or coarsely toothed, the stem leaves clasping. Petals white. Fruit borne on spreading pedicels, triangular, obcordate, 4–9mm long.

A cosmopolitan weed found in fields and gardens from 2200 to 3200m on the higher escarpment mountains and on the high plateau.

13. ANASTATICA *L.*

1. Anastatica hierochuntica *L.* Kāf, Jumays

Low-growing, grey stellate-hairy woody annual, persistent for many years after dying. Leaves petiolate, oblong, coarsely dentate. Flowers small, white, sessile. Fruiting plants larger with branches incurving when dry to form a fist-like ball. Fruit sessile, ovate with two broad ear-like appendages on either side of the persistent style.

This plant is strongly hygroscopic; the plant spreads out when wet and contracts into a fist when dry. This effect can be seen on dried specimens many years old. The Arabic word 'Kāf' means 'fist' but I have never heard the name 'Kāf Miriam' in Yemen. In Najd this name is widely used in allusion to Mary's clenched fist during her labour pains in the birth of Jesus.

A local plant of open gravelly plain between 1200 and 2300m in the Mashriq.

14. FARSETIA *Turra*

Perennial undershrubs, usually grey in colour from the numerous adpressed whitish hairs. Leaves simple, sessile, entire. Flowers in racemes with orange or purple petals that mostly open at night. Fruit a long pod-like capsule, flattened, shortly beaked, the seeds usually in two rows.

The following 3 species are all of similar habit and appearance and may be easily confused.

1.	Pod up to 2.5mm wide	*F. linearis*
	Pod 2.5–8mm wide	2
2.	Sepals less than 6mm long	*F. stylosa*
	Sepals 6–10mm long	*F. longisiliqua*

1. Farsetia linearis *Decne.*

A little-known plant differing from both the other species in its thick, leggy stems and long, narrow pods which are 3–4cm long and 1.5–2.5mm wide. The sepals are 4–7mm long and the petals c 10mm long and the style is usually long.

A rare plant of stony hillsides and plain found in a few localities between 1900 and 2500m on the high plateau and in the Mashriq: northeast of Dhamār (Acres); at the head of the Wadi Sirr, Bani Ḥushaysh); Majz; Marāshī (Steinberg).

2. Farsetia longisiliqua *Decne.*
Small, much branched, grey undershrub to 50cm. Leaves linear, 0.5–2cm long. Flowers orange or pinkish purple with sepals 6–10mm long and petals 10–15mm long. Pod 2–3cm long, 2.5–4mm wide.

Widespread throughout Yemen from sea level to 3000m growing in a variety of dry rocky habitats but rare in the Tihama and completely absent from the wetter parts of the escarpment and all alluvial deposits except for sand. It is one of the most typical species of the stony hills of the Mashriq.

3. Farsetia stylosa *R. Br.*
F. hamiltonii Royle
F. ramosissima Hochst. ex Fourn.
Similar to the two previous species but the sepals and petals less than 6mm long.

Known from a few scattered localities in the Tihama and in the Mashriq from sea level to 2300m: Mukhā (Wood); 15km south of Rada' (Popov); on the Mārib road northeast of Sana'a (Miller).

15. CARDAMINE *L.*

1. Cardamine africana *L.*
Low, delicate, rhizomatous perennial with scattered bristly hairs. Leaves trifoliolate, the leaflets ovate with crenate margins, the veins exserted as short mucros. Flowers white, inconspicuous, largely concealed by the conical pods which reach 20mm in length.

Found in a few places in damp, deeply shaded coffee groves in the Ja'farīyah around Ḥadīyah at 1100–1400m.

16. ARABIS *L.*

1. Arabis alpina *l.*
Mat-forming, stellate-hairy perennial with an erect stem to 25cm. Leaves obovate, coarsely toothed; basal leaves narrowed into a petiole-like base, stem leaves clasping. Flowers white. Fruit a pod c 20–30mm long borne on a pedicel 10–12mm long; seeds in one row.

An uncommon escarpment plant found on cliffs and broken stony slopes, often by mountain streams from 2700–3200m; on the Ibb mountains and from J. Nabī Shu'ayb to J. Maswar.

This is the only Arctic-alpine species found in Yemen. It occurs from Alaska, Iceland and Siberia south to the Himalayas and East African mountains.

17. NASTURTIUM *R. Br.*

1. Nasturtium microphyllum (*Boenn.*) *Reichenb.*
Water cress. Perennial with hollow glabrous shoots, rooting at the nodes but finally ascending. Leaves pinnate, leaflets sinuate-toothed. Flowers white. Pod borne on an ascending pedicel, cylindrical, curved, beaked, 13–18mm long; seeds in one row.

In swamp by the Wadi Banā between Saddah and Nādirah at 2500m.

18. RORIPPA *Scop.*

1. Rorippa palustris (*L.*) *All.*
R. islandica auct.
Glabrous, usually unbranched, erect annual to c 50cm, but commonly less. Leaves lyrate-pinnatifid, sinuate toothed, stalked below but sessile above. Flowers yellow. Pod borne on a spreading or slightly deflexed pedicel, 5–8mm long, narrowly oblong, slightly curved, with a persistent style; seeds in 2 rows.

A very local plant of mud at the edge of ponds and streams between 2000 and 2500m: near Suq al Jum'ah, J. Ash Shirq; Maḥwīt; by the Wadi Banā between Saddah and Nādirah.

19. NOTOCERAS *R. Br.*

1. Notoceras bicorne (*Ait.*) *Amo.*
Prostrate, appressed hairy perennial, with numerous, short stems about 10cm long. Leaves linear – narrowly obovate, entire, up to 20mm long. Flowers white. Pod subsessile, erect, rigid, bifid at the tip, 3–4 seeded.

Uncommon but scattered all over the drier parts of the high plateau and the western part of the Mashriq from Rada' north to Majz; usually in runoffs in sandstone country and apparently absent from limestone.

20. MORETTIA *DC.*

1. Morettia parviflora *Boiss.*
Low, grey, densely stellate-hairy, perennial, much-branched with branches entwined. Leaves petiolate, elliptic, entire, up to 25mm long and 15mm wide but commonly less. Flowers white. Pod borne on a short, rigid pedicel c 2mm long, 4-angled, straight or slightly curved, ± appressed against the stem.

This plant can be found in flower at almost any season.

Frequent on open sand and rocky plains from 1100–2300m throughout the Mashriq from Sana'a east to Mārib and from Rada' north to Majz.

21. EREMOBIUM *Boiss.*

1. Eremobium aegyptiacum (*Spreng.*) *Boiss.*
Malcolmia aegyptiaca Spreng.
Sprawling perennial with weakly ascending yellow-brown stems to 20cm. Leaves shortly petiolate, oblanceolate, entire, grey, stellate-hairy, c 25mm long. Flowers white or mauve. Pod borne on a pedicel c 5mm long, torulose, 15–25mm long.

A rare plant of sandy places in the Mashriq between 1000 and 1900m: Ghuraz near Ṣa'dah; Mārib (Common).

22. STERIGMOSTEMUM *M. Bieb.*

1. Sterigmostemum sulphureum (*Banks & Sol.*) *Bornm.*

Grey stellate-tomentose, branched annual to 80cm. Leaves oblanceolate, entire or obscurely toothed. Flowers yellow; filaments of the inner stamens joined; stigma conspicuously 2-lobed. Pod borne on ascending pedicels 5–10mm long, cylindrical, transversely jointed, 10–20mm long; seeds in one row.

A roadside casual at 2600m at Khidār.

23. SISYMBRIUM *L.*

Annual or perennial herbs. Leaves entire or pinnatifid. Flowers yellow. Pod long, slender, unbeaked; seeds in one row.

1. Petals 8–10mm long. Pods 5–10cm long — *S. orientale*
   Petals less than 6mm long. Pods less than 6cm long — **2**
2. Fruiting pedicels less than 5mm long. Pod tapering at the apex — *S. erysimoides*
   Fruiting pedicels 5–20mm long. Pod not tapering at the apex — *S. irio*

1. Sisymbrium erysimoides *Desf.*

Shortly-pubescent, sparingly-branched annual to 30cm. Leaves mostly basal, pinnatifid, serrate, the terminal lobe orbicular. Racemes 10–20cm long, the flowers small with petals 1–2.5mm long. Pods borne on pedicels less than 5mm long, spreading, tapering at the apex, 2.5–4cm long.

A locally frequent weed found in areas of plentiful mist between 1000 and 2000m in the Ḥujarīyah and the central escarpment; records from other areas have not been confirmed.

2. Sisymbrium irio *L.*

S. pinnatifidum Forssk.

Sparsely pubescent annual herb to 60cm, more branched and more vigorous than the previous species. Lower leaves pinnatifid, the terminal lobe much longer than broad; upper leaves simple. Racemes 20–40cm long, the flowers with petals 2.5–3.5mm long. Pods borne on pedicels up to 20mm long, ascending, not tapering at the apex, 3–6cm long.

S. pinnatifidum is sometimes treated as a separate species though the characters used to distinguish it seem tenuous. In any case the type has been lost and there is no means of knowing exactly what Forsskal collected.

The common weedy crucifer of altitudes between 2000 and 3200m. It is found all over the escarpment and high plateau, usually in villages and less commonly in fields.

3. Sisymbrium orientale *L.*

A sparsely pubescent, branched annual to 1m. Lower leaves pinnatifid, the terminal lobe ± lanceolate; upper leaves simple, often hastate. Racemes 40–60cm long, the flowers large with petals 8–10mm long. Pods borne on thick pedicels 3–5mm long, ascending, 4–10cm long.

A casual weed of waste ground in towns and villages between 2000 and 2900m: Turbah; Dhamār; Ḥaddah; Manākhah; Yāzil; Khamr.

Aubretia schweinfurthiana Muschl. was recorded from Yemen in error by Schwartz.

MORINGACEAE

1. MORINGA *Adans.*

1. Moringa peregrina (*Forssk.*) *Fiori* — Bān

M. aptera Gaertn.

Tree to about 10m with drooping branches. Leaves consisting of 2–5 pairs of stiff rush-like pinnae with small oblong leaflets (The leaflets are usually absent so the leaves look like virgate branchlets). Flowers in few-flowered axillary panicles, pink. Fruit a long ridged pod, up to 30cm long.

Found flowering in March in Yemen.

Originally collected at Bayt al Faqīh by Forsskal who described the species from there. It has never been re-collected in the Tihama and I only know it from stony plain north of Ḥarīb at 1100m near the edge of the Sands.

Oil can be extracted from the seeds.

Moringa oleifera *Lam.* (*M. pterygosperma* Gaertn.) is commonly cultivated in the Tihama particularly along the line of the Russian road and round the Russian Surdūd project. It is a small tree with tomentose branchlets, glaucous leaves, fragrant white flowers and long ridged pods about 30cm long.

RESEDACEAE

Herbs or shrubs with alternate stipulate leaves. Inflorescence a long terminal raceme of small yellow or cream flowers. The few Yemen species are usually found in dry habitats.

1. Shrub. Petals absent. Fruit a fleshy drupe — 4. *Ochradenus*
   Herb. Petals present. Fruit not a fleshy drupe — **2**
2. Flowers and fruit sessile — 3. *Oligomeris*
   Flowers and fruit pedicellate — **3**
3. Fruit a capsule — 2. *Reseda*
   Fruit of 6 free, 1-seeded carpels — 1. *Caylusea*

1. CAYLUSEA *A.St.-Hil.*

Herbs with undulate leaves. Flowers cream with 5 petals; fruit of 6 separate carpels.

Leaves c 6 × 45mm. Filaments c 4mm long — *C. abyssinica*
Leaves c 3 × 20mm. Filaments c 2mm long — *C. hexagyna*

1. Caylusea abyssinica (*Fresen*) *Fisch. & Mey.*

Perennial from a woody taproot with numerous ascending branches to 60cm. Leaves linear-lanceolate, very variable in size but at least 5mm wide. Flowers in racemes up to 25cm long.

Schweinfurth recorded this plant from Manākhah but almost certainly in error. Specimens of his collected in Eritrea in the 1890s are the next species which is common at Manākhah.

The only authentic record of this plant is at 1800m on the track from Al Jabin to Salafīyah. Jabal Raymah (Lyn Gillespie 90).

2. Caylusea hexagyna (*Forssk.*) *M.L. Green*
C. canescens (L.) A. St. Hil.
Very similar to the previous species from which it differs in its smaller leaves, shorter filaments and shorter racemes (up to 20cm long).

One of the most characteristic plants of the high plateau reaching 3400m on J. Nabī Shu'ayb. It is also found at Manākhah, in Khawlan Ash Sham and as a recent introduction along the Sana'a-Taiz road where it grows as low as 1700m near Qā'idah.

2. RESEDA *L.*

1. Reseda sphenocleoides *Deflers.*
Much-branched woody-based perennial to 1m. Leaves obovate, up to 5 × 2.5cm, dull green, coriaceous. Racemes of yellowish flowers up to about 6cm long, tapering gently upwards and so ± conical in shape.

Local on dry rocky hills between 1200 and 2300m. Frequent in the eastern part of the Ḥujarīyah; extending in an arc from Rada' northwest to Maḥābishah and Ṣa'dah, but absent from many apparently suitable areas.

This species is endemic to southwest Arabia.

3. OLIGOMERIS *Cambess.*

1. Oligomeris linifolia (*Vahl.*) *Macbr.*
O. subulata (Del.) Boiss.
Erect much-branched annual herb to 30cm. Leaves in clusters, glaucous, linear, c 2cm long. Flower spikes similar in general appearance to those of *Caylusea hexagyna* but flowers sessile. Fruit a deeply furrowed, sessile capsule.

An uncommon but perhaps overlooked plant of fields and drainage lines in gravelly country on the high plateau: Sana'a (Wissmann); near J. As Sama', Arḥab; Rada'; between Milāḥ and Sanabān.

4. OCHRADENUS *Del.*

1. Ochradenus baccata *Del.* Qurḍī
Yellow-green undershrub, typically 1m, high with numerous stiff, almost leafless branches, but rarely (especially in the protection of thorn bushes along wadis in the Mashriq) developing into a high climbing liana with soft green, leafy shoots. Leaves narrowly linear, up to 2cm long. Flowers yellow, in racemes up to 10cm long. Fruit a waxy white drupe.

Widespread but not very common throughout Yemen from about 200 to 3300m. It is absent from alluvial plain and all sand dune country whether in the Tihama or the east as well as areas enjoying rainfall above about 400mm per annum.

ERICACEAE

ERICA *L.*

1. Erica arborea *L.* Shandhaf
Small shrub to 1m in Yemen but reaching tree proportions elsewhere. Leaves similar to those of other heaths, simple, subulate, ascending, c 3–4mm long. Flowers white, clustered towards the tips of the branchlets; corolla bell-shaped, pendulous; style much exserted.

Unmistakable especially when in flower but very local in Yemen and restricted to areas of good rainfall with plenty of mist. It occupies three distinct areas: 1. Between 2800 and 3200m on most of the high peaks of the Ibb region (Ṣabir, Ta'kar, Manār, Sumārah) where it occurs as isolated bushes on cliffs and in short grassland. 2. By streams on sandstone rocks between 2700 and 3000m at Maḥjar and on Jabal Maswar. 3. Above 1900m on J. Raymah and J. Milḥān where it is locally abundant. It seems probable that the high peaks of J. Bārad and J. Ḥuzār in the Kusmah area originally supported communities dominated by this plant.

As with *Juniperus procera* the needle-like leaves of this plant are said to catch moisture from the winter mists. Interestingly its distribution is almost complementary to that of the Juniper which seems to avoid areas with rainfall above about 500mm a year.

SAPOTACEAE

Trees or shrubs often with latex. Leaves alternate or in apparent whorls, usually thick and leathery in texture, simple, entire, with or without stipules. Flowers in clusters, bisexual; calyx of free sepals, often in whorls; petals fused below. Fruit a berry.

1. Unarmed tree. Leaves long-petiolate, the petioles at least 2cm long — 3. *Mimusops*
   Armed or unarmed shrubs or small trees. Leaves shortly petiolate, the petioles less than 5mm long — 2
2. Leaves mostly in apparent whorls on small lateral branchlets. Pedicels 5–7mm long — 2. *Spiniluma*
   Leaves never in apparent whorls on small lateral branchlets. Flowers sessile — 1. *Monotheca*

1. MONOTHECA *A.DC.*

1. Monotheca buxifolia (*Falc.*) *A.DC.* Ẓulām
Reptonia muscatensis (A.DC.) Radlk. ex O. Schwartz
Sideroxylon buxifolia Hutch.
Shrub or small tree to 5mm, usually armed with spines but sometimes unarmed. Branchlets grey. Leaves alternate, shortly petiolate, oblanceolate, obtuse, 1–5cm long, ± glabrous and leathery, the undersurface paler than the upper surface, the veins indistinct. Flowers in dense subsessile axillary clusters, c 2.5mm long; sepals 5; corolla 5-lobed, yellow. Fruit a one-seeded berry 5–8mm in diameter.

Flowers in April.

A rare plant of dry hills between 1300 and 1900m in two distinct areas: south of Tawjān, Qābayṭah in the Ḥujarīyah;

in a rocky gully near Majz (Steinberg). Elsewhere it is a local tree of dry *Juniperus – Olea* woodland from Ethiopia through Arabia to Pakistan. In Yemen, though, it does not occur in this formation though *Juniperus* grows close by in Qābayṭah.

2. SPINILUMA (*Baill.*) *Aubrev.*

1. Spiniluma oxyacantha (*Baill.*) *Aubrev.* Jirūm
Sideroxylon oxyacantha Baill.
Spiny (or rarely unarmed) shrub to 4m with grey, finely pubescent branchlets. Leaves and flowers borne on very short specialised lateral branchlets from the base of which the spines develop. Leaves in apparent whorls on the branchlets or alternate towards the ends of the branches, shortly petiolate, oblanceolate, obtuse, 1–3cm long, ± glabrous and leathery, the undersurface paler and with distinct veins. Flowers solitary in the axils of the leaves, c 5mm long; pedicels 5–7mm long; sepals 5; corolla 5-lobed, white. Fruit a 1-seeded berry, c 7mm in diameter.

A local plant found between 1700 and 2600m around cultivation, in mountain gullies and in montane grassland in the high rainfall areas of Ibb province and the central escarpment: J. 'Ayfū; Mashwarah; between Suq As Sabt and J. Manār; at the head of the Wadi Banā above Saddah; on the western slopes of J. Raymah and J. Milḥān; Maḥwīt (F. Weber). Outside Yemen it is only known from Ethiopia.

3. MIMUSOPS *L.*

1. Mimusops laurifolia (*Forssk.*) *Friis* Libākh
M. schimperi A. Rich.
Tall tree to 40m with a rather narrow profile and a rounded crown; bark grey, vertically and horizontally fissured into small rectangles. Leaves clustered at the tips of the branchlets with petioles 2–3.5cm long, elliptic or oblong-elliptic, rather abruptly narrowed at both ends with a rounded or obtuse apex, 3–8cm long, glabrous, dark green, leathery. Flowers with pedicels c 1.5cm long, mostly 1–3 in axillary clusters, c 6mm long; sepals in 2 rows of 4, grey-pubescent; corolla 4-lobed, white. Fruit an ovoid berry, c 3cm long.

Probably the most attractive large tree in Yemen, but uncommon and found only between 500 and 1000m in the lusher valleys of the central escarpment: below Ḥadīyah (a single tree), above 'Alūjah and around Suq Ar Ribāṭ in J. Raymah; above Ḥillah and Suq As Sabt, J. Bura'; Wadi Ḥār and Rajamah near Ḥajjaylah in the Ḥarāz; above Walajah, J. Milḥān; Wadi Yūr where there are many fine specimens. I have heard rumours that it is in the Khabt area of Maḥwīt and in the J. Rāziḥ region of Khawlan Ash Sham but have been unable to confirm this.

As a native tree, *M. laurifolia* is only known outside Yemen in Saudi Arabia, Ethiopia and adjacent parts of Somalia but it has been widely cultivated. Indeed it was a cultivated tree that Forsskal described from Bayt al Faqīh and it grows well in arid plains elsewhere and notably in Egypt to which it was introduced in the second millennium B.C. In ancient Egypt the leaves were used for funeral garlands and they have been found in many tombs from the twelfth dynasty onwards, notably that of Tutankhamen.

EBENACEAE

Trees or shrubs with simple, entire leaves which lack stipules. Sepals and petals united. Corolla tubular, usually white. Fruit a berry.

Leaves opposite. Flowers in short racemes 1. *Euclea*
Leaves alternate. Flowers solitary or in clusters 2. *Diospyros*

1. EUCLEA *Murr.*

1. Euclea racemosa *Murr.* Bisa
E. kellau Hochst.
E. forskalii Schweinf.
Dioecious shrub reaching about 3m in Yemen. Leaves obovate to oblanceolate, up to about 7 × 2cm, glabrous, paler below, apex rounded, base cuneate, margin entire, undulate. Inflorescence a short raceme up to 3cm long; corolla white, glabrous, bell-shaped. Fruit a black, globose, 1-seeded berry c 5mm in diameter, the calyx not enlarging in fruit.

The Yemen plant is subspecies **schimperi** (A.DC.) F. White.

Widespread between 500 and 1500m in areas of good rainfall along the escarpment from the Wadi Rima' north to Khawlan Ash Sham. Apart from a single record from the Wadi Qusa' just south of Madīnat Al 'Abid it is unaccountably absent further south. It is found in both valley forest and on scrubby cliffs and hillsides.

2. DIOSPYROS *L.*

1. Diospyros mespiliformis *Hochst. ex A. DC.* Biṣaṣ
African Ebony. Dioecious. Usually a shrub to 2m but rarely (Wadi Yūr only?) a tree to 20m, the bark smooth, black. Leaves oblong to oblanceolate, up to c 14 × 4cm, with an acute apex, entire, glabrescent, shiny, coriaceous; undersurface with a prominent raised midrib, paler than the upper surface. Flowers white, pubescent, fragrant; male flowers in shortly pedunculate clusters; female flowers solitary, subsessile. Fruit yellowish, globose, c 2.5cm in diameter, the calyx enlarging in fruit.

An uncommon plant of valley scrub between 500 and 800m restricted to the western slopes of J. Raymah and J. Milḥān (unaccountably absent from J. Bura'), Wadi Yūr and Wadi Līyah, Khawlan Ash Sham.

MYRSINACEAE

A family of trees and shrubs with simple alternate leaves which lack stipules. Flowers small with the petals united in our species. Fruit a drupe.

Leaves 10–20cm long. Flowers in axillary panicles 1. *Maesa*
Leaves less than 2cm long. Flowers in axillary clusters 2. *Myrsine*

1. MAESA *Forssk.*

1. Maesa lanceolata *Forssk.* — Ma's

Scruffy looking shrub to about 3m. Bark red-brown. Leaves narrowly elliptic, up to 10 × 20cm long, glabrous, dark green above, paler beneath, apex acute, base cuneate, margin obscurely crenate, undersurface with prominent raised brownish veins. Inflorescence an axillary panicle with numerous small yellowish flowers, becoming pendent in fruit. Fruit globose, yellow-green, c 4mm wide, the style persistent.

Flowers throughout the year.

Frequent along the whole escarpment between 800 and 2800m but uncommon below about 1500m. It is usually found in scrub by semi-permanent streams in cultivated areas where it is often associated with *Debregeasia saenab*, *Senecio hadiensis* and *Pterolobium stellatum*. It is particularly common in areas where qat is grown without irrigation and is commonly used to wrap up bundles of qat. Although occurring in the Wadi Banā area as far east as Damt and in Wadi Aḥjar it is strikingly absent from the high plateau suggesting that it does not tolerate frost.

The generic name *Maesa* is based on the Arabic name of this plant.

2. MYRSINE *L.*

1. Myrsine africana *L.* — Qatam

Small dioecious shrub to 1m. Leaves sessile, glabrous, dark green above (paler beneath), variable in shape but commonly oblanceolate, up to 15 × 5mm, apex acute, base cuneate, margin serrate towards the apex. Flowers in small clusters in the leaf axils, subsessile, dull red, minute. Fruit globular, 1-seeded, purple, c 3mm wide, the style persistent.

Flowers in summer.

Frequent between 1800m and 3100m on nearly all the higher escarpment mountains from J. Ḥabashī north to J. Rāziḥ. It is most typical of grassland in areas of good rainfall above 2500m.

PRIMULACEAE

Annual or perennial herbs with opposite or alternate, usually simple leaves, lacking stipules. Sepals and petals 5, fused. Fruit a capsule.

1. Leaves whorled — 1. *Primula*
   Leaves alternate or in opposite pairs — **2**
2. Flowers in terminal racemes — **3**
   Flowers axillary — **4**
3. Leaves alternate. Corolla white — 5. *Samolus*
   Leaves mostly opposite. Corolla pinkish — 2. *Lysimachia*
4. Capsule opening horizontally by a cap. Flowers never yellow — 4. *Anagallis*
   Capsule opening by longitudinal valves. Flower yellow — 3. *Asterolinon*

1. PRIMULA *L.*

1. Primula verticillata *Forssk.*

Perennial herb with an erect stem to 40cm. Leaves whorled, forming a basal rosette and with 1–3 rather distant whorls on the stem, obovate to oblanceolate, 1–20cm long, diminishing regularly in size upwards, the margin dentate, the undersurface often white-mealy especially in dry conditions. Flowers in whorls in the axils of the upper leaves, pedicellate, the pedicels lengthening as the flowers mature; corolla with a long, often mealy tube up to 3cm long, much exceeding the calyx, the lobes spreading, yellow.

A conspicuous plant growing on damp shaded cliffs and by rocky mountain springs and ditches from 1500 to 3300m. Widely distributed on the escarpment from Ḥayfān (Deflers) and Turbah north to Kuḥlān, but absent from a number of apparently suitable areas on the outer escarpment, notably the Ta'kar – Mudhaykhirah mountains, J. Ḥubaysh, Waṣāb and J. Milḥān; in Khawlan Ash Sham known only from Wadi Rakwah (Steinberg); on the high plateau rare but found at Tin'am near J. Lawz, Ḥaddah and Yarīm. Outside southwest Arabia it is only known from Ethiopia.

2. LYSIMACHIA *L.*

1. Lysimachia ruhmeriana *Vatke*

Much-branched, decumbent glabrous herb, sometimes rooting at the nodes. Leaves opposite or nearly so, petiolate, lanceolate or elliptic, up to 8cm long, obscurely toothed, covered in small punctate glands on the undersurface. Flowers in long terminal racemes: corolla pinkish.

Only known from a streamside at 1900m in the grassland area of J. Kabūrah near Jirjir above Ḥadīyah.

3. ASTEROLINON *Hoffmanns & Link*

1. Asterolinon adoense *Kunze*

Small, erect, glabrous annual herb to 10cm. Leaves opposite, sessile, ovate c 1cm long. Flowers solitary in the leaf axils, the pedicels about as long as the leaves, becoming deflexed in fruit; corolla yellow, very small, the lobes about 1mm long. Capsule opening by longitudinal valves.

Only known at 1800m in shallow grassland on the lowest slopes of J. Ba'dān just north of Ibb where it was growing with *Striga asiatica* and *Wahlenbergia hirsuta*.

4. ANAGALLIS *L.*

Glabrous herbs with entire leaves. Flowers solitary in the leaf axils. Capsule opening horizontally by means of a lid.

1. Leaves alternate. Minute herb with a stem less than 5cm long *A. minima*
   Leaves opposite. Stems at least 5cm long **2**
2. Stems prostrate, rooting at the nodes *A. serpens*
   Stems erect, not rooting at the nodes **3**
3. Corolla lobes with numerous marginal glandular hairs. Fruiting pedicels usually exceeding the leaves *A. arvensis*
   Corolla lobes with few or no marginal glandular hairs. Fruiting pedicels equalling the leaves *A. foemina*

1. Anagallis arvensis *L.*

Pimpernel. Erect, much-branched annual to 30cm. Leaves opposite, sessile, ovate or lanceolate, up to 2cm long. Calyx 4–5mm long; corolla blue or red, very variable in size, the lobes always with prominent glandular hairs on the margins. Fruiting pedicels recurved just above the capsule, up to 3cm long, exceeding the leaves.

The typical plant with red flowers has not yet been found in Yemen and all specimens I know of belong to variety **caerulea** (*L.*) *Gouan* (var. *latifolia* (L.) Lange).

A common field weed found between 1000 and 3000m on the escarpment and high plateau but rare below 2000m and preferring irrigated land or high rainfall areas; often in wheat and barley fields.

2. Anagallis foemina *Miller*

Similar to the previous species but with a larger calyx up to 6mm long, corolla lobes largely without marginal glands and the fruiting pedicels irregularly recurved, often curled through more than 360°, up to 1.5cm long, equalling or shorter than the leaves.

The Yemen plant (Wood 72/67, 75/217; Miller 375, 465) differs from European *A. foemina* in its compact, usually perennial habit and dull mauve flowers, but seems in other respects identical despite its geographical isolation.

A rare plant of exposed mountain tops and loose rocky slopes, growing from 2600 to 3500m on a few of the higher escarpment mountains: J. Ṣabir; J. Sumārah; J. Bārad, Kusmah in Raymah; J. Shibām, Ḥarāz; J. Nabī Shu'ayb; J. Maḥdad; Kuḥlān.

3. Anagallis minima (*L.*) *E.H.L. Krause*

Minute erect annual to 5cm. Leaves alternate, shortly petiolate, elliptic or obovate, up to 5mm long. Flowers subsessile, the calyx c 2mm long, exceeding the minute white corolla.

Only known from short mossy grassland at c 3100m on J. Ta'kar.

4. Anagallis serpens *Hochst.* ex *DC.*

Prostrate herb (? perennial) creeping and rooting at the nodes. Leaves opposite, shortly petiolate, obovate or elliptic up to 8mm long. Flowers on erect pedicels up to 2.5cm long; calyx c 4mm long, much shorter than the pretty pink corolla.

A rare plant found in well-grazed, very wet marsh between 2400 and 3000m on and around the high plateau: J. Manār; J. Sumārah; by a pond 5km east of Dhamār; above Baw'ān.

5. SAMOLUS *L.*

1. Samolus valerandi *L.*

Brookweed. Glabrous herb with a basal rosette and leafy stems to c 30cm. Leaves obovate, entire, narrowed gradually into an indistinct petiole, up to 7cm long; stem leaves alternate. Flowers in a terminal raceme, the pedicels with a small bract in the middle; corolla white.

A local plant of streamsides between 1700 and 2500m which, like *Blumea bovei* and *Juniperus*, is absent from the escarpment except the southern and northern extremities: in the south it is known from the Turbah region, J. Ṣabir, Dhī Sufāl and Wadi Sūdān near Qā'idah; in Khawlan Ash Sham from Majz and Wadi Rakwah (Steinberg); otherwise known only from the high plateau near Dhamār and in the gorge above Tin'am, near J. Lawz.

PITTOSPORACEAE

PITTOSPORUM *Banks ex Soland.*

1. Pittosporum viridifolium *Sims*

P. abyssinicum auct.

Small tree to about 10m. Bark variable sometimes smooth, sometimes peeling. Leaves alternate to sub-opposite, without stipules, glabrous, oblanceolate, up to 8 × 3cm with an entire hyaline margin, base cuneate, apex mucronate. Inflorescence a panicle at the apex of branchlets; flowers with small yellow petals, fragrant. Fruit a globose black drupe with a persistent style.

Yemen plants are assigned to subspecies **arabicum** *Cuf.* This is endemic to Yemen and differs from African plants in its narrower leaves which have a mucronate tip.

Flowers April–May.

An uncommon tree of a few scattered localities between 1800 and 2700m where it may be a relic of mountain woodland: in several places between Attārah and Ḥajarah in the Ḥarāz; at the head of the Wadi Banā between Saddah and Ribāṭ; Maḥjar in Wadi Ahjar; Maḥwīt; around Aṣ Ṣa'bah and Al Markah, J. Milḥān which is the only place where it is at all frequent.

CRASSULACEAE

Succulent herbs or undershrubs. Leaves opposite or alternate, without stipules. Flowers regular, usually in cymes, petals and sepals 4 – many, free or united. Fruit a leathery follicle, often many-seeded.

1. Petals united into a conspicuous tubular corolla **2**
   Petals free, not tubular in appearance **4**
2. Leaves orbicular, peltate 4. *Umbilicus*
   Leaves not as above **3**

3. Flower parts in 4s, flowers erect (rarely the whole inflorescence nodding) — 2. *Kalanchoe*
   Flower parts in 5s, flowers nodding — 3. *Cotyledon*
4. Leaves crowded into dense rosettes. Flowers yellow — 6. *Aeonium*
   Leaves not crowded into dense rosettes (except in *Crassula alba*). Flowers white or pink — 5
5. Leaves alternate. Flower parts usually in 6s — 5. *Sedum*
   Leaves opposite. Flower parts in 4s and 5s — 1. *Crassula*

1. CRASSULA *L.*

Fleshy herbs with opposite leaves. Flower parts in 4s or 5s.

1. Leaves in a basal rosette (often dying when the plant flowers). Flowers in a corymb — *C. alba*
   Basal rosette of leaves absent. Flowers all axillary — 2
2. Leaves petiolate. Stem with a line of hairs — *C. alsinoides*
   Leaves sessile. Plant completely glabrous — 3
3. Aquatic herb with solitary axillary flowers — *C. hedbergii*
   Terrestial herb with flowers in axillary clusters — *C. schimperi*

1. Crassula alba *Forssk.* — Lisan Al Kabsh

An erect, probably biennial herb to 20cm with a ciliate often red-brown stem. Leaves sessile, lanceolate, fleshy, up to 4cm long, the margins ciliate; basal leaves in a rosette and often dying as the plant flowers; stem leaves opposite. Flowers small, white, in a terminal corymbs.

A common plant growing between 2000 and 3200m on the escarpment and extending east on the high plateau to Khamr, Ḥaddah near Sana'a and Yarīm. It is found on cliffs, terrace walls and around rock outcrops in high-altitude grassland.

2. Crassula alsinoides (*Hook. f.*) *Engl.*

A prostrate herb, glabrous except for a line of hairs on the stem. Leaves petiolate, very fleshy, elliptic, acute, c 1cm long. Flowers white, pedicellate, solitary in the leaf axils.

Only known from a terrace wall at Ḥadādah, J. Raymah at 1400m (Wood 1916).

3. Crassula hedbergii *Wickens & Bywater*

A glabrous trailing herb with short ascending stems. Leaves sessile, linear, acute, the base clasping the stem, c 6–8mm long. Flowers pink, almost sessile, solitary in the leaf axils. Follicles with 6–7 seeds in each carpel.

A rare plant found at c 3000m on mud by seasonal pools or, during the rainy season, floating on the pool surface at Kawkabān and on the ridge connecting J. Sumārah and J. Manār (Wood 1296, 1678); associated with *Limosella* spp, *Marsilea aegyptiaca* etc.

4. Crassula schimperi *Fisch. & C.A. Mey.*

C. pentandra (Royle ex Edgw.) Schoenl.
C. pharnaceoides sensu Deflers

A small glabrous trailing 'mosslike' herb with erect reddish stems c 5–8cm high. Leaves sessile, lanceolate to ± deltoid, acute, c 3–5mm long. Flowers white, sessile, in few-flowered axillary clusters.

Widespread and common between 2000 and 3200m on the escarpment from J. Ṣabir and J. Ḥabashī northwards and also occurring sporadically on the mountains round the high plateau east to the Jabal Lawz range. It is a plant of cliffs, rocks and especially terrace walls where it is associated with such species as *Pentas*, *Campanula edulis* etc.

I have in fact failed to find Deflers' specimens of *C. pharnaceoides*, but I think there is little doubt that the plant was *C. schimperi*, which Deflers did not record. There are at Paris some specimens of *C. schimperi* collected by Botta and named as *C. pharnaceoides* and Deflers probably took the name from these specimens.

2. KALANCHOE *Adans.*

Fleshy herbs with opposite leaves mostly clustered towards the base of the stem. Flowers in terminal cymes. Flower parts in 4s, the petals united to form a ± tubular corolla. Although superficially similar, the species are readily distinguished; important characters include indumentum, leaf shape, flower colour and flowering period, though the last two may not be constant outside Yemen.

Most species are called 'Khudar' or 'Khurūb'.

1. Plant pubescent or glandular at least on the inflorescence — 2
   Plant completely glabrous — 5
2. Plant pubescent, eglandular — *K. citrina*
   Plant glandular-hairy on the inflorescence — 3
3. Leaves ovate, petiolate — *K. deficiens*
   Leaves sessile, the stem leaves lanceolate to oblong — 4
4. Annual with pink flowers; glandular hairs extending nearly to the base of the stem — *K. lanceolata*
   Perennial with yellow flowers; glandular hairs on and just below the inflorescence only — *K. yemensis*
5. Flowers red, with short pedicels c 1–4mm long — *K. glaucescens*
   Flowers yellow-brown, pedicels usually exceeding 1cm long — *K. alternans*

1. Kalanchoe alternans (*Vahl.*) *Pers.*

K. rosulata E. Raadts

An erect, entirely glabrous perennial up to 40cm high. Leaves petiolate, entire, or crenate, variable in shape but typically bronze or blackish green in colour. Inflorescence lax, usually few-flowered and branched, sometimes nodding particularly in bud. Corolla yellow-brown, the tube 1.5–2cm long, the lobes linear 1–1.5cm long.

With the exception of one form mentioned below this species is always spring flowering.

A polymorphic species particularly with regard to leaf shape. The typical plant in the central escarpment (Wood 1756, 3185) has obovate to suborbicular leaves whereas plants in more arid conditions, particularly in Khawlan Ash Sham, have narrowly lanceolate leaves (Wood 2945, 3027). Intermediate forms (e.g. Wood 75/177) also occur. One form (Wood 954) from the wadi above 'Adūf near Turbah is outstanding in being autumn flowering and with very large flowers. It has lanceolate leaves.

Widespread but rather local on the escarpment and the western side of the high plateau from 1500 to 3000m. It usually grows on cliffs or in rather sparse stony grassland. Apparently endemic to southwest Arabia.

2. Kalanchoe citrina *Schweinf.*

An erect, pubescent, rather weak perennial c 30cm high. Leaves sessile, lanceolate or oblanceolate, dentate, dark green. Inflorescence small, dense. Corolla pubescent, yellow, the tube 1cm long, the lobes ovate, c 2mm long.

Flowers in spring.

A rare plant of stony hillsides in *Acacia – Commiphora* scrub between 1200 and 1500m: Wadi Ḥijān and Wadi Khufit near 'Usil, Ḥarāz and between Mafḥaq and Suq Samīr in Ḥaymat al Kharijīyah.

3. Kalanchoe deficiens *sensu Raadts*

K. crenata (Andr.) Haw. s.l.

Very similar to the next species from which it differs in having a densely glandular inflorescence.

Flowers in autumn.

A local plant found between 1600 and 2000m in the south-eastern mountains apparently with a strong preference for *Euphorbia* scrub on volcanic soils: J. Bayhān, Turbah; J. Ghayāmī, Ḥujarīyah (Deflers); J. Ḥabashī; Mirkiz As Salām; between Ḥawbān and Naqīl Ibl, Taiz; Qā'idah; near Khalāwah, J. Sawraq; between Rada' and Jubān (Westinga). (Wood 1020, 1530)

E. Raadts has identified this plant with *K. deficiens* (Forssk.) Aschers. & Schweinf., but unfortunately there is no type specimen and Forsskal's description could apply to any *Kalanchoe*. A manuscript note by Vahl says that *K. deficiens* has a hairy inflorescence but it is not certain that Vahl had seen the type and in any case this character is not unique to this plant. Since *K. deficiens* sensu Raadts does not grow anywhere near Forsskal's type locality (J. Milḥān) it seems unlikely that Raadts is correct and the name *K. deficiens* should be regarded as ambiguous.

4. Kalanchoe glaucescens *Britten*

K. laciniata sensu Schwartz

An erect, entirely glabrous perennial to 1m. Leaves petiolate, ovate to elliptic, crenate or entire, glaucous. Inflorescence dense, many-flowered; corolla red, the tube c 10–12mm long, the lobes ovate c 4mm long.

Flowers in autumn.

The Yemen plant is subspecies **arabica** *Cufod.* which has exserted stamens. It is apparently endemic.

Widespread and common between 1000 and 2500m throughout the escarpment from the Ḥujarīyah to Khawlan Ash Sham but more frequent in areas of good rainfall. It grows on cliffs, in scrub and in thorn fences round fields. (Wood 74/233)

5. Kalanchoe lanceolata (*Forssk.*) *Pers.*

An erect annual to 1½m, glandular – hairy in all parts except the lower leaves and stem base. Leaves sessile, lanceolate, entire, green. Inflorescence typically elongate with numerous lateral as well as terminal, many-flowered corymbs; corolla orange-red, the tube c 1cm long, the lobes ovate, c 5mm long.

Flowers in autumn.

Widespread between 600 and 1700m on the escarpment from J. Sawraq and the Ḥujarīyah north to Khawlan Ash Sham, but rather local and usually in areas of high rainfall, particularly the central escarpment and the lush middle-altitude wadis. (Wood 1000, 1452)

6. Kalanchoe yemensis (*Deflers*) *Schweinf.*

Similar to the previous species from which it differs in being a perennial of lower stature, reaching only 50cm. The corolla is bright yellow, the inflorescence more compact and the glandular hairs restricted to the inflorescence.

Flowers in autumn or, sometimes on the high plateau, in spring.

It is customary to regard this plant as a form of *K. lanceolata* which is extremely variable in Africa but the two species are quite distinct morphologically and ecologically in Yemen and are best treated as separate until they can be shown to be conspecific.

Widespread in Ibb province extending south to Shar'ab, east to the Wadi Banā and north along the line of the inner escarpment to Ma'bar and Qa' Bakīl in 'Anis. There are isolated colonies on the high plateau around Wa'lān and between Sana'a and Wadi Ḍahr. It is found between 1600 and 2500m usually on stony hills and rocky plain near cultivation. It is particularly abundant in the plain north of Ibb. (Wood 74/234, 977)

3. COTYLEDON *L.*

1. Cotyledon barbeyi *Schweinf.* Khudār, Tuflūq

An erect, many-stemmed undershrub to 1m. Leaves sessile, opposite, entire, fleshy, glaucous, oblanceolate, up to c 9 × 3.5cm. Inflorescence a terminal glandular cyme of large drooping red flowers; flower parts in 5s.

This attractive and very distinct species occurs on cliffs, rocky slopes and terrace walls between 1000 and 2800m. Its distribution is clearly centred on Ibb and Dhamār provinces, extending east to Damt, north to J. Bura', the Ḥarāz and the Yislah area and south to Qā'idah. In Taiz province there are two outlying locations in the Ḥujarīyah: the Naqil Jubirān, Turbah and the Naqil Wuslah, Qabayṭah. In general its occurrence is sporadic and unpredictable though it is widespread within the area of its distribution. It appears to avoid areas of very high rainfall and is most frequent along the inner escarpment particularly in 'Anis and around Iriyān.

4. UMBILICUS *DC.*

1. Umbilicus intermedius *Boiss.*

Perennial glabrous herb. Leaves alternate, peltate, orbicular, crenate, fleshy, c 2–8cm diameter. Flowers on an erect raceme 5–30cm long, shortly pedicellate, pale yellow, pendulous; bracts about as long as the flowers.

A species of the east Mediterranean region which in Yemen is a rare plant of shaded rocks and walls between 2600 and 3200m on a few high mountains: J. Ṣabir; J. Ta'kar; J. Sumārah; at the head of the Wadi Banā near Ribāṭ; J. Mada'; J. Maswar.

5. SEDUM *L.*

1. Sedum hispanicum *L.*

Small, erect, simple or branched, glandular, annual herb to 5cm. Leaves alternate, linear, glaucous about 1cm long. Inflorescence a few-flowered cyme; petals white with a pink mid-vein 5–7mm long.

Like *Umbilicus intermedius* this is a plant of the east Mediterranean region which in Yemen is a very rare but easily overlooked species of cliff ledges between 2600 and 3300m in Ibb province: J. Ta'kar; J. Manār; in a gorge near Ribāṭ at the head of the Wadi Banā. (Wood 1161, 1309)

6. AEONIUM *Webb & Berth.*

1. Aeonium sp.A — Khudār

A. chrysanthum sensu Schwartz

Undershrub to 50cm with a branched succulent stem and leaves crowded into dense rosettes at the ends of the branches. Leaves oblanceolate-cuneate, c 6 × 2cm, fleshy, entire, green (sometimes red-tinged), the margins ciliate. Inflorescence an erect leafy panicle c 15–25cm long; flowers yellow; sepals and petals 9–11.

A very local plant of inaccessible cliffs on the escarpment mountains between 1700 and 3300m, but sometimes abundant where it occurs: Akāḥilah near Turbah; near Suq Shirqi, Mudhaykhirah; Mashwarah between Ibb and 'Udayn; J. Ba'dān; on the Sumārah – Manār range; J. Shibām, Ḥarāz; Al Markah, J. Milḥān; Bayt Bashārī and Raḥbān, Maḥwīt.

The Yemen *Aeonium* is an undescribed species and the only two populations collected differ significantly from each other. The J. Shibām plant has a glandular-hairy inflorescence (Schweinfurth 1670, Wood 1684) while the J. Sumārah population has a glabrous inflorescence (Radcliffe-Smith & Henchie 4699, Lavranos 16227). It is important that further flowering plants are collected to assess the significance of this variation. Unfortunately *Aeonium* flowers irregularly and infrequently even under cultivation and in the wild flowering specimens are usually seen on the most inaccessible cliffs.

ROSACEAE

Trees, shrubs or herbs. Leaves usually alternate and with stipules. Flowers usually large and showy, regular, bisexual; sepals 5; petals 5, free; stamens often very numerous, borne on a cup-shaped receptacle. Fruit various but often fleshy and edible.

A mainly temperate family with few representatives in Yemen apart from a large number of fruit trees.

1. Shrubs armed with prickles — **2**
   Unarmed herbs or shrubs — **3**
2. Erect shrub. Seeds contained in the enlarged fleshy urceolate receptacle (hip) — 4. *Rosa*
   Decumbent or arching shrub. Fruit of numerous 1-seeded druplets (blackberry) — 1. *Rubus*
3. Shrub — 5. *Cotoneaster*
   Herbs — **4**
4. Petals present, yellow — 2. *Potentilla*
   Petals absent — 3. *Alchemilla*

1. RUBUS *L.*

Brambles. Spiny shrubs with arching, scrambling or decumbent stems and petiolate, digitate or pinnate leaves. Flowers in terminal panicles, the petals pink or white. Fruit of 1-seeded drupes (druplets) becoming at first red and in all our species eventually black. The fruit, known as a blackberry is edible, but in Yemen many of the carpels abort and the fruit is not as attractive as in temperate countries.

Most species of *Rubus* are apomictic and thus constant in their characters but do on occasion reproduce sexually giving rise to sterile hybrids and new 'species'. The number of such apomictic species in temperate countries is vast and *Rubus* has a good claim to be considered the most complex genus of flowering plants in the world. In Yemen *Rubus* is not particularly common and there are not many species. The following account is based on all the material I have been able to see, mostly my own collections, but many more collections need to be made before it is possible to assess the significance of the variation that can be observed. It is essential that healthy plants with good flowering and fruiting panicles are collected together with a piece from the central part of the first year stem as it is almost impossible to identify inadequate material. Many accounts of *Rubus* lump together dissimilar forms under one name, but I am reluctant to do this since this practice has almost invariably been mistaken in the past given the apomictic nature of most species. In particular I think it very doubtful whether our species are conspecific with African plants.

The following characters are of particular importance: the nature of the indumentum, the presence or absence of stalked glands, the number of leaflets and how they are arranged, the shape of the terminal leaflet, the shape of the inflorescence, the colour of the petals when the flowers open, the shape, size and frequency of the prickles. In this account all references to the stem are to the non-flowering first year stem.

In Yemen all *Rubus* species are found in damp rocky places, particularly mountain gullies and are most common in the wetter parts of the escarpment. They are called 'Ḥantafis', 'Ḥumays' or 'Nafas'.

1. All leaves with 3 leaflets. Prickles weak, up to 2mm long — *R. caesius*
   Some leaves pinnate with 5 leaflets. Prickles more than 2mm long — **2**
2. Stem and panicle rhachis with stalked glands — **3**
   Stem and panicle rhachis without stalked glands — **5**
3. Stem prickles mostly of the same size. Leaves, especially on the panicle, tomentose on the lower surface — *R. arabicus*
   Stem prickles of many different sizes. Leaves pubescent but never tomentose on the lower surface — **4**

4. Stem shortly tomentose with or without some long spreading hairs. Inflorescence dense, ± cylindrical *R. sp. D*
Stem thinly pilose, never tomentose. Inflorescence moderately dense, ± conical *R. sp. E*
5. Inflorescence with numerous stalked glands. Sepals conspicuously elongating in fruit and reaching 10–15mm in length. Carpels glabrous *R. sp. C*
Inflorescence without stalked glands or with stalked glands on the sepals only. Sepals not elongating much in fruit, less than 10mm long. Carpels hairy **6**
6. Petals absent. Leaves strongly grey-felted below. Stem sharply angled, deeply furrowed *R. sp. B*
Petals present. Leaves not grey-felted below. Stem at most weakly angled, never furrowed *R. sp. A*

1. Rubus arabicus (*Deflers*) *Schweinf.*
Stem vigorous, ± decumbent, rounded, sparsely to densely pilose with mostly large prickles and scattered stalked glands. Prickles falcate, mostly broad-based, up to 6mm long. Leaves pinnate with 3–5 leaflets, tomentose on the undersurface particularly on the panicle. Panicle short, lax, ± conical in shape; sepals with many stalked glands but without spines, erect and elongating slightly in fruit, up to 10mm long; petals pale pink (drying white), contrasting strongly with the reddish filaments; carpels glabrous.

Closely related to *R. sp. C* and possibly a derivative of it, perhaps by hybridisation with *R. sp. E*.

Quite frequent on J. Shibām and J. Masār in the Ḥarāz at c 2500m (Wood 2840, 3351), occurring also around Markah on J. Milḥān (Wood 3452, 3460) and at Kuḥlān (Miller 3222).

2. Rubus caesius *L.*
Stem weak, decumbent, rounded, sparsely pilose but without stalked glands. Prickles few, slender, recurved, up to 2mm long. Leaves digitate with 3 leaflets, sparsely pubescent below. Panicle lax, few-flowered; sepals without stalked glands, erect and elongating in fruit, up to 15mm long; petals pure white, large; filaments white; carpels glabrous.

A distinctive species widespread in the northern temperate regions but with an isolated population in Yemen.

Known from 2100 to 3000m in Bani Maṭar, Wadi Ḍahr and along the ridge from Aḥjar to Ṭawīlah and Maḥwīt, growing mostly on sandstone. (Wood 2399, 2557, 3110)

3. Rubus sp. A
Stem vigorous, arching, weakly angled, tomentose with or without spreading pilose hairs, stalked glands absent. Prickles falcate, broad-based, up to 5mm long. Leaves pinnate with 3–5 leaflets, ± tomentose below. Panicle long and lax with small cymes arising from quite low leaf axils, leafy to the very top; sepals usually with a few stalked glands, erect in fruit, up to 10mm long; petals pink; filaments red; carpels hairy.

Distinguished by its leafy inflorescence and eglandular stems.

Known from 1900 to 2100m in the central escarpment where it has been found on J. Ẓallamlam above Ḥadīyah and on J. Bura' between 'Ubr Sa'd and Al Qimaḥ (Wood 2117, 2812). An incomplete specimen from the experimental farm at Ibb may also belong here (Brunt 2457).

4. Rubus sp. B ? apetalus *Poir.*
Stem very vigorous, high arching, strongly angled and deeply furrowed, shortly pubescent but without stalked glands. Prickles straight or slightly recurved, up to 4mm long. Leaves pinnate with 3–5 leaflets, grey-felted and tomentellous on the lower surface. Panicle strikingly pyramidal, rather lax; sepals without stalked glands, erect in fruit, up to 5mm long; petals absent; filaments white except below; carpels densely hairy.

A distinctive species on account of its felted leaves, pyramidal panicle and petal-less flowers; it may be conspecific with *R. apetalus*, but I have not seen the type of that species.

Only known from swampy scrub at 1600m near Suq As Sabt, Ibb.

5. Rubus sp. C ? petitianus *A. Rich.*
Stem very vigorous, high arching, rounded, densely pilose, but without stalked glands. Prickles mostly narrow-based, straight or falcate up to 6mm long. Leaves pinnate with 3–5 leaflets, ± tomentose below. Panicle lax or rather dense, always short, ± cylindrical to conical in shape; sepals with numerous stalked glands, erect and elongating in fruit, up to 15mm long; petals pale pink (drying white) contrasting strongly with the red filaments; carpels glabrous.

Closely related to *R. arabicus* but readily distinguished by its eglandular stem. It is close to if not identical with specimens from Ethiopia identified by Gustafsson as *R. petitianus*.

Found on the central escarpment from 1800 to 2500m on J. Raymah, the Ḥarāz and Maḥwīt. (Wood 2839, 3109, 3186, 3306)

6. Rubus sp. D
Stem vigorous, arching or decumbent, rounded, grey-tomentose with some spreading hairs, numerous stalked glands and prickles of varying sizes. Prickles narrow-based, slightly curved, up to 6mm long. Leaves pinnate with 3–5 leaflets, pubescent below. Panicle dense, ± cylindrical with one or two axillary cymes below the main inflorescence; sepals covered in stalked glands and short spines, erect in fruit, up to 10mm long; petals pale pink; filaments white with pink bases; carpels glabrous.

Best recognised by its distinctive stem indumentum.

Known from between 1800 and 2200m at Dhī Sufāl (Hepper & Wood 5863), between Sayyānī and Ibb (Scott & Britton 379), Jiblah (Wood 74/246), 'Usil (Schweinfurth 1235), J. Marran in Khawlan Ash Shan (Wood 3034); Wood 2855 from J. Milḥān may belong here but is atypical in a number of respects.

7. Rubus sp. E
Stem vigorous, arching or decumbent, rounded, sparsely pilose, densely covered in prickles of varying sizes; stalked glands present, but often rather few. Prickles broad based, usually falcate, up to 6mm long. Leaves pinnate with 3–5 leaflets, pubescent below, usually very coarsely toothed. Panicle moderately dense, rather short, ± conical; sepals

covered in stalked glands and spines, erect and elongating in fruit, up to 13mm long; petals pale pink; filaments white with pink bases; carpels glabrous.

This is the common *Rubus* of the Ibb region found between 1800 and 2700m from Qā'idah (also J. Ṣabir ?) north to 'Utmah. There are particularly fine examples in the scrub on J. Ba'dān. (Wood 2650, 2882, 2991, 3116)

2. POTENTILLA *L.*

Perennial herbs. Flowers usually yellow; epicalyx present. Fruit a head of achenes.

Leaves pinnate	*P. dentata*
Leaves digitate	*P. reptans*

1. Potentilla dentata *Forssk.*

P. hispanica Zimm.

Erect perennial herb to c 60cm. Leaves pinnate; leaflets c 7–11, oblong, up to 3cm long, deeply serrate, brown tomentose below. Flowers in a rather dense terminal cyme; petals yellow, c 8mm long, slightly exceeding the sepals.

P. dentata is part of a complex of species around *P. pensylvanica* L. which is found throughout the northern hemisphere. Many populations of *P. pensylvanica* have been given separate names and it seems best to treat the isolated Yemen population under Forsskal's name until further study has been made of the whole complex.

A local plant found by springs and mountain ditches from 2300 to 3500m on the higher escarpment mountains: on nearly all the Ibb mountains from J. Ṣabir north to Waṣāb and J. Sumārah; along the inner escarpment from Baw'an, north to J. Maswar; Manākhah (Deflers).

2. Potentilla reptans *L.*

Cinquefoil. Slender procumbent perennial herb with long creeping stems which root at the nodes. Leaves digitate; leaflets usually 5, oblong-obovate, c 1–4cm long, coarsely serrate, sparsely pubescent below. Flowers solitary on long pedicels from the leaf axils; petals yellow, 8–10cm long, about twice as long as the sepals.

A local plant of grassy banks by streams and irrigation ditches found between 1500 and 2900m in Ibb province, along the inner escarpment, on the high plateau and in Khawlan Ash Sham. A widespread plant of the temperate northern hemisphere with its southernmost outpost in Yemen.

3. ALCHEMILLA *L.*

1. Alchemilla cryptantha *A. Rich.*

Lady's Mantle. Pilose herb with an ephemeral central basal rosette and spreading procumbent stems which root at the nodes. Leaves suborbicular in outline with a deep basal sinus, mostly 1–1.5cm broad, 5-lobed to about half way, the lobes deeply toothed. Flowers in subsessile clusters of 1–2 in the leaf axils, inconspicuous and without petals, ± concealed by the stipules.

A local plant found by streams and springs between 2400 and 3500m on the escarpment mountains; widespread in suitable places in Ibb province, but elsewhere rare and known only from J. Ṣabir, Kusmah in J. Raymah, Karāwan at Manākhah (Schweinfurth), Baw'an, J. Nabī Shu'ayb and Bayt 'Adhāqah.

4. ROSA *L.*

1. Rosa abyssinica *R. Br* Kilāb, 'Ashaq, Ḥawjam

R. indica sensu Forsskal

Wild rose. Erect shrub to 4m, the stem armed with numerous straight or slightly curved prickles. Leaves pinnate with up to 9 leaflets, the rhachis armed, leaflets glabrous, serrate. Flowers in terminal corymbs, fragrant; petals 5, white, up to 1.5cm long; sepals reflexed in fruit. The fruit consists of numerous seeds enclosed in an edible dull red fleshy urceolate container (hip), c 1cm long, formed from the enlarged receptacle.

Flowers throughout the year but especially after the spring rains.

Widespread on the escarpment and high plateau from 1500 to 3100m, but only common in areas of average to good rainfall; it is sometimes abundant notably at high altitudes on J. Ṣabir and in parts of the central escarpment. Outside southwest Arabia it is only known from Ethiopia.

Roses have been cultivated in Yemen for centuries and are generally called 'Warad'. Much the most common cultivated species is **Rosa x bifera** (*Poir.*) *Pers.* (*R. damascena* sensu Schwartz) which is widely grown in old gardens for its attractive red flowers and in the past was doubtless used to make rose water and perfumes. Modern cultivars of complex hybrid origin have been recently introduced to Sana'a and will doubtless be cultivated more widely in the future.

5. COTONEASTER *Medic.*

1. Cotoneaster racemiflorus (*Desf.*) *K. Koch*

C. nummularia Fisch. & C.A. Mey.

Small procumbent or ascending unarmed deciduous undershrub with dark red-brown branchlets which are ± tomentose when young. Leaves suborbicular, entire, mucronate, up to 2cm long, the upper surface dark green, the lower surface grey-tomentose. Flowers in small axillary cymes, white, the calyx tomentose. Fruit subglobose, red, fleshy, c 8mm long.

Only known from cliff ledges at c 2600m above Ḥiliyal, J. Sumārah.

Many members of the Rosaceae are cultivated for their edible fruit. The following are known to be grown in Yemen:

The Quince, **Cydonia oblonga** *Mill.* (*C. vulgaris* Pers.), known as 'Safarjal' is commonly cultivated on the high plateau particularly in the irrigated valleys around Sana'a. The fruit is ripe in late summer.

The Apple, **Malus domestica** *Borkh.* (*Pyrus malus* L.), known as 'Tufāḥ', is cultivated around Sana'a, but rather uncommonly. The local cultivar produces a small, rather unappetising fruit and it rarely appears in the markets since it cannot compete with imported apples.

The Apricot, **Prunus armeniaca** *L.*, known as 'Barqūq' or 'Mishmish' is cultivated sporadically and not very successfully on the escarpment, but in great abundance on the high plateau. Ḥaddah near Sana'a is renowned for its apricots, but production is large in all the valleys around Sana'a and also in districts like Al Hadā'. There are several local cultivars all with an excellent flavour. Apricots are ripe in May.

The Plum, **Prunus domestica** *L.*, known as ''Injāṣ', is rarely cultivated today and I only know of it around Shibam and Thilā. The local cultivar has a sweet red fruit which is ripe in May.

The Almond, **Prunus dulcis** (*Mill.*) D.A. Webb (*P. amygdalus* Batsch) known as 'Lawz' is commonly cultivated around the high plateau and grows at a higher altitude than any other fruit tree. It is the only cultivated fruit tree, for example in the villages in the aptly named Jabal Lawz region at 3000m. It is the first tree to flower, often beginning as early as December. The local almonds are more highly esteemed than the larger imported nuts.

The Peach, **Prunus persica** (*L.*) *Batsch*, known as 'Firsikh' or 'Khawkh', is commonly cultivated on both the escarpment and the high plateau, but good fruit is rarely if ever produced on the escarpment. The local cultivar is a white variety and has an excellent flavour though unfortunately it is usually picked unripe. Peaches are ripe in July.

The Pear, **Pyrus communis** *L.*, known as ''Ambarūt' is occasionally cultivated mostly on the high plateau. There is a large orchard at Shibām and at Saddah in the Wadi Banā and it is rumoured that many are grown in Al Ḥadā'. The local cultivar produces a small hard pear of good flavour which is ripe in June.

LEGUMINOSAE

A family of trees, shrubs or herbs which contains many species which climb either by twining or with tendrils. Leaves alternate, with stipules, usually compound, with the leaflets digitately or pinnately arranged. The flowers are usually bisexual and arranged in racemes. The sepals are ± joined, the calyx being regular or irregular. In the *Mimosoideae* (Genera 55–57) the flowers are regular, very small, and massed into dense heads or spikes. In the *Caesalpinioideae* (Genera 46–54) the flowers vary from almost regular in *Cassia* to highly irregular in *Tamarindus*. In the *Papilionoideae* the petals are arranged in a highly characteristic manner with the uppermost (the standard) enclosing the two laterals (the wings) which in turn enclose the two lowermost which are usually fused and known as the keel. The stamens are usually 10. The fruit is a pod which is dehiscent by two longitudinal valves.

The Legumes are the second largest family in the Yemen flora and are of great importance ecologically since huge tracts of country are dominated by members of this family notably the Acacia trees. Many species are of agricultural or horticultural importance. The family is surprisingly poor, though, in endemic species.

1 Corolla absent. Stamens 5. (Carob tree) 51. *Ceratonia*
Corolla present. Stamens usually 10 **2**

2. Flowers very small, regular, in dense spikes or heads (Mimosoideae) **3**
Flowers irregular, usually relatively large, not usually in dense heads or spikes **5**

3. Leaves with 1 pair of leaflets. Inflorescence of small heads arranged racemosely 57. *Pithecellobium*
Leaves with 2 or more pairs of leaflets. Inflorescence not arranged racemosely **4**

4. Branchlets spine-tipped. Inflorescence 2-coloured 55. *Dichrostachys*
Branchlets not spine-tipped. Inflorescence of 1 colour 56. *Acacia*

5. Petals similar, not obviously papilionaceous **6**
Petals dissimilar, papilionaceous **15**

6. Plant spiny **7**
Plant unarmed **9**

7. Flowers cream, very small, Acacia-like. Pod winged, usually red 47. *Pterolobium*
Flowers yellow, relatively large, not Acacia-like. Pod not winged nor red **8**

8. Pod linear, sub-torulose. Leaf petiole modified into a spine 49. *Parkinsonia*
Pod not linear or torulose. Leaf petiole not modified into a spine 46. *Caesalpinia*

9. Leaves simple, bilobed **10**
Leaves compound **11**

10. Tree with rusty indumentum. Pod indehiscent 53. *Piliostigma*
Shrub without rusty indumentum. Pod dehiscent 52. *Bauhinia*

11. Leaves bipinnate **12**
Leaves pinnate **13**

12. Shrub with foetid glandular inflorescence 46. *Caesalpinia*
Tree (rarely shrub) with eglandular inflorescence 48. *Delonix*

13. Calyx campanulate. Corolla pale to dark purple 45. *Cadia*
Calyx not campanulate. Corolla yellow or brown **14**

14. Tree, stamens 3. (Tamarind) 54. *Tamarindus*
Herbs, shrubs or trees. If trees, stamens 10 50. *Cassia*

15. Plant leafless when flowering **16**
Plant with leaves when flowering **17**

16. Spinescent undershrub. Pod inflated 3. *Crotalaria*
Unarmed herb. Pod not inflated 37. *Vigna*

17. Leaves simple **18**
Leaves 2 or more foliolate **23**

18. Leaves broadly ovate or orbicular, toothed 12. *Psoralea*
Leaves not broadly ovate to orbicular, entire **19**

19. Pod inflated. Flowers yellow 3. *Crotalaria*
Pod not inflated. Flowers mauve, pink or red 20

20. Calyx and stipules scarious 25. *Alysicarpus*
Calyx and stipules not scarious **21**

21. Leaves narrowly ovate. Pod jointed, constricted between the seeds 24. *Desmodium*
Leaves linear. Pod not jointed or constricted between the seeds **22**

22. Flowers solitary in the leaf axils. Pod elongate 16. *Tephrosia*
Flowers in racemes. Pod orbicular 14. *Indigofera*

23. Leaves terminating in a tendril **24**
Leaves terminating in a leaflet **26**
24. Leaflets 2. Stipules enlarged and leaflike 29. *Lathyrus*
Leaflets 4 or more. Stipules not enlarged and leaflike **25**
25. Calyx teeth about twice as long as the tube. Flowers less than 5mm long 28. *Lens*
Calyx teeth ± equalling the tube. Flowers c 10–15mm long 27. *Vicia*
26. Leaves 3 foliolate **27**
Leaves 5 or more foliolate **56**
27. Leaflets toothed **28**
Leaflets entire **32**
28. Plant with hairs attached at their middle. Leaflets more than 2cm broad. Pod winged 13. *Cyamopsis*
Plant lacking hairs attached at their middle. Leaflets less than 1cm broad. Pod not winged. **29**
29. Plant densely glandular pilose. Flowers in terminal leafy panicles, pink 5. *Ononis*
Plant largely lacking glandular hairs. Flowers in racemes or rarely in axillary clusters, never pink nor terminal **30**
30. Pod spirally twisted 7. *Medicago*
Pod not spirally twisted **31**
31. Pod ovoid. Flowers in elongated racemes 8. *Melilotus*
Pod elongate, ± linear. Flowers in head-like racemes or axillary clusters 6. *Trigonella*
32. Terminal leaflet with small stipels **33**
Terminal leaflet without stipels **45**
33. Pod jointed and constricted to more than half its width between the seeds, breaking up into 1-seeded units 24. *Desmodium*
Pod not jointed and constricted between the seeds, not breaking up into 1-seeded units **34**
34. Flowers red. Plant adpressed hairy with numerous white hairs attached at their centre (medifixed) 14. *Indigofera*
Flowers never red. Plant spreadingly hairy at least in some parts. (If ± adpressed hairy, hairs brownish on young growth, not medifixed) **35**
35. Underside of leaves and calyx dotted with glands **36**
Underside of leaves and calyx not dotted with glands **37**
36. Cultivated bush. Pod more than 4cm long 41. *Cajanus*
Undershrub or climber, not cultivated. Pod less than 3cm long 42. *Rhynchosia*
37. Flowers purple-black. Pod with irritant hairs 34. *Mucuna*
Flowers not purple-black. Pod without irritant hairs **38**
38. Flowers 2–3 in the leaf axils **39**
Flowers in 2 – many-flowered racemes **40**
39. Flowers yellow. Pod straight 40. *Macrotyloma*
Flowers mauve. Pod curved 38. *Dolichos*
40. Inflorescence of condensed umbel-like racemes at the apex of the usually long peduncle. 37. *Vigna*
Inflorescence various, but if long pedunculate clearly raceme-like and at least 5cm long **41**
41. Very robust plants. Leaflets more than 5cm broad. Flowers about 2–2.5cm long **42**
Plants not strikingly robust. Leaflets less than 2cm wide. Flowers less than 12mm long **43**
42. Leaflets rounded at base. Pods exceeding 15cm long **36.** *Canavalia*
Leaflets truncate or cuneate at base. Pods less than 8cm long 39. *Lablab*
43. Racemes dense, commonly with 50 or more flowers. Pods strongly curved near the apex 32. *Neonotonia*
Racemes lax with 15 or fewer flowers. Pods ± straight **44**
44. Flowers less than 5mm long. Pod up to 3mm wide 33. *Teramnus*
Flowers about 10mm long. Pod c 6–7mm wide 35. *Galactia*
45. Calyx and stipules scarious 25. *Alysicarpus*
Calyx and stipules not scarious **46**
46. Leaves gland-dotted below **47**
Leaves not gland-dotted below **48**
47. Shrub. Inflorescence a dense axillary cluster 44. *Flemingia*
Herb. Inflorescence a pedunculate raceme 43. *Eriosema*
48. Stipule base forming a sheath and enclosing the stem **49**
Stipule base not sheath-like nor enclosing the stem **50**
49. Leaflets obovate, toothed. Inflorescence a terminal leafless head 9. *Trifolium*
Leaflets elliptic, entire. Inflorescence a terminal leafy spike 23. *Stylosanthes*
50. Leaves sessile, fleshy. Pod with hooked bristles, jointed and constricted between the seeds 20. *Taverniera*
Leaves petiolate, not fleshy. Pod without hooked bristles, not constricted and jointed between the seeds **51**
51. Pod plicate (concertina-like) 15. *Ptycholobium*
Pod not plicate **52**
52. Flowers red. Hairs attached at the middle (medifixed) 14. *Indigofera*
Flowers yellow (sometimes blue-veined). Hairs not medifixed **53**
53. Pod inflated 3. *Crotalaria*
Pod not inflated **54**
54. Shrub 0.5–2m high 1. *Anagyris*
Low, sometimes woody herbs **55**
55. Pod flattened. Flowers in racemes or solitary in the leaf axils 4. *Argyrolobium*
Pod terete. Flowers in axillary clusters 2. *Lotononis*
56. Leaflets notched at the apex. Pod dentate on the margin 19. *Biserrula*
Leaflets entire or toothed but not notched at the apex. Pod entire **57**
57. Leaflets toothed 26. *Cicer*
Leaflets entire **58**
58. Leaflets 5, the basal pair stipule-like 10. *Lotus*
Leaflets 5 or more, the basal pair not stipule-like **59**

59. Shrub. Pod constricted between the seeds, jointed 22. *Ormocarpum*
Herbs, climbers or rarely shrubs. Pod not constricted between the seeds or jointed **60**
60. Pod spirally twisted 11. *Vermifrux*
Pod not spirally twisted **61**
61. Pod orbicular 21. *Onobrychis*
Pod elongate, never orbicular **62**
62. Some hairs attached by their middle. Flowers small, red 14. *Indigofera*
Hairs not attached by their middle. Flowers yellow, blue, purple or pink, but never red **63**
63. Corolla hairy. Leaves with 5 or more leaflets. Pod flattened. Leaves imparipinnate **64**
Corolla glabrous. Leaves with 7 or more leaflets. Pod terete or flattened. Leaves paripinnate or imparipinnate **65**
64. Standard large, blue, exceeding 2.5cm long 31. *Clitoria*
Standard small, pink or mauve 16. *Tephrosia*
65. Pod very long, exceeding 10cm. Standard yellow with dark purple dots 17. *Sesbania*
Pod up to 5cm. Flowers various but if yellow without dark dots **66**
66. Climbing shrubs. Pod 10–15mm broad 30. *Abrus*
Herbs. Pod 1–6mm broad 18. *Astragalus*

1. ANAGYRIS *L.*

1. Anagyris foetida *L.*

Foetid shrub to about 2m. Young branches white-tomentose. Leaves dark green, pubescent, 3-foliolate; the leaflets elliptic c 2–5 × 1–1.5 cm, apex acute to somewhat mucronate, base cuneate. Inflorescence a short axillary raceme of pale yellow flowers each about 2–2.5cm long. Pods 10–15cm long, up to 2cm broad, flattened, tapering towards the base.

A typical plant of the Mediterranean region which is a rare plant of high mountains between 2800 and 3200m in Yemen: J. Ṣabir; J. Ba'dān; J. Nuqum, Sana'a (Deflers); J. Nabī Shu'ayb; J. Mudhbur in J. Lawz.

2. LOTONONIS (*DC.*) *Eckl. & Zeyh.*

1. Lotononis platycarpos (*Viv.*) *Pic.-Serm.*

L. leobordea (Del.) Benth.

Prostrate mat-forming pubescent annual with minute digitately trifoliolate leaves. Flowers pale yellow in sessile axillary clusters. Pods straight, terete, c 5mm long, ± equalling the scarious persistent calyx.

Widespread but not very common on the high plateau between 2000 and 2500m; also occurring in Khawlan Ash Sham between Sāqayn and Ḥaydān and at Sharāwah. Characteristic of derelict fields and alluvial plain and commonly associated with *Althaea ludwigii.*

3. CROTALARIA *L.*

A distinctive genus of herbs and undershrubs with digitately 3-foliolate (rarely simple) leaves, yellow flowers and inflated pods. The keel is beaked and it is often helpful in identifying the species to see whether the beak is twisted or not. The pod is often distinctive and its indumentum, size and shape are important.

Many species are rare in Yemen and nearly all are restricted to the escarpment. *C. microphylla* and *C. senegalensis*, however, are only known from the Tihama and *C. aegyptiaca* from the Mashriq. *C. emarginella* is the only species occurring on the high plateau.

1. Leaves simple or absent **2**
Leaves 3-foliolate **4**
2. Spinescent undershrub with minute soon deciduous leaves up to 3mm long *C. aegyptiaca*
Herbs, not spinescent. Leaves persistent, exceeding 1.5cm long **3**
3. Leaves villous, 1.5–3.5cm long *C. hirta*
Leaves glabrous or adpressed pubescent, mostly 5–8cm long *C. retusa*
4. Plant spiny *C. spinosa*
Plant unarmed **5**
5. Flowers 1–3 in the leaf axils **6**
Flowers in racemes (rarely a few also in the leaf axils) **7**
6. Annual herb with strongly blue-veined flowers *C. microphylla*
Perennial shrub with orange-yellow flowers *C. squamigera*
7. Pod glabrous **8**
Pod pubescent to villous **9**
8. Stipules leaflike. Bracteoles absent *C. natalitia*
Stipules filiform. Bracteoles present *C. quartiniana*
9. Pod pilose. Stem with spreading hairs or not **10**
Pod pubescent. Stem never with spreading hairs **11**
10. Pod up to 1cm long. Leaves up to 1cm long, drying green *C. oöcarpa*
Pod 2.5–4cm long. Leaves 1–1.5cm long, drying black *C. incana*
11. Standard glabrous **12**
Standard pubescent **14**
12. Stipules absent. Racemes more than 5cm long *C. plowdenii*
Stipules present. Racemes very short usually less than 3cm long **13**
13. Leaflets cuneate to obovate. Racemes sessile *C. pycnostachya*
Leaflets linear-oblong. Racemes pedunculate *C. impressa*
14. Keel untwisted. Petioles commonly more than 2.5cm long *C. deflersii*
Keel twisted. Petioles usually less than 1.5cm long **15**

15. Leaflets obovate. Calyx lobes acuminate, becoming reflexed as the flowers mature *C. emarginella*
Leaflets oblong. Calyx lobes abruptly truncate, not becoming reflexed as the flowers mature *C. senegalensis*

1. Crotalaria aegyptiaca *Benth.*

Stiff spinescent grey-green undershrub with numerous intricate shortly pubescent branches. Leaves soon deciduous, minute, simple, sessile, 2–3mm long. Inflorescence a very lax terminal raceme of small glabrous yellow flowers; keel with a twisted beak. Pod glabrous, c 8mm long.

A very distinctive species found rarely between 1000 and 2000m in the Mashriq: Wadi Saba', Mā'rib; Khulāqah in Nihm.

2. Crotalaria deflersii *Schweinf.*

Vigorous perennial herb to 1m, shortly but densely pubescent in all parts. Leaves 3-foliolate; leaflets obovate 1.5–4cm long, apex rounded to emarginate. Inflorescence a very long terminal raceme often exceeding 40cm in length; flowers large, bright yellow, becoming wine-red with age; standard pubescent; keel without a twisted beak. Pods club-shaped, 2.5–4cm long, shortly pubescent.

A local but very attractive species, often on roadsides and characteristic of rainshadow areas in the central escarpment: around Ḥajjaylah and 'Usil, in the Ḥarāz (Deflers & Schweinfurth); Mawsanah, Al Qidam and Wadi Qaṣab, Ḥarāz; Wadi Kulābah and J. Masna'ah, Bilad Aṭ Ṭa'ām, J. Raymah (Wissmann); between Al Masbaḥ and Suq al Aḥad, Waṣab as Sāfal.

3. Crotalaria emarginella *Vatke*

C. rathjensiana Schwartz

Much-branched wiry perennial, often woody below, up to 20cm on the high plateau but twice that at low altitudes, usually sparsely adpressed pubescent in Yemen. Leaves 3-foliolate, often paired (one fully developed, one diminutive); leaflets obovate, very variable in size, 4–20mm long, apex truncate to emarginate. Inflorescence a lax terminal raceme of large yellow flowers, c 5–20cm long; standard pubescent; keel with a twisted beak. Pods ± club shaped, 1.5–2.5cm long, pubescent.

Very easily confused with *C. senegalensis* from which it is best distinguished by its acuminate calyx lobes which become recurved as the flowers mature.

Most characteristically a plant of gullies on rocky plain and mountain slopes on the high plateau and the Mashriq from Sana'a northwards reaching 2500m in the Bani Ḥushaysh. In this habitat it is locally common and bursts into flower after rain. It also occurs in a line through the Ḥarāz to the Tihama being found at Bani Manṣūr, Wadi Madfar, Ḥajjaylah (Schweinfurth) and as a field weed in the Tihama as low as 600m between Bājil and 'Ubāl. Elsewhere on the escarpment and in the Tihama it appears to be absent. Low-altitude plants (eg. Wood 2137) are much more vigorous.

4. Crotalaria hirta *Willd.*

Much-branched, densely pilose, decumbent, woody perennial herb. Leaves shortly petiolate, simple, oblong, ± rounded at both ends, 1.5–3.5cm long. Inflorescence of shortly pedunculate, few-flowered terminal racemes, 2–4cm long; flowers yellow; standard hairy; keel with a twisted beak. Pods glabrous, oblong-elliptic, 1–1.5cm long.

A rare plant known only from shallow grassland with scattered trees at 800m in the Wadi Ḥār, Ḥarāz.

C. hirta is an Indian species with its most westerly limits in Yemen.

5. Crotalaria impressa *Nees ex Walp.*

C. astragalina Hochst.

Prostrate or ascending pubescent annual herb. Leaves 3-foliolate; leaflets narrowly oblong c 25–30 × 3–6mm, apex acute, base cuneate. Inflorescence a dense, few-flowered terminal or axillary head-like raceme about 1–2cm long; standard, yellow; keel blue with an untwisted beak. Pods pubescent, c 12mm long.

Similar in general appearance to *C. microphylla* but readily distinguished by having flowers in racemes.

Still growing at c 1200m in open scrub at 'Usil in the Ḥarāz where Schweinfurth first found it in 1889.

6. Crotalaria incana *L.*

Erect usually rather bushy annual or perennial herb to 1m; stem pubescent with or without longer hairs. Leaves 3-foliolate, glabrous to thinly pilose; leaflets obovate, c 1.5–3cm long, apex rounded. Inflorescence a shortly pedunculate, terminal or axillary, lax raceme 2–20cm long; flowers yellow, hairy; the standard often red-veined; the keel with an untwisted beak. Pods pilose, c 2.5–4.5cm long.

The plant becomes black when dried.

Widespread on the escarpment from 400 to 2800m and from the Ḥujarīyah north to Khawlan Ash Sham, but scattered in its distribution and definitely preferring the high rainfall of the Ibb-'Udayn region and the central escarpment.

Two subspecies are known in Yemen, but their distribution has not been accurately worked out. Subspecies **incana**, which is without long spreading hairs on the stem and has bracts up to 3mm long, has been recorded at 1200m from Wadi Barakānī in the Ḥujarīyah (Wood 1410) and Taiz (Wood 74/56). Subspecies **purpurascens** (*Lam.*) *Milne-Redh.* with long spreading yellow-brown hairs on the stem and bracts 4–10mm long has been found at 2800m on J. Ba'dān (Wood 816) and at 1500m at 'Udayn (Hepper & Wood 5978).

7. Crotalaria microphylla *Vahl* Faja'ah

Prostrate, glabrous or sparsely pubescent annual. Leaves 3-foliolate; leaflets linear-oblong up to 33 × 5mm, apex acute, base narrowly cuneate, the terminal leaflet much longer than the laterals. Flowers solitary or paired in the leaf axils, small, very pale yellow with blue veins, glabrous; the keel with an untwisted beak. Pods pubescent, c 10–12mm long.

Abundant after rain on sand dunes in the Tihama from sea level to 300m. Commonly associated with *Indigofera argentea*, *Heliotropium pterocarpum* etc. The only *Crotalaria* found in this habitat in Yemen.

8. Crotalaria natalitia *Meissn.*

Vigorous bushy perennial to 1½m with pubescent stems, usually woody below. Leaves 3-foliolate; leaflets oblanceolate, glabrous above, sparsely adpressed pubescent below, up to c 3.5 × 1cm, apex truncate to emarginate, often apiculate. Inflorescence a pedunculate terminal raceme c 5–15cm long; flowers yellow; glabrous; keel with an untwisted beak. Pods glabrous, 2.5–4cm long.

Only variety **rutshuruensis** *De Wild.* with an adpressed pubescent calyx occurs in Yemen.

This is the only Yemen *Crotalaria* with leaf-like stipules.

A rare plant of grassland between 1200 and 2000m in the central escarpment: Bani Sa'īd, Ja'farīyah and near Ḥadādah, Al Jabīn, J. Raymah; between 'Ukaybir and Shaḥir, J. Milḥān.

9. Crotalaria oöcarpa *Bak.*

A weak, ascending, rather scruffy annual or perennial herb, grey-pilose in all parts. Leaves often clustered in the leaf axils, 3-foliolate; leaflets obovate to oblanceolate, c 10 × 5mm, apex rounded. Inflorescence of two kinds: short pedunculate terminal (rarely axillary) lax racemes c 5–8cm long and clusters of 1–4 flowers in the leaf axils. Flowers pale yellow, pubescent, the keel with a twisted beak. Pods densely pubescent c 8–10mm long.

The Yemen plant is subspecies **microcarpa** *Milne-Redh.*

Readily recognised by its distinctive indumentum and (where both are present) the two kinds of inflorescence.

Only known at 1800m on the ridge northeast of Markah, J. Milḥān, where it grows in exposed rocky cliff grassland. Outside Yemen it is widespread but apparently rare in east and central Africa.

10. Crotalaria plowdenii *Bak.*

Perennial herb with ascending pubescent stems which are prominently furrowed. Leaves 3-foliolate; leaflets glabrous above, adpressed pubescent below, elliptic, up to 5 × 2.5cm, apex rounded to emarginate. Inflorescence an almost sessile terminal raceme 5–20cm long; flowers yellow with purple veins, glabrous, the keel with an untwisted beak. Pods adpressed pubescent, c 10–12mm long.

Readily recognised from other species of *Crotalaria* in Yemen by the absence of stipules.

A rare plant of scrubby field borders between 1800 and 2500m in the Ibb area: between Ibb and Suq As Sabt; near Ra'isī, J. Ḥubaysh. Outside Yemen only known from Ethiopia.

11. Crotalaria pycnostachya *Benth.*

Procumbent to erect much-branched pubescent annual. Leaves 3-foliolate; leaflets glabrous above, thinly puberulous below, oblanceolate, c 2.5 × 1cm, apex rounded. Inflorescence of short sessile axillary racemes about 5cm long; flowers small, yellow, pubescent, the keel with an untwisted beak. Pods shortly adpressed pubescent, club-shaped, c 12mm long.

A common plant of stony hills and open scrub along the escarpment from 500 to 1800m, but infrequent in dry areas.

12. Crotalaria quartiniana *A. Rich.*

Much-branched, thinly pilose, weak ascending annual. Leaves 3-foliolate; leaflets glabrous above, pilose below, elliptic, c 2.5 × 1.5cm, apex acute, base rounded. Inflorescence of very short lax terminal racemes with 1–10 flowers; flowers yellow, glabrous, the keel with a twisted beak. Pods ± cylindrical but broadest at the apex, glabrous, c 3.5mm long.

Distinguished from other Yemen species by the presence of bracteoles.

Only known from grassy slopes at c 1800m near Aṣ Ṣa'bah, J. Milḥān. Outside Yemen it is widespread on the mountains of tropical Africa but very local in occurrence.

13. Crotalaria retusa *L.* Juljul

Erect, somewhat woody branched annual to c 1m, stem pubescent. Leaves simple, oblanceolate, up to c 8 × 2.5cm, glabrous above, adpressed pubescent below. Inflorescence a long terminal pedunculate raceme c 10–25cm long; flowers large, yellow, glabrous, the keel with a twisted beak. Pods glabrous, c 3–4cm long.

Like *C. incana*, this plant becomes black when dry.

Locally abundant as a field weed in the alluvial plains at the east of the Tihama around 'Abs, Bājil, Kadān etc. Common also in the foothills in areas of good rainfall becoming sporadic and mainly ruderal with altitude but reaching 1500m. Surprisingly very rare everywhere south of the Wadi Rīma'.

14. Crotalaria senegalensis (*Pers.*) *DC.*

Erect, much-branched, probably annual herb with shortly pubescent stems. Leaves in clusters, 3-foliolate; leaflets glabrous above, pubescent below, elliptic, c 30 × 15cm, apex rounded, the terminal leaflet much longer than the laterals. Inflorescence a long terminal pedunculate raceme up to about 25cm long; flowers pubescent, yellow, the standard turning wine red, the keel with a twisted beak. Pods pubescent, c 13mm long.

Easily confused with *C. emarginella* from which it can be distinguished by the truncate apiculate calyx lobes.

A rare field weed found at c 300–400m in the Tihama: Sa'id Misr east of Bayt al Faqīh; Qanāwis.

15. Crotalaria spinosa *Benth.*

Low bushy, usually annual plant with pubescent stems. Leaves solitary or appearing clustered, 3-foliolate; leaflets glabrous above, pubescent below, oblanceolate-cuneate, c 7 × 2mm, the apex usually emarginate. Inflorescence of short axillary racemes of 1 (–3) flowers, the rhachis spine-tipped; flowers small, pale yellow; standard pubescent; keel with a twisted beak. Pods pubescent, c 8–9mm long.

Widespread in areas of average to good rainfall between 1200 and 2300m in Taiz and Ibb province and abundant on grassy earth banks around fields on the Dhubhān plateau and in the Ibb valleys. Very rare further north and, only recorded near Suq al Jum'ah (J. Ash Shirq); 'Ukaybir, J. Milḥān; 'Usil, Ḥarāz; Shahārah.

16. Crotalaria squamigera *Deflers* Suraysah
Vigorous woody much-branched shrub to 1m, stems adpressed pubescent. Leaves 3-foliolate; leaflets glabrous above, minutely tomentellous below, elliptic, c 2–3.5 × 1–1.5cm, the apex obtuse and finely apiculate. Flowers solitary or paired in the leaf axils, pedicellate, large, orange-yellow with reddish veins, glabrous, the keel with an untwisted beak. Pods adpressed pubescent, club-shaped, c 4cm long.

Probably the most spectacular *Crotalaria* in Yemen and one of the most attractive and distinctive of endemic species. Its distribution is clearly centred on the Dhubhān plateau around Turbah where it is abundant on rocky slopes in open grassland and on the cliff tops. Elsewhere it is a rare and local plant of similar habitats between 1100 and 2300m: Tha'bat area, J. Ṣabir (Deflers, Scott & Britton); J. Sawraq; Al Khatwah, 'Aruq; Musalla, Aḥkum.

4. ARGYROLOBIUM *Eckl. & Zeyh.*

Perennial herbs or undershrubs with digitately 3-foliolate leaves, usually yellow flowers and a flattened, nearly straight, linear-oblong pod.

1. Pods less than 20mm long. Leaflets less than 5mm long — *A. confertum*
   Pods more than 20mm long. Leaflets more than 5mm long — 2
2. Flowers bright yellow, at least 10mm long. Leaflet margin with brown hairs — *A. rupestre*
   Flowers pale yellow to dull mauve, about 6–7mm long. Leaflets without brown hairs — *A. arabicum*

1. Argyrolobium arabicum (*Decne.*) *Jaub. & Spach*
A. abyssinicum Jaub. & Spach
Erect or spreading undershrub to about 30cm with adpressed hairy stems. Inflorescence a raceme of 1–4 pale yellow to dull mauve flowers, each about 7mm long and about equalling the calyx. Pods somewhat torulose, c 25–35 × 3mm.

Two distinct forms occur:
Form A (Wood 72/23, 2707) is a weak, scarcely woody, green perennial whose sparsely pubescent oblong leaves reach about 20 × 7mm. The flowers are dull mauve. This form is found between 1400 and 1900m in the central escarpment: Al Markah, J. Milḥān; Bayt Braz, Maḥwīt; Bani Sa'd, Ja'farīyah, J. Raymah; J. Sāfān.
Form B (Wood 75/155, Gillespie 54) is a wiry, not very leafy, woody perennial with pubescent linear-lanceolate to oblong leaves up to c 12 × 2mm. The flowers are pale yellow. This is common on bare rocky mountains around Sana'a and eastwards to Nihm and Bani Ḥushaysh, but is rare elsewhere and only known from J. Mirbish, Salafīyah (L. Gillespie); near Bayt 'Adhāqah and on J. Marran. It is found between 2000 and 2500m.

2. Argyrolobium confertum *Polhill*
Prostrate perennial with numerous straggling grey pilose stems sometimes forming 'carpets'. Leaflets small, oblanceolate, densely grey pilose, up to 5mm long. Flowers solitary, leaf opposed, shortly pedicellate, pale yellow, c 6mm long. Pods adpressed pilose, c 10–15 × 3mm.

An uncommon plant of cliffs and rocky slopes from 2400–3000m along the line of the inner escarpment from J. Ḍawrān north to the Kawkabān region thence to Bayt 'Adhāqah, Kuḥlān and finally the limestone hills between Khamr and Ḥūth.

3. Argyrolobium rupestre (*E. Mey.*) *Walp.*
A. virgatum Bak.
Low perennial with numerous prostrate to ascending pilose stems. Leaflets ± elliptic, c 16 × 7mm, pilose, the margins lined with long brown hairs. Inflorescence a head-like raceme of usually 1–2 bright yellow flowers each about 11mm long. Pods pilose, c 30–40 × 4mm.

The Yemen plant is subspecies **remotum** (*A. Rich.*) *Polhill.*

Frequent in short grassland and around rocks from 1900 to 3000m on the outer escarpment and J. Ash Shirq. Often associated with *Trifolium semipilosum* and showing a similar preference for mist-covered slopes.

5. ONONIS *L.*

1. Ononis reclinata *L.*
A low ascending sticky annual up to 10cm high, villous and glandular on its stem, leaves and fruit. Leaves pinnately 3-foliolate; leaflets obovate, cuneate, serrate at the apex. Flowers in leafy terminal panicles, pink. Pods deflexed, oblong – linear, c 1cm long.

Only known on cliff ledges at c 1800–2200m on J. Ẓallamlam and J. Kabūrah above Ḥadīyah, J. Raymah.

A common Mediterranean species with its southernmost outposts in Yemen and Ethiopia.

6. TRIGONELLA *L.*

Annual herbs with digitately 3-foliolate, dentate leaflets. Flowers pale yellow. Pod elongate, slightly compressed.

Flowers in racemes. Pod curved — *T. hamosa*
Flowers in axillary clusters. Pod 'wavy' (somewhat transversely plicate) — *T. anguina*

1. Trigonella anguina Del.
Prostrate herb with obovate leaflets c 6 × 6mm. Inflorescence an axillary cluster of 3–7 flowers, each c 3mm long. Pods wavy, c 7mm long.

Recorded once only, on mud by a cistern at Ḥūth (1900m).

2. Trigonella hamosa L.
Erect herb to 40cm with obovate, truncate leaflets c 15–16 × 8–12mm. Inflorescence a pedunculate head-like raceme of 5–15 flowers, each c 5mm long. Pods falcate, c 15mm long.

Recorded by Steinberg in the Majz area (c 2000m).

Fenugreek, **Trigonella foenum-graecum** *L.*, known in Yemen as 'Ḥulbah' is widely cultivated above 2000m on the escarpment and high plateau. It has usually solitary, pale yellow flowers and a curved pod. The dried seeds are ground into flour and, when mixed with meat broth and various vegetables, form the basis of one of the most characteristic Yemeni dishes, known simply as 'Hulbah'. It is extremely popular among all people living on the high plateau. Fenugreek is also esteemed for medicinal purposes.

7. MEDICAGO *L.*

Herbs with obovate to oblanceolate toothed pinnately trifoliolate leaves, head-like racemes of purple or yellow flowers and spirally twisted pods.

1. Racemes many flowered. Pod unarmed — *M. lupulina*
   Racemes 1–5 flowered. Pod spiny — 2
2. Whole plant, including fruit, densely pubescent — *M. minima*
   Fruit glabrous. Stems and leaves very sparsely hairy or subglabrous — *M. polymorpha*

1. Medicago lupulina *L.*

Procumbent or ascending pubescent annual. Leaflets obovate. Racemes long pedunculate (c 2–3cm), with 5–15 yellow flowers. Pods 1-seeded, reticulate, sparsely hairy, c 2mm diameter, coiled through 360°, becoming black when ripe.

Common as a field weed and on grassy banks by fields and on disturbed grassy mountain slopes on the escarpment from 2000 to 3400m on J. Nabī Shu'ayb. Also common on the high plateau but occurring there by irrigation ditches and in marsh.

2. Medicago minima (*L.*) *Bartal.*

Procumbent or ascending pubescent annual. Leaflets obovate (mostly narrowly so). Racemes shortly pedunculate (less than 1cm long) with 1–5 yellow flowers. Pods several seeded, spiny, pubescent, c 4–5mm diameter, coiled 4–5 turns, brown when ripe.

A rare plant of short grassland between 2000 and 3000m on a few mountains northwest of Sana'a: J. Shibām, Manākhah; Bayt al Bashāri, Maḥwīt; Shibām; Kawkabān.

3. Medicago polymorpha *L.*

M. hispida Gaertn.

Procumbent subglabrous or very sparsely pubescent annual. Leaflets cuneate-obovate. Racemes shortly pedunculate (less than 1cm long) with 1–5 yellow flowers. Pods several seeded, spiny, glabrous, c 5mm diameter, coiled for about three turns, brown when ripe.

In short grassland and as a field weed in scattered localities from 1800–3000m in the Ibb region, the central escarpment and the high mountains from Bani Maṭar to Kuḥlān.

Alfalfa or Lucerne, **Medicago sativa** *L.*, 'Qaḍb', is extensively cultivated on the high plateau and the escarpment above about 1800m. It is usually grown in irrigated fields and is on sale in almost all markets as animal fodder. It is sometimes found amongst cereal crops as a relic of former cultivation and is readily recognised by its attractive blue-purple flowers which can be seen throughout the year.

8. MELILOTUS *Mill.*

Annual or biennial sparsely pubescent herbs smelling strongly of coumarin especially when dry. Leaves pinnately trifoliolate, leaflets serrate. Flowers in axillary racemes. Pod small, ovoid, 1–2 seeded, glabrous in our species.

All three species in Yemen are probably introduced from the Mediterranean region.

1. Flowers white — *M. alba*
   Flowers yellow — 2
2. Flowers 2–3mm long. Pod 1.5–3mm long — *M. indica*
   Flowers 4–6mm long. Pod 3–5mm long — *M. officinalis*

1. Melilotus alba *Medic.*

Erect annual or biennial to about 1m. Leaflets linear-oblong, 1–2.5cm long. Flowers white, 4–5mm long. Pods reticulate veined, 3–5mm long.

It has only been found recently as a garden weed in Sana'a, Taiz and Yarīm (Hepper) but it appears to have been more common in the past and there are old records from sea level to 2600m: Luḥayyah (Ehrenberg); Ṭawīlah (Rathjens); J. Shibām, Manākhah (Schweinfurth).

2. Melilotus indica (*L.*) *All.*

Erect or in Yemen commonly decumbent annual. Leaflets obovate to oblong, c 1.5cm long. Flowers yellow, up to 3mm long. Pods strongly reticulate, 1.5–3mm long.

A field weed, often near irrigation ditches, found between 2000 and 2500m: Sana'a; Yarīm; between Bayḍā' and Dhī Nā'im.

3. Melilotus officinalis (*L.*) *Pall.*

Erect biennial very similar in general appearance to *M. alba*. Leaflets obovate to oblong, c 1.5cm long. Flowers yellow, usually about 4mm long. Pods rugose, 3–5mm long.

A rare casual of villages and roadsides 1900–2900m; J. Ta'kar; Mashwarah on the Ibb-'Udayn pass; Ḥūth.

9. TRIFOLIUM *L.*

Clovers. Low creeping perennials which root at the nodes. Leaves digitately trifoliolate with small obovate toothed leaflets and stipules forming a sheath at the base of the petiole. Inflorescence a pedunculate, dense, racemose head of small flowers. Pod very small, oblong, ± enclosed by the persistent calyx and flowers.

Flowers white. Calyx not accrescent in fruit — *T. semipilosum*
Flowers pink. Calyx strongly accrescent and reticulate in fruit — *T. fragiferum*

1. Trifolium fragiferum *L.*

Sparsely pubescent plant with long-petiolate leaves, c 10–15mm long. Inflorescence with peduncles 4–6cm long; flowers pink; calyx strongly accrescent and reticulate in fruit. Pods 1–2 seeded.

Common on grassy banks by streams and ponds on and around the high plateau, from 2300 to 3500m on J. Nabī Shu'ayb.

2. **Trifolium semipilosum** Fresen.

Pilose plant with petiolate leaves. Inflorescence with peduncles 1–5cm long; flowers white; calyx not accrescent in fruit. Pods 2–6 seeded.

Widespread on the escarpment from J. Sabir northwards growing in short grassland in areas that receive plentiful mist and good rains from 2000m on the outer escarpment to 3500m on J. Nabī Shu'ayb. It also occurs very rarely by ponds on the high plateau (Ribāṭ; Dhamār) and in similar conditions in Khawlan Ash Sham (Ḥaydān).

10. LOTUS *L.*

Herbs with 5-foliolate leaves, the lower pair of leaflets at the base of the short rhachis stipule-like. Inflorescence an axillary pedunculate umbel with a 1–3-foliolate bract. Pod straight, cylindrical.

1. Leaves glabrous. Flowers yellow — *L. corniculatus*
   Leaves hairy. Flowers pink or white — 2
2. Plant ± adpressed hairy. Leaflets acute — *L. quinatus*
   Plant densely pilose. Leaflets rounded to truncate, apiculate — *L. sp. A.*

1. **Lotus corniculatus** *L.*

L. arabicus sensu Schwartz (probably)

Prostrate or ascending perennial herb, mostly glabrous but with a few long hairs on the calyx, young leaves and especially the peduncles. Leaflets oblanceolate to elliptic c 14 × 4mm, acute, glabrous. Umbel mostly 2-flowered; calyx teeth c 4mm long, flowers yellow, the standard reddish, about 9mm long. Pods up to 3cm long.

A plant of marsh, springs and pond sides; common in suitable habitats on and around the high plateau; also on the Ibb mountains south to Wadi Sūdān, Qā'idah and in Wadi Rakhwah in Khawlan Ash Sham (Steinberg). Found from 1700–3100m. (Wood 74/38)

2. **Lotus quinatus** (*Forssk.*) *Gillett*

Dorycnium quinatum (Forssk.) C. Chr.

L. brachycarpus var. *menachensis* Schweinf.

L. arabicus sensu Deflers

Somewhat woody prostrate to weakly ascending perennial, sparsely adpressed pubescent in all parts. Leaflets oblanceolate, up to c 10 × 3mm, acute, adpressed hairy, but not obviously tuberculate. Umbels 2–4 flowered; calyx teeth c 3mm long, flowers white, the standard often pinkish, c 8mm long. Pods c 1.5cm long.

Frequent between 1500 and 2900m on the escarpment from J. Ṣabir and J. Ḥabashī north to Ḥajjah and Kuḥlān; in Khawlan Ash Sham at Ḥaydān and Madaq (Steinberg); on the high plateau only known from Yarīm and around Sana'a. Characteristic of sparsely vegetated crumbly volcanic rock in areas of reasonable rainfall where it is commonly associated with *Elionurus muticus* and *Becium serpyllifolium*. (Wood 72/44, 73/24)

Outside Yemen only known from Ethiopia.

3. **Lotus sp. A**

Ascending or erect annual to c 40cm, densely pilose in all parts. Leaflets obovate to oblanceolate, up to 15 × 9mm, apex rounded to truncate, apiculate, pilose, tuberculate. Umbels 3–7 flowered; calyx teeth c 4mm long; flowers white, the keel often pinkish, c 11mm long. Pods 1.5–2cm long.

A rare plant of roadsides and sandy plain at 2200–2400m on the high plateau; Wadi Rijām, Bani Ḥushaysh; Jiḥānah; Ma'bar. (Wood 1077, 2315) It is also known from Saudi Arabia where it was collected by Popov at Abha.

Lotus garcini DC. is listed by Schwartz from Yemen and the record is attributed to Deflers. I have been unable to trace the record in any of Deflers' writings and it must surely be a mistake.

11. VERMIFRUX *Gillett*

1. **Vermifrux abyssinica** (*A. Rich.*) *Gillett*

Helminthocarpon abyssinicum A. Rich.

Slender, wiry, procumbent perennial. Leaves 5–7-foliolate, grey-pubescent, very small; leaflets alternate, oblong, c 6mm long. Inflorescence an axillary pedunculate head with 6–8 very small orange-red flowers about 3mm long. Pods about 1m wide, spirally twisted in 1–2 coils, brown, reticulate.

A characteristic plant of mountains in Yemen found everywhere along the escarpment and around the high plateau from 2000 to 3400m. Usually found in rock crevices, around boulders and more rarely in short, rather thin grassland.

12. PSORALEA *L.*

1. **Psoralea corylifolia** *L.*

Erect much-branched perennial herb to about 80cm, sparsely pubescent but densely covered in dark red glands in all parts. Leaves simple, broadly ovate to sub-orbicular, c 2–5cm in diameter, the margin obscurely dentate. Inflorescence a long pedunculate dense head-like raceme of small mauve flowers. Pods dark brown, sub-reniform, c 4mm long, scarcely exceeding the calyx.

A very local plant of grassland by permanent streams between 1400 and 1800m: Wadi Ḥiḍrān, Taiz; Wadi Sūdān, Qā'idah.

13. CYAMOPSIS *DC.*

1. **Cyamopsis tetragonoloba** (*L.*) *Taub.*

C. psoralioides DC.

Erect annual with hairs attached at the middle. Stem somewhat winged. Leaves pinnately 3-foliolate, but without stipels; leaflets ovate to elliptic, c 3–5cm long, dentate. Inflorescence a sessile axillary raceme of small pale mauve flowers. Pods subsessile, erect, linear, flattened, 4–5cm long.

A rare cultivated plant grown under sorghum and appearing to be a weed. Known only from between 300 and 400m in the Tihama near Bājil and Khalīfah.

14. INDIGOFERA *L.*

Annual or perennial herbs almost always woody based and often distinctly shrubby, the stem usually strigose. Leaves very variable, but usually imparipinnate, small and grey-green in colour. Flowers red, usually in axillary racemes. These features usually make it easy to identify a plant as an *Indigofera* and the identification can be confirmed on close examination by the apiculate anthers and the presence of hairs attached at their middle.

Plants of this genus are the source of indigo ('Nil') and several species were formerly used and perhaps cultivated for this dye. They include *I. tinctoria*, *I arrecta*, *I. coerulea* but it seems that indigo is not produced nowadays in Yemen.

1. Leaves simple. Pod globular — *I. linifolia*
   Leaves compound. Pod elongate — 2
2. Plant spiny — 3
   Plant unarmed — 5
3. Inflorescence rhachis (the spine) up to 6cm long. Pod curved — *I. sp. A.*
   Inflorescence rhachis up to 4cm long. Pod straight — 4
4. Inflorescence rhachis orange with up to 5 flowers. Pod not torulose — *I. spinosa*
   Inflorescence green with up to 15 flowers. Pod strongly torulose — *I. spiniflora*
5. Inflorescence a sessile cluster in the leaf axils — *I. sessiliflora*
   Inflorescence an axillary raceme — 6
6. Leaves digitately 3-foliolate, leaflets obovate — *I. tritoides*
   Leaves pinnately 3 or more-foliolate; if 3-foliolate, leaflets not obovate — 7
7. Leaflets alternate — 8
   Leaflets opposite — 12
8. Leaflets usually 3 (rarely up to 7). Pod curved — *I. oblongifolia*
   Leaflets usually 4–7 (–11). Pod straight — 9
9. Plant densely white woolly. Pod torulose — *I. semitrijuga*
   Plant ± appressed hairy, never white woolly. Pod not torulose — 10
10. Calyx less than half the length of the flower, the lobes little longer than the tube. Pod more than 2cm long — *I. schimperi*
    Calyx more than half the length of the flower, the lobes much longer than the tube. Pod less than 2cm long — 11
11. Leaflets obovate. Stipules with a wide scarious base. — *I. spicata*
    Leaflets narrowly oblong. Stipules without a broad scarious base — *I. phillipsiae*
12. Standard glabrous outside — 13
    Standard hairy — 14
13. Inflorescence pedunculate, exceeding 5cm long — *I. costata*
    Inflorescence sessile, less than 3cm long — *I. parviflora*
14. Stem with spreading hairs. Pod with stalked glands — 15
    Stem adpressed hairy (rarely a few stalked glands present). Pod without stalked glands — 16
15. Stalked glands abundant on the stem. Calyx without brown hairs — *I. colutea*
    Stalked glands absent or very few on the stem. Calyx covered in brown hairs — *I. atriceps*
16. Leaflets with a translucent blister on either side and at the apex. Raceme 2–4 flowered, rhachis filiform, persistent — *I. brevicalyx*
    Leaflets without translucent blisters. Raceme usually with more than 5 flowers, rhachis stout, not persistent after the fruit falls — 17
17. Pod flattened. Raceme usually less than 1.5cm long. Leaflets usually 5 — 18
    Pod terete. Raceme variable but if short then leaflets more than 5 — 19
18. Pod straight. Leaflets up to c 1cm long — *I. arabica*
    Pod falcate. Leaflets c 1.5cm long — *I. hochstetteri*
19. Pod torulose. Leaflets obovate to orbicular often exceeding 1cm broad — 20
    Pod not torulose. Leaflets not usually obovate or orbicular but if so then less than 1cm broad — 21
20. Leaflets 3–7. Pod straight to slightly curved, usually less than 1cm long — *I. articulata*
    Leaflets 5–9. Pod strongly curved, about 1.5cm long — *I. coerulea*
21. Leaflets 3–9. Pod tetragonous. Stems distinctly ridged — *I. trita*
    Leaflets 7–25. Pod not tetragonous. Stems not distinctly ridged — 22
22. Pod exceeding 1.5cm long, deflexed — 23
    Pod about 1cm long or less, spreading — 24
23. Pod straight, about 6 seeded. Inflorescence hairs brown — *I. arrecta*
    Pod straight or slightly curved, 8–12 seeded. Inflorescence hairs whitish — *I. tinctoria*
24. Leaflets obovate, c 6mm long. Pod c 1.5mm broad — *I. argentea*
    Leaflets oblanceolate to oblong, c 11mm long. Pod c 3mm broad — *I. amorphoides*

1. Indigofera amorphoides *Jaub. & Spach*

Erect woody herb to 80cm. Leaflets opposite, 7–15, oblong to oblanceolate, c 11 × 4mm. Racemes sessile, many-flowered up to 3cm long. Pods thick, straight, spreading, 5 × 2.5mm.

Superficially very similar to *I. tinctoria* and *I. arrecta* but easily distinguished by the very short pod.

A plant of roadsides and open disturbed stony slopes near cultivation along the whole escarpment between 1100 and 2200m but never occurring in abundance. It has also been found in the Mashriq in the Ḥūth area and in the southeast mountains near Damt.

2. Indigofera arabica *Jaub. & Spach*

Small, usually procumbent woody herb. Leaflets opposite, usually 5, oblanceolate, c 6–9 × 2–3mm. Racemes sessile, many-flowered, 1–2cm long. Pods flat, straight, deflexed, c 7–10 × 2.5mm.

As noted by Schwartz two habitat forms occur. One of these is a decumbent plant with rather lax branches (Hepper and Wood 6058, Wood 468) found in protected places with relatively good rainfall at lower altitudes. The other is a procumbent, cushion-like plant with short stiff branches (e.g. Wood 75/132) usually found in exposed places on the high plateau. The second form is commonly associated with *I. phillipsiae* from which it can be easily distinguished even when sterile by its opposite leaflets.

Widespread on the escarpment, high plateau and south-east mountains from about 500m near Mishrāfah and Ḥajjaylah to 3100m on J. Lawz, but absent from areas of high rainfall in the central escarpment. It is often abundant on rocky plain and exposed hillsides and is particularly characteristic of rock crevices.

3. Indigofera argentea *Burm. f.* Draydah

I. burmanni Boiss.

Erect annual, commonly about 10cm high but reaching 1m after good rainfall. Stem with scattered stiff dark stalked glands amongst the white adpressed hairs. Leaflets opposite, 5–11, obovate, c 6 × 4mm. Racemes shortly pedunculate, 2–3cm long, up to 12-flowered. Pods straight, spreading, c 10 × 1mm.

One of the most characteristic plants of sand dunes in the Tihama occurring in great abundance after rain. It has been found from sea level to 250m from Mukhā north to Ḥaraḍ.

4. Indigofera arrecta *A. Rich.*

Erect undershrub to about 1m. Leaflets opposite, 7–15, oblanceolate to oblong, c 12 × 4mm. Racemes sessile, up to 5cm long but usually much less, many flowered. Pods straight, deflexed, c 16–20 × 2mm, up to 6-seeded.

Readily distinguished from similar species by the dark brown hairs found on the flowers and to a lesser extent on the calyx and young stems.

An uncommon plant of grassy banks in areas of high rainfall between 2000 and 2800m along the escarpment mostly in the Ibb area and the central escarpment and not found north of Kuḥlān. Schwartz has a number of high plateau records notably Wadi Ḍahr (Deflers) and Ghirās (Wissmann). It is not found there today and these records probably represent the relics of former cultivation.

5. Indigofera articulata *Gouan*

I. argentea sensu Schwartz p.p.

Erect silvery shrubby perennial to about 75cm. Leaflets opposite 3–5 (–7), obovate, large, c 1.5–3.5 × 0.5–2cm. Racemes sessile, up to 5cm long, with up to 15 pale red flowers. Pods slightly curved, spreading to deflexed, strongly torulose, c 10 × 3.5mm.

Easily confused with *I. coerulea* (as in Schwartz) but differing in its usually 3-seeded, short, very torulose, only slightly curved pods. The young shoots are a distinctive olive-green.

A plant of stony country with poor rainfall between 700 and 2300m. In Yemen apparently restricted to three areas: (a) the southwestern foothills from Waṣāb south to the Taiz area and thence east to Qabayṭah (b) around Madinat Al 'Abīd (c) a belt of country in the Mashriq from Khulāqah in Nihm and Bani Ḥārith northwest to Ḥarf Sufyān, Shahārah and Majz.

6. Indigofera atriceps *Hook.f.*

Erect woody perennial to about 1m with numerous white and brown hairs. Leaflets opposite, 5–13, elliptic-oblong c 8 × 3mm. Racemes long pedunculate, dense, about 3–5cm long. Inflorescence densely covered in brown hairs. Pods straight, spreading, covered in numerous ± spreading brown and white hairs and stalked glands, very short, c 8 × 2mm.

The Yemen plant belongs to subspecies **kaessneri** (*Bak.f.*) *Gillett*. It is easily recognised by the dark brown whiskery calyx lobes.

Only known from a steep grassy slope at 1900m between Markah and Sa'bah, J. Milḥān.

7. Indigofera brevicalyx *Bak.f.*

I. sparsa sensu Schwartz

Prostrate or ascending, delicate woody herb. Leaflets opposite, 5–11, oblong to oblanceolate, very small, c 4 × 1mm. Racemes pedunculate, about 3cm long, lax, 2–4 flowered. Pods straight, erect, c 18 × 1.5mm.

This species is unique among Yemen *Indigofera* in having leaflets with a translucent blister on either side and near the apex and in its delicate filiform but long persistent raceme rhachis.

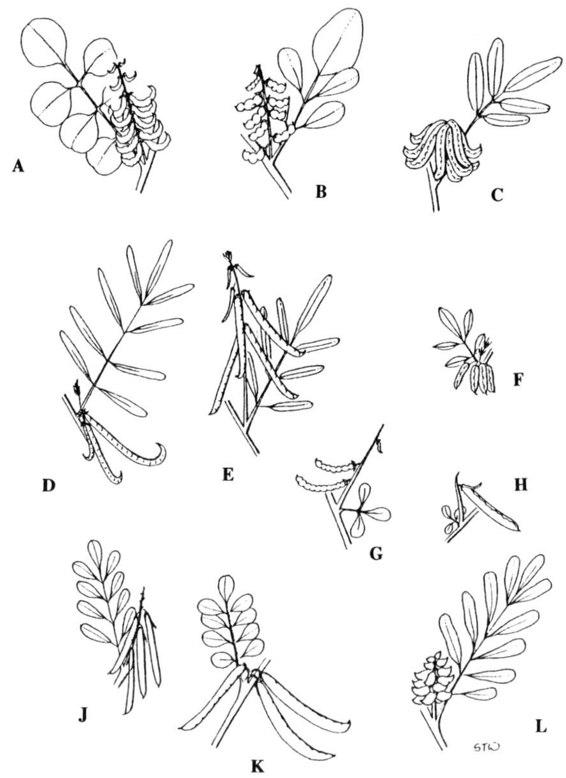

Fig. 11. Leaves and fruits of *Indigofera* species. **A** *I. coerulea*; **B** *I. articulata*; **C** *I. hochstetteri*; **D** *I parviflora*; **E** *I. costata*; **F** *I. arabica*; **G** *I. spiniflora*; **H** *I. spinosa*; **J** *I. arrecta*; **K** *I. tinctoria*; **L** *I. amorphoides*.

An uncommon plant of shallow stony grassland between 1500 and 2800m in areas of good rainfall; widespread in the central escarpment east to J. Shibām, Manākhah and Rujum; also at 'Udayn, Shahārah and 'Anis (J. Dawran (Wissmann) and Suq al Jum'ah).

8. Indigofera coerulea *Roxb.*
I. argentea sensu Schwartz p.p.
Erect shrubby perennial to about 75cm. Leaflets opposite, 5–9 (usually 7), obovate to suborbicular, c 1.5–3.5 × 1.5–3cm. Racemes sessile, up to 6cm long, many-flowered, rather dense. Pods strongly curved, deflexed, weakly torulose, c 15 × 2mm.

Yemen plants belong to variety **occidentalis** *Gillett & Ali* with leaflets hairy on the upper surface.

Very similar to *I. articulata* but usually with more leaflets and a longer, less torulose, strongly curved pod.

A weed of roadsides and fields in the silt plain at the east of the Tihama and on the escarpment; widespread but scattered, occurring between 300 and 1500m.

9. Indigofera colutea (*Burm. f.*) *Merr.*
I. viscosa Lam.
Erect or spreading annual to about 50cm with spreading hairs and numerous stalked glands. Leaflets opposite, 7–13, oblong – elliptic, c 8 × 3mm. Racemes pedunculate, 2–5cm long, rather lax with 5–10 (–15) flowers. Pods straight, spreading, c 20 × 1.5mm.

This is the only Yemen species whose stems are densely covered in stalked glands.

A weed of roadsides, fields and disturbed rocky ground between 500 and 1500m. Widespread on the escarpment north of the Wadi Lā'ah, but preferring areas of good rainfall and mainly appearing after heavy rains.

10. Indigofera costata *Guill. & Perr.*
I. goniodes Hochst.
Erect woody annual to about 50cm. Leaflets opposite, 5–9, linear-oblong, c 15–25 × 2–3mm. Racemes pedunculate, often exceeding 10cm long, lax, many-flowered. Pods slightly upturned towards the tip, deflexed, slightly flattened, c 25–35 × 2mm.

Yemen plants belong to subspecies **gonioides** (*Bak.*) *Gillett.*

Apart from the 3-foliolate *I. tritoides*, this species and the very similar *I. parviflora* are the only Yemen *Indigoferas* with glabrous flowers.

An uncommon weed of fields, roadsides and disturbed rocky grassland, found between 1400 and 2000m on the escarpment. It prefers well-drained stony ground in areas of average to good rainfall and has a strong southern tendency, not being found north of the Ḥarāz.

11. Indigofera hochstetteri *Bak.*
I. anabaptista Steud.
Usually decumbent annual. Leaflets opposite, usually 5, oblong to lanceolate, c 14–18 × 5mm. Racemes sessile, dense, many-flowered, up to 1.5cm long. Pods flattened, curved, deflexed, c 12–15 × 2.5mm.

Closely related to *I. arabica* from which it is distinguished by the larger leaflets and the longer, curved pod.

Very widespread occurring from 150 to 1500m in the eastern Tihama and on the escarpment north to Al 'Ashshah; usually found in short grassland on sand and well-drained stony ground (often on field borders), tolerant of high and low rainfall, but erratic in appearance and generally uncommon.

12. Indigofera linifolia *L.f.*
A very distinctive decumbent annual species with simple linear leaves c 8–30 × 1mm. Racemes sessile, dense, head-like, up to 1.5cm long with 5–12 flowers. Pods globular, apiculate, c 1.5mm long.

A rare plant of loose stony hillsides in areas of high rainfall between 600 and 1500m: Ḥadīyah and Suq Ar Ribāṭ, J. Raymah; Wadi Sukhnah; 'Udayn.

13. Indigofera oblongifolia *Forssk.* Ḥaṣār
A vigorous undershrub sometimes reaching 1½m high with distinctive flattened branches. Leaflets alternate, usually 3, but sometimes 1–2 or 4–7, oblong, c 8–20 × 3mm, the terminal rather larger than the lateral leaflets. Racemes shortly pedunculate, up to 7cm long, many-flowered. Pods torulose, slightly curved, spreading to deflexed, c 12–18 × 2mm.

A widespread plant in the Tihama and the escarpment foothills found from sea level to 1400m; abundant and locally dominant to the exclusion of almost all other species on silt and stony plain in the east of the Tihama and along drainage lines in the foothills.

14. Indigofera parviflora *Wight & Arn.*
Almost identical to *I. costata* except for the racemes which are sessile and up to 3cm long only.

Apparently common in short stony grassland and in fields between 1200 and 1500m from 'Udayn north to Waṣāb and thence northeast to Riḥāb and Mukhādir. Otherwise known only from Janad (Botta) and between Ḥammām Shamsān and Maḥwīt.

15. Indigofera phillipsiae *Bak. f.*
Procumbent perennial herb. Leaflets alternate, 3–6, oblong or lanceolate c 4–20 × 2–5mm, the terminal much larger than the lateral leaflets. Racemes long pedunculate, up to 12cm long, many-flowered. Pods straight, deflexed, c 10 × 2mm.

Plants from arid situations seem to have smaller more silvery leaves.

A plant of open stony plain found between 1700 and 2400m towards the east of the high plateau from Rada' north to the Sana'a area and thence through Nihm to Raydah, Shahārah and Ṣa'dah.

16. Indigofera schimperi *Jaub. & Spach*
Erect to decumbent undershrub to 50cm usually of noticeably untidy appearance. Leaflets alternate, 3–7, obovate to elliptic, rather large, c 9–20 × 6–14mm, the terminal usually larger than the lateral leaflets. Racemes pedunculate, up to 12cm long, many-flowered. Pods straight, weakly deflexed, c 22 × 1.5mm.

Distinguished from other similar species by the very short calyx which itself has very short triangular lobes (most easily seen on flowers about to open).

An uncommon but possibly overlooked species of open rocky hills in the escarpment foothills between 600 and 1500m: J. Jirdad, Bani 'Umar; between Ḥajdah and Barḥ; Ḥajjaylah (Schweinfurth); between Al Qidam and Khamīs Banī Sa'd; on the dip slope east of Maḥābishah.

17. Indiogofera semitrijuga *Forssk.*

Procumbent annual herb densely covered in white woolly hairs. Leaflets alternate to sub-opposite, usually 5, obovate, c 7 × 5mm. Racemes subsessile, c 2–4cm long, rather dense, many-flowered. Pods torulose, usually 2-seeded, straight, deflexed, c 5–6 × 2mm.

Readily identified by the distinctive woolly indumentum.

A rare plant of sand dunes near the sea found at Ṣalīf and between Luḥayyah and Khawbah.

18. Indigofera sessiliflora *DC.*

I. trigonelloides sensu Schwarz

Procumbent to ascending annual. Leaflets alternate to sub-opposite, 1–5, linear-oblong, c 9–20 × 2–4mm, the terminal much larger than the lateral leaflets. Flowers in sessile clusters in the leaf axils. Pods torulose, straight, spreading, c 12 × 1.5mm.

This is the only Yemen species with flowers clustered in the leaf axils.

A rare plant of sand dunes below 250m found between Hodeida and Ḍaḥī and between Qutay and Bājil.

19. Indigofera spicata *Forssk.*

I. endecaphylla Jacq.

Prostrate herb. Leaflets alternate, 5–10, obovate, c 5–12 × 4–8mm. Racemes pedunculate, 2–5cm long, dense, many-flowered. Pods torulose when immature, straight, deflexed, c 15 × 1.5mm.

Best distinguished from other species by the broad scarious margined stipules.

An uncommon plant of grassland in areas of good rainfall between 1000 and 1900m from Musalla, Aḥkum north to J. Milḥān and Wadi Sukhnah; particularly characteristic of grassy banks between irrigated fields as at Wadi Sudān near Qā'idah, Suq Ar Rubū', Waṣāb As Sāfal etc.

20. Indigofera spiniflora *Hochst. & Steud. ex Boiss.*

A spiny undershrub very similar in general appearance to *I. spinosa*. Leaflets opposite, 3, elliptic to oblanceolate c 6 × 2.5mm. Racemes of usually 10–15 flowers on a green spine-tipped rhachis 1–4cm long. Pods usually curved, ± spreading, strongly torulose, c 8–15 × 2mm.

Commonly confused with *I. spinosa* and as a result its distribution is unclear. It appears to be most characteristic of the plains towards the east of the Tihamah where it has been recorded from Bājil (Schweinfurth), Bayt al Faqīh and Al Barḥ. Inland it has been found at Ḥajjah and Madinat Al 'Abīd. It thus has an altitude range of 200 to 1400m. (Wood 73/37, 1361)

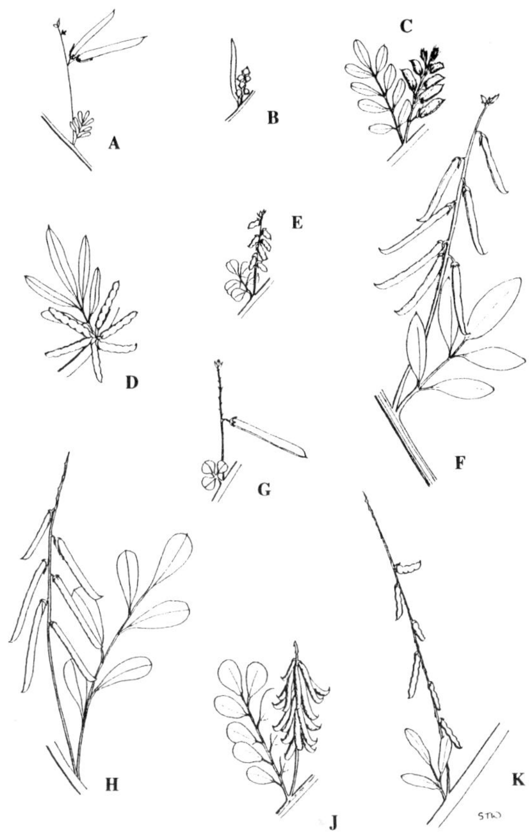

Fig. 12. Leaves and fruits of *Indigofera* species. **A**. *I. brevicalyx*; **B** *I. linifolia*; **C** *I. atriceps*; **D** *I. sessiliflora*; **E** *I. semitrijuga*; **F** *I. trita*; **G** *I. tritoides*; **H** *I. schimperi*; **J** *I. spicata*; **K** *I. oblongifolia*.

21. Indigofera spinosa *Forssk.* Qaṣār, Ḥil

A spiny undershrub with intricate branches. Leaflets opposite, 3, elliptic to oblanceolate, usually 4 × 2mm. Racemes consisting of 2–5 flowers usually towards the apex of an orange spine-tipped rhachis 10–15mm long. Pods somewhat tetragonal, never torulose, spreading to erect, c 12–22 × 2mm.

Abundant in Yemen from sea level to 1900m; although occurring on sand in the Tihama, it is more characteristic of stony hills occurring everywhere on the escarpment, the southeastern mountains and the Mashriq.

22. Indigofera tinctoria *L.* Huwir, Huwirwir

An erect undershrub similar in general appearance to *I. arrecta* but lacking the dark brown hairs found on the inflorescence of that species. The pod is usually 2–3cm long, 8–12 seeded and slightly curved.

Abundant around cultivation and on tracksides in the valleys of the escarpment foothills from the Wadi Rima' northwards; occasionally extending into the alluvial plains at the east of the Tihama and the valleys further inland; found between 200 and 1200m.

23. Indigofera trita *L.f.*

I. spachii Bak.

An erect or sprawling untidy annual or short-lived perennial herb. Leaflets opposite, 5, elliptic, up to 30 × 10mm. Racemes pedunculate, c 8–20cm long, rather lax but many-flowered. Pods tetragonal, straight or slightly curved, weakly deflexed, c 20–28 × 1.5mm.

The above description is of variety **scabra** (*Roth*) *Ali*, but variety **maffei** (*Chiov.*) *Ali* also occurs. This is rather distinct having 3-foliolate leaves, shorter subsessile racemes and the fruit pedicel bent sharply downwards.

The species as a whole is a plant of grassy scrub restricted to, but widely distributed in the Ibb-'Udayn region and the central escarpment north to Maḥābishah with an outlying record from J. Ash Shirq. Variety *scabra* is much the most common and is found from 1200 to 2000m. Variety *maffei* is restricted to J. Raymah (Ḥadīyah) and the Ḥarāz (Rajamah and Baḥal) and grows at the lower altitude range of 600 to 1200m.

24. Indigofera tritoides *Bak.*

Prostrate woody herb. Leaflets opposite, 3, obovate, c 4–7 × 3–5mm. Racemes sessile, 1–4cm long, many-flowered. Pods tetragonal, straight, erect, 20–29 × 2mm.

The only non-spiny Yemen species with digitately 3-foliolate leaves; also distinctive on account of its glabrous flowers.

A rare plant of gravelly plain in the Tihama found from 400 to 600m; near Suq al Kawlah, J. Ra's (Wadi Zabīd); Wadi Damīm near Ḥays.

25. Indigofera sp. A

A spiny undershrub very similar to *I. spiniflora* from which it differs in having subsessile leaves, a very long inflorescence rhachis up to 6cm long and a strongly arcuate pod which is somewhat compressed so the outline of the seeds is clearly visible.

This species is represented by two specimens at Kew collected by D. Pratt (Nos. 81 & 96) from between 250–500m in the Wadi Rima'. The specimens are very small and there are no collector's notes, but the pod is very distinctive and they probably represent an undescribed species.

15. PTYCHOLOBIUM *Harms*

1. Ptycholobium plicatum (*Oliv.*) *Harms*

A small, spreading, wiry, densely pubescent perennial similar in appearance to a *Tephrosia*. Leaves very shortly petiolate, mostly 3-foliolate, leaflets cuneate – oblanceolate, often emarginate at the apex, c 27 × 6mm. Inflorescence of few-flowered clusters in the leaf axils; flowers yellow, small, c 5mm long. Pods laterally plicate like a concertina, falcate, c 2cm long.

Yemen plants belong to subspecies **arabicum** *Brummitt*, but not to the extreme form of the species which is glabrous with unifoliolate leaves.

A rare plant found on open, well-drained, sparsely vegetated hillsides in areas of good rainfall between 600 and 1600m: in several places around Ḥadīyah; Ḥajjah; on the limestone dip slope east of Maḥābishah.

16. TEPHROSIA *Pers.*

Perennial or annual, often wiry herbs with simple adpressed or spreading silky hairs. Leaves simple or imparipinnate usually with oblanceolate leaflets. Flowers pink or purple, solitary or in racemes, always with a hairy standard. Pod linear – oblong, many-seeded, always beaked.

A 'difficult' genus. Many of the species are very widespread over the old world tropics and are very variable so that rather trivial characters often have to be used to separate species. *T. subtriflora* and *T. uniflora*, for example, are usually separable on leaflet and seed number but some specimens can only be distinguished by examination to see whether the style is twisted or not. It is perhaps worth mentioning that many species are more easily separable in the field than in the herbarium. *T. purpurea* and *T. pumila* for example never occur in the same habitat in Yemen.

1.	Leaves simple	*T. heterophylla*
	Leaves pinnate	**2**
2.	Flowers 2–3 in the leaf axils	**3**
	Flowers in racemes	**6**
3.	Leaves pubescent above	*T. quartiniana*
	Leaves glabrous above	**4**
4.	Leaf rhachis about half as long as the terminal leaflet; leaves strikingly white hairy below	*T. pentaphylla*
	Leaf rhachis about as long as the terminal leaflet or longer; leaves not strikingly white hairy below	**5**
5.	Pod 5–9 seeded. Style ribbon-shaped, not twisted (best seen on immature fruit)	*T. subtriflora*
	Pod 9–13 seeded. Style tapering, twisted	*T. uniflora*
6.	Pod ovoid, 1-seeded	*T. nubica*
	Pod linear-oblong, many-seeded	**7**
7.	Style pubescent	*T. elata*
	Style glabrous	**8**
8.	Calyx covered in yellow-brown spreading hairs	*T. villosa*
	Calyx adpressed hairy or if spreadingly hairy, hairs white	**9**
9.	Seeds 8–14, about 2.5mm apart	*T. pumila*
	Seeds 5–9, 4–8mm apart	**10**
10.	Leaflets 7–17, seeds 4–5mm apart	*T. purpurea*
	Leaflets 5–9, seeds 6–8mm apart	*T. apollinea*

1. Tephrosia apollinea *(Del.) DC.*

Erect bushy perennial to c 40 cm; stems densely but shortly pubescent. Leaflets 5–9, oblanceolate, adpressed pubescent on both surfaces. Inflorescence a short raceme, c 3–7 cm long, the flowers more dense than in related species; flowers dull pink or dirty white. Pods 5–8 seeded, the seeds 6–8 mm apart, up to 4.5 cm long.

Very often confused with *T. purpurea* of which it may be only a subspecies. Apart from the key characters it is of coarser, more robust appearance than *T. purpurea* and has a shorter, more crowded raceme. The distance between the seeds seems to me the most reliable character to use in separating the two species.

In Yemen only known at 1100 m from the edge of the Sands, south of Naqil Hajlah and at Harib. (Wood 3162)

2. Tephrosia elata *Deflers*

Erect bushy perennial to 1½m with a strongly ridged, sparsely adpressed hairy stem. Leaflets 7–13, oblanceolate, up to 4.5cm long. Inflorescence a terminal raceme c 5–10cm long; flowers purple, standard brown-velvety. Pods brown-tomentose, 7–9 seeded, c 5cm long.

This very attractive species is distinguished from all other species of *Tephrosia* in Yemen by its pubescent style.

An uncommon plant of stony grassland in areas of high rainfall between 1500 and 2500m: quite common in the area round Ibb west to Mashwarah and north to Mukhādir; Saddah, Wadi Banā; Suq al Jum'ah, J. Ash Shirq; J. Kabūrah, Ḥadīyah; Banī Dawān-Al Jabīn, J. Raymah; Ḥajarah, Ḥarāz; Aṣ Ṣa'bah, J. Milḥān.

3. Tephrosia heterophylla *Vatke.*

T. schweinfurthii Deflers.

Prostrate to erect perennial with a pubescent often somewhat zigzag stem. Leaves simple, usually oblanceolate or narrowly oblong below, becoming linear and longer above and reaching 5cm in length. Flowers solitary or more rarely paired in the leaf axils, deep pink. Pods ± 5 seeded, c 2.5–3.5cm long.

A widespread species with a very scattered distribution between 1000 and 2300m on open well-drained hillsides and rocky plain on the escarpment, high plateau and the Mashriq from Turbah north to Ṣa'dah. It seems to be particularly characteristic of rainshadow country in the central escarpment, of *Euphorbia* scrub around Taiz and Qā'idah and of limestone hills both in Banī 'Umar in the Ḥujarīyah and in the north around Ḥūth and Ṣa'dah, but it is by no means restricted to these vegetation types.

4. Tephrosia nubica (*Boiss.*) Bak.

T. pogonostigma Boiss.

Erect bushy annual or perennial to c 60cm with a densely pubescent stem. Leaflets 5–9, oblanceolate, ± tomentose on both surfaces. Inflorescence a raceme up to 20cm long; flowers pale pink. Pods densely long-white-tomentose, ovoid, 1-seeded, c 1cm long.

T. nubica is conventionally divided into two subspecies:

Subspecies **nubica** has 3–6 pairs of leaflets and a spreading indumentum on the stems. It is a rare plant of stony outcrops and gravel banks in the Tihama and the foothills from 100–500m: Bājil and Marāwi'ah (Schweinfurth); Wadi Rima' at 100m (Pratt YS65): Suq As Sabt, J. Bura'; between Bājil and Khamīs Banī Sa'd; Suq al Kawlah, Wadi Zabīd (Wood 1419); Ḥays.

Subspecies **arabica** (*Boiss.*) *Gillett* has 2–3 pairs of leaflets and an adpressed indumentum. It is only known from the edge of the Sands at Ḥarīb at 1100m (Wood 3061) and from Al Ḥazm, Jawf (J. King).

5. Tephrosia pentaphylla (*Roxb.*) *G. Don*

T. senticosa sensu Schwartz

Woody ascending perennial with ± silvery, adpressed pubescent stems. Leaves with a very short rhachis, always shorter than the terminal leaflet; leaflets usually 5, oblanceolate, glabrous above, silvery beneath. Inflorescence of 1–3 red flowers in the leaf axils. Pods somewhat silvery – pubescent, 7–8 seeded, c 4cm long, the style tapering but not twisted.

Easily recognised in the field by the silvery undersides of the leaves which contrast strongly with the green upper surface.

A rare plant of valley scrub between 500 and 1600m, mainly in the central escarpment: on the western slopes of J. Raymah and J. Bura'; Wadi Dūr, 'Udayn; on the limestone dip slope east of Maḥābishah.

6. Tephrosia pumila (*Lam.*) *Pers.*

T. diffusa sensu Schwartz.

Procumbent annual or perennial with densely pubescent stems. Leaflets 9–13, oblanceolate, pubescent on both surfaces. Inflorescence a short raceme c 3–7cm long, the flowers pale pink. Pods 8–14 seeded, the seeds c 2.5cm apart, 3–3.5cm long.

Very easily confused with forms of *T. purpurea* from which it can only be reliably separated by the more numerous seeds.

Widespread on the escarpment between 600 and 1500m from Wadi Ḍabab near Taiz north to J. Marran in Khawlan Ash Sham but rather local in occurrence and usually in partial shade on banks round fields or in scrub in areas of good rainfall.

7. Tephrosia purpurea (*L.*) *Pers.* Sanafī, Sanafa

Annual or perennial herb, often wiry and rather woody below; very variable in habit and indumentum. Leaflets 7–17, oblanceolate. Inflorescence a lax raceme c 5–20cm long; flowers pale pink to deep mauve. Pods 6–9 seeded, the seeds 4–5mm apart, up to 4.5cm long.

This notoriously variable species is divisible into two subspecies in Yemen, both restricted to the Tihama:

Subspecies **leptostachya** (*DC.*) *Brummitt*

T. leptostachya DC.

T. desertorum sensu Hepper

Procumbent, sparsely pubescent annual or perennial. Leaflets mostly 13–17, glabrous above. Racemes 10–20cm long; flowers deep mauve.

Readily recognised by the sparse indumentum, rich flower colour and very long racemes.

Apparently very local, but sometimes frequent on sand dunes after rain: Bayt al Faqīh; Manṣūrīyah; between Qutay and Bājil. (Wood 772, 1488)

Subspecies *Aff.* **dunensis** Brummitt

T. apollinea sensu Schweinfurth, Hepper etc.

Usually erect (rarely procumbent or ascending) perennial, almost always woody below, densely adpressed (less commonly spreadingly) pubescent, the young shoots appearing whitish. Leaflets 7–11, pubescent on both surfaces. Racemes c 5–12cm long; flowers usually pale pink.

Abundant on sand dunes, as a weed of sandy fields and along drainage lines in stony country throughout the Tihama up to 500m.

In its leaflet number this subspecies straddles the traditional concept of *T. purpurea* and *T. apollinea*; however, using the seed character which I have adopted here, the numerous examples I have examined all fall within *T. purpurea*. The Yemen plant is in fact very close to *T. purpurea subsp. dunensis* from the East African coast from which it seams to differ only in having fewer leaflets and a rather more erect habit. Examples of the typical Yemen plant include Wood 74/412, 75/98, Miller 3105, Hepper 5700, but some examples, notably Wood 1487 from Manṣūrīyah have a dense tomentose indumentum.

8. **Tephrosia quartiniana** *Cufod.*

T. vicioides A. Rich.

Procumbent perennial with sparsely pilose stems. Leaflets 5–9, obovate, the apex emarginate, pubescent on both surfaces. Inflorescence of 2–3 pale lilac flowers in the leaf axils. Pods pilose, 10–11 seeded, c 3–4cm long, the style ribbon-shaped, not twisted.

Only known from sand dunes at Bayt al Faqīh at 200m (Wood 75/774).

9. **Tephrosia subtriflora** *Bak.*

T. lathyrodes sensu Schwartz

Prostrate or ascending wiry annual with pubescent to villous stems. Leaflets usually 5–7 in Yemen plants, narrowly oblong to oblanceolate, glabrous above. Inflorescence of 2–3 pale pink flowers in the leaf axils. Pods with 5–9 seeds, mostly 2–3cm long.

A rare plant of stony plain in the Tihama and on rocky slopes, in the foothills between 200 and 1200m: Bani 'Umar, Ḥujarīyah (Wood 1399, rather untypical); between Mafraq and Nukaybah; south of Zabīd (Hepper); Rajamah near Ḥajjaylah; abundant near Khamīs Bani Sa'd. (Wood 1628)

10. **Tephrosia uniflora** *Pers.*

T. anthylloides Hochst.

A woody, branched, spreading to erect perennial, indumentum very variable. Leaflets 3–7, oblanceolate to obovate, glabrous above. Inflorescence of 2–3 pink flowers in the leaf axils. Pods (5–) 9–13 seeded, c 4–5.5cm long.

This species can be recognised by its twisted style.

Two subspecies are known in Yemen but their exact distribution is uncertain. Subspecies **uniflora** has a spreading indumentum, usually 7 leaflets and a pod commonly in excess of 5cm. It is represented by Wood 2290 from Mawsanah in the Ḥarāz and by Wood 2219 from between Ḥusaynīyah and Bayt al Faqīh. The latter example approaches subspecies **petrosa** *Gillett & Ali*. This has an adpressed indumentum, usually 5 leaflets and a pod less than 5cm long. It is represented by Wood 916 and Hepper & Wood 6029 both from Ḥajdah. It is probably the more common subspecies in Yemen.

A widespread plant with a scattered distribution between 200 and 1500m, usually occurring in small quantity; rare on sand and stony steppe in the east of the Tihama, more frequent on exposed rocky slopes in the foothills but avoiding areas of high rainfall.

11. **Tephrosia villosa** *(L.) Pers.*

Erect bushy perennial to 1m, stems tomentose. Leaflets 11–13, oblanceolate to oblong, tomentose on both surfaces. Inflorescence a densely brown-hairy raceme up to 15cm long; flowers purple, the standard brown-hairy outside. Pods densely tomentose, 6–12 seeded, c 3.5cm long.

The Yemen plant is subspecies **ehrenbergiana** (*Schweinf.*) *Brummitt.*

Only known from a grassy bank at 1300m on J. Marran, Khawlan Ash Sham.

17. **SESBANIA** Adans.

Erect woody annuals or rapid-growing shrubs. Leaves paripinnate with numerous leaflets. Inflorescence an axillary raceme. Pod usually very long, septate, beaked.

Style glabrous. Pod torulose	*S. leptocarpa*
Style pubescent. Pod not torulose	*S. pachycarpa*

1. **Sesbania leptocarpa** *DC.* Hawmir

S. arabica Hochst.

S. punctata sensu Deflers

Tall erect woody annual to 3m. Leaves glabrous, rhachis usually but not always unarmed, leaflets up to 15mm long. Flowers yellow with a purple spotted standard, c 10mm long; style glabrous. Pods c 12cm long, 15–20 seeded, strongly torulose.

Known with certainty from 'Abs (Wood 3060), Khalīfah (Schweinfurth 229, Wood 3429), Ḥajjaylah (Schweinfurth 1931), Wadi Sukhnah (Wood 3398) and Wadi Barakāni in the Ḥujarīyah (Wood 1409). It is very probable though that unconfirmed field records of *Sesbania* refer to this species and it is thus common as a field weed in the alluvial plains in the east of the Tihama from Zabīd northwards and in irrigated fields in the foothills and the southwest mountains of the Taiz area, being found from 200 to 1500m.

S. leptocarpa has been confused with *S. bispinosa* (Jacq.) W.F. Wight, a widespread tropical weed hitherto unrecorded from Yemen, which lacks the torulose pod of *S. leptocarpa*.

2. **Sesbania pachycarpa** *DC.*

Tall woody annual very similar to the previous species in general appearance but somewhat larger. Leaves glabrous, rhachis armed, leaflets up to 20mm long. Flowers yellow with a purple spotted standard c 15mm long, style pubescent (best seen on young fruit). Pods 14–20cm long, 20–30 seeded, not torulose.

Only known from a sorghum field at 1500m at Mukhādir near Ibb (Wood 2449).

Sesbania sesban (*L.*) *Merr.*, a small tree with yellow flowers, is occasionally cultivated and has been seen in a few places in the Ibb area. **Sesbania grandiflora** (*L.*) *Poir.*, a tree up to 5m tall with pale pink flowers, is one of the most widely planted trees in the Tihama and at Taiz. It is found in nearly all large Tihama towns and has been much used in recent tree planting projects. It is much less common in villages.

18. ASTRAGALUS *L.*

Annual or perennial herbs or woody undershrubs. Leaves imparipinnate with numerous leaflets. Pod elongate usually inflated, often ± narrowly conical with the upper margin inflexed so making it 2-roomed.

A huge genus centred on Turkey and Iran with few species in Yemen. One of these, *A. atropilosulus*, is the only species found in tropical Africa and the other four have a markedly northern distribution in Yemen.

1. Flowers in pedunculate racemes, purple — 2
   Flowers in ± sessile clusters, white or yellow — 3
2. Stipules leaf-like. Pod glabrous, not inflated — *A. atropilosulus*
   Stipules not leaf-like. Pod pubescent, inflated — *A. fatmensis*
3. Flowers white. Pod less than 1cm long — *A. tribuloides*
   Flowers yellow. Pod more than 1cm long — 4
4. Leaflets 40–50, less than 5mm long, leaf rhachis persistent, spine-like — *A. sieberi*
   Leaflets 15–25, more than 10mm long, leaf rhachis not persistent or spine-like — *A. sparsus*

1. Astragalus atropilosulus (*Hochst.*) *Bunge*

A. abnyssinicus Steud.

Erect, glabrous or sparsely hairy perennial to about 1m. Leaves with large ovate leaf-like stipules; leaflets usually 20–40, c 15mm long. Inflorescence a pedunculate, many-flowered raceme c 8–12cm long; flowers purple, c 12mm long. Pods flattened, ± oblong, acuminate at both ends, stipitate, c 3 × 0.5cm.

In Yemen there is only subspecies **burkeanus** (*Harv.*) *Gillett* with pod stipes more than 3mm long and the pod itself longer than 2.5cm.

Widespread but very thinly scattered between 1500 and 3100m on the escarpment and high plateau north to J. Minmār in Khawlan Ash Sham. It is found on roadsides, in grassland and most commonly on open stony slopes, but almost always as isolated individuals.

2. Astragalus fatmensis *Hochst.*

Erect or spreading annual herb to about 25cm. Leaflets c 10–20, 4–5mm long. Inflorescence a pedunculate head-like raceme of 5–15 purple flowers each about 3mm long. Pods inflated, pilose, narrowly conical, the upper margin strongly inflexed, c 10–13mm long.

Found on open rocky hillsides and sandy plain from sea level to 2400m. It is recorded from Mīdī (Ahmed Khattab) and the Majz area (Steinberg), but its main area of distribution is an arc from Rada' north to Sana'a and thence north to the Jawf (Zeller) and west to Ḥūth and the Ḥajjah area. Elsewhere it is unknown.

3. Astragalus sieberi *DC.*

Perennial undershrub with spreading or erect adpressed white-hairy stems to c 40cm. Leaves with the rhachis persistent after the leaflets have fallen and becoming spine-like; leaflets 40–50, 4–5mm long. Flowers yellow, about 2cm long, subsessile in clusters of 1–3 in the leaf axils. Pods inflated, adpressed hairy, narrowly conical, acuminate, c 4cm long.

Only known on sandy plain at c 2300m in Bani al Ḥārith and Bani Ḥushaysh, northeast of Sana'a.

4. Astragalus sparsus *Del. ex Decne.*

Spreading perennial herb with white woolly stems and leaves. Leaflets c 15–30, c 15mm long. Flowers yellow, about 2cm long, c 5–6 in subsessile axillary clusters. Pods inflated, pilose, narrowly conical, long acuminate, c 4–5 cm long.

Found only twice, on both occasions at about 2300m: on waste ground near Sana'a (Scott & Britton); on a stony hillside west of Sāqayn in Khawlan Ash Sham.

5. Astragalus tribuloides Del.

A usually prostrate and mat-like annual or short-lived perennial, sparsely pilose in all parts. Leaflets c 14–18, about 3mm long. Flowers white, about 2mm long, 1–4 in subsessile axillary clusters. Pods inflated, sparsely pilose, ± triangular in outline, curved, the upper margin strongly inflexed, c 7mm long.

Locally common and characteristic of derelict fields and stony plain on the high plateau from 2300 to 3200m on J. Nabī Shu'ayb. Also in similar conditions between Sāqayn and Ḥaydān in Khawlan Ash Sham.

19. BISERRULA *L.*

1. Biserrula palecinus *L.*

A delicate pubescent decumbent annual. Leaves imparipinnate; leaflets 15–20, oblanceolate-elliptic, about 6mm long with a notched apex and cuneate base. Inflorescence a long pedunculate raceme of usually 3 small pale blue flowers. Pods very distinctive, pendulous, c 20 × 7mm, with broad strongly dentate wings.

The Yemen plant is subspecies **palecinus**, whose distribution is centred on the Mediterranean.

Only known from c 2500–2700m on J. Shibām, Manākhah where it is quite common in sparse open grassland.

20. TAVERNIERA *DC.*

1. Taverniera lappacea (*Forssk.*) *DC.*

Usually procumbent, much-branched, grey-canescent undershrub. Leaves sessile, trifoliolate; leaflets fleshy, obovate, up to 5mm long but often less. Flowers solitary or paired in the leaf axils, shortly pedicellate, pink, c 8mm long. Pod 2–3 jointed, each joint c 7mm diameter, orbicular, covered in reddish spines.

A very rare plant of sand near the sea; collected by Forsskal at Luḥayyah but never found again in Yemen.

21. ONOBRYCHIS *Adans.*

1. Onobrychis ptolemaica (*Del.*) *DC.*

Low spreading or procumbent, softly villous perennial. Leaves imparipinnate with lanceolate to elliptic leaflets, each 1–2cm long. Inflorescence a long raceme with many pink, dark-veined flowers, each about 1.5cm long. Pods orbicular, spiny, woolly, c 1cm long.

A very rare plant of derelict fields on the high plateau. Known with accuracy only from 2200m northeast of Mayfīyah between Dhamār and Rada' (B. Goodman) but also collected without locality from the montane plains by B. Acres.

22. ORMOCARPUM *P. Beauv.*

Shrubs with imparipinnate leaves often clustered on short shoots. Leaflets ± alternate, oblong, ± truncate at the apex and the base. Flowers solitary or in racemes, yellow. Pod 1–6 seeded, constricted between the seeds, breaking up transversely.

Leaflets exceeding 10mm in length — *O. dhofarense*
Leaflets less than 10mm in length — *O. yemenense*

1. Ormocarpum dhofarense *Hillcoat & Gillett*

Very similar to the following species but differing in its larger leaflets (up to 15 × 7mm in Yemen), its more intricately branched racemes of flowers and the presence of swollen-based hairs at least on the pods.

Only known from scrub at 1500m just north of 'Udayn Suq (Wood 2012).

This plant is otherwise only known from the type collection in Dhofar. Wood 2012 is a fruiting specimen and flowers have never been collected in Yemen. There remains, therefore, some slight doubt as to the identity of the Yemen plant although no differences can be discerned when it is compared with the type.

2. Ormocarpum yemenense *Gillett* — Raḥdh

O. bibracteatum sensu Schwartz
O. spec. Schwartz p. 111

Shrub to 3m often with red-brown bark. Leaves with up to 10 leaflets; leaflets up to 6 × 3mm, adpressed pubescent on both surfaces, the midrib commonly reddish below. Flowers solitary or in few-flowered racemes, orange-yellow, the keel in particular becoming wine-red with age, almost glabrous, c 15mm long, much exceeding the somewhat campanulate calyx. Pods stipitate, reticulate, pubescent but lacking bulbous-based hairs, transversely jointed with up to 6 segments, each segment c 10 × 8mm.

Widespread on the escarpment and southeastern mountains from 500 to 1500m north to J. Sharafayn and Al 'Ashshah. An important constituent of *Grewia-Acacia* scrub which prefers areas of good rainfall and is absent from the more arid foothills. It bursts into flower immediately following heavy rain.

This species is endemic to southwest Arabia.

23. STYLOSANTHES *Swartz*

1. Stylosanthes fruticosa (*Retz.*) *Alston*

S. mucronata Willd.

Erect or ascending, pubescent wiry undershrub to about 30cm. Leaves trifoliolate with sheath-like stipules; leaflets elliptic, c 12 × 3mm. Inflorescence a dense leafy terminal spike of yellow flowers each about 7mm long. Pod jointed, compressed, c 8mm long, ± hidden by the bracts.

A plant of shallow stony grassland which is locally common between 800 and 1900m but restricted to the central escarpment including the Ḥarāz and the Maḥwīt region. It is often associated with *Indigofera brevicalyx*.

The Peanut, **Arachis hypogaea** *L.*, known in Yemen as 'Ḥab Al 'Azīz' has similar sheath-like stipules and yellow flowers. The well-known pods develop below the ground, hence the name 'Groundnut'. In Yemen I have only seen it cultivated in the 'Udayn region and on J. Milḥān. It is not naturalised.

24. DESMODIUM *Desv.*

Herbs with simple or 3-foliolate leaves, distinctive on account of their fruit which is transversely jointed and deeply constricted into one or more 'articles', which commonly break off separately.

1. Leaves simple — *D. gangeticum*
   Leaves 3-foliolate — 2
2. Creeping plant with obovate leaflets strongly resembling *Trifolium*. Inflorescence axillary, few-flowered — *D. triflorum*
   Plant not creeping, with rhomboid or ovate leaflets unlike *Trifolium*. Inflorescence of many-flowered axillary and terminal racemes — 3
3. Flowers exceeding 10mm long, red. Pod flat — *D. repandum*
   Flowers less than 4mm long, usually pale mauve. Pod with margins revolute — *D. ospriostreblum*

1. Desmodium gangeticum (*L.*) *DC.*

Perennial herb usually with weakly ascending stems. Leaves simple, ovate, c 3–5cm long. Inflorescence of long terminal and axillary racemes of mauve flowers each 3–5mm long. Pods not contorted or stipitate, about 2cm long, consisting usually of about 6 articles.

A very characteristic plant of damp shaded places in the central escarpment between 600 and 1200m including Maḥābishah, Wadi Yūr and Rajamah in the Ḥarāz.

2. Desmodium ospriostreblum *Chiov.*

Annual herb with weak, brittle, ascending stems and backward-directed 'clinging' hairs particularly on the inflorescence. Leaves 3-foliolate, terminal leaflet rhomboid, 3–5cm long, the laterals ovate 2–3cm long. Inflorescence of terminal and axillary racemes of pale mauve flowers each about 3mm long. Pods sometimes shortly stipitate, twisted with alternately revolute margins, about 1.5cm long and with 3–5 articles.

A rare plant on damp, partially shaded slopes associated with *Hyparrhenia coleotricha* by the Wadi Dūr, 'Udayn and Wadi Sukhnah.

3. Desmodium repandum (*Vahl*) *DC.*

D. scalpe DC.

Perennial herb with weakly ascending stems. Leaves 3-foliolate, terminal leaflet rhomboid 3–5cm long, the laterals ovate 2.5–4cm long. Inflorescence of long terminal and axillary racemes of red flowers each about 1cm long. Pods, long-stipitate, not contorted, 1–4cm long, commonly with 2–3 articles.

A plant of scrub-filled, partially shaded gullies, rarely grassland, almost entirely in the central escarpment: below Al Jir jir above Ḥadīyah, Kusmah, between Al Jabīn and Banī Dawān, all J. Raymah; J. Milḥān; 'Usil, Ḥarāz (Schweinfurth); Bayt Bashārī, Maḥwīt; in the gorge between Suq as Sabt and the market on the pass between J. Ba'dān and J. Manār, Ibb.

4. Desmodium triflorum (*L.*) *DC.*

Prostrate creeping perennial with 3-foliolate leaves whose broadly obovate, emarginate, 4–14mm long leaflets bear a strong resemblance to those of *Trifolium*. Inflorescence axillary, consisting of few-flowered racemes of blue flowers about 4mm long. Pods sessile, not stipitate or contorted, with up to 5 articles.

In short grass around rocks at 1100m between Ḥadīyah and Na'mah, Ja'farīyah.

25. ALYSICARPUS Desv.

Herbs with simple or rarely trifoliolate leaves easily recognised by the scarious stipules and calyx, and the jointed pods.

Pods not constricted between the seeds. Leaves mostly elliptic ... *A. vaginalis*
Pods constricted between the seeds. Leaves mostly linear-oblong or narrowly lanceolate ... *A. glumaceus*

1. Alysicarpus glumaceus (*Vahl*) *DC.*

A. rugosus sensu Schwartz.

Erect or sprawling adpressed-hairy herb. Leaves simple or trifoliolate, linear-oblong to lanceolate, the terminal leaflet 3.5–4.5cm long. Inflorescence a terminal or axillary raceme of red flowers about 5mm long. Pods constricted between the seeds, transversely ridged, 8–14mm long.

Locally common as a field weed on the silt plain in the east of the Tihama at places such as Bājil and 'Abs where it is commonly associated with *Crotalaria retusa*. It also occurs sporadically on field borders, terrace walls and stony banks on the outer escarpment up to 1600m.

2. Alysicarpus vaginalis (*L.*) *DC.*

Procumbent glabrescent perennial. Leaves simple, mostly elliptic, but becoming oblong on the flowering branches, 1–2cm long. Inflorescence a terminal or axillary raceme of mauve flowers about 5mm long. Pods not constricted between the seeds, reticulate but not ridged, about 1.5cm long.

A rare plant found between 500 and 1600m mostly in the central escarpment and very characteristic of stony wadi beds, liable to seasonal flood: Ḥadīyah, J. Raymah; above Ḥillah, J. Bura'; Walajah, J. Milḥān; near Rujum; Suq Ath Thalūth, Wadi Lā'ah; Wadi 'Ayn 'Ali, Ḥajjah.

26. CICER *L.*

1. Cicer arietinum *L.* — Ḥummaṣ

Chick Pea. Erect or spreading, grey-green, glandular pubescent annual. Leaves imparipinnate; leaflets about 12, elliptic, serrate, about 11 × 5mm. Flowers shortly pedicellate, solitary, white. Pods ± oblong, 2–3.5cm long.

Although recorded by Schweinfurth as a cultivated plant at Manākhah it does not seem to be cultivated in Yemen today. Recently it has only been found as a casual at Sana'a airport.

27. VICIA *L.*

Straggling herbs. Leaves with 3–8 pairs of linear – oblong leaflets, terminating in a tendril. Pods flattened. Both species appear to be native though the possibility of their being introductions cannot be ruled out.

Flowers subsessile or shortly pedicellate in the leaf axils ... *V. sativa*
Flowers in pedunculate racemes ... *V. villosa*

1. Vicia sativa *L.*

Sparsely pubescent annual. Leaves with 6–12 leaflets. Flowers usually solitary in the leaf axils, very shortly pedicellate, pink becoming violet as they dry, c 1–1.5cm long. Pods linear – oblong, 3–4 × 0.5cm, sparsely hairy.

All Yemen plants are variety **augustifolia** *L.*

A rare plant of grassy slopes in shade in areas of good rainfall on the escarpment from 1900 to 2900m: J. Ṣabir; J. Qarad (Ḥimyar), Mudhaykhirah; Jiblah (Hepper); J. Manār, Ibb; J. Ḥisn Madnan, Waṣāb al 'Ali; Kusmah, J. Raymah; J. Shibām, Manākhah; Bayt Bashārī, Maḥwīt.

2. Vicia villosa *L.*

Densely pubescent annual. Leaves with 12–20 leaflets. Inflorescence a long pedunculate raceme with up to 10 blue flowers. Pods oblong, 2–3.5 × 0.8cm, densely pubescent.

All Yemen plants are subspecies **eriocarpa** (*Hausskn.*) *P.W. Ball.*

A rare plant of montane grassland between 2600 and 3000m: near Dimarayn, J. Ṣabir; Kusmah, J. Raymah; near Bayt Yarīsh, Ḥaymat ad Dakhiliyah.

The Broad Bean, **Vicia faba** *L.*, known as 'Fūl' is commonly cultivated above 2000m on the escarpment and high plateau. Although dried broad beans are becoming a very popular breakfast dish, most local beans are sold at the roadside and in small markets to be eaten uncooked. The seeds are much smaller than the imported dried beans.

28. LENS *P. Mill.*

1. Lens culinaris *Medic.* — Bilsim, 'Adas

L. esculenta Moench.

Lentil. Erect annual up to about 30cm in Yemen. Leaves paripinnate, the upper terminating in a tendril; leaflets c 6–16. Inflorescence a pedunculate, usually 2-flowered, raceme; flowers pale blue. Pods 9–12mm long; seeds c 4mm diameter.

Widely cultivated on the escarpment and high plateau from 2000 to 3100m and occasionally found as a relic of cultivation, usually growing among other crops.

I have included all Yemen *Lens* under *L. culinaris* largely on the grounds that it is all of cultivated origin. *L. culinaris* is unknown as a wild plant and appears to have become distinct from *L. ervoides* (Brign.) Grande through selection and long cultivation. Most Yemen plants are in fact intermediate between the two species in terms of leaflet number, pod length and seed size (e.g. Wood 74/149, 1157) and this probably reflects Yemen farmers' reliance on their own seed stock which in turn is of very ancient origin.

29. LATHYRUS *L.*

1. Lathyrus pratensis *L.*

Low, scrambling, pubescent perennial with an angled, slightly winged stem. Leaves with 2 leaflets and an unbranched tendril; leaflets lanceolate, c 2–2.5cm long; stipules leaf-like, sagittate, c 1cm long. Inflorescence a long pedunculate raceme with the 3–10 yellow flowers clustered near the apex. Pods about 3cm long.

A very rare plant of damp grassy places in shade at 2900m: below the waterfall at Baw'an; in a cliff gully, J. Sumārah.

The Grass Pea, **Lathyrus sativus** *L.*, a small annual species with blue-purple flowers, was recorded as a cultivated plant by Deflers but has not been seen since.

The Pea, **Pisum sativum** *L.*, known in Yemen as ''Atar', is cultivated quite frequently on the escarpment above about 2500m. Although imported tinned peas are very popular, the local pea is always eaten uncooked and usually as a casual snack by field workers, travellers and people visiting markets. The plant cultivated in Yemen is subspecies **pumilio** *Meikle* with toothed stipules and leaflets, purple flowers and rather hard globose seeds.

30. ABRUS *Adans.*

Woody adpressed-hairy shrubs, usually climbing. Leaves paripinnate, usually terminating in a bristle, but occasionally in a diminutive leaflet; leaflets numerous, opposite, ± oblong, rounded at the base and the apex, c 10–15mm long. Inflorescence axillary and terminal, spike-like with flowers in sessile clusters on the rhachis. Flowers c 10–15mm long. Pods ± compressed.

Beak of pod hooked. Seeds black and red	*A. precatorius*
Beak of pod ± straight. Seeds dark brown	*A. bottae*

1. Abrus bottae *Deflers* Tinwāb

Erect spindly shrub to 3m, often supported by other bushes though not obviously climbing; stems brown. Flowers mauve, often very pale. Pods pale yellow-brown, ± smooth except for a few shallow pits, oblong, acuminate, c 4 × 1cm, the beak straight; seeds dark brown, flattened.

A plant of scrub and cliffs between 300 and 1500m from the Ḥujarīyah north to J. Sharafayn; generally common on the outer escarpment particularly at lower altitudes both in the wet areas of the central escarpment and the arid foothills; rare above 1200m and only recorded from the inner escarpment at Wadi Sukhnah and Al 'Ashshah.

This species, named after Botta, is endemic to Yemen, though it is closely related to *A. schimperi* Bak. from Ethiopia.

2. Abrus precatorius *L.*

Climbing shrub; stems green. Flowers dirty pink in Yemen. Pods grey, muricate, broadly oblong, truncate at base and apex, c 3 × 1.5cm, the beak hooked; seeds red and black, ovoid, poisonous.

The Yemen plant is subspecies **africanus** *Verdc.*

A local plant of scrub between 600 and 1500m in areas of good rainfall: frequent on the western slopes of the central escarpment (J. Raymah, J. Bura', J. Milḥān and Ḥarāz); elsewhere only known from 'Udayn and the lower slopes of J. Marran.

31. CLITORIA *L.*

1. Clitoria ternatia *L.*

An unmistakable and very attractive sparsely pubescent climbing herb. Leaves imparipinnate; leaflets usually 5, broadly elliptic, apex emarginate, base rounded to cuneate, very variable in size but commonly 1–3 × 1–1.5cm. Flowers axillary, usually solitary, the standard very large (up to 4cm long), blue except for a white area in the centre. Pods linear-oblong, flattened, c 6–10cm long.

Abundant in scrub in valleys between 500 and 1000m on the western slopes of the central escarpment (Raymah, Bura', Milḥān, Ḥarāz). Elsewhere only known from a thorn fence by a house along the Wadi Zabīd near J. Ra's, where it may have been introduced.

32. NEONOTONIA *Lackey*

1. Neonotonia wightii (*Wight & Arn.*) *Lackey*

Glycine wightii (Wight & Arn.) Verdc.
G. javanica L.

A perennial climber. Stem indumentum variable but usually densely ferruginous – hairy. Leaves 3-foliolate, the leaflets ovate to elliptic, acute, very variable in size but in Yemen commonly 3–6cm × 1.5–3cm. Inflorescence sessile, racemose, 10–20cm long, usually somewhat dense with 20–100 dirty white flowers 5–6mm long. Pods linear, curved, 1.5–3cm long, brown – tomentose.

All Yemen plants belong to subspecies **wightii**. They are mainly (? exclusively) autumn-flowering.

A locally common species found between 500 and 2000m along the escarpment from J., Ḥabashī north to J. Rāziḥ. It is most characteristic of scrub-filled gullies and overgrown fences near fields in the lush valleys around 1500m particularly in the Taiz – Ibb – 'Udayn region.

33. TERAMNUS *P. Br.*

Perennial climbing or trailing herbs. Leaves pinnately 3-foliolate. Inflorescence a lax few-flowered axillary raceme of small, 5mm long, pale pink flowers. Pod linear, nearly straight except for a right-angled hook at the apex formed by the style base.

Leaflets obovate to elliptic. Flowers 2–5. Pods c 3cm long *T. repens*
Leaflets oblong, elliptic or rhomboid, never obovate. Flowers 2–20. Pods c 4cm long. *T. labialis*

1. **Teramnus labialis** (*L.f.*) *Spreng.*

An extremely variable climbing or trailing herb, sometimes glabrous, sometimes adpressed hairy, sometimes spreadingly hairy. Leaflets variable in shape, but never obovate, usually with an acute apex, 1.5–5cm long. Inflorescence 2–10cm long with 2–20 flowers. Pods c 4cm long.

Almost every collection of this plant differs in some respects from every other but they have been conventionally placed in the following two subspecies:

Subsp. **labialis** with smooth seeds, pods up to 2.5mm wide and rather obtuse small leaves. This is widespread on the outer escarpment between 1300 and 2500m from Wadi Ḥidrān near Taiz north to J. Rāziḥ but is uncommon and very scattered in its occurrence, being most common in the lush 'hedgerows' just north of Ibb (Wood 1322, 2057, 2642).

Subsp. **arabicus** *Verdc.* with rough seeds, pods up to 3.5mm wide and rather larger acute leaves. This is apparently rare and restricted to low altitudes (500–1200m) in the central escarpment: above Suq As Sabt, J. Bura'; above Suq Ar Ribāṭ' J. Raymah (Wood 1895, Radcliffe-Smith & Henchie 4899); it is probable (on the grounds of altitude) that Schweinfurth's records of *T. labialis* from Ḥillah, J. Bura'; Walajah, J. Milḥān and Wadi Ḥijān, Ḥarāz, belong here.

2. **Teramnus repens** (*Taub.*) *Bak.f.*

Prostrate, trailing or mat-forming herb. Stems spreadingly hairy. Leaflets obovate to elliptic, ± rounded at the apex, 2–4cm long, thinly adpressed pilose on both surfaces. Inflorescence 2–3cm in length with 2–5 flowers. Pods c 3cm long, always spreadingly hairy.

All plants from Yemen belong to subspecies **gracilis** (*Chiov.*) *Verdc.*

A rare plant of thorn fences, grassland and bushy places around fields in wadis between 400 and 1000m, mostly, in the foothills: below Mishrāfah in the Wadi Rima'; below Ḥadīyah; above 'Alūjah, J. Raymah; Wadi Sharas; Wadi Līyah.

34. MUCUNA *Adans.*

1. **Mucuna pruriens** (*L.*) *DC.*

Vigorous short-lived perennial climbing herb. Stems pubescent. Leaves 3-foliolate; leaflets ovate, c 12 × 8cm, apex obtuse, apiculate, base rounded to truncate, the laterals very unequal-sided, adpressed pilose on both surfaces. Inflorescence racemose; rhachis and calyx silvery pubescent; flowers black-purple, 3–4cm long. Pods commonly *S*-shaped, 4–9cm long, covered with dense brown irritant hairs.

Only known at 600m in the wadi above 'Alūjah, J. Raymah where it scrambles over rocks on islands in the shaded wadi bed.

35. GALACTIA *P.Br.*

1. **Galactia tenuiflora** (*Willd.*) *Wight & Arn.*

Perennial climbing herb. Leaves 3-foliolate; leaflets oblong, c 4.5–6cm long, apex rounded or finely mucronate, base rounded, upper surface ± glabrous, undersurface paler and pubescent. Flowers 2–10 in lax racemes, pale lilac, c 10–12mm long. Pods linear-oblong, c 2–4cm × 6mm, straight, flattened, adpressed pubescent, 2–6 (–8) seeded.

Wood 2627 from J. Bura' appears to be identical to variety **biflora** *Schweinf.* described by Schweinfurth from the same locality. It has linear oblong leaves, 1–1.5cm wide and a 2-flowered raceme on a peduncle about 1cm long. Wood 1906 from J. Raymah is variety **villosa** (*Wight & Arn.*) *Benth.* It has oblong leaves c 2.5cm broad, a 2–7 flowered raceme on a peduncle at least 2cm long and is noticeably more pubescent on the undersurface of the leaf.

A rare plant of scrub between 700 and 1500m in the central escarpment: above Ḥillah, J. Bura'; near Bani Dawān, Al Jabīn, J. Raymah.

36. CANAVALIA *Adans.*

1. **Canavalia virosa** (*Roxb.*) *Wight & Arn.*

C. polystachya (Forssk.) Schweinf.

A very vigorous climbing perennial. Stem thinly pubescent. Leaves trifoliolate, the leaflets ovate (c 6–15 × 3–12cm), acute or shortly acuminate, base rounded, sparsely pubescent to ± glabrous on both surfaces. Inflorescence a long pedunculate axillary raceme, up to c 20cm long; flowers mauve, c 2.5–3cm long. Pods linear, pubescent, flattened, very long, up to c 15cm long and 3cm wide.

This species appears to have become rare since there are only two records this century: Bayt al Faqīh (Ahmed Khattab) and the wadi above 'Alūjah, J. Raymah. It seems to have been more frequent in the past in both the Tihama and the foothills of the central escarpment. Forsskal found it in the Wadis Mawr and Surdūd and at Ḥadīyah and Schweinfurth collected it at Bājil, Wadi Shawrī near Walajah and above Ḥillah on J. Bura'. These localities are between 200 and 800m.

37. VIGNA *Savi.*

A genus of twining or erect herbs usually easily recognised in Yemen by a combination of the following characters: pinnately trifoliolate leaves; inflorescence a raceme condensed at the apex of the peduncle (often into an umbel-like head) and commonly much shorter than the usually long peduncle; pod linear, terete, often long.

Several of the following species are cultivated, but either occur as casuals or can be easily mistaken for wild plants. They are all therefore keyed and described.

1. Leaves absent at flowering time — *V. sp. A*
   Leaves present at flowering time — 2
2. Leaflets deeply 3–5-lobed — *V. aconitifolia*
   Leaflets unlobed or shallowly 3-lobed — 3
3. Standard densely pubescent — *V. heterophylla*
   Standard glabrous — 4
4. Pod and commonly the stems also covered in brown hairs. Keel with a conical pocket on the left side — 5
   Pod and stems glabrous or hairy, but hairs never brown. Keel without a conical pocket on the left side — 6
5. Flowers mauve. Keel twisted — *V. vexillata*
   Flowers yellow. Keel not twisted — *V. radiata*
6. Stipules with a downward-directed spur. Stem ± glabrous — *V. unguiculata*
   Stipules without a downward-directed spur. Stem hairy — 7
7. Flowers yellow. Calyx teeth equalling the calyx — *V. luteola*
   Flowers mauve. Calyx teeth longer than the calyx — *V. membranacea*

1. **Vigna aconitifolia** (*Jacq.*) *Maréchal* — Quṭn

Phaseolus aconitifolius Jacq.

Erect annual, sparsely pilose-hispid on old growth but densely brown-pilose-hispid on young petioles and peduncles. Leaflets deeply 3–5 lobed, the segments linear – lanceolate. Flowers yellow, small, c 5mm long. Pods glabrous, 4–5cm long.

Cultivated in sandy fields in the Tihama and more rarely in the foothills up to 1000m. It is very drought-resistant and is one of the species planted in the sand dune country around Qanāwis, Bayt al Faqīh etc.

2. **Vigna heterophylla** *A. Rich.*

Climbing herb covered in spreading, mostly brown hairs. Leaflets unlobed, c 5 × 3.5cm, ovate, apex obtuse to rounded, base rounded, pubescent on both surfaces. Flowers about 1cm long consisting of a tomentose blue standard, glabrous mauve wings and a pale glabrous keel. Pods densely brown hispid, 2–4cm long.

Only known from a grassy bank among sorghum fields at 2300m on the Ibb – 'Udayn pass just below Mashwarah.

3. **Vigna luteola** (*Jacq.*) *Benth.*

Annual climbing herb, thinly pubescent in all parts. Leaflets unlobed, c 4–6 × 2–3cm, ovate to ovate – lanceolate, apex acute, base rounded. Flowers yellow, c 1.5–2.5cm long. Pods pubescent, 3–5cm long.

Only known from a single collection (Wood 2606) made in a stony millet field at c 1300m at Misrākh where it appeared to be cultivated.

4. **Vigna membranacea** *A. Rich.*

Climbing annual herb. Stem pilose. Leaflets ovate or triangular sometimes weakly 3-lobed, apex acuminate, c 5 × 3–4cm, base rounded or truncate, sparsely pilose. Flowers mauve, c 1–1.5cm long. Pods pubescent, c 5cm long.

Yemen plants all belong to subspecies **membranacea** which has the calyx teeth much longer than the calyx itself, a feature which distinguishes this plant from other species of *Vigna* in Yemen.

A rare plant of grassland and scrub in areas of high rainfall between 1600 and 2300m: in the gully below Al Jirgir, above Ḥadīyah, J. Raymah; on the Ibb – 'Udayn pass below Mashwarah.

5. **Vigna radiata** (*L.*) *Wilczek* — Qusharī

Phaseolus mungo L.

Mung Bean. Erect annual herb. Stem covered in brown bristly hairs. Leaflets unlobed, broadly ovate, c 8 × 7cm, apex shortly acuminate, base broadly cuneate to truncate, sparsely hispid on both surfaces. Flowers yellow, c 1cm long. Pods brown hispid, c 5–9cm long.

This plant has the marked pocket on the lefthand side of the keel also found in *V. vexillata* and the downward-directed spur to its stipules found in *V. unguiculata*.

Cultivated in sandy fields in the Tihama and more rarely in the foothills up to 1000m. Like *V. aconitifolia* it is planted opportunistically in the sand dunes, but seems to be less tolerant of really dry sandy conditions.

Variety **glabra** (*Roxb.*) *Verdc.* which is almost glabrous was recorded at the exceptional height of 2300m at Bayt An Nā'im by Rathjens.

6. **Vigna unguiculata** (*L.*) *Walp.* — Dijr

V. sinensis (L.) Endl.

Cow Bean. Erect or climbing annual. Stems ± glabrous. Leaflets unlobed, ovate to rhomboid, usually about 6 × 4cm, apex acute, base cuneate, glabrous. Flowers white, greenish or most commonly pale mauve, c 1.5–2.5cm long. Pods glabrous, 20–90cm long.

Subspecies **unguiculata** with pods 20–30cm long is the common form in Yemen, but subspecies **sesquipedalis** (*L.*) *Verdc.* with pods 30–90cm long has been recorded by Schweinfurth from Marāwi'ah and Bājil in the Tihama.

Cultivated from 100–2500m mainly in the silt plains in the east of the Tihama and on the escarpment and sometimes appearing to be naturalised on grassy banks and in scrub (Ḥadīyah, 'Udayn, Mashwarah). It is the most widely cultivated legume but seems to be of little economic importance. It is very widely planted as ground cover under sorghum, but neither the pods nor the seeds are sold in markets. They are eaten casually by people passing through the fields.

7. **Vigna vexillata** (*L.*) *A. Rich.*

Perennial climbing herb. Stem sparsely to densely covered in brown hairs. Leaflets unlobed, lanceolate to elliptic, c 6 × 2.5cm, apex acute to acuminate, often mucronate, base rounded to cuneate, thinly pubescent on both surfaces. Flowers often solitary, never more than 6 together, mauve, c 2–3cm long with the keel curled round and with a marked pocket on the left side. Pods covered in brown bristly hairs, 4–14cm long.

The typical plant, variety **vexillata** is a rare plant of scrubby slopes and gullies in high rainfall areas from 600 to 1200m mostly in the central escarpment: above Ḥadīyah, J. Raymah; above Ḥillah, J. Bura'; between Ḥabāsh and As Ṣa'bah, J. Milḥān; Wadi Līyah, Khawlan Ash Sham. Variety **angustifolia** (*Schumach. & Thonn..*) *Bak.*, a glabrescent plant with long, linear, oblong leaves up to 10 × 1.5cm was growing with the typical variety on J. Milḥān.

8. Vigna sp. A

Erect perennial herb. Leaves absent at flowering time. Flowers yellow with a mauve tip, 7–8mm long, the keel slightly curved. Pod c 4cm long.

Only known from banks around cultivated fields at 2400m at Mashwarah, near Ibb (Wood 1096).

This undescribed species with distinct smooth pollen grains is unrelated to any *Vigna* in Arabia or northeast Africa.

The related French Bean, **Phaseolus vulgaris** *L.*, known as 'Fasūliyah' is commonly cultivated around Sana'a, but I have no knowledge of its cultivation elsewhere in Yemen.

38. DOLICHOS *L.*

1. Dolichos trilobus *L.*

Perennial climbing herb. Stem pubescent, often deep violet in colour. Leaves 3-foliolate; leaflets ovate to ovate – rhomboid, usually c 3 × 2cm, apex rounded to shortly acuminate, base rounded, very sparsely pubescent on both surfaces. Flowers in clusters of 2–3 in the leaf axils, pedicellate, pinkish-mauve, c 8–10mm long. Pods ± oblong, slightly flattened, curved, c 3–4cm × 7–10mm, minutely puberulent. Flowers mostly in the autumn.

Found on the outer escarpment from 1500 to 2500m from the Turbah area north to J. Marran in Khawlan Ash Sham; common in grassland, on field borders and scrubby slopes in the high rainfall areas of Taiz and Ibb provinces becoming rarer further north and surprisingly uncommon in the central escarpment.

Dolichos sericeus E. Mey. subspecies *formosus* (A. Rich.) Verdc. has been recorded on several occasions (e.g. Schweinfurth, Hepper) but where it has been possible to examine the specimens the plants have turned out to be *D. trilobus*. The two species are very similar but *D. trilobus* has bracteoles up to 15mm long and acute stipules up to 4mm long whereas *D. sericeus* has bracteoles longer than 15mm and usually obtuse stipules more than 4mm long.

39. LABLAB *Adans.*

1. Lablab purpureus (*L.*) *Sweet* Kisht

L. vulgaris Savi

Vigorous climbing perennial. Stem sparsely pubescent. Leaves 3-foliolate; leaflets ovate – triangular, usually about 7 × 7cm, apex acuminate, base cuneate to truncate, sparsely pubescent. Inflorescence axillary, racemose, long pedunculate, c 10cm long; flowers purple, 1–2cm long. Pods ± oblong, c 4–8 × 2–3cm, flattened.

All Yemen plants appear to belong to the cultivated form, subspecies **purpureus**.

A cultivated 'bean' found from sea level to at least 2300m. It is cultivated mainly as an ornamental plant in the major towns, but was probably cultivated as a vegetable in places such as J. Milḥān and Ḥarāz where Schweinfurth collected it. It is very rarely naturalised but has been found on a roadside between Bāb Bāhal and Mawsanah (Ḥarāz).

40. MACROTYLOMA (*Wight & Arn.*) *Verdc.*

1. Macrotyloma axillare (*E. Mey.*) *Verdc.*

Perennial climbing herb. Stem with sparse spreading hairs. Leaves 3-foliolate; leaflets ovate-lanceolate, c 3–6 × 1.5–3cm, apex acuminate, base rounded, glabrous or thinly pubescent on both surfaces. Flowers in clusters of 2–3 in the leaf axils, very shortly pedicellate, greenish-yellow, c 1.5cm long, glabrous or thinly pubescent.

Rrestricted to the grassland areas of the western slopes of J. Raymah between 1500 and 2200m where it is locally common. It has been found in the Ja'farīyah, Kusmah and Al Jabīn districts.

41. CAJANUS *DC.*

1. Cajanus cajan (*L.*) *Millsp.*

C. indicus Spreng.

Pigeon Pea. Erect shrub to about 2½m with prominently ribbed dark green stems. Leaves 3-foliolate; leaflets elliptic to lanceolate, 4–6cm long, apex acute, base cuneate, undersurface densely pubescent, paler and more gland-dotted than the upper surface. Inflorescence a terminal panicle with numerous small axillary racemes; flowers bright yellow. Pods straight, densely glandular and hairy, grooved between the seeds, about 5cm long.

Sometimes cultivated and then characteristically on field borders and terrace edges in areas of good rainfall along the escarpment from 800 to 2200m; sometimes naturalised or appearing so as by the junction of Wadi Dūr and Wadi 'Annah at 'Udayn.

42. RHYNCHOSIA *Lour.*

Climbing or trailing perennial herbs or more rarely erect undershrubs with stipellate 3-foliolate leaves which are gland-dotted below, yellow flowers and 1–2 seeded pods which are usually oblong, apiculate and uninflated.

A complex genus which includes one species endemic to Yemen and several others which are very local in Africa and some whose exact status is unclear. In Yemen plants the leaf shape and indumentum are usually distinctive though often hard to define precisely.

1. Leaflets narrowly oblong, more than twice as long as broad — *R. totta*
   Leaflets rarely oblong, usually about as long as broad — 2
2. Pod strongly green-veined, laterally plicate and constricted between the seeds. Keel purple — *R. buramensis*
   Pod brown or weakly green-veined, never laterally plicate and constricted between the seeds. Keel yellow — 3
3. Flowers 10–20mm long — 4
   Flowers 5–10mm long — 6
4. Erect plant — *R. flava*
   Twining plant — 5
5. Pod inflated, 2–3cm long — *R. sublobata*
   Pod flat, c 1.5cm long — *R. malacophylla*
6. Pods up to about 2.5cm long with a mixed indumentum of long and short hairs. Leaflets usually with pale area along the midrib — 7
   Pods up to 2cm long, the hairs all of the same length. Leaflets without a pale area along the midrib — 8
7. Prostrate plant with rounded leaflets — *R. oblatifoliolata*
   Climbing plant with acute leaflets — *R. variegata*
8. Plant densely tomentose, grey or grey-green — 9
   Plant glabrous to pubescent, but never tomentose, always green — 12
9. Flowers solitary or paired in the leaf axils — 10
   Flowers in racemes — 11
10. Erect herb. Leaflets apiculate, commonly less than 1cm long — *R. sp. A*
    Twining herb. Leaflets emarginate more than 1cm long — *R. schimperi*
11. Corolla up to 5mm long. Leaflets grey-green with numerous very pale glands — *R. pulverulenta*
    Corolla c 9mm long. Leaflets grey with dark orange glands — *R. memnonia*
12. Racemes up to 5-flowered. Leaves less than 1cm long — *R. sp. B*
    Racemes more than 5-flowered. Leaves more than 1cm long — *R. minima*

1. Rhynchosia buramensis *Hutch. & Bruce*

Climbing herb with pubescent stems. Leaflets ovate, c 3–4cm long, apex acuminate, base truncate to ± cuneate, glabrescent on both surfaces except on the veins below, pleasantly scented. Inflorescence a few-flowered sticky-glandular axillary raceme; flowers about 1.5cm long, standard purple-veined, keel mostly dark purple, wings yellow. Pods c 2cm long, prominently green-veined, constricted between the seeds, somewhat laterally plicate with an indumentum of both long and short hairs.

Only known from scrubby field borders between Ibb and Suq As Sabt at c 1700m.

Outside Yemen only known from a few places on the Somali-Ethiopian border.

2. Rhynchosia flava (*Forssk.*) *Thulin*

Erect undershrub to about 40cm, stem with raised longitudinal lines, densely pubescent. Leaflets rhomboid, elliptic or obovate, 1–2cm long, apex rounded to acute and often apiculate, base rounded, shortly pubescent and glandular on both surfaces, margin often grey-canescent. Inflorescence a short axillary raceme of up to six flowers; flowers yellow except for the red-veined standard, c 10–13mm long. Pods c 1.5cm long, pilose.

This species is endemic to Yemen, but is very closely related to *R. erythraea* Schweinf. from which it differs in being more glandular and in having a sparser pubescence which is not silvery.

Widespread on the outer escarpment from J. Habashī north to J. Marran, extending to J. Ash Shirq on the inner escarpment, but eccentric in its occurrence being unaccountably absent from many suitable areas and showing a marked dislike of road verges and so rarely collected. It is found in short grassland and on stony, scrub-free slopes between 1800 and 2800m.

3. Rhynchosia malacophylla (*Spreng.*) *Boj.*

R. flavissima Hochst.

Climbing herb with pubescent stems. Leaflets ovate or rhomboid, c 2–3cm long, apex acute, base rounded to cuneate, glabrescent. Inflorescence a many-flowered axillary raceme, not sticky-glandular; flowers c 12–15mm long, yellow except for the standard which is usually purple-veined. Pods about 1.5cm long, shortly pubescent.

A local plant found between 600 and 1800m (almost always around 1500m) on the escarpment north to Khawlan Ash Sham. It is usually found in thorn fences and rough scrubby ground near cultivation and appears to be more frequent in years of good rainfall.

4. Rhynchosia memnonia (*Del.*) *DC.*

R. minima var. *Memnonia* (Del.) Cooke

Climbing herb, densely grey-tomentose in all parts. Leaflets 1–2cm long, base broadly cuneate to rounded, apex usually rounded, sometimes apiculate, the terminal leaflet transversely elliptic. Inflorescence a long axillary raceme of yellow flowers (standard dark-veined), each up to 8mm long. Pods c 2cm long, tomentose.

A locally abundant weed of roadsides and fields particularly in Taiz province and in the silt plains in the east of the Tihama; sometimes dominant in derelict fields. Very rare elsewhere but reaching 1800m at Ḥūth.

Schwartz cites J. Shibām, Manākhah and Ma'bar – Ḍawrān as localities for this plant (based on collections by von Wissmann). These are far above the altitude range I have observed for this plant and I suspect they are errors of identification.

5. Rhynchosia minima (*L.*) *DC.*
Climbing herb with glabrescent stems. Leaflets glabrous, rhomboid, c 2–3cm long, base cuneate, apex acute to acuminate. Inflorescence a long axillary raceme of yellow flowers (standard veined), similar to *R. malacophylla* but flowers only c 5–6mm long. Pods about 18mm long, glabrous or minutely puberulent.

A widespread but thinly scattered plant found on the escarpment north to Al 'Ashshah and rarely in the east of the Tihama, where it has only been recorded at Bājil (Schweinfurth) and between Nukaybah and Al Mafraq. It grows between 250 and 1500m usually in the protection of thorn bushes and hedges and, though occurring in such places as 'Alūjah, it appears to be more typical of rain-shadow areas.

In addition to the typical plant described above, variety **prostrata** (*Harv.*) *Meikle* is known from grassy slopes at 2700m on J. Ba'dān, Ibb (Wood 1855). This is a prostrate pubescent herb with rounded, reticulate leaflets, the terminal one transversely elliptic c 1–1.5cm long. The inflorescence is a short, rather dense raceme with flowers about 8mm long. The pod is shorter (c 12mm long) and densely pubescent.

6. Rhynchosia oblatifoliolata *Verdc.*
Prostrate herb with densely pubescent stems. Leaflets ovate – rhomboid, c 1–1.5cm long, pubescent and green on both surfaces but with a pale area along the midrib, rounded at the apex and the base. Inflorescence a usually 1–3 flowered axillary raceme; flowers c 8mm long, yellow except for the orange-red standard. Pods c 2–2.5cm long, covered in numerous long and short hairs.

Only known from bare sandy slopes among boulders below sandstone cliffs at Bāb al Aḥjar at 2600m.

A rare plant outside Yemen known from a few localities in Kenya and Ethiopia.

7. Rhynchosia pulverulenta *Stocks*
Woody climbing herb with tomentose stems. Leaflets distinctively grey-green-tomentose on both surfaces and densely covered in pale yellow glands below, terminal leaflet c 2cm long, markedly transversely elliptic to rhomboid with a cuneate base and rounded apiculate apex, lateral leaflets triangular-ovate. Inflorescence a short axillary raceme with pale yellow flowers about 5mm long. Pods about 1.5cm long, shortly but densely pubescent.

Only known on a scrub-covered roadside bank at Mandib in the Wadi Qaṣab, Ḥarāz (700m).

8. Rhynchosia schimperi (*Hochst. & Steud.*) *Boiss.*
R. pulverulenta sensu Deflers
Very similar in general appearance to *R. memnonia* from which it differs in having solitary or paired axillary flowers about 6mm long. The leaflets are very distinctive being obovate with a cuneate base and a rounded almost always emarginate apex. They are 1–2cm long. Pods about 15mm long, grey-tomentose.

A rare plant of sand dunes at c 100–200m near Bayt al Faqīh (Wood 773) between Bājil and Qutay and at Marāwi'ah (Deflers 43).

This is a good example of a plant which extends from Yemen northwest to the Indian subcontinent but is absent from Africa.

9. Rhynchosia sublobata (*Schumach.*) *Meikle*
R. viscosa DC.
Climbing herb, stem and leaves usually pubescent. Leaflets broadly ovate to rhomboid, c 2–4cm long, rounded at apex and base, often slightly lobed. Inflorescence an axillary raceme; flowers yellow except for the standard which is often dark-veined, 1–2cm long. Pods 2–3cm long, inflated when ripe, pubescent.

Recorded by Deflers at 700m in the Wadi Hawayat at Ḥajjaylah, but not found since.

10. Rhynchosia totta (*Thunb.*) *DC.*
Slender climbing herb with very sparsely pubescent stems. Leaflets narrowly oblong, usually about 15–20 × 7mm, apex obtuse, base rounded to subcordate, glabrescent and reticulate on both surfaces. Flowers solitary or in 2–3 flowered axillary racemes, yellow, about 1cm long. Pods about 1.5cm long, with an indumentum of both long and short hairs.

Found in sparse shallow grassland over rocks and on broken cliffs exposed to the sun on the escarpment and high plateau between 900 and 2400m. Not recorded from Taiz province and Khawlan Ash Sham.

11. Rhynchosia variegata (*Deflers*) *J.R.I. Wood* comb. nov.
Vigna variegata Deflers, Voyage au Yemen: 132 (1889)
Rhynchosia mensensis Schweinf.
Climbing herb with pubescent stems. Leaflets pubescent and reticulate on both surfaces, ovate to elliptic, c 3–5cm long, apex acute, base rounded, the upper surface dark green with a distinctive pale area along the midrib. Inflorescence a lax few-flowered axillary raceme, flowers c 1cm long, yellow except for the purplish standard. Pods c 2.5cm long, green-veined with an indumentum of both long and short hairs.

Found between 1600 and 2500m in scrub and grassland in areas of good rainfall on the outer escarpment from J. Raymah north to J. Rāziḥ. Elsewhere very rare and only known from J. Ḥubaysh near Ibb and on the inner escarpment at Ṭawīlah and Kuḥlān.

Outside Yemen it is only known from four collections from Ethiopia (fide Mats Thulin). It is similar to the widespread African species *R. usambarensis* Taub. from which it differs mainly in its white retrorse indumentum.

12. Rhynchosia sp. A
Erect woody herb, softly grey-tomentose in all parts. Leaves very shortly petiolate (petioles often less than 5mm), leaflets very small (5–10mm long), apex rounded, apiculate, base broadly cuneate, terminal leaflet obovate, lateral leaflets ovate to elliptic. Flowers solitary or paired in the leaf axils, pure yellow, c 9mm long. Pods about 15mm long, grey-tomentose.

Only known from a sandy gully below sandstone hills at 2000m just north of Rada' (Wood 3215).

13. Rhynchosia sp. B

Prostrate woody plant with adpressed pubescent stems. Leaflets obovate, rhomboid or suborbicular, 5–10mm long, rounded but with the apex often apiculate, pubescent especially on the margins. Inflorescence of solitary axillary flowers or of lax few-flowered racemes. Flowers yellow except for the veined standard, 7–8mm long. Pods about 13mm long, very shortly tomentose, glandular.

A plant of exposed stony plain and hillsides found between 1800 and 2500m in scattered localities mostly on the high plateau but also, though rarely, on the escarpment from Naqīl Jubirān near Turbah north to Shahārah. It is commonly a chewed-down and inconspicuous plant except after plentiful rain (Wood 1082, 1949).

Not found in Africa and restricted in Arabia to the southwest, but very similar to plants of the *R. minima* complex from India which were formerly separated as *R. microphylla* Wall. The Yemen plant is more sparsely pubescent, but more densely glandular, especially on the pods.

43. ERIOSEMA (*DC.*) *Desv.*

1. Eriosema longipedunculatum *A. Rich.*

Erect or more commonly sprawling, pubescent perennial herb, often with reddish-brown hairs on young growth. Leaves pinnately 3-foliolate with distinctive red-brown stipules; leaflets narrowly elliptic, c 2–6 × 1–1.5cm, apex acute, base rounded, dotted with glands on the undersurface. Inflorescence a long pedunculate dense head-like raceme; flowers usually yellow but the standard often purple-black. Pods erect, elliptic, flattened, c 14 × 9mm.

The Yemen plant is variety **hirsutum** *Verdc.*

A locally common plant of grassland in areas of good rainfall between 1500 and 2500m: J. 'Ayfū and J. Qarad, Mudhaykhirah; the western slopes of J. Raymah.

44. FLEMINGIA *Ait. f.*

1. Flemingia sp. — Waras

F. rhodocarpa sensu Schwartz

Vigorous erect shrubby herb about 1m high. Stem densely covered in brown hairs. Leaves 3-foliolate, pubescent, glandular and scaly beneath; leaflets ± elliptic, apex acute, base cuneate, the terminal leaflet about 11cm long. Inflorescence of clustered racemes of yellow-green flowers. Pods c 1cm long, pubescent.

Known only from Ḥadīyah (Forsskal) and at c 1800m on J. 'Ayfū in the Mudhaykhirah region which was renowned for the cultivation of 'Waras' in medieval times. *Flemingia* appears to be an introduction and was used as a dye but is now extremely rare.

Both my collection and Forsskal's are sterile and in consequence it is impossible to say whether Yemen plants are the mainly African *F. grahamiana* Wight & Arn. or the wholly Asiatic *F. macrophylla* (*Willd.*) *Merr.* The former has rigid bracts, a woolly calyx and pods with glands which secrete an orange dye. The latter lacks rigid bracts and glandular fruit but has a silky calyx. *F. grahamiana* has been found in South Yemen and this is most likely to be our species, though the indumentum of both North Yemen collections resembles that of *F. macrophylla*. If the correct identity of the North Yemen plants is ever discovered, It will be necessary to adopt the name *F. hadiensis* for whichever species it turns out to be.

45. CADIA *Forssk.*

1. Cadia purpurea (*Picc.*) *Ait.* — Hawmir, Humar, Kadhī

C. varia L'Herit.

Shrub to about 3m with rather slender branches. Leaves imparipinnate with about 30–40 sessile, linear leaflets, each about 1–1.5cm long and resembling those of the Tamarind. Flowers pedunculate, usually solitary but sometimes in a cyme of up to 4 flowers, pendulous, pale to dark purple, the petals equal, not papilionaceous; calyx distinctly campanulate. Pods straight, stipitate, linear, c 12 × 1cm.

Cadia was named by Forsskal after one of the Arabic names for this species, 'Kadhī'.

A common plant on the escarpment and southeastern mountains from 800–3500m; at low altitudes characteristic of areas of good rainfall such as the western slopes of the central escarpment but at higher altitudes more typical of rainshadow mountains where it is commonly associated with *Acacia etbaica*, *Dodonaea*, *Psiadia* etc. It often becomes abundant on heavily eroded slopes where the terracing has collapsed.

Two trees from related genera are known to be cultivated in Yemen. **Calpurnia aurea** (*Ait.*) *Benth.*, a small tree with pendulous racemes of yellow flowers has been recorded from Ibb Forestry Nursery (Henry). **Sophora japonica** *L.*, a tree with large terminal panicles of cream or pink flowers was found by Rathjens in Sana'a. Neither has been reported by other observers.

46. CAESALPINIA *L.*

Usually spiny shrubs with bipinnate leaves, the leaf opposite. Inflorescence of terminal or axillary racemes, pubescent (*except C. pulcherrima*). Pod various, but always dehiscent in our species.

All species of *Caesalpinia* are introduced to Yemen but the following three are ± naturalised.

1.	Plant unarmed	*C. gilliesii*
	Plant spiny	2
2.	Leaflets less than 1cm wide. Pod not prickly	*C. decapetala*
	Leaflets usually exceeding 1cm wide. Pod prickly	*C. bonduc*

1. Caesalpinia bonduc (*L.*) *Roxb.* — Kilab

C. crista sensu Schwartz

Broad spreading bush to about 2m, stem and leaf rhachis pubescent, yellow-brown, covered in numerous falcate spines. Leaflets pubescent, 6–9 pairs, oblong-elliptic, c 1.5–3cm long, the apex obtuse, apiculate. Racemes long, axillary; petals yellow, c 12mm long. Pod densely prickly, c 5–6 × 4cm; seeds globose, large, c 1.5–2cm diameter.

Common by almost all the flowing wadis between 500 and 1500m in Taiz province, extending through 'Udayn and Waṣāb north to 'Alūjah in J. Raymah. Found as long ago as 1762 at Ḥadīyah (Forsskal) but showing no signs of spreading northwards.

2. Caesalpinia decapetala (*Roth.*) *Alston*
Climbing or straggling shrub to about 3m in Yemen, stem pubescent with straight spines. Leaf rhachis with falcate spines; leaflets pubescent, 8–12 pairs, oblong-elliptic, c 1.5cm long, the apex rounded. Racemes at the end of branches, very long, often exceeding 25cm; petals yellow, c 15mm long. Pod flattened, unarmed, c 6–9 × 2.5cm, with a long slender beak; seeds ellipsoid, up to 1cm long.

Only known from two localities in areas of high rainfall: at 1700m on a roadside between Ibb and Suq As Sabt; in some abundance between 2000 and 2500m on the north side of J. Ṣabir where it is used as a hedge around the qat fields.

3. Caesalpinia gilliesii (*Hook.*) *Dietr.*
Erect straggling foetid unarmed shrub to 1½m. Leaflets glabrous, 8–12 pairs, oblong c 4mm long, gland-dotted below. Racemes terminal, glandular, up to 25cm long; petals yellow, c 2cm long; filaments very long, red. Pod flattened, glandular and pubescent, c 6–8 × 1.5cm; seeds flattened, c 1cm broad.

Recorded in the 1930s by Rathjens in Sana'a and by Scott and Britton in Taiz, where it still persists on rocky roadsides.

The Barbados Pride, **Caesalpinia pulcherrima** (*L.*) *Sw.* a glabrous sparsely spiny shrub with attractive red or orange flowers is cultivated in Taiz.

47. PTEROLOBIUM *Wight & Arn.*

1. Pterolobium stellatum (*Forssk.*) *Brenan* Hawjam, Kilab
P. lacerans R. Br.
Scrambling shrub, viciously spiny on the stem and leaf rhachis, the prickles falcate, 1–3 together. Leaves bipinnate; leaflets glabrous, 4–9 pairs, oblong, c 8–15mm long, rounded at the apex and the base. Inflorescence of long Acacia-like terminal and axillary racemes, often massed together into a panicle; flowers very small, cream, with petals c 3mm long. Pods usually brick red, flattened, winged, 1-seeded, c 3–4 × 1cm.

Flowers throughout the year but mostly in the summer. The red pods are a conspicuous feature in some areas on the autumn, notably around Ibb.

Widespread and locally abundant between 1000 and 2200m on the outer escarpments from Aḥkum north to Maḥābishah; particularly characteristic of scrubby gullies and ditches in the Ibb – 'Udayn region, showing a southern tendency in its distribution. Sometimes encouraged as a hedge particularly around qat fields.

48. DELONIX *Raf.*

1. Delonix elata (*L.*) *Gamble* Ranf
Poinciana elata L.
Tree (rarely a shrub) to c 12m with a rounded crown and smooth grey bark. Leaves bipinnate; leaflets adpressed puberulous, 11–25 pairs, linear – oblong, c 1cm long, rounded. Inflorescence of racemes near the ends of branchlets; petals large, pale yellow and white becoming orange with age; filaments very long, red. Pods woody, flattened, dehiscent, c 12–20 × 3cm.

Flowers throughout the year.

This very attractive species is widespread but never very abundant in the foothills of the escarpment from 300 to 1100m. It is in fact very rare above 600m and only occurs above that in areas of low rainfall. A very drought-resistant species, it is also found in the Mashriq by the Wadi Nuwāsh north of Rada' and on the Naqīl Ḥajlah between Ḥarīb and Jūbah. It usually grows on cliffs or steep rocky slopes above wadis.

The Flamboyant, **Delonix regia** (*Hook.*) *Raf.*, is commonly grown in Taiz and Ibb. It is a deciduous tree whose brilliant scarlet flowers appear in April and May. It has been much planted in recent years and lines some of the streets in Taiz.

49. PARKINSONIA *L.*

1. Parkinsonia aculeata *L.* Saysībān
Small glabrous tree or bush to c 4m with a smooth green trunk. Leaves bipinnate with 2–4 pinnae, the rhachis of the pinnae very long, spine-tipped with numerous very small obovate leaflets. Inflorescence of long racemes of attractive yellow flowers. Pod torulose, c 6–12cm long.

Commonly cultivated in the Tihama and on the escarpment up to 1500m and a familiar sight around villages where it appears to regenerate spontaneously.

50. CASSIA *L.*

Annual or perennial herbs, shrubs or trees. Leaves paripinnate. Inflorescence racemose, but racemes often much reduced so the flowers appear solitary and axillary. Petals usually yellow. Pod variable.

It is often of value to look for glands on the leaves. These may be present or absent, stalked or sessile, on the leaf petiole or between the leaflets. *C. fallacina* is the only species in Yemen with a stalked petiolar gland *C. obtusifolia* has glands between the leaflets. *C. absus* is glandular hairy. Several species are eglandular (Nos. 2, 4, 6, 7, 12). The remainder have sessile glands on the leaf petiole.

1. Tree with persistent pendulous cylindrical pods often exceeding 30cm long *C. fistula*
   Shrubs or herbs. Pod flattened or if ± cylindrical up to c 20cm long, not pendulous **2**
2. Sepals obtuse. Leaflets more than 5mm broad, rarely less and if so exceeding 3cm long **3**
   Sepals acute. Leaflets less than 5mm broad **10**

3. Plant with glandular hairs. Leaves with 2 pairs of leaflets *C. absus*
   Plant without glandular hairs. Leaves with more than 2 pairs of leaflets **4**
4. Pod flattened, exceeding 1cm broad. Leaves eglandular **5**
   Pod terete or flattened, less than 1cm broad. Leaves glandular **8**
5. Stipules broadly ovate-cordate. Leaves with more than 10 pairs of leaflets *C. didymobotrya*
   Stipules narrow, linear to triangular. Leaves with less than 10 pairs of leaflets **6**
6. Leaflets acuminate, exceeding 3cm long. Pod almost straight *C. senna*
   Leaflets obtuse to rounded, usually not exceeding 2.5cm. Pod falcate **7**
7. Plant almost glabrous. Pod with a crest along the centre *C. italica*
   Plant shortly tomentose. Pod without a central ridge *C. holosericea*
8. Leaflets rounded, usually 3 pairs *C. obtusifolia*
   Leaflets acute, more than 3 pairs **9**
9. Leaflets 1–2cm wide. Peduncles 8–20mm long *C. sophera*
   Leaflets 2–3cm wide. Peduncles 3–5mm long *C. occidentalis*
10. Vigorous erect plant to 1m; flowers usually 3–5 together. Pedicels c 3mm long *C. nigricans*
    Prostrate plants; flowers 1–2 (–3) together. Pedicels usually more than 5mm long **11**
11. Gland on petiole stalked *C. fallacina*
    Gland on petiole sessile **12**
12. Rhachis of leaves crenate (best seen when leaflets are pulled off). Pedicels exceeding 10mm long *C. mimosoides*
    Rhachis of leaves not crenate. Pedicels about 5–6mm long *C. hochstetteri*

1. Cassia absus *L.*

Procumbent or ascending, sometimes woody annual herb, the whole plant except the leaves covered in stalked glands. Leaves with 2 pairs of asymmetric, ± rounded, obovate leaflets, c 1–4 × 0.5–2.5cm. Racemes usually short, few-flowered; petals 5–7mm long, yellow to pink. Pods straight, flattened, glandular-hairy, c 50 × 7mm.

A rare plant of tracksides and field borders in a few areas of high rainfall between 1000 and 1500m: above Ḥillah, J. Bura' (Schweinfurth); in a number of places around 'Udayn.

2. Cassia didymobotrya *Fres.*

Shrub to 3m with pubescent stems. Leaves eglandular, with c 12–18 pairs of oblong – elliptic, rounded, mucronate leaflets c 2–4 × 1cm. Racemes long, many-flowered; petals 1.5–2.5cm long, bright yellow. Pods almost straight, flattened, transversely ridged, pubescent c 8–12 × 2cm.

Readily recognised by its large cordate stipules, numerous leaflets and large flowers.

A recent introduction naturalised on roadsides in Taiz and near Nashamah in the Ḥujarīyah (1300–1500m).

3. Cassia fallacina *Chiov.*

C. pumila Lam. var. *yemensis* Deflers

Prostrate, sparsely pubescent perennial. Leaves with 10–25 pairs of strongly asymmetric, oblong, mucronate leaflets c 7 × 1mm. Flowers solitary (rarely paired), supra-axillary; pedicels c 5mm long; petals yellow, 5–8mm long. Pods straight, flattened, with a few crisped hairs, 20–40 × 3mm.

Readily distinguished from other similar species in Yemen by the stalked gland on the leaf petiole.

An uncommon plant of open sparse grassland found between 700 and 1700m at 'Udayn and in the central escarpment: Ḥadīyah and Suq Ar Ribāṭ, J. Raymah; Wadi Yūr; Ḥillah and Maghrabah, J. Bura'; 'Usil, Ḥarāz; Maḥābishah. (Wood 1647, 1898, 1993, 3268)

4. Cassia fistula *L.* *Shambār*

Small tree to 5m. Leaves with 4–10 pairs of large, ovate, acute leaflets c 5–12 × 3–6cm, the undersurface rather glaucous, minutely puberulent. Racemes very long, pendulous; petals yellow, c 1.5–2.5cm long. Pods straight, cylindrical, indehiscent, pendulous, c 25–50 × 2cm, long persistent after the tree has ceased flowering.

This very attractive species is probably an ancient introduction from India. It grows in great abundance between 500 and 800m in scrub around Ḥadīyah where it was first found by Forsskal. It also grows, though rarely, in scrub below Suq Ar Ribāṭ, J. Raymah and near Suq al Aḥad, Waṣāb As Sāfal.

5. Cassia ? hochstetteri *Ghesq.*

Prostrate or weakly ascending pubescent annual. Leaves with 15–30 pairs of asymmetric, oblong, mucronate leaflets c 6 × 1mm. Flowers 1–3, supra-axillary; pedicels c 5–6mm long; petals yellow, 4–5mm long. Pods straight, flattened, pubescent, c 25–40 × 4mm.

Some doubt exists over the identity of the Yemen plant (Wood 1224, 1262, 1834, 1951) because it does not constantly have the 4–5 stamens characteristic of *C. hochstetteri*. The number has varied between 5 and 8 in the flowers that have been examined.

A rare plant of grassland between 1800 and 2400m in high rainfall areas: near Suq Al Jum'ah, J. Ash Shirq; Mashwarah near Ibb; between Ibb and Suq As Sabt; J. Milḥān.

6. Cassia holosericea *Fres.*

Prostrate to erect, softly tomentose, glaucous woody perennial. Leaves with 3–8 pairs of oblong – elliptic, rounded leaflets c 10–20 × 5–8mm. Racemes rather short, about 10-flowered; petals dull yellow, c 5–7mm long. Pods falcate, flattened, tomentellous c 3 × 1.5cm.

Superficially very similar to the more common *C. italica* but readily distinguished by its softly tomentose indumentum.

Common from sea level to 400m on sandy and stony plain in the Tihama, favouring depressions where water accumulates and frequently near irrigation channels; largely absent from sand dunes.

7. Cassia italica (*Mill.*) *Lam.* 'Ashraq

C. obovata Collad.

Usually prostrate, glaucous woody perennial with finely puberulent stems. Leaves with 3–6 pairs of obovate, rounded leaflets c 1–4 × 0.5–2cm. Racemes c 10–15 flowered; petals pale yellow, 8–20mm long. Pods falcate, flattened, minutely puberulous, c 3–5 × 1.5–2cm.

In Yemen two subspecies occur: subspecies **italica** (e.g. Wood 75/7 from Taiz) with petals more than 1cm in length and racemes longer than the leaves and subspecies **micrantha** *Brenan* (e.g. Hepper 5717 from the Wadi Rima' area of the Tihama) with petals 8.5–9mm long and racemes shorter than the subtending leaves. The exact distribution of the two subspecies is unknown.

C. italica can be distinguished from all other Cassias in Yemen by the crest running along the centre of the pod.

One of the most widespread plants in Yemen, occurring in all areas from sea level to 2400m east to Mārib and Ḥarīb; completely absent only from the Sands and the mountain tops. It is usually found on open stony and sandy plain and is generally common though rarely occurring in great abundance.

8. Cassia mimosoides *L.*

Prostrate annual or short-lived perennial herb with pubescent stems. Leaves with c 15–30 pairs of slightly asymmetric, oblong, acute leaflets c 5 × 1mm. Flowers mostly solitary, ± axillary; pedicels c 12–14mm long; petals yellow, 5–7mm long. Pods straight, flattened, sparsely adpressed pubescent, c 50 × 4–5mm.

Distinguished from *C. hochstetteri* by the long pedicels and the crenate leaf rhachis which can be clearly seen if the leaflets are carefully removed.

Only known from northwest-facing grassland at c 1300m at Bani Sa'id, Ja'farīyah.

9. Cassia nigricans *Vahl*

Erect annual to 1m with a short pubescent and long pilose indumentum mixed together on the stem. Leaves with c 10–15 pairs of oblong, rounded, mucronate leaflets c 15–20 × 4–5mm. Flowers mostly 3–5, supra-axillary; pedicels c 3mm long; petals yellow, c 4mm long. Pods erect, straight, flattened, pubescent, c 2–3 × 0.5cm.

A common field weed between 200 and 500m in the alluvial plains in the east of the Tihama. Inland found once only at 1100m between Riḥāb and Waṣāb al 'Ali.

10. Cassia obtusifolia *L.*

C. tora sensu Schwartz

Erect, usually annual herb to c 1m with pubescent stems. Leaves with 3 pairs of obovate, rounded, mucronate leaflets c 2–4 × 1–2.5cm. Racemes 1–2 flowered, very short; petals yellow, c 1.5cm long. Pods linear, curved, ± square in cross-section, ± glabrous, c 12–20 × 0.5cm.

The only species with glands between the leaflets.

A plant of waste ground, roadsides and field borders between 300 and 1700m; it is found on the eastern plains of the Tihama and is widespread on the escarpment though with a southern tendency and is particularly abundant in high rainfall areas of Taiz and Ibb provinces.

11. Cassia occidentalis *L.* Juljul

Erect, usually annual herb to c 1m with subglabrous stems. Leaves with 4–5 pairs of ovate-elliptic, acute, leaflets c 2.5–8 × 2–3cm. Racemes shortly pedunculate, few-flowered; petals yellow, c 1–1.5cm long. Pods linear, slightly curved, flattened, glabrous, c 10–11cm × 6–7mm.

Abundant on waste ground, field borders, in wadi beds and along irrigation channels from 300–1700m; found in the eastern Tihama alluvial plains and on the escarpment and often associated with the previous species.

12. Cassia senna *L.* Sana

C. angustifolia Vahl

Erect perennial undershrub to c 1½m, the stems subglabrous. Leaves with 4–9 pairs of narrowly elliptic, acute leaflets c 2–5 × 0.5–0.8cm. Racemes very long, many flowered; petals yellow, c 10–15mm long. Pods straight, flattened, sparsely puberulous, c 4–6 × 1.5–2cm.

This species is remarkable for its ability to continue flowering after prolonged drought.

A plant of arid conditions from sea level to 1200m, found in the Tihama, the drier areas of the escarpment and the Mashriq. It grows characteristically along sandy and gravelly drainage lines including the larger wadis.

The dried leaves are sold in markets as a remedy for constipation.

13. Cassia sophera *L.*

Very similar to *C. occidentalis* from which it differs in the following particulars: leaf petiole with a prominent club-shaped gland (in *C. occidentalis* the gland is globular); leaves with up to 10 pairs of leaflets which do not exceed 2cm in width; raceme peduncles c 8–20mm long (best seen on fruiting specimens); pods subcylindrical.

A plant of roadsides, waste ground and wadi beds between 300 and 1700m in the eastern Tihama plains and the escarpment but virtually restricted to the south and particularly common around Taiz. North of J. Raymah it has only been found on J. Marran and by Forsskal in the Wadi Surdūd and Wadi Mawr.

Cassia tomentosa *L.f.* a shrub or small tree with grey tomentose leaves and attractive yellow flowers is commonly planted in gardens in Sana'a.

51. CERATONIA *L.*

1. Ceratonia siliqua *L.* Kiranīt, Khurūb

Carob Tree. Shrub or small tree to 10m, dioecious or bisexual. Leaves dark green and shiny above, coriacious, paripinnate; leaflets glabrous, 2–5 pairs, orate to elliptic, up to 7cm long. Inflorescence a short axillary catkin-like raceme of green flowers on the older wood; petals absent; stamens 5. Pods indehiscent, leathery, flattened, curved, c 10–20 × 1.5–2cm.

Flowers mostly in spring.

A common tree on open hillsides, especially near villages, from 1000 to 2500m throughout Taiz and Ibb provinces; elsewhere very rare and only known from J. Bura', J. Milḥān and Maḥwīt in the central escarpment. The pods are sold in the suqs for animal food.

Since the recent discovery of *Ceratonia oreothauma* Hillcoat, Lewis & Verdc. in Oman and Somalia, it has been suggested that the carob tree might also have originated in Arabia. Deflers (1889) noted *Ceratonia siliqua* on J. Ṣabir and stated his opinion that it was native in Yemen. It is impossible to disprove Deflers' opinion, but I have noted (as did Deflers himself) that the tree usually grows near houses. Yemen is, of course, densely populated so this in itself is not particularly significant but it does appear to be absent from those hills and scrub-filled gullies which are remote from hosues. I would place it with *Iris albicans*, *Myrtus communis* and *Ruta chalepensis*. All four species are essentially Mediterranean in their known distribution, all have some economic or at least aesthetic value, all have been claimed as indigenous to Yemen, but all have decidedly eccentric distribution patterns within the country. *C. siliqua* is virtually restricted to the Shafei tribal areas and it may have been introduced by one of the Shafei dynasties interested in agriculture such as the Rasulids.

52. BAUHINIA *L.*

1. Bauhinia tomentosa *L.*

B. inermis Forssk.

Shrub to 2m with shortly puberulous branchlets which are never rust-coloured. Leaves simple, bilobed, 1–5cm long, about as broad as long, base truncate. Inflorescence of mostly terminal racemes of few, rather large, pale yellow bisexual flowers. Pods not stipitate, flattened, dehiscent, slightly woody, tomentellous, c 6–8 × 1.5cm.

An uncommon plant of riverine scrub between 500 and 900m in the central escarpment: Ḥadīyah, 'Alūjah and Suq Ar Ribāṭ, J. Raymah; near Walajah, J. Milḥān (Schweinfurth, Forsskal); Wadi Yūr; Wadi Ḥijān, Ḥarāz. One of several species unaccountably absent from J. Bura'.

53. PILIOSTIGMA *Hochst.*

1. Piliostigma thonningii (*Schumach.*) *Milne-Redh.* *Rukāb*

Tree to 10m with dark brown to black fissured bark; branchlets, leaf veins, inflorescence and fruit all ± densely rusty-tomentellous when young. Leaves simple, bilobed, reticulate below, c 5–15cm long, about as broad as long. Inflorescence of axillary panicles of pale pink to white, dioecious flowers. Pods stipitate, woody, flattened, indehiscent, c 12–25 × 3–6cm.

Flowers in late summer.

A rare plant found between 400 and 700m in the central escarpment: in valley forest above Suq al Sabt, J. Bura'; in scrub below Suq Ar Ribāṭ, J. Raymah.

54. TAMARINDUS *L.*

1. Tamarindus indica *L.* Ḥumar

Tamarind. Tree to 20m with a broad spreading crown. Leaves paripinnate; leaflets 10–18 pairs, glabrous, oblong, c 2.5cm long, rounded at the base and the apex. Inflorescence of terminal and lateral racemes of red-veined, yellow-brown flowers; petals 5, but 2 rudimentary and much smaller than the others, fertile stamens 3. Pods indehiscent, sausage-shaped, usually curved, covered in red-brown scurf, c 7–12 × 2–3cm.

Flowers April–June.

Abundant along the escarpment from 400 to 1500m, giving every appearance of being indigenous. Usually found by or at least near wadis but extending onto open hillsides in high rainfall areas. Much valued for the shade it gives, but the pulpy part of the fruit used elsewhere for flavouring drinks and preserves is little used in Yemen though it is sold in the suqs.

55. DICHROSTACHYS (*DC.*) *Wight & Arn.*

1. Dichrostachys cinerea (*L.*) *Wight & Arn.* Jizīr

D. nutans Benth.

Shrub to 3m, often growing in groups and apparently suckering, armed with spine-tipped, often leafy, branchlets. Leaves bipinnate, the leaflets numerous and small (c 5mm long). Inflorescence a spike, the upper part yellow, the lower part mauve, usually ± pendent. Pods densely clustered, contorted, indehiscent, about 4cm long.

Widespread on the escarpment between 700 and 2000m, growing on roadsides, field borders, in valley forest and even on open stony hills, but only frequent in overgrazed grassland in areas of good rainfall.

56. ACACIA *Mill.*

Trees or shrubs, the native species almost always armed with spines and prickles. Leaves bipinnate with (usually) numerous small leaflets. Inflorescence usually axillary, the flowers in heads or spikes; stamens numerous. Pods very variable but usually flat, dehiscent.

Acacias are very important in the Yemen flora, occurring in almost all communities except the high mountains and the Sands. They are usually the most important species in scrub, often becoming dominant. The various communities they occur in are discussed under the species below. Their most important value in the traditional economy is for firewood and a good deal of Acacia scrub clearance has taken place particularly in the Mashriq. Some species were tapped for gum but this practice has fallen into disuse.

The different species are much easier to identify in the field than in the herbarium. This is because dried specimens usually show only leaf, flower and pod characters, whereas in the field it is usually easy to observe whether the plant is a tree or a shrub, whether it has a distinct shape and what its bark is like. The bark is a particularly useful character but it is best to look at healthy mature trees as immature, very old or damaged trees may not show their typical bark. In addition most species have definite ecological requirements particularly with regard to altitude, and in Yemen flower at a particular season. It is often possible to name an Acacia simply by knowing the date and location where it was collected. Flowering period though must be used as a character with some caution since isolated trees are often found in flower outside the normal flowering period.

The following key is designed to be used with dried specimens and generally does not include the field characters mentioned above. These are, however, given in the descriptions.

Most species are rather variable in many smaller, more technical characters and have been divided into subspecies or varieties. These are mentioned where appropriate but, with the exception of *Acacia nilotica*, do not affect the general appearance and ecology of the species at least within Yemen. Hybrids are known but seem to occur very rarely.

There are a number of points about the distribution of *Acacia* in Yemen which deserve some comment. *A. gerrardii*, *A. etbaica* and *A. hockii* are generally common in Yemen but are strikingly absent from the central escarpment though all occur in similar situations around Ibb. *A. mellifera*, *A. ehrenbergiana*, *A. hamulosa* and *A. oerfota* occur at much higher altitudes in the Mashriq than on the escarpment even though the Mashriq is both drier and has lower winter temperatures than the escarpment. Several species, notably *A. hockii*, *A. nilotica* ssp. *kraussiana* and *A. seyal*, have strangely disjunct distributions. There are no obvious explanations for these facts and the ecology of *Acacia* in Yemen would be an interesting study.

1. Flowers in spikes. Spines all curved, less than 7mm long — **2**
Flowers in ± globular heads. Some or all spines straight and exceeding 10mm long — **5**

2. Leaves and pinnae rhachis with scattered prickles — *A. hamulosa*
Leaves and pinnae rhachis unarmed — **3**

3. Pinnae with 8–20 pairs of leaflets — *A. asak*
Pinnae with 1–4 pairs of leaflets — **4**

4. Leaflets 1–2 pairs. Calyx 0.5–1cm long — *A. mellifera*
Leaflets 3–4 pairs. Calyx 1–2cm long — *A. laeta*

5. Flowers yellow. Mature pod usually constricted between the seeds (except in *A. hockii*), straight or falcate — **6**
Flowers white or cream. Mature pod not constricted between the seeds or, if so, pod strongly curled or twisted — **9**

6. Pinnae (1–) 2 pairs. Pod less than 5mm wide — *A. ehrenbergiana*
Pinnae 2–11 pairs. Pod more than 7mm wide — **7**

7. Branchlets pubescent, ± eglandular. Pod straight or nearly so — *A. nilotica*
Branchlets glabrous, densely glandular. Pod falcate — **8**

8. Epidermis of branchlets peeling off to reveal a yellow-green powdery undersurface. Tree — *A. seyal*
Epidermis of branchlets not peeling off to reveal a powdery undersurface. Shrub. (Very rarely a tree) — *A. hockii*

9. Pod ± glabrous; straight or nearly so. — **10**
Pod distinctly hairy; straight, falcate, twisted or coiled — **12**

10. Branchlets glabrous or finely puberulent. Pods usually exceeding 6cm long. Bark never fissured — **11**
Branchlets pubescent. Pods less than 6cm long. Bark fissured — *A. etbaica*

11. Leaflets c 4mm long. Pinnae usually less than 10 pairs. Highland tree with distinctive peeling, papery bark on branches — *A. origena*
Leaflets c 2.5mm long. Pinnae usually more than 10 pairs. Lowland tree with smooth bark — *A. abyssinica*

12. Pods less than 6mm wide, usually twisted or at least curled. Flat-topped trees or bushes — **13**
Pods 6–20mm wide, straight or falcate (rarely coiled). Trees or shrubs, not obviously flat-topped — **14**

13. Pods spirally twisted, pubescent, not markedly constricted between the seeds (Tihama and escarpment) — *A. tortilis*
Pods sinuate, falcate or curled to form a circle, pilose, strongly constricted between the seeds (Mashriq) — *A. campoptila*

14. Pods ± straight. Spines rarely exceeding 3cm long. Shrubs — **15**
Pods falcate or curled to form a circle (very rarely straight). Some spines up to 6cm long. Tree — *A. gerrardii*

15. Branchlets whitish. Pod flattened, veined, quickly dehiscent — *A. oerfota*
Branchlets red-brown. Pod sub-cylindrical, woody, not veined, tardily dehiscent — *A. edgeworthii*

1. Acacia abyssinica *Hochst. ex Benth.*

Shrub or small tree to 6m; bark smooth, dark grey. Spines paired, straight, very variable in length, up to 4cm long, sometimes absent. Branchlets green when young, glabrous or puberulent. Pinnae and leaflets very numerous both often with more than 30 pairs; leaflets very small. Flowers in heads, cream. Pods glabrous, straight or nearly so, veined, c 8–12 × 1–1.5cm.

Flowers May–September.

The Yemen plant is variety **macroloba** *Schweinf.* with erect 10–11 seeded pods and is probably a distinct species.

Local between 300 and 1700m along the escarpment from Shar'ab and Maqbanah northwards, but largely restricted to the foothills of the central escarpment where it is locally common between 400 and 600m.

2. Acacia asak (*Forssk.*) *Willd.* 'Asaq, Daḥī

A. glaucophylla Steud.

A. senegalensis sensu Schwartz

Shrub or small tree to 4m; bark yellow-green, papery, peeling. Spines usually in 3s (rarely absent), straight or slightly curved, up to 7mm long. Branchlets purple-brown, ± glabrous. Pinnae 3–6 (usually 4) pairs, leaflets c 7–18 pairs, glaucous. Flowers in spikes, cream. Pods glabrous, straight, slightly veined, often purplish, c 6–12 × 1–1.6cm.

Flowers mostly in summer following the spring rains.

Plants collected in high rainfall areas, notably in the valley forest above Suq As Sabt, J. Bura', tend to be unarmed and to have larger leaflets and pods.

Abundant on steep well-drained mountain slopes between 600 and 1700m along the whole escarpment extending southeast to J. Sawraq. Sometimes almost the only constituent of relatively dense scrub, usually on very loose scree-like volcanic slopes (e.g. Naqīl Hasus, Nashimah).

3. Acacia campoptila *Schweinf.* Sumur

Shrub or rarely a small tree to 4m, ± flat-topped; bark smooth, brown. Spines paired, usually short and hooked on young wood, straight on older wood and reaching 5cm in length. Branchlets red-brown, densely pubescent. Pinnae 2–6 pairs; leaflets 5–8 pairs, very small. Flowers in heads, cream. Pods pale red-brown, white-pilose, sinuate, falcate or curled into circle (not spirally coiled), strongly constricted between the seeds, 3–6 × 0.5cm.

Apparently flowering after substantial spring rain.

This plant is very closely related to the African *Acacia paolii* Chiov. as well as to *Acacia tortilis*. Apart from the pod characters it differs from *A. tortilis* in having ± glabrous leaflets and a black (not greeny-brown) seed which is often pitted along the edge. It will probably turn out to be a subspecies of *A. tortilis*; one intermediate specimen (Wood 3134 from the Wadi Mawr between Ḥajjah and Maḥābishah) has been found.

Known with certainty from Al 'Ashshah (Wood 3248), Wadi Madhab, near Marāshī (D. Wood Y 1220), between Mārib and the Fardah pass and Wadi Nuwāsh north of Rada' (Wood 3090) but it is probably abundant on stony plain and along drainage lines between 1100 and 1900m in the Mashriq from Ḥarīb north to Saudi Arabia.

4. Acacia edgeworthii *T. Anders.*

Shrub to 1½m. Spines paired, straight, up to c 3cm long. Branchlets red or purple brown, pubescent becoming glabrous with age. Pinnae 4–6 (–10) pairs; leaflets 6 (–15) pairs. Flowers in heads, white. Pods purple-brown with a white tomentellous indumentum with red glands, woody and subcylindrical, straight or slightly curved, 7–12 × 1–2cm.

Found flowering in winter.

A rare plant of stony steppe between 200 and 400m in the Tihama extending from Nukaybah south to Mawza' and doubtless thence to the South Yemen border.

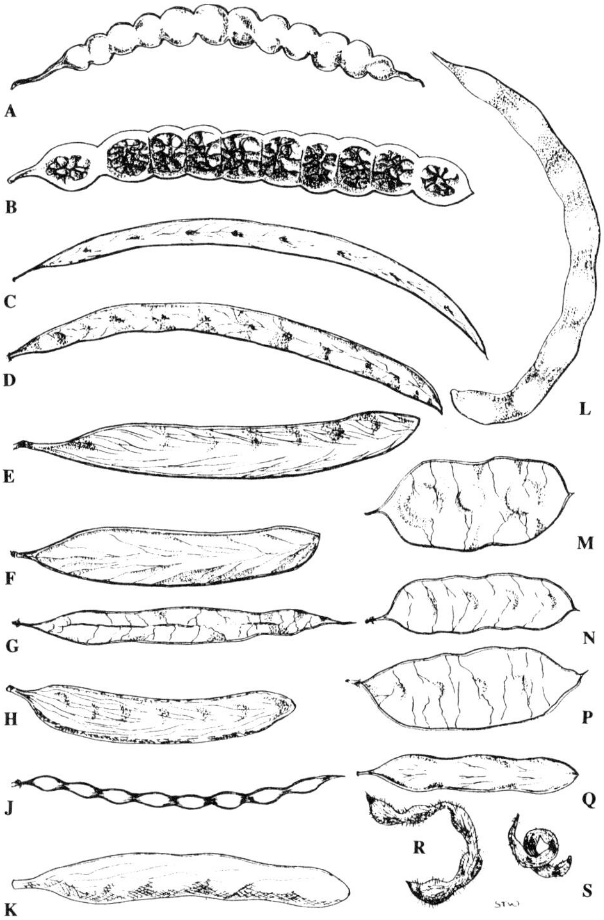

Fig. 13. Pods of *Acacia* species. **A.** *A. nilotica* subsp. *indica*; **B** *A. nilotica* subsp. *kraussiana*; **C** *A. seyal*; **D** *A. hockii*; **E** *A. abyssinica*; **F** *A. origena*; **G** *A. asak*; **H** *A. oerfota*; **J** *A. ehrenbergiana*; **K** *A. edgeworthii*; **L** *A. gerrardii*; **M** *A. hamulosa*; **N** *A. mellifera*; **P** *A. laeta*; **Q** *A. etbaica*; **R** *A. campoptila*; **S** *A. tortilis*.

5. Acacia ehrenbergiana *Hayne* Silam

Shrub or small tree to 5m, usually branched near the base and with many ascending branches; bark brown, papery, peeling. Spines paired, straight, up to 3cm long. Branchlets red-brown, glabrous but with a few reddish glands. Pinnae (1–) 2 pairs; leaflets 8–12 pairs. Flowers in heads, yellow. Pods glabrous, falcate, constricted between the seeds at maturity, 5–10 × 0.3–0.5cm.

Flowers throughout the year even after prolonged drought.

A widespread and locally abundant species found from sea level to 1900m. In the Tihama it occurs almost everywhere and is the only Acacia likely to be encountered by the sea. Locally on sand it forms dense scrub woodland in which it is almost the only species. Such woodland is a feature of the Ḍaḥī – Zaydīyah area but can be found in many other places in the Tihama north of the Wadi Zabīd. On the escarpment it is mainly found along wadi lines in dry country, but has penetrated Ḥaymah to 1900m along the road built by the Chinese. In the Mashriq it penetrates far into the mountains along the wadis reaching Ṣa'dah, Watadah and the Rada' area. At the edge of the Sands around Mārib and Ḥarīb it is abundant, but does not appear to grow on the Sands themselves.

6. Acacia etbaica *Schweinf.* Qaraḍ

Tree or rarely a shrub to 10m, often flat-topped; bark brown, deeply fissured. Spines paired, usually short and hooked, mixed with long and straight, up to c 4cm in length. Branchlets pale brown, densely grey puberulous when young, becoming glabrous, the bark often splitting to reveal the red brown undersurface. Pinnae 4–9 pairs; leaflets 10–30 pairs. Flowers in heads, cream. Pods glabrous, straight, rather shiny, veined, 4–6 (–9) × 1cm.

Flowers in early summer following the spring rains.

Most Yemen plants belong to subspecies **uncinata** *Brenan*, in which the rhachis of the leaves and the pinnae are densely puberulent below and the spines curved as well as straight. A collection from Majz (Steinberg 381) has glabrous leaves and mostly straight spines and should be assigned to subspecies **etbaica** but is remarkable for its short pods (less than 3cm long) containing only 1–2 seeds.

Widespread between 1200 and 2300m. On the escarpment and southeast mountains it is characteristic of stony uncultivated mountains in the rainshadow areas closely paralleling the distribution of *Juniperus*, being particularly abundant in Khawlan Ash Sham and the Ḥujarīyah and often occupying an altitude zone intermediate between *Acacia asak* and the *Juniperus*. It was probably dominant in prehistoric times in woodland around 1500–1800m in such areas. Few pure stands remain though small examples may be seen on Jabal Jirdad, Bani 'Umar, Ḥujarīyah and west of Suq al Jumlī, Khawlan Ash Sham. Although quite common in Waṣāb, 'Utmah and Ibb province it is strikingly absent from the central escarpment with the exception of a single small colony near Manākhah. In the Mashriq it is local and apparently restricted to 1600–1900m in the Ḥūth – Ḥarf Sufyān region and the area to the east of Al Ḥadā' and north of Rada'.

7. Acacia gerrardii *Benth.* Ṭalḥ

A. pachyceras O. Schwartz
A. iraquensis Rech. f.

Tree or rarely a shrub to 10m; bark grey, smooth. Spines straight or slightly curved, very variable in size but reaching 5cm long. Branchlets grey-brown, densely pubescent. Pinnae 2–10 pairs; leaflets 12–20 (–30) pairs. Flowers in heads, cream. Pods pubescent, usually falcate (rarely straight or coiled into a circle), veined, yellow to purple-brown, c 5–12 × 0.6–1.5cm.

Flowers after the spring rains and then throughout the summer.

This species is very variable particularly in its indumentum (plants from the Ibb area are notably hairy), the number of pinnae (much reduced in low rainfall areas) and the length of the stipe of the pod. It has been customary to regard Arabian *A. gerrardii* as a separate subspecies *negevensis* Zoh. or even as a separate species *A. iraquensis* Rech. f., but the welter of forms in Yemen appear to straddle the features on which these taxa are based and I think it is best to place all Yemen plants within the very variable *A. gerrardii.*

Locally abundant from 1500 to 2500m. Common in Ibb province extending southwards to the Taiz area, but otherwise completely absent from the escarpment. Everywhere the commonest Acacia on the high plateau and entering the Mashriq where, however, it does not seem to occur lower than about 1800m though it is widespread, being found from Al Bayḍā' north to the Majz area. It is rare on open hillsides and is usually found on alluvial plains and around cultivation, being particularly abundant in the Ṣa'dah plain and in the valleys north of Ibb. This preference for alluvium may partially explain its curious absence from much of the escarpment and the Mashriq.

8. Acacia hamulosa *Benth.* Qitat

A. asak sensu Schwartz.

Shrub to 1½m. Spines paired or in 3s, straight or curved, up to 7mm long. Branchlets grey or dark red-brown, glabrous or pubescent. Pinnae 2–3 pairs; leaflets 5–8 pairs. Flowers in spikes, cream. Pod finely puberulent, straight, yellow-brown, veined, c 3–6 × 2–2.5cm.

Distinguished from all other species by the small hooked prickle(s) on the leaf rhachis.

Flowers after any considerable rain.

Probably divisible into two subspecies: Tihama plants (e.g. Wood 75/104) appear to be ± glabrous but plants from the Mashriq (e.g. Wood 75/648, Steinberg 1170) are pubescent on the branchlets, leaflets, pedicels and calyx.

Locally abundant on stony plain in dry country; frequent on the stony steppe in the southern Tihama from Ḥays southwards where it is found between 200 and 600m; in the Mashriq widespread between 1200 and 1800m from the Rada' area through the Jawf to Majz.

9. Acacia hockii *De Wild.*

A. orfota sensu Hepper
A. seyal sensu Schwartz

Shrub to 1½m; bark red-brown, peeling, papery. Spines straight, up to 3.5cm long. Branchlets red-brown, glabrous but densely glandular. Pinnae 1–5 pairs; leaflets 7–11 pairs.

Flowers in heads, bright yellow. Pods glabrous, falcate, sometimes slightly constricted between the seeds, veined, red-brown, 9–11 × 0.7–1cm.

Flowers from April to October.

Rather variable, plants from dry country having narrower pods and fewer pinnae. Wood 3247 from Raydah is remarkable for developing into a tree. Yemen plants may be separated from African plants as a distinct subspecies since they are always glabrous and usually have wider pods than African plants.

Locally abundant on the escarpment between 1600 and 2500m but restricted to two distinct areas: Taiz and Ibb provinces north to the Sumārah pass; from Ḥajjah, Kuḥlān and Raydah north to Maḥābishah, Shahārah and Ḥūth. It is a plant of stony hills and, in the Ibb area, rock outcrops and grassy banks around cultivation.

10. **Acacia laeta** *R. Br. ex Benth.*

Shrub or small tree to 6m; bark smooth, grey-black. Spines paired, hooked, up to 5mm long. Branchlets grey, glabrous. Pinnae 2–4 pairs; leaflets 3–5 pairs, glaucous, rather large. Flowers in racemes, cream. Pods glabrous, straight, veined, pale brown, c 6 × 2cm.

Flowers in winter.

Very closely related to *A. mellifera* (possibly derived from it) and growing as scattered individuals throughout its range; widespread but very local between 500 and 1900m on the escarpment and in the Mashriq: Wadi Rima' presumably near Mishrāfah (Pratt); Ḥajjaylah (Deflers); Mafḥaq; Wadi Mawr between Ḥajjah and Maḥābishah; between Ḥajdah and Ḥays; Naqil Ḥasūs near Nashimah; north of Rada' (Westinga); Naqīl Ḥajlah between Jūbah and Ḥarīb.

11. **Acacia mellifera** (*Vahl*) *Benth.*

Ḍay'an, Ẓubah

Shrub or small tree to c 4m; bark grey-brown, smooth. Spines paired, hooked, up to 5mm long. Branchlets pale brown, glabrous. Pinnae 2 pairs; leaflets 1 (–2) pairs, rather large. Flowers in spikes, white. Pods glabrous, straight, veined, yellow-brown, c 3–6 × 1.5cm.

Flowers mainly in winter, often when the plant is leafless.

Widespread and often abundant on the escarpment from 300 to 1700m though rarely found above 1500m. In many areas, particularly between 500 and 1000m in the foothills of the central escarpment, it is the dominant species in scrub on stony uncultivated hillsides. It is also very important in scrub in rainshadow areas of the central escarpment and in parts of the Ḥujarīyah, but is noticeably less common in *Euphorbia* scrub around Taiz. In the Mashriq it is widespread between 1100 and 1900m but apparently never dominant and not recorded north of Al Ḥarf.

12. **Acacia nilotica** (*L.*) *Del.*

Tree to 12m high; bark dark brown to black, fissured. Spines paired, straight, often deflexed, up to 5cm long but sometimes absent. Branchlets grey or brown, pubescent. Pinnae 2–11 pairs; leaflets 10–25 pairs. Flowers in heads, yellow. Pods very variable but always indehiscent, straight or slightly curved.

Two very distinct subspecies occur in Yemen:

Subspecies **indica** (*Benth.*) *Brenan* Suyal

A. arabica (Lam.) Willd.

Mimosa sejal Forssk. (perhaps)

A broad spreading tree to 12m with glaucous leaflets, often unarmed. Pods grey tomentose, c 7–12 × 1–1.5cm but very variable in shape even on the same pod – sometimes necklace-like with rounded seed segments connected by a small joint, sometimes deeply and regularly constricted between the seeds, sometimes coarsely and irregularly crenate, sometimes ± entire or any combination of these on the same tree. (Wood 74/448, Radcliffe-Smith & Henchie 4529, Henry 14/76).

Flowers throughout the year.

Possibly introduced and with a strong southern tendency in its distribution. Widespread from 100–1400m in Taiz province and the Tihama north to Ḥadīyah and Bayt al Faqīh; absent further north apart from a single record by Forsskal from the Wadi Mawr near Luḥayyah. Usually occurring as an isolated tree by roadsides and in villages, but abundant in the Zabīd area.

Subspecies **kraussiana** (*Benth.*) *Brenan* Qaraḍ

Small tree to c 4m with green leaflets. Pods pubescent at first, but the raised parts over the seeds becoming shiny, glabrous and eventually almost black, not necklace-like but with slightly crenate margins, c 8–12 × 1–1.7cm.

Flowers in summer.

A local tree around fields and on stony plain in five areas between 1200 and 2000m: on the plain bounded by Nashimah, Misrākh, and Yufrūs; Taiz west to Ramādah and north to Mikhlāf; from Mawīyah to the border east of Raḥidah; between 'Utmah and Madīnat Al 'Abīd; from Rada' east to Dhī Nā'im and south to Damt. (Wood 74/370, Hepper & Wood 5836)

13. **Acacia oerfota** (*Forssk.*) *Schweinf.* 'Urfūṭ

A. nubica Benth.

Shrub to 2½m. Spines paired, straight or slightly curved, up to 2cm long, but commonly much less. Branchlets white, pubescent. Pinnae 2–7 pairs; leaflets 3–16 pairs, grey-green. Flowers in heads, white. Pods puberulent, ± straight, woody, veined, grey-green, often with a wing-like margin, c 4–10 × 1–1.5cm. The plant is strongly foetid when cut.

Flowers after rain.

Two quite distinct varieties occur in Yemen: the typical plant has short, often slightly curved spines up to c 8mm long, leaves relatively long with a rhachis exceeding 2cm, c 5 pairs of pinnae and at least 10 pairs of leaflets, each leaflet less than 1mm wide. This is the Tihama plant which is abundant between 200 and 700m on alluvial and stony plain in the Tihama and the escarpment foothills north of the Wadi Zabīd. Further south it is replaced by *A. hamulosa*. (Radcliffe-Smith & Henchie 4932, Wood 74/417).

The other variety has straight spines exceeding 1cm in length, short leaves with a rhachis less than 2cm in length, 2 pairs of pinnae and 4–6 pairs of leaflets, each leaflet more than 1mm wide. Locally common on gravel and stony plain in the Mashriq between 100 and 1900m from the Wadi Nuwāsh area and the Naqil Ḥajlah near Ḥarīb north to Ṣa'dah (Wood 75/605, 3081).

14. Acacia origena *Hunde* Ṭalḥ

A. negrii auct.

A. menachensis Schweinf. ined.

A. abyssinica sensu Schwartz & Deflers

Tree to c 12m, ± flat-topped; bark yellow-brown, papery, peeling off in layers. Spines paired, straight, up to 6cm long. Branchlets yellow-brown, glabrous or puberulent. Pinnae 5–10 pairs; leaflets 10–25 pairs. Flowers in heads, cream. Pods glabrous, ± straight, veined, greeny-brown, c 5–8 × 1.5–2cm.

Flowers in summer.

Although the Yemen plant is quite constant it is doubtful whether it is the same as the type from Ethiopia.

Widespread and common on the escarpment above 1900m and on the mountains around the high plateau from 2300m to 2900m thus growing higher than any other *Acacia* in Yemen. It is found around cultivation and on open hillsides and was probably the main constituent of highland forest on the escarpment in prehistoric times. In the Ḥarāz on J. Shibām and J. Masār there are a number of small relic woodland areas in which *A. origena* is dominant and it often forms scrub on steep slopes in other highland areas.

15. Acacia seyal *Del.*

Tree to 9m; bark powdery, yellow-green. Spines paired, straight, up to 6cm long. Branchlets glabrous, red-brown with many reddish glands, the epidermis peeling off to reveal the powdery, yellow-green undersurface. Pinnae 3–7 pairs; leaflets 11–20 pairs. Flowers in heads, yellow. Pods glabrous, falcate, constricted between the seeds at maturity, veined, c 7–15 × 0.7–0.9cm.

Flowers from January to May.

Apart from a single record by Schweinfurth from Al Khalīfah in the Tihama at 250m, this tree is restricted to a number of scattered areas between 1300–1900m, mostly but not exclusively on alluvial plain: locally abundant on the plains north of Taiz extending in an arc southeast to Raḥidah and even climbing the mountains to Ḥayfān; in the plain bounded by Yifrūs, Misrākh and Nashimah; in the plain between Riḥāb and Waṣāb; in the valley between Ḥūth and Khaywan; abundant in the plain between Shahārah and Al 'Ashshah.

16. Acacia tortilis (*Forssk.*) *Hayne* Sumur, Ḥaras

A. spirocarpa Hochst. ex A. Rich.

Tree to 15m, rarely a bush, characteristically flat-topped. Bark grey to black, smooth. Spines paired, usually short and hooked on young wood, straight on older wood, up to c 5cm long. Branchlets brown, pubescent. Pinnae 2–10 pairs; leaflets 6–19 pairs, very small. Flowers in heads, cream. Pod pubescent, spirally twisted usually for more than a single coil, slightly constricted between the seeds, veined, c 3–6mm broad.

Flowers mostly April–June.

The Yemen plant is probably all subspecies **tortilis** with narrow eglandular pods (e.g. Wood 75/41, Pratt 75). Intermediates with *A. campoptila* occur in the Wadi Mawr.

Widespread in the Tihama, penetrating the escarpment along the broader valleys up to 1000m (very rarely to 1500m); rather local in the north and generally near the foothills but becoming more common further south and locally dominant in open Acacia 'parkland'; in Taiz province where it grows from the coast right into the mountains up to 1200m and is by no means restricted to wadis.

Acacia lahai Steud. & Hochst. ex Benth. was recorded by Schweinfurth from Ḥillah, J. Bura' and Manākhah but almost certainly in error. The plant has not been refound, Schweinfurth's specimens were sterile and he himself was doubtful about their identity.

At least two species of *Acacia* are known to be cultivated in Yemen. **Acacia albida** *Del.* is a tall tree with whitish branchlets, spikes of white flowers, and a strongly curled indehiscent pod. It has been found on the Qa' Shirāh (B. Acres). **Acacia cyanophylla** *Lindl.* is an unarmed tree with heads of yellow flowers and differs from all native species in having simple leaves (technically phyllodes). It is commonly cultivated in Sana'a.

Several species from related genera are cultivated in the Tihama. The following three are commonly found around towns and projects. All have bipinnate leaves and flowers in round heads. **Albizia lebbeck** (*L.*) *Benth.* is an unarmed tree whose flowers have distinctive green filaments up to 3cm long. The pod is 3–5cm broad and thus wider than in other related species. It has been cultivated in Yemen since at least the eighteenth century. **Leucaena leucocephala** (*Lam.*) *De Wit* (*L. glauca* (L.) Benth.) is also an unarmed tree. It has greenish flowers and a long stipitate pod up to 2cm wide. Its hairy anthers distinguish it from all related species in Yemen. **Prosopsis juliflora** (*Sw.*) *DC.* is an armed shrub with cream flowers. It is commonly used as a hedge plant round 'projects' as at Al Kadan.

57. PITHECELLOBIUM *Hart.*

1. Pithecellobium dulce (*Roxb.*) *Benth.* Daymān

Small tree to 10m with a pair of spines in the place of stipules. Leaves bipinnate with 1 pair of leaflets; leaflets asymmetric, elliptic, large, c 2.5 × 1.5cm. Flowers creamy in small heads arranged paniculately in leafless inflorescences. Pods spirally twisted.

An introduced tree commonly planted in the Tihama and the drier escarpment foothills up to 1500m. Occasionally appearing naturalised as near Raḥidah and at Bathayjah, both in the Ḥujarīyah.

ELAEAGNACEAE

Elaeagnus angustifolia *L.* has been found in village gardens on Jabal Ṣabir and since it has been recommended for planting in Yemen it is likely to become more frequent. It is an ornamental slightly spiny shrub which has silvery scales on the twigs, undersurfaces of the leaves and fruit. The yellow flowers appear in spring.

HALORAGACEAE

MYRIOPHYLLUM *L.*

1. Myriophyllum spicatum *L.*

Dark green perennial aquatic herb entirely submerged except for the flowering spike. Leaves in whorls of 4, pinnate with numerous capillary segments. Inflorescence a terminal spike with the flowers in sessile whorls in the axils of the bracts, female flowers below, male flowers above, the latter with small red petals.

Superficially similar to *Ceratophyllum demersum* which differs in having the leaves once or twice forked, not pinnate.

A rare plant found in ponds between 2400 and 3000m: Kawkabān; Bayt Badī near 'Amrān.

LYTHRACEAE

A family of herbs and shrubs with simple, entire, usually opposite leaves. Flowers variously arranged. Fruit a capsule.

1.	Shrubs	**2**
	Herbs	**3**
2.	Leaves glabrous. Flowers white, in many-flowered panicles	5. *Lawsonia*
	Leaves hairy, gland-dotted below. Flowers red, axillary, solitary or in few-flowered racemes	3. *Woodfordia*
3.	Flowers solitary in the leaf axils	4. *Lythrum*
	Flowers in axillary cymes or clusters	**4**
4.	Flowers in sessile axillary clusters. Capsule opening by valves at the apex	2. *Rotala*
	Flowers pedicellate in pedunculate axillary cymes. Capsule opening transversely	1. *Ammania*

1. AMMANIA *L.*

Annual herbs with opposite, sessile leaves. Flowers small, in axillary cymes, the calyx campanulate. Capsule opening transversely and irregularly.

Leaf base auriculate. Style 2–3mm long	*A. auriculata*
Leaf base attenuate. Style less than 0.5mm long	*A. baccifera*

1. Ammania baccifera *L.*

A. attenuata Hochst. ex A. Rich.

Erect, branched, glabrous herb to 50cm. Leaves oblong, attenuate at both ends. Cymes shortly pedunculate (rarely sessile); flowers pedicellate. Capsule reddish with a very short persistent style 0.5mm long.

Only known from irrigation ditches at c 1000m near Suq al Aḥad, Waṣāb As Sāfal.

2. Ammania auriculata *Willd.*

Very similar to the previous species from which it differs in having leaves with cordate-auriculate bases and a capsule with a persistent style 2–3mm long.

A rare plant of streamsides between 500 and 600m in the central escarpment: Walajah, J. Milḥān (Schweinfurth); by the wadi below Ḥadīyah, J. Raymah.

2. ROTALA *L.*

1. Rotala ? serpiculoides *Hiern*

Erect glabrous annual, often very short (less than 10cm high) and sparsely branched. Leaves sessile, opposite, oblanceolate, the base ± attenuate, c 8–17 × 2–4mm. Flowers very small in sessile clusters in the leaf axils, calyx campanulate. Capsule opening by valves from the apex.

There is some doubt about ascribing the sole Yemen collection (Wood 992) to this species. The specimen is not very typical of *R. serpiculoides* and in particular has less strikingly attenuate leaves than in the type, but it fits this species better than any other I have seen.

Known only from a flooded field at 1400m in Wadi Ḍabāb, Taiz where it was growing with *Guizotia scabra*, *Oldenlandia corymbosa* and *Pycreus flavescens*.

3. WOODFORDIA *Salisb.*

1. Woodfordia uniflora (*A. Rich.*) *Koehne.*

Small branched shrub to 3m but commonly less in Yemen. Leaves opposite, lanceolate, acuminate, c 3–7cm long, the undersurface sparsely pubescent and dotted with black glands. Flowers large (c 2cm long), red, axillary, solitary or in short, few-flowered racemes; flower parts in 6s. Fruit an ovoid capsule c 8mm long.

Frequent but easily overlooked between 700 and 2100m on the western slopes of the central escarpment (Bura', Milḥān, Ḥarāz, Raymah); elsewhere only known from Shahārah and east of Rujum. It is found in valley scrub and on steep, scrub-covered slopes where grazing is difficult.

4. LYTHRUM *L.*

1. Lythrum hyssopifolia *L.*

Erect or ascending, glabrous, branched annual herb. Leaves sessile, alternate, ± erect, linear-oblong, c 10–15 × 2–3mm. Flowers numerous, small, solitary in the leaf axils; calyx cylindrical with 6 apical teeth, c 5mm long; petals mauve.

A common plant of streamsides, ditches and marshes between 2200 and 3000m on the high plateau extending southwards in similar situations to the Wadi Banā, J. Ba'dān and Dhī Sufāl. Also at Sāqayn in Khawlan Ash Sham.

5. LAWSONIA *L.*

1. Lawsonia inermis *L.* — Ḥinah

Henna. Shrub to 3m, sometimes spiny especially on old wood. Leaves opposite, glabrous, elliptic or obovate, acute, c 1–2cm × 4–7mm. Flowers white, fragrant, in many-flowered panicles; flower parts in 4s. Fruit a small globose capsule c 5mm diameter, the style persistent.

Originally introduced but now well naturalised near villages in the east of the Tihama and along the wadis in the foothills of the escarpment up to about 1400m; mostly occurring in small quantity and often as isolated bushes.

The leaves are the source of the well-known dye, Henna.

THYMELAEACEAE

GNIDIA *L.*

1. Gnidia somalensis *Franch.*

Lasiosiphon sphaerocephalus (Bak.) Domke.

Erect bushy perennial to 50cm, woody below and with many branches arising from the base; stems pilose especially above. Leaves glaucous, shortly petiolate, oblong-linear, c 2–3cm long, with an acute apex. Flowers in attractive, dense, bracteate, terminal heads; calyx tube cylindrical, pilose with long silky hairs at the base and with 5 orange-yellow petaloid lobes. Fruit consisting of dry seeds enclosed in the persistent base of the calyx tube.

A common and conspicuous plant of stony hillsides and rocky plain on the high plateau and in similar areas of Khawlan Ash Sham extending southeast to Damt and Bayḏā' and west to Maḥwīt. There is also a colony at Shahārah and another on the Dhubhān plateau around Turbah which extends west to Bathayjah and east to Akāḥilah. It grows between 1800 and 3000m.

MYRTACEAE

A distinctive family of trees and shrubs with entire, coriaceous, gland-dotted and usually aromatic leaves that lack stipules. Flowers with a conspicuous hollow receptacle united to the ovary, inconspicuous petals and numerous prominent stamens which are cream in all our species.

Flowers solitary, axillary	1. *Myrtus*
Flowers in terminal panicles	2. *Syzygium*

1. MYRTUS *L.*

1. Myrtus communis *L.* Hadas

Myrtle. Erect, much-branched shrub reaching 2m in Yemen. Leaves opposite, lanceolate to narrowly elliptic, up to about 3 × 1.5cm, acute, coriaceous, glabrous, strongly aromatic when crushed. Flowers solitary, axillary, long pedicellate, fragrant; petals white. Fruit a subglobose berry c 8mm diameter, blue-black when ripe.

Apart from a single plant in *Dodonaea* scrub by the path from Midāhit to Aḥkūm in the Ḥujarīyah, this species is restricted to the central escarpment. It is rare on J. Raymah (J. Ẕallamlam), J. Bura' (near Maghrabah) and J. Sāfān, Ḥarāz, but is abundant in the grassland around 'Ukaybir on J. Milḥān.

The status of this plant in Yemen is rather doubtful. Its somewhat eccentric distribution and its attractive fragrance suggest it is an introduction, but it is found elsewhere in Arabia so it is possible to regard the Yemen populations as isolated relics.

2. SYZYGIUM *Gaertn.*

1. Syzygium guineense (*Willd.*) *DC.*
Riya', Athalamah

Tree to about 10m. Bark dark grey, smooth. Leaves opposite, glabrous, elliptic, reaching 12 × 5cm, coriaceous, waxy-green, not markedly fragrant; apex shortly acuminate; base attenuate. Inflorescence a terminal panicle of fragrant white flowers in subsessile clusters at the apex of the panicle branches. Fruit an ovoid, juicy, purple-black berry up to about 1.5cm long.

Yemen plants all seem to be referable to subspecies **guineense**.

It flowers from April to June when the whole tree is covered in a mass of flowers.

Very widespread along the Yemen escarpment, occurring from the Ḥujarīyah north to Khawlan Ash Sham and from about 800 to 2700m, but very uncommon. It is found most typically as a riparian species with *Ficus* spp. along the lusher wadis at middle altitudes, such as the Wadis Līyah, Yūr, Ḥadabah ('Udayn) and Sukhnah. It also occurs sporadically by permanent streams at higher altitudes such as Wadi Aḥjar and above Dhī Sufāl where it is associated with *Myrica salicifolia* and *Debregeasia saenab*.

The Guava, **Psidium guajava** *L.*, known in Yemen as 'Zaytūn' is sporadically cultivated by wadis along the escarpment below about 1500m. It is occasionally seen in markets especially in Taiz. It is a shrub or small tree with elliptic leaves up to 5cm broad and pubescent below.

Various species of *Eucalyptus* have been introduced to Yemen and they are now a conspicuous part of the scenery especially around towns where they have been planted to surround buildings and to line roads. Although they set seed freely, there is little sign of spontaneous regeneration at present. All are rapid growing trees with pendent, lanceolate, aromatic mature leaves. They are of course renowned for their drought resistance and so have been favourites in the government's various tree-planting schemes. They are known locally as 'Kāfūr'.

Eucalyptus is a very large and complex genus whose species are very difficult to identify. It is necessary to collect mature and juvenile leaves, flowers and fruit, and to take photographs of the habit and bark to ensure that a particular tree can be correctly identified. In view of the difficulties of identifying these introduced species, the following account must be regarded as tentative.

There appear to be only three species which are at present found as mature trees in Yemen. Of these much the most common is **Eucalyptus camaldulensis** *Dehnh.* which is planted throughout almost the whole of Yemen including the Tihama. Its mature leaves are very variable in size but normally exceed 8cm in length and 1.5cm in width. The umbels of 5–10 flowers are subtended by terete peduncles which do not normally exceed 12mm in length. A very similar species which may be **Eucalyptus marginata** *Sm.*, was probably the first *Eucalyptus* to be introduced to Yemen. It is found mainly in the Sana'a area where there are a number of large trees. It differs from *E. camaldulensis* in having smaller leaves up to about 8cm long but only 1cm wide or less. The peduncle, which exceeds 15cm in length, is sharply angled almost winged. The third species, **Eucalyptus globulus** *Labill*, is mainly found on deep soils in areas of good rainfall particularly in Ibb and the Wadi Banā. It is immediately distinguished from the other species by its broadly lanceolate juvenile leaves which reach 11 × 7cm and are conspicuously different from the mature leaves which are themselves broader than those of the previous two species (about 5cm wide).

In recent years experimental plantings have been made of many other species and it is likely that many more species will be found as mature trees in the future.

PUNICACEAE

PUNICA *L.*

1. Punica granatum *L.* — Rumān

Pomegranate. A familiar small spiny shrub to about 3m. Leaves opposite, oblong, lanceolate or obovate, entire, glabrous, dark green and shiny. Flowers with 4 showy, scarlet, crinkled petals; calyx funnel-shaped, fleshy, persistent on the globular red or dull yellow fruit which may reach 10cm in diameter.

Flowers in spring. Fruit ripe July to October.

The Pomegranate is cultivated in all parts of Yemen except the Tihama, but it is only really productive in the irrigated valleys around Sana'a, Ṣa'dah and further east. The pomegranates of Ṣa'dah are particularly esteemed and are among the best I have ever eaten. Along the escarpment it is often found naturalised around villages and, particularly in the wetter areas of the central escarpment, it has become a constituent of scrub around 1500–2000m. There is no real evidence to suggest it is native in Yemen but since the only other member of the family is a Socotran endemic it is at least possible that the Pomegranate originated in Arabia.

ONAGRACEAE

A family of herbs with simple leaves.

Leaves alternate. Flowers yellow. Seeds without hairs — 1. *Ludwigia*
Leaves opposite. Flowers pink. Seeds silky hairy — 2. *Epilobium*

1. LUDWIGIA *L.*

1. Ludwigia octovalvis (*Jacq.*) *Raven*

Erect branched herb to 50cm (–3m in Africa), stem sparsely pilose. Leaves alternate, entire, linear-lanceolate. Flowers solitary in the leaf axils, shortly pedicellate, yellow; sepals 4, persistent as teeth on the apex of the 2.5cm long cylindrical capsule.

The Yemen plant is subspecies **brevisepala** (*Brenan*) *Raven*.

Only known from swamp at 400m at Dayr 'Urāj below Ḥillah. J. Bura'.

2. EPILOBIUM *L.*

1. Epilobium hirsutum *L.*

Willow herb. Erect herb to 1m spreading by rhizomes and often forming dense colonies, stem villous. Leaves opposite, serrate, lanceolate. Flowers solitary in the leaf axils, pink, the pedicels c 1cm long. Fruit a capsule, c 5cm long; seeds with long silky hairs.

Common in ditches and by streams between 1800 and 3000m in Ibb province; elsewhere rare and recorded only from Kusmah, J. Raymah, the Manākhah area (Wissmann and Schweinfurth), the valleys on the southwest of J. Nabī Shu'ayb, Ṭawīlah (Rathjens).

COMBRETACEAE

A family of trees with simple entire leaves which lack stipules. Flowers in racemes, heads or spikes. Fruit winged.

Petals absent. Fruit 2-winged — 2. *Terminalia*
Petals present. Fruit 4-winged — 1. *Combretum*

1. COMBRETUM *Loefl.*

1. Combretum molle *R. Br. ex G. Don* — Dhurūb, 'Asam

C. trichanthum Fresen.
C. lepidotum A. Rich.
Small tree to about 10m with a spreading crown. Bark smooth, grey. Leaves opposite, oblong to oblanceolate, up to 11 × 5cm, base rounded, apex acute or acuminate; under-surface strongly reticulate, covered in numerous small white scales, tomentose when young but becoming glabrous with maturity except on the veins. Flowers produced when the tree is leafless or has only small immature leaves, small, yellow, in dense spikes up to 5cm long. Fruit ovoid with 4 well-developed, golden-yellow wings about 2cm long.

Flowers in spring.

Widespread on the escarpment from 500–1800m from the Ḥujarīyah north to Khawlan Ash Sham. In drier areas it is mostly found in gullies and along small seasonal watercourses. In the central escarpment it is not restricted to drainage lines and grows in abundance on scrubby slopes. In the best-developed valley forest in Yemen above Suq As Sabt on J. Bura' it is the most important tree and locally dominant. It was probably the dominant tree in valleys at low altitudes in the central escarpment in prehistoric times.

2. TERMINALIA *L.*

1. Terminalia brownii *Fresen.* — Qa', Thū'ab

Tree to about 15m, typically of pagoda-like outline. Bark pale grey. Leaves alternate and somewhat clustered at the ends of the branches, long petiolate, broadly obovate to elliptic, up to 8 × 5cm; apex rounded to acute or shortly apiculate; base cuneate to rounded; upper surface dull dark green, undersurface paler and strongly pubescent. Flowers white, in dense spikes up to 10cm long. Fruit elliptic with 2 purple wings, c 4cm long.

Flowers in spring after *Combretum molle*.

Along the whole length of the escarpment from the Ḥujarīyah to Khawlan Ash Sham occurring between 500 and 1500m. It is, however, much less common than *Combretum molle* and unlike that species is restricted to the outer escarpment and is not a plant of drainage lines. It is usually found on hills and banks surrounding cultivated fields and less commonly in scrub away from cultivation. It seems to require a greater depth of soil than many Yemen tree species.

In addition **Conocarpus lancifolius** *Engl.* has been introduced from Somalia where it is native. It is a drought-resistant tree which has alternate lanceolate leaves and a racemose inflorescence with heads of yellow-green flowers. An avenue of this tree has been planted at the entrance to Mukhā, but I have only noticed it there and at Hodeida.

RHIZOPHORACEAE

RHIZOPHORA *L.*

1. Rhizophora mucronata *Lam.*

A mangrove which can reach 20m or more in height. Leaves entire, elliptic, up to 15 × 7cm with a cuneate base and mucronate apex, dark green, coriaceous. Inflorescence a cyme of a few fleshy cream flowers. Fruit an indehiscent berry.

Only once recorded with certainty in Yemen from Hodeida by Bové. It has not been seen since, and though it may have disappeared from Hodeida it probably exists in the extensive mangrove swamps near Mīdī and Luḥayyah since it is found elsewhere along the Red Sea. When not in flower it is best distinguished by its leaves which are cork dotted below whereas in our other mangrove, *Avicenna marina*, they are shortly white-tomentose.

SANTALACEAE

Shrubs or herbs, usually semi-parasitic. Leaves simple, entire, alternate, without stipules. Flowers small, greenish, consisting of a 3–6 lobed perianth. Fruit a one-seeded drupe.

Shrub	2. *Osyris*
Herb	1. *Thesium*

THESIUM *L.*

Semi-parasitic herbs with a wiry stem and small alternate leaves. Flowers bisexual, the perianth yellow-green, 5-lobed.

Erect plant with scale-like leaves c 2mm long	*T. stuhlmanii*
Prostrate plant with linear leaves c 10mm long	*T. radicans*

1. Thesium radicans *Hochst. ex A. Rich.*

Minute, glabrous, dark green, probably perennial herb with several short prostrate stems arising from a woody rootstock. Leaves linear, c 10mm long. Flowers solitary in the leaf axils, shortly pedunculate, subtended by two small bracteoles. Fruit a globose orange drupe, c 2.5mm in diameter.

An inconspicuous plant of short montane grassland usually only noticed because of its bright orange fruit. Widespread on the higher escarpment mountains from the Ibb region north to J. Maswār, mostly between 2500 and 3100m but descending in the central escarpment to about 2000m near Markah, J. Milḥān, probably because of the influence of the mist; rare on the high plateau but found at Matnah (Deflers), Al Zīl, Ḥadā' and on the Qa'al Ḥaql.

2. Thesium stuhlmanii *Engl.*

Glabrous, dark green herb with numerous erect stems to 30cm arising from a woody (? annual) rootstock. Leaves scale-like, c 2mm long. Flowers sessile, yellowish, crowded into short axillary spikes. Fruit globose, c 1.5mm in diameter.

Only known from montane grassland at c 1900m on the northwest of J. Ẓallamlam, near Nā'mah, above Ḥadīyah.

2. OSYRIS *L.*

1. Osyris abyssinica *Hochst. ex A. Rich.*

Erect, branched, glabrous, monoecious shrub to 2m, the stem and branchlets with green ridged bark. Leaves alternate, leathery, subsessile, oblong-lanceolate, up to 5cm long, with very obscure side veins and a distinct pale border. Male flowers in small cymes; female flowers solitary with two prominent bracts. Fruit a red globose drupe, c 5mm diameter.

A plant of grassland and steep scrubby slopes from 900 to 2300m on the escarpment north of the Wadi Rima'; widespread and quite frequent throughout the central escarpment on the west-facing slopes, elsewhere very rare and only known from Shahārah and J. 'Alab in Khawlan Ash Sham (Steinberg). It thus has a strong northern tendency in its distribution which closely parallels that of *Euclea schimperi*.

LORANTHACEAE

Woody parasites which are attached to the host (usually trees) by means of suckers. Leaves green, leathery, entire.

All Yemen species belong to *Loranthus* in the broad sense. The flowers are usually in cymes which are reduced to umbels; they consist of a small rim-like receptacle from which the calyx arises. The petals are usually joined to form a tubular corolla. The fruit is a sticky berry.

1.	Filaments straight	**2**
	Filaments coiled	**3**
2.	Flowers in racemes	1. *Helixanthera*
	Flowers in umbels	2. *Plicosepalus*
3.	Corolla lobes equal to and slightly longer than the tube	3. *Oncocalyx*
	Corolla lobes shorter than the tube	**4**
4.	Corolla glabrous	5. *Tapinanthus*
	Corolla hairy	4. *Phragmanthera*

1. HELIXANTHERA *Lour.*

1. Helixanthera thomsonii (*Sprague*) *Danser*

Loranthus thomsonii Sprague

Glabrous plant with small obovate-spathulate leaves c 1cm long arranged in alternate clusters. Flowers up to 15 in short racemes; calyx cup-shaped, c 1.5mm long; petals free to the base, orange, filiform; filaments red, almost straight at maturity.

A very rare parasite on *Commiphora* found between 1100 and 1300m in Taiz province near Bir Bāsha and at Barakānī.

2. PLICOSEPALUS *Van Tiegh.*

Plants with a small cup-shaped calyx c 5mm long and conspicuous red flowers in small umbels.

Corolla formed of free petals which remain erect as the flowers grow — *P. curviflorus*
Corolla formed of petals which are united for their length, the free part becoming reflexed as the flowers grow — *P. acaciae*

1. Plicosepalus acaciae (*Zucc.*) *Wiens & Polhill*

Loranthus acaciae Zucc.
L. arabicus Deflers.

Glabrous plant creeping over the surface of the host by means of rhizomes. Leaves mostly alternate, solitary or in clusters, oblong to oblanceolate up to 4cm long. Flowers 2–3, in solitary or clustered, pedunculate umbels; corolla dark red (with some green patches), often warted below, the lobes about equalling the tube and becoming reflexed at maturity.

Flowers mainly in autumn.

Occurs in 3 quite distinct areas: 1. In the Tihama at c 150m on *Acacia edgeworthii* west of Al Mafraq. 2. Locally abundant in *Euphorbia* scrub on the escarpment from 1400 to 2000m growing on *Euphorbia ammak, E. cactus and E. inarticulata:* Sayrah near Turbah; around Taiz and Qā'idah; Suq as Sabt near Ibb; between 'Utmah and Madīnat Al 'Abid. 3. Abundant in the Mashriq, at 1200–2000m around 'Ashshah, Ḥūth, and Al Ḥarf where it grows on various Acacias, especially *A. mellifera*, in open *Acacia* scrub on dry limestone hills.

2. Plicosepalus curviflorus (*Benth. ex Oliv.*) *Danser*

Loranthus curviflorus Benth. ex Oliv.
L. faurotii Franch.

Glabrous plant with alternate, linear or oblanceolate leaves up to 6cm long. Flowers 3–6, in solitary pedunculate umbels; corolla sharply bent above the calyx, orange below, red above; petals free to the base.

Mostly autumn and winter flowering in Yemen.

A parasite on *Acacia* (rarely *Zizyphus* fide Schweinfurth) and usually on *A. gerrardii* and *A. mellifera*; locally abundant in dry country from sea level to 2000m: in the Tihama on Kamaran Island (Faurot), at Sayyīd Sulayman and between Bājil and Ḥajjaylah (Schweinfurth); common in the Ḥujarīyah; common between Rada' and Jūbah; Maqsalah in Khawlan Ash Sham (Steinberg), Wadi Amlah (Fleurentin).

3. ONCOCALYX *Van Tiegh.*

Plants with a tubular calyx c 4mm long and yellow-brown flowers. Petals united to form a corolla which has a V-shaped split on one side, the lobes slightly longer than the tube. Filaments coiled at maturity.

Leaves glabrous, elliptic or obovate — *O. schimperi*
Leaves tomentose, ovate or elliptic — *O. doberae*

1. Oncocalyx doberae *(Schweinf.) A.G.Miller & J.A.Nyberg* comb. nov.

Loranthus doberae Schweinf.

Plant with tomentose branchlets and alternate, grey-tomentose, ovate leaves up to 2.5cm long. Flowers 2–5 in solitary, sessile, axillary umbels. Receptacle tomentose but calyx and corolla almost glabrous.

Flowers in early summer.

One of the rarest and most threatened plants in the Yemen flora. It was found by Schweinfurth growing on *Dobera* at Khalīfah (300m) and on *Tamarix* at Ḥajjaylah (600m), but despite careful search I have failed to find it at either locality. I only know it in the open *Dobera* woodland in the Wadi Yūr where it is quite common, but any attempt to fell the *Dobera* for firewood would make the *Loranthus* extinct as it is known nowhere else in the world.

2. Oncocalyx schimperi (*Hochst. ex A. Rich.*) *M.G.Gilbert*

Loranthus schimperi Hochst. ex A. Rich.

Similar to *O. doberae* but with tomentellous branchlets and alternate dull green, obovate or oblong leaves up to 4cm long. Flowers 1–5 in solitary, sessile, axillary umbels; receptacle tomentellous but calyx and corolla glabrous.

Uncommon but widespread and very tolerant ecologically being found between 600 and 2600m in all regions of Yemen apart from the Tihama and the "Sands" and on a great variety of hosts including *Tamarix*, *Salvadora*, *Acacia*, *Zizyphus*, *Ficus*, *Punica*, *Citrus* and *Segeretia*. Its recorded localities include such diverse places as Jiblah, J. Minmār in Khawlan Ash Sham, Wadi Yūr, Ḥamīd near Raḥidah, the Yislah Pass and the wadis north of Rada'.

4. PHRAGMANTHERA *Van Tiegh.*

1. Phragmanthera regularis *(Steud. ex Sprague) M.G.Gilbert* — Kabsh, Kurad

Loranthus regularis Stend. ex Sprague
Loranthus rufescens DC.

Robust plant with stellate-hairy branchlets. Leaves large, opposite, ovate, up to 12cm long, glabrous except for the petiole. Flowers mostly 4–10 in clustered, subsessile umbels; calyx cup-shaped, c 2mm long; corolla shortly villous, split down one side, the lobes spathulate, much shorter than the tube, remaining erect at maturity; filaments coiled.

Flowers mostly in early summer.

The commonest "*Loranthus*" in Yemen; widespread and frequent on the escarpment and south eastern mountains from 1200 to 3000m with an isolated record from Wa'lan on the high plateau and Wadi Amlah (Fleurentin) in the Mashriq. It is particularly common in the Ḥujarīyah. It has been found on a wide variety of hosts including *Ceratonia* and *Olea*, but is most common on *Zizyphus spina-christi*, *Ficus* (especially *F. salicifolia*) and *Acacia* (especially *A. gerrardii* and *A. origena*).

5. TAPINANTHUS *Blume*

1. Tapinanthus globiferus (*A. Rich.*) *Van Tiegh.*
Loranthus globiferus A. Rich.
Glabrous plant with opposite (rarely alternate), oblong or oblanceolate leaves up to 6cm long. Flowers 3–6 in clustered, subsessile umbels; calyx cup-shaped, c 2mm long; corolla glabrous, strongly swollen above the calyx, split down one side the lobes linear, much shorter than the tube, strongly deflexed at maturity; filaments coiled.

Flowers in winter.

A local plant found from 500 to 1200 (–2200) on the escarpment from Samsarah in the Ḥujarīyah north to Maḥābishah, but far more common in the central escarpment than elsewhere. It grows on a variety of hosts including *Citrus*, *Commiphora*, *Ficus*, *Jatropha spinosa* and *Zizyphus* but shows a marked preference for *Acacia asak*.

HYDNORACEAE

HYDNORA *Thunb.*

1. Hydnora abyssinica *A. Br. ex Schweinf.*
H. johannis Becc.
A very distinctive fungus-like parasite with an underground, warted cylindrical rhizome attached usually to *Acacia* roots. Flowers arising singly on the rhizome, sessile, foetid, consisting of a tubular 4-lobed fleshy calyx up to 15cm long, the inside buff in colour, the lobes pink; stamens and stigma on the inside of the tube.

Found by Lavranos between Taiz and Turbah but the exact locality was not noted.

CELASTRACEAE

A family of trees and shrubs with small stipules which are often deciduous. Inflorescence cymose; flowers with 5 free petals and usually a conspicuous disk.

MAYTENUS *Molina*

Shrubs with simple, alternate leaves; often spiny. Flowers very small, cream coloured. Capsules 2–3 locular.

1. Undersurface of leaves distinctly paler than upper surface. Inflorescence an axillary cluster. Plant unarmed — *M. undata*
   Undersurface of leaves not much paler than upper surface. Inflorescence distinctly pedunculate. Plant usually spiny — **2**
2. Leaves obovate with rounded marginal teeth. Capsule 2-locular — *M. senegalensis*
   Leaves ovate to elliptic with pointed marginal teeth. Capsule 3-locular — **3**
3. Leaves hairy — *M. forsskaoliana*
   Leaves completely glabrous — **4**
4. Branches glabrous or pubescent. Inflorescence 1.5–8.5cm long. Leaves equally reticulate on both surfaces — *M. parviflora*
   Branches pubescent when young. Inflorescence less than 2.5cm long. Leaves reticulate on one side only — *M. arbutifolia*

1. Maytenus arbutifolia (*Hochst. ex A. Rich.*) *Wilczek*
A small spiny shrub to 3m with branches hairy when young. Leaves ovate to elliptic, up to 4cm long, glabrous in Yemen. Inflorescence very short, less than 2.5cm long. Capsule 3-locular, 3–8mm long, the seed covering purple, rugose.

Known only from Yarīm, 2600m (Wood 72/54) but doubtless elsewhere in Yemen.

2. Maytenus forsskaoliana *Sebsebe*
Gymnosporia serrata variety *pubescens* Schweinf.
A very variable plant sometimes chewed down and ± procumbent, sometimes reaching 3m in scrub, usually spiny but sometimes unarmed. Spines usually slender, 2–4cm long. Leaves ovate to elliptic, up to 4cm long, pubescent. Inflorescence pedunculate, up to 7cm long. Fruit 3-locular, 4–8mm long, the seed covering white, smooth.

The exact distribution of this newly described endemic species is not accurately known. It appears to be common at low altitudes in the central escarpment.

3. Maytenus parviflora (*Vahl*) *Sebsebe*
Catha spinosa Forssk.
Gymnosporia serrata sensu Schwartz
Gymnosporia arbutifolia sensu Schwartz
A variable plant similar to the two previous species but having glabrous or hairy branches, glabrous leaves and an inflorescence up to 8.5cm long. Seed covering purple, rugose.

Common along the whole escarpment from 500–2800m in scrub, on cliffs, terrace walls, rocky slopes and field borders. It also occurs rarely on the western side of the high plateau. It is one of the commonest low thorny bushes in the Yemen flora but, curiously, seems to lack a well-established Arabic name.

4. Maytenus senegalensis (*Lam.*) *Exell*
Gymnosporia senegalensis (Lam.) Loes.
Very variable erect shrub or small tree, in Yemen typically with short stiff spreading branches and reaching 3m high; bark grey. Spines usually present, usually stout, reaching 4cm long. Leaves obovate, serrulate, 3–12cm long, glabrous, pale green; base long-cuneate. Flowers not often seen in Yemen. Inflorescence a cyme with short peduncles 1–2cm long. Fruit 2-locular, reddish.

Essentially a plant of the foothills between 300 and 800m, widespread but not very common along the whole escarpment from Bathayjah (Ḥujarīyah) to near Malāḥiṭ (Khawlān Ash Sham); very drought tolerant and also occurring in the stony steppe in the southern Tihama; very rarely occurring in the mountains but quite common at Sūq Shirqī, Mudhaykhirah at c 2400m.

5. Maytenus undata (*Thunb.*) *Blakelock*

Unarmed shrub or small tree not exceeding 3m in Yemen. Leaves elliptic, c 5–7cm long, greeny-white below with coarsely dentate margins; base long-cuneate. Inflorescence an axillary cluster without a common peduncle; pedicels 1cm long. Fruit 2–3-locular, greenish.

Only known from dense scrub at c 1500–1900m around Al Mirkah and Aṣ Ṣa'bah, J. Milḥān.

Qāt, **Catha edulis** (*Vahl*) *Forssk. ex Endl.*, known everywhere in Yemen as 'Qāt' is the most important cash crop in the Yemen economy. The Qāt plant is an unarmed shrub or small tree usually 1–3m in height with shiny-green leaves. The small yellowish-green flowers are rarely seen.

Like coffee, Qāt was introduced to Yemen from Ethiopia, possibly in the 15th century since it is not mentioned in literature until the 16th century. Its use as a stimulant drug led to the rapid spread of its cultivation so that by the 18th century at the latest it was being grown along the whole escarpment, principally for local consumption. The advent of motor transport this century has given a strong stimulus to its production as it allows the transport of the young shoots at speed to all parts of Yemen so that they can be consumed within 24 hours of being picked. Great care has to be taken that the shoots retain their freshness so they are carefully wrapped in leaves before despatch.

Today Qāt is grown all along the escarpment on hillside terraces between about 1700 and 2700m wherever rainfall is sufficient. Particularly famous areas of production include J. Ṣabir above Taiz and the Maḥābishah region but nearly everyone has their own favoured region. On the high plateau and in the Mashriq, Qāt can only be cultivated on irrigated land. It is especially common in the valleys around Sana'a such as the Banī Ḥushaysh but can be found in most areas east to Al Baydā' and Ḥarīb.

The cultivation of Qāt has been blamed for the decline in Yemen coffee production but this accusation does not seem to be justified. Coffee is grown between 1000 and 2000m, at a lower altitude than that favoured by Qāt. It is true that coffee terraces have been planted with Qāt in some areas such as Shar'ab but coffee is unlikely to have thrived in these areas in any case. The cause of the decline of Yemen coffee production should be looked for elsewhere.

Qāt is normally chewed socially after lunch. Its effect is the subject of wild claims. Many foreigners denounce it as dangerously addictive while many Yemenis claim amazing hallucinatory, aphrodisiac and stimulant effects for the drug. The truth seems more mundane. A large quantity has to be chewed for several hours to get much effect. A Qāt party is very stimulated in its early stages but later becomes quiet and subdued. It prevents sleep, dulls appetite and sexual potency, but is in no way clinically addictive or intoxicating. Nothing has been proved against it medically, but there is some suggestion that it may cause gastric troubles and especially constipation. Its main danger seems to be economic since its cost is astronomical and its cultivation on the best lands makes Yemen more dependent on imported cereals than would otherwise be the case.

SALVADORACEAE

A small family of trees and shrubs with simple opposite entire leaves which have minute stipules. Flowers small, in panicles; flower parts in 4s. Fruit a drupe in our species.

Tree. Flowers about 4mm long. Fruit 10–15mm long .... 1. *Dobera*
Shrub. Flowers about 1.5mm long. Fruit less than 5mm long .... 2. *Salvadora*

1. DOBERA *Jüss.*

1. Dobera glabra (*Forssk.*) *R. Br.* Ḍabr, Ḍubar

Tree to c 8m with a relatively thick trunk and small rounded crown (? due to pollarding), reminiscent of a child's drawing of a tree; bark smooth, dark. Leaves opposite, leathery, dark green, variable in shape but usually elliptic or obovate, apex acute to emarginate, up to 10cm long but usually half this. Inflorescence of axillary and terminal panicles of cream flowers, each about 4mm long. Fruit an ellipsoid drupe 1–1.5cm long.

Flowers in spring.

This tree is one of the characteristic species of the alluvial plains in the east of the Tihama and the broader wadis in the foothills. It usually covers the plains as scattered individuals giving the country a park-like appearance. Such 'parkland' can be found almost everywhere north of the Wadi Zabīd and I have noticed particularly fine examples around Sa'īd Miṣr, in the Wadi Yūr and between Dayr Khushm and Ṭur. Such communities probably represent the remains of prehistoric woodland dominated by *Dobera*. *Dobera* also occurs sparsely on rocky slopes throughout the escarpment region up to 1500m.

2. SALVADORA *L.*

1. Salvadora persica *L.* Rāk

Much-branched, pale stemmed, usually glabrous shrub to 4m, often forming broad impenetrable bushes. Leaves opposite, leathery, slightly succulent, broadly ovate to elliptic, acute, usually c 5–7 × 2–3cm. Flowers in large ± leafless axillary and terminal panicles, white, c 1.5mm long. Fruit an orbicular, pale red drupe, c 2–5mm long.

The above description is of the typical variety represented by Wood 73/81 and 75/6, but the very distinctive variety **angustifolia** *Verdc.* occurs abundantly in the south-west around Mukhā and Mawza'. This has small, narrow, linear-oblong leaves, usually c 2–4cm × 2–4mm (e.g. Wood 1839) and smaller more leafy panicles. I have not observed any intermediates between the two varieties.

In the Tihama *Salvadora* is abundant and sometimes dominant on uncultivated, perhaps slightly saline alluvial deposits notably between Qutay and Khalīfah and east of Mukhā. It is surprisingly uncommon on sand dunes. On the escarpment it is restricted to drier areas, growing on rock outcrops and along wadis reaching 1400m. It is absent from the high plateau but is widespread throughout the Mashriq up to 1900m growing mostly along wadis. In the Sands it is very abundant, growing on hummocks as do *Rhazya stricta* and *Calligonum crinitum*.

Salvadora branches are used for the traditional toothbrushes known as 'Miswaq'.

EUPHORBIACEAE

Plants of varying habit, but never with bisexual flowers. Leaves alternate or opposite, usually simple and with stipules. Flowers monoecious or dioecious, usually small, either without petals or with small greenish ones. Fruit usually a capsule.

An important family in Yemen with a large number of endemic species.

1. Inflorescence consisting of a glandular cup containing numerous male flowers and a solitary ovary. Plants with latex. — 1. *Euphorbia*
Inflorescence consisting of separate male and female flowers. Plants without latex — **2**
2. Leaves palmatifid or palmitisect — **3**
Leaves simple — **6**
3. Climbing plant. Inflorescence enclosed in 2 large bracts — 16. *Dalechampia*
Erect plant. Inflorescence not enclosed in 2 large bracts — **4**
4. Shrub or small tree with softly spiny fruit and peltate leaves — 14. *Ricinus*
Shrubs or herbs without spiny fruit or peltate leaves — **5**
5. Annual herb with spicate inflorescence — 8. *Croton*
Shrubs with inflorescence of axillary and terminal cymes — 7. *Jatropha*
6. Plant stellate-hairy, especially on the undersurface of the leaves — **7**
Plant without stellate hairs — **8**
7. Tree — 8. *Croton*
Woody herb — 10. *Chrozophora*
8. Leaves serrate, dentate or crenate — **9**
Leaves entire or very obscurely crenate, the apex sometimes emarginate — **12**
9. Dioecious shrub, not aromatic — 11. *Erythrococca*
Monoecious herbs, or shrubs; if a shrub, strongly aromatic — **10**
10. Climbing, rarely erect, plants with stinging hairs — 15. *Tragia*
Erect plants without stinging hairs — **11**
11. Raceme very interrupted with clusters of male and female flowers together along its length — 12. *Micrococca*
Raceme usually with male flowers above and female below, the male flowers often dense and catkin-like — 13. *Acalypha*
12. Undershrub with spiny stipules — 7. *Jatropha*
Spiny stipules absent — **13**
13. Tree or shrub with large elliptic leaves exceeding 10cm long which are grey and reticulate below — 2. *Bridelia*
Herbs or shrubs. Leaves usually less than 3cm long but, if longer, green below and not reticulate — **14**
14. Dioecious plants; shrubs or vigorous undershrubs with herbaceous branches — **15**
Monoecious plants, herbs or shrubs — **16**
15. Leaves obovate, up to 2cm long. Fruit a white drupe — 6. *Flueggea*
Leaves elliptic, 3–9cm long. Fruit a capsule — 9. *Clutia*
16. Prostrate perennial herbs, rarely erect and then spreadingly hairy. Petals sometimes present — 3. *Andrachne*
Erect or ascending annual herbs or shrubs (if a perennial herb, ± glabrous). Petals always absent — **17**
17. Shrub. Leaves more than 1cm broad — 4. *Meineckia*
Herbs or shrubs. Leaves less than 1cm broad — 5. *Phyllanthus*

1. EUPHORBIA *L.*

The largest and much the most diverse genus in habit in the Yemen flora. Trees, shrubs or herbs, often succulent; leaves opposite or alternate, with or without stipules. Flowers monoecious, much reduced and crowded into a lobed involucre called a cyathium, which bears 4–5 'petal-like' glands. Fruit a capsule. All species have a caustic white latex which can cause intense irritation especially to the eyes.

This genus is of outstanding interest in Yemen on account of the numerous succulent species, most of which are endemic. In several areas, notably around Taiz and Qā'idah and in the foothills below the central escarpment, these succulent species are dominant over huge areas of otherwise barren stony hills and are among the most conspicuous features of the Yemen landscape.

1. Plant succulent. Leaves absent or, if present, few and quickly deciduous — **2**
Plant not succulent. Leaves present at least when the plant is in flower — **13**
2. Plant spiny — **3**
Plant unarmed — **11**
3. Spines paired, ± equal in length — **4**
Spines in 3s, the central much longer than the two lateral spines — *E. triaculeata*
4. Stem with 7–10 wing-like angles — **5**
Stem 3–5 angled or subcylindrical — **6**
5. Stems 1–5 together, usually exceeding 30cm in height — *E. fruticosa*
Plant ± tufted with numerous stems, usually less than 10cm high — *E. sp. aff. fruticosa*
6. Angles of branches winged, deeply lobed — *E. cactus*
Angles of branches, winged or not, unlobed or nearly so — **7**
7. Branches weakly 5 angled, subcylindrical — *E. qarad*
Branches strongly 3–5 angled, sometimes winged, never subcylindrical — **8**

8. Branches 3 angled, strongly winged, the wings bent backwards and forwards in zig-zag fashion *E. fractiflexa*
Branches 3–5 angled, usually only weakly winged, the wings never bent in zig-zag fashion **9**
9. Tree with rigid erect branches. Capsule exceeding 1cm broad *E. ammak*
Shrub, or if a tree branches erect, but eventually spreading. Capsule less than 5mm broad **10**
10. Shrub. Branches 3–5 angled. Capsule purple-black when ripe *E. inarticulata*
Tree. Branches 5 angled. Capsule red when ripe *E. parciramulosa*
11. Small tree with a distinct trunk. Capsule on a recurved pedicel *E. uzmuk*
Shrub without a distinct trunk. Capsule erect **12**
12. Stem green, c 7mm in diameter. Flowering shoots often unbranched; glands c 3.5mm broad *E. schimperi*
Stem grey-green c 5mm in diameter. Flowering shoots usually branched; glands c 2mm broad *E. sp. aff. schimperi*
13. Shrubs with woody stems **14**
Herbs **15**
14. Branchlets spine-tipped. Leaves truncate or emarginate at the apex *E. cuneata*
Branchlets unarmed. Leaves acute, ± mucronate *E. balsamifera*
15. All leaves opposite. Plants mostly small, prostrate and with stipules **16**
Leaves alternate at least below. Plants mostly erect, without stipules **23**
16. Plant entirely glabrous **17**
Plant hairy at least on the capsule and/or stems **18**
17. Leaves linear, entire *E. arabica*
Leaves oblong, serrate *E. inaequilatera*
18. Leaves 1–3 (–5)mm broad **19**
Leaves 5–15mm broad **21**
19. Stem with a line of hairs on the upper surface *E. prostrata*
Stem hairy all round, sometimes very thinly so **20**
20. Leaves ± entire, up to c 7mm long *E. granulata*
Leaves distinctly serrate, mostly exceeding 10mm in length *E. sp. aff. aegyptiaca*
21. Cyathia 1–3 in the leaf axils, mostly on short raceme-like branchlets *E. scordifolia*
Cyathia mostly 5–15 in pedunculate clusters arising from the main stem **22**
22. Plant with scattered long yellowish hairs. Leaves acute *E. hirta*
Plant without yellowish hairs. Leaves rounded or obtuse *E. indica*
23. Cyathia with a single gland. Leaves dark green, very large, often exceeding 10cm in length *E. heterophylla*
Cyathia with 4–5 glands. Leaves usually much smaller, rarely exceeding 6cm, usually yellow-green **24**
24. Cyathia in densely bracteate umbels. Capsule glabrous **25**
Cyathia in small cymes at the ends of the umbel branches, bracts few. Capsule pubescent **27**
25. Leaves serrate *E. helioscopia*
Leaves entire **26**
26. Glands entire. Leaves petiolate *E. peplus*
Glands with 2 small horns. Leaves sessile *E. schimperiana*
27. Leaves linear *E. pseudoholstii*
Leaves ovate or elliptic *E. agowensis*

1. Euphorbia sp. aff. aegyptiaca *Boiss.*

Procumbent or ascending annual herb, the stem sparsely pilose, reddish. Leaves opposite, lanceolate, c 10–15mm long, base unequal, margin serrate. Cyathia on short, axillary, leafy, raceme-like branchlets. Capsule sparsely pilose.

This plant is very similar to *E. inaequilatera* from which it differs in being hairy. It differs from typical examples of *E. aegyptiaca* in having consistently smaller leaves, but is otherwise identical.

Known only from disturbed ground near paths between 800 and 1600m near Ḥadīyah, J. Raymah and between Suq Ar Rubū' and Dan, Waṣāb al'Ali. (Wood 2459, 2797)

2. Euphorbia agowensis *Hochst. ex Boiss.*

Erect annual herb to 30cm, usually rather woody below, stems glabrous. Leaves alternate, petiolate, ovate to elliptic, c 2–6cm long, apex acute, base cuneate, the margin obscurely serrate. Inflorescence consisting of terminal umbels with small cymes of cyathia subtended by reduced leaves at the end of the umbel branches. Capsule pubescent.

Known only from valley scrub at 700m at Rajamah in the Ḥarāz.

3. Euphorbia ammak *Schweinf.* 'Amq

Spiny succulent tree to 6m with a distinct trunk. Branches erect, dark green, 4 (–5) angled, the angles winged but not lobed. Spines paired, 5–10mm long. Cyathia in sessile cymes at the top of the branches. Capsule glabrous, trigonous, sharply angled, each side c 12–15mm wide, pale brown when ripe.

Readily distinguished from all other species by its habit and the distinctive size, shape and colour of the capsule. Flowers erratically throughout the year.

Common along the escarpment between 1500 and 2500m from the Ḥujarīyah to Khawlan Ash Sham. Its natural habitat appears to be cliffs (for example at Shahārah) and open stony hills (as around Qā'idah and Ibb), but it is also widely planted as a fence and in graveyards and so is particularly common around villages. It is the characteristic 'Tree' *Euphorbia* of higher altitudes. It is endemic to southwest Arabia but is clearly closely related to *E. abyssinica* J.F. Gmel. from Ethiopia and Sudan.

4. Euphorbia arabica *Hochst. & Steud. ex Boiss.*

Erect or ascending glabrous perennial herb, usually rather woody below. Leaves opposite, linear, c 1–2cm long, not more than 2mm wide, entire. Cyathia on short axillary leafy raceme-like branches.

A local but sometimes abundant plant of dry stony hills on the rainshadow areas of the escarpment between 400 and 1200m: Bājil (Schweinfurth); around Khamīs Bani

Sa'd; Mawsanah, Ḥarāz; southwest of Madīnat Al 'Abīd; Bani Miftāḥ between Riḥāb and Waṣāb al 'Ali.

5. Euphorbia balsamifera *Ait.*

Much-branched, unarmed, grey-barked undershrub to 50cm. Leaves densely clustered at the tips of the branchlets, obovate, 1–1.5cm long, apex acute to mucronate, glabrous, slightly fleshy, often glaucous. Inflorescence consisting of solitary, terminal cyathia. Capsule pilose.

The Yemen plant is subspecies **adenensis** (*Deflers*) *Bally* which is endemic to southwest Arabia. *E. balsamifera* is one of a small group of plants showing a connection between the Canary Islands and the Red Sea area.

Locally abundant between 1600 and 2500m in a narrow belt of country on the border of the Mashriq and the high plateau, stretching from Bayḍā' and Dhī Nā'im northwest to the Arḥab, Ḥūth, Ṣa'dah and Majz. It grows on open rocky hills showing a preference for limestone.

6. Euphorbia cactus *Ehrenb. ex Boiss.*

Kirath, Killaḥ, Qaṣāṣ, Ikrith

Spiny succulent shrub to 2m, lacking a trunk and much branched at ground level with numerous ascending branches. Branches 3–4 (–5) angled, the angles broadly winged and the wings deeply and irregularly lobed. Spines paired, very variable in length, usually 5–10mm long but in the Taiz area up to 4cm long, the horny grey spine shields ± continuous along the angles. Cyathia in sessile cymes at the top of the branches. Capsule similar to that of *E. ammak*, glabrous, trigonous, sharply angled, each side c 12–15mm wide, but always dull red when mature.

A distinctive succulent species on account of its deeply lobed wings and ± continuous spine shields. It flowers in autumn beginning shortly before *E. inarticulata*.

Widespread between 300 and 2000m on the escarpment but becoming rather less common northwards. It is often abundant and sometimes dominant on stony hillsides particularly around Taiz and in the foothills of the central escarpment, but is unaccountably rare in some areas. Outside southwest Arabia it is only known from the Eritrean coast.

7. Euphorbia cuneata *Vahl* — Shinzib

Non-succulent deciduous grey-barked shrub to 4m, somewhat similar to *Commiphora* when leafless. Branchlets spine-tipped. Leaves mostly in clusters, entire, glabrous, obovate with an emarginate apex, c 1–3cm long. Cyathia mostly axillary, solitary or in small umbels. Capsule pubescent.

Widely distributed on stony steppe in the Tihama and on rocky hills in low rainfall areas on the escarpment from 200 to 1500m. It is most characteristic of rainshadow areas and the drier hills of Taiz province and is strikingly absent from the western slopes of the central escarpment and from 'Udayn. It is also widespread on the southeastern mountains and in the Mashriq, reaching 2000m at Damt and near Rada'. It is extremely drought-resistant, occurring on hills near Jubah, and is commonly associated with *Commiphora*, being an important constituent of drier *Commiphora-Acacia* scrub.

8. Euphorbia fractiflexa *S. Carter & J.R.I. Wood*

Spiny succulent shrub to 2½m, lacking a trunk and much branched at ground level with numerous at first decumbent but finally ascending branches. Branches 3-angled, the angles broadly winged, the wings not lobed but bent backwards and forwards in zig-zag fashion. Spines paired, 10–15mm long, the spine shields decurrent. Cyathia in pedunculate cymes at the top of the branches. Capsule glabrous, trigonous, rounded, c 7mm diameter, red when ripe.

Flowers in spring.

An uncommon plant usually growing in colonies on gravel mounds and rocky outcrops between 150 and 500m in the east of the Tihama from Suq Al Khamīs north to the Wadi Līyah. Outside Yemen it is only known from the adjacent area of Saudi Arabia.

9. Euphorbia fruticosa *Forssk.* — Shurūr

Spiny succulent shrub to c 40cm with 1–5, mostly unbranched stems arising from an underground base. Stems with 7–10 (–12) winged angles. Spines paired, 5–15mm long, very numerous. Cyathia in sessile cymes at the top of the stems. Capsule on a decurved pedicel, glabrous, trigonous, rounded, c 3mm diameter, becoming reddish with maturity.

Mostly winter-flowering.

A local plant found between 1300 and 2200m mostly on cliffs in the central escarpment extending north to Maḥābishah, Ḥajjah and Kuḥlān. It also occurs on cliffs in Mikhlāf and on stony plain between Khamr and Ḥūth and between Nashimah and Barakānī in the Ḥujarīyah. Plants from the last locality have rather more stems than usual. (Wood 75/18, Schweinfurth 1361) It is endemic to Yemen.

10. Euphorbia sp. aff. fruticosa *Forssk.*

Wakad, Jumays, Fa'

E. officinalis sensu Forssk.

Very similar to the previous species with which it may be conspecific, but differing in being tufted with numerous very short stems arising from an underground base and rarely exceeding 10cm in height. The low stature is not maintained in cultivation and little reliance should be placed on it.

A local but sometimes abundant plant of rocky plains and stony hillsides most typical of plateau country: Dhubḥān plateau; between Najd al Aḥmar and Jiblah; J. Ba'dān; J. Ḥiṣn Madnān, Waṣāb al 'Ali; J. Ḍawrān; from Sana'a westwards to Ḥaddah and Thilah; Khamr; Maḥwīt; Shahārah; Khawlan Ash Sham (common). (Wood 75/14, 75/235)

11. Euphorbia granulata *Forssk.* — Libanah

Procumbent or rarely ascending, usually grey-green annual herb, sparsely or densely pubescent throughout. Leaves opposite, obovate to oblong, c 3–7mm long, base unequal, margin entire. Cyathia axillary, usually on short raceme-like branchlets, rarely solitary. Capsule pubescent.

A rather variable plant particularly in its indumentum. The typical plant is densely pubescent with crowded green leaves (e.g. Wood 75/321). In variety **glabrata** (*Gay*) *Boiss.* the whole plant is sparsely pilose and the leaves rather

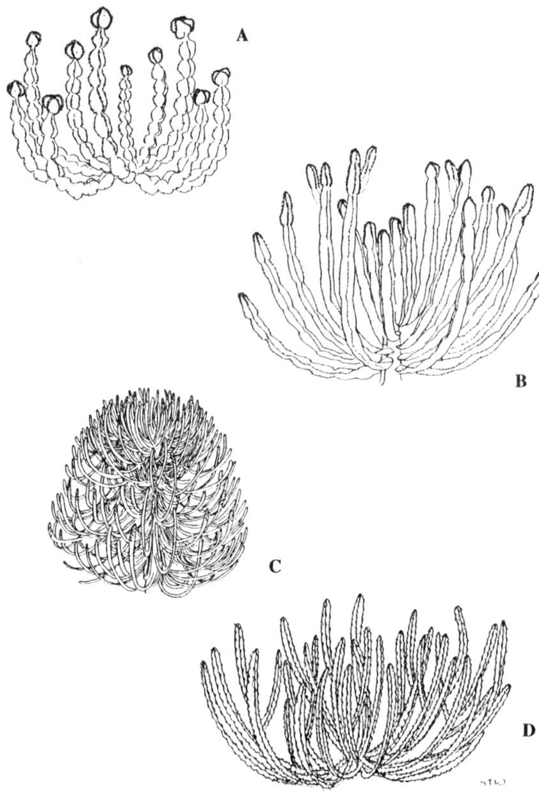

Fig. 14. Habits of *Euphorbia* species. **A** *E. cactus*; **B** *E. inarticulata*; **C** *E. qarad*; **D** *E. fractiflexa*.

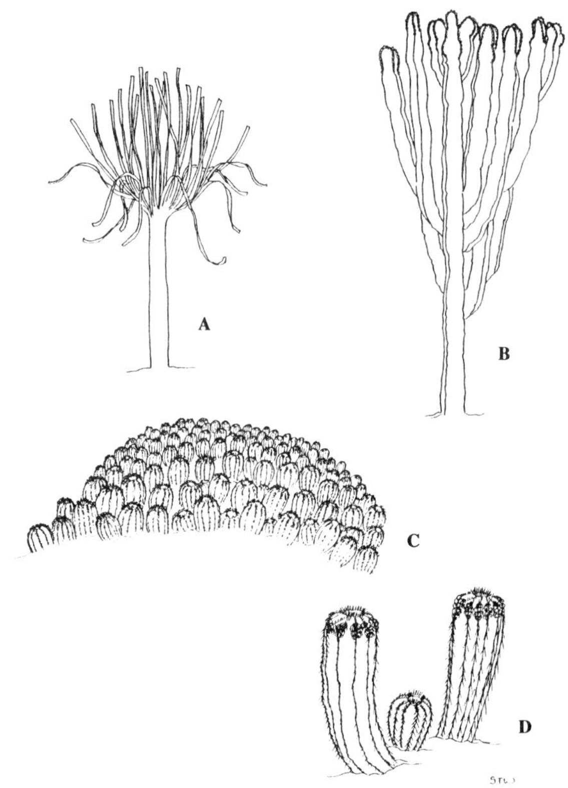

Fig. 15. Habits of *Euphorbia* species. **A** *E. parciramulosa*; **B** *E. ammak*; **C** *E. sp.* aff. *fruticosa*; **D** *E. fruticosa*.

distant so giving a lax green appearance (e.g. Wood 2594 from Raḥidah).

The species as a whole is an abundant weed in the Tihama, growing in fields, waste places and depressions in sand dunes where water accumulates. It rarely penetrates the escarpment but reaches 1200m in dry areas such as Raḥidah and Al 'Ashshah. It also grows at Marashī (Heckel and D. Wood), Mārib and elsewhere in the Mashriq.

12. **Euphorbia helioscopia** *L.*

Erect, glabrous or pilose, usually single-stemmed annual to 40cm. Leaves alternate at least below, sessile, obovate-spathulate, c 2–5cm long, the margin serrulate. Cyathia in terminal, densely bracteate, compound umbels. Capsule glabrous; seeds reticulate.

The Yemen plant is usually (? always) sparsely but persistently pilose.

A locally common weed of low-growing crops between 2000 and 3000m on the escarpment and the western part of the high plateau.

13. **Euphorbia heterophylla** *L.*

E. geniculata Ortega
E. prunifolia Jacq.

An erect, sparsely branched, glabrescent annual to ¾m. Leaves alternate, petiolate, entire, elliptic, c 6–12cm long, dark green above, paler and sparsely pubescent below. Cyathia in terminal cymes, each one with a single gland. Capsule glabrescent, rather sharply trigonous.

A rare introduced weed found in irrigated fields between 500 and 1500m along the Wadi Rima' at Mishrāfah and above Madīnat Al 'Abīd.

14. **Euphorbia hirta** *L.*

Erect or decumbent annual herb to c 20cm, the stem sparsely pilose with yellow, large-celled hairs. Leaves opposite, shortly petiolate, pubescent, serrate, lanceolate to elliptic, acute, c 1–4cm long, very unequal sided, usually purplish above, but greyish below. Cyathia in dense, axillary pedunculate heads. Capsule thinly pubescent.

Widely distributed and generally common in the east of the Tihama and on the escarpment from 300 to 1700m. It grows on open scrubby hills, on banks around cultivation and as a field weed. It is strange that such a common plant was not found by the earlier botanists and it may be a comparatively recent introduction.

15. Euphorbia inaequilatera *Sond.* Libanah
E. sanguinea Hochst. & Steud. ex Boiss.
E. polycnemoides Hochst ex Boiss?
Prostrate, entirely glabrous annual herb. Leaves opposite, lanceolate to oblong, c 3–10mm long, base unequal, margin serrate. Cyathia on short, leafy, axillary, raceme-like branchlets.

Rather variable in leaf size and shape: plants from arid high-altitude situations tend to have shorter almost elliptic leaves whereas plants from lusher situations have larger clearly oblong leaves. I have placed *E. polycnemoides*, recorded by Deflers from J. Munīf under this species since I have seen no Arabian material that agrees with the type of this species which may in any case only be a form of *E. inaequilatera* with linear-lanceolate leaves.

Widely distributed and common from 500 to 3000m on the escarpment, high plateau and southeastern mountains extending east to Bayḍā'. It grows as a field and garden weed, in derelict fields, on alluvial plain and roadsides, and more rarely on open stony hills.

16. Euphorbia inarticulata *Schweinf.* Suyayb, Qaṣāṣ
Much-branched, spiny succulent shrub to 2½m, usually with a short trunk but often trunkless and with branches arising at ground level. Branches ascending, 3–5 angled, the angles slightly winged but not lobed. Spines paired, 5–15mm long, the spine shields long, decurrent. Cyathia in sessile cymes at the top of the branches. Capsule glabrous, trigonous, rounded, c 3–4mm diameter, purple-black when ripe.

Flowers in autumn and early winter usually beginning a week or two after *E. cactus*.

A very variable plant possibly divisible into two subspecies. Plants from the foothills have strictly 3-angled branches and the inflorescence develops at the top of mature branchlets with well-developed spines (Wood 74/415). Plants from upland areas, notably around Taiz, have 3–5 angled branches and the inflorescence develops on new growth at the top of the branchlets, often marked by a distinct joint at the base and usually with poorly developed spines (Wood 74/61). Both forms are readily distinguished from all other species in Yemen by their purple-black fruit.

This is the commonest succulent *Euphorbia* in Yemen, growing from 300 to 2000m at the east of the Tihama, on the escarpment and on all the southeastern mountains just extending into the Mashriq around Rada' and Ḥūth. It grows on open stony ground and is commonly associated with *E. cactus*. Like that species, it is often abundant and sometimes dominant around Taiz and in the foothills, but is unaccountably rare or absent from some areas. It is endemic to southwest Arabia but it is clearly closely related to the Ethiopian *E. polyacantha* Boiss.

17. Euphorbia indica *Lam.*
E. hypericifolia sensu Schwartz
Prostrate or ascending annual herb with brown, thinly pubescent stems. Leaves opposite, shortly petiolate, pubescent, entire or sparsely dentate, elliptic, obtuse, c 1.5–3cm long, unequal sided, dark green above, glaucous below. Cyathia in dense axillary, pedunculate cymes. Capsule pubescent.

A common field weed in the eastern Tihama and on the escarpment from 200 to 1500m. Usually in higher rainfall areas or irrigated land.

18. Euphorbia parciramulosa *Schweinf.*
Spiny succulent tree to 5m with a distinct trunk. Branches arising in a mass at the top of the trunk, at first ascending, but becoming long and pendulous with maturity, 3–4 angled, the angles slightly winged but not lobed. Spines paired, mostly 5–6mm long, the spine shields decurrent. Cyathia in sessile cymes on new growth with poorly developed spines at the top of the branches. Capsule glabrous, trigonous, rounded, c 4–6mm diameter, red when ripe.

Flowers in spring.

This species has been much confused with *E. inarticulata* and it is in fact impossible to identify specimens reliably without fruit. In the field it is immediately recognisable by its distinct habit.

Very local but sometimes abundant on open stony hillsides between 1700 and 2300m: abundant between Qā'idah and Sayyānī; a few trees between Ibb and Mashwarah; abundant in the Wadi Banā between Nādirah and Damt; a single large colony between Dhamār and Shirqabi; a few trees, Suq al Jum'ah, J. Ash Shirq; a few trees, Hajarah, Ḥarāz; abundant, J. Masār, Ḥarāz.

19. Euphorbia peplus *L.*
Erect, branched, glabrous annual herb to c 25cm. Leaves alternate below, petiolate, entire, obovate, c 1–2cm long. Cyathia in densely bracteate, terminal compound umbels. Capsule glabrous; seeds pitted.

A locally common weed of fields, gardens and waste ground in villages between 1500 and 3000m on the escarpment and western part of the high plateau. Often associated with *E. helioscopia*.

20. Euphorbia prostrata *Ait.*
Prostrate annual herb; stem glabrous except for a line of hairs on the upper surface. Leaves dark green, opposite, entire, obovate to elliptic, c 5mm long, sometimes sparsely pubescent below, base slightly unequal. Cyathia axillary, usually solitary, sometimes on raceme-like branches. Capsule glabrous except for the angles which are pilose.

An introduced weed of fields and waste places between 200 and 1200m, at present rare but likely to become frequent: Suq al Mirkiz, Ḥujarīyah; Mishrāfah, Wadi Rima'; Al Kadān; near 'Ukaybir, J. Milḥān.

21. Euphorbia pseudoholstii *Pax*
Very similar to *E. agowensis*, of which it may only be a variety, differing only in its long, linear leaves which reach 8cm in length.

Only known from scrub in the valley just above Suq As Sabt on J. Bura' at 500m.

22. Euphorbia qarad *Deflers*
E. aculeata Forssk. (probably)
Spiny succulent shrub to 3m with a distinct trunk. Branches very numerous, at first spreading horizontally or even slightly pendulous from the trunk, eventually ascending or erect, subcylindrical with 5 very weak angles. Spines

paired, 5–10mm long, the spine shields decurrent. Cyathia in pedunculate cymes at the top of the branches. Capsule on a recurved pedicel, glabrous, trigonous, rounded, c 3–3.5mm diameter, colour not known.

Flowering period not known, but it was found with a few flowers in August.

Readily distinguished from all other succulent 2-spined species by its ± cylindrical branches.

Known only from sandy plain at Mawza' (250m) where it is abundant and from the foothills south of J. Ghayāmī in the Ḥujarīyah where it waş found by Deflers. It is endemic to southwest Arabia but is much more common in South Yemen.

23. Euphorbia schimperi *Presl* Dahan

E. bottae Boiss.

Unarmed succulent shrub to 2m, without a distinct trunk but with numerous long decumbent or ascending branches and often forming thickets. Branches cylindrical, green, c 7mm in diameter. Leaves and bracts ovate to ensiform, quickly deciduous. Cyathia in terminal umbels usually at the end of long undivided branches, glands c 3.5mm broad. Capsule erect, glabrous.

Widely distributed and often abundant on open stony hills between 1000 and 2200m in Taiz province extending north to J. Ra's (Botta), J. 'Ayfū, Najd Al Aḥmar and J. Sawraq; particularly common in the *Euphorbia* scrub around Taiz. Elsewhere extremely rare and known only from between Suq Ar Rubū' and Dan, Waṣāb Al 'Ali; J. Mirbish, Salafīyah in Raymah (Gillespie); at the head of the Wadi Ḥijān near Mujwarah, Ḥarāz (Wood 74/199, Hepper 5666). Outside southwest Arabia this species is only known from Somalia.

24. Euphorbia sp. aff. schimperi

Similar to the previous species but usually a low, erect branched undershrub to ½m, though sometimes smaller and almost unbranched. Branches grey-green, c 5mm in diameter. Flowering branches usually short and divided, the glands c 2mm broad and often reddish tinged.

In the field this species is readily distinguished from *E. schimperi* by its low stature, but under cultivation it grows vigorously and attains a similar height though it never seems to produce so many branches.

Widely distributed between 2200 and 3100m on the mountains on and around the high plateau and extending into the Mashriq and south to Damt; also in similar habitats in Khawlan Ash Sham. It usually grows in rock crevices and more rarely on rocky plain. It appears to be endemic to southwest Arabia.

25. Euphorbia schimperiana *Scheele*

Rumayd, Sawsab, Sabaysab

E. petitiana A. Rich.
E. yemenica Boiss.
E. terracina sensu Khattab & Hadidi

Very variable, erect, branched, glabrous or thinly pubescent annual or perennial herb to 50cm. Leaves alternate (at least below), sessile, entire, linear-lanceolate, c 3–7cm long. Cyathia in terminal, densely bracteate, compound umbels; bracts deltoid, strikingly dissimilar to the leaves; glands with 2 small horns attached. Capsule usually glabrous; seeds smooth.

Abundant from 2000 to 3000m on the escarpment and high plateau, growing on roadsides, as a field weed, in grassland, on rocky hills and even in marsh. It may well be the commonest plant at this altitude. Although it descends to 400m it is much rarer at lower altitudes where it is usually found in irrigated fields. It is absent from the Mashriq.

26. Euphorbia scordifolia *Jacq.*

Prostrate or ascending, tomentose perenial herb with a red stem which is often woody below. Leaves opposite, shortly petiolate, serrate, ovate to oblong, c 5–20mm long, very unequal sided. Cyathia axillary, solitary or in clusters on small leafy raceme-like branchlets; glands often white when fresh but drying pink. Capsule pubescent.

Locally common in the Tihama from sea level to 400m from Bayt Al Faqīh northwards. It is one of the characteristic plants of sand dunes associated with *Indigofera argentea* and *Heliotropium pterocarpum*.

27. Euphorbia triaculeata *Forssk.* Suyayb

Much-branched spiny succulent shrub to 80cm, lacking a trunk. Branches decumbent or ascending, dark green with paler blotches, ± cylindrical. Spines in 3s, the central spine c 1.5–2cm long, the two lateral above the central spine and only 4–5mm long, spine shields slightly decurrent. Cyathia in sessile cymes at the top of the branches. Fruit rare; capsule glabrous, trigonous, rounded, c 3mm diameter, reddish when mature.

Flowers erratically throughout the year.

Abundant on stony and sandy plain between 100 and 400m around Mawza', extending north to Nukaybah. Elsewhere very rare in similar habitats and known to me only at Ḥaraḍ, but also recorded by Deflers between Bājil and Ḥajjaylah. Outside Yemen it is quite common on the coastal plains on both sides of the Red Sea and the Gulf of Aden.

28. Euphorbia uzmuk *S. Carter & J.R.I. Wood*

'Uzmuq

Unarmed succulent tree to 2½m with a distinct dull-green barked trunk about 1m high, from the top of which arise numerous erect or ascending, glaucous, cylindrical branches. Leaves quickly deciduous. Cyathia in small clusters at the top of the branches, pubescent; glands very small, c 1mm broad. Capsule on a recurved pedicel, thinly pubescent, but rarely seen.

A local species usually growing in small colonies between 800 and 1300m on the escarpment north to the Wadi Rima'. It is found in gullies and on hills with loose volcanic rocks: between Mishrāfah and Al Masbaḥ and near Suq Ar Rubū', Waṣāb As Sāfal; Hajdah; Wadi Hiḍrān near Taiz; in several places centred on Nishamah, Ḥujarīyah; Bathayjah and Bani 'Umar, Ḥujarīyah. It is endemic to Yemen.

Schwartz records *Euphorbia abortiva* Forssk. but since there is neither a specimen nor a description it is impossible to know what the plant is and it cannot be included in the Yemen flora.

Two species of *Euphorbia* are cultivated in Taiz. The Poinsettia, **Euphorbia pulcherrima** *Willd.* is an ornamental shrub with attractive red bracts. **Euphorbia cyathophora** *Murr.* is an annual herb also with red bracts. It self-seeds readily in gardens and may well become naturalised as has happened in other countries.

2. BRIDELIA *Willd.*

1. Bridelia scleroneura *Muell.-Arg.*

B. tomentosa Blume var. *glabrata* Schweinf.
Monoecious shrub or small tree to 3m with furrowed bark. Branchlets red-brown, pubescent. Leaves ± glabrous, dark green above, grey and reticulate below, oblong-elliptic, up to 12 × 5cm, apex obtuse, margin obscurely crenate. Flowers sessile in the leaf axils, the male much more numerous than the female. Fruit a black globose drupe c 5mm in diameter.

A very rare plant of scrub between 600 and 900m in the central escarpment: above Walajah, J. Milḥān (Schweinfurth); above Ḥillah, J. Bura'; above Suq Ar Ribāṭ, J. Raymah. At both the last two localities I found a single plant while I have failed to refind it on J. Milḥān.

3. ANDRACHNE *L.*

Monoecious perennial herbs, woody below and usually procumbent in habit. Leaves stipulate. Flowers solitary or in clusters in the leaf axils; petals present, yellow-green. Fruit a capsule.

Leaves green, elliptic with a cuneate base — *A. telephioides*
Leaves blue-green, orbicular with a cordate base — *A. aspera*

1. Andrachne aspera *Spreng.*

A. telephioides sensu Hepper
Prostrate, glabrous, sparsely scabrid or rarely glandular-hairy, blue-green perennial with long slender branches. Leaves ± orbicular, base cordate, apex emarginate or rounded, c 3–8mm long; stipules entire.

The typical plant described above grows from 2000 to 2900m and is a widespread and characteristic species of the high plateau and similar regions in Khawlan Ash Sham. It is extremly rare on the escarpment but is known from Hajjah (Rathjens), the Manakhah area and the east side of J. Raymah. It grows on rocky hillsides and gravelly plain, showing a preference for volcanic soils and particularly recent lavas. It is quite variable in indumentum, sometimes being completely glabrous (Wood 1612), most commonly sparsely scabrid (Wood 74/362), and very rarely glandular hairy. The last has been described as variety **glandulosa** *Hochst.* and is only known from between Sana'a and Wa'lān (Wood 2156).

A very distinctive, erect, spreadingly-hairy variety grows on volcanic rocks between 800 and 1100m on the escarpment, but is only known from the Ramādah-Ḥajdah area and in the Wadi Ḥillah, J. Milḥān (Wood 917, 2845).

Finally there are some puzzling forms which are intermediate with the next species (Wood 2207 and D. Wood Y 1273) and may be hybrids between the two species.

2. Andrachne telephioides *L.*

Prostrate glabrous green perennial, often with reddish stipules. Leaves obovate or elliptic, base cuneate, apex acute or obtuse, never rounded or emarginate, usually c 7mm long; stipules fimbriate, the teeth whitish.

Only known from rock crevices above the village of Maḥalayn on J. Madhbur in the J. Lawz region at 3000m.

4. MEINECKIA *Baill.*

1. Meineckia phyllanthoides *Baill.*

Securinega phyllanthoides Muell. Arg.
Erect, glabrous, monoecious undershrub to 50cm, somewhat resembling *Phyllanthus*, but the leaves not arranged in two rows on specialised leafy shoots and the stamens united into a column. Leaves entire, elliptic, rounded, c 1–2.5cm long; stipules present. Flowers solitary or in small clusters in the leaf axils; sepals 5, with a broad hyaline margin. Fruit pedicels c 1–1.5cm long.

A rare plant of valley scrub between 500 and 900m in the central escarpment: near the suq, Ḥadīyah, J. Raymah; above Ḥillah (Schweinfurth) and Suq As Sabt, J. Bura'; Rajamah, Ḥarāz; Walajah, J. Milḥān (Schweinfurth); Wadi Yūr.

5. PHYLLANTHUS *L.*

In Yemen monoecious, ± glabrous shrubs or herbs. Leaves usually arranged in two rows on specialised leafy shoots and absent from the main stem, entire, usually rather small and in our species not exceeding 1.5cm in length; stipules present. Flowers solitary or in clusters in the leaf axils, usually pedicellate and ± pendulous from the rhachis of the leafy shoots; petals absent. Fruit a capsule.

A difficult genus containing a number of interesting species.

1. Woody stemmed shrub up to 2m high — *P. ovalifolius*
   Herbs, sometimes slightly woody below, not exceeding 50cm in height — 2
2. Perennial herb. Sepals obovate, green, with a narrow hyaline margin — *P. maderaspatensis*
   Annual herbs, sometimes briefly perennial. Sepals ovate or oblong, always with a broad hyaline margin — 3
3. Sepals 4–5 — 4
   Sepals 6 — 6
4. Fruiting pedicels 5–8mm long, filiform. Plant almost always bronze tinged — *P. tenellus*
   Fruiting pedicels less than 5mm long, not filiform. Plant usually green — 5
5. Leafy shoots c 5–22cm long — *P. hodjelensis*
   Leafy shoots usually less than 5cm long, rarely reaching 10cm — *P. amarus*
6. Leaves obovate to orbicular, green or glaucous. Stems often rough — *P. rotundifolius*
   Leaves oblong to elliptic, never glaucous. Stem smooth — *P. fraternus*

1. Phyllanthus amarus *Schum. & Thonn.*
Erect annual herb to 30cm. Leafy shoots short, usually less than 5cm long. Leaves oblong. Sepals 4–5. Fruit pedicels c 1.5mm long.

Very easily confused with *P. fraternus* from which it differs in having 5 not 6 sepals.

Known only as a weed of irrigated fields between 500 and 1500m in Wadi Sukhnah and Mishrāfah in the Wadi Rima', but quite possibly more frequent as it is easily overlooked (Wood 2359, 3178).

2. Phyllanthus fraternus *Webster*
P. niruri sensu Schwartz.
Erect or ascending annual herb to 30cm. Leafy shoots short, usually less than 5cm long, but sometimes reaching 10cm. Leaves oblong or elliptic. Sepals 6. Fruit pedicels c 1.5–2mm long.

Distinguished from *P. rotundifolius* by its oblong leaves and from *P. amarus* by its 6 sepals.

Widely distributed between 300 and 1500m as a field weed on rich alluvial soils in the east of the Tihama and in irrigated fields along wadis on the escarpment. Its exact frequency is uncertain. (Wood 2295, 2716, 2967)

3. Phyllanthus hodjelensis *Schweinf.*
Erect annual herb to 30cm, sometimes woody below. Leafy shoots very long, reaching 22cm in length. Leaves elliptic. Sepals 5. Fruit pedicels c 1.5–2mm long.

Distinguished from all other species, except *P. maderaspatensis*, by the very long leafy shoots.

This species was found by Schweinfurth at 900m at Rajamah near Ḥajjaylah, after which village it was named. I searched the coffee terraces at Rajamah thoroughly and failed to refind it, but subsequently discovered it under coffee trees at the head of the Wadi Ḥijān in the same general area (Wood 2582). I also found a single plant at Ḥabash on J. Milḥān at 1200m in the coffee zone (Wood 2847). This endemic species is therefore extremely rare and has good claims to be considered the rarest plant in the Yemen flora. It is curious that it appears to be entirely restricted to artificial habitats – on or by coffee terraces.

4. Phyllanthus maderaspatensis *L.*
Erect or spreading perennial herb to 30cm, usually woody below. Leafy shoots very long, reaching 40cm. Leaves oblanceolate below, oblong or linear above. Sepals 6, differing from those in other species in being obovate and almost entirely green. Fruit sessile or nearly so.

A widespread species growing from 200 to 2000m; uncommon in the Tihama, growing occasionally as a field weed on the alluvial plains in the east; locally abundant on the escarpment up to 1500m, growing mostly in the foothills, in rainshadow areas and in the more arid hills in Taiz province; apparently rare in the Mashriq, occurring at c 2000m near Ṣa'dah and east of Rada'.

5. Phyllanthus ovalifolius *Forssk.* Ḥawmir
P. lalambensis Schweinf.
P. guineensis Pax
Erect, many-stemmed shrub to 2m with woody branches. Fertile leafy shoots up to 10cm long, sterile shoots much longer. Leaves oblong, dark green above, contrasting strongly with the paler lower surface. Sepals 4–5, often pinkish. Fruit pedicels up to 4mm long.

A very local plant of shaded rocky gullies between 1200 and 1700m: common in the Ja'farīyah but not seen elsewhere on J. Raymah except at Bani Dawan; near 'Ukaybir and between Ḥabash and Aṣ Ṣa'bah, J. Milḥañ; below Shamsān, Maḥābishah.

6. Phyllanthus rotundifolius *Klein ex Willd.*
Erect annual or short-lived perennial, sometimes woody below, reaching 40cm in height. Leafy shoots up to 7cm long. Leaves obovate to ± orbicular. Sepals 6. Fruit pedicels c 1.5mm long.

This species occurs in two rather distinct forms:

The first of these (Wood 2170, 2527, 2714, 2896) is somewhat woody below and has dark green suborbicular leaves. It grows in open scrub and grassland between 500 and 1500m in high rainfall areas from 'Udayn north to the Wadi Lā'ah, but is mainly found in the central escarpment. It gives every appearance of being part of the natural vegetation.

The second form (Wood 2291, 2359a) is obviously annual with pale green, obovate leaves. It grows as a weed of irrigated fields in rich wadis (e.g. Sukhnah, Qaṣab), the same habitat as *P. fraternus*. Its exact distribution is uncertain but it is probably widespread, though uncommon.

7. Phyllanthus tenellus *Muell. Arg.*
Delicate, erect, bronze-green annual to 30cm. Leafy shoots up to 10cm long. Leaves elliptic. Sepals 5. Fruit pedicels, 5–8mm (–10mm) long.

The Yemen plant is the endemic variety **arabicus** *Muell. Arg.* which differs from the type in having minutely scabrid stems. It is largely restricted to artificial habitats but it may grow naturally on shaded cliff gullies where it occasionally occurs. The type is known as a native species in the Indian Ocean islands and as an introduction in tropical America and the existence of an endemic variety in Yemen is something of a botanical puzzle.

Distinguished from all other Yemen species by the long delicate filiform pedicels.

Locally abundant between 1000 and 1800m as a weed of coffee terraces and of shaded paths in the central escarpment: Ja'farīyah district, J. Raymah (abundant); around Mirzin and Ṭuba', J. Bura' (abundant); between Ḥabash and As Sa'bah, J. Milḥān.

6. FLUEGGEA *Willd.*

1. Flueggia virosa (*Roxb. ex Willd.*) *Voight* Naḥaf
Securinega virosa (Roxb. ex Willd.) Baill.
Glabrous dioecious shrub to 2m. Branchlets red-brown when young, becoming rigid and spine-like with age. Leaves entire, obovate, usually 1–1.5cm long, dark green above, ± glaucous below. Flowers in axillary clusters, shortly pedicellate, yellow or cream, fragrant. Fruit a white globose drupe, c 5mm in diameter.

Flowers in spring, the flowers very short-lived.

Widely distributed and common between 500 and 2100m in the southeastern mountains and on the escarpment north to Shahārah. It grows in scrub, on cliffs and in open stony ground and is tolerant of both high and low rainfall. It is of rather nondescript appearance when not in flower or fruit and is easily overlooked.

7. JATROPHA *L.*

Small monoecious shrubs usually with palmatifid leaves, often with spiny or at least filiform stipules. Flowers in terminal or axillary cymes with small petals which are often slightly red tinged. Fruit a capsule.

All the Yemen species contain a sticky latex which is highly reputed for staunching and healing cuts and sores.

1. Stipules spiny or filiform, 5–15mm long — 2
   Stipules very small, never spiny or filiform — 4
2. Leaves palmately lobed — 3
   Leaves unlobed — *J. variegata*
3. Stipules transformed into spines, never gland-tipped — *J. spinosa*
   Stipules filiform, gland tipped — *J. glauca*
4. Stipules transformed into a mass of sticky glands. Leaves tomentose on both surfaces — *J. pelargoniifolia*
   Stipules very small, not sticky. Leaves ± glabrous or sparsely pubescent below — *J. curcas*

1. Jatropha curcas *L.* — Sharib

Physic-nut. Vigorous, erect shrub to 3m. Leaves dark green and glabrescent above, paler and often pubescent on the veins below, very shallowly 5-lobed but otherwise entire, base cordate, c 7–10cm long. Stipules minute, neither glandular nor spiny. Cymes mostly axillary, many-flowered. Capsule glabrous, large, c 22mm long.

Introduced from America and originally cultivated for its seeds which are used as a purgative, but now widely naturalised in the Tihama and on the escarpment between 200 and 1500m, but not recorded north of the Wadi Mawr. It is usually found growing by wadis (sometimes planted in lines) and less commonly by irrigation ditches, on field borders and around villages.

2. Jatropha glauca *Vahl*

J. lobata (Forssk.) Muell.-Arg.

Erect undershrub to ½m, usually branched at the base. Leaves bright green, glabrous above, glabrous or pubescent on the veins below, palmately 3–5 lobed, c 4–7cm long; base variable, cordate, truncate or cuneate; margin dentate. Stipules branched, filiform, gland-tipped. Cymes mostly axillary, long-peduncled, few-flowered. Capsule glabrous, c 8mm long.

Abundant on the alluvial plains in the east of the Tihama extending along the broader valleys into the escarpment and reaching 1000m in Wadi Sharas. Like *Acacia oerfota* and *Aloe vera*, with which it is commonly associated, it is unknown south of the Ḥays region. It grows on open plain, in derelict fields and on uncultivated stony hillocks.

3. Jatropha pelargoniifolia *Courb.* — 'Ubab

J. villosa (Forssk.) Muell.-Arg.

J. glandulosa Vahl

Erect or often decumbent, branched undershrub with tomentose stems to 1m. Leaves dull green, tomentose on both sides, palmately 3–5 lobed, shallowly cordate, serrate, c 2–4cm long. Stipules transformed into sticky sessile glands. Cymes mostly terminal, few-flowered. Capsule tomentellous, c 8mm long.

All plants I have seen have tomentose leaves, but variety **glabra** *Muell.-Arg.* with glabrous leaves was collected by Botta from Yemen, but without locality.

Abundant on sand throughout the Tihama sometimes becoming dominant on sand dunes as around Marāwi'ah. More local on rock but quite common on the stony steppe south of Ḥays and found in a few places inland in the Ḥujarīyah where it reaches 1200m. In the Mashriq it grows at the edge of the sands around Ḥarīb, Jūbah and Mārib and doubtless further north.

4. Jatropha spinosa *Vahl* — Ibkī, Baka, Duma'

Erect, entirely glabrous branched undershrub to 1½m. Leaves shiny green, ± deeply palmately 3–5 lobed, with shallowly cordate base, c 2–5cm long. Stipules transformed into rigid spines c 5–10mm long. Cymes mostly terminal, few-flowered. Capsule c 8mm long.

A rather variable species, the spines sometimes abundant but often infrequent or completely absent. Leaf size and lobing vary from population to population; sometimes the leaves are entire apart from the main lobes but sometimes the margin is crenate (variety **crenulata** *Pax*).

A local but sometimes abundant species of dry rocky hills between 500 and 1900m mostly in the south of the country. Generally common in Taiz province extending northwest through Maqbanah to Waṣāb As Sāfal and northeast to Qa'ṭabah and thence northwards through the Mashriq north to the Wadi Amlah. Elsewhere it is only known from the hills around Khamīs Bani Sa'd and Mafḥaq. Outside southwest Arabia it is restricted to Somalia.

5. Jatropha variegata *Vahl* — Dundul, Ibkī

Erect, entirely glabrous branched undershrub to 1½m. Leaves shiny green, unlobed, entire, oblanceolate, c 5–12cm long. Stipules transformed into weak filiform spines c 5–10cm long. Cymes mostly terminal, few-flowered. Capsule c 10cm long.

A local but sometimes abundant endemic species of open stony hills, field borders and rocky plain between 1000 and 2000m: widespread in the Ḥujarīyah extending north to J. Ḥabashī and Wadi Ḍabāb in the west and to J. Sawraq in the east; frequent from 'Udayn north to the valleys between Riḥāb and Waṣāb; Wadi Sukhnah (uncommon); between Sharas and Kuḥlān (rare).

In parts of the Ḥujarīyah where *J. variegata* and *J. spinosa* grow together apparent hybrid swarms develop. Such plants (e.g. Wood 1038) retain the stipular thorns of *J. variegata* but have at least some leaves 3-lobed as in *J. spinosa*. This hybrid has been seen at Badhayjah and nearby at Bani 'Umar and in the 'Abūs region at Mushawis and in the Wadi Marah below Najd Dhamrān.

Fig. 16. **A** *Acalypha racemosa*; **B** *Jatropha pelargoniifolia*; **C** *Meineckia phyllanthoides*; **D** *Acalypha fruticosa*.

8. CROTON *L.*

A large genus poorly represented in Yemen, consisting mostly of trees.

Herb. Leaves palmately 3–5 lobed — *C. lobatus*
Tree. Leaves unlobed, ovate — *C. macrostachyus*

1. Croton lobatus *L.*

Erect, branched, monoecious annual herb to c 80cm, stellate-hairy on young growth. Leaves palmately 3–5 lobed almost to the base, c 4–8cm long, glabrescent except on the nerves below, the margin serrate. Inflorescence axillary, long pedunculate, spicate, the female flowers below and the male above. Capsule c 6–7mm long, stellate-hairy.

A common weed of irrigated fields and stream beds from 300 to 1500m on the escarpment and more rarely in the east of the Tihama north to the Wadi Mawr. In higher rainfall areas it also grows in rain-fed fields, coffee terraces and on tracksides.

2. Croton macrostachyus *Hochst. ex Del.*

A usually dioecious tree 6–10m high. Leaves ovate, acuminate, 3–10cm long, stellate-tomentose particularly below, with 2 conspicuous brown glands at the base. Inflorescence of lateral and terminal racemes, 10–20cm long; flowers whitish. Capsule c 10mm long, tomentose.

Known only from J. Bura' (Muller-Hohenstein).

9. CLUTIA *L.*

Dioecious undershrubs to 1½m, woody below and usually with long, leggy, branched, erect herbaceous stems, often almost leafless except at the top. Leaves alternate, oblanceolate to elliptic, obtuse, c 3–9cm long, turning orange as they age. Flowers pedicellate, solitary or in small clusters in the leaf axils. Petals 5, yellow-green. Capsule wrinkled, c 5mm long.

The two following species are very close and male plants are almost impossible to identify. They differ essentially in their indumentum but this character is not entirely constant. Other characters such as how the style is forked and the arrangement of the glands on the sepals of the male flowers do not seem to correlate well with the presence or absence of hairs on the fruit. It is quite possible that there is only a single variable species in Yemen.

***Clutia* is widespread and often abundant on the escarpment and high plateau from 1700 to 3100m growing on field borders, grassy slopes and open rocky hills. It is often associated with *Cadia purpurea*, *Maytenus parviflora* and other plants on heavily eroded hillsides where the terrace system has broken down.**

Capsule glabrous — *C. lanceolata*
Capsule pubescent — *C. richardiana*

1. Clutia lanceolata *Forssk.*

C. myricoides Jaub. & Spach

Female plants can be identified by the glabrous ovary and capsule. Usually both sexes can be identified by the glabrous mature leaves which are typically darker green and more leathery than in the following species, but sometimes the leaves are sparsely hairy.

Distribution imperfectly known but occurring on both the escarpment and the high plateau where it appears to be the more common species. (Wood 72/19, 1320, 1347, 1484)

2. Clutia richardiana *Muell.-Arg.*

C. jaubertiana Muell.-Arg.

Female plants can be identified by the pubescent ovary and capsule. The leaves are pubescent, usually on both surfaces and are green and of a thinner texture than in the previous species.

The distribution of this species is not accurately known, but it occurs both on the high plateau and the escarpment, where it appears to be the more common species. (Wood 74/325, 1450, 1794; Smith and Henchie 4534, 4568).

10. CHROZOPHORA *Neck.*

1. Chrozophora oblongifolia *(Del.) A. Juss. ex Spreng.*

Much-branched, stellate-hairy, monoecious perennial to 1m, usually woody below. Leaves grey, ovate-rhomboid, c 3–6cm long, base cuneate to truncate, the margin ± incised or sinuate. Flowers in small axillary racemes; male flowers above, sessile, rather numerous; female flowers below, fewer, pedicellate. Capsule on a reflexed pedicel, c 7mm long, purple-brown, covered in white scales.

Widely distributed in the Tihama and on the escarpment from sea level to 1500m, but absent from areas of high rainfall. It also occurs at the edge of the sands at Mārib and Ḥarīb. It grows on roadsides, field borders, in derelict fields and more rarely, amongst growing crops, but is seldom in large quantity.

11. ERYTHROCOCCA *Benth.*

1. Erythrococca abyssinica *Pax*

Dioecious shrub to 3m, the branchlets grey-green, lenticulate. Leaves dark green, lanceolate, acuminate, c 3–5cm long, margin crenate, ± glabrous except for the pubescent petiole. Flowers in small axillary racemes. Capsule deeply 3-lobed (though commonly with one or two of the ovules aborting), glabrous, c 4mm long.

Widely distributed on the escarpment between 800 and 2400m from Turbah north to Shahārah, but uncommon and mainly in the high rainfall areas. It usually grows on shaded stream sides and in scrubby gullies, often on cliffs. It is usually a low plant but very vigorous specimens occur sometimes, notably at Jiblah and north of Ṣa'bah, J. Milḥān.

12. MICROCOCCA *Benth.*

1. Micrococca mercurialis *(L.) Benth.*

Erect, sparingly branched monoecious annual herb to 30cm, stem pubescent. Leaves subglabrous, ovate to elliptic, c 3–4cm long, apex acuminate, margin crenate. Flowers in slender interrupted axillary racemes with clusters of male and female flowers together. Capsule very deeply 3-lobed, pilose, c 2–3mm long.

A field weed, widely distributed but scattered in distribution growing from 300 to 1000m (–1500m) in the east of the Tihama and on the escarpment north to the Wadi Mawr.

13. ACALYPHA *L.*

Monoecious shrubs or herbs with alternate, thinly pubescent, serrate leaves. Inflorescence variable, but mostly axillary and spicate with the male flowers in catkins above the female flowers which are often surrounded by a large bract which is of importance in identifying the different species. Fruit a small capsule.

1. Shrub. Leaves aromatic and dotted with yellowish glands — *A. fruticosa*
   Annual herb. Leaves neither aromatic nor dotted with yellowish glands — 2
2. Inflorescence spicate, the female flowers sessile at the base of the male catkin — 3
   Female flowers in terminal panicles, separate from the axillary male catkins — *A. racemosa*
3. Bracts of female flowers with a fringed margin — *A. ciliata*
   Bracts of female flowers variously toothed but not fringed — 4
4. Female flowers somewhat distant in fruit, not overlapping, the bracts c 5mm long — *A. indica*
   Female flowers rather crowded in fruit, ± overlapping, the bracts c 2–3mm long — 5
5. Bracts of female flowers with long glandular hairs — *A. lanceolata*
   Bracts of female flowers ciliate but largely or entirely without glandular hairs — *A. crenata*

1. Acalypha ciliata *Forssk.*
Erect pubescent annual herb to c 30cm. Leaves ovate, acuminate, the base broadly cuneate, c 4–9cm long, often somewhat glaucous. Flowers axillary, spicate, the male catkin above the rather crowded female flowers; female bracts fringed, c 6mm long, the teeth half this length.

A widespread but local weed of irrigated fields between 500 and 1500m on the escarpment, usually found in great abundance where it occurs and tolerating shade from very tall cereals.

2. Acalypha crenata *A. Rich.*
Erect pubescent annual herb to 30cm. Leaves ovate, acute, often with a rounded base, c 3–7cm long. Flowers similar to *A. ciliata*, but the female bracts c 2–3mm long, crenate, ciliate, very rarely with a few glandular hairs.

Like the other annual species, abnormal female flowers sometimes develop above the male catkin. In *A. crenata* these are muricate (not ± smooth) and this character can be used to distinguish *A. crenata* from related species.

Recorded by Schweinfurth at 600m at Walajah, J. Milḥān, but its occurrence in Yemen must be regarded as uncertain. I have not traced Schweinfurth's specimen, but other *Acalypha* specimens of his have sometimes been misnamed and it is strange that this species has not been refound. Nevertheless, its occurrence is quite probable.

3. Acalypha fruticosa *Forssk.* Difrān, 'Inshāṭ
Much-branched erect shrub to 1m. Leaves ovate, shortly acuminate with a cuneate, rounded or truncate base, c 3–6cm long, strongly aromatic and dotted with numerous yellow glands on the undersurface. Flowers axillary, spicate, the male catkin above the female flowers; female bracts enlarging in fruit, up to 6mm long, dentate.

Rather variable in indumentum; the typical plant is pubescent, but variety **villosa** *Pax ex Hutch.* with villous leaves is known from low altitudes in the central escarpment at Ḥadīyah (Wood 2235), Ḥillah, J. Bura' and Walajah, J. Milḥān (Schweinfurth).

Abundant everywhere between 300 and 1600m on the escarpment and southeastern mountains east to Qa'ṭabah, being the most characteristic undershrub below *Acacia* and *Euphorbia*. It is absent from the Tihama but appears in abundance on the first stony outcrops of the foothills.

4. Acalypha indica *L.*
Erect pubescent annual herb to 40cm. Leaves ovate, acute base rounded or truncate, c 3–6cm long. Flowers axillary, spicate, the male catkin above the rather distant female flowers; female bracts enlarging in fruit, up to 5mm long, dentate.

The commonest annual *Acalypha* in Yemen, occurring widely on the escarpment between 500 and 1500m; unlike the other annuals, it does not occur as a field weed but grows in disturbed stony ground in open scrub particularly by tracks, on field borders and near villages.

5. Acalypha lanceolata *Willd.*
A. glomerata Hutch.
Yellow-green annual herb similar in general appearance to *A. crenata* but the leaf base distinctly cuneate, the female bracts with numerous long glandular hairs and the abnormal female flowers not muricate.

A characteristic weed of the coffee terraces in the central escarpment (Raymah, Bura', Milḥān, Ḥarāz) growing between 1000 and 2000m; elsewhere only known from the Wadi Līyah in Khawlan Ash Sham.

6. Acalypha racemosa *Wall. ex Baill.*
A. paniculata Miq.
Erect, often much-branched, villous annual herb to c 60cm. Leaves ovate to suborbicular, shortly acuminate with a ± truncate base, c 3–5cm long. Male flowers in slender pedunculate catkins; female flowers in terminal and axillary panicles, the stigmas often bright red.

Found between 600 and 1700m in the central escarpment from J. Raymah north to Ḥajjah, but not recorded from the Ḥarāz. It is locally frequent in disturbed ground by paths, in villages and as a weed of coffee terraces. Like the previous species it is essentially a plant of the coffee regions.

Another annual species, *Acalypha brachystachya* Hornem. with digitately tripartite female bracts, is listed by Schwartz but this cannot yet be included in the Yemen flora. Schwartz's record is based on *Acalypha supera* Forssk. but since Forsskal's description is inadequate and there is no specimen at Copenhagen it is impossible to know what Forsskal's plant was.

14. RICINUS *L.*

1. Ricinus communis *L.* Tubshah, Jir
Castor oil tree. Monoecious tree or shrub to 8m. Leaves dark green, long petiolate, peltate, ± orbicular in outline, but deeply palmately lobed, c 5–20cm in diameter, the margin dentate with gland-tipped teeth. Flowers in large erect terminal panicles, the male below the female. Capsule ± globose, covered in soft spines, red or green when mature, c 10–15mm in diameter.

The bean-like seeds are the source of castor oil and they (and indeed the whole plant) are extremely poisonous. Since the plant is commonly found in gardens, it is a likely cause of poisoning to children who may find the seeds attractive. It is said that as few as six seeds can be fatal to a small child.

Although presumably an ancient introduction, this plant is found throughout Yemen but almost always near human habitations. It grows in gardens, on waste ground and roadsides and by streams and irrigation ditches.

15. TRAGIA *L.*

1. Tragia pungens (Forssk.) Muell.-Arg. Ḥumayt
T. arabica (Baill. ex Muell.-Arg.) Prain
T. maamarensis Baill.
Erect, or more usually climbing, perennial herb, often woody below and always covered in stinging hairs. Leaves ovate, acuminate, 3–5cm long, the base slightly cordate, margin serrate. Flowers in axillary and terminal racemes, the 1–3 female flowers below the numerous male flowers. Capsule 3-lobed, ± enclosed by the 5–6 lobed calyx which enlarges greatly in fruit.

Widespread and frequent between 1000 and 2800m on the escarpment, high plateau and the western parts of the Mashriq, growing in rough grassland, scrub and on open stony hills.

16. DALECHAMPIA *L.*

1. Dalechampia scandens *L.*
Slender pubescent monoecious climbing perennial. Leaves dentate, palmately 3–5 lobed almost to the base, c 3–6cm long. Flowers axillary, long pedunculate, ± enclosed by a pair of acutely 3-lobed leaf-like bracts which enlarge and become yellow as the fruit matures. Fruit a deeply 3-lobed capsule.

The Yemen plant is variety **cordofana** (*Hochst.*) *Muell.-Arg.*

A widespread plant of open scrub on the escarpment from 300 to 1400m but avoiding high rainfall areas. It is characteristic of the foothills, the rainshadow areas of the central escarpment and the western parts of Taiz province but is rather scattered in occurrence.

RHAMNACEAE

Trees, shrubs or climbers with simple, alternate or opposite, stipulate leaves. Flowers small, yellow-green, regular with 4–5 petals. Fruit often an edible drupe.

1. Climbing plant with tendrils — 5. *Helinus*
   Trees or shrubs without tendrils — 2
2. Some leaves opposite — 3
   All leaves alternate — 4
3. Flowers in axillary clusters — 2. *Berchemia*
   Flowers in terminal panicles — 4. *Segeretia*
4. Thorns stipulate, some curved, rarely absent. Leaves 3–5 nerved from the base — 1. *Zizyphus*
   Plants unarmed or with straight spines terminating the branchlets. Leaves pinnately nerved — 3. *Rhamnus*

1. ZIZYPHUS *Mill.*

Trees or shrubs with leaves 3–5 nerved from the base and paired stipular spines, one of which is curved and the other straight.

Leaves acuminate. Inflorescence glabrous or nearly so — *Z. mucronata*
Leaves obtuse. Inflorescence especially the pedicels pubescent — *Z. spina-christi*

1. Zizyphus mucronata *Willd.* 'Ulab
Tree or shrub to 10m with a broad, spreading crown. Branchlets brown. Leaves glabrous, green on both sides, ovate, c 7 × 4cm with an unequal, rounded to cordate base, an acuminate, mucronate tip and crenate margin. Flowers yellow-green in dense, glabrous or minutely puberulent axillary cymes. Fruit a dark red-brown, spherical (1cm) drupe of rather bitter taste.

Flowers in the summer rainy season.

Scattered along the western escarpment from 'Aruq in the Ḥujarīyah to Wadi Līyah in Khawlan Ash Sham from about 500 to 1700m. It is more frequent in areas of high rainfall and is usually found in riverine scrub, of which it may be an important constituent.

2. Zizyphus spina-christi (*L.*) *Willd.* 'Ulb, 'Arj, Sidr
Tree or shrub to about 12m with a rounded crown. Branchlets white. Leaves pubescent below, green on both sides, elliptic, c 4.5 × 2cm with an equal cuneate base, obtuse, mucronate tip and finely serrate margin. Flowers whitish-green in few-flowered pubescent axillary cymes. Fruit a dark red-brown spherical (1cm) drupe of sweetish taste.

Flowers in the summer rainy season.

Probably the most widespread plant in the whole Yemen flora, being absent only from the coastal fringe, the eastern sands and mountains above about 2500m. It is common almost everywhere but is especially abundant around fields and on cultivated alluvial plain.

This is the most useful of all trees in the traditional Yemeni economy. Its timber is highly valued for roof beams and has a reputation for durability; its leafy branches provide a reserve of food for goats and camels in the dry season; the dead branches are widely used as 'zerb' to make thorn bush hedges and the fruit are sometimes sold in country markets but are mainly eaten by children, being rather dry and juiceless.

Two species of *Zizyphus* are doubtfully recorded from Yemen: *Z.lotus* (L.) Lam. was recorded by Schwartz on the basis of a number of collections by von Wissmann and Rathjens. The plant has not been refound and the records are almost certainly errors for *Zizyphus spina-christi* from which *Z. lotus* differs in always being a bush with strongly zigzag branchlets and very small leaves which do not exceed 15mm in length. *Z. jujuba* Miller (*Z.vulgaris* Lam.) was recorded by Deflers as a cultivated plant. The Arabic name given ('Arj) is commonly used for *Z. spina-christi* and the record is probably an error for that species.

2. BERCHEMIA *DC.*

1. Berchemia discolor (*Klotzch*) *Hemsl.* Tumār

Phyllogeiton discolor (Klotzsch) Herzog

Glabrous shrub or tree with a rounded crown, to 15m. Leaves opposite towards the tip of the branchlets, becoming alternate towards the base, variable in shape and size but commonly elliptic, with a cuneate base and obtuse apex, about 4cm long, shiny above, markedly paler and glaucous below, with prominent pinnate veining. Flowers in few-flowered axillary clusters, pale yellow. Fruit a yellow, fleshy, edible drupe about 1cm long.

Flowers in spring.

Widespread along the whole escarpment from 400 to 1500m, occurring sporadically in areas of low rainfall but becoming an important constituent of valley scrub in the foothills of the central escarpment.

3. RHAMNUS *L.*

Shrubs with alternate leaves.

Unarmed shrub. Leaves serrate, up to 9cm long *R. prinoides*
Branches terminating in a spine. Leaves entire, c 1cm long *R. staddo*

1. Rhamnus prinoides *L'Herit.*

Low unarmed shrub, probably never exceeding 1m in Yemen. Branchlets pubescent when young becoming darker and glabrous with age. Leaves alternate, up to 9cm long, ovate to lanceolate with a rounded to weakly cuneate base and a long acuminate tip, serrate, dark shiny green above but markedly paler beneath. Flowers in few-flowered axillary clusters, yellow-green. Fruit a globular drupe, turning from red to purple-black with age, c 5mm long.

Flowers in summer.

In grassland and scrub between 2000 and 2900m on a few of the wettest escarpment mountains: on the southeast summit of J. Ḥabashī; J. Ba'dān; in some frequency in the mountains between Kusmah and Ḥadīyah, J. Raymah; near Al Markah, J. Milḥān.

2. Rhamnus staddo *A. Rich.* Jidb

Dioecious shrub to 2m. Branchlets terminating in a spine. Leaves alternate, about 1cm long, elliptic, lanceolate or obovate, entire with a cuneate base and a rounded mucronate apex, green on both sides. Flowers in axillary clusters, yellow-green. Fruit a 1–3-lobed drupe becoming pale red at maturity, c 3mm long.

This plant flowers very early in the year (January–March) and is commonly ± leafless at flowering time.

Yemen material is usually included in variety **deflersii** (*Schweinf.*) *Chiov.*

A common but nondescript and hence easily overlooked shrub of cliffs, tracksides and scrubby slopes on the escarpment and around the high plateau from 1500 to 3100m.

4. SEGERETIA *Brongn.*

1. Segeretia thea (*Osbeck*) *Johnst.*

Hinah Rubāḥ, Sharas, Limūm, Numnum

S. brandrethiana Aitch.
S. spiciflora (A. Rich.) Hutch. & Bruce
Berchemia yemensis Deflers

Small, erect shrub to 2m, usually with distinctive long, narrow, tapering, somewhat flattened branches. Leaves glabrous, dark green and shiny above but paler below, opposite, shortly petiolate, ovate to elliptic, up to 20 × 10mm with a rounded base, acute or shortly mucronate apex and serrate margin. Flowers dull yellow in a paniculate inflorescence consisting of numerous short spikes. Fruit a small blackish globular drupe.

Flowers throughout the summer rainy season.

Frequent on cliffs and scrubby hill slopes especially along drainage lines from 1000 to 3100m along the whole western escarpment and in the high plateau region.

5. HELINUS *E. Mey.*

1. Helinus integrifolius (*Lam.*) *Kuntze*

H. arabicus Jaub. & Spach

Climbing plant with woody stems and coiled tendrils. Leaves alternate, dull glaucous green, ovate, weakly cordate with a mucronate tip. Flowers small, yellow-green in glabrous umbels. Fruit ovoid, capsular.

Apparently winter-flowering.

Originally collected by Botta without precise locality, this plant is now known from only three localities between 1400 and 1500m in rocky scrub in areas of good rainfall: Hasabān Asfal, Misrākh at the southwest foot of J. Ṣabir; Bir Ghūbaḥ, Shar'ab and just north of 'Udayn market.

VITACEAE

Perennial herbs or shrubs, climbing by means of tendrils. Leaves alternate, usually palmate, with stipules. Flowers regular, flower parts in 4s or 6s; petals usually small, yellow-green. Fruit a berry with 1–4 seeds.

A mainly tropical family restricted to the Tihama and the escarpment.

1. Leaves simple 2. *Cissus*
   Leaves compound **2**

2. Petals 5–6. Woody climbing shrubs with trifoliolate leaves — 1. *Rhoicissus*
Petals 4. Climbing perennials, sometimes woody below. Leaves 3–5 foliolate — **3**
3. Fruit 1-seeded. Flower buds constricted in the middle — 4. *Cyphostemma*
Fruit 2–4-seeded. Flower buds not constricted in the middle — 3. *Cayratia*

1. RHOICISSUS *Planch.*

Woody climbing shrubs with tendrils. Leaves 3-foliolate. Inflorescence of umbel-like cymes arising opposite the leaves. Petals 5–6. Berry with 1 or 2 seeds, ± globose.

Leaflets entire, glabrous below — *R. revoilii*
Leaflets coarsely dentate, pubescent below — *R. tridentata*

1. **Rhoicissus revoilii** Planch. — Mirū

Climbing shrub with woody pubescent stems. Leaflets ± lanceolate, entire, glabrous on both surfaces, dark green above, up to c 7 × 2.5cm, the lateral leaflets very unequal sided. Cymes on short lateral branches arising opposite the leaves, usually densely dark red-tomentose when young.

A characteristic plant of riverine scrub between 600 and 1600m in the central escarpment including Maḥwīt and J. Sharafayn; it also grows abundantly at 'Udayn. It is found climbing over streamside bushes and is very rare away from water.

2. **Rhoicissus tridentata** (*L.f.*) *Wild & Drummond*

R. erythroides (Fresen.) Planch.

Rather similar to the previous species from which it is best distinguished by the larger leaflets (up to 10 × 6cm) which are obovate, coarsely dentate and tomentose below. In addition the inflorescence is typically pale-brown tomentose when young and arises on much longer lateral branches.

Locally common between 1500 and 2800m on most of the higher rainfall escarpment mountains from J. Ṣabir and J. Sawraq north to Maḥwīt. Unlike the previous species it rarely climbs over other bushes, but is usually found sprawling over rocks or in grassland on open hillsides.

2. CISSUS *L.*

Climbing fleshy undershrubs with tendrils and simple leaves. Flowers in cymes. Fruit a red, fleshy, usually 1-seeded berry.

Stem ± rounded. Leaves ovate to orbicular, never lobed — *C. rotundifolia*
Stem strongly quadrangular. Leaves ovate to triangular, commonly 3-lobed. — *C. quadrangularis*

1. **Cissus quadrangularis** *L.* — Sila', Khudām

Glabrous climbing plant with stout quadrangular, slightly winged succulent stems. Leaves fleshy, ovate – triangular, often 3-lobed; margin dentate, base truncate or rounded. Flowers in small, pedunculate, umbel-like cymes arising opposite the leaves.

Abundant throughout the Tihama and the escarpment, occurring in a wide variety of habitats including stony and sandy steppe, scrub, stream sides and rocky hillslopes, becoming less common above 1000m and not recorded above 1500m. Absent only from salt marsh, sand dunes and cultivated land.

2. **Cissus rotundifolia** (*Forssk.*) *Vahl* — Halaṣ

Climbing plant with shortly pubescent, ± rounded woody stems. Leaves fleshy, often folded, ovate to orbicular, never 3-lobed; margin dentate; base cordate. Flowers in panicled umbel-like cymes arising in the leaf axils.

Abundant on the escarpment between 400 and 1800m and one of the characteristic plants of *Euphorbia* scrub where it is the most common species after *Euphorbia* both in the central escarpment foothills and around Taiz. It is strikingly less common in *Acacia-Commiphora* bushland.

The leaves are boiled and eaten as a vegetable by poorer families.

3. CAYRATIA *Juss.*

1. **Cayratia gracilis** (*Guill. & Perr.*) *Suesseng*

Glabrous to sparsely pubescent, scarcely succulent climbing herb. Leaves 3–5 foliolate, the lateral leaflets pedate. Leaflets ovate, serrate, base variable, rounded to cordate. Flowers in small pedunculate lateral cymes, arising irregularly but most commonly from the leaf axils. Fruit black, globose, 2–4 seeded.

A rare plant found between 1500 and 1800m in a few high rainfall areas of Ibb province: in riverine scrub on Jabalayn and by the Wadi Dūr, 'Udayn; on scrubby field borders between Ibb and Suq As Sabt.

4. CYPHOSTEMMA (*Planch.*) *Alston*

Perennial, climbing, succulent undershrubs with compound fleshy leaves and tendrils. Flowers in cymes. Fruit a 1-seeded berry. Very similar to the previous genera but readily recognised by the flower buds which are strongly constricted in the middle.

Glabrous plant. Leaves 3-foliolate, sessile — *C. ternatum*
Plant pubescent at least on young growth. Leaves 3–5-foliolate, petiolate — *C. digitatum*

1. **Cyphostemma digitatum** (*Forssk.*) *Desc.* — Ḥalqa

Cissus digitata (Forssk.) Lam.

Climbing plant, pubescent on young growth. Leaves fleshy, petiolate, digitately 3–5-foliolate, usually with 5 leaflets below and 3 above; leaflets ovate, dentate. Flowers in pedunculate axillary cymes. Berry red.

Widespread and common between 1400 and 2500m on the escarpment, often on cliffs and with a marked preference for shaded stony places such as terrace walls and gullies where it is often associated with *Clematis spp.* and *Senecio hadiensis*.

2. **Cyphostemma ternatum** (*Forssk.*) *Desc.*

Cissus ternata (Forssk.) J.F. Gmel.

Robust glabrous climbing plant. Leaves fleshy, ± sessile,

3-foliolate; leaflets ovate, dentate. Flowers in long pedunculate axillary cymes.

An uncommon but easily overlooked plant usually growing amongst *Salvadora* between sea level and 500m in the Tihama: around Ḥays; by the Wadi Sihām between Qutay and Khalīfah; between Bayt al Faqīh and Manṣūrīyah; at 21km between Hodeida and Manṣūrīyah; between 'Abs and Mīdī; near Suq al Jammah.

The Grape, **Vitis vinifera** *L.* known as ''Inab', is cultivated on the high plateau and in the Mashriq. It is grown on irrigated land in rather dry country notably around Ṣa'dah and in the valleys northeast of Sana'a, particularly the Bani Ḥushaysh and Wadi Ḍahr but, in fact, few settlements in the Mashriq are without vineyards. Cultivation in the high rainfall, more humid areas is generally unsuccessful.

Grapes are ripe in Yemen from June to December. Many varieties are grown. The season usually begins with small seedless grapes. The main season varieties are a black very sweet ellipsoid grape and a large globular red, very crisp-textured grape which are found throughout autumn in all markets. Other varieties are grown but are seen less commonly. Although a large part of the crop is sold fresh, considerable quantities are dried and sold as raisins, sultanas etc. Traditionally the Jews made wine, but since their departure wine making has largely ceased.

LINACEAE

LINUM *L.*

Herbs with alternate, sessile, simple leaves. Inflorescence a terminal cyme; petals 5, clawed. Fruit a capsule.

1. Flowers blue. Capsule 6–8mm long — *L. usitatissimum*
   Flowers yellow. Capsule 2–2.5mm long — 2
2. Upper leaves with a pair of dark glands at the base — *L. volkensii*
   Leaves without glands at the base — 3
3. Leaf margins smooth, entire. Sepals slightly exceeding the capsule — *L. trigynum*
   Leaf margins rough, finely serrulate. Sepals about twice as long as the capsule — *L. strictum*

1. Linum strictum *L.*

Erect, sparingly branched annual to 20cm, sparsely puberulous below. Leaves crowded, erect, narrowly lanceolate, acute, 1.5–2cm long, the margins rough, finely serrulate, often inrolled. Inflorescence dense; petals yellow, 6–8mm long; sepals erect, twice as long as the 2mm long capsule.

A rare plant of short grassland between 2400 and 3200m on a few high rainfall southern mountains: J. Kubūrah and J. Bārad, Kusmah, Raymah; J. Ḥiṣn Madnān, Waṣāb al 'Ālī; J. Qarad, Mudhaykhirah; J. Ta'kar; J. Ba'dān, J. Sumārah (Radcliffe-Smith and Henchie). (Wood 75/525)

2. Linum trigynum *L.*

L. gallicum L.

Spreading, much-branched, glabrous annual. Leaves distant, spreading, narrowly elliptic, acute, c 1cm long, the margins smooth, entire, not inrolled. Inflorescence lax; petals yellow, 4–6mm long; sepals erect, slightly overtopping the 2mm long capsule.

A very rare plant of short grassland at high altitudes (2700–2900m): J. Shibām, Manākhah; J. Ḥiṣn Madnān, Waṣāb al 'Ālī. (Wood 1690)

3. Linum usitatissimum *L.* — Mawmir

L. humile Miller

L. brachypetalum Schwartz.

Flax. Robust, erect, annual herb to c 40cm, usually branched near the base. Leaves crowded, spreading, linear, acute c 2–3cm long, the margin smooth, entire, not inrolled. Petals blue, c 1cm long; sepals erect, shorter than the 6–8mm long capsule.

Commonly cultivated between 2000 and 3000m on the high plateau and the escarpment and often found as a relic of cultivation. It is presumably cultivated for its oil and its fibre but I have no information on how they are extracted. It is also well reputed medically as a cure for kidney stones.

4. Linum volkensii *Engl.*

Erect annual herb to 35cm, much branched above, puberulent below. Leaves crowded, erect, narrowly lanceolate, c 2cm long; apex aristate; margins rough, finely serrulate, not inrolled. Inflorescence lax; petals yellow, 6–8mm long; sepals erect, slightly overtopping the 2mm long capsule.

Readily distinguished from other species by the pair of conspicuous dark glands at the base of the leaves.

Locally abundant in shallow grassland between 1400 and 2400m on the western slopes of J. Raymah; also between Al Markah and Aṣ Ṣa'bah on J. Milḥān and southwest of Bayt Shamrān, J. Masār, Ḥarāz. (Wood 2705)

MALPHIGIACEAE

CAUCANTHUS *Forssk.*

1. Caucanthus edulis Forssk. — Qawqa'

Small climbing bush to about 2½m. Branchlets red-brown with soft appressed white hairs especially when young. Leaves with reddish petioles, opposite, rounded, entire except for the retuse or shortly mucronate apex, up to 3cm long. Flowers on reddish pedicels in many-flowered corymbs, small, cream-coloured. Fruit up to 6cm long with 6 prominent reddish wings.

Flowers mainly in early summer.

An attractive plant of cliffs and steep rocky slopes on the western escarpment from 400 to 1300m. It occurs from Bathayjah in the Ḥujarīyah north to the Wadi Sara' near Khamīs Banī Sa'd and Bāb Baḥal in the Ḥarāz. Despite its great abundance in the valley forest communities of J. Raymah and J. Bura' it is totally absent from similar habitats further north.

This plant is virtually restricted to Yemen and is only known elsewhere from Somalia. The name of the genus, *Caucanthus*, is based on the Arabic name, Qawqa'.

POLYGALACEAE

POLYGALA *L.*

Herbs or undershrubs. Leaves alternate. Flowers in terminal racemes or less commonly, axillary; sepals 5, the inner (wings) much larger than the outer; petals 3, joined, the lower (keel) large and often with a fimbriate crest, ± enclosing the stamens whose filaments are joined into a tube. Fruit a compressed winged capsule containing 2 silky-hairy seeds.

1. Inflorescence axillary only **2**
   Inflorescence terminal, sometimes axillary as well **3**
2. Undershrub. Wings with numerous minute veins, 5–10mm long *P. senensis*
   Herbs or small undershrubs. Wings with a prominent green midrib 4–5mm long *P. erioptera*
3. Terminal racemes very short, equalled or overtopped by the leaves. Leaves 7–20mm broad **4**
   Terminal racemes much longer than the leaves. Leaves less than 5mm broad **5**
4. Annual. Wings glabrous *P. persicariifolia*
   Perennial undershrub. Wings pubescent *P. tinctoria*
5. Wings pubescent. Corolla yellow *P. irregularis*
   Wings glabrous. Corolla pink or mauve **6**
6. Prostrate plant. Racemes up to 2cm long. Leaves 5–9mm long *P. rupicola*
   Erect plant. Racemes 1–15cm long. Leaves 10–40mm long *P. abyssinica*

1. Polygala abyssinica *R. Br. ex Fresen.*

Erect branched, glabrous or sparsely pubescent woody perennial with long rod-like branches reaching 1m high. Leaves linear, 1–4cm long, 2–3mm broad. Flowers in terminal racemes 1–15cm long, wings glabrous with 3 principal veins from the base, 6–9mm long; corolla mauve.

In addition to the typical plant there also occurs variety **subaphylla** (*O. Schwartz*) *J.R.I. Wood* stat. nov. (*Polygala subaphylla* O. Schwartz, 'Flora des tropischen Arabien' in *Mitt. Inst. Bot. Hamburg* 10:132 (1939). This is a very vigorous plant with stout, almost leafless stems and long racemes of large flowers. Although readily recognised in the field by its large size, it does not seem to differ from variety *abyssinica* in anything except its slightly larger dimensions.

Variety *abyssinica* is widespread and common on the escarpment and high plateau growing 1200 to 2900m on cliffs, terrace walls and steep rocky banks. Variety *subaphylla* occurs on dry stony hills in small scattered colonies mostly in the Mashriq: near Majz (Steinberg 638, 989, 1177); Aqabah Bani Dawd in Bani Ḥushaysh (Wood 1083); 20km east of Rada'; Mafḥaq (Wood 718); Wasit (Miller).

2. Polygala erioptera *DC.* Junayd

P. thurmanniana Chodat

Prostrate or erect, pubescent annual or perennial herb, sometimes woody below. Leaves linear, 10–35mm long, 2–3mm broad. Flowers in axillary clusters; wings pubescent with a prominent green midrib but no lateral veins, 4–5mm long; corolla mauve.

An extremely variable plant with intergrading forms. Forms from cultivated land, sand dunes and similar disturbed places (e.g. Wood 74/42) are obviously annual, usually with long erect herbaceous sparingly branched stems, 50cm or more in length. The almost glabrous leaves are usually more than 15mm long. Forms from rocky situations (e.g. Wood 75/129) are perennial and woody based usually with short prostrate, much-branched stems less than 10cm in length. The leaves are densely pubescent and mostly 10–15mm long. An extreme perennial form with woody branches and grey-pubescent apparently fleshy leaves up to 8mm long has been treated as a separate species (*P. thurmanniana* Chodat) but I agree with Schweinfurth in regarding this as merely variety **perennis** *Schweinf.* of *P. erioptera*. This was described from Aden but has been recorded from North Yemen at Qa'ṭabah (Wissmann) and on Ḥanish Island (Slade).

Widespread and common from sea level to 2500m in all areas except the eastern part of the Mashriq and the Sands. It occurs in a variety of well-drained habitats including sand dunes, sandy fields, open stony *Acacia* scrub, rock outcrops and exposed mountain ridges.

3. Polygala irregularis *Boiss.*

Sparsely puberulent perennial herb, much branched from the woody base with spreading or ascending branches to 30cm. Leaves linear-oblanceolate, 2–3cm long, 2–6mm broad. Flowers partly in the leaf axils and partly in terminal racemes 1–10cm long; wings shortly pubescent, prominently veined with a white membranous margin, 5–6mm long; corolla yellow.

Widespread in the Tihama on sand dunes and in sandy fields from sea level to 400m but very local: Mukhā; between Mafraq and Barḥ; Bayt al Faqīh; between 'Abs and Mīdī.

4. Polygala persicariifolia *DC.*

Pubescent annual herb with erect, sparingly branched stems to 50cm. Leaves narrowly elliptic, 3–8cm long, 7–20mm broad. Flowers partly in clusters in the leaf axils, and partly in short lateral and terminal racemes, 3–5cm long; wings glabrous with 3–5 obscure looping veins from the base, 6–7mm long; corolla cream or pale pink.

A rare plant only known from a grassy bank by a sorghum field between Ibb and Suq As Sabt at 1600m.

5. Polygala rupicola *Hochst. & Steud. ex A. Rich.*

P. aethiopica Chodat

Small glabrous perennial, much branched from a woody rootstock with short prostrate stems. Leaves linear, 5–9mm long, 1–1.5mm broad. Flowers in short terminal racemes 1–2cm long; wings glabrous with a prominent branching midrib, 5–7mm long; corolla mauve.

An inconspicuous plant of rock crevices on exposed mountain ridges growing from 2600 to 3300m on the higher mountains of the escarpment and high plateau from J. Ṣabir north to Aḥjar. It is, however, of very local occurrence though found somewhere on nearly all the main mountain groups. Outside Yemen it is only known from the Ethiopian highlands.

6. Polygala senensis *Klotzsch.*

P. obtusissima Hochst.

Small pubescent undershrub with spreading or erect, wiry branches to c 60cm. Leaves oblong or oblanceolate, often emarginate, 10–15mm long, 3–5mm broad. Flowers in axillary clusters or very short axillary racemes less than 1cm long; wings pubescent, pale yellow with numerous minute veins; corolla mauve.

Widespread and frequent from 500 to 1600m on the escarpment and more rarely up to 2000m in the Mashriq where it has been found between Khaywān and Ḥarf and near Khulāqah in Nihm. It is a plant of open *Acacia* scrub and similar stony bushland areas in dry country and is often associated with *Vernonia cinerascens*, *Barleria proxima* and *Solanum cordatum*.

7. Polygala tinctoria *Vahl*

P. yemenica Chodat ?

Much-branched, pubescent perennial undershrub with erect or spreading woody branches to c 1m. Leaves oblanceolate or oblong, ± tomentose below, 20–40mm long, 7–15mm broad. Flowers in short lateral or terminal racemes 1–3cm long; wings pubescent, obscurely veined, suborbicular, 9–11mm broad; corolla mauve.

Rather variable in terms of leaf size and indumentum and it seems probable that *P. yemenica* was only a large-leaved, hirsute form of this species. Unfortunately it is impossible to be certain as the type specimen was probably destroyed in a fire in the last century.

An attractive and frequent plant of damp scrubby hillsides particularly in areas of winter mist, growing from 800 to 2000m on the outer escarpment from Turbah north to Shahārah. It is often associated with *Hybanthus enneaspermus* and *Pelargonium alchemilloides*. It is endemic to southwest Arabia.

MELIANTHACEAE

BERSAMA *Fresen.*

1. Bersama abyssinica *Fresen.*

Small straggling tree to 8m. Leaves with stipules, alternate, imparipinnate with 9–13 leaflets, up to 50cm long; leaflets entire or somewhat dentate, glabrous when mature, very variable in size and shape but commonly oblanceolate, about 8cm long with an acuminate apex. Inflorescence a dense pubescent axillary raceme to 10cm long; petals cream, recurved. Fruit a hard, globose capsule c 2cm long.

Flowers and fruit never observed in Yemen.

Known only from two collections by Schweinfurth: at 1000m above Ḥillah, J. Bura' and at 900m near 'Ukaybir, J. Milḥān. It has never been recollected.

SAPINDACEAE

A family of trees, shrubs and climbers. Leaves alternate, simple or compound, usually without stipules. Flowers bisexual, small and inconspicuous with parts in 4s or 5s. Seeds often winged.

1. Herbaceous climbing plants with coiled tendrils on the peduncle — 1. *Cardiospermum*
   Trees or shrubs — **2**
2. Leaves simple — 3. *Dodonaea*
   Leaves trifoliolate — 2. *Allophylus*

1. CARDIOSPERMUM *L.*

Annual climbing plants with biternate leaves. Inflorescence an axillary umbel or corymb with a pair of curled tendrils on the peduncle. Flowers pedicellate with 4 cream petals. Fruit a pale green bladder-like inflated capsule.

Petals 5–6mm long. Inflorescence with 4 or more branches — *C. corindum*
Petals 2–3mm long. Inflorescence with 3 branches. — *C. halicacabum*

1. Cardiospermum corindum *L.*

Densely pubescent dull green climber. Leaflets deeply cut with weakly acute or rounded lobes. Inflorescence a corymb usually with 4–6 branches.

An uncommon plant of open stony hills between 500 and 1200m in the central escarpment, but strikingly absent from J. Raymah. It is usually found in very open *Acacia – Commiphora* bushland: 'Ukaybir and Walajah, J. Milḥān; Wadi Yūr; Ḥillah, J. Bura'; 'Usil, Rajamah and Mawsanah in the Ḥarāz.

2. Cardiospermum halicacabum *L.*

Glabrous to thinly hairy, green plant very similar to the last species, but differing in its colour, indumentum, more acutely lobed leaves, its 3-branched umbellate inflorescence and petals which only reach about 3mm in length.

A frequent plant of scrub and field borders between 300 to 1800m along the whole escarpment from the Ḥujarīyah north to Khawlan Ash Sham. It is often but not exclusively found by ditches and irrigation channels in irrigated valleys.

2. ALLOPHYLUS *L.*

1. Allophylus rubifolius (*Hochst.*) *Engler*

Low, weak shrub to 1½m with finely pubescent branchlets and petioles. Leaves 3-foliolate, the leaflets shortly petiolate, obovate, sparsely pubescent, coarsely serrate towards the acute apex, the base attenuate especially on the terminal leaflet which is about twice the size of the two laterals reaching 9 × 5cm. Flowers in simple, or few-branched, lateral racemes, small, cream. Fruit a blackish turbinate drupe.

There is some doubt as to whether the Yemen plant is really conspecific with the East African species, which usually has leaves densely tomentose below.

A rare plant of low-altitude scrub between 500–800m in

the central escarpment: Rajamah near Ḥajjaylah and J. Sāfān, Ḥarāz; Walajah, J. Milḥān; along the western foothills of J. Raymah; one of several species unaccountably absent from J. Bura'.

3. DODONAEA *Mill*

1. Dodonaea viscosa (*L.*) *Jacq.* Shaḥth, Shaḥs
Shrub to about 3m. Branchlets red-brown, glabrous, somewhat sticky. Leaves c 9 × 1.5cm, subsessile, glabrous, sticky, oblanceolate with a long attenuate base, acute apex and entire margin. Flowers mostly in small terminal racemes, sometimes of one sex only, sometimes mixed, without petals, usually reddish-green in colour. Fruit small but with 2–3 large reddish membraneous wings.

Flowers in spring.

Scattered along the whole escarpment from 1000 to 3100m, but only common in Khawlan Ash Sham and the southeast mountains, especially parts of the Ḥujarīyah where it is an important constituent of scrub in the drier rainshadow hills and is associated with *Juniperus*, *Psiadia*, *Solanum schimperianum* and *Lavandula* species.

BURSERACEAE

Trees and shrubs which are commonly resinous. Frankincense and myrrh come from this family but, contrary to popular belief, frankincense itself does not come from Yemen but from Dhofar and Somalia.

COMMIPHORA *Jacq.*

The species of *Commiphora* found in Yemen share many features in common. They are typically small trees (less commonly shrubs) about 2–4m high with relatively stout dark green trunks and thin, papery, peeling bark. They have numerous branches towards the top of the trunk which form a dense rounded entwined mass. The glabrous (rarely finely pubescent) leaves are clustered at the end of short lateral branchlets which are often spine-tipped. The flowers are clustered or in cymes, with usually 4 very small, red or yellow-green petals and commonly appear during the dry season when the plant is leafless. The fruit is a glabrous drupe containing a hard stone.

***Commiphora* is an important genus as, together with *Acacia* and *Grewia*, it provides the scrub cover on most dry stony hills up to about 1500m. Most species are very drought-resistant.**

Several species were traditionally tapped for resin (e.g. *C. gileadensis*) but, as with many features of the traditional economy, this practice has fallen into disuse. I have never seen or heard of resin collection in my time in Yemen and the resins sold in the suqs, generally under the name of Lubān, are of Indian or Somali origin. Unfortunately, *Commiphora* is still of economic importance in one respect: it is the main source of firewood and it seems to be much less frequent near main roads than on more remote hills.

***Commiphora* is a very complex genus but there are not too many species in Yemen. Aberrant forms and intermediates, perhaps hybrids, occur, for example: Wood 3023 from near Suq al Markaz in the Ḥujarīyah is intermediate between *C. habessinica* and *C. schimperi*. The following key which is based mainly on leaf characteristics should therefore be used with caution.**

1.	All leaves simple	2
	Some leaves 3–5-foliolate	4
2.	Leaves sessile or nearly so	*C. sp.A*
	Leaves petiolate, the petioles exceeding 5mm long	3
3.	Leaves broadly ovate to suborbicular, ± entire	*C. quadrincincta*
	Leaves obovate, dentate	*C. habessinica*
4.	Some leaves 5-foliolate, entire	*C. gileadensis*
	All leaves 1–3-foliolate, the terminal leaflets dentate	5
5.	Lateral leaflets exceeding 1cm long. Flowers in cymes, the peduncle commonly 1.5–4cm long.	*C. kataf*
	Lateral leaflets very small usually less than 5mm long. Flowers in subsessile clusters	6
6.	Terminal leaflet 1.5–4.5cm long	*C. habessinica*
	Terminal leaflet up to 1.5cm long	7
7.	Lateral leaflets rudimentary, entire	*C. myrrha*
	Lateral leaflets about half the size of the terminal leaflet, dentate	*C. schimperi*

1. Commiphora gileadensis (*L.*) *C. Chr.* Bishām, Risha'
C. opobalsamum (Forssk.) Engl.
Branchlets without spines, commonly long and slender. Leaves 3–5-foliolate, terminal leaflets obovate (rarely elliptic) about 1cm long, base attenuate, apex rounded to emarginate, margin entire, lateral leaflets fully developed, about the same size as the terminal leaflets. Flowers in clusters. Drupe apiculate.

The crushed leaves of this species are strongly fragrant. It is the source of Mecca Balsam and the Balm of Gilead.

A rather local plant often associated with *C. myrrha* on dry stony hills in the Tihama foothills up to 1200m and on the stony steppe south of Ḥays. It has not been found north of J. Ash Sharafayn.

2. Commiphora habessinica (*Berg.*) *Engl.* Khadash
Branchlets spine-tipped. Leaves 3-foliolate, terminal leaflets obovate to oblanceolate, c 2–4.5cm long, apex obtuse to rounded, base attenuate, margin dentate, lateral leaflets much reduced, 2–3mm long, quickly deciduous so the leaves sometimes appear to be simple. Flowers in clusters. Drupe apiculate.

All Yemen specimens examined by J.B. Gillett have 1-celled fruits except for Wood 75/80, which has some fruits 1-celled and some 2-celled, thus approaching the Indian species *C. wightii* (Arn.) Bhaudari.

A local but perhaps overlooked species found between 1400 and 1900m on the escarpment from the Ḥujarīyah north to Bayt 'Adhāqah.

3. Commiphora kataf (*Forssk.*) *Engl.* Qaṭaf
Branchlets without spines. Leaves 3-foliolate, terminal leaflets obovate, 2–4cm long, base attenuate, apex usually rounded, margin weakly dentate, rarely entire, lateral leaflets fully developed. Flowers in cymes, the peduncle 1.5–4cm long. Drupe somewhat flattened, rounded.

This species is endemic but is very closely related to *C. erythraea* (Ehrenb.) Engl. found on the west of the Red Sea which seems to be little more than a tomentose form of *C. kataf*.

Widespread on the escarpment up to 1500m from the Ḥujarīyah north to Malāḥiṭ, but only frequent in areas of good rainfall in the central escarpment where it is common but never dominant on scrubby hillsides. It appears to root readily since it is often grown as corner posts of shops in the foothill markets and more rarely in the Tihama towns such as Bayt al Faqīh.

4. Commiphora myrrha (*Nees*) *Engl.* Qafal

Amyris kafal Forssk. nom. nud.

Branchlets spine-tipped, numerous. Leaves 3-foliolate, terminal leaflets obovate to elliptic, up to 15mm long but commonly 5–10mm, variable in shape and dentation on the same plant, some leaves entire, others dentate, others ± sinuate lobed, apex rounded to acute, base rounded to attenuate, lateral leaflets not fully developed, commonly only 2–3mm long, always entire. Flowers in clusters. Fruit apiculate, the drupe opening more readily than in other species, always (?) by two valves.

Despite the name of this plant and its bitter taste, there is no evidence that it is a source of myrrh.

This is the most drought-resistant *Commiphora*. It occurs in abundance along the Tihama foothills from Malāḥiṭ southwards and covers vast areas of the stony steppe south of Ḥays. It is not found on sand or alluvium but appears on the first hills arising out of the Tihama plain. It occurs more rarely in the drier rainshadow areas of the escarpment up to about 1500m, such as Al 'Ashshah, Mafḥaq and between Waṣāb and Riḥāb. In the Mashriq it is the only *Commiphora* and is widely distributed from the stony plains bordering the Jawf to the arid hills east of Jūbah.

5. Commiphora quadricincta *Schweinf. ex Engl.*

Branchlets spine-tipped. Leaves simple, entire, broadly ovate to orbicular, commonly about 2.5 × 2.5cm, apex obtuse to rounded, base truncate. Flowers in clusters. Drupe apiculate, the stone of the fruit with four longitudinal wings, hence the name of the species.

All plants I have seen have simple entire leaves but Schweinfurth's description states that 3-foliolate forms occur in which the leaves are dentate. The orbicular leaf shape would readily distinguish these forms from other species.

A very rare plant of sand and alluvial plain in the Tihama: it is found as an isolated tree in a number of places near Bājil including Sayyid Sulayman (Schweinfurth) and amongst *Acacia ehrenbergiana* in a village between Bājil and Qutay. Otherwise it is only known between Malāḥiṭ and Suq al Jammah.

6. Commiphora schimperi (*Berg*) *Engl.* Khadash

Branchlets spine-tipped. Leaves 3-foliolate, terminal leaflets obovate, c 10–15mm long, dentate, apex rounded, base attenuate. Flowers in clusters.

Very similar to some forms of *C. habessinica* but the lateral leaflets are fully developed and clearly dentate, the terminal leaflet does not normally exceed 15m and the drupe is more apiculate.

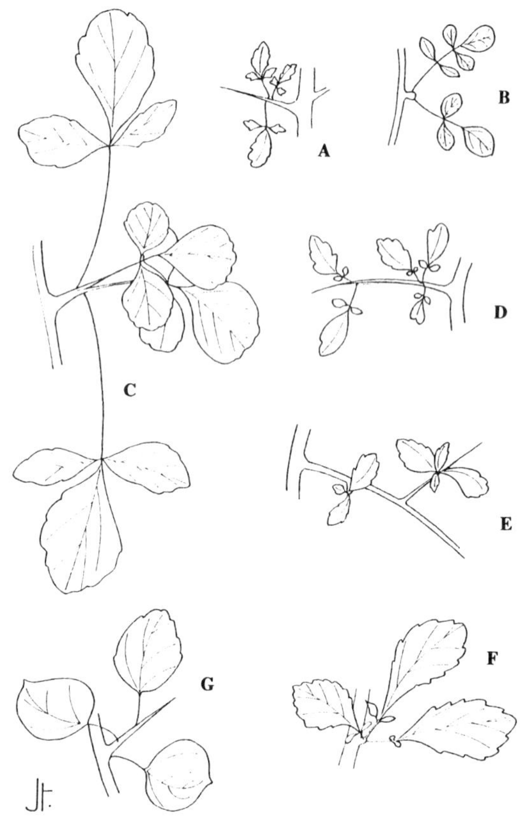

Fig. 17. Leaves and branchlets of *Commiphora* species. **A** *C. schimperi*; **B** *C. gileadensis*; **C** *C. kataf*; **D** *C. myrrha*; **E** *C. sp. A*; **F** *C. quadricincta*; **G** *C. habessinica*.

A sterile shoot (Wood 2115), collected between Suq As Sabt and Al 'Ajūr, J. Bura', has leaves about twice the size of typical *C. schimperi*, but is probably only a shade form. Some examples (e.g. Wood 2311 from between Raḥidah and Ghalaybah, 'Abūs) appear to lack spines.

In scattered localities along the escarpment between 1000 and 2300m from the Ḥujarīyah north to Maḥābishah. It rarely occurs in abundance and is usually found in small groups near cultivation. It grows at higher altitudes and prefers a higher rainfall than other species and is thus the only *Commiphora* found in the Ibb-Jiblah area.

7. Commiphora sp.A Khadash

C. habessinica (Berg) Engl. var. *simplicifolia* Schweinf.

Branchlets spine-tipped. Leaves simple, sessile, obovate to oblanceolate, up to 2.5cm long, apex rounded, base cuneate, margin serrate especially towards the apex. Flowers in clusters. Drupe abruptly apiculate.

This species is often regarded as part of *C. habessinica* but is readily distinguished by its smaller, sessile, simple leaves. I have called it *Species A* as *simplicifolia* cannot be used at a specific level and it is not clear at present what its correct name should be. Examples include Wood 514, 1099 and 2065.

Widespread on the escarpment from the Wadi Badr south to the Ḥujarīyah and thence eastwards through the southeast mountains to Bayḍā. It is an important constituent of scrub particularly in the rainshadow areas of the central escarpment e.g. Madīnat Al 'Abīd, Mafḥaq, but is absent from the low-altitude dry areas favoured by *C. myrrha*. Recorded between 600 and 2000m, but rarely above 1500m except in the southeast mountains.

ANACARDIACEAE

Trees and shrubs often with resinous bark. Leaves lacking stipules, alternate, usually compound. Flowers small, bisexual, regular; petals free, 5. Fruit a drupe.

1. Leaves simple — 2. *Ozoroa*
   Leaves compound — **2**
2. Leaves 3-foliolate — 1. *Rhus*
   Leaves pinnate — **3**
3. Leaflets ± equal sided. Flowers in dense spikes or racemes — 4. *Lannea*
   Leaflets unequal sided. Flowers in lax panicles — 3. *Pistacia*

1. RHUS *L.*

Small shrubs up to 3m high with alternate trifoliolate leaves. Flowers in panicles. Fruit small.

1. Leaves glabrous when mature — **2**
   Leaves pubescent — **3**
2. Leaflets sessile, obtuse — *R. natalensis*
   Leaflets petiolate, acuminate — *R. retinorrhaea*
3. Leaflets up to 3cm long — *R. flexicaulis*
   Leaflets at least 6cm long and commonly more — *R. abyssinica*

1. Rhus abyssinica *Hochst. ex Oliv.*
Leaves including the petiole densely pubescent; leaflets sessile, crenate or entire, often brownish-tomentose below, usually acute; the central leaflet 6–15cm long usually ovate to lanceolate; lateral leaflets rather smaller, elliptic. Flowers brownish-white.

A very rare plant found between 'Usil and 'Attārah in the Ḥarāz at c 1400 to 1800m by Schweinfurth and von Wissmann, but not refound. A sterile plant collected on J. Um Laylah (Steinberg 496) may well be this species.

2. Rhus flexicaulis *Baker*
Leaves densely tomentose on both surfaces; leaflets sessile, entire, oblanceolate, with a very variable apex (acute, obtuse, apiculate or mucronate), not exceeding 3cm long, the terminal leaflet slightly larger than the laterals. Flowers not known.

Known only from the Ḥujarīyah region; between Najd Dhamrān and Al Qabayn, Qabayṭah; between Wadi Rujaymah and Ḥayfān; between Naqil Jubirān and Saqūl, Aḥkum; Bani Shaybah; J. Jirdad, Bani 'Umar. In all these localities it occurs as a single bush except in the last where it is frequent. It was found in open scrub on exposed rocky hills between 1200 and 1800m.

This bush is also known from South Yemen, but appears to be endemic to southwest Arabia since African plants given this name are probably a different species.

3. Rhus natalensis *Bernh. ex Krauss.* Tālab, Tawlab
R. glaucescens A. Rich.
Leaves dull grey-green, glabrous; leaflets sessile, crenate, narrowly ovate to oblong lanceolate with an attenuate base and obtuse, sometimes mucronate apex; the terminal leaflet about twice as long as the laterals, reaching up to 9cm long. Flowers yellow-green.

Common in valley forest and on scrubby hillslopes from 500 to 1500m in the central escarpment including Maḥwīt. Also known from Wadi Ḥadabah, 'Udayn; Wadi Qūsa', Madīnat al 'Abīd; Wadi Sukhnah.

4. Rhus retinorrhaea *Steud. ex Oliv.* Tālab, Tawlab
Leaves yellow-green, somewhat resinous and viscid, glabrous; leaflets shortly petiolate, denticulate, narrowly elliptic with an attenuate base and acuminate tip, the terminal leaflet slightly larger than the laterals, reaching 13cm long. Flowers yellow-green.

Immature plants sometimes have pubescent leaves and may be confused with *R. abyssinica*.

Common especially on cliffs and in rocky gullies along the whole escarpment from 1500 to 2800m.

2. OZOROA *Del.*

1. Ozoroa insignis (*Del.*) *O. Kuntze* Ra'd
Heeria insignis (Del.) O. Kuntze
Shrub or small tree to 4m. Leaves simple, often opposite, oblong-lanceolate with a cuneate base and an acute mucronate apex, up to 15 × 4cm; undersurface shortly tomentose, whitish and markedly paler than the upper surface, with prominent parallel lateral veins. Inflorescence a hairy terminal panicle with insignificant white flowers. Fruit a black, currant-like drupe.

Flowers in summer.

A gregarious but very local shrub occurring on scrubby slopes in a few high rainfall areas on the escarpment between 600 and 1500m; north of 'Udayn Suq; above 'Alujah, J. Raymah; on the western slopes of J. Bura' above Ḥillah and Suq As Sabt; Wadi Sukhnah.

3. PISTACIA *L.*

1. Pistacia falcata *Becc. ex Martelli*
Shrub to 3m, somewhat resinous scented. Leaves with a reddish petiole, usually paripinnate, but sometimes with a diminutive terminal leaflet, glabrous, about 15cm long. Leaflets in 4–7 pairs, subsessile, narrowly lanceolate with a long acuminate apex and cuneate base, entire, very unequal sided, falcate. Inflorescence consisting of panicles of small yellow-green flowers, the panicles clustered at the end of branchlets. Fruit a small reddish drupe.

Flowers in spring.

An uncommon plant of cliffs and scrubby slopes along the escarpment: J. Ash Shirq; between Suq Ar Rubū' and

Ad Dan, Waṣāb al 'Ālī; in various localities on J. Bura', Raymah and Milḥān; Maḥābishah; Shahārah. It is found between 1500 and 2500m, usually as isolated bushes and the only place it is at all frequent is on the northeast of J. Milḥān between Markah and Maghrabah.

4. LANNEA *A. Rich.*

1. Lannea fruticosa (*Hochst.*) *Engl.*
Shrub or small tree reaching about 3m in Yemen. Branchlets with numerous scar-like pores in the bark. Leaves clustered at the branch tips, imparipinnate, glabrous when mature, up to 25cm long; leaflets in 5–7 pairs, sessile, lanceolate with an obtuse apex, entire, almost equal sided. Inflorescence consisting of racemes clustered at the ends of the branchlets, each raceme c 10cm long; flowers small, yellow. Fruit a hard, sessile, black, capstan-shaped drupe c 8mm long.

Flowers in autumn.

Only known from 1400 to 1500m on scrubby slopes at 'Udayn (Wadi Dūr and Wadi 'Annah) and Wadi Sukhnah.

In addition to the native species two introduced *Anacardiaceae* have become prominent features of the Yemen scenery:

The Mango, **Mangifera indica** *L.*, known locally as 'Manja' or ''Amb', is a broad spreading tree with simple, glabrous dark green leaves. The flowers appear in winter and the characteristic Mango fruit is ripe in May and June. It is widely cultivated in almost all the wadis with permanent flowing water from 500 to 1500m along the escarpment. In many areas, notably the central escarpment, it appears to grow spontaneously and shows few signs of systematic cultivation. In Yemen the fruit is small (about 6cm long) with both orange skin and orange flesh. It has an excellent flavour. It is only easily available in the markets along the foothills of the escarpment.

Schinus molle *L.*, the Pepper tree, known in Yemen as 'Filfil', is a small aromatic tree 3–4m high, though sometimes attaining 12m or more. The branches, leaves, flowers and fruit are all pendulous. The leaves are (im-)paripinnate with at least 10 pairs of leaflets. The small yellow-green flowers are in panicles. The fruit becomes dull red as it ripens and bears a strong visual resemblance to pepper seed but has a sweet aromatic flavour even though smelling of pepper. It is very drought resistant and has been widely planted in recent years on the escarpment and high plateau. Mature trees, though, are rare and virtually restricted to Sana'a.

SIMAROUBACEAE

BRUCEA *J.F. Mill.*

1. Brucea antidysenterica *J.F. Mill.*
Small tree with a smooth trunk to 6m; leafy shoots clustered towards the end of branches so the tree recalls *Hyphaene thebaica* from a distance. Leaves imparipinnate, up to 5cm long, with about 9 pairs of sub-opposite leaflets which are oblong-lanceolate, 5–15cm long, densely brown-pubescent below and with unequal bases. Flowers small, yellow-green, borne in a spike-like panicle c 15cm long. Fruit an oval, bright red drupe c 10mm long.

Flowers and fruit have never been collected in Yemen.

Known from two trees in the Ja'farīyah region of Jabal Raymah: one at 1500m between 'Alujah and Sha'af and one at 1800m below Na'mah (Wood 2886).

MELIACEAE

Trees or shrubs usually with alternate pinnate leaves, lacking stipules. Flowers usually in panicles; petals free. Fruit a capsule. Usually easily recognised by the stamens which are united into a tube.

1.	Leaves simple	1. *Turraea*
	Leaves compound	**2**
2.	Leaves pinnate, hairy	3. *Trichelia*
	Leaves bipinnate, glabrous	2. *Melia*

1. TURRAEA *L.*

Trees or shrubs differing from other genera in *Meliaceae* by having simple leaves.

Leaves to 12cm long. Flowers orange	*T. holstii*
Leaves 5–15mm long. Flowers cream	*T. parvifolia*

1. Turraea holstii *Gurke*
Small tree to 3m high. Leaves shortly petiolate, up to 12 × 6cm, elliptic with a cuneate base and acute apex, glabrous, entire. Flowers pale orange, 3cm long, 1–3 in axillary cymes on a slender peduncle c 2cm long.

Only known from the scrub-filled gullies from 1500–1800m around Aṣ Ṣa'bah, J. Milḥān.

2. Turraea parvifolia *Deflers*
Shrub to 3m high often with long slender branches. Leaves almost sessile, up to 15 × 4mm, linear to oblanceolate, obtuse, glabrous, entire. Flowers cream, c 12mm long, in clusters of 1–3 on short lateral branchlets.

Locally abundant in scrub between 300 and 1200m in the southwestern foothills mostly in Maqbanah and in the Barakānī to Bani 'Umar region of the Ḥujarīyah.

Outside Yemen known from Somalia.

2. MELIA *L.*

1. Melia azedarach *L.* Muraymirah
Persian Lilac. Shrub or small tree to 6m developing with age long drooping branches which become leafless except at the apex. Leaves bipinnate, glabrous; leaflets 2.5–8cm long, coarsely serrate, acuminate. Flowers in large panicles, very fragrant, pale lilac. Fruit a hard brown drupe whose seeds are poisonous.

Commonly planted by villages from sea level to 1500m, sometimes self-sown or at least appearing to be so.

3. TRICHELIA *P. Browne*

1. Trichelia emetica *Vahl* Ruqa'

T. roka (Forssk.) Chiov.

Tree to 12m with a broad dark green spreading crown, densely pubescent in almost all parts except the old wood. Leaves up to 35cm long, pinnate with 4–5 pairs of leaflets, tomentose below; leaflets subsessile, oblong, up to 7 × 3cm, but the terminal leaflet commonly obovate. Flowers in short axillary racemes, white, c 8mm long. Capsule globose, very hard, c 15mm long; seeds red.

Flowers in early spring.

The bark contains an emetic and the wood is both flexible and strong.

A mostly riparian species which is frequent along the whole western escarpment between 500 and 1500m, growing by permanently flowing wadis, and more rarely in alluvial plain as around Al 'Ashshah.

In addition **Azadirachta indica** *A. Juss.* is commonly planted around the larger Tihama towns notably Hodeida, Zabīd and Bayt al Faqīh. It has pinnate leaves with serrate acuminate leaflets similar to those of *Melia azedarach*. Its flowers are white and fragrant, but are rarely observed in Yemen.

RUTACEAE

Trees, shrubs or herbs with gland-dotted aromatic leaves without stipules. Flowers in terminal inflorescences, regular, bisexual, greenish-yellow. Fruit various.

1.	Tree or shrub	3. *Teclea*
	Herbs	2
2.	Leaves simple. Petals entire	2. *Haplophyllum*
	Leaves pinnatisect. Petals ciliate fringed	1. *Ruta*

1. RUTA *L.*

1. Ruta chalepensis *L.* Shadhab

Rue. Erect, glabrous, foetid, branched, glaucous perennial to about 50cm. Leaves alternate, bipinnatisect, the segments oblanceolate. Inflorescence a terminal cyme of yellow flowers; petals 4, ciliate margined. Fruit a 4-lobed glabrous capsule about 7mm long.

Cultivated throughout Yemen as a potherb. It is much esteemed for its fragrance by women who place small bunches of the plant behind their ears. The smell of rue is very typical of women's parties. It also appears to be naturalised or perhaps native on rocks on a few high mountains between 2500 and 2800m notably on J. Ṣabir, J. Sāmi' (Clissold) and on J. Ḥuzār and J. Bārad near Kusmah.

2. HAPLOPHYLLUM *A. Juss.*

1. Haplophyllum tuberculatum (*Forssk.*) *A. Juss.*

Ruta tuberculata Forssk.

R. obovata (Steud.) O. Schwartz

Erect or ascending, foetid, branched, glaucous perennial up to 70cm, covered in numerous wart-like glands. Leaves simple, crenate, variable in shape but in Yemen narrowly oblanceolate, up to 3cm long. Inflorescence terminal, many-flowered, consisting of forked cymes; petals usually small, yellow, entire. Capsule 5-lobed, c 1.5mm long.

Widespread but scattered and in small quantity on the high plateau, growing on roadsides, derelict fields and sandy plains; it also occurs in the Mashriq but is apparently rare and known only from Ḥarf Sufyān, the Fardah pass in Nihm and Mārib. It grows between 1100 and 2600m.

3. TECLEA *Del.*

1. Teclea nobilis *Del.* Ḍarīm, Dharīm

In Yemen a shrub to 2m but in Africa often a tall forest tree; bark grey. Leaves alternate, petiolate, dark green, ± glabrous, gland-dotted and slightly aromatic when crushed, 3-foliolate; leaflets entire, lanceolate to narrowly elliptic, acute, 8–15 × 2.5–3.5cm. Flowers small, yellow, in large erect panicles. Fruit an ovoid drupe, eventually becoming red, c 5–7mm long.

Flowers in spring.

Frequent between 700 and 2100m in areas of average to good rainfall along the escarpment from Ḥayfān and Turbah north to the Wadi Lā'ah; unrecorded further north. It is usually found in small quantities on cliffs, along watercourses and on open scrubby hills and is most frequent in the Ibb-'Udayn region and the central escarpment.

Various species of *Citrus* are grown in Yemen but *Citrus* cultivation is not very widespread possibly because of zinc deficiency in the soil. The taxonomy of *Citrus* is very complicated and the following account is a tentative outline only. All *Citrus* species in Yemen are shrubs with dark leathery fragrant leaves, often spiny at least when young. The flowers are fragrant, white or partly suffused with pink. They are most readily identified by their fruit.

Much the commonest species in Yemen is the Lime, **Citrus aurantifolia** (*Christm.*) *Swingle*, known as 'Līm Ḥāmidh'. Within this species I include the plant called by Schwartz *C. limonia* Osbeck var. *pusilla* Risso. Schweinfurth (1895–99) has a very interesting discussion on this plant which has been neglected in later monographic treatments of *Citrus*. He considered it to be the form from which the cultivated lemon was developed. Forsskal claimed that it grew wild in Yemen, but there has been no subsequent confirmation of this. The small orbicular fruit has a green rind and white flesh tasting of lemon. It is widely cultivated on the escarpment and particularly in irrigated valleys in the foothills. The juice is used for making drinks and for flavouring salads etc.

None of the other species cultivated are important economically:

The Citron, **Citrus medica** *L.* known as 'Trūnj' has a large fruit with a thick orange rind. It is sold in most of the larger markets, the pith is eaten and the juice made into drinks. The Lemon, **Citrus limon** (*L.*) *Burm. f.*, known as 'Līmūn', and the Sweet Lemon, **Citrus limetta** *Risso* (*C. limonia* Osbeck var. *dulcis* Moris.) appear rarely in Taiz market. The latter was also collected by Schweinfurth on J. Milḥān. The Orange, **Citrus sinensis** (*L.*) *Osbeck* known as 'Līm Ḥālī' or 'Burtaqal' is occasionally cultivated. The rind of the Yemen orange is mottled green and it has a sharper taste than imported fruit. The Tangerine, **Citrus deliciosa** *Ten.*, known as 'Yūsafī', is cultivated in Wadi Ḍabāb near Taiz and perhaps elsewhere. The Yemen plant has a green rind when ripe. The Seville Orange, **Citrus aurantium** *L.* is sporadically cultivated and is reputedly used for making drinks. Large quantities of this species appear in Raḥidah market every autumn, but I have never seen anyone use them and their price always seems unrealistically high.

BALANITACEAE

BALANITES *Del.*

1. Balanites aegyptiaca (*L.*) *Del.* *Ḥalaj*

Tree to 6m with a rounded crown composed of a tangled mass of spiny branches. Bark grey. Branchlets green with straight spines up to 2cm long. Leaves shortly petiolate, grey-green, glabrous, obovate, up to 4 × 3cm, but commonly much less. Flowers in small axillary clusters, yellow-green. Fruit an edible drupe which becomes yellow when ripe.

The most conspicuous and abundant shade tree in the Tihama between the Wadi Zabīd and the Wadi Surdūd but unaccountably rare or absent outside this area though occurring sporadically in the escarpment foothills and reaching 1200m in the Ḥujarīyah; at higher altitudes it always grows as a bush.

The seeds of *Balanites* are remarkable in that they contain a poison that can kill the snails that act as a host for the Bilharzia parasite.

ZYGOPHYLLACEAE

A family of herbs and undershrubs usually with opposite leaves, stipules present. Flowers regular usually with 5 free petals and 10 stamens. Fruit a capsule or consisting of hard mericarps.

The family is characteristic of desert conditions and most species are rare or absent on the escarpment.

1. Stipules transformed into spines. Flowers pink — 3. *Fagonia*
   Stipules not transformed into spines. Flowers white or yellow — 2
2. Leaves pinnate. Fruit consisting of hard, commonly spiny or winged mericarps — 1. *Tribulus*
   Leaves various but not pinnate. Fruit an unarmed capsule — 3
3. Leaves alternate, mostly irregularly forked, not fleshy — 4. *Peganum*
   Leaves opposite, simple or 1–2-foliolate, terete, succulent — 2. *Zygophyllum*

1. TRIBULUS *L.*

Herbs with opposite, pinnate leaves, usually prostrate. Flowers solitary in the leaf axils. Petals 5, yellow. Fruit consisting of 4–5 hard mericarps.

***Tribulus* is a difficult genus and a lot more collecting needs to be done before all problems can be settled. Species with winged fruit are virtually restricted to mobile sands in Yemen and are little known. Species with unwinged fruit are weeds of arable land and waste places. In this group there is some correlation between the species' vigour, the number of leaflets, stamens and spines and the length of the petals and pedicels. Most plants fall into one of the described species with little difficulty but intermediates do occur, notably the aptly named *Tribulus intermedius* Kralik, collected by Botta at Janad near Taiz which appears to combine characters of *T. terrestris* and *T. bimucronatus*. The status of such intermediates is not at present known.**

Most species are characteristic desert plants: annuals, with a long taproot which appear in great abundance after the rains. They are widely valued in traditional medicine as a cure for kidney stones.

1. Mericarps distinctly winged — 2
   Mericarps without wings — 4
2. Petals small, c 4mm long. Fruit with sharply dentate spinescent wings — *T. pentandrus*
   Petals large, c 10–16mm long. Fruit with crenate but not spinescent wings — 3
3. Erect plant. Fruit ± globose, c 8mm across, borne on pedicels 10–15mm long — *T. arabicus*
   Prostrate plant. Fruit obliquely ellipsoid, much broader than long, borne on pedicels mostly 15–22mm long — *T. omanensis*
4. Stamens 10 — 5
   Stamens 5 — 6
5. Mericarps armed with 4 spines — *T. terrestris*
   Mericarps unarmed but muricate — *T. mollis*
6. Mericarps armed with 4 spines — *T. parvispinus*
   Mericarps unarmed or with 2 weak spines — *T. bimucronatus*

1. Tribulus arabicus *H. Hosni* *Zaḥar*

Erect, densely villous herb (? annual) to 1m. Leaves with 7 pairs of rather crowded leaflets. Flowers large; pedicels c 12mm long; petals 10–12mm (–16)mm long; stamens 10. Fruit with broad crenate-margined wings, spineless, ± globose, c 8mm diameter.

Known in Yemen only from the type collection made in the lower Wadi Najrān (Zeller 19401) and said by the collector to be within Yemen.

2. Tribulus bimucronatus *Viv.* Quṭayb

T. spurius Kralik

Prostrate, grey-green annual with mixed short and long pubescent hairs. Leaves with 5–6 pairs of leaflets. Flowers small; pedicels 2–5mm long; petals c 4mm long; stamens 5. Fruit villous, unwinged, pyramidal, c 5mm long.

This species has two subspecies. In subspecies **bimucronatus** each mericarp has 2, often rather soft spines. This is probably the common weedy *Tribulus* in the Tihama and has been found at Zabīd (Wood 74/439), between Khalīfah and J. Bura' (Wood 3430, 3433) and at Zuḥrah (Miller 313). In subspecies **inermis** (*Kralik*) *H. Hosni*, the mericarps are unarmed. This subspecies has not yet been recorded in Yemen, but grows at Jīzān and will almost certainly be found in the Tihama in the future.

3. Tribulus mollis *Ehrenb. ex Schweinf.* Quṭayb
T. ochroleucus Maire
Green, prostrate, usually softly hairy annual. Leaves with 5–7 pairs of leaflets. Flowers medium sized; pedicels c 6–9mm long; petals rather broad, c 6mm long; stamens 10. Fruit villous, muricate but without wings or spines, pyramidal, c 7mm long.

Only known from Ḥays where it was found by Ahmed Khattab.

4. Tribulus omanensis *H. Hosni ex Hadidi* Zaḥar
Prostrate green annual with very long villous stems. Leaves mostly with 5–6 pairs of leaflets. Flowers large; pedicels very long, up to 22mm; petals c 15mm long; stamens 10. Fruit sparsely pilose with broad, ± crenate wings, obliquely ellipsoid, c 16mm long and 10mm broad.

Restricted to the Sands in Yemen and known definitely only from the lower Wadi Najrān (Zeller 19004) and the Mārib area (Wood 75/327), but probably common. It occurred in enormous abundance after the heavy autumn rains of 1977 in the deserts south of Mārib.

5. Tribulus parvispinus *Presl* Quṭayb
T. pentandrus sensu Schwartz
Prostrate, green, pubescent annual similar to *T. terrestris*, but smaller and more delicate in all parts. Leaves usually with 5 pairs of leaflets. Flowers small; pedicels c 1–2mm long; petals c 3mm long; stamens 5. Fruit sparsely pubescent, unwinged, ± globose, 4–6mm diameter, the mericarps each with 4 spines.

The exact distribution is not known but it is probably frequent in exposed habitats such as denuded hillsides, earth banks and gravel plains throughout Yemen from 200 to 2300m, but appears to be absent from sand. (Wood 75/85, 3471, Miller and Long 3514)

6. Tribulus pentandrus *Forssk.* Quṭayb
T. alatus Del.
T. longipetalus Viv.
Prostrate grey annual densely covered in long and short hairs. Leaves with 5–6 pairs of leaflets. Flowers small; pedicels 1–3mm long; petals c 4mm long; stamens variable in number, 5–8. Fruit densely pubescent, with sharply dentate wings, the teeth spinescent, ± globose, 5–6mm diameter.

Known with certainty only from Luḥayyah (Forsskal), Ḥays (Miller and Long 3514) and Mukhā (Wood 931), but probably widespread on sand in the Tihama.

7. Tribulus terrestris *L.* Quṭayb
Prostrate, green, pubescent annual with numerous long stems. Leaves with 5–6 pairs of widely spaced leaflets. Flowers medium sized; pedicels 5–6mm long; petals c 6mm long; stamens 10. Fruit variable in indumentum but usually sparsely pubescent, unwinged, obliquely ellipsoid, about 6mm long and 8–10mm broad, each mericarp armed with 4 spines.

Widespread and often abundant between 200 and 2500m in the Tihama, on the escarpment and the high plateau. It is most commonly found on paths, in gardens and villages but less commonly as a field weed. It is frequently dominant on the mud roofs of old houses.

2. ZYGOPHYLLUM *L.*

Herbs or undershrubs with fleshy, ± cylindrical, opposite leaves. Flowers white or yellow, solitary in the leaf axils. Fruit a 5-valved capsule.

1. Leaves simple. Flowers yellow ... *Z. simplex*
   Leaves compound, 1–2 foliolate. Flowers white ... 2
2. Herb. Leaves with a single leaflet at the apex of the petiole ... *Z. hamiense*
   Shrub. Leaves with a pair of opposite leaflets at the apex of the petiole ... 3
3. Plant glabrous. Fruit cylindrical ... *Z. coccineum*
   Plant tomentose. Fruit pear-shaped, the lobes divergent above ... *Z. album*

1. Zygophyllum album *L.f.* Ḥumaym
Grey-green, tomentose, ascending or decumbent, much-branched undershrub to 1m. Leaves terete, fleshy with 2, c 5–7mm long, opposite leaflets at the apex of the fleshy petiole. Flowers white, small, c 5mm long. Fruit c 8mm long, deeply 5-lobed, the lobes divergent.

The Yemen plant is variety **amblyocarpum** (*Baker*) *Hadidi* which differs from the type in having the fruit very deeply and acutely lobed.

Only known from sandy beaches around Mawshij and Khawkah where it is abundant. (Wood 75/228, 1524).

2. Zygophyllum coccineum *L.*
Very similar in habit to the previous species but green in colour and completely glabrous. The leaves are slightly larger with the leaflets about 10mm long. The fruit is an unlobed, angled but scarcely winged cylindrical capsule about 6–10mm long.

Only known from coastal sands at Mīdī where it is abundant.

3. Zygophyllum hamiense *Schweinf.*
Grey-green, tomentose, procumbent perennial herb. Leaves terete, fleshy with a single leaflet about 5–10mm long at the apex of the fleshy petiole. Flowers white, c 8mm long. Fruit very similar to that of the last species – an unlobed, angled, cylindrical capsule about 10mm long.

Locally abundant between 1100 and 2000m on stony and gravelly plain in Nihm and Sirwāḥ districts extending north to the Jawf and 'Ujli, south of Ṣa'dah (Ellenberg). (Wood 2328).

4. Zygophyllum simplex *L.* Kirmil
Green, glabrous, very succulent, procumbent (? annual) herb. Leaves simple, cylindrical, sessile, c 1cm long. Flowers yellow, c 5mm long, usually very numerous and conspicuous.

Fruit a small, deeply 5-lobed capsule c 3–4mm long.

Abundant on fixed sand and gravel near the sea in the Tihama (also on Ḥanish Island (Slade)), apparently very tolerant of salt spray. Widespread but much less common inland in the Tihama, decreasing in frequency eastwards. It also occurs rarely in the Mashriq up to 2300m, perhaps indicating the presence of salt, and has been recorded at the head of the Wadi Khārid, in Bani al Ḥarith, at Marāshī (Heckel and D. Wood) and at Mārib.

3. FAGONIA *L.*

Perennial or annual herbs typically dark green in colour. Leaves opposite, 1- or 3-foliolate, the stipules transformed into spines. Flowers pink. Fruit a deeply ridged, 5-carpelled capsule.

This genus has been confused taxonomically in the past and I am not convinced that the situation is stable yet. The species are superficially very similar and are distinguished from each other in the main by a single character.

All species are known as 'Ḥil', 'Siḥah', 'Shuwayk' or 'Ḥilwah'.

1.	Stem ± quadrangular, the sides deeply sulcate. Spines and internodes very short, usually less than 1cm long	*F. bruguieri*
	Stem rounded, striate. Spines and internodes usually exceeding 1cm in length	2
2.	Leaves 3-foliolate	*F. schweinfurthii*
	Leaves simple	3
3.	Leaves linear, c 1.5mm wide	*F. indica*
	Leaves ovate, c 5mm wide	*F. ovalifolia*

1. Fagonia bruguieri *DC.*

Low, often ± procumbent, densely glandular-hairy perennial up to c 10cm high. Stem quadrangular, deeply sulcate. Leaves 3-foliolate below, 1-foliolate above. Spines and internodes short, usually less than 1cm long. Petals c 4mm long. Capsule shortly pedicellate, c 3mm broad, 4mm long.

Usually easily recognised by its dense appearance caused by the short internodes.

Locally common on exposed stony plain and hillsides, apparently with a preference for volcanic lavas and thus common around Rada' and in Ḥamdān-Arḥab. It has been found between 2000 and 2500m mainly along the border of the high plateau and the Mashriq from Dhī Nā'im north to the Ṣa'dah region. (Wood 74/348)

2. Fagonia indica *Burm. f.*

F. parviflora Boiss.
F. arabica sensu Schwartz
? *F. bruguieri* sensu Schweinfurth

Low, sparsely glandular-hairy perennial to c 20cm, varying in habit from procumbent to ascending. Stem cylindrical, striate. Leaves unifoliolate, usually with very short petioles about 1mm long. Spines and internodes mostly 1–2.5cm long. Petals c 5–6mm long. Capsule on a short recurved pedicel, c 4 × 4mm.

There is a good deal of variation particularly in petiole length and indumentum. Most plants are quite glandular (e.g. Wood 2678), but some, notably Wood 2763, are very sparsely glandular giving the plant a different, more green appearance.

Widespread and often common from sea level to 3000m in the Tihama, the drier parts of the escarpment, the high plateau and the Mashriq. It usually grows on stony steppe, rocky hillsides and alluvial plain and, though occurring on sand at Mīdī, Qanāwis etc., it is not a feature of the sand dune flora either in the Tihama or the Sands.

3. Fagonia ovalifolia *Hadidi*

Very similar to the previous species from which it differs principally in having ovate leaves. In addition the leaves are always long petiolate and the flowers and fruit are usually somewhat larger. It may be annual.

Only known from sandy plain at Jūbah at 1100m (Wood 3156). This species is otherwise known from eastern Arabia and it is only likely to occur in Yemen at the edge of the Sands.

4. Fagonia schweinfurthii (*Hadidi*) *Hadidi*

? *F. bruguieri* DC. var. *ehrenbergii* Schweinf.

Very similar to *F. indica* from which it differs in having trifoliolate leaves.

This species has the same distribution as *F. indica* with which it often grows and is in my opinion doubtfully distinct (Wood 74/349, 2056).

Fagonia khattabii Hadidi is listed in Khattab and Hadidi's *Results of a Botanic Expedition to Arabia in 1944–5* from Mīdī. The plant was never actually described and Professor Hadidi tells me that the specimen has been mislaid. The species cannot therefore be accepted as part of the Yemen flora.

4. PEGANUM *L.*

1. Peganum harmala *L.* Ḥarmil, Ḥurumil

Erect, glabrous, dark green perennial to 50cm, much branched near the base. Leaves 1–3 together in alternate clusters, simple or irregularly forked, the lobes linear – lanceolate, acute. Flowers solitary, terminal, large, white, c 15mm long, overtopped by the linear sepals. Fruit a 3-lobed capsule.

One of the most characteristic plants of the high plateau and higher parts of the Mashriq growing between 1700 and 2500m but absent south of the Dhamār – Al Bayḍā' road. There is also an isolated record from the escarpment on the limestone dip slope on the east of J. Sharafayn. It grows on roadsides, field borders and waste ground, on alluvial plain and, more rarely, on rocky hillsides and is abundant around Sana'a, Ṣa'dah and the other towns of the interior.

OXALIDACEAE

OXALIS *L.*

Delicate herbs with trifoliolate leaves which are sensitive to light.

The following two species are very similar and have been confused so their exact distribution is not known. In an aggregate sense they are widespread and common between 1000 and 3000m on the escarpment and high plateau, growing in open rocky ground such as wadi beds, terrace walls, tracksides and stony hills.

Stipules present, joined to the base of the petiole *O. corniculata*
Stipules absent *O. radicosa*

1. Oxalis corniculata *L.* Ḥamaḍmaḍ

Pubescent creeping perennial herb with long procumbent brown stems which often root at the nodes. Leaves alternate, stipulate, 3-foliolate; leaflets obcordate, up to 1cm long. Flowers solitary or in small umbels arising from the leaf axils; petals pale yellow, 4–7mm long. Capsule pod-like, 5–15mm long.

Known with certainty only from the escarpment and probably less common than the following species, occurring mostly as a weed of cultivation. (Wood 74/115)

2. Oxalis radicosa *A. Rich.* Ḥamaḍmaḍ

Differs from the last species in always having very slender stems which do not root at the nodes. The leaves are without stipules.

Known with certainty only from the high plateau but probably the commonest species in Yemen growing in dry stony habitats (Acres 374, 403).

GERANIACEAE

A family of herbs with stipulate, usually lobed or divided leaves. Flowers in umbels or cymes with rather delicate white or pinkish-mauve petals. Sepals and petals 5, free. Fruit of mericarps attached to a long beak formed from the persistent accrescent style.

1. Leaves palmately divided 2
   Leaves pinnately lobed or, if unlobed, then pinnately veined 3
2. Flowers regular. Sepals without spurs 2. *Geranium*
   Flowers irregular. Posterior sepal with a spur fused to the pedicel 4. *Pelargonium*
3. Stamens 15. Leaves unlobed. Beak of fruit plumose 1. *Monsonia*
   Stamens 5. Leaves lobed. Beak of fruit not plumose 3. *Erodium*

1. MONSONIA *L.*

Perennial herbs with unlobed pinnately veined leaves. Our species grow in semi-desert conditions and can be readily recognised by the untwisted plumose beak to the fruit.

Main veins deeply impressed. Beak 4–5cm long *M. nivea*
Main veins not impressed. Beak 6–8cm long *M. heliotropoides*

1. Monsonia heliotropoides (*Cav.*) *Boiss.*

Erect herb of grey-green appearance, stem stout and woody below. Leaves long-petiolate, broadly ovate, c 2.5–3 × 2cm, veins not impressed, margin coarsely dentate but not lobed; the petiole and undersurface of leaf villous but not silvery canescent. Flowers in long pedunculate umbels, pink. Fruit beak 6–8cm long, plumose.

A rare plant of sandy and stony plain found between 1100 and 2500m in the Mashriq, usually appearing only after heavy rains; between Ṣa'dah and J. Minmār; Al 'Ashshah; between Ma'mar and 'Amrān. (Wood 75/612)

2. Monsonia nivea (*Decne.*) *Decne. ex Webb*

Very similar to the previous species with which it is easily confused but silvery not grey-green in appearance, the leaves narrowly ovate with the main veins deeply impressed and with the beak of the fruit only 4–5cm long.

Only known from Maqsalah near Majz at 1900m (Steinberg).

2. GERANIUM *L.*

Slender herbs with palmately divided leaves which are ± orbicular in outline. Flowers in cymes, each with 10 fertile stamens. Beak of fruit neither twisted nor plumose.

1. Leaves palmately divided to the base. Petals clawed. *G. robertianum*
   Leaves deeply palmately lobed but never to the base. Petals without a claw 2
2. Inflorescence glandular. Flowers mauve with a dark centre. Mericarps glabrous 3
   Inflorescence eglandular. Flowers white. Mericarps pilose *G. arabicum*
3. Mericarps grey, 1.5–2mm long, reticulate but not winged *G. ocellatum*
   Mericarps orange, 4–5mm long, with dentate wings 4
4. Mericarps with 2 recurved horn-like appendages at the apex *G. biuncinatum*
   Mericarps without horn-like appendages at the apex *G. trilophum*

1. Geranium arabicum *Forssk.*

G. simense Hochst. ex A. Rich.

Perennial herb with procumbent reddish stoloniferous stems. Leaves orbicular in outline, c 3cm diameter, deeply palmately 5-lobed, the lobes also divided. Inflorescence pubescent, but not glandular; petals c 7mm long, claw absent, white with pink veins. Mericarps c 3–4mm long, smooth, pilose, brown, remaining attached to the beak-like style at maturity.

Widespread on the escarpment between 1600 and 3000m, generally common and showing a marked preference for damp, shaded places such as cliff gullies, stream sides and terrace walls.

2. Geranium biuncinatum *Kokwaro*

G. yemense sensu Schwartz

Annual herb with erect or ascending stems. Leaves orbicular in outline, c 4–5cm diameter, deeply palmately 5-lobed, the lobes also divided. Inflorescence glandular-pilose; flowers mauve with a dark centre, the petals c 10mm long, claw absent. Mericarps c 4–5mm long, orange, glabrous with strongly dentate wings and 2 curved horn-like appendages at the apex, becoming detached from the beak-like style at maturity.

A characteristic weed of coffee terraces in the central escarpment which occurs rarely elsewhere on the escarpment between 900 and 2500m.

3. Geranium ocellatum *Camb.*

G. muscatense Boiss.

Very similar to *G. biuncinatum* from which it differs in having grey, strongly reticulate mericarps, c 2mm long, which lack both dentate wings and apical horns.

A local field weed growing between 2000 and 2900m on the escarpment and high plateau; its occurrence appears to be very sporadic except in the Ibb area where it is a constant and frequent weed of sorghum fields.

4. Geranium robertianum *L.*

Herb Robert. Strongly aromatic annual or biennial herb with procumbent or ascending reddish stems. Leaves divided to the base and appearing compound, the 3 (–5) divisions 2-pinnatisect. Inflorescence glandular; petals including the distinct pale claw, 10–12mm long, pink. Mericarps c 2.5mm long, weakly reticulate, brown, glabrous except for the silky threads by which alone they remain attached to the beak-like style at maturity.

A very rare plant of shaded gullies at 2800–2900m on the escarpment: J. Bārad, Kusmah in Raymah; Shirnamah, J. Maswār.

This widespread plant in the temperate regions of the Old World appears to have its southernmost outpost in Yemen.

5. Geranium trilophum *Boiss.*

G. yemense Deflers

Identical to *G. biuncinatum* except for the mericarps which lack the two apical horns found in that species.

Like *G. biuncinatum*, with which it often grows, it is a characteristic weed of coffee terraces but also occurs in a few other scattered localities on the escarpment between 1200 and 1800m.

3. ERODIUM *L.Herit.*

Herbs with pinnate or pinnatisect leaves. Flowers in umbels each with 5 stamens. Beak of fruit not plumose but becoming spirally twisted as the fruit ripens.

Leaves pinnate — *E. cicutarium*
Leaves simple, usually shallowly pinnatisect — *E. malacoides*

1. Erodium cicutarium (*L.*) *L'Herit. in Aiton*

Usually prostrate annual herb with several procumbent stems. Leaves pinnate with deeply pinnatifid leaflets. Umbels usually with 4–5 pink flowers. Fruit beak c 3–5cm long.

A local plant growing as a weed of damp fields and tracks, between 2300 and 3100m on the higher mountains of the escarpment and high plateau region.

2. Erodium malacoides (*L.*) *L'Herit. in Aiton*

Similar in general appearance to the previous species but more vigorous and sometimes erect, the leaves simple, coarsely dentate to pinnatisect and the fruit beak only 1–2cm long.

Like *E. cicutarium* this species grows on and around the higher mountains where it has been found between 2100 and 3000m. It is also a weed of fields and waste places but perhaps shows a preference for nitrogen-rich habitats in villages.

4. PELARGONIUM *L'Herit.*

1. Pelargonium alchemilloides (*L.*) *Ait.*

P. multibracteatum A. Rich.

Vigorous pubescent herb to c 50cm. Leaves orbicular in outline, palmately 5–7 lobed. Flowers in long pedunculate 5–8 flowered umbels; posterior sepal with a spur fused to the pedicel; petals white, c 2cm long; fertile stamens usually 5 or 7. Fruit beak twisting spirally when ripe, c 3cm long.

The Yemen plant is subspecies **multibracteatum** (*Hochst. ex A. Rich.*) *Kokwaro*.

Widespread on the escarpment and high plateau between 1400 and 2900m occurring frequently but sporadically on steep slopes in mountain gullies, on terrace walls and in scrubby grassland. It appears to be most common on the western slopes of the central escarpment where there is plentiful mist.

Several other species of 'Geranium' are cultivated, particularly in Sana'a, but much the most frequent is **Pelargonium hybridum** (*L.*) *L'Herit.*, a vigorous species with red flowers and mottled leaves. This is occasionally found in village gardens as well as in the major towns above 2000m.

TROPAEOLACEAE

The Nasturtium, **Tropaeolum majus** *L.* is commonly cultivated in Sana'a. It is a glabrous herb with orbicular, peltate leaves and attractive red, orange or yellow spurred flowers.

ARALIACEAE

CUSSONIA *Thunb.*

1. Cussonia holstii *Harms ex Engl.*

Khawnab, 'Uām, 'Ujām

Small tree to about 12m but usually much less in Yemen. Leaves digitate, clustered at the end of branchlets so giving the tree a characteristic appearance rather like that of a branched pawpaw. Leaflets usually 5, petiolate, glabrous, dark green above but markedly paler beneath, broadly ovate, up to c 12 × 6cm; apex long-attenuate; margin coarsely crenate; base unequal, truncate. Flowers minute, yellow-green, massed in 15–25cm long, thick, flower spikes clustered at the tips of the branchlets. Fruit small, ovoid, dull yellow.

The tree is usually covered in ants who probably come in search of a resin which is freely exuded from the branchlets.

A very local plant of a few places in high rainfall areas along the escarpment from 1600m to 2500m on the very summit of J. Milḥān: below Hasabān, J. Ṣabir; in gullies at the western foot of J. Ba'dān, just north of Ibb; on the Ibb-'Udayn pass; Bani Mukhtār, above Ḥadīyah; Wadi Sukhnah; Bayt Shamrān, J. Masār, Ḥarāz; between Bayt Adhāqah and Suq Ath Thalūth, Wadi Lā'ah. It appears to have been an important constituent of highland forest in the central escarpment together with species such as *Acacia origena*, *Pittosporum* etc., as testified by its relative abundance around Al Markah, J. Milḥān.

UMBELLIFERAE

Herbs or rarely shrubs, often aromatic. Leaves alternate, usually without stipules, mostly pinnately or palmately divided. Flowers usually bisexual, in umbels, which are often compound, the primary rays generally subtended by bracts and the secondary by bracteoles. Petals free, usually white or yellow. Fruit dry, consisting of two mericarps joined to a common axis, often ribbed or winged.

A very distinctive family, centred on the temperate regions of the world. Many of the species in Yemen are widely distributed throughout Europe (*Conium*, *Anthriscus*, *Apium* etc.) and there are comparatively few plants with an African distribution. One or two species are endemic but they belong to critical genera.

1. Leaves orbicular in outline. Flowers in small head-like umbels **2**
   Leaves various but not orbicular in outline. Umbels rarely head-like **3**
2. Leaves lobed 1. *Hydrocotyle*
   Leaves unlobed 2. *Centella*
3. Shrub 6. *Heteromorpha*
   Herbs **4**
4. Leaves palmately lobed to or nearly to the base 3. *Sanicula*
   Leaves ± pinnately lobed, never palmately lobed **5**
5. Flowers yellow **6**
   Flowers white, sometimes pink tinged or blackish **8**
6. Leaf segments capillary, hair-like **7**
   Leaf segments flattened, not hair-like 15. *Peucedanum*
7. Robust plant to 2½m. Fruit winged 14. *Ferula*
   Erect plant to 1m. Fruit unwinged 13. *Foeniculum*
8. Basal leaves (and commonly stem leaves) simply pinnate **9**
   Basal leaves 2 or more pinnate **11**
9. Umbels with conspicuous bracts 9. *Berula*
   Umbels without bracts **10**
10. Umbels mostly leaf opposed, sessile or on very short peduncles c 1cm long 7. *Apium*
    Umbels mostly terminal, long pedunculate, the peduncles 2cm long or commonly much more 10. *Pimpinella*
11. Fruits spiny or hairy or both. Stems hairy **12**
    Fruits glabrous. Stems glabrous or nearly so **14**
12. Leaves linear, the margins involute. Fruit not spiny 4. *Pycnocycla*
    Leaves with flattened segments. Fruit spiny **13**
13. Umbels head-like. Petals blackish 16. *Agrocharis*
    Umbels not head-like. Petals pinkish 17. *Torilis*
14. Bracts present (sometimes very small) **15**
    Bracts absent 12. *Anthriscus*
15. Fruit ridged, not warty. Bracts without a broad scarious sheathing base, simple or if forked c 2cm long **16**
    Fruit unridged, covered in numerous warty protuberances. Bracts forked or simple c 1cm long, always with a broad scarious base 11. *Oreoschimperella*
16. Bracts forked c 2cm long 8. *Ammi*
    Bracts undivided c 5mm long 5. *Conium*

1. HYDROCOTYLE *L.*

Creeping perennial herbs growing in aquatic habitats. Leaves simple, orbicular, lobed with (in our species) a deep basal sinus. Flowers in small, pedunculate, ± capitate umbels. Fruit ridged. Compressed.

Leaf petiole stout, c 2–3mm broad and commonly exceeding 20cm long, quite glabrous *H. ranunculoides*
Leaf petiole slender, less than 1mm broad, rather short, not exceeding 5cm long, ± pilose at the top *H. sibthorpioides*

1. Hydrocotyle ranunculoides *L.f.*

Glabrous plant with creeping stems, rooting at the nodes. Leaves with fleshy erect petioles 2–3mm broad and 15–35cm long, dark shiny green, ± orbicular, lobed, c 5–6cm in diameter, the margins crenate. Umbels 5–10 flowered, on 2–3cm long peduncles arising in the leaf axils.

An uncommon plant growing in swamp between 1500 and 2600m, mostly on the high plateau; Suq As Sabt, Ibb; between Yarīm and Raḍmah; near Al Zīl, Al Ḥadā'; Wadi Ḥabab between Madinat Al 'Abid and Hammām Ali; Ribāṭ, south of Yarīm. It is usually found growing in water between taller rushes and sedges.

2. Hydrocotyle sibthorpioides *Lam.*
Very similar to the previous species but more slender in all parts. Leaves less than 1cm in diameter, the petiole less than 1mm broad and only about 4–5cm long. Umbels 3–8 flowered.

A very rare plant growing in swamp at the edge of seasonal pools from 2600 to 3000m: J. Manār, Ibb; at the head of the Wadi Banā near Ribāṭ; Kawkabān.

2. CENTELLA *L.*

1. Centella asiatica (*L.*) *Urb.*
Thinly pubescent, prostrate, stoloniferous perennial herb. Leaves ± orbicular with a deep basal sinus c 1.5–2cm diameter, the margin crenate. Umbels head-like with 1–5 flowers, several arising on short peduncles c 3–10mm long from each leaf axil. Fruit reddish, faintly ridged, compressed.

Essentially a plant of springs and marsh on the high plateau where it is common in suitable habitats. It occurs rarely on the escarpment in Wadi Ḥabab (Madīnat Al 'Abīd – Hammām 'Ali), at Ibb and around Turbah. At Turbah, as also in its two known locations in the Mashriq ('Ujli, south of Ṣa'dah and at the head of the Wadi Khārid, Arḥab) it grows by springs in association with *Blumea bovei*. It is found between 1500 and 2700m.

3. SANICULA *L.*

1. Sanicula elata *Buch.-Ham. ex D. Don*
S. europaea sensu Schwartz.
Erect glabrous herb to c 30cm. Leaves rosetted, deeply palmately divided almost to the base, c 2–4cm long, the petiole long and reddish. Flowers white or pinkish in small few-flowered umbels; bracts 2, usually 3-lobed; bracteoles present. Fruit ovoid, covered in reddish hooked bristles.

A rare plant of stream sides between 1900 and 2800m in high rainfall areas on the escarpment: J. Ba'dān, Ibb; Karāwin spring near Manākhah (Schweinfurth); Al Jabin, Kusmah and Jirjir, J. Raymah.

4. PYCNOCYCLA *Lindl.*

1. Pycnocycla glauca *Lindl.*
Prostrate, ascending or erect perennial to 30cm. Leaves glabrous, glaucous when fresh, bipinnate, the segments linear with involute margins, often forked. Umbels with very short rays, often ± capitate, on a 10–20cm long, unbranched, pubescent peduncle, which is leafless except below; petals white or pinkish; bracts and bracteoles subulate. Fruit cylindrical, villous.

Widely distributed on the escarpment from 1500 to 3000m from J. Ṣabir north to Shahārah, but erratic in occurrence, apparently avoiding areas of very high rainfall and being most typical of the inner escarpment, its distribution closely parallelling that of *Aloe tomentosa*. It is found on open rocky slopes and in sparse stony grassland.

5. CONIUM *L.*

1. Conium maculatum *L.*
Hemlock. Erect foetid biennial to c 1½m in Yemen, stem glabrous, often purple spotted. Leaves bright green, glabrous, rather large, the lower leaves often reaching 50cm long, bi-quadripinnate, the segments coarsely serrate. Umbels large, terminal, usually puberulent; petals white, bracts narrowly triangular. Fruit glabrous, cylindrical, slightly compressed, ridged.

The whole plant is extremely poisonous.

A local plant occurring in organically rich habitats such as sewage outlets around villages on and near high-altitude escarpment mountains. It grows on J. Ṣabir and the mountains round Ibb and from J. Nabī Shu'ayb north to J. Maswār. It descends to 2100m at Ibb but reaches 3600m on J. Nabī Shu'ayb.

6. HETEROMORPHA *Cham. & Schlecht.*

1. Heteromorpha arborescens (*Spreng.*) *Cham. & Schlecht.*
Glabrous shrub to 3m, bark red-brown. Leaves simple or more commonly pinnately 4(–5)-foliolate, the leaflets lanceolate, c 4–7cm long. Umbels mostly terminal, large; petals pale yellow; bracts subulate. Fruit, elliptic, winged, glabrous.

The only shrubby Umbellifer in the Yemen flora.

A characteristic species of the escarpment mountains growing from 2100 to 3100m in areas of good rainfall from Turbah north to Shahārah. It grows on cliffs and in rocky gullies.

7. APIUM *L.*

Herbs with simply pinnate, serrate leaves and mostly leaf-opposed, shortly pedunculate umbels. Bracts absent. Fruit ovoid.

Bracteoles present. Leaflets unlobed, ovate, sessile ... *A. nodiflorum*
Bracteoles absent. Leaflets usually 3-lobed, deltoid, petiolate ... *A. graveolens*

1. Apium graveolens *L.*
Wild Celery. Ascending or erect, glabrous biennial herb to c 60cm, stem deeply grooved. Leaves dark green, pinnate, the leaflets petiolate, deltoid or rhomboid and often 3-lobed. Umbels sessile or shortly pedunculate; petals greenish-white. Fruit glabrous, ridged.

The whole plant smells strongly of celery.

An uncommon plant found growing by small streams and ditches in a number of scattered localities between 1700 and 3000m: on the cliffs at Turbah; Wadi Sudān, Qā'idah; Dhī Sufāl; Rada'; Wadi Ḍahr; Kawkabān; Manākhah.

2. Apium nodiflorum (*L.*) *Lag.*
Fool's watercress. Usually procumbent, glabrous perennial herb, stems finely furrowed. Leaves shiny green, pinnate, the leaflets sessile, ovate to lanceolate, very rarely lobed. Umbels on short peduncles, 1cm or less in length; petals

white. Fruit glabrous, ridged.

A common and characteristic plant of ditches, ponds and streams in Khawlan Ash Sham, on the high plateau and the inner escarpment extending to the Ḥarāz and south through the Ibb region to Wadi Ḍabāb and Hiḍrān near Taiz; absent from other areas including the Ḥujarīyah and the outer escarpment. Found from 1300 to 3000m.

8. AMMI *L.*

1. Ammi majus *L.*

Erect glabrous annual herb to c 80cm. Leaves dull green, bi-tripinnate, usually rather small and not exceeding 5cm in length, the segments usually serrate. Umbels mostly terminal; petals white; bracts forked, up to 2cm long; bracteoles present. Fruit oblong, glabrous, ridged.

An abundant field weed on the rich alluvial plains at c 1600–2300m around Ibb, also occurring at Dhī Sufāl. Unknown elsewhere in Yemen.

9. BERULA *Koch*

1. Berula erecta (*Huds.*) *Coville*

B. thunbergii (DC.) Wolff

An erect glabrous perennial to 1m. Stem ridged. Leaves dark green, pinnate, the leaflets sessile, ovate to lanceolate, serrate and very similar in appearance to those of *Apium nodiflorum*. Umbels leaf-opposed, the peduncles 1–3cm long; petals white; bracts lanceolate, sometimes trifid, up to 8mm long, usually strongly reflexed; bracteoles present. Fruit globose, glabrous, ridged.

A local plant growing in ponds and less commonly streams between 2300 and 2900m, virtually restricted to the high plateau: Aḥjar; Ẓulā' (Deflers); above Baw'an; Al Zīl, Al Ḥadā'; below Wa'lān; common around Yarīm extending towards Raḍmah and also in the Wadi Banā between Saddah and Nādirah. It is a stoloniferous plant often spreading to completely cover shallow ponds.

10. PIMPINELLA *L.*

Short-lived erect herbs with pinnate basal leaves. Petals white. Bracts and bracteoles absent. Fruit ovoid.

Fruit glabrous. Basal leaves up to 40cm long ... *P. woodii*
Fruit hairy. Basal leaves not exceeding 10cm in length ... *P. menachensis*

1. Pimpinella menachensis *Schweinf. ex Wolff*

P. hirtella sensu Schwartz

Erect biennial herb to 60cm, stem ± glabrous. Basal leaves pinnate with 5 leaflets; the leaflets very variable in size, 8–35mm long, ovate usually pubescent below, serrate, the terminal leaflet cordate. Stem leaves pinnate, but becoming much reduced in size upwards and ultimately with linear, entire leaflets. Umbels terminal; petals white. Fruit not ridged, densely covered in rather stiff hairs.

The basal leaves are present when the plant first flowers, usually in early spring, but die before it fruits. It is difficult therefore to collect fruiting specimens with basal leaves.

A widespread and characteristic plant of short grassland between 1300 and 3200m on the escarpment from J. Ḥabashī north to J. Rāziḥ. Its distribution seems to be related to the presence of mist and it occurs at much lower altitudes in the central escarpment than in the Ibb region where it is not found below about 2300m.

This species is endemic, but is part of a complex of forms centred on *P. peregrina* L. and is doubtfully distinct from plants in Ethiopia and Sudan.

2. Pimpinella woodii *C. C. Townsend*

Erect biennial herb to 2½m, stem thinly pubescent. Basal leaves yellow-green, pubescent, pinnate with 5 leaflets; the leaflets variable in size, 4–12cm long, serrate, ovate, cordate; stem leaves much reduced in size, the leaflets becoming lanceolate and very coarsely serrate above. Umbels terminal; petals white. Fruit unridged, glabrous.

Flowers from February to April.

Only known with certainty from a scrub-filled gully with ± permanent water near Jabin, J. Ba'dān at 2700m (Wood 2731), but leaves probably belonging to this species have been seen on J. Manār and near al Qimah, J. Bura'.

11. OREOSCHIMPERELLA Rauschert.

1. Oreoschimperella arabiae-felicis *C.C. Townsend*

Trachyspermum arabiae-felicis C.C. Townsend

Glabrous ascending perennial herb to 30cm, smelling slightly of carrot. Leaves rather small, lanceolate, bipinnate, the segments deeply dissected. Umbels terminal; petals white; bracts linear, sometimes forked, with a broad, scarious, sheathing base. Fruit ovoid, glabrous, but densely covered in wart-like protuberances, black.

Flowers in late summer.

A rare plant of high-altitude grassland found between 2600 and 3200m, occurring on all the high peaks around Ibb, extending south to J. Ṣabir and north through Waṣāb (J. Ḥisn Madnān) and 'Utmah (J. Al 'Amāl) to the Kusmah area of J. Raymah.

12. ANTHRISCUS *Pers.*

1. Anthriscus sylvestris (*L.*) *Hoffm.*

Cow Parsley. An erect biennial to 1½m, apparently always glabrous in Yemen. Leaves very large, especially below and reaching 40cm in length, 2–3 pinnate, the segments ± pinnatifid and coarsely serrate. Umbels terminal; petals white; bracts absent; bracteoles prominent, strongly reflexed. Fruit oblong, glabrous, ridged, c 7mm long and thus longer than in other white-flowered umbellifers in Yemen.

An uncommon species growing in shaded gullies and along hedges between 2600 and 3200m on the Ibb mountains: J. Ṣabir; J. Qarad, Mudhaykhirah; J. Ta'kar; J. Nu'mān, J. Ba'dān; J. Manār.

13. FOENICULUM *Miller*

1. Foeniculum vulgare *Miller* — Shamar

Fennel. Erect, glabrous, glaucous, short-lived, aromatic perennial to 1½m. Leaves mainly basal, 2–4 pinnate, the segments capillary. Umbels terminal; petals yellow; bracts absent. Fruit oblong, glabrous, ridged.

Cultivated in Yemen for its seeds and freely escaping. It is widely distributed on the escarpment and high plateau from 1000 to 2700m and is common near cultivation, growing on grassy field borders and tracksides.

14. FERULA *L.*

1. Ferula communis *L.* Munīsh

Massive erect glabrous herb to 3m. Basal leaves green, very large, 3–4 pinnate, the segments capillary, upper leaves progressively reduced to conspicuous sheathing bases which subtend the umbels. Umbels lateral and terminal; petals yellow; bracts absent. Fruit elliptic, glabrous, winged, ridged, c 1.5cm long.

Common on stony hillsides, terrace walls and cliffs between 2200 and 3200m on the Ibb mountains extending west to J. Qarad, Mudhaykhirah and south to J. Sawraq, J. Ṣabir and J. Habashī. It has not been found north of the Sumarah pass.

15. PEUCEDANUM *L.*

1. Peucedanum inaccessum *C.C. Townsend*

Slightly aromatic, erect, glabrous, probably biennial herb to 50cm, stem glaucous. Leaves green, tri-pinnate, the segments pinnatifid with ± triangular lobes; petiole base sheathing. Umbels terminal; petals yellow; bracts and bracteoles few, linear. Fruit prominently 4-winged, glabrous.

A very rare endemic species known from grassy slopes between 1600 and 2000m at Maḥābishah (below Shamsān) and on J. Milḥān (between Markah and Aṣ Ṣa'bah and between Markah and Maghrabah).

16. AGROCHARIS *Hochst.*

1. Agrocharis melananthera *Hochst.*

Caucalis melanantha (Hochst.) Benth. & Hook.
Daucus yemensis Deflers

Slightly aromatic, procumbent or ascending perennial herb, stem yellow, sparsely pilose. Leaves green, pilose, lanceolate, 2–3 pinnate, the segments linear. Umbels terminal, capitate; petals purple-black; bracts linear. Fruit oblong, ridged, the primary ridges spiny and the secondary ridges pilose.

A characteristic plant of grassland and derelict fields on the higher escarpment mountains from J. Ṣabir north to J. Maswār. This is another species which appears to require mist, occurring as low as 1900m on the outer escarpment though unknown below 2600m on the inner escarpment. It reaches 3500m on J. Nabī Shu'ayb.

17. TORILIS *Adans.*

1. Torilis arvensis (*Huds.*) *Link*

An erect annual herb to 40cm, stem sparsely adpressed hairy. Leaves usually purplish, thinly pilose, 2(–3) pinnate, the segments deeply pinnatisect and remotely serrate. Umbels terminal, rather small; petals white or pinkish; bracts usually absent; bracteoles minute. Fruit ridged, densely spiny.

A local plant found between 2400 and 3200m on the higher escarpment mountains from J. Ṣabir north to Kawkabān. It usually grows in disturbed stony ground such as tracksides and terrace walls.

Several umbellifers are cultivated for their aromatic leaves or fruit: the most common is Coriander, **Coriandrum sativum** *L*, 'Kabsarah', whose leaves are widely used in salads, soups and other dishes. The fruit is less commonly used but is widely available in markets. It is grown almost everywhere as a pot herb.

Parsley, **Petroselinum crispum** (*Mill.*) *A.W. Hill*, 'Baqdūnis', is also very common. The form traditionally grown in Yemen is an annual with flat leaves. The crisped leaved biennial forms are a recent introduction.

Dill, **Anethum graveolens** *L.*, was recorded from Manākhah (Deflers) and Sana'a (Rathjens) but I have not seen it outside the gardens of Europeans and it is certainly rare if it is still cultivated.

The Carrot, **Daucus carota** *L.*, 'Jazar' is cultivated in two forms. A red-rooted variety of recent introduction is cultivated around Sana'a, whereas a pale orange-rooted variety of more bland flavour has long been cultivated in the Ibb area.

LOGANIACEAE

Trees with simple, opposite leaves. Stipules present, inconspicuous in *Nuxia* but auriculate in *Buddleja*. Petals united into a corolla.

Leaves glabrous. Inflorescence a cyme-like panicle — 1. *Nuxia*
Leaves tomentose below. Inflorescence a spike-like panicle — 2. *Buddleja*

1. NUXIA *Lam.*

Trees. Young leaves and calyx sticky. Flowers in terminal panicles. Fruit a capsule.

Leaves in pairs, margins serrate — *N. oppositifolia*
Leaves in 3s, margins entire — *N. congesta*

1. Nuxia congesta *R. Br. ex Fresen.* Qaṭīn

Tree to c 8m with rough brown bark. branchlets angled, the bark peeling in strips. Leaves in 3s, entire, elliptic to oblanceolate, obtuse, up to 8 × 2cm. Flowers cream in dense sticky orange-brown panicles.

A very local tree of some high rainfall mountains in the Ibb area (J. Ṣabir; J. Manār; J. Ba'dān; J. Sumārah) and the central escarpment (Kusmah area, J. Raymah; near Shaḥir, J. Milhān). It is found between 2300 and 2900m and is particularly common on J. Ṣabir and J. Ba'dān.

2. Nuxia oppositifolia (*Hochst.*) *Benth.* Rumah

N. dentata R. Br.

Tree to c 8m, similar to the previous species but branchlets less strongly angled and frequently drooping. Leaves opposite, paired, coarsely serrate, obtuse, up to c 10 × 2.5cm. Flowers cream in a relatively lax, sticky panicle.

Found in riverine scrub between 700 and 1800m in a few scattered localities along the escarpment: 'Ukaybir, J. Milḥān (Schweinfurth); Wadi Yūr; 'Usil (Schweinfurth) and Rajamah, Ḥarāz; Wadi Sukhnah; Musalla, Aḥkūm.

2. BUDDLEJA *L.*

1. Buddleja polystachya *Fresen.* 'Afār

Shrub to about 4m with pale brown tomentose branchlets. Leaves lanceolate, up to c 7 × 4cm, dull green above, grey tomentose below, serrate, apex shortly acuminate, base rounded. Inflorescence an elongated spike-like panicle at the end of the branchlets. Flowers orange-red, sweetly but not very pleasantly scented.

Flowers mainly in winter.

Widespread and locally common on the escarpment between 1500 and 3000m. It is restricted to areas of moderate or high rainfall but is very tolerant of exposure and frost occurring at altitudes where the only other large shrubs or trees are *Acacia origena* and *Rosa abyssinica*. It thus occurs in Sāqayn, J. Ḍawrān and Banī Maṭar as well as more favoured places.

GENTIANACEAE

Herbs with opposite, entire leaves, without stipules. Flowers often in diachasial cymes; sepals free or joined; corolla of 4 or 5 joined petals. Fruit a capsule.

1.	Flowers white, sometimes with a dark blue central line	2
	Flowers pink or yellow	3
2.	Flowers in axillary clusters	2. *Enicostemma*
	Flowers in terminal cymes	4. *Swertia*
3.	Flowers pink	3. *Centaurium*
	Flowers yellow	1. *Sebaea*

1. SEBAEA *R. Br.*

Annual or perennial herbs. Flowers in terminal cymes: sepals free, corolla yellow.

Basal rosette present. Flowers several on each branch	*S. pentandra*
Basal rosette absent. Flowers usually solitary on each branch	*S. microphylla*

1. Sebaea microphylla (*Edgw.*) *Knobl.*

Small rhizomatous perennial with simple or sparingly branched erect stems to 5cm, basal rosette absent. Leaves few, subulate or narrowly triangular, 1–3mm long. Flowers usually solitary on each branch, yellow.

Only known from bare patches in short grassland at 2700m on J. Khudrā' on J. Ḥubaysh. (Wood 1779)

2. Sebaea pentandra *E. Mey.*

S. barbeyiana Schinz

Annual herb with a basal rosette from which arise several erect stems to 10cm. Leaves ovate, acute, 5–10mm long. Inflorescence plentifully branched with well-developed diachasial cymes of yellow flowers.

The Yemen plant is variety **burchellii** (*Gilg.*) *Marais.*

Known with certainty only from marsh in the gorge above Tin'am at 2500m (Wood 2341) but field records elsewhere on the high plateau by a pond near Dhamār and in marsh at the head of the Wadi Banā near Ribāṭ are probably of this species.

S. pentandra is widely distributed from Kenya to South Africa but has not yet been found in Ethiopia or Somalia.

2. ENICOSTEMMA *Blume*

1. Enicostemma verticillare (*Retz.*) *Baill.*

E. littorale Blume

Erect glabrous perennial to 30cm, usually branched at the base; stems winged. Leaves sessile, oblong-lanceolate, up to 8cm long. Flowers in dense sessile, axillary clusters; calyx much shorter than the corolla tube; corolla white, the lobes c 2mm long.

Widespread on rocky outcrops and in sparse stony grassland in areas of average to good rainfall on the escarpment and in suitable places in the east of the Tihama growing from 300 to 2300m.

3. CENTAURIUM *Hill.*

1. Centaurium erythraea *Rafn.*

Biennial herb with an erect, often solitary stem 30cm from a basal rosette. Rosette leaves elliptic, 1–5cm long; stems leaves, oblong, acute, usually less than 2.5cm long. Flowers pedicellate in a lax terminal diachasial cyme; calyx slightly shorter than the corolla tube; corolla pink, the lobes c 7mm long.

A rare plant found between 2400 and 3000m by springs and in other damp rocky places: above Baw'an, Bāb al Aḥjar; in the gorge above Tin'am; Wadi Rakwah, Khawlan Ash Sham (Steinberg).

4. SWERTIA *L.*

Glabrous herbs. Inflorescence of few-flowered terminal cymes; sepals free; petals white. usually with a dark blue central line.

Leaves linear, less than 2mm broad; the stems usually with at least ten pairs of leaves	*S. polynectaria*
Leaves spathulate, oblanceolate or elliptic, at least 3mm broad and commonly much more; stems rarely with more than 5 pairs of leaves	*S. woodii*

1. Swertia polynectaria (*Forssk.*) *Aschers. ex C. Chr.*

S. decumbens Vahl

Woody perennial, usually much-branched below with numerous decumbent twiggy branches, each with many pairs of leaves. Basal rosette absent, all leaves similar, linear, 8–16mm long. Flowers in a terminal monachasial cyme; calyx lobes acute about a third as long as the corolla; corolla c 12mm long, white, often with a dark blue central line.

A characteristic species of rocky slopes and open montane grassland between 2000 and 3100m on the higher mountains of the escarpment and high plateau from J. Ṣabir north to J. Maswār and east to J. Lawz; in Khawlan Ash Sham on J. Rāziḥ only. It appears to be endemic to Yemen but should occur in the Asir mountains.

2. Swertia woodii *Shah*

Slender herb, usually branched at the base with several decumbent or ascending branches each with up to about 5 pairs of leaves, probably perennial but dying back at the end of the rainy season. Basal leaves ± forming a rosette, obovate or spathulate, 2–3cm long, dying back as the plant flowers; stem leaves oblanceolate-oblong, up to 2.5cm long. Flowers in a terminal diachasial cyme; calyx lobes acute, about half as long as the corolla; corolla 12–15mm long, the lobes white, often with dark blue veins and a thick blue central line.

There is some variation within the Yemen plant. The most common form (Wood 73/160, 1147, 1316) is 10–15cm high and is clearly at most a short-lived perennial. One collection (Wood 2909) from J. Nabī Shu'ayb is a dwarf, obviously perennial plant but in all other respects identical to the common Yemen form and probably represents an adaptation to a higher, more arid situation than usual. The presence and absence of the dark blue central line on the corolla lobes does not seem to be taxonomically significant in our plants. There is a further collection made by R. Spellenberg at Ibb which probably represents another species.

A local plant of stream sides, rocky slopes and high-altitude grassland found between 2300 and 3400m on all the main mountains around Ibb and on the mountains from Baw'an north to J. Maswār.

APOCYNACEAE

A family mostly of trees and shrubs usually readily recognized by the presence of latex and by the simple, entire, usually glossy green leaves without stipules. Petals united; seeds winged or with a tuft of hairs.

1.	Flowers pink. Trunk massively swollen	4. *Adenium*
	Flowers white or cream. Stem or trunk not swollen	**2**
2.	Leaves alternate	3. *Rhazya*
	Leaves opposite	**3**
3.	Plant spiny	1. *Carissa*
	Plant unarmed	2. *Acokanthera*

1. CARISSA *L.*

1. Carissa edulis (*Forssk.*) *Vahl*

'Utūr, 'Untūr, 'Arm, Lūḍ, Lahas

Much-branched spiny shrub to c 3m, with green bark. Leaves opposite, broadly ovate or elliptic, acute or ± mucronate, c 2–4cm long. Flowers in terminal cymes, fragrant; corolla small, white (pinkish outside), about 1cm long. Fruit a red, eventually black, globular edible berry about 5mm long.

Flowers throughout the rainy season and sporadically at other times.

The typical plant has glabrescent leaves, but variety **tomentosa** *Stapf* was found at Al Jakhrah northwest of Manākhah by Schweinfurth.

Widespread and often abundant on the outer escarpment from 500 to 2500m from Turbah north to Khawlan Ash Sham. It is an important constituent of valley scrub in the central escarpment and is also common in scrub on uncultivated hillocks in the Ibb-Udayn region, where it is associated with *Euphorbia ammak* and *Acacia hockii*.

2. ACOKANTHERA *G. Don*

1. Acokanthera schimperi (*A. DC.*) *Oliv.*

'Imrikh, Athbīr, Tumāq

A. deflersii Schweinf.

Completely glabrous, unarmed shrub or small tree to c 6m with rough deeply fissured bark. Leaves opposite, leathery, elliptic, c 2–5cm long, the apex very variable but usually rounded, mucronate, less commonly acute or emarginate. Flowers in small axillary clusters or corymbs; corolla white, fragrant, c 1cm long. Fruit an eventually black, ovoid berry, c 1.5cm long.

Flowers in March and April.

Locally abundant on open hillsides and around cultivation between 700 and 2000m on the escarpment but apparently absent from the Ḥujarīyah though present on J. Sawraq. A gregarious species, it is often found in small colonies usually in areas of above average rainfall and is thus most common in Ibb province and the central escarpment.

3. RHAZYA *Decne.*

1. Rhazya stricta *Decne.*

Ḥarmil

Vigorous, spreading to ascending, much-branched, glabrous undershrub or woody herb reaching 1m in height. Leaves alternate, oblanceolate, acute. Inflorescence a terminal cyme. Corolla (including the tube) cream, about 1cm long. Fruit a linear pod-like capsule 5–10cm long.

One of the most characteristic desert plants in the Yemen flora which is abundant in the Sands east of Mārib and penetrates the Mashriq along sandy wadi beds climbing to 2400m but just failing to reach Rada', Sana'a and Ṣa'dah.

An extraordinarily drought-resistant plant that flowers throughout the year. It is said to be poisonous and does not seem to be grazed by stock and this may partly explain its great abundance in desert areas.

4. ADENIUM *Roem. & Schult.*

1. Adenium obesum (*Forssk.*) *Roem. & Schult.*

'Adan, Tibḥī

A. arabicum Balf. f.
A. micranthum Stapf.

An unmistakable shrub to 3m in height, often flowering when the plant is leafless. Stems grossly swollen and succulent, bark purple-brown, smooth. Leaves spirally arranged at the branchlet tips, obovate to oblanceolate, rounded or mucronate c 5–10cm long, pubescent or glabrous. Flowers in terminal cymes; corolla large, pink, 2–5cm long. Fruit a pod-like capsule up to 10cm long; seeds woolly.

Two distinct forms occur. One has a narrowly conical stem 1–3m high and is found at lower altitudes. The other has a very low but massive stem of nondescript shape up to c 50cm high. This is found at higher altitudes and is common, for example, around Taiz.

Widespread and common on the escarpment from the first foothills at 300 to 2000m; frequent in the southeast mountains but rare in the Mashriq and known only from Rada', Ṣa'dah to Baraṭ (Fleurentin) and Majz (Steinberg). There is some evidence to suggest that it occurs at higher altitudes in low rainfall areas. In the foothills it is locally in great abundance and is occasionally ± dominant on stony slopes, notably on the western approaches to J. Milḥān.

A number of plants from this family are cultivated in Yemen. The most commonly seen is the Oleander, **Nerium oleander** *L.*, a shrub with coriacious, lanceolate, dark green leaves and corymbs of fragrant pink or rarely white flowers. It is found in gardens throughout Yemen but especially in Taiz and Sana'a. **Thevetia peruviana** (*Pers.*) *Merr.* a shrub with glabrous linear leaves and large yellow funnel-shaped flowers is planted mainly round Taiz. **Plumeria acuminata** *Ait.*, the Frangipani, a much-branched succulent shrub with fragrant cream flowers is grown in Taiz. Its leaves are deciduous in winter. **Catharanthus roseus** (*L.*) *G. Don*, a perennial, slightly woody herb with somewhat fleshy leaves and pink (rarely white) flowers is very commonly grown in Taiz. It self-seeds readily and can be expected to become naturalised in waste places fairly soon.

ASCLEPIADACEAE

Herbs, shrubs or climbers, often succulent and with milky juice. Leaves opposite, entire, sometimes vestigial or absent; stipules minute or absent. Flowers usually in cymes or umbels, regular, bisexual; sepals 5, joined at the base; petals 5, joined at least at the base; corona usually present, 5-lobed, attached to the corolla or the stamens; stamens 5, the filaments fused or free. Fruit a pair of pod-like follicles (often solitary as one follicle may fail to develop), bearing numerous silky-haired seeds.

The formal generic distinctions are usually very hard to observe since they are often based on small features of coronal structure and on the very complex structure of the anthers and pollen transfer system which is a distinctive feature of the family. The key below relies therefore on very artificial but easily observed characters.

The *Asclepiadaceae* are of outstanding interest in the Yemen flora partly because of the exotic nature of the flowers in some of the succulent genera notably *Ceropegia* and *Caralluma* and partly because of the high proportion of endemic species. Endemism here as in other families such as the *Euphorbiaceae* and *Liliaceae* is much more marked in the succulent genera. The *Asclepiadaceae* are also poorly known taxonomically and there are great difficulties in several genera, notably *Huernia* and *Sarcostemma*.

1. Leafless succulents. Stems angled or if ± cylindrical, then never climbing nor long, smooth and rod-like **2**
Leafy or leafless plants; if leafless succulents stems either climbing or long, smooth and rod-like **7**

2. Stem decumbent, covered in sharp, brown, conical spines. Flowers very large 10–12cm in diameter 19. *Edithcolea*
Stem erect or decumbent, unarmed or with soft, fleshy teeth. Flowers less than 4cm in diameter **3**

3. Stems ± cylindrical, without fleshy teeth or wings **4**
Stems angled, often winged or with fleshy teeth **5**

4. Stem trailing, dull green, densely covered with a network of obtuse tubercles 18. *Echidnopsis*
Stems erect, grey, with obscure elongate protuberances 20. *Rhytidocaulon*

5. Flowers in terminal racemes or in clusters of 1–100 arising at or near the apex of the stem 21. *Caralluma*
Flowers borne in small clusters on a very short, thickened peduncle arising near the base of the stem **6**

6. Pedicels at flowering time less than 15mm long. Corolla with small teeth in the sinuses between the lobes 23. *Huernia*
Pedicels more than 20mm long. Corolla without small teeth in the sinuses between the lobes 22. *Duvalia*

7. Erect leafy plants **8**
Climbing plants or, if erect, with long, rod-like, leafless, succulent stems **14**

8. Leaves linear, gradually narrowed into an attenuate base **9**
Leaves broadly lanceolate to ± orbicular, abruptly narrowed into a cuneate or truncate base **11**

9. Fruits inflated, bristly. Corolla lobes reflexed 4. *Gomphocarpus*
Fruits not inflated, smooth. Corolla lobes erect or spreading **10**

10. Corolla glabrous. Follicles usually paired 1. *Periploca*
Corolla hairy. Follicles solitary 6. *Kanahia*

11. Large shrub with leaves more than 10cm long. Fruit inflated, subglobose to ellipsoid, c 6–12cm long 5. *Calotropis*
Small shrubs or herbs with leaves less than 3cm long. Fruit not inflated, less than 6cm long **12**

12. Small shrub. Leaves less than 1cm long. Flowers solitary 7. *Blyttia*
Herbs. Leaves 1–3cm long. Flowers not solitary **13**

13. Flowers in sessile clusters, ± rotate 3. *Glossonema*
Flowers in pedunculate umbels, ± tubular 2. *Odontanthera*

14. Flowers with a long flask-shaped tube at least 1cm long. Corolla lobes usually joined at the apex but free for most of their length 17. *Ceropegia*
Flowers not as above; tube if present, less than 5mm long, corolla lobes never joined at the apex **15**

15. Plants leafless or with minute, reduced leaves **16**
Plants leafy with fully developed leaves **19**

16. Stamens conspicuously woolly 1. *Periploca*
Stamens glabrous **17**

17. Corolla and calyx pubescent 16. *Leptadenia*
Flowers glabrous **18**

18. Flowers more than 5mm long, petals erect. Corona double, fleshy 9. *Sarcostemma*
Flowers less than 3mm long, petals spreading or reflexed. Corona single 8. *Cynanchum*

19. Leaves mucronate. Flowers glabrous with acuminate corolla lobes — 10. *Pentatropis*
    Leaves not mucronate. Flowers not as above — **20**
20. Leaves cordate — **21**
    Leaves cunate to truncate — **22**
21. Flowers in sessile umbels. Follicles smooth — 8. *Cynanchum*
    Flowers in pedunculate cymes. Follicles bristly or muricate — 11. *Pergularia*
22. Corolla 2–3mm long — **23**
    Corolla more than 5mm long — **24**
23. Leaves glaucous, minutely puberulent — 16. *Leptadenia*
    Leaves dark green, densely pubescent — 13. *Gymnema*
24. Leaves lanceolate, less than 2cm wide. Flowers yellow, tinged with maroon; corona obscure — 12. *Tylophoropsis*
    Leaves ovate, more than 3cm wide. Flowers yellowish or white, corona prominent — **25**
25. Corolla more than 10mm long, densely woolly in the mouth — 14. *Telosma*
    Corolla 5–8mm long, not woolly in the mouth — 15. *Dregea*

1. PERIPLOCA *L.*

Shrubs with white latex but of varying habit. Leaves present or absent. Flowers in cymes; corolla ± rotate with a short, 1–2mm long, tube; corona present. Follicles almost always in pairs, diverging from each other at 180°, linear, smooth.

1. Plants leafless or with small, reduced leaves less than 6mm long. Anthers conspicuously woolly — 2
   Plant with well-developed leaves at least 2cm long. Anthers glabrous — *P. brevicoronata*
2. Corolla lobes ± glabrous, cream, yellow or flesh-coloured. Leaves subulate. Plant usually decumbent — *P. visciformis*
   Corolla lobes villous-margined, dark purple. Leaves ovate. Erect shrub — *P. aphylla*

1. Periploca aphylla *Decne.*

Much-branched shrub to 2½m with long, stiff, erect, glaucous rod-like branches which are glabrous and leafless when mature but pubescent with minute, ovate leaves, c 2mm long when young. Cymes on short, lateral branches; corolla c 7mm long with dark purple, woolly-margined, lanceolate lobes; anthers woolly. Follicles 7–9mm long.

Flowerless plants bear a strong superficial resemblance to *Leptadenia pyrotechnica* and to a lesser extent to *Euphorbia schimperi*, but the former always grows in sand and the latter is quite glabrous.

In scattered localities on dry, stony plain between 1000 and 2000m, extending from the western Ḥujarīyah and the Taiz area through the southeast mountains to Damt and northwards through the Mashriq to Ḥarf Sufyān.

P. aphylla is a plant of Sahara-Sindian distribution being found from the eastern Sahara to northwest India.

2. Periploca brevicoronata *D. Goyder and L. Boulos*

Erect, much-branched, glabrous, leafy shrub with woody, red-brown branches. Leaves linear-oblanceolate, 4–6cm long. Cymes axillary, shortly pedunculate, compact; corolla c 3mm long, with pale yellow, oblong lobes; anthers glabrous. Follicles 6–7cm long.

Only known from a rocky slope at 2000m between Iriyān and Riḥāb (Wood 1041).

3. Periploca visciformis (*Vatke*) *K. Schum.*

Socotora visciformis (Vatke) Bullock

Glabrous undershrub, much branched above a woody base, the branches usually decumbent, and flexuous but sometimes stout and stiff. Leaves minute, subulate, c 5mm long, soon deciduous. Cymes few-flowered, axillary, pedunculate; corolla c 6mm long, with oblong, glabrous, cream, yellow or flesh-coloured lobes; anthers woolly. Follicles 5–6.5cm long.

A local plant of cliffs and steep rocky slopes between 1000 and 2600m: widespread on the escarpment from the Wadi Zabīd north to J. Minmār but only frequent in the central escarpment; in the southeast mountains at Turbah and J. Sawraq; in the Mashriq on the Fardah pass, Wadi 'Abdayn near Ṣa'dah and Wadi Rakwah near Majz (Steinberg); absent from the high plateau and presumably frost sensitive.

Outside southwest Arabia it is known only from Somalia and Socotra.

2. ODONTANTHERA *Wight*

1. Odontanthera radians (*Forssk.*) *D. Field*

Steinheilia radians (Forssk.) Decne.

Erect, unbranched, densely pubescent, annual herb, 5–15cm high. Leaves 2–2.5cm long, broadly ovate, mucronate, shallowly cordate at the base, ± tomentose, the veins highlighted with white hairs. Inflorescence of long pedunculate axillary umbels; corolla 8–9mm long, glabrous, tubular the lobes pinkish, erect. Follicles usually solitary, 4–5cm long, ellipsoid, acuminate, covered in soft prickles.

Widespread on sand in the Tihama up to 400m but only appearing after heavy rain and apparently uncommon being recorded from few, widely scattered localities: Mafraq; Ḥays; Khawkhah (Miller); Bayt al Faqīh; Marāwi'ah (Schweinfurth); Dayr As Suhayl, west of Bājil (Wissmann).

3. GLOSSONEMA *Decne.*

Annual or perennial herbs. Flowers in sessile, axillary umbels; corolla with a very short tube and 5 ± rotate lobes. Follicles usually solitary, ovoid or ellipsoid, acuminate, covered in soft, fleshy prickles.

Leaves suborbicular. Style truncate — *G. varians*
Leaves broadly lanceolate to oblong. Style conical — *G. boveanum*

1. Glossonema boveanum (*Decne.*) *Decne.*

Perennial pubescent woody herb, much branched at the base and with ascending stems c 20cm high. Leaves 1.5–2cm long, broadly lanceolate to oblong, minutely mucronate, cuneate or truncate at the base, thinly white pubescent and ± green in colour, the margin often crisped. Corolla c 4mm long, whitish-yellow, glabrous. Follicles 3–5cm long; style tip conical.

A rare plant in Yemen known only from stony plain at 600m near Ḥajjaylah.

2. Glossonema varians (*Stocks*) *Benth. ex Hook. f.*
G. edule N.E. Br.
G. arabicum Deflers
Similar to the previous species but probably annual, the leaves sub-orbicular, usually grey in colour because of the dense white-hairy indumentum and the style tip truncate.

A local plant of rocky plain which comes up after heavy rain; it is found between 1100 and 2600m in dry areas of the escarpment, high plateau and southeast mountains: Barakānī, Ḥujarīyah; east of Rada'; between Milāḥ and Sanabān; Yislah pass; Wadi Ḍahr; Mafḥaq.

4. GOMPHOCARPUS *R. Br.*

Erect perennials with milky latex, much branched, especially above. Leaves linear, gradually narrowed into an attenuate, sessile base. Flowers in pedunculate, lateral umbels, which are usually few in number and towards the apex of the stem; corolla divided almost to the base, the lobes obovate, reflexed; corona attached to the stamens. Follicles ovoid, acuminate, inflated, covered in soft fleshy bristles.

The treatment of this genus in Schwartz is very misleading.

Follicles and the outside of the corolla pubescent. Corona without teeth *G. sinaicus*
Follicles and outside of the corolla glabrous. Corona with recurved teeth *G. fruticosus*

1. Gomphocarpus fruticosus (*L.*) *Ait. f.*
Branched undershrub to 1m with grey stems. Leaves up to 9cm long. Flowers pale yellow, suffused with purple, ± glabrous except for the pedicels; corona with recurved teeth. Follicles 2–5cm long, brown, ± glabrous.

Most, if not all, Yemen plants belong to variety **setosus** (*Forssk.*) *O. Schwartz* which differs from the type in its very narrow leaves (less than 5mm broad) and smaller corolla and corona. These differences, though, are probably not very significant.

Widespread and frequent from 1000 to 2900m on the escarpment and high plateau growing on roadsides and field borders, in grassland and damp scrubby places, often near streams, particularly at lower altitudes. It is the commonest *Gomphocarpus* on the escarpment.

2. Gomphocarpus sinaicus *Boiss.*
Undershrub similar in appearnce to the last species but hairier and often with pinkish stems. Leaves dull green, up to 6cm long. Flowers cream, the corolla pubescent on the outside and the corona without teeth. Follicles 2–5cm long, purple-brown, pubescent.

Widespread on the escarpment and high plateau from 1900 to 2700m growing on roadsides, in derelict fields, on open stony plain and in rock crevices. It is more frequent on the high plateau and in the north than *G. fruticosus* but definitely less common on the escarpment, where it is absent at low altitudes.

5. CALOTROPIS *R. Br.*

1. Calotropis procera (*Ait.*) *Ait. f.* 'Ushār, Qushār
Soft-wooded shrub, up to 4m high, with plentiful white latex and rather weak spreading branches; bark fissured. Leaves sessile, weakly cordate, obovate or elliptic, 10–25cm long. Flowers in dense, pedunculate, lateral cymes; corolla deeply 5-lobed, the lobes triangular, c 10mm long, white on the outside, purple inside; corona prominent, white in the centre. Follicles usually solitary, inflated, globose or ovoid, smooth, 5–15cm long.

One of the most conspicuous plants in Yemen which is widespread up to 2300m though at higher altitudes mostly occurring as scattered individuals on roadsides and in waste ground. It is abundant along sandy wadis in the Tihama, the escarpment foothills and particularly in the Mashriq where with *Tamarix* it is the most characteristic species of wadi margins.

Calotropis procera is common throughout the Sahara-Sindian regions.

6. KANAHIA *R. Br.*

1. Kanahia laniflora (*Forssk.*) *R. Br.* Kanāḥ
Undershrub to 1½m with many, erect, sparingly branched, leafy, glabrous stems arising in a tuft from a perennial rootstock; white latex present. Leaves petiolate, linear-lanceolate, acuminate, glabrous, 10–15cm long. Flowers in pedunculate, lateral cymes; corolla c 10mm long, cream, deeply 5-lobed, the lobes woolly-margined; corona attached to the stamens. Follicles usually solitary, boat-shaped, smooth, 4–5.5cm long.

Frequent from 500 to 2500m along the margins of almost all escarpment wadis where there is at least a fairly constant flow of water in the rainy season; much rarer on the high plateau and in the Mashriq but recorded from the wadis east of Jiḥānah, Nihm, Wadi Shalālah (Hepper), Wadi Ḍahr and the Wadi Thay'ān between 'Amrān and Thilā.

The name *Kanahia* is derived from the name of this species used in Yemen.

7. BLYTTIA *Arn.*

1. Blyttia spiralis (*Forssk.*) *D. Field & J.R.I. Wood*
Much-branched, bushy undershrub to 80cm, the branches with spinescent tips, whitish-grey, pubescent; white latex absent. Leaves petiolate, ovate-oblong, obtuse, pubescent, 0.5–1cm long. Flowers arising beside the leaves, solitary, shortly pedicellate; corolla 2–3mm long, cream, deeply 5-lobed, the lobes erect, spirally twisted, hairy on the inner surface; corona single, very small. Follicles smooth, narrowly cylindrical, tapering at the apex, c 4cm long.

Abundant on stony and sandy plain between 50 and 500m in the southern Tihama extending north to the Wadi Mawr but rare north of the Wadi Zabīd.

This species is known outside southwest Arabia only from one place in Eritrea.

8. CYNANCHUM *L.*

Climbing herbs or shrubs, sometimes succulent. Flowers in sessile umbels; corolla with a very short tube and spreading or reflexed lobes; corona single, tubular, toothed or laciniate. Follicles smooth, narrowly cylindrical, tapering to the apex.

Glabrous, leafless, succulent climber — *C. tetrapterum*
Hairy, leafy, woody climber — *C. hastifolium*

1. Cynanchum hastifolium *N.E. Br.*

Small, climbing shrub, densely pubescent at the leaf nodes, glabrescent elsewhere. Leaves petiolate, sometimes clustered, ovate, acuminate, hastate or cordate at the base, 1–2cm long. Flowers in sessile umbels arising opposite the leaves; corolla lobes yellow-green, spreading, c 4mm long; corona cream, with 5 subulate lobes alternating with 5 teeth.

A rare plant known from dry open *Acacia* bushland at c 1600m: between Mafḥaq and Suq Samīr; Marāshī (D. Wood).

2. Cynanchum tetrapterum (*Turcz.*) *R.A. Dyer*

C. sarcostemmatoides K. Schum.

Glabrous, glaucous, leafless, succulent climber. Flowers in sessile umbels arising at the nodes; corolla lobes yellow-green, reflexed, c 1.5mm long; corona orange, with many short, blunt teeth.

An uncommon but easily overlooked plant of *Acacia* scrub found in scattered localities between 400 and 1700m: near Taiz airport; J. Sawraq: 'Alūjah and Ḥadīyah, J. Raymah; Madīnat al 'Abīd; Suq As Sabt, J. Bura'; Ḥajjaylah and 'Usil, Ḥarāz; Wadi Sukhnah.

9. SARCOSTEMMA *R. Br.*

Decumbent, erect or climbing succulents with fleshy cylindrical stems and plentiful white latex. Leaves reduced, quickly deciduous so the plant is commonly leafless. Flowers in clusters; corolla lobes almost free, erect; corona double, joined to the stamens. Follicles linear, smooth.

***Sarcostemma* is possibly the least understood genus in the Yemen flora. It makes very unsatisfactory herbarium specimens but most plants are in any case similar in floral characters though differing considerably vegetatively. Habit photographs, specimens of the root system and flowers are needed from individual plants before the genus can be worked out satisfactorily. There may well be other species in addition to the following three.**

1. Corolla pink. Stems trailing, 3–4mm diameter — *S. vanlessenii*
   Corolla white or cream. Stems erect, climbing or (rarely) trailing, 2–6mm diameter — 2
2. Stems 2–3mm diameter. Corolla c 4mm long — *S. arabicum*
   Stems 3–6mm diameter. Corolla 5–6mm long — *S. viminale*

1. Sarcostemma arabicum *Bruyns & Forster*

Stems trailing over rocks or climbing over scrub vegetation, rooting infrequently, green, 2–3mm in diameter. Flowers in dense terminal and axillary clusters; corolla cream, the lobes c 4mm long, the outer corona forming a continuous ring around the stamens.

Known only from Al Barḥ (Lavranos & Newton), Shibām and Suq As Sabt, J. Bura' (Wood 3431) but probably quite common.

2. Sarcostemma vanlessenii *Lavranos*

Stems trailing, frequently rooting at the nodes, terete, green, 3–4mm in diameter. Flowers shortly pedicellate, in few-flowered clusters; corolla dark pink, the lobes c 5mm long.

An uncommon plant of cliff gullies and ledges where there is a reasonable accumulation of soil and no grazing; it is found between 2000 and 2400m mostly on the high plateau but possibly overlooked elsewhere; near Riṣābah (Radcliffe-Smith and Henchie); between Sana'a and Ḥaddah; Shamlān; Wadi Ḍahr (Lavranos); by a waterfall below Manākhah to the north west; between Ḥadādah and Al Jabin, J. Raymah.

A distinctive species on account of its flower colour; endemic to southwest Arabia.

3. Sarcostemma viminale (*L.*) *R. Br.*

Raḍiḥ, Rīd, Miru'ad

S. stipitaceum (Forssk.) R. Br.

A very variable plant, much branched at the base, usually with short erect rod-like or twisted stems in open habitats but developing long, trailing or climbing stems in protected scrub vegetation. Stems rooting occasionally, terete, waxy-grey except in wet conditions, 3–6cm in diameter. Flowers in dense terminal and axillary clusters, fragrant; corolla whitish, the lobes 5–6mm long; the outer corona with lobes joined at the base only.

Locally abundant in dry country in the Tihama, on the escarpment and in the Mashriq from 100–2000m; it grows on rock outcrops, cliffs, dry hills, stony and gravelly plain and in acacia bushland; most common in Taiz province and along the dry foothills of the escarpment further north; in the Mashriq uncommon, known from Damt, Rada' and Nihm but doubtless occurring elsewhere.

10. PENTATROPIS *Wight & Arn.*

1. Pentatropis nivalis (*J.F. Gmel.*) *D. Field and J. R. I. Wood*

'Arkis

P. spiralis auct.

Asclepias forskalei Roem. & Schult.

Slender twiner, glabrous except on the petioles and the young inflorescences. Leaves petiolate, oblong, mucronate, shallowly cordate at the base, 1.5–3.5cm long. Flowers in subsessile, 3–6-flowered, axillary umbels; corolla campanulate, the lobes linear with a broad base, whitish-yellow, 6–7mm long. Follicles usually paired, narrowly ovoid, acuminate, smooth, 5–6cm long.

Widespread and frequent in the Tihama and on the drier parts of the escarpment up to 1500m, growing wherever there are suitable bushes to climb on; it is one of the few species found in the *Acacia ehrenbergiana* thickets in the Tihama and is also strikingly common in Taiz province. In the Mashriq it is a rare plant found at Damt and between Mārib and the Fardah pass.

11. PERGULARIA *L.*

Hairy twining plants with ovate-cordate leaves and plentiful white latex. Flowers in pedunculate corymb-like racemes; corolla with a short narrow tube and spreading or erect lobes; corona double, prominent. Follicles usually paired, both erect and pressed against each other, lanceolate in outline, bristly or muricate.

Plant grey canescent. Follicles muricate — *P. tomentosa*
Plant green, usually pilose (rarely glabrous). Follicles bristly — *P. daemia*

1. Pergularia daemia (*Forssk.*) *Chiov.*

P. extensa (Jacq.) N.E. Br.
P. glabra (Forssk.) O. Schwartz?
Twining or less commonly trailing herb, almost always with pilose, purplish stems. Leaves petiolate, ovate, cordate, shortly acuminate, green, nearly glabrous except on the veins below, 3–5cm long. Racemes with peduncles 5–10cm long; corolla lobes greeny-purple, c 5mm long; corona white. Follicles linear-lanceolate, 4–5cm long, densely covered in fleshy bristles, usually ± hairy.

P. daemia is rather variable in indumentum and it seems probable that *P. glabra* was merely a glabrous form, though it is impossible to be certain as there is no type specimen.

Widespread and frequent on the escarpment from 1000 to 1900m growing on fences around fields, in scrub and in damp overgrown gullies; it also occurs in the southeast mountains to Al Bayḍā' and has been seen at 2200m in the Wadi Ḍahr on the high plateau.

2. Pergularia tomentosa *L.*

Similar to *P. daemia* but softly grey-canescent in all parts and though potentially a climber it usually grows in open places and is ± erect in habit. In addition the peduncles are much shorter, being 1–5cm long and the follicles are broadly lanceolate and sparsely covered with small fleshy teeth.

A locally abundant plant of open rocky plain in dry country from 500 to 2300m. It is most abundant in Taiz province extending north along the escarpment to Waṣāb and through the southeast mountains to the Mashriq here it extends north to Ṣa'dah, east to Mārib and west to the Sana'a area; there is also an isolated record from Mafḥaq. This distribution is strikingly similar to that of *Jatropha spinosa*, with which *P. tomentosa* often grows.

12. TYLOPHOROPSIS *N.E. Br.*

1. Tylophoropsis heterophylla (*A. Rich.*) *N.E. Br.*

T. yemensis (Deflers) N.E. Br.
Herbaceous twining plant with 2 lines of pubescence along the stem, white latex absent. Leaves petiolate, broadly lanceolate, acuminate, rounded or shallowly cordate at the base, glabrous except on the veins below, 2.5–4cm long. Flowers in shortly pedunculate, lateral umbels; corolla with a short tube, the lobes c 5mm long, ovate, pubescent, yellow often strongly tinged with maroon; corona absent. Follicles oblong, acuminate, smooth, ± glabrous.

A local plant of bushy places between 2500 and 3200m on the higher mountains of the escarpment and high plateau: on J. Ṣabir and the Ibb mountains; J. Shibām, Ḥarāz; on the mountains from Baw'ān north to J. Maswār; J. Lawz. It often grows on *Rosa abyssinica* and in association with *Zehneria scabra*.

13. GYMNEMA *R. Br.*

1. Gymnema sylvestre (*Retz.*) *R. Br. ex Schult.*

Softly pubescent, woody climber without white latex. Leaves petiolate, ovate or elliptic, acute, rounded to cuneate at the base, 4–7cm long. Flowers in shortly pedunculate lateral umbels; corolla yellow, c 3mm long, tubular for nearly half its length, the lobes ovate, minutely hairy; corona present. Folicles linear-lanceolate in outline, acuminate, smooth, glabrous, 6–8cm long.

Only known at 600m in the valley forest above Suq As Sabt, J. Bura'.

14. TELOSMA *Coville*

1. Telosma africana (*N.E. Br.*) *N.E. Br.*

Glabrous or sparsely pubescent woody climber without white latex. Leaves petiolate, ovate, shortly acuminate, cuneate or rounded at the base, glabrous except on the veins below, 5–9cm long. Flowers in subsessile, lateral umbels; corolla 13–20mm long, yellow-green, tubular in the lower third, the tube inflated with a villous mouth, the lobes linear-lanceolate, sinuate, hairy on the inside. Follicles unknown.

Known only from scrubby banks around irrigated coffee fields at 1400m in Wadi Ḍabāb near Taiz.

Widespread in Africa but rare and scattered in occurrence and also remarkable because the fruit has apparently never been collected.

15. DREGEA *E. Mey.*

Woody, pubescent climbers without white latex. Inflorescence umbellate or cymose; corolla with a short tube and spreading, cream-coloured lobes; corona attached to the stamens. Follicles usually solitary.

Corolla lobes with villous margins. Follicles with large wings — *D. abyssinica*
Corolla lobes glabrous. Follicles wrinkled but not winged — *D. schimperi*

1. Dregea abyssinica (*Hochst.*) *K. Schum.*

Marsdenia abyssinica (Hochst.) Schlecht.
Leaves petiolate, ovate, acuminate, truncate or ± cuneate at the base, nearly glabrous, 5–12cm long. Flowers in shortly pedunculate lateral umbels; corolla 6–8mm long, the lobes oblong, villous-margined; stigmas not exserted. Follicles 6–8cm long, narrowly ovoid, acuminate, with prominent crenate wings.

A rare plant found in dense scrub in high rainfall areas of the escarpment between 1000 and 1500m: north of the suq,

'Udayn; between Suq Ar Ribāṭ and Bani Daw'ān, J. Raymah.

2. Dregea schimperi (*Decne.*) *Bullock*

Marsdenia schimperi Decne.

Vegetatively very similar to the last species but pubescent on the petiole and leaf veins. Flowers in lateral, pedunculate, compound cymes; corolla 5–8mm long, the lobes narrowly lanceolate, glabrous; stigmas exserted above the staminal disc. Follicles similar to those of *D. abyssinica* but longitudinally wrinkled and without wings.

An uncommon plant found on the escarpment between 1500 and 2000m, usually climbing over trees along wadi margins: Wadi Sūdān, Qā'idah; between Ibb and Suq As Sabt; near Dalīl; Wadi Ajlūb near Damt; between Madīnat Al 'Abīd and Ḥammām 'Alī; J. Ḍawrān (Wissmann); Wadi Sukhnah.

16. LEPTADENIA *R. Br.*

Shrubs or climbers with glaucous, glabrous woody stems. Leaves present or absent. Flowers in shortly pedunculate or subsessile, lateral umbels; corolla pubescent, with a very short tube and small, dirty-white lobes; corona hairy at the tips. Follicles usually solitary, smooth, glabrous.

Erect leafless shrub — *L. pyrotechnica*
Leafy twiner — *L. arborea*

1. Leptadenia arborea (*Forssk.*) *Schweinf.*

L. heterophylla (Del.) Decne.

Woody climber. Leaves petiolate, very variable in shape, sometimes linear, sometimes hastate, most commonly ovate, acuminate, truncate at the base, 3–7cm long. Corolla 2–2.5cm long, pubescent. Follicles lanceolate, acuminate, 6–8cm long.

Widespread and frequent from the first rocky outcrops in the Tihama at 300 to 1400m on the escarpment; it is almost always found growing over shrubs and is particularly common on bushes along wadi margins.

2. Leptadenia pyrotechnica (*Forssk.*) *Decne.*

Markh

Leafless shrub to 3m with numerous erect, whip-like, glaucous branches. Umbels densely pubescent, very compact, the flowers with short pedicels; corolla c 2mm long. Follicles borne on thickened peduncules, deflexed, linear, acuminate, 8–12cm long.

Widespread and locally frequent on sand in the Tihama, spreading to the dry stony plains and sandy wadis of the southwest escarpment foothills where it reaches 1000m; also common on the Sands spreading westwards into the Mashriq along sandy wadis and onto rocky plain to reach about 1700m on the Fardah pass.

17. CEROPEGIA *L.*

Twining, often succulent perennial herbs. Leaves, petiolate or sessile, commonly deciduous. Flowers usually in pedunculate, lateral cymes or umbels; corolla elongate, slightly inflated at the base, tubular in the lower ⅔, the lobes free for most of their length but usually coherent at the tips; corona double. Follicles smooth.

A fascinating genus rich in endemic and poorly understood species. Ceropegias typically occur singly and are easily overlooked. The species appear to be very variable and more collections are needed so the variation can be better understood.

1. Stems 1–2mm thick, not succulent. Leaves fully developed, not fleshy or deciduous at flowering time — 2
   Stems at least 2.5mm thick, succulent. Leaves absent or, if present, few, fleshy and/or quickly deciduous — 4
2. Leaves glabrous. Stems arising from a corm-like tuber — *C. bulbosa*
   Leaves hairy. Stems arising from elongated thickened tuberous roots — 3
3. Corolla uniformly shortly pubescent, lacking long ciliate hairs at the tip — *C. sepium*
   Corolla nearly glabrous except for the prominent purple hairs at the tip — *C. foliosa*
4. Stems at least 6mm thick — 5
   Stems less than 5mm thick — 6
5. Corolla with two separate inflations near the base, the lobes with prominent lateral projections 0.5–3.5cm long — *C. variegata*
   Corolla with a single basal inflation, the lobes without prominent lateral projections — *C. rupicola*
6. Corolla glabrous outside (except for long hairs at the tip) — 7
   Corolla uniformly finely pubescent outside — *C. somalensis*
7. Corolla 2–3cm long — 8
   Corolla 4–7cm long — 9
8. Stems arising from a thickened corm-like lump. Leaves (if present) linear. Corolla lobes joined near the apex to form a knob-like structure — *C. tihamana*
   Stems arising from a cluster of fibrous roots. Leaves (if present) ovate. Corolla lobes joined at the apex but not forming a knob-like structure — *C. aristolochioides*
9. Stem arising from a thickened irregular lump. Corolla lobes strongly convergent before diverging to the apex, having a large purple blotch below the point of convergence — *C. botrys*
   Stem arising from elongate tuberous roots. Corolla lobes not strongly convergent and without a prominent purple blotch — *C. arabica*

1. Ceropegia arabica *Huber*

Leafless twining succulent with stems 2–4mm thick, roots tuberous. Flowers shortly pedicellate, usually solitary, borne on short, fleshy peduncles less than 5mm long; corolla 5–7cm long, greenish-yellow with dark veins and spots particularly on the swollen base, the lobes long and linear, ciliate on the margins, often entirely free at the apex, the tips sometimes becoming spirally twisted together. This is the only species combining elongate tuberous roots and leafless stems.

C. arabica is divisible into three varieties based on the structure of the corolla lobes. In variety **abbreviata** *Bruyns* the lobes are only about 1cm long and the sepals correspondingly very short (2mm long). In the other two varieties the corolla lobes are at least 1.5cm long and the sepals exceed 2.5mm in length. However in variety **superba** (*Field & Collenette*) *Bruyns* (*Ceropegia superba* Field & Collenette) the corolla lobes are rigid and dark green with a prominent keel while in variety *arabica* the prominent dark green keel is absent.

Variety *arabica* is widespread on the escarpment, south-east mountains and in the Mashriq growing from 500 to 2000m in open, rocky country, particularly in *Euphorbia* scrub; it is easily overlooked and very scattered in occurrence: between Nashimah and Barakānī; around Taiz; between Qā'idah and Sayyānī; Ḥammām Damt; Wadi Ḥijān, Ḥarāz; near 'Ukaybir, J. Milḥān; Maghrabah at the head of the Wadi Yūr; Mafḥaq; between Khaywān and Al Ḥarf; Wadi Lā'ah (Müller-Hohenstein & Deil). Variety *abbreviata* is a rare plant restricted to J. Raymah (Ḥadīyah and Banī Dawān). Variety *superba* is known from Maḥwīt, Manākhah (Müller-Hohenstein) and Misrākh.

2. **Ceropegia aristolochioides** *Decne.*

Glabrous twining succulent with stems 2–3mm thick; roots fibrous. Leaves usually present at flowering time, ovate, acute or mucronate, base cordate or truncate, 2–4cm long. Flowers pedicellate in pedunculate lateral cymes; corolla 2.5–4cm long, yellow or green, often purple at the base and the apex or speckled throughout, glabrous apart from the margins of the lobes which are sparsely white-pilose.

The Yemen plant is subspecies **deflersiana** *Bruyns* which is restricted to SW Arabia.

An uncommon endemic plant of ungrazed, partially shaded, rocky slopes including cliffs found from 700 to 2000m in areas of high rainfall on the escarpment: J. Ḥawbān, Taiz (Butler), Wadi Tha'bat, J. Ṣabir; Dhī Sufāl, on river cliffs just above Dalīl; between Ẓulmah and Ra'isī, J. Ḥubaysh; Wadi Dūr, 'Udayn; between Ibb and 'Udayn (Noltee); between 'Alūjah and Kusmah, J. Raymah; Aṭ Ṭūr (Müller-Hohenstein & Deil), Wadi Līyah (Haig-Thomas).

3. **Ceropegia botrys** *K. Schumann*

C. mansouriana Chaudhary & Lavranos

Similar to *Ceropegia arabica* from which it differs in its top-shaped rootstock and different corolla. The corolla lobes converge tightly at their middle before expanding again to the apex. There is a prominent purple blotch just below the point of convergence.

Only known from Aṭ Tūr at 350m (Müller-Hohenstein & Deil).

4. **Ceropegia bulbosa** *Roxb.*

C. vignaldiana A. Rich.

Slender, non-succulent, glabrous, leafy twiner from a corm-like tuber. Leaves very variable in shape, 2–4cm long, linear to ovate, ± truncate at the base, acuminate to cuspidate at the apex, usually (always?) finely dark veined. Flowers pedicellate in lateral pedunculate cymes; corolla 1–2cm long, the tube yellow (tinged purple to a greater or lesser extent), the lobes purple, remaining joined at the apex which is densely covered in conspicuous, long, purple hairs.

A rare plant found in scrub between 1300 and 1900m in high rainfall areas of the escarpment: between Qā'idah and Dhī Sufāl; on the lower slopes of J. Ba'dān just north of Ibb; 'Udayn; just below Ḥadādah, J. Raymah; Al Maḥwīt (Noltee).

A rather widespread species being found from Ethiopia to India.

5. **Ceropegia foliosa** *Bruyns*

Non-succulent, leafy twiner with pilose stems; roots of thin, elongate tubers. Leaves ovate, acuminate, cordate at the base, pubescent, 4–8cm long. Flowers long-pedicellate, borne in lateral pedunculate umbels; corolla 2–3cm long, narrowly obovoid, yellow with obscure green veins, a swollen purple base, and greenish purple tips to the lobes which are fused at the apex and densely covered in prominent, long, purple and white hairs.

A rare plant found in bushy places between 1500 and 1900m around 'Udayn, on the lower slopes of J. Ba'dān just north of Ibb and on J. Mabyān near Ḥajjah (Müller-Hohenstein & Deil). It is endemic to Yemen.

6. **Ceropegia rupicola** *Deflers*

Rather variable in appearance, sometimes pendulous on cliffs but more usually a succulent bush c 1m high, much branched at the base with many stout, erect, cylindrical, glaucous stems 5–10mm thick, these becoming thinner and twining in character towards their tips; roots fibrous. Leaves usually present at flowering time, variable in shape, lanceolate, ovate or orbicular, cuneate to rounded at the base, acute, mucronate or emarginate at the apex, 2–6cm long. Flowers pedicellate, in lateral cymes borne on a short, fleshy peduncle; corolla 3–4.7cm long, variable in colouring but usually maroon, completely glabrous, the lobes separating only slightly and never becoming free at the apex.

Two varieties are known from Yemen. Variety **rupicola** has a uniformly maroon corolla which is c 4.5cm long. In variety **stictantha** *N. Taylor* (probably the hybrid (*Ceropegia rupicola x aristolochioides*) the corolla is slightly shorter (c 4cm long) and creamy-white, spotted dark maroon.

An uncommon but conspicuous endemic plant found in scattered colonies on cliffs and on rocky hillslopes on the escarpment between 1700 and 2600m. Variety *stictantha* is only known from the *Euphorbia* scrub north of Qā'idah. Variety *rupicola* is much more common: on the hills from Ḥayfān east to J. Ghayāmī; Wadi Tha'bat, J. Ṣabir; Ḥajdah (Noltee); Qā'idah; J. 'Ayfū; J. Ḥubaysh; near Dan, Waṣāb al 'Ālī; J. Milḥān; J. Masār, 'Attārah and Manākhah in the Ḥarāz; Maḥwīt (D. Wood); Aḥjar (Choudhary).

7. **Ceropegia sepium** *Deflers*

Leafy, non-succulent twiner with glabrous or sparsely pilose stems; rootstock of thin, elongate tubers. Leaves broadly-lanceolate, acuminate, weakly cordate at the base, 4–5cm long, ± pilose on the petiole and margins. Flowers pedicellate in pedunculate, lateral umbels; corolla 2–2.5cm long, purplish in the lower half, yellow in the upper half, finely pubescent but without long hairs, the lobes joined at the apex.

A rare endemic species found in bushy places in high rainfall areas of the escarpment between 1900 and 2500m: on the lower slopes of J. Ba'dān and J. Sumārah, north of Ibb; near Jirjir, above Ḥadīyah; in a number of places around J. Shibām, Ḥarāz.

8. Ceropegia somalensis *Chiov.*

Glabrous twining succulent with stems 3–4mm thick; roots fibrous. Leaves usually present at flowering time, ovate, acuminate, base cuneate, 1–2cm long. Flowers pedicellate in few-flowered pedunculate cymes; corolla grey-green with purplish patches, shortly pubescent, c 5cm long, the lobes distinctly longer than the tube, connivent around their middle as in *C. botrys*.

Between Khamīs Banī Sa'd and Manākhah at 1050m (Noltee).

9. Ceropegia tihamana *Chaudhary & Lavranos*

Glabrous twining succulent resembling *C. arabica* and *C. botrys* in its stem, leaves and general appearance but differing from the former in its lump-shaped rootstock and from both in its very short corolla (about 2cm long) whose lobes join at the apex to form a distinct rounded knob.

Known only from Mawza' at 300m (Lavranos).

10. Ceropegia variegata *Decne.*

C. tubulifera Deflers
C. de-vecchii Chiov.

Leafless climbing succulent with fibrous roots; stems stout, much-branched, often rather scarred and knobbly, variegated green and violet, c 6–10mm thick. Flowers shortly pedicellate, in small clusters terminal on short lateral branches; corolla 3–6cm long, with a distinct inflation just below the lobes as well as at the base of the tube, white with purple spots and markings particularly at the apex, the lobes with numerous long, white or purple hairs along the margins, the tips joined.

Two distinct varieties are found in Yemen characterised by the manner in which the free parts of the corolla lobes are extended outwards. In variety **variegata** they are extended into short deltoid auricles c 5mm long but in variety **cornigera** *Huber* (*C. de-vecchii*) they are extended into conspicuous, narrow, horn-like channels 1–3.5cm long. Intermediate plants are sometimes found.

A characteristic plant of open *Acacia* scrub in the dry escarpment foothills, rarely occurring further inland but reaching about 1500m at Mafḥaq and Mirkiz As Salām, Shar'ab. The exact frequency of the two varieties is not known but variety *cornigera* appears to be more frequent and is known from between 'Abs and Suq al Khamīs, Khalīfah, Shar'ab and Misrākh. Variety *variegata* is known from Wadi Ḥaraḍ, Walajah on J. Milḥān (Schweinfurth) and J. Ra's (Botta).

Ceropegia deflersii Schweinf. is recorded in Schwartz from 'Usil in the Ḥarāz where it was collected by Schweinfurth. Unfortunately Schweinfurth never published the name and the specimen has disappeared. It was most probably *C. arabica* which was not described at that time but is known today from near 'Usil.

18. ECHIDNOPSIS *Hook. f.*

Distinctive leafless succulent herbs with cylindrical, trailing stems which are covered in a network of obtuse, square tubercles. Flowers lateral, solitary or in clusters; corolla with a short or long tube; outer corona present or absent; inner corona of 5 fleshy lobes bent over the anthers. Follicles smooth.

All our species are similar vegetatively and cannot be reliably identified without flowers.

1.	Corolla 10–15mm long, tubular with very short lobes	*E. squamulata*
	Corolla less than 5mm long, lobed to at least half way	2
2.	Outer corona cup-shaped	*E. scutellata*
	Outer corona absent	*E. cereiformis*

1. Echidnopsis cereiformis *Hook. f.*

E. nubica N. E. Br.

Stems usually with 8 rows of tubercles. Flowers in clusters near the tips, subsessile; corolla yellow, ± rotate, 5-lobed to more than half way, the lobes broadly ovate, 3–4mm long; outer corona absent; inner corona with horizontally spreading ovate-deltoid lobes.

Known only at 1400m near 'Usil in the Ḥarāz (Schweinfurth).

2. Echidnopsis scutellata *Deflers*

Stems usually with 6 rows of tubercles. Flowers in small clusters, the pedicels 1–2mm long; corolla reddish-brown becoming yellow towards the tips of the lobes, lobed to about half way, the lobes triangular-ovate, c 2mm long; outer corona ± cup-shaped; inner corona with erect, lanceolate lobes.

Apparently frequent from 400 to 1700m in the central escarpment growing around the bases of large succulents and on earth banks among rocks: Mishrāfah, Wadi Rima' Usil (Schweinfurth), Bājil (Lavranos); Wadi Sara' (Miller); between Ḥammām Shamsān and Maḥwīt; Wadi Sharas (Miller); Maḥābishah.

3. Echidnopsis squamulata (*Decne.*) *Bally*

Ceropegia squamulata Decne.

Similar in habit to the last two species but the flowers borne on pedicels up to 5mm long and the corolla dark red, tubular, 10–15mm long, with minute lobes. The outer corona is cup-shaped, the inner corona has subulate ascending lobes forming a ± erect column.

An uncommon plant of well-drained earthy places particularly around the bases of *Aloe* and *Euphorbia* where there has been some soil accumulation; it is known definitely from J. Ra's (Miller), Taiz, Aḥkūm and Raḥidah but field records from J. Ḥabashī and Bir Ghubāḥ in Shar'ab are probably of this species. It is apparently endemic to southwest Arabia.

Echidnopsis multangula (Forssk.) Chiov. though probably an *Echidnopsis* cannot be used as a name since it is impossible to know what species Forsskal intended. No specimen has survived and in any case it seems that Forsskal found a sterile plant.

19. EDITHCOLEA *N.E. Br.*

1. Edithcolea grandis *N.E. Br.*

Much-branched leafless succulent herb forming a mat of decumbent, ± cylindrical stems which are densely covered in sharp, brown, conical spines. Flowers solitary, borne close to the branch tips on short peduncles c 1.5cm long; corolla almost rotate, very large, deep maroon and yellow, 10–12cm in diameter, the tube short, the lobes ovate with deep maroon margins which are densely covered with purple vibratile hairs.

A spectacular succulent which is widespread in East Africa but in Arabia is known only from Al Barḥ at 500m (Lavranos and Newton).

20. RHYTIDOCAULON *Bally*

1. Rhytidocaulon macrolobum *Lavranos*

Erect, leafless succulent, up to 25cm high. Stem sparingly branched, ± cylindrical, waxy-grey, rugose, covered in obscure elongate protuberances but without fleshy wings or prickles. Flowers scattered over the younger parts of the stem, solitary, borne on pedicels 2–3mm long; corolla without a tube, c 15mm in diameter, the lobes ovate-deltoid, spreading, up to 6.5mm long, the lower part whitish-yellow with purple bands, the apical part uniformly dark purple, the apex with several dark purple, vibratile hairs; outer corona dark purple with deeply bifid lobes.

A rare and easily overlooked succulent found in dry rocky country from 500 to 1900m in the south: Barḥ and Wadi Ḍabāb (Lavranos and Newton); Al Ḥaradah, Damt.

21. CARALLUMA *R. Br.*

Decumbent or erect, leafless succulent herbs. Stems 4–6 angled, the angles toothed and sometimes winged. Flowers in terminal racemes or in clusters at or near the tips of the branches, rarely solitary; corolla 5-lobed, often rotate, sometimes with the fused part forming a campanulate tube, glabrous or hairy, sometimes with club-shaped vibratile hairs; corona usually double. Follicles usually paired, cylindrical but narrowed at both ends, smooth.

A fascinating genus rich in endemic species which are poorly understood taxonomically. There are undoubted taxonomic problems but a lot of the difficulties are caused by the poor quality of dried specimens and the fact that many species seem only to flower for a very short time after heavy rain. Certainly many of the more interesting species have been found during the high rainfall years of 1975, 1977 and 1981. It seems probable that more species will be found particularly in the southeast of the country. Flowers should be collected in spirit and good photographs can be a great aid to identification as dried specimens are almost useless unless very carefully made.

The species of *Caralluma* in Yemen fall into three groups which can be recognised vegetatively. *Caralluma* proper (*C. edulis*, *C. sinaica*, *C. subulata*) are erect plants with a tapering stem which forms the rhachis of the racemose inflorescence. The so-called 'Ango' group (*C. chrysostephana*, *C. commutata*, *C. deflersiana*, *C. wissmannii* and *C. sp.B*) have decumbent, variegated stems with prominent fleshy teeth and an inflorescence of one or a few flowers in a cluster near the branch tips. The remaining species are sometimes treated as a separate genus, *Desmidorchis* Decne. and are a rather heterogenous group which does not have the distinctive features of the other two groups. Apart from the vegetative characters the following descriptions concentrate on the shape, size, colour and indumentum of the corolla and this is sufficient to identify the species found in Yemen. Coronal characters often support distinctions based on the corolla but are only mentioned where they are particularly obvious.

1. Stem tapering in the upper half and forming the rhachis of the inflorescence which is clearly a raceme — 2
   Stem not tapering. Inflorescence of clusters of 1–many flowers at or near the branch tips — 4
2. Corolla completely glabrous, the fused part campanulate, about as long as the lobes. Pedicels 4–10mm long — *C. edulis*
   Corolla hairy, the fused part saucer-shaped, much shorter than the lobes. Pedicels 2–4mm long — 3
3. Corolla red-brown, glabrous except for the vibratile hairs at the tips of the lobes — *C. subulata*
   Corolla yellow-green or mauve, spotted red, uniformly covered in long purple hairs but without vibratile hairs at the tips of the lobes — *C. sinaica*
4. Stem decumbent, variegated, armed with fleshy teeth 5–15mm long — 5
   Stem erect or decumbent, uniformly glaucous or grey-green, the teeth usually reduced to small blunt projections on the angles of the stem, always less than 5mm long — 9
5. Corolla with a distinct campanulate tube almost as long as the lobes. Flowers always solitary — *C. deflersiana*
   Corolla ± rotate, usually lobed to near the base. Flowers usually 2–4 together, rarely solitary — 6
6. Corona brown. Corolla lobes broadly triangular, about as wide as long, the margins not recurved — *C. commutata*
   Corona yellow. Corolla lobes narrowly triangular, the margins recurved — 7
7. Corolla yellow, pubescent — *C. sp.B*
   Corolla reddish or red-spotted or yellow, glabrous or with white vibratile hairs, never pubescent — 8
8. Corolla with numerous white vibratile hairs. Pedicels 14–18mm long — *C. chrysostephana*
   Corolla completely glabrous. Pedicels 5–6mm long — *C. wissmannii*
9. Corolla tube distinctly longer than the lobes. — 10
   Corolla tube slightly to much shorter than the lobes — 12
10. Corolla lobes with vibratile hairs at the tips — 11
    Corolla completely glabrous — *C. sp.D*

11. Corolla dark red-brown, the lobes 5–7mm long, distinctly longer than broad *C. petraea*
Corolla dark purple-black, the lobes 2–3mm long, distinctly broader than long *C. sp. C*
12. Flowers 1–2 together at the branch tips **13**
Flowers in clusters of 5–100 at the branch tips **14**
13. Corolla and corona uniformly yellow *C. quadrangula*
Corolla lobes dark red-brown, the corona and fused parts of the corolla yellow *C. cicatricosa*
14. Corolla shortly tubular; the margins of the lobes strongly recurved **15**
Corolla ± rotate; the lobes, the margins not recurved **17**
15. Corolla pubescent, the tips of the lobes with vibratile hairs *C. plicatiloba*
Corolla completely glabrous **16**
16. Corolla dark red-brown on the inside and outside. Pedicels less than 3mm long *C. sp. E*
Corolla pale yellow with dark purple spots on the inside, greenish on the outside. Pedicels 5–7mm long *C. awdeliana*
17. Vigorous erect plants usually exceeding 40cm in height **18**
Decumbent plants **20**
18. Corolla yellow with purple-brown spots, 10–12mm in diameter, the lobes always tipped with vibratile hairs *C. penicillata*
Corolla dark reddish-purple, 14–25mm in diameter, with or without vibratile hairs **19**
19. Corolla hairy. Plant usually c 1m high *C. acutangula*
Corolla glabrous except for vibratile hairs which are sometimes present at the tips of the lobes. Plant less than 70cm high *C. sp. A*
20. Corolla tuberculate, the lobes about as broad as long. Pedicels less than 3mm long *C. hexagona*
Corolla smooth, the lobes longer than broad. Pedicels more than 5mm long *C. shadhbana*

1. Caralluma acutangula (*Decne.*) *N. E. Br.*
Ghulāf, Ghulāth

C. retrospiciens (Ehrenb.) N.E.Br.
C. hirtiflora N.E.Br.
C. russelliana (Courb. ex Brogn.) Cuf.
A massive, erect, branched succulent to 1½m. Stems grey or glaucous, quadrangular, the angles with broad wings and small recurved teeth. Flowers in a dense terminal, globose head 5–10cm in diameter, foetid; pedicels 2–5cm long; corolla 20–30mm in diameter, not fleshy, blackish-purple, hairy, ± rotate, lobed to more than half way, the lobes spreading, ovate, slightly longer than broad; corona pubescent, dark purple.

Much the largest *Caralluma* in Yemen, but uncommon being found in scattered localities on dry stony plain up to 1800m in the Tihama and the Mashriq: Ḥanish Island (Slade); between Suq al Jammah and Malāḥiṭ (flowers not seen); Ḥarf Sufyān; Dhī Bīn (Lavranos); Wadi Ḥabab, Sirwāḥ; Jūbah.

2. Caralluma awdeliana (*Deflers*) *Berger*
Erect, branched succulent to 30cm. Stems grey-green, quadrangular, obscurely toothed on the angles. Flowers in clusters at the tips of the branches; pedicels 5–7mm long; corolla 8–12mm in diameter, glabrous, pale yellow spotted dark purple on the inside, greenish on the outside, divided to slightly more than half its length, the fused part campanulate, the lobes lanceolate with revolute margins, 6–7mm long, 3–4mm wide.

Only known from a silty wadi bed at 1100m at Mārib (Miller 3441).

3. Caralluma chrysostephana (*Deflers*) *Berger*
Decumbent, branched succulent with ascending flowering stems. Stems 4-angled, grey-green with many dark spots of different shape, bearing prominent, patent, fleshy teeth 8–12mm long. Flowers 1–2 together near the tips of the branches; pedicels 14–18mm long; corolla 2.5–4.5cm in diameter, maroon, glabrous except for the prominent white vibratile hairs found towards the centre, deeply lobed almost to the base, the lobes spreading, narrowly triangular, 15–21mm long, c 5mm wide, the margins recurved; outer corona golden-yellow, with pouch-shaped lobes; inner corona with yellow lobes.

Known only from J. Ra's (Miller and Long). A rare plant found also in South Yemen.

4. Caralluma cicatricosa (*Deflers*) *N. E. Br.*
Much-branched, erect, succulent similar in general appearance to *C. quadrangula* but taller, reaching 80cm in height. Stems grey, quadrangular, the angles with small blunt teeth and numerous scars particularly towards the tips. Flowers solitary or in few-flowered clusters at the tips of the branches; pedicels 2–3mm long; corolla 15–20mm in diameter, glabrous, divided almost to the base with narrowly triangular, spreading, red-brown lobes 6–8mm long, the fused part of the corolla with a raised yellow disc covered in numerous papillae; corona apparently single, yellow, 3-lobed, the middle lobe bent over the anthers, the laterals erect.

A local plant of rocky outcrops in areas of good rainfall between 1000 and 2300m, mostly in Ibb province and the central escarpment but with an isolated record from the high plateau near Ma'bar (Radcliffe-Smith and Henchie). Apparently endemic to southwest Arabia and probably requiring a higher rainfall than any other species in Arabia.

J. Lavranos has recently questioned whether the traditional interpretation of this species adopted here is correct, pointing out that it differs in several crucial ways from Deflers' description. The matter can only be settled by examination of the type but there is a good deal of circumstantial evidence to support the usual interpretation.

5. Caralluma commutata *Berger*
Similar in vegetative habit to *C. chrysostephana* but the stems green, marbled dull red. Flowers 1–2 together near the tips of the branches; pedicels 3–4mm long; corolla 2–3cm in diameter, reddish-brown, glabrous, rotate, lobed to rather more than half way, the lobes ovate-triangular, about 7mm long and broad; outer corona bowl-shaped, unlobed, dark brown; inner corona paler.

Known only from the Manākhah area (Schweinfurth). Outside Yemen it has been found in Saudi Arabia and perhaps also in Ethiopia.

6. **Caralluma deflersiana** *Lavranos*

Similar in vegetative habit to *C. chrysostephana*. Flowers solitary at the tips of the branches; pedicels c 10mm long; corolla 2–3cm in diameter, dark purple-brown, glabrous, campanulate with a tube 10–13mm long and spreading, triangular lobes, each 10–18mm long and 7–9mm wide; corona dome-shaped, purple-brown.

A widespread and quite frequent plant of rocky plain in dry *Acacia* bushland; in suitable habitats throughout Taiz province, on rock outcrops in the eastern Tihama and in the foothills of the escarpment up to c 1300m extending north to Muḥarraq.

7. **Caralluma edulis** (*Edgw.*) *Benth.*

C. vittata N.E.Br.

Erect succulent to 20cm. Stems quadrangular with small blunt teeth on the angles, the upper half sometimes with reduced leaves, tapering and forming the rhachis of the racemose inflorescence. Pedicels filiform, 4–12mm long; corolla 10–12mm in diameter, glabrous, with a rounded campanulate tube and dark purple, spreading, subulate lobes 4–5mm long; corona dark purple.

A rare plant known only at 1900m at Majz (Steinberg) but widespread in the Sahara-Sindian region from North Africa to India.

8. **Caralluma hexagona** *Lavranos*

Prostrate, branched succulent. Stems grey-green, 4–6 angled, the angles slightly winged and with obscure blunt teeth. Flowers in clusters at the tips of the branches; pedicels 1–3mm long; corolla 15–18mm in diameter, glabrous, pale yellow with numerous red-brown spots, tuberculate, lobed to about half way, the fused part bowl-shaped, the lobes triangular, c 5mm long and wide; corona red-brown, the outer with bifid, shortly recurved lobes.

The Yemen plant is variety **septentrionalis** *Lavranos and Newton*.

A plant of rock crevices, usually in sandstone country found in scattered localities between 1700 and 2400m on the high plateau and in the west of the Mashriq: 20km east of Rada'; Wadi Ḍahr; Naqīl Ghaylān; Khulāqah in Nihm; Dhī Bīn (Lavranos).

9. **Caralluma penicillata** (*Deflers*) *N.E.Br.*

Ghulāf, Ghulāth

Erect, branched, succulent to c 1m. Stems grey-green, quadrangular, the angles slightly winged and with small recurved teeth. Flowers in dense clusters at the tips of the branches, slightly foetid; pedicels 8–10mm long; corolla 10–12mm in diameter, greenish-yellow with numerous small purple-brown spots, glabrous except for the tips of the lobes which each have a small tuft of purple-brown vibratile hairs, rotate, divided almost to the base, the lobes spreading, ovate, c 4mm long and 3mm broad; outer corona cup-shaped, hairy, pale brown.

Widespread and abundant in *Acacia* and *Euphorbia* scrub from the first rocky outcrops in the east of the Tihama at about 200m to about 1600m on the escarpment; it is also common in similar country in the southeast mountains where it grows as high as 2000m around Damt, Rada' and Bayḍā'.

Found on the mountains on either side of the Red Sea.

10. **Caralluma petraea** *Lavranos*

Ascending or erect, much-branched succulent to 30cm, related to *C. awdeliana* (Defl.) Berg. Stems grey, quadrangular, the angles rather blunt, obscurely toothed, much scarred. Flowers in dense clusters at the tips of the branches; pedicels 6–10mm long; corolla 9–12mm in diameter, dark red-brown, tuberculate, glabrous except for the tips of the lobes which bear a tuft of brown vibratile hairs, divided to about ⅓ of its length, the fused part cylindrical c 10–13mm long, the lobes spreading, triangular 5–7mm long and 4mm wide; outer corona glabrous, cup-shaped, the lobes filiform, erect.

Known only at 2300m on sandstone hills near the Sana'a-Mārib road, c 20km east of Sana'a. (Miller 3403)

11. **Caralluma plicatiloba** *Lavranos*

Much-branched, erect succulent usually forming dense clumps 10–25cm high. Stems grey-green or glaucous, sometimes pubescent when young, usually bluntly 4-angled, the angles with numerous, almost imbricate blunt teeth, each capped with a small scar. Flowers in dense clusters towards the tips of the branches, foetid; pedicels c 2mm long; corolla 15–20mm in diameter, shortly pubescent, lobed for about ⅔ of its length, the fused part ± campanulate, yellow with purple spots and lines, the lobes spreading, purple, broad at the base but abruptly narrowed into a long slender tip, 7–8mm long, c 3mm wide, the tips with white, purple-tipped vibratile hairs, the margins bent strongly downwards so the lobes are ± tubular; corona pale purple.

With *C. quadrangula*, the common *Caralluma* of the high plateau, where it is found almost everywhere on rocky plain and stony hillsides between 2000 and 2500m extending south to Bayḍā' and Damt. In Khawlan Ash Sham it is probably common in equivalent areas and has been found at Sāqayn and west of Majz.

C. plicatiloba is widely distributed in western Arabia from Jordan southwards.

12. **Caralluma quadrangula** (*Forssk.*) *N.E.Br.*

Much-branched, clump-forming, erect, grey succulent to 50cm. Stems quadrangular, the angles often scarred and with small blunt teeth. Flowers in few-flowered clusters (sometimes solitary) at the tips of the branches; pedicels 2–3mm long; corolla 15–20mm in diameter, very fleshy, yellow, glabrous, divided almost to the base with narrowly triangular spreading lobes 6–8mm long; corona yellow.

Two colour forms are found. In the south particularly Taiz province, the corolla and corona are uniformly pale yellow while further north the corolla and corona are a rich yellow.

The most widespread *Caralluma* in Yemen, being found on dry rocky hills from 500 to 2500m on the escarpment, high plateau and western part of the Mashriq. It is absent only from the wettest parts of Ibb province and the central escarpment.

Endemic to southwest Arabia.

13. Caralluma shadhbana *Lavranos*
Prostrate, branched succulent. Stems grey-green, quadrangular, the angles slightly winged and with small teeth. Flowers in dense clusters at the tips of the branches; pedicels 6–8mm long; corolla 11–14mm in diameter, purple-brown, ± rotate, lobed to slightly more than half way, the lobes broadly triangular, c 5mm long and 4mm wide, vibratile hairs present on the tips (the type) or absent (variety **barhana** *Lavranos and Newton*); outer corona purple, the lobes bifid with recurved tips; inner corona dark purple.

Widespread and apparently frequent on the escarpment from the Taiz area north to Khawlan Ash Sham, growing on stony hills and on rock outcrops between 500 and 1500m, usually in open *Acacia* bushland. All specimens seen in Yemen have been var. *barhana* except the type which was collected in Wadi Shadhbān, east of Manākhah.

Outside Yemen it is only known from Saudi Arabia.

14. Caralluma sinaica (*Decne.*) *Berger*
C. maris-mortui Zohary
Erect succulent to 18cm. Stems quadrangular with small blunt teeth on the angles, the upper half tapering and forming the rhachis of the racemose inflorescence. Pedicels 2–4mm long; corolla 10–12mm in diameter, yellow-green with scattered red spots, densely covered with long pale purple hairs, deeply lobed almost to the base, the lobes spreading, narrowly triangular 5–6mm long; outer corona reddish-purple; inner corona hairy with erect subulate lobes.

Only known from Al Barḥ at 500m (Lavranos and Newton).

15. Caralluma subulata (*Forssk.*) *Decne.*
Much-branched, decumbent or erect succulent to 30cm. Stems quadrangular with small blunt teeth on the angles, the upper half usually with reduced leaves tapering and forming the rhachis of the racemose inflorescence. Flowers pendulous; pedicels 2–4mm long; corolla 11–14mm in diameter, deeply lobed nearly to the base, the fused part saucer-shaped, white with red-brown spots, the lobes c 6mm long, linear, spreading, rich red-brown, with many red-brown vibratile hairs near the tips; outer corona white; inner corona red-brown.

Much the most common *Caralluma* in the Tihama, being found everywhere except on the coastal fringe, usually growing on rock outcrops or on hard alluvial banks at the bases of trees and in fences around villages. It also grows in the drier parts of the escarpment, particularly in Taiz province, and reaches 1300m in Wadi Ḥidrān near Taiz. It is endemic to southwest Arabia.

16. Caralluma wissmannii *O. Schwartz* Kusmah
C. meintjesiana Lavranos
C. eremastrum O. Schwartz
Rather similar to *C. chrysostephana* but differing in the much shorter pedicels (5–6mm long), the entirely glabrous corolla and the reddish inner corona lobes.

As understood here, *C. wissmannii* is a very variable species. The corolla shows variation in texture and colour. The inner surface may be wholly or partical tuberculate or completely smooth. The basic corolla colour is dull yellow with numerous small reddish spots but in most specimens the red is so prominent that the yellow is not obvious without a lens. In other specimens the red is almost absent and the tips of the lobes in particular may be wholly yellow. There is also considerable variation in the inner corona. In the type of *C. wissmannii* and in most recent collections, the inner corona bears a short erect bifid appendage while the lobes are bent over the anthers, ± concealing them. Another form represented in Yemen only by the type collection of *C. eremastrum* from Ḥajjah (Rathjens) has very short tooth-like lobes which leave the anthers uncovered. A third form known only from Qā'idah (Wood 1125, Radcliffe-Smith and Henchie 4424) is like the type of *C. wissmannii* but the appendages on the inner corona are very prominent, being 2–3mm long. Further investigation of all these forms is desirable.

Widespread and frequent in rock crevices and on terrace walls from 1800 to 3000m growing throughout the high plateau and on the escarpment from J. Bura' and the Ḥarāz northwards; unknown in the south of the country except at Qā'idah.

17. Caralluma sp.A
Erect, much-branched, clump-forming succulent to 70cm. Stems glaucous or grey-green, 3–4 angled, the angles slightly winged and with small, recurved teeth. Flowers in a terminal head, foetid; pedicels 7–10mm long; corolla 14–15mm in diameter, dark reddish-purple, glabrous, rotate, lobed to more than half way, the fused part of the corolla with a distinct paler, raised disc-like area, the lobes ovate, slightly longer than broad, with or without purplish vibratile hairs at the tips; corona glabrous, dark reddish-purple.

Locally common on dry stony ground in the foothills of the southwest escarpment between 700 and 1200m: J. Ra's (Miller 3499); over a wide area of Maqbanah centred on Ḥajdah (Wood 915, 2640).

18. Caralluma sp.B
Similar in vegetative habit to *C. chrysostephana* but the teeth rather shorter and usually 5mm long. Flowers in small clusters near the tips of the branches; pedicels c 3mm long; corolla 2–3cm in diameter, pale yellow, pubescent, deeply lobed almost to the base, the lobes spreading, narrowly triangular, 10–11mm long, 3–4mm wide, the margins recurved; corona cup-shaped, bright yellow.

An uncommon species found on cliffs and steep grassy slopes on the escarpment between 1500 and 2100m: Jiblah (Miller and Long 3256); between Ḥadādah and Al Jabin, J. Raymah (Wood 1913); Mafḥaq (Wood 694).

19. Caralluma sp.C
Similar to *C. petraea* in vegetative growth, corolla size and in the presence of vibratile hairs on the tips of the lobes but differing in the blackish-purple colour of the corolla and in the short, broadly triangular corolla lobes which are only 2–3mm long but 4–5mm in width.

Known only from Wasit, 20km north of Ṣa'dah (Miller 3176).

20. Caralluma sp. D

Apparently very similar to *C. petraea*, differing only in the corolla which is yellow with numerous red-brown spots and not tuberculate, the fused part ± campanulate and the lobes without vibratile hairs at the tips.

The only material of this species I have available is a photograph so it is not possible to be certain of its identity, though it is clearly not conspecific with any species at present known from North Yemen. It is close to *C. kalmbachiana* Lavranos from South Yemen with which it may be conspecific.

Known only from stony plain about 10km north of Al Ḥarf (Miller 3159).

21. Caralluma sp. E

Rather similar to *C. plicatiloba*, having the same type of stem and a similar inflorescence with red-brown flowers borne on short pedicels 2–3mm in length but differing in the completely glabrous, predominantly red-brown corolla whose narrowly triangular lobes taper evenly to the apex and are only about 6mm long.

A rare plant known only at 2300m about 20km northeast of Sana'a on the road to Mārib (Miller 3402).

Three species of *Caralluma* recorded in Schwartz have been excluded. *C. dentata* (Forssk.) Blatter is rejected because there is no way of knowing what the plant Forsskal collected was without either a type specimen or a description of the flowers. It may well have been a *Huernia* in any case. The record of *C. sprengeri* (Schweinf. ex Hort. Damm.), N.E.Br. is based on the assumption by Berger that a sterile specimen of an 'Ango' group *Caralluma* collected by Schweinfurth on J. Milḥān must be that species. Since *C. sprengeri* has not been collected subsequently in Arabia but several other members of the 'Ango' group have, the record cannot be accepted. *C. arabica* N.E.Br. is excluded because it is only otherwise known from the type collection from Hadramaut and Schwartz would not have seen the type. It is impossible to guess what plant Schwartz actually saw.

22. DUVALIA *Haw.*

Leafless succulent herbs with the variegated stems and large fleshy teeth of *Huernia* and the *Caralluma ango* group but differing vegetatively in that new shoots are produced from the upper part of the old stem. Inflorescence lateral, consisting of 1–4 flowers borne on a short, fleshy peduncle arising from near the base of the stem; pedicels 2–5cm long; corolla ± rotate, large; corona double. Technically the genus is distinguished by its stipitate corona.

1. Corolla completely glabrous — *D. somaliensis*
   Corolla variously hairy, either shortly pubescent or with vibratile hairs — 2
2. Corolla with vibratile hairs on the margins of the lobes — *D. sulcata*
   Corolla shortly pubescent on the inside but without vibratile hairs — *D. velutina*

1. Duvalia somaliensis *Lavranos*

Stem prostrate, quadrangular with patent fleshy teeth 8–10mm long, glaucous, mottled purple. Inflorescence with a short fleshy peduncle from which arises 1–4 flowers borne on pedicels c 3cm long; corolla 3–3.5cm in diameter, glabrous, fleshy, yellow with large red-brown spots and lines, ± rotate, deeply lobed to near the base, the fused part with a prominent raised disc (annulus), the lobes with strongly recurved margins, deltoid, 15mm long and 7mm wide; outer corona disc-like, yellow with red-brown spots; inner corona white with five erect horn-like appendages.

Known only from stony plain at about 300m in the Mafraq area (Lavranos).

2. Duvalia sulcata *N.E.Br.*

Similar to the last species from which it is best distinguished by the presence of purple vibratile hairs along the margins of the corolla lobes. In addition the corolla is uniformly dark purple and is hardly fleshy so dried specimens of this plant are much more satisfactory than are those of most other succulent Asclepiads; the margins of the corolla lobes are at most weakly recurved.

There is much variation in the hairiness of the annulus, the type being pilose while the only collection from Yemen has an almost glabrous annulus and is placed in variety **seminuda** *Lavranos*.

Only known on sandy plain at 300m west of Aṭ Ṭūr (Muller-Hohenstein and Diel 200).

3. Duvalia velutina *Lavranos*

Similar to the previous two species but the inner surface of the corolla shortly but densely pubescent.

This species appears to be very variable in size, the pedicels varying from 2–5cm long and the corolla 1.5–4cm in diameter. The corolla is always fleshy as in *D. somaliensis*.

A rare plant found on alluvial plain in the northern Tihama: in the protection of *Capparis decidua* at Ḥaraḍ (Wood 2682) and of *Aloe vera* at Suq al Jammah (Wood 3055).

23. HUERNIA *R. Br.*

All of our species are vegetatively alike and share the following characters. They are small succulent herbs, often spreading by horizontal rhizomes to form large patches. Flowering stems similar to the *Caralluma ango* group, short, erect, 4–5-angled, variegated, with fleshy teeth 5–10mm long. Flowers in clusters of 1–5 borne on a short fleshy peduncle arising from near the base of the stem by which the flowers are often ± concealed; pedicels 5–10mm long; corolla large, fused for rather more than half its length, the lobes spreading, triangular, acuminate; outer corona lying flat against the back of the corolla, reddish-purple, 5-lobed; inner corona with fine subulate lobes.

***Huernia* can be distinguished from all other succulent *Asclepiadaceae* occuring in Yemen by the presence of small teeth in the sinuses between the lobes. All species are called 'Kusmah'.**

1. Corolla tuberculate or with papillae on the inner surface, the fused part campanulate, without a flat disc-like area **2**
   Corolla entirely smooth, both inside and outside, the fused parts with a short funnel-shaped tube surrounded by a flat disc-like area *H. laevis*
2. Inner surface of the corolla covered in long papillae c 1mm in length. Inner corona lobes erect, not convergent *H. lodarensis*
   Inner surface of the corolla tuberculate. Inner corona lobes bent over the stigma and convergent on each other *H. macrocarpa*

1. Huernia laevis *J.R.I. Wood*

Corolla c 15mm long, the fused part with a short funnel-shaped tube c 4mm long surrounded by a flat disc-like area, the lobes triangular, c 10mm wide and 8mm long, both surfaces quite smooth, the inner pale yellow with elongate maroon markings. Inner corona with subulate lobes bent over the stigma, convergent.

A rare species known only from between 1800 and 2600m in Khawlan Ash Sham: Madaq (Steinberg 1101); J. Marran (Wood 3037).

2. Huernia lodarensis *Lavranos*

Corolla 12–18mm long, the fused part campanulate, the lobes 5–9mm long and wide, the inner surface cream with numerous purple spots, covered in papillae about 1mm long, the outer surface minutely tuberculate. Inner corona lobes subulate, erect, not convergent, c 3.5mm long.

The two collections known from Yemen differ somewhat in their corolla dimensions and further collections are much needed to assess this variation and to clarify the relationship between this species and *H. saudi-arabica* D.V. Field.

Known only on open stony hillsides between 1000 and 1600m near Dalīl (Wood 2779) and on the escarpment below Maḥābishah (Wood 3271).

3. Huernia macrocarpa (*A. Rich.*) *Spreng.*

H. penzigii N.E.Br.
H. arabica N.E.Br.

An extremely variable plant, almost every collection differing in some respects from every other. Corolla 10–20mm long, the fused part campanulate, the lobes 5–9mm wide, 7–12mm long, sometimes broader than long, sometimes longer than broad, the inner surface always tuberculate but variable in colour, the outer surface varying from tuberculate to almost smooth (as in Wood 1980 which approaches subspecies *concinna* (N.E.Br.) M.G. Gilbert). Inner corona lobes subulate, the tips bent over the stigma and ± convergent.

At least three forms can be distinguished in Yemen, based mainly on colour. The most distinct is variety **arabica** (N.E.Br.) White and Sloan. This has a small uniformly blackish-purple corolla 10(–15)mm long and is frequent between 500 and 1500m on J. Bura', J. Milḥān, the Ḥarāz and in the Maḥwīt region (Wood 1202, 2524, 2581, 2622). The second variety has a uniformly red-brown corolla and is only known from 1500–2000m at Maḥwīt (Miller 3359) and Wadi Sukhnah (Wood 2969). In the third variety, which is closest to the type of *H. macrocarpa*, the corolla is cream with purple spots and may be up to 20mm long. This is quite frequent between 1000 and 1800m in areas of good rainfall in the southern escarpment: J. Raymah, 'Udayn, Ibb and the Taiz area (Wood 1700, 1980, 2236).

SOLANACEAE

Herbs and shrubs with alternate leaves; stipules absent. Flowers solitary, in racemes or cymes; sepals united; corolla 5-lobed, usually regular; stamens 5. Fruit a berry or a capsule.

An economically very important family containing such plants as the potato, tomato, tobacco and chilli, all of which are cultivated in Yemen. Many species contain alkaloids and have been much used in traditional medicine, though when used in quantity the plants are, of course, poisonous.

1. Fruit enclosed by the unarmed calyx **2**
   Fruit not enclosed by the calyx, if nearly so, calyx spiny **5**
2. Fruit a berry. Flowers axillary **3**
   Fruit a capsule. Flowers in terminal cymes, racemes or panicles **4**
3. Flowers solitary 2. *Physalis*
   Flowers in clusters 3. *Withania*
4. Inflorescence a terminal raceme-like scorpioid cyme. Leaves deeply incised 6. *Hyoscyamus*
   Inflorescence a terminal panicle. Leaves entire 7. *Nicotiana*
5. Corolla saucer-shaped, tube very short 1. *Solanum*
   Corolla with a tube at least 1 cm long **6**
6. Stem spiny. Fruit an unarmed berry 4. *Lycium*
   Stem unarmed. Fruit a spiny capsule 5. *Datura*

1. SOLANUM *L.*

Herbs or shrubs, armed or unarmed, often stellate-hairy. Flowers usually in leaf-opposed cymes, sometimes reduced to umbels, less commonly solitary or in racemes; calyx 5-lobed, not tubular; corolla saucer-shaped with 5 spreading lobes and a very short tube, violet or white in colour. Fruit a globose berry.

A complex genus in which nearly all the species are very variable, particularly in indumentum.

1. Plant without spines 2
   Plant armed with spines 6
2. Climbing plant. Inflorescence strictly terminal *S. terminale*
   Erect or decumbent herbs or shrubs. Inflorescence lateral, leaf-opposed 3
3. Herbs, usually annual, glabrous or with simple hairs 4
   Undershrubs, always perennial, stellate hairs always present at least on new growth at the top of the stem 5
4. Berry orange (often drying black) *S. villosum*
   Berry black when fresh *S. nigrum*
5. Flowers in racemes, white. Inflorescences numerous arising opposite most leaves on the stem *S. platacanthum*
   Flowers in cymes or corymbs, violet. Inflorescences usually 1 or 2 at the top of the stem only *S. schimperianum*
6. Herbs. Spines straight 7
   Shrubs. Spines usually falcate, rarely straight 8
7. Calyx enlarging and becoming very spiny in fruit. Spines mostly less than 5mm long *S. coagulans*
   Calyx not enlarging in fruit, sparsely spiny. Some spines more than 10mm long *S. surattense*
8. Leaves ovate, ± entire, less than 5cm long 9
   Leaves varied, but if less than 5cm long, neither ovate nor entire 10
9. Flowers solitary *S. cordatum*
   Flowers in cymes *S. forskalii*
10. Ripe fruit less than 1cm in diameter 11
    Ripe fruit more than 2cm in diameter *S. incanum*
11. Stem stellate-hairy. Flowers white *S. platacanthum*
    Stem sub-glabrous. Flowers blue *S. glabratum*

1. Solanum coagulans *Forssk.* Baqām, Buqīm

S. dubium Fresen.

Erect, spiny, woody annual to 1m, the stems shortly white-tomentose. Spines straight, white, up to 5mm long. Leaves ovate or oblong, 3–9cm long, stellate-hairy on both surfaces and grey below, sinuate-lobed, spiny on the petiole and main veins. Flowers in small cymes; corolla violet. Berry c 1cm in diameter, partially enclosed by the calyx lobes which enlarge and become spiny in fruit.

An abundant field weed in the alluvial plains in the east of the Tihama and in the main valleys extending into the escarpment up to about 1000m; also occurring as a rare casual at higher altitudes and found up to c 2200m near Khamir.

2. Solanum cordatum *Forssk.* Ḥadaq

S. gracilipes Decne.
S. hadaq Deflers

Decumbent or ascending, spiny undershrub to 30cm, the stems glabrescent, usually scurfy-stellate-hairy when young. Spines straight or slightly curved, 4–5mm long. Leaves ovate, rounded, 1–2(–3)cm long, completely unarmed, margin usually entire, very variable in indumentum, sometimes subglabrous, sometimes ± pilose, sometimes glandular-hairy (Wood 2727), most commonly thinly stellate-hairy (Wood 2765). Flowers solitary, long pedicellate, the pedicel often bent; corolla violet. Berry orange-red, 6–7mm diameter.

A characteristic plant of stony hillsides and open *Acacia* scrub in dry country: rather local on the escarpment and mostly found between 1000 and 1500m in Taiz province or in rainshadow areas elsewhere; in the Mashriq it is widespread though easily overlooked in the dry season and grows up to 2400m.

S. cordatum is a good example of an Indian species which extends west to Yemen but is only found in Africa in Somalia.

3. Solanum forskalii *Dunal*

S. albicaule Kotschy ex Dunal

Decumbent or erect, spiny undershrub to 30cm, the young stems densely white-stellate-hairy. Spines usually strongly recurved, 4–8mm long. Leaves ovate, obtuse, 2–3cm long, the margin repand (rarely entire), both surfaces ± densely stellate-hairy, unarmed except for a few obscure spines on the petiole. Flowers in small cymes; corolla pale violet. Berry orange, 7–8mm in diameter.

An uncommon but probably overlooked plant of dry *Acacia* scrub in the escarpment foothills between 400 and 1000m: J. Ra's (Miller); Ḥajjaylah (Schweinfurth); Surdūd (Forsskal); Walajah, J. Milḥān (Schweinfurth); Wadi Mawr (Ehrenberg), between 'Abs and Maḥābishah (Wood 3260, Miller 3494).

4. Solanum glabratum *Dunal* Ḥadaq, Ḥuraysh

S. palmetorum Dunal
S. sepicula Dunal

Erect undershrub to 70cm, stems ± glabrous, very spiny. Spines hooked, yellow, 4–7mm long. Leaves very variable in shape and size, lanceolate to oblong, 2–10cm long, entire, coarsely-sinuate or sub-hastate, subglabrous or thinly stellate-hairy on both surfaces, the petiole and the midrib spiny. Flowers 1–5 in small cymes; corolla violet. Berry orange-red, c 8mm in diameter.

The two following varieties can generally be distinguished in Yemen, though intermediates occur:

Variety **glabratum** is relatively sparingly branched, with fewer spines and subglabrous leaves 4–10cm long (Wood 2456, 2564, 3437; Schweinfurth 204, 663).

Variety **sepicula** (*Dunal*) *J.R.I. Wood* stat. nov. (*Solanum sepicula* Dunal in DC. Prod. XIII (1) 283 (1852)) is intricately branched with entwined stems, numerous spines and sparsely stellate-hairy leaves mostly less than 3cm long (Wood 2485, 3222, 3341; Deflers 455, 493).

The species as a whole is widespread and abundant in the east of the Tihama, on the escarpment, high plateau and southeastern mountains, growing from 200 to 2800m on roadsides, field borders, disturbed rocky slopes and abandoned fields. Variety *glabratum* occurs mainly (exclusively ?) in the Tihama and the foothills; variety *sepicula* is most common on the high plateau.

5. **Solanum incanum** *L.* Nuqum, 'Irsim

Erect, much-branched, spiny shrub to 2m, the stems densely stellate-hairy. Spines broad-based, curved, 5–10mm long. Leaves ovate, obtuse, 5–15mm long, the base very unequal, the margin sinuate, both surfaces stellate-tomentose, the petiole and midrib usually spiny. Flowers solitary or in small racemes of up to 6; corolla violet, large, the lobes c 1cm long. Berry yellow, 3–4cm in diameter.

One of the most conspicuous and common plants in Yemen found in disturbed places such as roadsides, field borders and stony flood beds and absent only from the Tihama, the Sands and mountains above 2600m since it is very sensitive to severe frosts. Known as 'Sodom Apple' in English, it is widely known for its poisonous properties in Yemen.

Three collections, all from single bushes (Wood 1939 from Ḥiliyal, J. Sumārah, Wood 2554 from below Ṭawīlah by the bridge on the Maḥwīt footpath and Wood 3223 from Manākhah School) show characters intermediate between *S. incanum* and *S. platacanthum*. The three plants are not identical but all have the violet flowers and large spines of *S. incanum* but the corolla lobes are shorter (7–8mm long), the fruit smaller (1–1.5cm diameter) and the flowers always in racemes. Since both *S. incanum* and *S. platacanthum* grow in the immediate neighbourhood of these plants it seems probable that they are hybrids.

6. **Solanum nigrum** *L.*

Erect, glabrous or pubescent annual herb to c 50cm, unarmed and without stellate hairs. Leaves ovate, acute, up to 8cm long, the margin entire or sinuate-lobed. Flowers in small lateral cymes or umbels; corolla white, rather small. Berry black, c 8mm diameter.

Although there are numerous old records of this species from the escarpment and high plateau, I very much doubt whether it occurs in Yemen. All plants I have seen have orange fruit and are thus referable to *S. villosum*. Herbarium specimens may have black fruit but this cannot be regarded as diagnostic as the orange fruit of *S. villosum* frequently becomes black in the drying process.

7. **Solanum platacanthum** *Dunal*

Erect, much-branched spiny shrub to 1m (unarmed in Wood 2857 from Ṣa'bah, J. Milḥān), stem stellate-hairy. Spines strongly curved, 4–6mm long. Leaves ovate, obtuse, 2–12cm long, base unequal, margin entire or sinuate, both surfaces stellate-hairy and grey, the petiole and midrib usually spiny. Flowers 3–10 in small lateral racemes; corolla white, the lobes 5–10mm long. Berry orange, c 8–9mm diameter.

Widespread on the escarpment between 1500 and 2800m from Turbah north to J. Rāziḥ growing on roadsides, field borders, steep eroded hillsides and scrub-filled cliff gullies, often associated with *Vernonia leopoldii*, *Acanthus arboreus* and *Pennisetum unisetum*. It is endemic to Yemen but is closely related to several other species including *S. adoense* Dunal from Ethiopia.

8. **Solanum schimperianum** *Hochst. ex A. Rich.*

S. polyanthemum Hochst. ex A. Rich.
S. carense Dunal

An erect, branched, unarmed shrub to 1½m, stems stellate-hairy above, glabrous below. Leaves ovate, acute, ± entire, 3–12cm long, glabrous or stellate hairy on both surfaces. Flowers 5–25 in lateral, often forked cymes, which are always few in number and near the top of the stem; corolla violet (rarely white). Berry orange-red, c 7mm diameter.

The common Yemen plant is variety **carense** (*Dunal*) *J.R.I. Wood* stat. nov. (*Solanum carense* Dunal in DC. Prod XIII (1), 105 (1852)). This has nearly glabrous leaves and is common on rocky hills between 1500 and 2900m in the southeastern mountains, on the high plateau and in Khawlan Ash Sham and is often associated with *Psiadia*, *Euryops* and *Clutia*. (Wood 2652, 2944)

It is very probable that variety *schimperianum* with densely stellate-hairy leaves does not occur in Arabia, but plants with a heavier indumentum than typical variety *carense* occur on the escarpment and have been called variety **subglabrum** *Bitter*. This is an uncommon plant known from J. Bura' and Manākhah (Schweinfurth), J. Milḥān (Wood 3449), Suq al Jum'ah, J. Ash Shirq (Radcliffe-Smith and Henchie 5021) and at Majz (Steinberg 146).

9. **Solanum surattense** *Burm. f.*

S. xanthocarpum Schrad.
S. armatum Forssk.

Prostrate branched herb, stems spiny, ± glabrous except on new growth which is sparsely stellate-hairy. Spines straight, yellow, 7–15mm long. Leaves oblong, deeply pinnately lobed, base ± truncate, thinly stellate-hairy on both surfaces, spines present on the petiole and the main veins. Flowers 3–8 in small lateral racemes or cymes; corolla violet, large. Berry yellow, c 15mm diameter.

An uncommon plant of sandy and gravelly flood beds found between 100 and 1200m in the area bounded by the Wadi Rima', Suq Ar Rūbū' in Waṣāb al'Ālī, Ḥays and Zabīd.

10. **Solanum terminale** *Forssk.*

Unarmed climber, stems sparsely stellate-hairy above, glabrous below. Leaves elliptic, narrowed at both ends, 4–8cm long, ± entire, stellate-hairy especially below. Flowers many, in terminal cymes or umbels, corolla white or very pale violet. Berry orange, 5–6mm diameter.

A very local plant of deep-shaded cliff gullies in the mist zone of the escarpment mountains from 1800 to 2800m: above Jabin, J. Ba'dān; Manākhah; in a gully below Al Jirjir, above Ḥadīyah; in a gully north of Ṣa'bah, J. Milḥān; Bayt Bashārī, Maḥwīt; between Bayt 'Adhāqah and Suq Ath Thalūth, Wadi Lā'ah.

S. terminale is often recorded by herbarium botanists in error for *S. schimperianum* var. *carense*, but although the latter may have nearly terminal inflorescences, there is always a leaf bud above the inflorescence and some clearly lateral inflorescences can usually be found.

11. Solanum villosum (*L.*) *Lam.* 'Anamnamah

S. hirsutum Dunal

An extremely variable decumbent or erect annual (or short-lived perennial ?) herb to 1m, glabrous or villous but always unarmed and without stellate hairs. Leaves ovate or elliptic, acute, entire or sinuate-lobed, up to 10cm long. Flowers 2–8 in lateral cymes or umbels; corolla white or violet, very variable in size. Berry orange-red when ripe, c 5–8mm diameter.

This species has been divided recently into the two following subspecies but they do not really do justice to the welter of forms found in Yemen:

Subspecies **villosum** is a villous plant with glandular hairs and smooth stems that lack dentate ridges. The most common form in Yemen (Wood 74/265) has an erect woody but annual stem and a white corolla whose lobes are 5–6mm long. It is common in nitrogen-rich habitats in villages between 1500 and 2800m on the escarpment and high plateau. Another common form (Wood 75/742) is perhaps perennial and has a procumbent stem and a violet or white corolla. It is a plant of terrace walls on the mountains around the high plateau reaching 3400m on J. Nabī Shu'ayb. In addition to these two predominant forms, plants occur with a very large corolla (lobes 1cm long);, with various forms of leaf lobing and with characters intermediate between the more extreme forms. Plants with strongly sinnate-lobed leaves have been called *S. grosse dentatum* A. Rich but this seems only to be part of the variation in this species.

Subspecies **puniceum** (*Kirschleger*) *Edmonds* is a glabrescent plant, whose stem has dentate longitudinal ridges. The Yemen form of this subspecies is quite uniform, having a very small white corolla whose lobes are about 2mm long. It is almost always found as a weed, particularly favouring irrigated fields or high rainfall areas. It is widely distributed between 800 and 2500m on the escarpment, but is much less common on the high plateau, being recorded only from Sana'a, Dhamār, Rada' and Raḍmah.

The potato, **Solanum tuberosum** *L.*, 'Batātā' in Arabic, is an increasingly important crop. The main centres of production are in Ibb province, especially near Dhī Sufāl and around Kitāb but there is a huge potato project at Dhamār and production should increase significantly in the future. Potatoes are still marginal to most Yemenis' diet being eaten in side dishes with lunch or as a snack food in the markets.

The aubergine, **Solanum melongena** *L.*, 'Badhinjān' in Arabic, is cultivated for the markets in Hodeida, Taiz and Sana'a but is not an important crop.

The tomato, **Lycopersicum esculentum** *Mill.* (*Solanum lycopersicum* L.), 'Tumatim' or 'Tumatah', is in a closely related genus. It is one of the most popular vegetables in the country and is cultivated in various varieties everywhere. In general supplies come from the high plateau in summer and from the Tihama in winter. Production in the Ibb region is unexpectedly low.

2. PHYSALIS *L.*

Herbs. Flowers solitary in the leaf axils; calyx enlarging and becoming loose and papery in fruit. Fruit a globose berry, completely enclosed by the calyx.

The following two species, both introductions from America, are extremely closely related and may only be varieties of the same species. Their individual distribution is not known accurately, but they are field weeds found on the escarpment between 500 and 1800m, commonly in Taiz province, but elsewhere mostly in irrigated fields in the foothills.

Mature fruiting calyx more than 2cm long	*P. angulata*
Mature fruiting calyx less than 2cm long	*P. micrantha*

1. Physalis angulata *L.*

Glabrous or pubescent, erect or decumbent annual herb. Leaves ovate, dentate, up to 6cm long, base rounded to cuneate. Corolla very pale yellow, the lobes about 4mm long. Fruiting calyx c 2.5–3cm long.

Known with certainty from the wadi above 'Alūjah in J. Raymah (Wood 1369) and from Mishrāfah in the Wadi Rima' (Wood 74/423).

2. Physalis micrantha *Link*

P. minima L.

Very similar to the previous species but usually pubescent and the fruiting calyx slightly less than 2cm long.

Known with certainty from Ḥajjaylah (Schweinfurth) and Suq Ar Rūbū', Waṣāb Al 'Ālī (Wood 2453).

3. WITHANIA *Pauq.*

1. Withania somnifera (*L.*) *Dunal* 'Ubab

Erect, woody, pubescent perennial herb to 1m, much branched from the base. Leaves dull green, elliptic, usually up to 12cm long. Flowers 2–6, subsessile in axillary clusters; corolla campanulate, greenish, c 6mm long; calyx enlarging, becoming papery and enclosing the globose berry.

Abundant from 100 to 2900m on roadsides, field borders, ditch sides and along wadis everywhere in Yemen except for the coastal fringe, the Sands and the mountain peaks.

4. LYCIUM *L.*

1. Lycium shawii *Roem. & Schult.* 'Awsaj

L. persicum Miers

L. arabicum Schweinf. ex Boiss.

Small spiny shrub to 1m with many grey, pubescent or glabrous branches often forming a spiny entwined mass. Spines straight. Leaves in small clusters at the base of the spines, oblanceolate, up to 3cm long, pubescent or glabrous, entire. Flowers usually solitary, on fine pedicels arising with the leaves; corolla tubular, c 10–12mm long, blue-violet. Fruit a red globose berry.

A common plant of field borders, wadi sides and other disturbed bushy places between 800 and 3000m in the Mashriq, high plateau and southeastern mountains; absent from the escarpment.

5. DATURA *L.*

Vigorous herbs or undershrubs, often foetid and poisonous. Flowers large, solitary; calyx tubular; corolla funnel-shaped with a long tube and 5 very short lobes. Fruit a capsule, armed with spines or blunt protuberances.

All species are called 'Manj' in Arabic.

1. Corolla more than 10cm long. Fruit pedicel recurved **2**
   Corolla up to 8cm long. Fruit pedicel erect *D. stramonium*
2. Stem and leaves softly grey-pubescent. Fruit spiny *D. innoxia*
   Stem and leaves sparsely pubescent, never grey. Fruit covered in blunt protuberances *D. metel*

1. Datura innoxia *Mill.*

? D. metel sensu Deflers

Erect, softly grey-pubescent, sparingly branched annual herb to 1m. Leaves unequal-based, ovate, up to 15cm long, the veins failing to reach the repand margin. Calyx c 9–10cm long with narrowly triangular lobes; corolla white, c 12cm long. Capsule on a deflexed pedicel, covered in numerous spines about 1cm long.

A common weed of fields and waste ground in villages in the southwest. It grows from 100 to 1900m but though occurring in the Tihama is most characteristic of dry cultivated hills in the Ḥujarīyah, Maqbanah and Waṣāb; not known in Ibb province or north of the Wadi Rima'. Like *Argemone* this species was probably introduced through Aden and is slowly spreading northwards.

2. Datura metel *L.*

D. fastuosa var *alba* (Nees) Clarke

Erect, thinly pubescent, dark green bushy annual to 1½m with woody, often purple brown, stems. Leaves long petioled, equal-based, ovate, up to c 10cm long, the veins reaching the margin which is entire or with a few large teeth. Calyx 6–8cm long with broadly triangular lobes; corolla white, 12–15cm long. Capsule on a deflexed pedicel, covered in short blunt protuberances.

A very poisonous plant, originally introduced from India and now locally frequent between 400 and 1200m along the escarpment from Maqbanah to Shahārah. It usually grows by streams and on wadi margins but sometimes occurs as a weed of waste ground in villages.

3. Datura stramonium *L.*

Erect, thinly pubescent, dark green annual, rarely exceeding 60cm in Yemen. Leaves equal-based, ovate, up to c 10cm long, the veins reaching the coarsely dentate margin. Calyx 3–4cm long with broadly triangular lobes; corolla white, 5–8cm long. Capsule on an erect pedicel, covered in spines 3–5mm long.

Originally from America, this famous poisonous plant, known as Jimson Weed or Thorn Apple, is widely used in traditional medicine as a sedative and an anaesthetic since it can readily induce coma.

A widespread weed of fields and waste places found from 500 to 2500m wherever there is cultivated land. It is rare below 1000m and since it is also very susceptible to frost it is most common around 1500–2000m.

6. HYOSCYAMUS *L.*

1. Hyoscyamus albus *L.*

Erect, sticky glandular-hairy annual or biennial herb to 40cm. Leaves petiolate, ovate, coarsely incised, up to 5cm long (sometimes much larger below). Flowers solitary in the leaf axils forming a leafy terminal raceme-like scorpioid cyme; corolla funnel-shaped, irregular, yellow; calyx urceolate, enclosing the capsule and becoming papery in fruit.

A rare plant of waste ground in villages between 2300 and 2700m: Dhamār (Miller); Thilā; Bāb al Aḥjar; Shahārah.

7. NICOTIANA *L.*

1. Nicotiana rustica *L.* — Titūn, Tumbac

Erect, sticky glandular-hairy annual herb to 1m. Leaves petiolate, ovate or elliptic, obtuse, 4–16cm long. Flowers in a terminal panicle; corolla tubular, yellow, 12–20mm long; calyx enlarging in fruit to ± enclose the globular capsule.

Formerly cultivated (according to Deflers) and now locally naturalised on terrace walls in villages on the escarpment and high plateau between 1500 and 3000m, mostly in small quantity and in areas of high rainfall.

Nicotiana tabacum *L.* also known as 'Titūn' or 'Tumbac' is occasionally cultivated in the Tihama and on the escarpment up to c 2000m, but does not appear to be naturalised. It has sessile, oblanceolate acuminate leaves which are much larger and reach 50cm in length. The corolla is often pinkish and is about 4–5cm in length.

Two species of *Capsicum* are cultivated in Yemen. **Capsicum annuum** *L.*, the Green Pepper, known as 'Baybah', is cultivated for sale in the large towns but is not much used in traditional cooking. **Capsicum frutescens** *L.*, the chilli, which is called 'Bisbās' in Yemen, is an almost essential item in Yemeni cooking and is cultivated as a potherb and around villages throughout the country. It is used with tomato and onion to make a sauce and with milk to make a dip for bread and also to flavour soups, vegetables and meat.

CONVOLVULACEAE

A distinctive family of herbs, shrubs or parasites, often with a twining habit. Leaves alternate, simple, without stipules, entire or lobed. Flowers bracteate, regular, bisexual (except in *Hildebrandtia*); sepals 4–5, free or joined; corolla usually funnel-shaped, 5-lobed or entire. Fruit a capsule.

1. Leafless parasites 1. *Cuscuta*
   Leafy, green, non-parasitic plants **2**
2. Capsule distinctly lobed. Small prostrate plants **3**
   Capsule unlobed **4**
3. Capsule 2-lobed. Corolla shorter than the calyx, insignificant 2. *Dichondra*
   Capsule 4-lobed. Corolla longer than the calyx, prominent 3. *Falkia*

4. Dioecious shrubs. 2 sepals of the female flowers enlarging considerably in fruit — 5. *Hildebrandtia*
Bisexual herbs or shrubs. Sepals not much enlarging in fruit — **5**
5. Styles 2 — **6**
Style 1 — **8**
6. Corolla blue. Stigmas 4, i.e. each style forked — 4. *Evolvulus*
Corolla white. Stigmas 2, i.e. styles not forked — **7**
7. Styles and stamens exserted. Maritime plant — 7. *Cressa*
Styles and stamens included. Inland plants — 6. *Seddera*
8. Stigmas linear. Corolla pale pink or white — 8. *Convolvulus*
Stigmas oblong to globose. Corolla variously coloured — **9**
9. Stigmas ovate-oblong. Rare plant with yellow flowers in bracteate inflorescences — 9. *Hewittia*
Stigmas globose. Flowers, if yellow, not in bracteate inflorescences — 10. *Ipomoea*

1. CUSCUTA *L.*

1. Cuscuta planiflora *Tenore*

Slender, leafless, twining parasite with dark red stems. Flowers subsessile in globose clusters; calyx and corolla cream, slightly fleshy, c 2mm long; stigmas linear.

Widespread on the escarpment and high plateau growing from low altitude bushland at 1000m to exposed mountain tops at 3200m but seldom occurring in abundance. It grows on a variety of hosts, most commonly *Thymus laevigatus*, *Acalypha fruticosa* and *Hypoestes forskalei*, but occurs on many others including *Crepis ruepellii* and *Barleria prionitis*.

2. DICHONDRA *J.R. & G. Forst.*

1. Dichondra repens *J.A. & G. Forst.*

Prostrate, pubescent perennial herb, the stem trailing and rooting at the nodes. Leaves petiolate, cordate, reniform, 1–2cm long. Flowers solitary, axillary, shortly pedunculate; corolla whitish, c 2.5cm long, ± equalling the calyx; styles 2. Capsule 2-lobed, pilose.

Widespread on the escarpment between 1000 and 2700m, always growing in grassy places and usually on grassy banks by irrigation ditches, streams or springs.

3. FALKIA *L.f.*

1. Falkia oblonga *Krauss*

Sparsely pubescent perennial herb spreading by underground rhizomes from which arise aerial stems a few centimetres high. Leaves petiolate, ovate–oblong or oblong, truncate at the base, 1–1.5cm long. Flowers solitary, axillary, shortly pedunculate; corolla white, c 1cm long, much exceeding the calyx; styles 2. Capsule 4-lobed, pilose.

The Yemen plant is variety **minor** *C.H. Wright* with pedicels less than 1cm long.

A gregarious, locally abundant plant of seasonally flooded grassland between 2300 and 2800m on the high plateau; particularly common in the Yarīm area.

4. EVOLVULUS *L.*

1. Evolvulus alsinoides (*L.*) *L.*

E. yemensis Deflers

Decumbent or erect, silky-hairy, woody herb. Leaves subsessile, narrowly elliptic, 1–2.5cm long. Flowers in 1–5-flowered axillary cymes borne on slender peduncles 1–4cm long; corolla blue, broadly funnel-shaped, scarcely lobed, c 6mm long; styles 2, deeply bifid, with 4 stigmas. Capsule globose.

Frequent between 500 and 1600m in *Acacia* scrub in the central escarpment; rare in similar habitats elsewhere and known only from Mafḩaq, near Suq Ar Rubū' in Waṣāb al 'Alī, Suq As Sabt north of Ibb and Bir Ghubāh in Shar'ab.

5. HILDEBRANDTIA *Vatke*

Much-branched dioecious shrubs. Flower parts in 4s; sepals equal in the male flowers but the outer 2 much larger and accrescent in female flowers; corolla narrowly funnel-shaped, yellow-green; styles 2. Capsule enclosed in the enlarged sepals.

Both the species in Yemen come into flower about a week after the first substantial rains of the year. The flowers last only for a few days.

Leaves adpressed pilose when young. Enlarged sepals rounded to cuneate at the base — *H. africana*
Leaves scabrid. Enlarged sepals cordate based — *H. sp.A*

1. Hildebrandtia africana *Vatke*

Much-branched shrub 1–2m high, the branchlets purple-brown, glabrous, spinescent at the tips. Leaves borne in clusters on short, pubescent lateral shoots, oblanceolate-obovate, adpressed-pilose when young, glabrescent when mature, 5–15mm long. Flowers shortly pedicelled, borne in small clusters of 1–4 on the lateral shoots, yellow-green, 6–12mm long, the female slightly larger and longer pedicelled than the male. Enlarged sepals elliptic, rounded to cuneate at the base and decurrent onto the pedicel, c 2cm long.

Very local but sometimes abundant in dry bushland on the southern parts of the escarpment between 900 and 1400m: Barakānī, Ḩujarīyah; between Wadi Hiḑrān and Shar'ab; Taiz; between Riḩāb and Waṣāb; near Madīnat Al 'Abīd (D. Wood); between Mishrāfah and Masbaḩ, Waṣāb As Sāfal.

A plant known from southwest Arabia and the Horn of Africa.

2. Hildebrandtia sp.A

Similar to *H. africana* but the branchlets white, scabrid and the leaves also scabrid. Enlarged sepals orbicular, connate in the lower third, cordate at the base, not decurrent onto the pedicel, c 1.5cm long. Male flowers not known.

A distinctive undescribed species known only from a single collection made in open *Acacia* bushland at 1900m about 20kms east of Rada'. (Wood 2752)

6. SEDDERA *Hochst.*

Small, intricately branched undershrubs with prostrate or erect branches which sometimes become spinescent when old. Flowers axillary, solitary or in few-flowered cymes; flower parts in 5s; corolla funnel-shaped, weakly lobed, small, white; styles 2. Capsule ovoid to globose.

A genus of dry country shrublets mostly distinguished by few rather imprecise characters but nonetheless with distinct ecological requirements.

1. Flowers ± sessile — *S. latifolia*
   Flowers borne on distinct pedicels or in pedunculate cymes — 2
2. All pedicels less than 8mm long. Leaves linear or lanceolate — 3
   All or most pedicels or peduncles (in cymose inflorescences) more than 8mm long. Leaves elliptic to suborbicular — 4
3. Leaves linear, sparsely adpressed pilose, green. Capsule with an apical tuft of hairs — *S. virgata*
   Leaves lanceolate, densely adpressed silky-pilose, silvery. Capsule silky — *S. secundiflora*
4. Leaves elliptic to suborbicular, up to 18mm broad — *S. sp.A*
   Leaves narrowly elliptic, up to 10mm broad but usually much less — *S. arabica*

1. Seddera arabica (*Forssk.*) *Choisy*

S. somalensis (Vatke) Hall. f.
S. bottae Jaub. & Spach

Low, much-branched undershrub with decumbent or ascending branches, pubescent in all parts with adpressed or spreading hairs. Leaves shortly petiolate, narrowly elliptic to oblong, obtuse and mucronate at the apex, 10–20mm long, 5–10mm wide. Flowers borne in 1–4-flowered, pedunculate axillary cymes, the peduncle usually 8–25mm long; corolla 5–6mm long. Capsule with an apical tuft of hairs.

A very variable plant depending on the conditions it grows in. Plants from arid, grazed situations (e.g. Wood 75/67) tend to be bush-like with short entwined spinescent branches. The leaves are small and the peduncles invariably one-flowered. In contrast plants from high rainfall areas (e.g. Wood 1653) are spreading plants with long herbaceous branches. The leaves are relatively large and the inflorescence clearly cymose with up to 4 flowers on each peduncle. Although the extremes described above are very distinct all types of intermediate can be found.

Widespread and often abundant between 1000 and 2000m on the escarpment, growing in open stony ground mostly in *Euphorbia* or *Acacia* scrub; also in the southeast mountains and the Mashriq where it reaches 2400m but much less common and often found as chewed-down remnants in very exposed situations.

2. Seddera latifolia *Hochst. & Steud.*

Low, much-branched, bushy undershrub to 30cm, the whole plant covered in a silky-grey indumentum. Leaves shortly petiolate, oblong-elliptic, ± obtuse at the apex, mostly 6–12mm long, 3–5mm wide. Flowers solitary, axillary, sessile; corolla 4–5mm long. Capsule with an apical tuft of hairs.

A local but sometimes abundant plant of very dry bushland between 200 and 1200m on the escarpment; particularly characteristic of the Maqbanah area, the plains round al 'Ashshah and the valleys between 'Ubāl and Madīnat Al 'Abīd. In the Mashriq rare and found at c 1900m in the Rada' area and near Khulāqah in Nihm.

3. Seddera secundiflora *Jaub. & Spach*

S. hallieri Engl., & Pilg.

Soft, silvery, silky-hairy undershrub with prostrate or decumbent stems. Leaves subsessile, lanceolate or oblong-lanceolate, acute, 8–16mm long, 2–6mm wide. Flowers solitary, axillary, borne on pedicels 2–8mm long; corolla c 7mm long. Capsule silky.

As the name of this plant implies, the flowering branches are strikingly one-sided.

Widespread on rocky plain, cliff tops and similar habitats between 1900 and 2500m on the high plateau and at Majz in Khawlan Ash Sham (Steinberg 1026).

This little-known species appears to be restricted to the highlands of Yemen and Ethiopia.

4. Seddera virgata *Hochst. & Steud.*

Much-branched, wiry undershrub with long, slender, sparsely adpressed-pilose ascending branches. Leaves subsessile, linear, 11–20mm long, 1–2.5mm wide. Flowers solitary, axillary, borne on pedicels 1–4mm long; corolla c 7mm long. Capsule with an apical tuft of hairs.

An uncommon plant of arid bushland in the southern escarpment between 200 and 1000m: in several places in the Maqbanah area; Suq al Kawlah, Wadi Zabīd; Khalīfah; Bājil (Schweinfurth); Khamīs Bani Sa'd.

5. Seddera sp.A

Similar to *S. arabica* but the leaves broadly elliptic to suborbicular, rounded and finely mucronate at the apex, 10–30mm long, 8–18mm broad. The peduncles are also strikingly long, reaching as much as 5cm.

Only known from scrubby margins of the wadi at Rajamah near Ḥajjaylah (Wood 2141).

It is not impossible that this plant is merely an outsize version of *S. arabica* but it occurs at a lower altitude than *S. arabica* and I have been unable to find any specimen of *S. arabica* from Africa or Arabia which in any way approximates to the large dimensions of this plant.

Seddera intermedia Hochst. & Steud. is recorded from Mafḥaq by Schwartz but almost certainly in error. This plant seems only to be known from the type collection which looks like a broad-leaved form of *S. virgata*. Both species were described from the same locality.

7. CRESSA *L.*

1. Cressa cretica *L.*

Erect, much-branched, leafy, pilose annual or briefly perennial woody herb to 30cm. Leaves sessile, ovate or lanceolate, c 7mm long. Flowers subsessile in dense bracteate spikes at the tips of the branches; corolla white, deeply lobed, c 5mm long; stamens and styles exserted; styles 2. Capsule ovoid.

A plant of scattered localities on sand dunes along the Red Sea coast: Khawkhah; Hodeida; Ṣalīf; between Khawbah and Ibn 'Abbās; Luḥayyah.

8. CONVOLVULUS *L.*

Annual or perennial herbs (rarely undershrubs) with erect, prostrate or twining stems. Leaves often hastate or sagittate. Flowers axillary in 1 – few-flowered cymes or heads; corolla usually funnel-shaped and unlobed; style 1 with 2 filiform stigmas. Capsule usually glabrous, globose.

1. Spinescent undershrub. Flowers sessile in the leaf axils *C. hystrix*
   Herbs, never spinescent. Flowers pedicelled or in pedunculate cymes 2
2. Leaves sessile, cuneate or rounded at the base *C. prostratus*
   Leaves petiolate, some or all of them hastate, sagittate or truncate at the base 3
3. Flowers in dense pedunculate heads, the sepals densely brown-villous *C. glomeratus*
   Flowers solitary or in lax pedunculate cymes, the sepals glabrous to pilose 4
4. Leaf margin deeply crenate *C. fatmensis*
   Leaf margin entire to weakly crenate 5
5. Annual. Corolla c 6mm long. Leaf base mostly truncate or cuneate *C. siculus*
   Perennial. Corolla at least 8mm long. Leaf base cordate, hastate or sagittate 6
6. Leaves and peduncles glabrescent *C. arvensis*
   Leaves and peduncles distinctly hairy 7
7. Corolla c 16mm long. Small prostrate plant with leaves 10mm or less in length *C. thymoides*
   Corolla 9–15mm long. Twining or prostrate plants with leaves mostly more than 15mm long 8
8. Leaves ovate to triangular. Twining plant *C. farinosus*
   Leaves oblong. Prostate plant (in Yemen) *C. sagittatus*

1. Convolvulus arvensis *L.* Faja'ah, 'Ujal

Bindweed. Perennial herb with prostrate or twining stems. Leaves petiolate, lanceolate or oblong, hastate or sagittate at the base, glabrescent, mostly 2–4cm long, 1cm wide. Flowers usually solitary, borne on a glabrescent peduncle with linear bracts c 3mm long; corolla white or pink, 15–20mm long.

Usually recognised by its large flowers and glabrescent leaves.

A common weed of fields and gardens occurring in suitable habitats throughout Yemen between 1000 and 3300m, though only on irrigated land in dry country.

2. Convolvulus farinosus *L.*

Perennial twining herb (rarely prostrate), white-silky on young growth, softly long-pubescent when older. Leaves petiolate, ovate-triangular, cordate with a broad sinus, 4–7cm long, 2–4cm wide, the margin obscurely crenate. Inflorescence a 1–6-flowered pedunculate cyme, bracts minute 1–2mm long; corolla white or very pale pink, c 11mm long.

Very close to *C. sagittatus* but usually twining in habit and with softly pubescent distinctly triangular leaves.

A rare plant found in bushy places between 2300 and 2600m, often near streams: near Mirkiz 'Utmah; near Bayt Mughlad, J. Shibām, Ḥarāz; Dār al Ḥajjah, Wadi Ḍahr. (Wood 2990, 3200, 3281)

3. Convolvulus fatmensis *Kuntze*

Slender, sparsely pubescent, twining perennial herb. Leaves petiolate, ovate, cordate with a broad sinus, 1–3cm long, 1–1.5cm wide, the margin deeply crenate. Inflorescence a shortly pedunculate 1–2-flowered cyme, the peduncles c 1cm long, bracts minute; corolla pale pink, 9–10mm long.

The short peduncles and crenate leaves make this a distinctive species.

Only known at 2300m on sandy roadside in Banī Hārith northeast of Sana'a.

4. Convolvulus glomeratus *Choisy*

C. arabicus Hochst.

Sparsely adpressed pilose, twining perennial herb. Leaves petiolate, lanceolate or oblong, truncate, hastate or shallowly cordate at the base, 3–6cm long, 1–2cm wide. Inflorescence a dense many-flowered pedunculate head subtended by prominent leafy bracts 10–18mm long; bracts and sepals densely covered in long, brownish hairs; corolla usually white, 8–10mm long.

Widespread and quite frequent in the Tihama and on the foothills up to 1000m growing as a field weed and in bushy places around cultivation.

It is rather surprising that the closely allied *C. rhyniospermus* Hochst. ex Choisy has not been found in Yemen. It can be recognised by the subsessile inflorescence and deeply lobed corolla but is otherwise similar to *C. glomeratus*.

5. Convolvulus hystrix *Vahl*

Grey-green pilose undershrub to 80cm with long, decumbent primary branches and short spinescent secondary branches. Leaves sessile, oblong, truncate at the base, 0.5–1cm long, 2–4mm wide. Flowers 1–3 in sessile, axillary clusters; bracts oblanceolate, c 5mm long; sepals densely silky pilose; corolla white or pale blue, c 5mm long.

An uncommon plant found on sandy plain in the Tihama: between Mukhā and Mafraq; Zabīd; Ḍaḥī; Ṣalīf; Zuhrah (Miller).

6. Convolvulus prostratus *Forssk.*

C. microphyllus Sieb.

Perennial herb, much branched at the base with many long, ascending branches to 80cm, the indumentum pilose with some hairs spreading and some adpressed and those on young growth ± brownish. Leaves sessile, linear, oblong to oblanceolate, cuneate or rounded at the base, 1–6cm long. Inflorescence a 1–3-flowered pedunculate cyme, subtended by linear bracts 3–10mm long; corolla pale pink, 6–12mm long.

C. prostratus belongs to a poorly understood group many of which have been given specific names. I have included all forms found in Yemen under this name since it is the oldest available and is based on a Yemen plant. The actual type appears to be a low plant less than 30cm high with small leaves and very shortly pedunculate inflorescences and is similar to Wood 1185. Most plants collected in modern times (e.g. Wood 2097, 3427) are more robust with leaves up to 6cm long and with a distinctly pedunculate inflorescence. I am not convinced that the two extremes are specifically distinct and suspect they may be adaptations to dry and wet seasons respectively. Certainly considerable variation in leaf size and peduncle length has been observed on individual plants and from collection to collection and similar variation has been noted in other species of *Convolvulaceae*. A curious feature of recent collections in Yemen is that the plants appear to be sterile. Further investigation is needed into all these forms.

Locally abundant on sand dunes and in sandy fields in the Tihama from Manṣūrīyah north to the Wadi Mawr.

7. Convolvulus sagittatus *Thunb.*

Very variable dull green perennial herb, stems usually prostrate and radiating from a central rootstock, rarely (? never in Yemen) twining, ± densely white-pubescent, the hairs often brownish on young growth. Leaves petiolate, ± oblong, sagittate at the base, obscurely crenate, usually 3–4cm long, 0.5–1cm wide. Inflorescence a long pedunculate 1–2-flowered axillary cyme, bracts minute, 1–2mm long; corolla pale pink or white, c 12mm long.

A rare plant found on gravelly tracksides and banks around fields in scattered localities between 2200 and 2700m: J. 'Ayfū; above Saddah in the Wadi Banā; J. Hisn Madnān, Waṣāb al 'Ālī; Manākhah (Schweinfurth); Aḥjar. (Wood 1796, 2165, 3340)

8. Convolvulus siculus *L.*

C. agrestis Schweinf.

Annual herb with pubescent, usually simple, decumbent or ascending stems to c 50cm. Leaves petiolate, oblong-lanceolate, truncate or abruptly cuneate at the base, sparsely pilose on the margins, 2.5–5cm long, 0.5–2cm wide. Inflorescence a 1–3-flowered, pedunculate axillary cyme; bracts subulate, up to 10mm long; corolla pale lilac, c 6mm long.

The Yemen plant is subspecies **agrestis** (*Schweinf.*) *Verdc.*

A plant of disturbed grassy places, particularly field borders and tracksides, found between 2400 and 3300m in scattered localities on the escarpment from J. Ṣabir north to Shahārah but only at all frequent in the Ibb region.

9. Convolvulus thymoides *O. Schwartz*

C. sagittatus var. *grandiflorus* sensu Schwartz

Perennial herb with short, densely pubescent, prostrate stems less than 10cm long. Leaves shortly petiolate, ovate-oblong, truncate to ± auriculate-hastate at the base, pilose on the margins, 0.5–1.5cm long, 0.2–0.6cm wide. Flowers solitary in the leaf axils, shortly pedunculate, bracts minute; corolla pink, 15–17mm long.

A plant of grassy banks around fields almost always associated with *Cichorium bottae* and *Dianthus uniflorus*; quite frequent between 2400 and 3300m on the inner escarpment and western parts of the high plateau from J. Sumārah north to J. Maswār but very rare on the outer escarpment when it is only known on J. Nu'mān and J. Shibām in the Ḥarāz.

This endemic species is closely related to the variable *C. sagittatus* and is not as easily separated from Ethiopian forms of that species as Schwartz suggested. It is very uniform in Yemen and seems best treated as a distinct species until it can be clearly demonstrated that it is part of *C. sagittatus*.

9. HEWITTIA *Wight & Arn.*

1. Hewittia scandens (*Milne*) *Mabberley*

H. sublobata (L.f.) O. Kuntze

H. bicolor (Vahl) Wight & Arn.

Perennial herb with long, prostrate or twining, pubescent stems. Leaves petiolate, ovate, usually cordate, ± pilose especially below, c 5–10cm long. Inflorescence a 1–5-flowered, axillary, pedunculate, bracteate cyme with a long pubescent peduncle and very short pedicels; corolla funnel-shaped, yellow with a purple centre, densely pilose, 2–2.5cm long; style 1, filiform, stigma ovate-oblong. Capsule pilose.

A rare plant of overgrown cliff gullies and wadi beds between 600 and 1200m in the central escarpment: in the wadi above Ḥillah, J. Bura'; near 'Ukaybir (Schweinfurth) and between Ḥabāsh and Ṣa'bah, J. Milḥān.

10. IPOMOEA *L.*

Morning Glory. Annual or perennial herbs with woody or herbaceous, twining or prostrate stems. Leaves petiolate, entire or lobed. Flowers axillary, solitary or in cymes; corolla funnel-shaped, inconspicuously lobed, often brightly coloured; style 1, stigmas 2, globose. Capsule ovoid or globose.

1. Leaves palmately divided to the base into 5–7 lobes — *I. cairica*
   Leaves entire or lobed but never palmately lobed to the base — **2**
2. Leaves bilobed. — *I. pes-caprae*
   Leaves entire or 3–5-lobed — **3**
3. Leaves covered in dense matted white indumentum on the undersurface — **4**
   Leaves green below, never with a matted indumentum — **5**
4. Corolla less than 2cm long — *I. dichroa*
   Corolla 2–5cm long — *I. ficifolia*
5. Corolla yellow, with or without a dark centre — *I. obscura*
   Corolla white or pinkish — **6**
6. Corolla more than 4cm long — **7**
   Corolla less than 2.5cm long — **10**
7. Leaves and sepals ± glabrous. Leaves always entire — **8**
   Leaves and sepals pilose or bristly. Leaves often lobed — **9**
8. Woody twiner, often leafless when in flower — *I. triflora*
   Prostrate marsh plant with succulent stems. Plant leafy when in flower — *I. aquatica*
9. Sepals long acuminate, more than 2cm long. Leaves always lobed — *I. nil*
   Sepals acute, 1.3–1.6cm long. Leaves usually entire — *I. purpurea*

10. Flowers in subsessile clusters. Capsule pilose *I. eriocarpa*
Flowers in pedunculate cymes. Capsule glabrous **11**
11. Sepals distinctly widened to a cordate or ± hastate base. Corolla more than 1cm long *I. sinensis*
Sepals not widened at the base which is ± cuneate. Corolla less than 1cm long *I. verticillata*

1. Ipomoea aquatica *Forssk.*
I. reptans sensu Schwartz
Glabrous annual or perennial herb with long, prostrate or floating, semi-succulent stems. Leaves entire but otherwise variable, oblong to triangular, truncate, cordate or hastate at the base, 3–8cm long. Inflorescence pedunculate, 1–few-flowered; corolla purple, pink or white with a darker centre, 4–7.5cm long; sepals ovate, obtuse, 6–12mm long.

Collected by Forsskal in the Wadi Zabīd but never found again in Arabia. It is a pantropical plant of swamps and streamsides and in the Far East is popular as a vegetable.

2. Ipomoea cairica (*L.*) *Sweet*
I. palmata Forssk.
Glabrous, twining or prostrate, perennial herb, the stems often tuberculate. Leaves palmately divided to the base into 5–7 elliptic lobes, each lobe tapered at the ends, 3–6cm long. Inflorescence pedunculate, 1–many-flowered; corolla usually purple, 4.5–6cm long; sepals ovate, obtuse, 4–6mm long.

A rare plant found on scrubby slopes by wadis at c 800m between Aṭ Ṭūr and Suq al Amān.

3. Ipomoea dichroa *Choisy*
I. arachnosperma Welw.
Annual herb with prostrate or twining stems, almost the whole plant covered in long, white or yellowish bristly hairs. Leaves usually entire in Yemen (often 3-lobed elsewhere), ovate, cordate, covered in matted white-cottony indumentum below, 5–10cm long. Inflorescence pedunculate, few-flowered; corolla mauve, 1.2–1.7cm long; sepals lanceolate, 6–18cm long. Capsule pubescent.

A rare plant of bushy places in the middle-altitude wadis between 1200 and 1500m, often growing in overgrown thorn fences around irrigated fields: Misrākh; 'Udayn; Bani Miftaḥ between Riḥāb and Waṣāb; between Suq Ath Thalūth and Hammām Shamsān in the Wadi Lā'ah.

4. Ipomoea eriocarpa *R. Br.*
I. hispida (Vahl) Roem. & Schult.
Pubescent or slightly hispid annual herb with twining stems. Leaves entire, ovate or oblong, subhastate at the base, 2.5–6cm long. Inflorescence an axillary cluster of 1–5 subsessile flowers; corolla mauve, pink or white, 6–9mm long; sepals ovate, acuminate, spreading in young fruit, c 9mm long. Capsule pubescent.

A weed of cultivated and disturbed places between 200 and 1500m in the eastern Tihama and on the lower slopes of the escarpment; particularly frequent in irrigated land.

5. Ipomoea ficifolia *Lindl.*
Similar to *I. dichroa* but less densely hairy and usually with 3-lobed leaves (in Yemen) which are less strongly white-cottony below. The corolla is much larger, being about 5cm long.

A rare plant restricted to damp northwest-facing cliff gullies between 1400 and 1800m in the central escarpment where it is part of a highly specialised fern-rich community: in the gully below Al Jirjir above Ḥadīyah; in the gullies around Ṣa'bah, J. Milḥān; below the main suq, Maḥābishah.

6. Ipomoea nil (*L.*) *Roth*
I. scabra Forssk.
Annual twining herb, bristly-pilose in most parts. Leaves ovate, cordate, 3-lobed, 6–10cm long. Inflorescence pedunculate, 1–few-flowered; corolla magenta with a pale tube, 6–7.5cm long; sepals linear-lanceolate, long-acuminate, 22–27mm long. Capsule glabrous.

Originally from America, this plant was established as long ago as 1762 and is now locally frequent from 600 to 1500m in scrub along wadis and irrigation ditches on the escarpment from Aḥkūm north to J. Bura'; it is most frequent in the Wadi Ḍabāb area, at 'Udayn and in the central escarpment.

7. Ipomoea obscura (*L.*) *Ker-Gawl.*
Glabrous or pubescent herb with slender twining stems. Leaves entire, ovate, cordate, 3–7cm long. Inflorescence pedunculate, 1–few-flowered; corolla pale yellow usually with a dark purple centre, 1.5–2.5cm long; sepals lanceolate to ovate, acute, 4–8mm long. Capsule glabrous.

Widespread and frequent on the escarpment between 500 and 2000m, growing in bushland and on overgrown fences particularly in wadis, around cultivated land and near irrigation ditches.

8. Ipomoea pes-caprae (*L.*) *R. Br.*
Glabrous perennial with long, thick, hollow, creeping stems, often forming extensive mats. Leaves fleshy, suborbicular to elliptic, bilobed at the apex, ± rounded at the base, 3–8cm long. Inflorescence pedunculate, 1–many-flowered; corolla pink with a darkish centre, 4–5cm long; sepals ovate, obtuse, 5–12mm long.

A salt-tolerant plant occurring on waste ground and in gardens near the Red Sea, notably in Hodeida; rarer inland and restricted to scattered localities in the Tihama apart from an isolated occurrence near the salt springs at Ḥawbān near Taiz at 1500m. In many of its localities its status is unclear and it is often either cultivated or an escape but is probably native along the coast.

9. Ipomoea purpurea (*L.*) *Roth*
Bristly-pubescent annual herb with trailing or twining stems. Leaves ovate, cordate, entire (rarely 3-lobed), 6–10cm long. Inflorescence pedunculate, 1–few-flowered; corolla magenta with a white tube, 4–5cm long; sepals lanceolate, acute, 13–16mm long. Capsule glabrous.

Originally from America, this species is now well naturalised in roadside vegetation at 2000–2400m around Ibb and near Nādirah in the Wadi Banā. However it appears to have disappeared from Sana'a and Ḥajjah where Rathjens collected it in the 1930s.

10. **Ipomoea sinensis** (*Desr.*) *Choisy*

I. calycina sensu Schwartz

Pilose annual herb with prostrate or twining stems. Leaves entire, ovate to oblong, broadly cordate to almost sagittate at the base, 2–8cm long. Inflorescence pedunculate, 1–3-flowered; corolla white or pink, 1.3–2.2cm long; sepals accrescent in fruit, up to 21mm long, the outer ovate with a conspicuous, broad, cordate or hastate base. Capsule glabrous.

Two rather distinct subspecies occur in Yemen. Subspecies **sinensis** is a relatively robust plant, both the fruiting sepals and the pure white corolla being about 2cm long. In subspecies **blepharosepala** (*A. Rich.*) *Meeuse* the corolla is pink and only c 1.3cm long while the fruiting sepals are c 8mm long.

Subspecies *sinensis* is a rare plant known only from overgrown fences just west of 'Udayn suq at 1500m. Subspecies *blepharosepala* occurs in scattered localities on the escarpment between 1100 and 2300m from J. Ṣabir north to Al 'Ashshah growing in and around cultivated land, particularly on fences around fields.

11. **Ipomoea triflora** *Forssk.*

I. gossypina Deflers

Robust glabrous twiner with woody, roughened stems. Leaves entire, ovate, shallowly cordate to ± truncate at the base, 4–6cm long. Inflorescence very shortly pedunculate, 1–few-flowered; corolla white with a reddish purple centre, 3.5–4.5cm long; sepals oblong, obtuse, pallid-margined, 10–12mm long.

The most remarkable *Ipomoea* in the Yemen flora. It is the only species with a woody stem, it is strictly winter-flowering and is usually completely leafless when in flower.

A local plant of rough scrubby banks along wadis between 600 and 1100m on the escarpment, often associated with *Celosia polystachia*: locally frequent in Waṣāb As Sāfal; around 'Alūjah and Ḥadīyah in J. Raymah; frequent around Ḥajjaylah in the Ḥarāz; Khamīs Bani Sa'd; Wadi Sharas.

13. **Ipomoea verticillata** *Forssk.*

Prostrate or twining annual herb similar to *I. sinensis* subsp. *blepharosepala* from which it is best distinguished by its very short corolla (less than 5mm long) and by the usually conspicuously bristly sepals which are cuneate at the base and never cordate or hastate. The peduncles are always (?) 1–2-flowered.

A little-known species of northeast tropical Africa and Yemen, where it is only known from Forsskal's collection at Kudmīyah in the Wadi Mawr.

The sweet potato, **Ipomoea batatas** (*L.*) *Lam.*, 'Jazar', has prostrate stems, entire or palmately lobed, cordate leaves and lilac flowers. It is cultivated on a small scale for its edible tubers and is found below about 2000m on the escarpment, usually as isolated plants along the edge of terraces.

BORAGINACEAE

A family of trees, shrubs and herbs with alternate, simple, usually scabrid leaves without stipules. Flowers commonly in spike-like scorpioid cymes, usually regular; sepals united; corolla with a tube. Fruit usually of 4 nutlets, but sometimes less or in our tree species a drupe.

Most species except the trees are found in dry country.

1.	Trees or shrubs. Fruit a drupe	**2**
	Herbs or small undershrubs. Fruit of nutlets	**3**
2.	Leaves glabrous below. Style once divided	2. *Ehretia*
	Leaves hairy below. Style twice divided	1. *Cordia*
3.	Flowers and fruit sessile in leafless scorpioid cymes, small bracts rarely present and if so corolla white	3. *Heliotropium*
	Flowers and fruit distinctly pedicellate, sometimes sessile but if so inflorescence always bracteate and the corolla yellow or mauve	**4**
4.	Stamens forming a cone-like appendix exserted above the corolla. Flowers in panicles	**5**
	Stamens not forming a cone-like appendix exserted above the corolla. Flowers in scorpioid cymes or apparent racemes	**6**
5.	Staminal appendix hairy. Calyx enlarging in fruit	5. *Trichodesma*
	Staminal appendix glabrous. Calyx not enlarging in fruit	4. *Cystostemon*
6.	Corolla blue, white or mauve	**7**
	Corolla yellow	**10**
7.	Corolla irregular, mauve-pink. Filaments long-exserted	11. *Echium*
	Corolla regular, white or blue. Filaments not exserted	**8**
8.	Nutlets with hooked spines	6. *Cynoglossum*
	Nutlets without hooked spines	**9**
9.	Inflorescence densely bracteate	7. *Anchusa*
	Inflorescence without bracts	9. *Myosotis*
10.	Plant sticky-glandular	8. *Alkanna*
	Plant hispid	10. *Arnebia*

1. CORDIA *L.*

Trees or shrubs with white or cream flowers in panicled cymes. Fruit a drupe surrounded by a prominent, persistent, shallowly lobed calyx.

1.	Leaves opposite	*C. sinensis*
	Leaves alternate	**2**
2.	Leaves scabrid above. Corolla less than 1cm long	*C. ovalis*
	Leaves smooth above. Corolla c 1.5cm long	*C. africana*

1. **Cordia africana** *Lam.* — Tānib

C. abyssinica R. Br.

Tree to c 10m with pale brown bark. Branchlets pale brown, pubescent. Leaves alternate, entire or obscurely dentate, ovate to suborbicular, up to 12 × 9cm, base ± rounded, upper surface glabrous and ± smooth, undersurface shortly tomentose. Flowers large and attractive, cream, sessile in large lax terminal panicles; calyx ribbed, brown-tomentose; corolla c 15cm long. Drupe ovoid, yellow, apiculate, c 15mm long.

Flowers mostly in autumn.

Widespread and common between 500 and 2600m on the escarpment but seldom in large quantity and usually found in villages or around cultivation; much less common away from human activity. It is commonly planted as it is valued as a shade tree for coffee and other crops and its wood is much esteemed for carpentry and joinery.

2. Cordia ovalis *R. Br. ex DC.* Gharāf

C. dioica Boj.

Shrub or (rarely) a small tree to 6m. Branchlets pale brown, ± scabrid when very young, becoming glabrous and smooth. Leaves alternate, entire or obscurely dentate, ovate to suborbicular, up to c 8 × 6cm, base rounded, upper surface scabrid, undersurface pubescent. Flowers cream, subsessile in small compact axillary panicles; calyx brown-tomentellous, not ribbed; corolla c 5mm long. Drupe ovoid, yellow, apiculate, c 15mm long.

Flowers in spring.

Found between 500 and 1700m on the escarpment but preferring high rainfall areas and only really common in valley scrub in the central escarpment and the 'Udayn region. It appears to be commoner in the south where it is less exacting in its rainfall requirements, occurring in arid areas such as Ḥamīd, southeast of Raḥidah; it is unrecorded north of the Wadi Mawr.

3. Cordia sinensis *Lam.* Sakhil, Miskhil, 'Inab, Gharāf

C. gharaf Ehrenb. ex Aschers.

C. rothii Roem. & Schult.

Shrub or small tree to 6m with peeling bark. Branchlets pale brown, glabrescent. Leaves opposite, entire, oblanceolate, up to c 9 × 3cm, base cuneate, upper surface scabrid, undersurface pubescent, grey-green. Flowers cream, pedicellate in small lax cymes; calyx thinly pubescent, unribbed, ± unlobed; corolla inconspicuous, c 4mm long. Drupe ovoid, yellow, apiculate, c 15mm long, edible.

Flowers mostly in spring.

A locally common shrub of dry *Acacia* scrub in the escarpment foothills which occurs rarely in similar situations elsewhere, notably on the dip slope east of Maḥābishah and at Ḥamīd southeast of Raḥidah, where it reaches 1200m. It is also cultivated (for its edible fruit?) and is found in many of the Tihama towns (Mawza', Zabīd, Luḥayyah etc.) and more rarely in old escarpment towns such as Yifrūs.

2. EHRETIA *Browne*

Trees and shrubs with white flowers in panicles. Calyx deeply lobed, not particularly prominent. Fruit a four-seeded, obliquely ellipsoidal, nearly black drupe.

Leaves acute, commonly exceeding 10 × 6cm *E. cymosa*
Leaves obtuse, c 5 × 3cm *E. obtusifolia*

1. Ehretia cymosa *Thonn.* 'Asār, Waraf

E. abyssinica R.Br.

Shrub or tree to c 10m. Branchlets brown to grey, glabrous. Leaves alternate, entire, ovate to elliptic, up to c 13 × 8cm, base rounded, apex acute, both surfaces glabrous and smooth. Flowers white, sessile in terminal panicles; corolla c 6mm long. Drupe c 4 × 6mm.

Flowers sporadically throughout the year.

Widespread on the escarpment between 700 and 2600m with a very similar distribution to *Cordia abyssinica* with which it commonly grows. Like that species it is rarely found in large quantity and usually grows in villages and around fields and other areas of human activity.

2. Ehretia obtusifolia *Hochst. ex DC.* Misūq

Shrub to 3m. Branchlets grey, glabrous. Leaves alternate, entire, obovate, up to c 5 × 3cm, base cuneate, apex rounded, both surfaces glabrous and smooth. Flowers white, sessile in terminal and axillary panicles; corolla c 8mm long. Drupe c 5 × 7mm.

Flowers sporadically throughout the year.

A plant with a strikingly similar distribution to *Cordia ovalis*; found between 500 and 1500m on the escarpment, but with a distinct southern tendency, being found in relatively dry country southwest of Taiz and not recorded north of the Wadi Mawr. It is most typical of valley scrub in the central escarpment where it is common in suitable habitats.

3. HELIOTROPIUM *L.*

Undershrubs, perennial or annual herbs. Inflorescence of terminal scorpioid cymes, usually leafless but bracteate in some species; flowers sessile or nearly so; corolla with a cylindrical tube and five spreading lobes, small, white or pale yellow. Fruit with 4 (rarely 2) nutlets, the style terminal on top.

1. Inflorescence with scattered bracts between the flowers 2
   Inflorescence without bracts 4
2. Leaves elliptic, hispid. Stigma sessile *H. sessilistigma*
   Leaves linear, adpressed pubescent. Stigma with a distinct style 3
3. Perennial undershrub with white peeling bark. Stigma ± globose *H. rariflorum*
   Annual (rarely perennial) herbs, bark brown. Stigma conical with a distinct 'saucer-like' ring below it 4 *H. strigosum*
4. One sepal much larger than the others. Decumbent grey plant of damp habitats *H. ovalifolium*
   All sepals equal. Erect plant of dry habitats 5

5. Corolla yellow, the lobes long acuminate *H. zeylanicum*
Corolla white or yellow, the lobes rounded, not acuminate 6
6. Desert shrub with subsessile leaves, white stems and a yellow corolla *H. digynum*
Not as above 7
7. Plants with adpressed strigose indumentum even on the inflorescence 8
Plants variously hairy, but usually hispid and the inflorescence always villous with spreading hairs 9
8. Leaves less than 4cm long. Nutlets shortly pubescent *H. bottae*
Leaves often more than 5cm long. Nutlets glabrous *H. longiflorum*
9. Leaves with an undulate-dentate margin and a gradually narrowed base 10
Leaves entire; base abruptly rounded 11
10. Nutlets 4, pubescent. Calyx c 2mm long *H. ramosissimum*
Nutlets 2, glabrous. Calyx 3–4mm long *H. pterocarpum*
11. Nutlets glabrous *H. pallens*
Nutlets hairy *H. europaeum*

1. Heliotropium bottae *Deflers*

Perennial strigose undershrub to 40cm, usually much branched and woody below, the bark grey. Leaves entire, green, very variable in size and shape, generally ovate and up to 4cm long in wetter regions but linear lanceolate and only 1.5cm long in dry country. Cymes long-pedunculate, weakly scorpioid, without bracts, up to 15cm long; corolla white, rarely yellow; calyx strigose, 2mm long. Nutlets 4, shortly pubescent, but infrequent – the plant does not appear to be very fertile.

Apart from *H. rariflorum* and *H. digynum* this is the only definitely shrubby species in Yemen.

A plant of open stony hillsides and field borders between 1100 and 2300m; abundant in the southeastern mountains from Taiz to Baydā' becoming steadily less common north and west but reaching Shahārah.

H. bottae is endemic to Yemen but is closely related to the African *H. steudneri* Vatke.

2. Heliotropium digynum (*Forssk.*) *Aschers. ex C. Chr.*

H. luteum Poir.

Perennial scabrid undershrub to 40cm, usually much branched and with many white-glabrescent woody stems. Leaves subsessile, yellow-green with prominent veins, ovate or elliptic, 1–2cm long, the margin obscurely undulate. Cymes pedunculate, weakly scorpioid, without bracts, 1–2cm long; corolla yellow; calyx scabrid-hairy, 3–4mm long. Nutlets 4, yellow, rather swollen, usually ciliate but sometimes glabrous.

A characteristic desert plant only known in Yemen from sand dunes just north of Mārib (Miller) at 1100m.

3. Heliotropium europaeum *L.*

Erect, much-branched, densely hairy, grey annual to 50cm, stems densely crisped-hairy. Leaves entire, ovate or elliptic, 1–3cm long, the base abruptly cuneate. Cymes shortly pedunculate, weakly scorpioid, without bracts, up to 8cm long; corolla white; calyx villous 2–3mm long. Nutlets 4, densely pubescent.

Easily confused with *H. pallens* but more hairy, with shorter cymes and hairy nutlets.

A local field weed of 'good' agricultural land between 1500 and 2500m mostly on the high plateau but also in the Ibb and Turbah regions.

4. Heliotropium longiflorum (*A.DC.*) *Steud. & Hochst. ex Bunge*

Erect strigose perennial herb to 50cm usually much branched and slightly woody below. Leaves entire, green, lanceolate or linear, 3–8cm long. Cymes pedunculate, weakly scorpioid, without bracts, lengthening dramatically in fruit to reach 25cm long; corolla white; calyx strigose, c 1mm long. Nutlets 4, glabrous.

Easily recognised by its minute calyx and thinly strigose indumentum but variable and with two distinct varieties in Yemen. Variety **longiflorum** (Wood 75/412) has lanceolate leaves and very long fruiting cymes. It is a common field weed in the Tihama and foothills. Variety **stenophyllum** *O. Schwartz* (Wood 1092) has linear-lanceolate leaves less than 1cm broad and fruiting cymes up to 7cm long only. This is common in waste places at higher altitudes.

The species as a whole is widespread and common throughout the Tihama, escarpment and high plateau reaching 2300m. It grows in a great variety of disturbed open habitats including fields, roadsides, wadi beds and rough grassland.

5. Heliotropium ovalifolium *Forssk.*

Prostrate, much-branched, silky-hairy, grey annual. Leaves entire, oblanceolate, 1–2.5cm long. Cymes long pedunculate, strongly scorpioid, without bracts, up to 6cm long; corolla white; calyx ciliate, c 3mm long, one lobe much longer than the others. Nutlets 4, silky-hairy.

A distinctive species which is the only *Heliotropium* to grow in aquatic habitats in Yemen. It is common both by the larger wadis and by small streams between 500 and 1500m, particularly in the central escarpment but is only known between the Wadi Zabīd and the Wadi Lā'ah.

6. Heliotropium pallens *Del.*

Erect, branched, dull green annual to 60cm, the stems adpressed-hairy. Leaves entire, ovate to rhomboid up to 6cm long, the base truncate or abruptly cuneate. Cymes pedunculate, scorpioid, without bracts, up to 14cm long; corolla white; calyx villous, c 3mm long. Nutlets 4, glabrous.

Quite frequent in sandy fields in the Tihama from the Wadi Sihām northwards but possibly overlooked further south.

7. Heliotropium pterocarpum (*DC.*) *Steud. & Hochst. ex Bunge*

Erect or ascending, grey-hispid annual to 50cm, though commonly much less in dry years. Leaves lanceolate to oblong, 3–5cm long, gradually narrowed into the petiole, the margin undulate and appearing crenate. Cymes shortly pedunculate, strongly scorpioid, without bracts, 1–2.5cm long; corolla white; calyx villous 3–4mm long. Nutlets 2, usually winged, glabrous.

This is the only Yemen species whose fruit has 2 nutlets only. The development of wings on the nutlets is very variable. The typical plant (Wood 76/1186, 2111) has rudimentary wings c 1mm wide. In the region around Quṭay there is another form whose nutlets have wings 3mm broad and this is associated with a larger corolla (Wood 1495). These two forms have sometimes been separated as distinct species but intermediate forms occur.

Common on sand in the Tihama from Zabīd northwards, growing from sea level to 400m. The species occurs in neighbouring countries but is restricted to the coastal regions of the Red Sea.

8. **Heliotropium ramosissimum** (*Lehm.*) *Sieb. ex DC.*

H. persicum Boiss.

H. undulatum sensu Schwartz

Erect, hispid, grey perennial to 50cm, much branched from the woody base. Leaves oblong or lanceolate, up to 6cm long, the margin undulate and appearing dentate. Cymes shortly pedunculate, weakly scorpioid, without bracts, 2–3cm long; corolla white; calyx villous, c 2mm long. Nutlets 4, glabrous or thinly pilose.

Superficially very similar to *H. pterocarpum* but distinguished by the shorter calyx and 4 nutlets.

Widespread in Yemen, growing from 200 to 2400m in all regions east to Mārib, but rare in the Tihama and absent from the wetter parts of the escarpment. It is most frequent in Taiz province and on the high plateau growing in disturbed stony ground such as fields, roadsides and drainage lines. It shows a strong preference for volcanic rock of recent origin and is often associated with *Forsskalea tenacissima*.

9. **Heliotropium rariflorum** *Stocks*

Perennial strigose undershrub to 40cm, much branched below and with numerous ascending woody branches with distinctive peeling grey bark. Leaves entire, green, linear-oblanceolate, 1–2.5cm long. Cymes sessile, weakly scorpioid, lax, with scattered bracts in between the flowers, up to 5cm long; corolla white; calyx strigose, 1.5mm long. Nutlets 4, strigose, the stigma globose borne on a distinct persistent style.

Easily confused with *H. strigosum* variety *cordofanum* and best distinguished by its shrubby habit and globose stigma.

Widespread but very scattered, growing between 600 and 2500m in the Mashriq and on the drier parts of the escarpment particularly the rainshadow areas of the central escarpment; not recorded either from the Ḥujarīyah or Khawlan Ash Sham.

A characteristic Sahara-Sindian species growing from Senegal to India in semi-desert conditions.

10. **Heliotropium sessilistigma** *Hutch. & E.A. Bruce*

Perennial grey hispid herb, rather woody below and with many short decumbent branches. Leaves entire, elliptic, 1–2cm long. Cymes sessile, weakly scorpioid, with numerous leafy bracts, up to 4cm long; corolla white; calyx hispid, c 1.5mm long. Nutlets 4, hispid, the stigma sessile and ± hidden.

A rare plant of stony hillsides on the escarpment from 1100 to 1900m showing a preference for limestone: Bani 'Umar, Ḥujarīyah; J. Ra's (Miller); Maghrabah, J. Bura'; between Ḥūth and 'Ashshah; on the dip slope east of Maḥābishah.

11. **Heliotropium strigosum** *Willd.*

Procumbent or erect annual herb to 40cm, very variable in branching and indumentum but always strigose. Leaves entire, linear, 1–2.5cm long. Cymes sessile, scarcely scorpioid, with scattered bracts in between the flowers, up to 8cm long; corolla white; calyx strigose, c 1.5mm long. Nutlets 4, strigose, the stigma conical with a saucer-like ring below, borne on a distinct style.

In Yemen there are two distinct varieties that have sometimes been treated as different species.

Variety **bicolor** (*Hochst. & Steud.*) *O. Schwartz* (Wood 74/400) is a low usually prostrate grey plant, densely strigose throughout. The leaves are usually less than 1cm long and the inflorescence is relatively lax and long. It is usually found on alluvial plains in and around cultivation and is particularly common in the Tihama.

Variety **cordofanum** (*Hochst.*) *Schweinf.* (Wood 75/295) is an erect green plant, sparsely strigose throughout. The leaves are mostly longer than 1.5cm, the inflorescence is comparatively dense with few cymes longer than 4cm. It is frequent in open stony grassland (absent from alluvial deposits) and grows mostly on the escarpment.

H. strigosum as a whole is widespread and common in the Tihama and on the escarpment up to 1500m (rarely 2000m) in bare, open, often disturbed habitats.

12. **Heliotropium zeylanicum** (*Burm.f.*) *Lam.*

H. subulatum (DC.) Martelli

Erect, branched, yellow-green, hispid and often glandular perennial to 50cm, the stem usually woody below. Leaves entire, oblong or elliptic, 3–6cm long. Cymes pedunculate, weakly scorpioid, without bracts, up to 18cm long; corolla yellow, the lobes long-acuminate; calyx glandular, ± hispid, c 2.5mm long. Nutlets 4, glabrous, rugose.

Variable in indumentum; most commonly the plant is covered in sessile and stalked glands and stiff spreading hairs of varying length (eg Wood 75/695) but plants with few glands and adpressed hairs also occur (Wood 74/172).

Widely distributed from 500 to 1500 (–2000)m on the escarpment and southeastern mountains but local, usually in small quantity and avoiding areas with low rainfall. It grows in disturbed grassy habitats such as field borders and track sides.

Heliotropium bacciferum Forssk. was described from Luḥayyah. There is some doubt as to what the type of this plant is and the name has been used for different species. It seems most probable that it is *H. pterocarpum* but the matter has not yet been cleared up.

4. CYSTOSTEMON *Balf. f.*

1. Cystostemon heliocharis (*S. Moore*) *A. Miller & Reidl*

Vaupelia heliocharis (S. Moore) Brand

Bristly undershrub with numerous ascending branches to 25cm. Leaves shortly petiolate, narrowly oblong, c 1–2cm long, entire. Flowers in terminal panicles; calyx lobes linear, c 7mm long, not accrescent in fruit; corolla white, the staminal appendix long exserted, narrowly conical, glabrous. Nutlets whitish-brown, slightly warted, 2–3mm long.

A rare plant of limestone pavement known only from near Raydah (Hepper) and Ḥūth.

5. TRICHODESMA *R. Br.*

Erect herbs, usually very scabrid and often becoming woody below. Flowers in panicles; calyx enlarging in fruit; corolla usually blue with a short tube and 5 spreading lobes; stamens forming a hairy cone-like appendix above the corolla.

1. Undersurface of leaves grey-tomentose — 2
   Undersurface of leaves not tomentose, at most thinly strigose or sparsely bristly — 3
2. Annual. Leaves subsessile. Nutlets smooth, c 4mm long — *T. zeylanicum*
   Perennial. Leaves petiolate. Nutlets with an obscurely dentate margin, c 6mm long — *T. sp. A*
3. Calyx 1–2cm long. Nutlets c 6mm long — *T. trichodesmoides*
   Calyx less than 1cm long. Nutlets less than 4mm long — 4
4. Leaves elliptic, gradually narrowed into the petiole, less than 2cm broad — *T. africanum*
   Leaves ovate, abruptly narrowed into the petiole, some more than 2cm broad — *T. ehrenbergii*

1. Trichodesma africanum (*L.*) *Lehm.*

Erect, much-branched woody herb to 50cm, the stem and leaves scabrid with numerous bulbous-based bristles, though otherwise glabrous. Leaves petiolate, elliptic, 3–4cm long, 1.5cm broad, gradually narrowed at both ends. Flowers pale blue; calyx becoming ovate and reaching 8–9mm long in fruit. Nutlets slightly scabrid with a spiny margin, c 3.5mm long.

Quite frequent on volcanic hills on the high plateau around Sana'a in the area bounded by 'Amrān, Wa'lān and Jihānah. Elsewhere rare and only recorded from Shahārah and 'Ujlī, south of Ṣa'dah. Found between 1800 and 2400m.

2. Trichodesma ehrenbergii *Schweinf.*

Erect, much-branched scabrid annual to 1m with numerous bulbous-based bristles. Leaves petiolate, ovate, up to 7 × 5cm though much less on the inflorescence, abruptly narrowed into the petiole and often ± truncate. Flowers pale blue; calyx lanceolate in fruit, reaching 8mm long. Nutlets very scabrid, with a spiny margin, c 3.5mm long.

Very similar to *T. africanum* from which it can be distinguished by its taller habit, larger, almost truncate leaves and less harsh indumentum.

An uncommon plant of gravelly banks in volcanic regions of the escarpment from 800 to 1600m, usually on roadsides or in wadis: Ḥajdah; Mawsanah and Wadi Ḥijān, Ḥarāz; Wadi Yūr; Wadi Sukhnah; below Kuḥlān; Ḥajjah; Shahārah; Maḥābishah. It has also been found in the Mashriq at Wasīt, north of Ṣa'dah at c 1900m (Miller).

3. Trichodesma trichodesmoides (Bunge) Gurke

T. calathiforme Hochst.

Much-branched, ascending perennial herb to 50cm, the stem scabrid-pubescent, woody below. Leaves petiolate, ovate or lanceolate, 3–10cm long, 1.5–3.5cm broad, gradually narrowed at both ends, ± strigose with bulbous-based hairs. Flowers bright blue; calyx ovate-cordate, reaching nearly 2cm long in fruit. Nutlets with a spiny margin, c 6mm long.

A local plant of stony volcanic hillsides between 1500 and 2500m in four distinct areas: Wadi Hijān and J. Masār, Ḥarāz; Wadi Ḍahr (Choudhary) and J. Nuqum (Rathjens); Wadi Shalālah (Hepper), Raḍmah and Wadi Sabalī in Maghrib 'Ans; J. Ṣabir (Scott and Britton), Misrakh, J. Ḥabashī and Rubaysah near Turbah.

4. Trichodesma zeylanicum (*L.*) *R. Br.*

Erect, much-branched woody annual herb to 1m, the stem pubescent and with scattered bulbous-based scabrid hairs. Leaves subsessile, oblong-lanceolate, up to 8 × 2cm, grey-tomentose below with scabrid hairs on the main veins only. Flowers pale blue; calyx broadly lanceolate, reaching 15mm long in fruit. Nutlets smooth, c 4mm long.

An uncommon field weed found at 200–400m in the alluvial plains in the east of the Tihama: between Khalīfah and J. Bura'; Kadān; Jirābī; Dayr Khushm.

5. Trichodesma sp. A

Much-branched, ascending perennial herb to 50cm, scarcely scabrid except on the upper surface of the leaves, stem grey-canescent. Leaves petiolate, lanceolate or elliptic, up to 9 × 4cm, gradually narrowed at both ends, grey-tomentose below. Flowers white or flesh-coloured; calyx ovate-cordate reaching 2cm in fruit. Nutlets with an obscurely spiny margin, 6–7mm long.

An uncommon plant of roadsides and field borders between 1500 and 2400m on the escarpment: J. Mirbish, Salafīyah (Gillespie); Suq Ar Ribāṭ, J. Raymah; between Suq al Jum'ah and Ma'bar (D. Wood); Khamīs Madyūr, Ḥaymat al Kharijīyah; Maḥwīt; Ḥajjah; Shahārah. (Wood 866; 2097A) The plant has a northern tendency and is only known outside Yemen in Saudi Arabia.

6. CYNOGLOSSUM *L.*

Erect herbs with ± scabrid leaves and blue or white flowers in raceme-like scorpioid cymes; corolla with a short tube and 5 spreading lobes. Fruit of four nutlets, each with short hooked spines.

In Arabia *Cynoglossum* is virtually restricted to the Yemen escarpment where three of the four species are endemic. The fourth, *C. lanceolatum*, is widely distributed in Asia and Africa.

1. Nutlets less than 2.5mm broad. Corolla less than 2mm diameter, usually white — *C. lanceolatum*
   Nutlets more than 3mm broad. Corolla more than 2mm diameter, blue or at least with a blue throat — 2
2. Nutlets with few spines on the upper surface. Young inflorescence ± enclosed by the imbricate bracts — *C. sabirense*
   Nutlets densely covered with spines on the upper surface. Young inflorescence not bracteate — 3
3. Corolla blue. Spines equally distributed over the nutlets — *C. yemenense*
   Corolla white with a blue throat. Spines more numerous at the margin than at the centre of the nutlets — *C. bottae*

1. Cynoglossum bottae *Deflers*

An erect, sparingly branched, scabrid biennial to 50cm. Leaves lanceolate to elliptic, up to 8cm long. Cymes mostly long and lax; corolla white with a blue throat, c 4mm broad. Fruit on recurved pedicels up to 10mm long, the nutlets 3–5mm wide, spines numerous but unevenly distributed forming a distinct marginal ring.

Frequent between 2000 and 3100m on the escarpment mountains from 'Utmah and Ḍawrān northwards; very rare on the high plateau but recorded from Nakhl al Ḥamrā in Ḥadā'. It grows in villages, on terrace walls and in disturbed rocky places. Outside Yemen it is only known from the Asir region of Saudi Arabia.

2. Cynoglossum lanceolatum *Forssk.*

An erect, scabrid annual or biennial to 60cm with numerous branches spreading at a wide angle and often forming an entwined mass. Leaves ovate or lanceolate, up to 8cm long. Cymes usually in pairs, dense-flowered; corolla white (rarely pale blue), very small, up to 2mm diameter. Fruit on recurved pedicels less than 3mm long, the nutlets 2–2.5mm wide, the spines numerous and evenly distributed.

The commonest *Cynoglossum* in Yemen; widely distributed on the escarpment (there is also an isolated record from a streamside in Nihm in the Mashriq) growing in disturbed rocky ground and on field borders from 1000 to 3000m though rare at low altitudes except in the central escarpment.

3. Cynoglossum sabirense (*R. Mill & A. G. Miller*) *J.R.I. Wood* comb. nov.

Paracynoglossum sabirense R. Mill & A.G. Miller in Notes Royal Bot. Gard. Edin. 41:481 (1984)

A decumbent, sparingly branched, scabrid biennial herb. Leaves lanceolate, up to 20cm long. Cymes lax; corolla sky-blue, c 5–6mm diameter. Fruit with nutlets 4–5mm broad and with few spines on the upper surface.

A little-known species with large leaves, distinctive young inflorescences which are enclosed by the numerous imbricate bracts, sky-blue flowers and few-spined nutlets.

A very rare endemic plant of shaded gullies between 2500 and 2900m on a few escarpment mountains where it is associated with *Arisaema bottae*: J. Ṣabir (Lavranos); J. Manār; in the gully below Shirnamah, J. Maswār.

4. Cynoglossum yemenense (*R. Mill & A.G. Miller*) *J.R.I. Wood* comb. nov.

Paracynoglossum yemenense R. Mill & A.G. Miller in Notes Royal Bot. Gard. Edin. 41:478 (1984)

Very similar in general appearance to *C. bottae* from which it differs in being more freely branched, in having deep blue flowers and in the nutlets which have the spines evenly distributed over their surface.

Quite frequent as a weed of fields and track sides in the high altitude southern mountains from J. Ṣabir north to Yarīm and Kusmah; elsewhere only known from Shibām and J. Nabī Shu'ayb. It grows from 2500 to 3300m and is endemic to Yemen.

7. ANCHUSA *L.*

1. Anchusa arvensis (*L.*) *Bieb.*

Lycopsis orientalis L.

Decumbent or ascending, bristly annual herb to 30cm, stem sparingly branched. Leaves sessile, oblanceolate, up to 6cm long, obscurely sinuate, green. Flowers pedicelled in lax, bracteate, branched, terminal, raceme-like cymes; calyx lobes narrowly lanceolate, c 10mm long; corolla bright blue. Nutlets wrinkled.

The Yemen plant is subspecies **orientalis** (*L.*) *Nordh.*

A local field weed occuring in scattered localities between 2000 and 3100m mainly on the wetter mountains of the escarpment but also on the high plateau where it usually grows in irrigated fields.

8. ALKANNA *Tausch*

1. Alkanna orientalis (*L.*) *Boiss.* — Azawa'rān

An erect, much-branched, densely sticky glandular-hairy, yellow-green perennial to 50cm. Leaves sessile, lanceolate or oblong, up to 8cm long, entire. Flowers shortly pedicellate in very leafy raceme-like inflorescences; calyx lobes triangular, c 8mm long; corolla funnel-shaped, yellow. Nutlets pitted.

A characteristic and common plant of open stony roadsides, hills and plains on the high plateau between 2000 and 3000m; also common in similar areas in Khawlan Ash Sham. Otherwise known only from the Manākhah area.

Outside Yemen this species is abundant in the East Mediterranean basin.

9. MYOSOTIS *L.*

Forget-me-not. Herbs with entire, subsessile leaves and small blue flowers in leafless raceme-like inflorescences; corolla with a short tube and with 5 spreading, rounded lobes. Nutlets smooth and glossy.

Stem, leaves and calyx densely spreadingly hairy *M. abyssinica*
Stem, leaves and calyx sparsely adpressed hairy *M. laxa*

1. Myosotis abyssinica *Boiss. & Reut.*

An erect, sparsely branched, pubescent annual to c 10cm. Leaves oblong-elliptic, c 2cm long. Calyx c 2mm long, the lobes narrowly triangular; corolla deep blue, very small, c 1mm diameter.

Only known from shaded cliff ledges at 3200m just below the summit of J. Ta'kar.

2. Myosotis laxa *Lehm.*

A decumbent or ascending, branched, sparsely adpressed pubescent biennial. Leaves oblong-elliptic, c 4–6cm long. Calyx c 5mm long, the lobes broadly triangular; corolla pale blue, c 3–4mm diameter.

The Yemen plant is subspecies **caespitosa** (*O.F. Schultz*) *Hyl. ex Nordl.*

A very local plant of marsh and streamsides between 2200 and 3000m: in a number of places at the head of the Wadi Banā in the Yarīm region; in a gorge between J. Manār and Suq As Sabt, Ibb; Qa' Bosan (Balasan?) (Acres); above Baw'an.

10. ARNEBIA *Forssk.*

1. Arnebia hispidissima (*Lehm.*) *DC.*

Much-branched, erect perennial herb from a woody base to 20cm, grey-green in colour because of the numerous white bristly hairs. Leaves sessile, linear, up to 5cm long, entire. Flowers sessile in bracteate terminal scorpioid cymes; calyx lobes linear, c 8mm long; corolla funnel-shaped, yellow, drying violet. Nutlets pitted.

The roots produce a purple dye.

A local species of field borders and stony hillsides virtually restricted to the high plateau where it is found in the Yarīm, Ḍawrān, Sana'a, Aḥjar and Khamīs Madyūr areas; elsewhere only known from a flood bed near Banī Manṣūr (doubtless washed down) and from the Majz area (Steinberg). It grows from (1500–) 2000 to 2700m and is most typical of sandstone areas.

11. ECHIUM *L.*

1. Echium rauwalfii *Del.* Lisān Ath Thūr

E. longifolium auct.
E. sericeum auct.

Much-branched, erect or ascending probably biennial herb, often grey-green in colour because of the numerous white bristly hairs. Leaves sessile, entire, narrowly lanceolate, up to 4cm long (though sometimes much longer in the first-year basal rosette). Flowers sessile in bracteate terminal scorpioid cymes; calyx lobes narrowly lanceolate, 10–15mm long; corolla irregular, mauve but drying blue. Nutlets wrinkled.

Widely scattered but infrequent between 1300 and 2400m on the escarpment and high plateau, usually growing on roadsides, stony flood beds or in disturbed rocky places, possibly favouring volcanic rocks; completely absent from the central escarpment.

E. rauwalfii is widely valued in Yemen for its medical properties, being used as a diuretic, tranquilliser and aphrodisiac.

VERBENACEAE

Trees, shrubs or herbs. Leaves opposite, without stipules. Flowers usually arranged in racemes or cymes, bisexual, irregular; calyx 4–5-lobed; corolla tubular. Fruit drupaceous or of 2–4 hard mericarps, rarely a capsule.

1. Coastal mangrove tree 9. *Avicennia*
   Herbs or shrubs, not coastal **2**
2. Inflorescence of elongate spikes or racemes, 4–40cm long **3**
   Inflorescence of cymes or of dense head-like spikes not exceeding 3cm in length **5**
3. Leaves pinnatifid. Corolla pale blue 5. *Verbena*
   Leaves not pinnatifid. Corolla white or cream **4**
4. Fruit elongate, erect, adpressed to the rhachis of the inflorescence 3. *Chascanum*
   Fruit subglobose, pendulous 4. *Priva*
5. Prostrate herb, rooting at the nodes 2. *Phyla*
   Erect shrubs **6**
6. Flowers in short head-like spikes borne on peduncles from the leaf axils. Scabrid shrub 1. *Lantana*
   Flowers in terminal cymes. Plants not scabrid **7**
7. Leaves less than 2cm long. Flowers cream, small 6. *Premna*
   Leaves more than 3cm long. Flowers blue and white, at least 1cm long 7. *Clerodendrum*

1. LANTANA *L.*

1. Lantana viburnoides (*Forssk.*) *Vahl*

L. salvifolia sensu Schwartz

Untidily erect or prostrate and heavily grazed, aromatic branched shrub to 2m with scabrid, quadrangular, yellow-brown stems. Leaves petiolate, lanceolate to ovate, cuneate at the base, crenate, scabrid above, pubescent below, very variable in size, 1–12cm long. Flowers in long pedunculate, bracteate, head-like axillary spikes which elongate to c 3cm in fruit; corolla with a short tube and a pink or mauve limb 2–4mm in diameter. Fruit an edible black drupe.

Very variable in leaf size and often separated into two species in consequence. Plants from the wettest areas of the escarpment (eg Wood 1373) have leaves 5–12cm long and agree well with the type of *L. viburnoides*. Plants from drier situations have leaves less than 3cm long and are probably a distinct variety though all sorts of intermediates can be found.

Widespread between 500 and 2500m in a variety of well-drained, protected habitats on the escarpment, high plateau and western part of the Mashriq; it is found on cliffs, in valley scrub, around larger thorny shrubs in *Acacia* bushland and in thorn fences around fields.

The Curse of India, **Lantana camara** *L.* is cultivated as a hedge plant in Taiz and readily self-seeds though it does not appear to have become naturalised yet. It is like a robust form of *L. viburnoides* but has spiny stems and a larger orange corolla which turns yellow as the flowers mature.

2. PHYLA *Lour.*

1. Phyla nodiflora (*L.*) *Greene*

Lippia nodiflora (L.) A. Rich.

Perennial herb with a trailing, sparsely strigose, purplish stem which roots at the nodes. Leaves oblanceolate, dentate near the apex, 1–1.5cm long. Inflorescence of dense, axillary, pedunculate, purplish spikes, mostly 3–12mm long; corolla small, pinkish or cream. Fruit of 2 small mericarps.

Characteristic of the sandy and gravelly margins of permanent flowing water in the Tihama and on the escarpment up to 2000m; rare in the Mashriq and only known from Rada' and the Wadi Khārid. It is often associated with *Eclipta prostrata*.

3. CHASCANUM *E. Mey.*

Small undershrubs with simple leaves. Flowers in terminal spikes; corolla with a slender tube, 2-lipped, cream but drying brown or purplish; calyx narrow, cylindrical, enclosing the fruit which is released when the side furthest from the inflorescence rhachis splits open. Fruit erect, consisting of 2 dry cylindrical mericarps.

1.	Leaves cuneate at the base, tapering to the stem without a petiole	*C. sessilifolium*
	Leaves truncate at the base, distinctly petiolate	2
2.	Calyx c 6mm long. Leaves grey below	*C. marrubifolium*
	Calyx c 8mm long. Leaves green below	*C. laetum*

1. Chascanum laetum *Fenzl ex Walp.*

C. arabicum Moldenke
Svensonia laeta (Fenzl ex Walp.) Moldenke
Bouchea pterygocarpa Schau.

Erect, branched, pubescent, leggy undershrub to 1½m. Leaves petiolate, ovate, truncate at the base, crenate, green on both surfaces, 1–3cm long. Flowers in terminal spikes 10–25cm long; calyx c 8mm long; corolla c 10mm long.

A local plant of roadsides and disturbed rocky slopes between 300 and 1500m mostly in the central escarpment: on the western slopes of J. Bura'; in the Ḥarāz around Ḥajjaylah, 'Usil and 'Mawsanah; Khamīs Banī Sa'd; between Bājil and Kadān; Mafḥaq; between Madīnat al Abīd and Ḥammām 'Alī.

2. Chascanum marrubifolium *Fenzl ex Walp.*

Bouchea marrubifolia (Fenzl) Schau.

Erect, much-branched, pubescent undershrub to 70cm. Leaves petiolate, ovate, crenate, truncate at the base, green above, grey-pubescent below, 1–3cm long. Flower spikes 3–7cm long; calyx and corolla both c 6mm long.

A local plant of arid stony plain between 100 and 1300m: abundant in open *Acacia-Commiphora* scrub in the stony steppe of the southern Tihama and the Maqbanah region extending almost to Taiz; in the protection of *Capparis decidua*, Ḥaraḍ; on the lava flows north of Mārib.

3. Chascanum sessilifolium (*Vatke*) *Moldenke*

Prostrate or ascending, much-branched, pubescent undershrub less than 10cm high. Leaves sessile, obovate to oblanceolate, crenate at the apex, green on both surfaces, 1–1.5cm long. Flower spikes 5–10cm long; calyx c 6mm long, often purplish; corolla c 11mm long, fragrant.

A local plant of cliff tops and similar open stony places betwen 1500 and 2300m, always on sandstone; near the hospital, Turbah; east of Rada'; Suq al Jum'ah, J. Ash Shirq; around Sana'a and on the cliffs south of Wadi Ḍahr; east of Rujum.

Known only from Yemen and Somalia.

4. PRIVA *Adans.*

Herbs with scabrid leaves. Flowers in long simple racemes; corolla tubular with 5 ± equal, white lobes; calyx tubular in flower, enclosing the fruit and becoming globular as the mericarps enlarge, densely covered in hooked hairs and clinging to clothing and passing animals. Fruit pendulous, consisting of 2 hard, spiny mericarps.

Mericarps densely pubescent	*P. tenax*
Mericarps glabrous or nearly so	*P. adhaerens*

1. Priva cordifolia (*L.f.*) *Druce* Ḥamshad

Erect, sparingly branched, perennial herb with quadrangular stems to 80cm, the whole plant rough with longish hooked hairs. Leaves petiolate, ovate, truncate at the base, crenate, 2–8cm long. Racemes up to 50cm long but often much less; corolla small, white; calyx c 6mm long; mericarps hard, woody, few spined, shiny, glabrous.

Widespread on the escarpment from 500 to 1900m growing on waste ground, terrace walls and stony banks with a marked preference for the mist zones of the Ḥujarīyah and the central escarpment.

2. Priva tenax Verdc.

P. adhaerens auct.

Very similar to the previous species and only certainly distinguished by the mericarps which are many-spined and densely pubescent. It is a clearly annual plant, greyer in appearance and with longer racemes.

A weed of tracksides, waste ground and irrigated fields in the alluvial plains in the east of the Tihama and on the escarpment up to 1500m. Perhaps the only *Priva* in the Tihama but very scattered in occurrence.

5. VERBENA *L.*

1. Verbena officinalis *L.*

Vervain. Glabrous perennial herb, usually branched at the base with erect, square stems to 50cm. Leaves pinnatifid, 2–7cm long. Flowers in slender branched terminal spikes 6–15cm long; corolla pale lilac, ± 2-lipped, the tube c 4mm long, exceeding the calyx. Fruit of 4 nutlets.

Frequent on the grassy margins of streams and ditches between 1300 and 2900m on the escarpment and high plateau.

Verbena x hybrida *Voss* is commonly cultivated in gardens in Sana'a.

6. PREMNA *L.*

1. Premna resinosa (*Hochst.*) *Schau.* Shaqab

Erect aromatic shrub to 3m, the branches pubescent and brownish when young but becoming grey and glabrescent at maturity. Leaves petiolate, ovate, entire or obscurely crenate, ± glabrous, 1–2cm long. Flowers in small, terminal panicled cymes; corolla yellow-green, minute; calyx dotted with yellow glands. Fruit a small black, obovoid drupe, c 5mm long.

Locally abundant in *Acacia-Commiphora* bushland on rocky outcrops in the Tihama and on the escarpment from 200 to 1200m; particularly characteristic of the first foothills and drier, rainshadow areas inland.

7. CLERODENDRUM *L.*

1. Clerodendrum myricoides (*Hochst.*) *Vatke*

Untidy, erect, branched shrub to 2m, the branchlets brown, pubescent. Leaves petiolate, oblong or elliptic, obscurely or strongly crenate, 3–8cm long. Flowers in terminal panicled cymes; corolla c 15mm long, shortly tubular, ± 2-lipped, the upper lobes pale blue, the lower lobe darker blue, concave, much longer; calyx cup-shaped; filaments and style long-exserted, pale blue. Fruit a black drupe.

Two distinct forms occur in Yemen. Populations from Taiz province (Wood 1446) are strongly aromatic with broadly elliptic to suborbicular leaves which are glabrescent below. Populations from Ibb (Wood 1988) are scentless with narrowly elliptic leaves which are densely tomentose below. *C. myricoides* is a very variable species in Africa and it is not clear how these forms relate to those in Africa.

Frequent on terrace walls, cliffs and bushy places near cultivation between 1200 and 2000m in the Ḥujarīyah north to the Wadi Ḍabāb area; elsewhere known only from scrubby field borders between Ibb and Suq As Sabt.

Clerodendron inerme (*L.*) *Gaertn*, a Jasmine-like climbing or creeping plant with white flowers, is cultivated in Hodeida and Taiz.

9. AVICENNIA *L.*

1. Avicennia marina (*Forssk.*) *Vierh.* Shūrah

White Mangrove. An unmistakable small tree or shrub growing in coastal mud. Leaves shortly petiolate, ovate or elliptic, entire, shiny and dark-green above, white-tomentellous below, 3–8cm long. Inflorescence a small terminal panicle with the flowers arranged in dense heads; corolla small, yellow. Fruit a pubescent, ovoid capsule, c 2.5cm long.

Frequent and locally dominant on mudflats and in small creeks along the Red Sea from Hodeida northwards with the best-developed mangrove swamps found around Mīdī and Luḥayyah. South of Hodeida it is very rare and only known from between Yakhtul and Mawshij.

The Lemon Verbena, **Lippia triphylla** (*L'Hérit) Kunze* (*L. citriodora Kunth*), has been cultivated at Sana'a (Rathjens) but it has not been seen recently.

LABIATAE

A very distinctive family of herbs or undershrubs. Stems usually quadrangular. Leaves opposite, without stipules, usually simple, diminishing regularly in size upwards and changing imperceptibly into bracts. Flowers bisexual, typically in cymose whorls in the axils of the bracts forming an interrupted raceme but becoming condensed and spicate at the apex of the stem; calyx tubular or bell-shaped, usually 5-lobed; corolla tubular, the lobes forming two distinct lips. Fruit of 4 nutlets.

Nearly all species are strongly aromatic, a particular aroma often being characteristic of a particular genus. Several species are cultivated as potherbs because of their attractive smell including Mint, Basil and Marjoram.

The family is rich in endemic species both in tropical genera like *Plectranthus* and *Becium* and in temperate genera like *Nepeta* and *Thymus*.

In the following key (as also in some generic keys) attention is focused on the calyx. Calyx characters should be observed when the plant is in fruit.

1. Leaves pinnatisect — 8. *Lavandula*
   Leaves entire or shallowly toothed, never pinnatisect — **2**
2. Calyx with a broad spreading scarious lower lip at least 3 times as long as the other lobes — 19. *Otostegia*
   Calyx without a broad spreading lower lip — **3**
3. Calyx 2-lipped, both lips rounded, entire, the upper lip with a flat appendage attached to it — 15. *Scutellaria*
   Calyx 2-lipped or regular, but if 2-lipped at least one lip terminating in teeth or bristles — **4**
4. Calyx 2-lipped, the upper lip formed by the single upper calyx lobe which is rounded, not bristle-like, and much larger than the other bristle-like calyx lobes — **5**
   Calyx with 4–10 similar teeth or if 2-lipped, the upper lip formed of 3 joined teeth — **10**
5. Corolla tube straight. Flowers cream, white, pink or mauve. (Violet only in *Orthosiphon thymifolius.*) Plant never fleshy — **6**
   Corolla tube bent at the mouth of the calyx. Flowers blue or violet. Plants often fleshy — **9**
6. Stamens included in the corolla tube. Leaves oblanceolate — 3. *Endostemon*
   Stamens exserted from the corolla tube. Leaves obovate, ovate, elliptic or lanceolate — **7**

7. Filaments longer than the corolla. Leaf base cuneate **8**
Filaments shorter than the lower lip of the corolla. Leaf base truncate 4. *Orthosiphon*
8. Bracts of inflorescence whorls transformed into glands. Lateral calyx lobes obscure so calyx apparently 3-lobed 2. *Becium*
Bracts of inflorescence whorls not transformed into glands. Lateral calyx lobes, tooth-like, similar to the two lower teeth 1. *Ocimum*
9. Lateral calyx lobes rounded. Annual plant 6. *Solenostemon*
Lateral calyx lobes acute, bristle-like. Annual or perennial plants 5. *Plectranthus*
10. Corolla with 4 ± equal lobes 9. *Mentha*
Corolla distinctly 2-lipped (rarely the upper lip much reduced) **11**
11. Calyx bladder-like, distinctly inflated in fruit, soft **12**
Calyx not inflated or bladder-like, usually stiff **13**
12. Corolla pink. Calyx with 5 teeth 21. *Teucrium*
Corolla white. Calyx with 10 teeth 20. *Leucas*
13. Calyx with 10 teeth. Corolla white, usually villous **14**
Calyx with 5 teeth. Corolla rarely white, never villous **15**
14. Stems white-woolly 16. *Marrubium*
Stems glabrous or pubescent, never woolly 20. *Leucas*
15. Flowers in dense heads. Prostrate creeping plant 10. *Thymus*
Flowers not in dense heads. Plants ± erect **16**
16. Flowers in dense uninterrupted spikes **17**
Inflorescence various but, if spike-like, interrupted **18**
17. Spikes consisting of shortly pedunculate cymes. Leaves ovate 7. *Hyptis*
Spikes consisting of opposite, sessile flowers. Leaves linear or lanceolate 8. *Lavandula*
18. Leaves entire **19**
Leaves toothed **20**
19. Leaves less than 1.5cm long, green below 11. *Micromeria*
Leaves more than 2cm long, white below 17. *Stachys*
20. Flowers in axillary spike-like racemes 7. *Hyptis*
Flowers whorled in the axils of the bracts **21**
21. Fertile stamens 2 **22**
Fertile stamens 4 **23**
22. Shrub. Calyx woolly 12. *Meriandra*
Herb or undershrub. Calyx at most densely pubescent 13. *Salvia*
23. Calyx oblique with a protruding lower lip. Flowers white 20. *Leucas*
Calyx with 5 ± equal teeth. Flowers variously coloured **24**
24. Bracts sessile, all much longer than the calyx **25**
Bracts petiolate, mostly shorter than the calyx **26**
25. Corolla red. Leaves ovate or orbicular 18. *Lamium*
Corolla blue. Leaves oblanceolate 22. *Ajuga*
26. Flowers up to 10 in lax whorls. Corolla pink 11. *Micromeria*
Flowers numerous in dense whorls. Corolla blue or white 14. *Nepeta*

1. OCIMUM *L.*

Herbs or undershrubs. Flowers shortly pedicellate, in axillary whorls, forming terminal, sometimes panicled racemes, the bracts small and inconspicuous; calyx 5-lobed, the upper lobe broad, rounded, decurrent into the calyx tube, forming the upper lip of the calyx, the four other lobes bristle-like, similar, the lower 2 attached to the lower lip of the calyx; corolla 2-lipped with an obconical tube; filaments long exserted.

1. Lower lip of the calyx bent sharply upwards in fruit tightly closing the calyx mouth **2**
Calyx remaining open in fruit **4**
2. Inflorescence congested, spike-like, the rhachis and pedicels both concealed. Plant stellate-hairy *O. spicatum*
Inflorescence of panicled racemes, the rhachis and pedicels both clearly visible. Hairs simple **3**
3. Leaves narrowly elliptic, sparsely pubescent below *O. gratissimum*
Leaves ovate or broadly lanceolate, grey tomentose below *O. suave*
4. Leaves entire or obscurely toothed. Stem with crisped hairs **5**
Leaves strongly crenate. Stem with long spreading large-celled hairs *O. tenuiflorum*
5. Corolla less than 5mm long *O. canum*
Corolla more than 7mm long *O. forskolei*

1. Ocimum canum *Sims*

O. americanum auct.

Similar to the much more common *O. forskolei* but annual with green, more densely crisped-hairy stems which only become woody when old and leaves which are always pubescent below. The racemes are rather dense, the lowest whorls less than 1cm apart, with small flowers whose cream corolla is only 3–5mm long.

Strongly aromatic, the aroma different from that of *O. forskolei* but similar to that of *O. basilicum*, the cultivated Basil which may be a derivative of *O. canum*.

Locally abundant by irrigation channels in the alluvial plains below Mishrāfah in the Wadi Rima' area (Wood 1385). All other records I have been able to trace are in fact *O. forskolei*.

2. Ocimum forskolei *Benth.* Ḥubaqbaq, Dawmir

O. menthifolium Hochst. ex Benth.
O. hadiense sensu E. A. Bruce
O. dichotomum sensu Deflers
O. canum sensu Schweinf.

Erect branched perennial to 50cm, the stems crisped-pubescent, brown, rather woody, the leaves deciduous in the dry season when the plant becomes a small leafless twiggy bush. Leaves lanceolate, ovate or elliptic, up to 4cm long, glabrous or nearly so, the margin entire or obscurely repand. Racemes with rather distant whorls, the lower more than 1cm apart; calyx often purplish, open in fruit; corolla pale mauve, rarely cream, c 7–10mm long.

The plant is strongly aromatic.

Widespread and often abundant in the east of the Tihama and on the escarpment up to c 1200m, occurring more rarely at higher altitudes but reaching 2400m and penetrating the high plateau at Riṣābah (Hepper) and Sana'a (Rathjens); also in the southeastern mountains, but not known east of Qa'ṭabah (Wissmann). It grows in open *Acacia* scrub, on tracksides, in stony wadi beds, on field margins, on open alluvial plains and by irrigation ditches. (Wood 546, 1387)

3. Ocimum gratissimum *L.*

O. viride Willd.
Similar to the much more common *O. suave*, but larger and more obviously shrubby reaching 2m in height. The leaves are narrowly elliptic up to 12 × 3cm, dark green and at most sparsely pubescent below. The racemes are comparatively lax and long often exceeding 15cm in length.

Only known from the margins of the permanently flowing Wadi Sukhnah at 1500m (Wood 3394.)

4. Ocimum spicatum *Deflers*

O. cylindrostachys Schweinf. nom nud.
Erect, branched undershrub to 1m, the stem brown, woody and glabrous below, but densely white stellate-hairy above. Leaves ± ovate, up to 5cm long, densely white stellate-tomentose below, dark green and thinly stellate-hairy above, the margin weakly crenate. Racemes solitary, dense, spike-like, the whorls undifferentiated; calyx densely hairy, the lower lip closing the calyx mouth; corolla yellow-green, very short, ± enclosed in the calyx, c 3–4mm long.

The plant is weakly aromatic.

A local plant of rocky slopes in *Acacia* scrub found between 700 and 1400m in the Ḥujarīyah (J. Masanah (Deflers) and at Bani 'Umar); the rainshadow areas of the central escarpment (around Madīnat Al 'Abīd; behind Sukhnah in Wadi Kulabah (Wissmann) and below Nasham on J. Bura'; in the Ḥarāz in Wadi Khufīt near 'Usil (Schweinfurth) and at Baḥal; Wadi Yūr; on the dip slope east of Maḥābishah.) Outside southwest Arabia it is only known from Somalia.

5. Ocimum suave *Willd.*

Erect, branched perennial to 1m, stems crisped-pubescent, woody below. Leaves ovate or broadly lanceolate up to 10 × 4cm, ± grey tomentose below, dark green and pubescent above, the margin serrate. Racemes panicled or in 3s, rarely more than 10cm long, rather dense, the lowest whorls about 5mm apart; calyx green, the lower lip closing the calyx mouth; corolla cream, c 3mm long, scarcely exceeding the calyx.

The plant smells strongly of cloves.

Widely distributed on the escarpment between 1000 and 2400m from Wadi Ḍabāb near Taiz north to the Wadi Lā'ah, but rather local, occurring on the margins of the lush middle-altitude wadis and also on open grassy mountainsides in areas of good rainfall such as 'Utmah and J. Masār in the Ḥarāz where it may be a relic of highland forest. (Wood 75/268, 3353; Miller 3072)

6. Ocimum tenuiflorum *L.* Qurunful

O. sanctum L.
Erect, branched perennial herb to 50cm, the stems becoming woody with age, pilose with spreading, white, large-celled hairs. Leaves ovate, up to 7cm long, thinly pubescent on both surfaces, the margin crenate. Racemes panicled or in 3s, rather dense, the lowest whorls less than 1cm apart; calyx olive-green, open in fruit; corolla cream, c 3mm long, scarcely exceeding the calyx.

The plant smells strongly of cloves.

Often confused with *O. suave*, but smaller in habit with a distinctive stem indumentum and the calyx remaining open in fruit.

Quite frequent between 200 and 800m in the alluvial plain in the east of the Tihama and in the wadis penetrating the foothills from the Wadi Zabīd north to the Wadi Mawr, usually growing by irrigation ditches, in seasonally swampy land or wadi beds (Wood 1386, 3436). It is widespread in tropical Asia, but unknown in Africa.

Basil, **Ocimum basilicum** *L.*, is grown as a potherb throughout Yemen and is known as 'Rīḥān' or 'Ḥabaq'. It is not used in cooking, but fresh sprigs are worn by both men and women for decoration in their turbans. The typical plant has green racemes, but variety **purpurascens** *Benth.* with purple racemes is also widely grown.

Ocimum lamiifolium Hochst. ex Benth. was recorded by Deflers from Manākhah but the record cannot be accepted. The specimen is not at Paris nor has the plant been recollected at Manākhah (or anywhere else in Arabia) and Deflers himself stated that the identification was uncertain.

2. BECIUM *Lindl.*

Similar to *Ocimum* but differing in the bracts which fall early leaving the enlarged glandular bases and in the calyx where the two lateral lobes are missing and replaced by a toothed or ciliate margin.

1. Leaves serrate, glabrescent — *B. filamentosum*
   Leaves usually entire, pubescent on both surfaces — 2
2. Flowers in apparent heads, the racemes not elongating till the plant is in fruit. Leaves subsessile, more than 2cm long — *B. obovatum*
   Flowers in interrupted racemes. Leaves petiolate, less than 2cm long — *B. serpyllifolium*

1. Becium filamentosum (*Forssk.*) *Chiov.*

Ocimum filamentosum Forssk.
O. eu-filamentosum Schweinf. nom. nud.
Erect, branched, woody perennial to 50cm, rather variable in habit, usually shrubby, lax and straggling in scrub but compact, rounded and scarcely 15cm high on exposed plain; stem crisped-pubescent. Leaves petiolate, ovate or oblong, up to 4cm long, glabrescent on both surfaces, the margin serrate. Racemes with few rather distant whorls, the lower more than 1cm apart, calyx green or purplish, glabrescent except on the veins and pedicels, the lateral lobes replaced by minute teeth and a line of white hairs,

remaining open in fruit; corolla pale violet, rarely white, 9–12mm long; filaments and style long exserted, often as much as 2cm.

B. filamentosum is commonly confused with *O. forskolei* but apart from the generic differences, it is not aromatic and usually has a rather unhealthy appearance, the racemes usually with only 3–4 whorls thus contrasting with the strongly aromatic *O. forskolei* which regularly has 10 or more whorls in each raceme.

Widespread and often common on the escarpment from 600 to 2400m, usually growing in scrub, sometimes in very dry conditions, extending sporadically eastwards across the high plateau to the Bayḏā' – Nihm – Majz line and there usually growing on open stony plain or in crevices in limestone pavement. (Wood 460, 1438, 1575)

2. **Becium obovatum** (*E. Mey.*) *N.E. Br.*

Erect, branched perennial herb to 40cm; stems woody, crisped-pubescent. Leaves subsessile, oblong or obovate, up to c 5cm long, thinly pubescent on both surfaces, margin entire. Racemes head-like, not elongating till the plant is in fruit; calyx with prominent green veins, becoming purplish towards the margin, pubescent except for the prominent white-ciliate margin which replaces the lateral lobes, remaining open in fruit; corolla pink, 12–15mm long.

On an open hillside on loose volcanic rock adjoining cultivated terraces at 2600m just south of Mashwarah between Ibb and 'Udayn. (Wood 2736)

3. **Becium serpyllifolium** (*Forssk.*) *J.R.I. Wood*

Ocimum serpyllifolium Forssk.

Small, weakly aromatic, erect, much-branched, woody undershrub to 50cm; stems crisped-pubescent. Leaves petiolate, oblong or obovate, rarely exceeding 1cm long, thinly pubescent on both surfaces, margin entire. Racemes with numerous rather distant whorls, the lower more than 1cm apart; calyx purplish, pubescent, the lateral lobes replaced by a row of teeth, remaining open in fruit; corolla pink, 9–10mm long.

Widespread and often abundant on the escarpment and high plateau between 1500 and 2700m but completely absent from Taiz province apart from J. Sawraq. It grows on bare rocky slopes and mountain ridges, often on crumbly volcanic rock and is sometimes locally dominant. It is often associated with *Lotus quinatus*, *Indigofera arabica* and *Elionurus muticus*. It is apparently endemic to Yemen, but probably also occurs in the Asir region of Saudi Arabia. (Wood 75/154, 2838)

3. ENDOSTEMON *N.E. Br.*

Similar to *Ocimum* but the filaments short and the stamens included in the corolla tube. In addition the corolla is weakly 2-lipped, the whorls have only 2–6 flowers and our species all have oblanceolate leaves.

1. Whorls consisting of 2 flowers only. Leaves sessile or nearly so — *E. tenuiflorus*
   Whorls consisting of 2–6 flowers. Leaves distinctly petiolate even above — **2**
2. Calyx remaining open in fruit. Whorls with up to 4 flowers only — *E. gracilis*
   Lower lip of calyx bent sharply upwards in fruit, tightly closing the calyx mouth. Whorls with up to 6 flowers — *E. tereticaulis*

1. **Endostemon gracilis** (*Benth.*) *M. Ashby*

Orthosiphon melhanensis Schweinf. nom. nud.

Weak, slightly aromatic, ± erect, branched annual or short-lived perennial herb to 20cm, stem densely pubescent with numerous spreading glandular hairs. Leaves petiolate, oblanceolate, up to 4cm long, pubescent, the margin crenate. Racemes rarely exceeding 5cm long with up to 5 whorls, each with 2–4 flowers only; calyx greenish, densely covered in glandular hairs, remaining open in fruit so the white-villous throat is clearly visible; corolla cream, c 2mm long.

A very local plant found in open bushland between 500 and 1000m in the central escarpment: 'Ukaybir, J. Milḥān (Schweinfurth); Suq As Sabt, J. Bura'; Suq Ar Ribāṭ, J. Raymah. (Wood 1929, 2428)

2. **Endostemon tenuiflorus** (*Benth.*) *M. Ashby*

Aromatic perennial with a woody rootstock, much branched at the base and with numerous, erect, scabrid-pubescent stems to 15cm. Leaves subsessile, linear-oblanceolate up to 2cm long, crisped-pubescent, grey, the margin crenate. Racemes up to c 10cm long with up to 15 whorls, each with 2 flowers, the flowers quickly deciduous at maturity leaving the persistent pedicel; calyx with a distinctive purple upper lip, sparsely scabrid-pubescent, remaining open in fruit so the white-villous throat is clearly visible; corolla mauve, c 7mm long.

Widely distributed on dry rocky hills and stony plain in very open *Acacia* scrub between 500 and 1500m on the escarpment and southeastern mountains east to Dhī Nā'im; absent from high rainfall areas.

3. **Endostemon tereticaulis** (*Poir.*) *M. Ashby*

Weakly aromatic, erect branched annual or perennial herb to 40cm, the stem crisped-pubescent, becoming woody when old. Leaves petiolate, oblanceolate, up to 4cm long, pubescent, the margin crenate. Racemes often more than 10cm long with up to 15 whorls, each with 4–6 flowers; calyx greenish, densely glandular-hairy, the lower lip bent sharply upwards in fruit and closing the mouth of the calyx; corolla mauve, c 3mm long.

A rare, but possibly overlooked plant of open gravelly slopes in *Acacia* scrub between 600 and 1500m in the northern escarpment: Wadi Sukhnah; on the foothills below Maḥābishah; Malāḥiṭ; on the lower slopes of J. Marran above the Wadi Līyah. (Wood 2966, 3053)

4. ORTHOSIPHON *Benth.*

Similar to *Ocimum* but the filaments short, ± enclosed in the corolla tube with just the stamens showing. In addition in our species there are only 6 flowers in each whorl and the leaf base is truncate (not cuneate as in related genera). Our plants are not aromatic.

Stem glabrous. Corolla violet *O. glabratus*
Stem pubescent. Corolla cream *O. pallidus*

1. Orthosiphon pallidus *Royle ex Benth.*

Erect, branched, scentless annual or short-lived perennial to 30cm, stem crisped-pubescent becoming woody with age. Leaves ovate, ± triangular in outline with a truncate base, up to 4cm long, ± glabrous on both surfaces but paler beneath, margin serrate. Racemes short, usually with 4–5 rather distant whorls; calyx green, sparsely pubescent, remaining open in fruit; corolla cream, small, 2–3mm long.

A widespread but infrequent plant tolerating quite arid conditions and growing from 200 to 1500m on rocky outcrops in the Tihama and on the escarpment, usually in small quantity. It is found in open *Acacia* scrub, on field borders and stony tracksides, being most characteristic of the escarpment foothills and dry areas such as Waṣāb and Maqbanah.

2. Orthosiphon thymiflorus (*Roth*) *Sleesen*

O. glabratus Benth.

Erect, branched undershrub to 30cm, the stem glabrous, woody. Leaves ovate, ± triangular in outline with a truncate base, up to c 4cm long, ± glabrous on both surfaces but paler beneath, margin serrate. Racemes long, delicate, lax, the lower whorls more than 1cm apart; calyx deep violet, glabrous, remaining open in fruit; corolla violet with a white tube, c 7mm long.

A scentless plant which bursts into flower immediately after the onset of the spring rains.

Locally frequent in the protection of thorn fences and spiny shrubs between 500 and 1800m in the wetter parts of the escarpment from Wadi Ḍabāb near Taiz north to Maḥābishah. (Wood 75/62, 2429)

5. PLECTRANTHUS *L'Herit.*

Herbs or undershrubs, usually aromatic and in varying degrees fleshy. Flowers pedicellate, in axillary whorls, forming terminal, sometimes panicled racemes; bracts leaf-like and persistent or more commonly membranous and deciduous; calyx ± 2-lipped with a large upper lobe and 4 bristle-like lower lobes, remaining open in fruit; corolla usually blue or violet, the tube bent downwards at the mouth of the calyx, 2-lipped, the upper lip often short and flat and the lower lip deeply concave.

A complex genus which has been much confused largely because of the poor quality of most herbarium specimens. The species are, in fact, distinct and are readily identified, particularly in the field. The following characters are listed as they make identification easier:

Shrubs at least 1m high: *P. asirensis, P. cicatricosus, P. hyemalis, P. ignarius.*
Strongly aromatic plants with very succulent brittle leaves: *P. arabicus, P. cylindraceus, P. tenuiflorus.*
Leaves ± clustered at the ends of the branches: *P. coerulescens, P. hadiensis, P. hyemalis.*
Plant leafless when flowering: *P. cicatricosus, P. ignarius, P. tenuiflorus.*
Leaves sessile: *P. arabicus, P. coerulescens.*
Fruiting calyx villous in the throat: *P. barbatus, P. hyemalis.*
Leaves sticky: *P. cicatricosus.*
Plants winter-flowering: *P. asirensis, P. cicatricosus, P. coerulescens, P. hyemalis, P. ignarius.*
Plants flowering in early summer: *P. ovatus.*
Plants flowering in late summer: *P. barbatus, P. cylindraceus, P. lanuginosus.*
Plants never flowering in Yemen: *P. amboinicus.*

1. Leaves sessile, oblanceolate, up to 2cm long. racemes with persistent fleshy bracts similar to the leaves *P. arabicus*
Leaves petiolate or, if sessile, elliptic and more than 2cm long. Racemes without bracts or, if bracts present, dissimilar to the leaves **2**
2. Inflorescence of panicled spike-like racemes; corolla less than 5mm long *P. cylindraceus*
Inflorescence not spike-like; corolla more than 10mm long **3**
3. Fruiting calyx white-villous in the throat **4**
Fruiting calyx glabrous or with a few hairs in the throad, never white-villous **5**
4. Leaves obovate. Racemes with up to 12 whorls *P. hyemalis*
Leaves lanceolate, ovate or elliptic. Racemes with 10–25 whorls *P. barbatus*
5. Leaves glabrous or minutely tomentellous, usually entire. Calyx scarious *P. ignarius*
Leaves pubescent, villous or tomentose, distinctly toothed. Calyx not scarious **6**
6. Leaves distinctly longer than broad, the base cuneate, very rarely truncate **7**
Leaves about as long as broad, the base cordate or truncate, rarely cuneate **10**
7. Leaves sessile. Racemes simple. Corolla nearly glabrous on the outside *P. coerulescens*
Leaves petiolate. Inflorescence usually paniculate. Corolla densely hairy on the outside **8**
8. Erect shrub to 1½m. Calyx not enlarging in fruit *P. asirensis*
Herbs to 1m (usually less). Calyx enlarging in fruit **9**
9. Inflorescence glandular-hairy. Bracts deciduous before the flowers open *P. lanuginosus*
Inflorescence largely without glandular hairs. Bracts deciduous as the flowers open *P. ovatus*
10. Lower lip of corolla deeply concave, at least 2mm deep. Inflorescence and stem densely covered in long white glandular hairs **11**
Lower lip of corolla shallowly concave, less than 2mm deep. Inflorescence and stem with few or no long white glandular hairs **12**

11. Leaves sticky. Flowers in short, apparently lateral racemes up to 4cm long *P. cicatricosus*
Leaves not sticky. Flowers in terminal panicles, the individual racemes 5–15cm long *P. lanuginosus*
12. Leaves very succulent, brittle. Pedicels mostly 5–8mm long. Stems glandular *P. tenuiflorus*
Leaves scarcely succulent. Pedicels less than 5mm long. Stems eglandular *P. hadiensis*

1. **Plectranthus arabicus** *E. A. Bruce*

P. quadridentatus sensu Schwartz
Ocimum β zatarhendi Forssk.
Small, strongly aromatic, very succulent, decumbent annual herb to 10cm. Leaves oblanceolate, gradually narrowed to the cuneate sessile base, 1–2cm long, pubescent, each side with 4 prominent teeth. Flowers in terminal bracteate racemes, the bracts similar to the leaves, but up to 4mm long only; corolla 3–4mm long, the upper lip white, the lower lip violet.

A very distinctive species readily identified by its slender decumbent habit, small, very succulent, sessile leaves and racemes with green leaf-like bracts. It flowers after any substantial rainfall.

Widespread but local and easily overlooked, growing from 500 to 1400m on the escarpment from Bani 'Umar in the Ḥujarīyah north to Maḥābishah. It is found on cliffs and bare stony slopes in *Acacia* scrub where it is often partially hidden around the roots of spiny shrubs.

P. arabicus is endemic to Yemen but is very closely related to *P. prostratus* Guerke from East Africa.

2. **Plectranthus asirensis** *J.R.I. Wood* Sha'r

Coleus arabicus Benth.
Weakly aromatic, slightly succulent, branched undershrub to 1½m. Leaves petiolate, oblong or elliptic with a cuneate base, up to 10cm long, densely pubescent or ± tomentose on both surfaces, margin crenate. Flowers in simple or sparingly branched, densely pubescent terminal racemes, bracts colourless, deciduous as the flowers open; corolla, violet c 15mm long, the lower lip deeply concave; calyx not enlarging in fruit.

Recognized by its shrubby habit, thick ± tomentose leaves and small calyx which, unlike that of other species, does not enlarge in fruit. It flowers mainly in winter, but erratically and only a few plants are usually found in flower at any one time.

A characteristic plant of the northern escarpment found on stony mountain slopes from 1500 to 2600m, but absent from places with winter frost: between Rujm and Ṭawīlah; between Bayt 'Adhāqah and the Wadi Lā'ah; below Kuḥlān; abundant on J. Sharafayn; J. Marran; J. Minmār; J. Rāziḥ. Also cultivated at Sana'a. It is endemic to the western mountains of Arabia with its distribution centred on the Asir.

3. **Plectranthus barbatus** *Andr.* Sha'r

Coleus barbatus (Andr.) Benth.
Weakly aromatic, slightly succulent perennial with decumbent woody stems and erect simple terminal racemes to 50cm (often a larger, more shrubby plant in other countries). Leaves petiolate, narrowly ovate or elliptic with a cuneate base up to 8cm long, pubescent on both surfaces, the margin crenate. Racemes simple, very long, with up to 25 whorls of shortly pedicellate flowers, the bracts large, ovate, colourless, deciduous as the flowers open; corolla violet, very sharply bent downwards, c 10–12mm long, the lower lip deeply concave.

A distinctive species on account of its many whorls of flowers and fruiting calyx whose mouth is closed by a mass of white hairs. It is essentially summer-flowering but is occasionally found in flower in the central escarpment in winter.

Widespread and often abundant between 1200 and 3100m in areas of average to good rainfall on the escarpment; particularly abundant around Ibb. It grows on cliffs, stony hillsides and in shallow grassland.

4. **Plectranthus cicatricosus** (*Hutch. & E. A. Bruce*) *J.R.I. Wood* comb. nov.

Coleus cicatricosus Hutch & E. A. Bruce in Kew Bull. 1941: 180
Aromatic, slightly succulent, branched undershrub to 1½m, the young stems sticky and covered in long white glandular hairs. Leaves long-petiolate, broadly ovate with a truncate to weakly cordate base, up to 9cm long, glandular pubescent and rather sticky especially below, the margin coarsely crenate. Plant apparently always leafless when flowering; flowers in small, mostly lateral racemes, 3–4cm long, forming a long narrow panicle; corolla dark blue, c 12mm long, the lower lip deeply concave.

Readily identified by its soft, ovate-cordate, sticky leaves and distinctive inflorescence. It flowers in winter.

Only known at 1900m by the Wadi Sabalī in Maghrib Ans (Wood 2988). This is an extremely rare plant known elsewhere only from the type collection in Somalia.

5. **Plectranthus coerulescens** (*Guerke*) *Willemse*

Strongly aromatic, slightly succulent perennial with decumbent woody stems and erect, simple, terminal racemes to 30cm. Leaves ± clustered at the ends of the branches, sessile, elliptic with a cuneate base, up to 4cm long, ± tomentose on both surfaces, the margin obscurely crenate. Racemes simple, rather short, the rhachis, pedicels and calyx all deep violet, bracts colourless, deciduous as the flowers open; corolla dark blue, 12–14mm long, the lower lip deeply concave.

Distinguished by the distinctive, clustered, sessile, elliptic leaves, which have a very thick succulent texture, and by the dark violet-stained inflorescence. It flowers in winter only.

Restricted to the Ḥujarīyah where it is found on open stony hilltops between 1700 and 2100m: Ḥayfān; abundant on the Turbah plateau. (Wood 75/21) Outside Yemen, known only from the highlands of northern Somalia.

6. **Plectranthus cylindraceus** *Hochst. ex Benth.* Khurūb

Strongly aromatic, succulent, branched, erect perennial to 1m, stems woody below. Leaves shortly petiolate, elliptic or obovate with a broadly cuneate base, up to 5cm long, the margin crenate to nearly entire. Flowers numerous in short, spike-like lateral and terminal racemes forming a long, narrow panicle; bracts ovate, green, c 5mm long, persistent; corolla violet, 3–4mm long.

Readily identified by its fleshy leaves, spike-like racemes and very small corolla. It flowers in late summer and autumn.

Widespread and frequent on stony, often very eroded hillsides, in open scrub and on cliffs, growing between 1000 and 2600m on the escarpment and southeastern mountains as far east as Al Bayḍā'.

7. Plectranthus hadiensis (*Forssk.*) *Schweinf. ex Sprenger* Midan, Bajdah

P. zatarhendi sensu E.A. Bruce
P. madagascariensis sensu Deflers

Aromatic, slightly succulent perennial with decumbent or prostrate woody stems and erect terminal racemes to 30cm. Leaves ± clustered at the ends of the branches, shortly petiolate, ± orbicular with a broadly cuneate or truncate base, up to 3cm long, densely pubescent, the margin crenate. Racemes simple or sparingly branched, rather lax, bracts apparently absent; corolla blue or violet, c 15mm long, the lower lip shallowly concave.

Easily identified by the small, clustered, orbicular leaves and the long lax racemes of rather delicate, narrow flowers. In Africa this is a very variable plant, in which *P. madagascariensis* Benth. should probably be included, and there are two distinct races in Yemen. Plants from the Ḥujarīyah are more robust, more strongly aromatic and with more deeply toothed leaves than those in the central escarpment. It flowers erratically throughout the year, though mainly after the rains.

Locally common on cliffs and terrace walls from 1200 to 2500m, apparently restricted to areas of prolonged winter mist on the escarpment: frequent in the Ḥujarīyah from Turbah to Najd Dhamrān; J. Sawraq; J. Ṣabir above Misrākh; J. Ra's (Miller); on all the central escarpment mountains north to the Wadi Lā'ah area, but particularly abundant above Ḥadīyah after which it is appropriately named. (Wood 75/504, 1645)

8. Plectranthus hyemalis *J.R.I. Wood* 'Irdhib

Weakly aromatic, slightly succulent, erect branched shrub to 1½m. Leaves ± clustered at the ends of the branches, obovate, narrowed gradually into the petiole, up to 4cm long, finely pubescent, crenate near the apex only. Racemes simple, short, bracts apparently absent; corolla violet, c 1cm long, the lower lip deeply concave.

Readily identified by the shrubby habit, clustered obovate leaves and fruiting calyx which is closed by a mass of white hairs. It flowers in winter.

Very local but sometimes abundant on open stony hillsides and rocky plain between 1000 and 2000m in the southeastern mountains; near Saqūl, Ahkum and between Yusafīn and Qabayṭah in the Ḥujarīyah; abundant around Damt. It is sometimes cultivated in Sana'a but is otherwise known elsewhere only from South Yemen. (Wood 1532)

9. Plectranthus ignarius (*Schweinf.*) *Agnew*

Coleus ignarius Schweinf.

Weakly aromatic, scarcely succulent, erect undershrub to 2m, leafless when in flower. Leaves petiolate, ovate or orbicular with a ± truncate base, up to 6cm long, glabrous or tomentellous, the margin entire. Flowers in short, simple racemes; bracts apparently absent; calyx dry, scarious; corolla violet, strongly bent, c 12mm long.

A distinctive plant on account of its shrubby habit, nearly glabrous, entire leaves and scarious calyx. Outside Yemen the leaves are often crenate and the indumentum glandular-pubescent, but the significance of this variation is difficult to assess. It flowers in winter.

Only known from rocky hills at 600m around Khamīs Banī Sa'd.

10. Plectranthus lanuginosus (*Hochst. ex Benth.*) *Agnew*

Coleus arabicus sensu Deflers

Weakly aromatic, slightly succulent, much-branched, erect or decumbent herb to c 60cm. Leaves petiolate, broadly ovate to ± elliptic with a cuneate, truncate or weakly cordate base, up to 7cm long, densely pubescent, the margin crenate. Racemes much branched, forming a terminal panicle, densely glandular with numerous white, spreading glandular hairs, bracts apparently absent; corolla blue, sharply bent, c 15mm long, the lower lip very deeply concave.

Recognised by its pure blue flowers and very glandular-hairy paniculate inflorescence. It flowers erratically throughout the year.

A local plant, very characteristic of damp mist-filled cliff gullies growing between 1800 and 3100m on the escarpment from Akāḥilah near Turbah north to Shirnamah on J. Maswar; often associated with *Solanum terminale*, *Geranium arabicum* and uncommon fern species.

11. Plectranthus ovatus *Benth.*

Erect, branched, aromatic, slightly succulent perennial herb to c 80cm growing from a tuberous rootstock each spring and dying back in the autumn. Leaves shortly petiolate, ovate or elliptic with a broadly cuneate or rounded base, up to 8cm long, pubescent, the margin crenate. Racemes much branched, forming a terminal panicle, the broadly lanceolate white bracts falling as the flowers open; corolla dark violet, c 12mm long, the lower lip deeply concave.

This is the only Yemen species with a tuberous rootstock. The inflorescence is similar to *P. lanuginosus* but lacks the glandular hairs of that species. It flowers from June to August.

Locally abundant in shallow stony grassland between 2100 and 2700m on the Ta'kar – Mudhaykhirah range west to J. 'Ayfū and east to J. Khadrā; also at Mashwarah and on J. Ba'dān. It is endemic to Yemen. (Wood 75/236, 3334)

12. Plectranthus tenuiflorus (*Vatke*) *Agnew* Midan

Coleus zatarhendi (Forssk.) Benth.
Ocimum α zatarhendi Forssk.?

Strongly aromatic, very succulent perennial with long decumbent woody stems spreading creeper-like to form dense clums. Leaves petiolate, broadly ovate-elliptic or orbicular with a ± truncate base, up to 5cm long, pubescent, margin crenate. Flowers in simple terminal racemes, often growing on leafless shoots and appearing to be partly dried up, each whorl with very few open flowers; bracts small, lanceolate, deflexed, whitish, persistent till the plant is in fruit; pedicels up to 8mm long, longer than in our other species; corolla pale violet, c 1cm long, the lower lip very narrow.

Readily identified by its creeping habit, very aromatic, succulent leaves and long pedicellate, narrow flowers. It flowers sporadically throughout the year though mainly in winter.

Widespread but local and often in small quantity between 500 and 1600m on the escarpment from Taiz northwards, growing on rock outcrops and stony hillsides, usually in *Acacia* scrub. (Wood 1033, 3259)

Plectranthus amboinicus (*Lour.*) *Spreng.* (*Ocimum vaalae* Forssk. ?), known as 'Wālah', is cultivated in a few places at low altitudes along the escarpment notably on J. Bura' (Schweinfurth), J. Milḥān and J. Marran. It is a strongly aromatic, densely pubescent succulent herb with ovate, crenate leaves. It is not known to flower in Yemen.

6. SOLENOSTEMON *Schum.*

1. Solenostemon latifolius (*Hochst. ex Benth.*) *J. K. Morton*

Coleus latifolius Hochst. ex Benth.

Erect, slightly succulent, sparingly branched, scentless annual herb to 50cm. Leaves ovate with a truncate base, up to 9cm long, glabrescent, the margin crenate. Flowers in long, simple or sparingly branched terminal racemes; bracts ovate, whitish, falling as the flowers open; calyx 2-lipped, the lateral lobes absent, the lower lobes reduced to two small teeth at the apex of the lower lip; corolla small, blue, c 6mm long.

A rare plant found between 700 and 1600m on terrace walls and in wadi beds in a few high rainfall areas of the escarpment: near Suq As Sabt, Ibb; Wadi Dūr, 'Udayn; among the coffee terraces above Ḥadīyah.

7. HYPTIS *Jacq.*

1. Hyptis pectinata (*L.*) *Poit.*

An erect or decumbent, straggling, pubescent annual herb to 1m. Leaves ovate, weakly cordate, up to 6cm long, much paler on the undersurface. Inflorescence a narrow terminal panicle consisting of opposite pairs of spike-like racemes which in turn are composed of small, shortly pedunculate cymes; bracts leaf-like, shorter than the racemes; calyx c 4mm long, with 5 equal teeth slightly exceeded by the small, 2-lipped, cream or pale purple corolla.

A rare plant found near water in scrubby wadis between 500 and 1200m, mostly in the central escarpment: Ḥillah, J. Bura' (Schweinfurth); Walajah, J. Milḥān (Schweinfurth); Wadi Ḥijān, Ḥarāz; Wadi Līyah, J. Marran.

8. LAVANDULA *L.*

Lavenders. A distinctive genus of aromatic undershrubs. Leaves simple or deeply pinnatisect. Inflorescence of long, usually slender and crowded, terminal spikes; bracts not leaf-like; calyx with 5 equal teeth; corolla 2-lipped, usually blue.

All our species have their own distinct smell and show a strong preference for volcanic soils.

1.	Leaves entire. Flowers yellow	*L. atriplicifolia*
	Leaves crenate or pinnatisect. Flowers blue	2
2.	Leaves pinnatisect. Spikes in terminal panicles	3
	Leaves crenate. Spikes solitary, terminal	*L. dentata*
3.	Bracts 1.5–2mm long	*L. coronopifolia*
	Bracts 4–5mm long	4
4.	Stems spreading pubescent. Leaf segments usually toothed or pinnatifid	*L. pubescens*
	Stems shortly puberulent or sub-glabrous. Leaf segments mostly entire	*L. citriodora*

1. Lavandula atriplicifolia *Benth.*

L. canescens Deflers

Erect, aromatic, grey-canescent undershrub to 1m, usually much branched especially below. Leaves oblanceolate, entire, up to 4cm long. Inflorescence very variable, the spikes sometimes solitary and terminal on short lateral branches, more usually paired in lax terminal panicles; spikes always short, up to 5cm long, flowers 2 in each whorl; bracts c 5mm long; corolla dull yellow.

A local plant of volcanic (rarely sandstone) cliffs and rock outcrops between 2000 and 3000m on the high plateau and surrounding mountains; also occurring in Khawlan Ash Sham at Ḥaydān and more surprisingly near Ad Dan in Waṣāb Al 'Āli and on J. Ḥawah in Salafīyah on J. Raymah (Gillespie). It is sometimes cultivated as a potherb and sprigs are mixed with Rue and Basil and used by women as a decoration in their turbans.

2. Lavandula citriodora *A. Miller*

Very similar to *L. coronopifolia* and *L. pubescens*, differing from the former in its bracts which are 4–5mm long, from the latter in its puberulent or sub-glabrous indumentum and from both in its pinnatisect leaves whose segments are broader (c 2–3mm wide) and almost always entire. The whole plant usually smells strongly of lemon.

A local plant of open rocky hills between 2300 and 3000m on the high plateau and in Khawlan Ash Sham often associated with *Juniperus*, *Clutia*, *Psiadia* and *Solanum schimperianum* var. *carense*: Yislah pass; on hills near Ḥaddah; near Tinam; J. Lawz; J. Minmār; J. Marran; between Majz and Jifān. (Wood 74/363, 613, 1970)

3. Lavandula coronopifolia *Poir.*

L. stricta Del.

Similar to *L. pubescens* but glabrous or thinly pubescent, the spikes very long and slender reaching 10cm in length, often interrupted below and the bracts only 1.5–2mm long.

Like *L. atriplicifolia* essentially a local plant of volcanic hillsides between 2000 and 2500m on the high plateau but extending south to Damt, occurring in Khawlan Ash Sham at Majz (Steinberg) and near Ṣa'dah and on the escarpment at Mafḥaq and between Ḥajjah and Sharas where it descends to 1200m. (Wood 75/162, 611, 770)

4. Lavandula dentata *L.*
Erect, aromatic, much-branched, densely pubescent, bushy undershrub to 1½m. Leaves sessile, linear, up to 2cm long, the margin deeply and regularly crenate. Inflorescence of stout, solitary, terminal spikes, up to 5cm long; flowers in whorls of 6–10; bracts 8–15mm long, whitish; corolla blue.

Locally common on stony hillsides above 2300m in Khawlan Ash Sham (Sāqayn, Ḥaydān, Jifān, Madaq); elsewhere very rare and known only from J. Maswār and J. Jalāl (Scott & Britton) where it reaches 2900m.

5. Lavandula pubescens *Decne.* Khawah, Sinayb
Erect, strongly aromatic, densely pubescent undershrub to 1½m, much branched below. Leaves deeply pinnatisect, the segments usually dentate, sometimes pinnatifid 1–2mm wide. Flowers in terminal panicles; the spikes long and slender, flowers 2 in each whorl; bracts 4–5mm long; corolla blue.

Widespread on old volcanic cones, lava flows and other open rocky volcanic places from 1000 to 2700 on the escarpment, high plateau and southeastern mountains; abundant on the high plateau but more local in other areas where it is apparently more demanding in its ecological requirements.

The cultivated Lavender, **Lavandula angustifolia** *Miller* (*L. vera* DC.) was recorded by Deflers as a cultivated plant, but has not been seen since.

9. MENTHA *L.*

1. Mentha longifolia (*L.*) *Hudson* Na'ud
M. sylvestris L.
Erect, aromatic perennial herb to 50cm, spreading by creeping rhizomes and usually growing in dense colonies. Stem villous. Leaves sessile, lanceolate, obscurely serrate, up to 8cm long, the undersurface grey and more densely tomentose than the upper. Flowers in terminal, usually branched spikes, 3–10cm long; calyx with 5 teeth; corolla regular, 4-lobed, lilac.

The Yemen plant is subspecies **schimperi** (*Briq.*) *Briq.*

Widespread and often abundant by permanent water from 500 to 3200m on the escarpment and high plateau extending rarely into the Mashriq where suitable habitats are rare. At low altitudes it is local, growing mainly by the larger wadis, but at higher altitudes it occurs wherever there are suitable wet land habitats.

The cultivated garden mint, known as 'Na'Na'', is probably one of the hybirds of **Mentha spicata** *L.* (such as *M. x piperita* L.). It is cultivated as a potherb throughout Yemen and is used mainly to flavour tea.

10. THYMUS *L.*

1. Thymus laevigatus *Vahl* Za'tar, Sa'tar
T. serpyllum sensu Schwartz
Wild Thyme. A distinctive low, aromatic, mat-forming perennial with woody procumbent non-flowering stems and short erect pubescent flowering stems less than 5cm long. Leaves sessile, obovate-cuneate, entire, c 5mm long. Flowers in a dense terminal head; calyx purple, 2-lipped, the 3 upper pubescent teeth broader than the 2 lower ciliate teeth; corolla white, 2-lipped.

A characteristic plant of bare stony mountain-sides and summit ridges between 2400 and 3500m on and around the high plateau, extending west to Ṭawīlah and J. Maswār and onto all the Ibb mountains south to the Ta'kar – Mudhaykhirah range; also occurring in similar habitats in Khawlan Ash Sham at Madaq (Steinberg) and between Ḥaydān and Sāqayn but completely absent from the outer escarpment. It is endemic to southwest Arabia.

The leaves are used medically and in cooking.

Thymus pulegioides L. is listed by Schwartz but this species cannot be accepted as part of the Yemen flora. It was recorded by Forsskal with doubt, there is no description or specimen and the occurrence of this European plant in the Tihama is extremely unlikely.

11. MICROMERIA *Benth.*

Perennial herbs. Flowers in long, lax, mostly terminal racemes. Bracts leaf-like, mostly shorter than the flowers; calyx tubular with (in our species) 5 ± equal teeth; corolla pink, 2-lipped.

Leaves entire with a thickened margin	*M. imbricata*
Leaves serrate, margin not thickened	*M. abyssinica*

1. Micromeria abyssinica *Benth.*
Satureja abyssinica (Benth.) Briq.
Pubescent perennial to 50cm with many ascending unbranched stems from a woody base. Leaves petiolate, ovate, 1–4cm long, serrate. Calyx 5–6cm long; corolla pink.

The plant smells of Calamint.

Locally abundant between 1800 and 2500m in the grassland areas of the central escarpment on J. Raymah, J. Milḥān, Maḥwīt (Miller) and J. Sāfān in the Ḥarāz; elsewhere very rare and known only from J. Ṣabir (Radcliffe-Smith and Henchie) and near Khamīs Madyūr, Ḥaymat al Kharijīyah (D. Wood).

2. Micromeria imbricata (*Forssk.*) *C.Chr.*
M. biflora (Buch.-Ham. ex D.Don) Benth.
M. punctata Benth.
Satureja biflora (Buch.-Ham. ex D.Don) Benth.
Erect, woody-based, variably aromatic, glabrous to villous, much-branched perennial to 40cm. Leaves subsessile or shortly petiolate, elliptic, lanceolate or ovate, up to 13mm long, the margin always entire and thickened. Inflorescence of leafy terminal racemes, the whorls made up of opposite, sessile or shortly pedunculate cymes with up to 20 flowers; calyx c 3mm long; corolla pink or mauve.

As understood here, an extremly variable plant divisible into at least three forms worthy of some sort of taxonomic recognition. The common form in Yemen (Wood 74/143, Miller 272) is an erect, often robust, plant with crisped pubescent stems, lanceolate or elliptic leaves and whorls of 4–16 flowers. The second form is characterised by its villous indumentum and dense, ± sessile flower clusters. It is only known from Sharāwah in Khawlan Ash Sham (Steinberg 963) but is similar to, if not identical with, forms known from the Asir. The third form is a low, cushion-like plant with short stems, broadly ovate leaves and whorls of only 1–3 flowers. This is a rare plant of high altitudes known only from J. Ṣabir (e.g. Hepper 5946) and J. Sumārah (e.g. Lavranos and Newton 13078) where it has been collected by a number of people perhaps because it grows by the car park at the summit.

M. imbricata as a whole is widespread and abundant from 1000 to 3300m on the escarpment, high plateau and southeastern mountains growing on open stony hillsides, on rocky outcrops and on terrace walls.

12. MERIANDRA *Benth.*

1. Meriandra benghalensis (*Roxb.*) *Benth.*

Much-branched aromatic shrub to 2m. Leaves lanceolate, up to 7cm long, the larger leaves with a strongly truncate base, margin finely crenate; upper surface green, strongly rugose; lower surface white-tomentellous. Flowers in terminal inflorescences, sessile, whorled, the whorls dense; bracts ovate, short, becoming concealed as the whorls develop; calyx woolly, obconical, weakly 2-lipped, longer than the calyx; stamens 2.

The plant smells strongly of Vick.

A characteristic but local plant of rocky hills between 2000 and 2800m on the high plateau, extending south to Damt; also in Khawlan Ash Sham at Sharāwah (Steinberg).

13. SALVIA *L.*

A distinctive genus of aromatic herbs or undershrubs. Flowers whorled (not cymose) in terminal racemes; bracts shorter than the calyx; calyx strongly 2-lipped, the upper lip consisting of 3 teeth (rarely sub-entire); corolla 2-lipped, stamens 2.

1. Corolla more than 2cm long, white. Leaves woolly below — *S. schimperi*
   Corolla less than 2cm long, usually coloured. Leaves not woolly below — 2
2. Corolla less than 5mm long. Leaves less than 2cm long — *S. aegyptiaca*
   Corolla more than 7mm long. Leaves more than 2cm long — 3
3. Upper lip of calyx sub-entire with 3 minute teeth. Leaves rugose above — *S. merjamie*
   Upper lip of calyx with 3 distinct teeth at least 1mm long. Leaves not rugose above — *S. nilotica*

1. Salvia aegyptiaca *L.*

Small, scabrid-pubescent, much-branched woody undershrub with stiff ascending whitish stems to c 30cm. Leaves sessile above, shortly petiolate below, linear-lanceolate, up to 2cm long, the margin ± crenate, recurved. Racemes short, the whorls with 2–4 flowers; corolla lilac, 3–4mm long.

A common plant of stony plains and hillsides on the high plateau extending west to the Ḥarāz, Maḥwīt and Hajjah. It also grows around Ṣa'dah and in the mountains of Khawlan Ash Sham and occurs sporadically in the southeastern mountains having been found near Bayḍā' and at Bathayjah in the Ḥujarīyah. It grows from 1300 to 2800m.

2. Salvia merjamie *Forssk.* — Rīḥān Al Baqr

S. nudicaulis Vahl

Erect, much-branched, pilose perennial from a basal rosette of leaves, up to 40cm high. Leaves petiolate, oblong-lanceolate, up to 6cm long, the margin crenate, the upper surface strongly rugose. Racemes branched, rather long, the whorls with 5–8 flowers; calyx with a broad, purple, obscurely toothed upper lip; corolla purple or white, 7–10mm long.

Rather variable, some plants like Wood 75/219 from J. Ṣabir have white flowers and a very glandular-pilose indumentum whereas the common plant is eglandular with purple flowers.

Widespread and frequent in short, often disturbed grassland around fields and by tracks from 2200 to 3600m on the escarpment and high plateau; not recorded from the Ḥujarīyah and Khawlan Ash Sham.

3. Salvia nilotica *Jacq.*

Erect, much-branched, coarsely pubescent perennial herb to 40cm. Basal leaves petiolate, obovate, crenate, often pinnatifid, up to 8cm long; stem leaves sessile, ovate, crenate, 2–4cm long. Racemes lax, often branched, rather long, the whorls with 4–5 flowers; corolla usually pink, c 8mm long.

A rare plant of shaded disturbed ground such as pathsides and cliffs, found between 1900 and 2300m in the central escarpment: J. Kabūrah and J. Ẓallamlam above Ḥadīyah in J. Raymah; between Shaḥir and 'Ukaybir, J. Milḥān; on cliffs Bayt Bashārī, Maḥwīt.

4. Salvia schimperi *Benth.*

S. spinosa sensu Deflers

Erect perennial to 50cm, branched at the base, the stem woolly below. Leaves clustered towards the base of the stem but not forming a rosette, petiolate, ovate, dentate, woolly, up to 10cm long, the upper surface rugose. Racemes long, branched, glandular-pubescent; the whorls subtended by a broadly ovate scarious bract, usually 4-flowered; corolla white, 2.5–3cm long.

Widespread and common on open stony hills, roadsides and field borders between 2000 and 3500m on the high plateau and in similar areas of Khawlan Ash Sham. Elsewhere known only from Manākhah (Wissmann) and Bayḍā'.

14. NEPETA *L.*

Catmints. Perennial herbs. Inflorescence terminal, the flowers in dense axillary whorls formed of opposite, subsessile, many-flowered cymes; calyx cylindrical with 5 ± equal teeth; corolla 2-lipped.

The plants have a distinctive smell which is very attractive to cats.

Flowers blue. Plant shortly pubescent — *N. deflersiana*
Flowers white. Plant densely covered in spreading glandular hairs — *N. woodiana*

1. **Nepeta deflersiana** *Schweinf. ex Hedge* Firuwāk

N. mussini sensu Schwartz
N. rugosa Benth. p.p.
Shortly pubescent, woody-based perennial, much branched below with many erect, usually simple stems to 80cm. Leaves shortly petiolate, ovate, up to 4cm long, crenate, the undersurface grey-tomentellous. Corolla dark blue, c 1cm long, about twice as long as the bluish calyx. Nutlets brown, finely muricate.

Widespread and frequent from 2000 to 3600m on the escarpment and high plateau from Akāḥilah near Turbah north to J. Rāziḥ. It grows on rocky outcrops, disturbed eroded hillslopes, terrace walls and cliffs. It is endemic to southwest Arabia.

2. **Nepeta woodiana** *Hedge*

Similar in habit and appearance to the previous species but differing in its spreading glandular-pilose indumentum, larger leaves (to 6cm) which are glandular-pubescent below, white flowers and grey nutlets.

Only known from rocky slopes between 2500 and 3000m along the Sumārah-Manār ridge and mostly on the latter mountain where it is quite frequent, though scattered in occurrence. Endemic to Yemen.

15. SCUTELLARIA *L.*

1. **Scutellaria arabica** *Jaub. & Spach.*

S. peregrina sensu Schwartz
Erect or decumbent, much-branched perennial to 40cm, the purple stems woody and pubescent below, glandular-pilose above. Leaves ovate, crenate, pubescent, up to 2.5cm long. Inflorescence a terminal raceme, the flowers in axillary pairs; bracts leaf-like, slightly exceeding the calyx; calyx with 2 broad entire lips, the upper lip with a large, flat, usually purple appendage, closed in fruit; corolla blue or (rarely) mauve, 2-lipped, 10–13mm long.

Rather variable in the density of the glandular indumentum on the inflorescence.

A characteristic plant of rock crevices, terrace walls and stony mountain slopes between 2000 and 3500m on all suitable escarpment mountains. Apparently endemic to southwest Arabia but closely related to *S. rubicunda* from the Mediterranean region and to an undescribed African species.

16. MARRUBIUM *L.*

1. **Marrubium vulgare** *L.*

Perennial herb to 40cm, branched at the base; stems white-woolly, at least below. Leaves broadly ovate or orbicular, up to 3cm long, crenate, the upper surface dark green and thinly tomentose, the lower surface densely white-tomentose. Flowers sessile, in dense many-flowered axillary whorls, forming a terminal inflorescence; bracts leaf-like, longer than the whorls; calyx with 10 teeth; corolla white, exceeding the calyx.

Widespread and frequent between 2000 and 3200m on the high plateau, growing on roadsides and on wasteground, showing a preference for organically rich soil in villages; more local in similar situations on the escarpment. This species is common round the Mediterranean and in Central Asia and reaches its southernmost limit in Yemen.

17. STACHYS *L.*

1. **Stachys yemenensis** *Hedge* Ḥakl

S. hypoleuca sensu Schwartz
Small, spreading, much-branched, aromatic undershrub to 50cm, stems white-tomentose above. Leaves subsessile, oblong-lanceolate, entire, up to 5cm long, the upper surface green, the lower surface white-tomentose. Flowers shortly pedicellate in axillary whorls of 4–6, forming leafy terminal racemes; bracts leaf-like, longer than the flowers; calyx white, with 5 equal teeth; corolla pink, 2-lipped.

Widespread, frequent but not usually in large quantity, growing from 1700 to 3300m on open stony hills along the escarpment from J. Sawraq, Najd Dhamrān and Turbah north to J. Maswār; apparently very rare on the high plateau and only recorded from J. 'Aybān (Acres), J. Nuqum (Deflers) and J. Lawz. Endemic to Yemen.

Stachys aegyptiaca Pers. was recorded for Yemen by Schwartz on the basis of a record by Botta, but Botta's collection is unlocalised and was probably made near Taif in Saudi Arabia.

18. LAMIUM *L.*

1. **Lamium amplexicaule** *L.*

Dead-nettle. Erect, pubescent annual herb to 30cm, usually branched at the base. Leaves shortly petiolate, ovate or orbicular, deeply crenate, up to 3cm long. Flowers sessile in axillary whorls of 2–10, forming a leafy terminal inflorescence; bracts leaf-like, sessile, much longer than the calyx; calyx with 5 equal teeth; corolla pink or purple, 2-lipped, c 15mm long.

A rare field weed found between 2600 and 3000m on J. Ṣabir, J. Manār and near Baw'ān.

19. OTOSTEGIA *Benth.*

1. **Otostegia fruticosa** (*Forssk.*) *Briq.* Shaqab

Decumbent or erect, much-branched, aromatic undershrub to 2m, the branches usually long and slender. Leaves petiolate, ovate, crenate, up to 4cm long, the upper surface dark green, pubescent, the undersurface white-tomentose.

Inflorescence a leafy panicle made up of short branches, each with 1–4 axillary whorls of flowers; bracts leaf-like, longer than the flowers; calyx 2-lipped, the lower lip broad, spreading, scarious, 3 or 4 times as long as the small upper lip; corolla white, 2-lipped, the upper lip with a densely villous margin.

A variable plant divided up into separate species, subspecies or varieties by different authors. The most recent treatment divides our plant into two subspecies which are themselves variable and also appear to intergrade with each other.

Subspecies **fruticosa** (*O. repanda* Benth; *O. scariosa* Benth.) is characterised by the relatively large leaves which are closely spaced (the internodes less than twice as long as the leaves). The plant thus has a rather dense, leafy appearance. Within this subspecies are extremely vigorous plants such as Wood 75/402 and Hepper 6151 which attain 2m in height, have leaves about 4cm long and a huge scarious lower lip to the calyx in excess of 1cm in length. Other forms of this subspecies are much more modest in their dimensions (Wood 1512).

Subspecies **schimperi** (*Benth.*) *Sebald* (*O. arabica* Jaub. & Spach; *O. benthamiana* Jaub. & Spach; *O. microphylla* (Del.) Aschers. & Schweinf.) has long internodes, more than twice the length of the smaller leaves which never exceed 3cm in length. The plant has a very lax, sparsely leafy appearance (Wood 74/280, 3371).

The species as a whole is widespread on the escarpment, high plateau and south eastern mountains and even extends into the Mashriq; it grows on cliffs, steep rocky slopes and rock outcrops from 800 to 3000m. Subspecies *fruticosa* is found throughout the range of the species except in the Mashriq but is commoner in high rainfall areas and is probably the only subspecies in the wetter parts of the Ibb area and the central escarpment. Subspecies *schimperi* is also widespread but is characteristic of drier areas.

20. LEUCAS *R.Br.*

Herbs or undershrubs. Inflorescence terminal, the flowers subsessile in whorls in the axils of the large leaf-like bracts; bracteoles usually present; calyx usually with 10 teeth, sometimes reduced to 5, regular or ± oblique due to the lengthening of one side; corolla white, 2-lipped, the upper lip densely villous.

1. Annual herbs — 2
   Perennial undershrubs, always woody below — 3
2. Calyx rigid, the bristle at the top much longer than the others — *L. martinicensis*
   Calyx soft, the mouth very oblique with the lower part extended to form a lower lip — *L. urticifolia*
3. Calyx soft, white, inflated in fruit. Leaves subsessile — *L. inflata*
   Calyx rigid, green, not inflated in fruit. Leaves petiolate — 4
4. Calyx with 10 teeth. Bracteoles less than 2mm long — *L. glabrata*
   Calyx with 5–7 teeth. Bracteoles 2–4mm long — *L. alba*

1. Leucas alba (*Forssk.*) *Sebald* — Shaqab

Leucas royleoides (Benth.) Guerke
Ballota forskahlei Benth.

Perennial with numerous erect or ascending stems to 1½m from a woody base. Leaves petiolate, ovate, up to 8cm long, the apex often ± rounded, the margin crenate. Whorls with 5–30 flowers; bracteoles filiform, 3–4mm long; calyx usually with 5 teeth, 6–8mm long, very oblique, the lower two teeth drawn forward, green, rigid, not inflated in fruit.

This species is divisible into two distinct forms in Yemen which appear to be geographically separated and require further study. The first form, which is probably the type, is a vigorous plant to 1½m in height, apparently restricted to 500–700m in the valley scrub communities of the central escarpment (Wood 1638, 1891, 2110). It is sparsely pubescent with large leaves which are never tomentose below. One specimen, Wood 1638, from J. Bura' has a calyx with 7 teeth.

The second form is frequent from 1200 to 2000m on open rocky ground on the escarpment and southeastern mountains. It rarely exceeds 50cm in height, is densely pubescent and has leaves up to 3cm long which are ± tomentose below, (Wood 74/144, 1870).

L. alba is endemic to southwest Arabia.

2. Leucas glabrata (*Vahl*) *R.Br.*

L. glaberrima Jaub. & Spach
L. trachyphylla Jaub. & Spach

Nearly glabrous perennial with erect or ascending virgate stems to 1½m from a woody base. Leaves petiolate, ovate, up to 4cm long, the margin serrate, apex acute. Whorls with up to 12(–20) flowers; bracteoles filiform, inconspicuous, 1–2mm long; calyx with 10 teeth, 6–8mm long, very oblique, the lower 3 teeth drawn forward, green, rigid, not inflated in fruit.

The whole plant usually becomes blackish in fruit.

A local plant of thorn fences, bushy field borders, cliffs and other ungrazed places between 1600 and 2900m on the escarpment and high plateau; most common in the Ibb region, there being few records elsewhere and none north of Kuḥlān. (Wood 1132, 1909; Hepper 5811)

3. Leucas inflata *Benth.*

Physoleucas inflata (Benth.) Jaub. & Spach

Perennial with numerous erect stems to 30cm from a woody base, stems tomentose above. Leaves subsessile, ovate or orbicular, up to 2cm long, grey-tomentose, crenate at the apex only. Whorls with 6–10 flowers; bracteoles minute; calyx with 10 teeth, c 5mm long, not oblique at the mouth, soft, papery, white, villous, inflated in fruit.

Widespread and frequent on open rocky plain and stony hills on the high plateau extending through the plains around Shahārah to Khawlan Ash Sham; also in the southeastern mountains where it is less frequent but extends southwards to Qa'ṭabah (Wissmann) and J. Masanah in the Ḥujarīyah (Deflers); grows from 1100 to 2500m.

4. Leucas martinicensis (*Jacq.*) *Ait. f.*

Erect, branched, pubescent annual herb to 50cm. Leaves petiolate, ovate, crenate, up to 5cm long. Flowers numerous in very dense whorls; bracteoles subulate, 4–5mm long; calyx with 10 teeth, c 5mm long, oblique, the upper margin drawn forward and terminating in a bristle more than twice as long as the other teeth, green-veined, rather rigid, not inflated in fruit.

A weed of fields, roadsides and waste places in the richer agricultural areas of the escarpment between 400 and 2000m; rather local, particularly in the north, but abundant in some areas notably the Dhubhān plateau and Ibb province where it extends as far east as Damt.

5. **Leucas urticifolia** (*Vahl*) *R.Br.*

Erect or decumbent, branched, puberulent annual herb to 50cm. Leaves petiolate, ovate, crenate, up to 6cm long. Flowers numerous in rather dense whorls (less dense than in *L. martinicensis*); bracteoles linear, ciliate, c 8mm long; calyx with 10 teeth, c 9mm long, very oblique, the lower 3 teeth drawn forward, dirty-white, soft, not inflated in fruit.

Widespread but very scattered in occurrence, growing usually as a field weed or more rarely in gravelly wadi beds on the escarpment from 600 to 2000m.

21. TEUCRIUM *L.*

1. **Teucrium yemense** *Deflers* Istaqūtas

Decumbent or ascending, grey-tomentose perennial, much branched from a woody base. Leaves sessile, oblanceolate, serrate, up to 1cm long. Flowers sessile in dense terminal heads; bracts leaf-like, about as long as the calyx; calyx soft, inflated, bladder-like with 5 ± equal teeth; corolla pink (rarely white), slightly longer than the calyx, 2-lipped, the upper lip very short.

The plant smells strongly of incense.

A plant of open rocky hills between 1500 and 3000m with a similar distribution to *Salvia aegyptiaca*. Widespread and generally common all over the high plateau extending westwards on the escarpment to Madīnat Al 'Abīd, Manākhah, Rujum, Ḥajjah, Maḥābishah and Shahārah, but absent from the high rainfall areas of Ibb and the central escarpment; also widespread in Khawlan Ash Sham and the southeastern mountains. It is endemic to southwest Arabia.

Schwartz includes *Teucrium leucocladum* Boiss. as part of the Yemen flora on the basis of a record by Botta. This cannot be accepted, since Botta's plant was collected at Taif in Saudi Arabia.

22. AJUGA *L.*

1. **Ajuga bracteosa** *Wall. ex Benth.*

Decumbent, pubescent, perennial herb, much branched at the base. Leaves oblanceolate, crenate, gradually narrowed into the petiole, up to 6cm long. Flowers subsessile in axillary whorls forming a leafy terminal inflorescence; bracts sessile, leaf-like, much longer than the flowers; calyx campanulate, with 5 ± equal teeth; corolla pale blue, about twice as long as the calyx, 2-lipped, the upper lip very short.

Widespread on the escarpment between 1800 and 3000m, but infrequent, growing in damp shaded rocky places such as cliff gullies and the beds of seasonal mountain streams; most common in the Ibb region.

Marjoram, **Origanum majorana** *L.*, known as 'Lizab' or ''Uzab', is cultivated as a potherb. The flowers are in panicled globular spikes. It is used both medicinally and decoratively, being one of a number of aromatic plants used by women in sprigs in their turbans.

PLANTAGINACEAE

PLANTAGO *L.*

Small herbs, usually scapose and with the leaves in a rosette. Flowers sessile in a usually dense spike or head; petals united, corolla lobes 4, pale brown. Fruit a capsule.

1.	Leaves in a rosette. Stem absent or very short	**2**
	Leaves opposite on a distinct ± erect stem	*P. afra*
2.	Peduncle and leaves glabrous or very thinly hairy	**3**
	Peduncle and leaves villous	**5**
3.	Leaves ovate. Flower spike rather lax, usually exceeding 3cm in length	*P. major*
	Leaves linear to lanceolate. Flower spike dense, up to 2.5cm long	**4**
4.	Peduncles mostly more than 10cm in length. Bracts with an indistinct brown midrib	*P. lanceolata*
	Peduncles less than 10cm in length. Bracts with a pronounced dark brown midrib	*P. amplexicaulis*
5.	Spike ovate, dense, 1–2cm long	*P. ovata*
	Spike elongated, usually interrupted, more than 2cm long	*P. albicans*

1. **Plantago afra** *L.*

Erect, hispid, branched annual with a distinct stem to 20cm. Leaves opposite or whorled, narrowly linear, up to c 5 × 0.2cm. Peduncles c 3–5cm long, arising from the leaf axils; flowers in a dense globose or ovate head c 1cm long, greeny brown in colour.

Only known from a rocky slope near Maghrabah, J. Bura' at 1800m.

2. **Plantago albicans** *L.*

Softly grey-pilose, usually perennial herb, stem very short or absent. Leaves in a basal rosette, linear-lanceolate, up to 15 × 6mm. Peduncles c 3–8cm long; flowers in an elongate, interrupted, pale brown spike 3–15cm long.

An extremely variable plant ranging from definitely perennial plants with spikes about 15cm long (Wood 2403) to probably annual plants with spikes barely 3cm long (Wood 598). The annual plant has been separated as *P. boissieri* Hausskn. & Bornm. but annual growth forms may only be the result of desert conditions which do not allow the plant to perenniate. The distinction between *P. albicans* s.l. and *P. ovata* likewise seems unsatisfactory and all three 'species' seem to merge into each other.

Apparently a rare plant of rocky plain, gullies and terrace walls between 2400 and 3100m on the mountains around the high plateau from Yarīm north to 'Amrān and from near Baw'an east to J. Lawz; always (?) on volcanic soils.

Fig. 18. **A** *Plantago major*; **B** *P. ovata*; **C** *P. amplexicaulis*; **D** *P. albicans*; **E** *P. afra*; **F** *P. lanceolata*.

3. Plantago amplexicaulis *Cav.*

Sparsely pilose annual herb, usually with a very short stem. Leaves in a basal rosette, linear-lanceolate with a distinctly amplexicaul base, up to 8 × 0.5cm. Peduncles 3–8cm long; flowers in a dense, ovate or very shortly cylindrical head 1–2cm long, shiny-white in colour, the bracts with a very dark midrib.

A rare plant known with accuracy only from a field border near Shamlān at 2300m, but also found by Rathjens in the Sana'a area.

4. Plantago lanceolata *L.*

Ribwort Plantain. Thinly pubescent perennial herb, stem absent. Leaves in a basal rosette, lanceolate, 5–15 × 2–3cm. Peduncles 10–20cm long; flowers in a dense, cylindrical, brown spike, 1–2.5cm long.

Widely distributed and frequent between 1500 and 2900m on the escarpment and high plateau but not recorded north of Maḥābishah and Kuḥlan; it grows on grassy banks by ditches and ponds and more rarely as a weed of irrigated fields and gardens.

5. Plantago major *L.*

Ratstail Plantain. Thinly pubescent annual or perennial herb, stem absent. Leaves in a basal rosette, ovate, 8–12 × 2.5–4cm. Peduncles c 5–10cm long; flowers in a lax greenish spike, often interrupted below, c 2–6cm long.

Widely distributed and frequent between 1400 and 2900m on the escarpment and high plateau, growing in disturbed wet places, where it is often associated with *P. lanceolata*.

6. Plantago ovata *Forssk.*

Softly grey-pilose annual or perennial very similar to *P., albicans* but much smaller in all parts. Leaves linear, up to 2mm wide. Peduncles less than 5cm long; flowers in dense ovate or slightly elongated spikes 1–2cm long.

Essentially a plant of the inner escarpment from J. Ḍawrān north to Ḥūth, but also occurring in similar places in Khawlan Ash Sham (Aqabah Jifān, Sāqayn), on J. Lawz and on J. Ḥawah, Salafīyah (Gillespie). It thus shows a preference for higher rainfall areas than *P. albicans* and also strongly favours limestone and sandstone hills being particularly common in the Kawkabān – Ṭawīlah region. It has been found between 2000 and 3300m.

OLEACEAE

Trees or shrubs with opposite entire leaves without stipules. Flowers regular; petals and sepals united. Stamens 2 only.

Climbing shrubs. Corolla with a long tube — 1. *Jasminum*
Tree. Corolla with a very short tube — 2. *Olea*

1. JASMINUM *Dum.*

Jasmines. Climbing shrubs with dark green opposite leaves and small cymes of large, white, fragrant flowers. Fruit a berry, often deeply lobed.

1. Leaves simple — *J. sambac*
   Leaves compound — **2**
2. Leaves usually pinnate with 5 leaflets. Calyx lobes 5–8mm long — *J. grandiflorum*
   Leaves trifoliolate. Calyx truncate or with very short lobes less than 3mm long — *J. fluminense*

1. Jasminum fluminense *Vell.* — Yasmin

J. mauritianum Boj.
J. gratissimum Deflers

Climbing shrub, glabrous except on the inflorescence and young branchlets. Leaves trifoliolate, the leaflets ovate, acuminate, c 3–6cm long. Cymes usually on short ± leafless branches; calyx ± truncate or with very short lobes, c 1(–3)mm long. Corolla white, the tube 3–3.5cm long, very fragrant.

The Arabian plant is subsp. **gratissimum** (*Deflers*) *P.S. Green.*

Locally common on the escarpment between 500 and 1500m from 'Udayn northwards, usually growing in valley scrub but spreading upwards onto scrub-covered hillsides in high rainfall areas, where it is more common. It is also cultivated in Taiz and other towns.

2. Jasminum grandiflorum *L.* — Qa'in, Khab

J. floribundum R. Br. ex Fres.
J. officinale sensu Schwartz

Glabrous or pubescent climbing shrub similar to the previous species in general appearance. Leaves pinnate usually with 5 leaflets, rarely up to 7 or reduced to 3, the leaflets ovate, acuminate, c 2–5cm long. Cymes often on longer, more leafy shoots than in the previous species; calyx with linear teeth c 5–8mm long; corolla white but often reddish outside, the tube c 1.5–2cm long.

The Arabian plant is subspecies **floribundum** (R. Br. ex Fres.) P. S. Green.

Widespread and often common between 1500 and 3000m on the escarpment and high plateau. It grows in scrub and, particularly at high altitudes, on cliffs. On the high plateau it is virtually restricted to cliffs where it roots in rock crevices and is virtually immune from grazing.

3. Jasminum sambac (*L.*) *Ait.* — Ful

Climbing shrub with pubescent stems. Leaves simple, ovate to elliptic, shortly acuminate, c 3–8cm long. Cymes mostly terminal; calyx with linear teeth c 1cm long; corolla very fragrant, white, the tube c 1–1.5cm long.

Commonly cultivated in the Tihama towns and more rarely on the escarpment up to 1500m at Taiz; well naturalised in scrub in the wadi below Ḥadīyah at 600m. The flowers are sold in markets and by street salesmen and are much esteemed for their fragrance.

2. OLEA *L.*

1. Olea europaea *L.* — 'Utum

O. africana Mill.
O. chrysophylla Lam.

Tree to 10m but usually much less in Yemen; bark finely fissured, grey. Leaves linear-lanceolate, narrowly attenuate at the base and apex, 6–10cm long, dark olive green above, pale and minutely scaly below. Flowers in small axillary panicles; corolla cream, c 1mm long. Fruit a subglobose purple drupe about 12mm long.

The Wild Olive of Yemen is subspecies **cuspidata** (*DC.*) *Ciffieri* which differs from the cultivated olive, subspecies *europaea*, in little more than the size of its fruit and its country of origin. It is a curious fact that the cultivated olive, so popular in the Mediterranean world, has not been grown in Yemen except as an experimental introduction and its Arabic name, 'Zaytūn', is applied in Yemen to the Guava.

The Wild Olive is widespread on the escarpment between 1200 and 2500m, but is generally infrequent and usually grows in small numbers or as isolated trees. It is probably an indicator of the former presence of highland forest. This tree was doubtless much more common in the past as testified by the existence of place names such as 'Utmah.

SCROPHULARIACEAE

Herbs (rarely shrubs) with opposite or alternate leaves, lacking stipules. Flowers usually in racemes but sometimes axillary or in panicles; calyx and corolla usually with a tube, the latter 5-lobed and often 2-lipped; stamens usually 4; style long. Fruit a capsule containing numerous seeds.

A large and rather interesting family in Yemen containing, on the one hand, a number of endemic species, though mostly in critical genera, and on the other a number of semi-parasitic plants, containing little chlorophyll and drying black. The family is of little economic importance and few species have Arabic names.

1. Woody shrub 2–3m tall — 8. *Halleria*
Herbs or low undershrubs — **2**
2. All leaves in a basal rosette — **3**
Leaves on an elongated stem (very short in *Aptosimum*) sometimes also in a basal rosette — **4**
3. Plants stoloniferous. Fruit pedicels recurved — 14. *Limosella*
Plants not stoloniferous. Fruit pedicels not recurved — 13. *Craterostigma*
4. Glabrous herb of aquatic habitats, creeping and rooting at the nodes — 12. *Bacopa*
Erect herbs or, if creeping, neither glabrous nor rooting at the nodes — **5**
5. Corolla yellow (sometimes violet tinged) — **6**
Corolla white, pink, red, blue or mauve, never yellow — **11**
6. Corolla regular, tube very short or absent — **7**
Corolla ± 2-lipped with a distinct tube — **8**
7. Plant with a basal rosette. Pedicels longer than the calyx — 3. *Verbascum*
Plant without a basal rosette. Pedicels shorter than the calyx — 18. *Alectra*
8. Flowers spurred. (Spur rarely absent and if so leaves entire) — 4. *Kickxia*
Flowers without a spur. Leaves dentate — **9**
9. Leaves about as broad as long. Calyx 5-lobed — 11. *Lindenbergia*
Leaves at least 3 times as long as broad, Calyx 4-lobed — **10**
10. Leaves glandular-hairy — 24. *Bellardia*
Leaves eglandular — 23. *Bartsia*
11. Flowers in a terminal spike — **12**
Flowers axillary or, if terminal, inflorescence not spike-like — **15**
12. Plant densely glandular — 19. Harveya
Plant eglandular — **13**
13. Non-parasitic glabrous undershrub. Flowers white — 5. *Anarrhinum*
Semi-parasitic, usually scabrid annual. Flowers red, pink, mauve or white — **14**
14. Corolla tube straight — 20. *Buchnera*
Corolla tube strongly bent — 21. *Striga*
15. Corolla with a distinct tube — **16**
Corolla tube absent or nearly so — **23**
16. Grey, commonly leafless undershrubs. Leaves, if present, linear, less than 1cm long — 16. *Campylanthus*
Herbs, very rarely undershrubs. Leaves always present, more than 1cm long — **17**
17. Flowers in terminal panicles or axillary cymes — 9. *Scrophularia*
Flowers solitary in the leaf axils, sometimes forming a leafy terminal raceme — **18**
18. Flowers and fruit sessile or almost so — **19**
Flowers and fruit distinctly pedicellate, the pedicels at least 5mm long — **21**
19. Prostrate trailing herb. Flowers white — 22. *Cycniopsis*
Plant not trailing. Flowers blue or mauve — **20**
20. Low tufted plant. Corolla regular — 1. *Aptosimum*
Erect plant. Corolla 2-lipped — 7. *Misopates*
21. Plant densely glandular-hairy. Corolla regular — 2. *Anticharis*
Plant eglandular. Corolla 2-lipped — **22**
22. Leaves opposite, sessile — 10. *Mimulus*
Leaves alternate, petiolate — 6. *Schweinfurthia*
23. Stamens 2. Leaves in opposite pairs — 17. *Veronica*
Stamens 4. Leaves commonly in whorls of 3–5 — 15. *Scoparia*

1. APTOSIMUM *Burch.*

1. Aptosimum pumilum *Benth.*

Densely tufted, wiry annual herb; stem very short or apparently absent, never (?) exceeding 3cm long. Leaves alternate (appearing whorled), linear, up to 5cm long, conspicuously ciliate at the base, margin entire but scabrid. Flowers sessile in the leaf axils, crowded, nearly concealed by the leaves; corolla pale blue, the tube funnel-shaped, the lobes regular, spreading.

A local, easily overlooked but characteristic species of alluvial plain between 1900 and 2400m on the high plateau and on the plains around Ṣa'dah and Majz.

2. ANTICHARIS *Endl.*

Glandular hairy annual herbs with alternate entire leaves. Flowers long pedicellate, axillary, forming terminal leafy racemes; corolla regular, with a dilated tube, funnel-shaped.

Leaves lanceolate, c 2cm long	*A. arabica*
Leaves linear, 2–4cm long	*A. linearis*

1. Anticharis arabica *Endl.*

Erect, branched annual to 10cm. Leaves distinctly petiolate, lanceolate, c 2cm long. Pedicels 5–8mm long; corolla white or pink.

Only found once on stony plain in open *Commiphora* scrub between Al Mafraq and Nukaybah at c 200m.

2. Anticharis linearis (*Benth.*) *Hochst. ex Aschers.*

Very similar to *A. arabica* but usually more vigorous and more branched, becoming blue-black when dry. Leaves linear without a distinct petiole, c 3–4cm long. Pedicels 10–15mm long; corolla pale blue.

A locally common field weed in the Tihama and drier foothills but not recorded north of the Wadi Surdūd. Elsewhere rare and scattered in occurrence and recorded only on the plain below Shahārah, on an old volcano at Shamlān near Sana'a and in a number of places around Mārib; found between 100 and 2300m.

3. VERBASCUM *L.*

Mullein. Usually biennial herbs with leaves in a basal rosette and with tall, erect stems, the stem leaves alternate. Flowers in terminal racemes; corolla regular, rotate, yellow, without a tube. Fruit an ovate capsule.

1.	Plant stellate-hairy. Flowers in small subsessile clusters, the pedicels equalling the calyx or shorter	2
	Plant with simple hairs. Flowers solitary, pedicellate, the pedicels much longer than the calyx	3
2.	Basal leaves ovate, at least 5cm broad. Calyx lobes triangular	*V. sinaiticum*
	Basal leaves lanceolate, 2–3cm broad. Calyx lobes linear	*V. yemense*
3.	Style 18–22mm long. Inflorescence sparsely glandular	*V. melhanense*
	Style 12–15mm long. Inflorescence densely glandular	*V. bottae*

1. Verbascum bottae (Deflers) *Huber-Mor.*

Celsia bottae Deflers

An erect, unbranched or sparingly branched herb to c 80cm; stem purple, sparsely pilose with simple hairs. Rosette leaves long petiolate, lanceolate, up to 15 × 4cm, crenate, hairy on the veins only. Inflorescence long, rather lax, densely glandular-hairy; flowers solitary; pedicels 1–3cm long; calyx lobes linear; corolla yellow; stamens 4, the filaments pilose with mauve hairs for more than half their length; style 12–15mm long.

Frequent on cliffs, terrace walls and grassy banks between 2000 and 3200m in the wetter parts of the escarpment from Turbah north to Maḥwīt and Kuḥlān; very rare on the high plateau where it usually grows near water as at Tin'am. Apparently endemic to Yemen. (Wood 1707, 2251, 3108)

2. Verbascum melhanense (Murb.) *Huber-Mor.*

Celsia melhanensis Murb.

Very similar to the previous species with which it is easily confused, but differing in its almost eglandular inflorescence, filaments which are pilose at the base only and the longer style which is 18–22mm long.

In similar habitats to *V. bottae* but less common, growing at slightly lower altitudes (1400–2900m) and only on the northern escarpment: Manākhah, 'Usil (Schweinfurth) and J. Masār, Harāz; J. Milḥān; Shahārah; J. Marran; J. Rāziḥ. (Wood 2835, 2864, 3384)

V. melhanense is named after J. Milḥān and outside Yemen is only known from the Asir region of Saudi Arabia.

3. Verbascum sinaiticum *Benth.*

An erect, much-branched, woolly, densely stellate-hairy herb to 2m, the stem usually green. Rosette leaves shortly petiolate, ovate, up to 25 × 10cm, crenate, woolly, grey. Inflorescence long, dense, densely stellate-hairy; flowers in clusters; pedicels c 1–3mm long; calyx lobes triangular; corolla yellow; stamens 5, the filaments white, woolly; style c 6–9mm long.

A recent introduction (first recorded by Rathjens in 1937), now a widespread but infrequent weed of stony roadsides between 1600 and 2400m from Qā'idah north to Maḥwīt extending eastwards to Sana'a and Raḍmah.

4. Verbascum yemense *Deflers*

Similar to the previous species but less densely stellate-hairy in all parts. An erect herb to c 50cm with a simple or sparingly branched stem. Rosette leaves long-petiolate, lanceolate, up to 10 × 3cm, the upper surface green rather than grey. Inflorescence parts all slightly smaller than in *V. sinaiticum* and the calyx lobes linear.

A characteristic plant of the high plateau which grows on stony hills and as a weed of roadsides and fields from 2000 to 3000m. It is endemic to southwest Arabia.

4. KICKXIA *Dumort.*

Herbs or undershrubs with alternate, entire or dentate leaves. Flowers solitary in the leaf axils; corolla yellow, 2-lipped, the lower lip 3-lobed with a projecting 'palate' closing the mouth of the corolla, tube with a bulge at the base, usually spurred.

The distribution of *Kickxia* in Yemen is of considerable interest. The species are generally restricted to particular regions in which they are abundant. Neighbouring regions are either devoid of *Kickxia* altogether or have a different species. Most of the species are little known, but are probably endemic.

1. Pedicels very short, mostly 1–2mm long. Plant glandular-hairy — *K. acerbiana*
   Pedicels more than 2.5mm long. Plant glabrous to villous but not glandular-hairy — 2
2. Leaves hastate or sagittate — 3
   Leaves gradually narrowed into the petiole — 5
3. Pedicels villous, less than 10mm long — *K. elatine*
   Pedicels glabrous, more than 10mm long — 4
4. Stem villous. Leaves ovate-triangular — *K. woodii*
   Stem scabrid, sometimes pubescent. Leaves lanceolate — *K. scalarum*
5. Flowers without a spur — *K. saccata*
   Flowers with a spur — 6
6. Spur 8–15mm long — *K. sabaea*
   Spur less than 8mm long — 7
7. Leaves linear, with a scarcely perceptible petiole. Pedicels 2.5–6mm long — *K. asparagoides*
   Leaves linear-lanceolate or lanceolate with a distinct petiole. Pedicels mostly 15–20mm long — *K. petiolata*

1. Kickxia acerbiana (*Boiss.*) *Taeckh. & Boulos*

Erect, much-branched, grey, glandular-pilose, woody-based perennial to 30cm. Petioles short, 1–2mm long; leaves ovate with a rounded to subcordate base, 0.3–0.8cm long. Pedicels 1–2mm long; corolla small, villous; spur c 3mm long.

Only known at 2300m on a volcanic cone just south of Ma'mar. (Wood 74/357)

2. Kickxia asparagoides (*Schweinf.*) *Cufod.*

Linaria asparagoides Schweinf.

Erect or ascending, much-branched, glabrous, woody-based perennial to 30cm. Leaves linear with a scarcely perceptible petiole, 1.5–3cm long. Pedicels 2.5–6mm long; corolla small, glabrous, 7–10mm long; spur 2–4.5mm long.

Known only from Madān (D. Wood) and Shahārah where it is common on terrace walls and rocky slopes c 1700–2400m. Outside Yemen known from Ethiopia only. (Wood 2509, 3380)

3. Kickxia elatine (*L.*) *Dum.*

Elatinoides elatine (L.) Wettst.

Villous grey annual, much branched at the base with numerous long procumbent stems. Petioles short, 1–3mm long; leaves ovate with a hastate base, 1–2cm long. Pedicels 4–5mm long, villous; corolla villous 5–7mm long; spur slender, c 5mm long.

The Yemen plant is subspecies **crinita**(*Mabille*) *W. Greuter* with a densely villous indumentum.

A local field weed found mostly from 2200 to 2600m and restricted to the Ibb region and the high plateau apart from an isolated record from Wadi Ḍabāb where it descends to 1300m.

4. Kickxia petiolata *D. Sutton*

Much-branched, thinly pubescent decumbent perennial. Petioles of very varied length, 2–20mm long; leaves lanceolate or linear-lanceolate with a rounded to cuneate base, 1–3.5cm long. Pedicels mostly 1.5–2cm long; corolla pubescent; spur short, 2–5mm long.

A local endemic plant of stony hillsides between 1300 and 1800m found between Taiz and 'Udayn: between Ḥuqayl and Mirkiz Maqbabah; Rawnah, Shar'ab; J. 'Ayfū; Suq Shirqī, Mudhaykhirah; Taiz. (Wood 74/336, 2223)

5. Kickxia sabaea *D. Sutton*

Linaria sagittata sensu Schwartz p.p.

Glabrous woody-based perennial with numerous erect or ascending branches to c 50cm. Basal leaves, where present, with 5mm long petioles, oblanceolate with a cuneate base, 6–18mm long; stem leaves linear 0.5–4cm long with a scarcely perceptible petiole. Pedicels 4–5.5mm long; corolla glabrous or pubescent; spur very long, 8–15mm long.

An uncommon plant of rocky slopes, stony fields and cliffs found between 2000 and 2700m mostly on the high plateau: J. Ḍawrān; Yislah Pass; in several places in Bani Ḥushaysh; about 20kms south of Ṣa'dah. Known certainly only from Yemen but perhaps also in Saudi Arabia. (Wood 638, 1084, 2448, 3238)

6. Kickxia saccata *D. Sutton*

Similar to the previous species in appearance but entirely glabrous and with linear leaves narrowed imperceptibly into the petioles. Pedicels only about 1cm long. The glabrous corolla is very distinctive since the tube is without a spur.

Only known from Jabal Ra's. (Miller 3493)

7. Kickxia scalarum *Schweinf ex D. Sutton*

Much-branched, scabrid, sometimes pubescent, decumbent perennial. Petioles 2–11mm long; leaves lanceolate with a sagittate base, 0.5–2.5cm long. Pedicels mostly 1–1.5cm long; corolla large, pubescent, often violet tinged; spur narrowly triangular, 5–9mm long.

Abundant from 1200 to 2600m on terrace walls and rocky slopes in the Ḥarāz (J. Shibām; J. Masār; J. Sāfān; 'Usil); J. Milḥān; Wadi Yūr; Maḥwīt. (Wood 670, 3350)

8. Kickxia woodii *D. Sutton*

Very similar to the previous species but more robust in all parts, the stems and corolla villous, the leaves ovate-triangular with a hastate base (often also with a single lateral tooth), the pedicels 1.8–2.4cm long.

Frequent on cliffs and rocky banks between 1200 and 2100m in the Ḥujarīyah from Turbah east to Qabayṭah. Unknown elsewhere. (Wood 74/133)

5. ANARRHINUM *Desf.*

1. Anarrhinum forskahlii (*J.F. Gmel.*) *Cufod.*

A. orientale Benth.

Glabrous perennial with a woody base from which arise numerous slender stems to 30cm. Leaves alternate, sessile, linear-oblong 1–2cm long, entire. Flowers in a terminal leafless raceme, shortly pedicellate, small, numerous; corolla small, tubular, white, 2-lipped.

A widespread and common plant of open, stony mountain slopes between 2000 and 3200m on the escarpment and high plateau but apparently absent from all the southeastern mountains.

6. SCHWEINFURTHIA *A. Br.*

Herbs with alternate, petiolate, entire leaves. Flowers solitary in the leaf axils, finely pedicellate; corolla mauve, pink or white, 2-lipped with a short swollen tube.

Leaves linear-oblong *S. pedicellata*
Leaves oblanceolate-spathulate *S. pterosperma*

1. Schweinfurthia pedicellata (*T. Anders.*) *Balf. f.*

Erect, glabrous, much-branched, twiggy annual herb to 40cm, the branches spreading at a rather wide angle. Leaves linear-oblong, 2–3cm long. Fruiting pedicels 10–17mm long, sharply reflexed.

On gravel plain in the Tihama from 100 to 300m in the Mafraq region and probably at Barāqish in the Jawf at 1100m.

2. Schweinfurthia pterosperma (*A. Rich.*) *A. Br.*

Very similar to the previous species from which it differs in being definitely herbaceous with slightly fleshy branches spreading at a narrow angle, and in having oblanceolate-spathulate leaves and the fruiting pedicels weakly recurved.

Abundant in the Tihama north of the Wadi Mawr around 'Abs, Mīdī and Ḩaraḑ growing on sandy plain from 0–400m.

7. MISOPATES *Raf.*

1. Misopates orontium (*L.*) *Raf.*

Antirrhinum orontium L.

An erect, sparingly branched annual herb to 30cm, glabrous below but sparsely glandular-pubescent above. Leaves alternate, petiolate, linear, c 3–4cm long, entire. Flowers subsessile, solitary in the leaf axils; calyx pilose; corolla mauve, 2-lipped, with a swollen tube.

A local field weed found between 1000 and 2900m on the escarpment, rarely in abundance.

8. HALLERIA *L.*

1. Halleria lucida L. — Shibrad

An erect branched glabrous shrub to 3m. Leaves opposite, petiolate, ovate, acuminate, 4–7cm long, the margins finely dentate. Flowers pedicellate, solitary or in clusters in the leaf axils or developing on the stem below the leaves; calyx very small; corolla red, 2-lipped with a long tube, 2.5–3cm long; stamens exserted. Fruit a berry.

Only known from wooded slopes at 2700m by the waterfall at Maḩjar in Wadi Aḩjar.

9. SCROPHULARIA *L.*

Figwort. Herbs with square stems and opposite petiolate leaves. Flowers in cymes in the axils of the upper leaves or in terminal panicles. Corolla red-brown, 2-lipped with a short almost globular tube.

Flowers in a terminal panicle. Plant glabrous *S. deserti*
Flowers in axillary cymes. Plant glandular-pubescent *S. arguta*

1. Scrophularia arguta *Ait.*

Erect, glandular-pubescent annual herb to 70cm. Leaves ovate to elliptic with a cuneate or cordate base, 4–5cm long, the margin serrate. Flowers in small, rather rigid, often leafy axillary cymes; corolla c 6mm long. Capsule ovoid, beaked.

An infrequent plant of terrace walls found between 800 and 2000m in the central escarpment (Bura', Milḩān, Ḩarāz and Maḩwīt) and in the Ḩujarīyah (Turbah, Aḩkūm, 'Abūs). These rather contrasting areas both have plentiful winter mist and this may be the climatic feature which determines the distribution of *S. arguta*.

2. Scrophularia deserti *Del.*

Glabrous perennial herb to 40cm with numerous erect stems arising from the woody base. Leaves dark green, oblanceolate in outline, deeply pinnatisect, 2–3cm long, the margins often white. Flowers in leafless terminal panicles; corolla c 4mm long. Capsule globose, not beaked.

A rare plant found on cindery volcanic hillsides on the high plateau between 2200 and 3100m: J. Nabī Shu'ayb above Matnah; near Ma'mar; Sanabān; Shamlān.

10. MIMULUS *L.*

1. Mimulus gracilis *R. Br.*

Erect, glabrous, sparingly branched herb to 40m, stems quadrangular. Leaves opposite, sessile, often clasping, 2–5cm long, entire. Flowers long-pedicellate, solitary in the leaf axils; calyx almost as long as the 2-lipped, tubular, white corolla.

A very rare plant of pond sides between 2700 and 3000m in the Ibb mountains; J. Ḩubaysh; J. Sumārah.

11. LINDENBERGIA *Lehm. ex Link & Otto*

1. Lindenbergia indica (*L.*) *Kuntze*

L. abyssinica Hochst. ex Benth.
L. sinaica (Decne.) Benth.

Much-branched, decumbent or ascending perennial herb, stems pubescent. Leaves opposite, petiolate, up to 3cm long, the margin dentate. Flowers shortly pedicellate in leafy, sometimes glandular-hairy, terminal racemes; corolla orange-yellow, 2-lipped, with a long tube, much longer than the 5-lobed calyx.

A rare plant of cliffs between 1100 and 1600m on the northern escarpment: Wadi Sukhnah; Wadi Sharas; Maḩābishah (eastern side); Shahārah (lower slopes).

12. BACOPA *Aubl.*

1. Bacopa monnieri (*L.*) *Pennell*

Herpestes monniera (L.) H.B.K.

Procumbent glabrous herb, creeping and rooting at the nodes. Leaves opposite, shortly petiolate, oblanceolate, c 1cm long, slightly fleshy, entire. Flowers with 1cm long pedicels, solitary in the leaf axils; corolla pale blue or white, campanulate, weakly 2-lipped, c 1cm long, slightly longer than the calyx.

A plant of stream sides and gravelly wadi beds where there is ± permanent flowing water from 200 to 2300m, often associated with *Phyla nodiflora*. Abundant in suitable habitats in the Tihama and on the escarpment; completely absent from the high plateau (sensitive to frost ?); in the Mashriq at 'Ujlī only.

13. CRATEROSTIGMA *Hochst.*

Small rosette herbs. Flowers in a terminal raceme; calyx tubular; corolla violet with a yellow throat, strongly 2-lipped, the upper lip small, the lower broad and spreading.

Both the following plants, which superficially recall the violet of temperate regions, are characteristic of sandy and gravelly depressions between rocks where water accumulates. They are often associated with *Ophioglossum polyphyllum* and *Microchloa kunthii* and, like those species, come into flower rapidly after the onset of the rains.

Peduncle shorter than the pedicels ... *C. pumilum*
Peduncle longer than the pedicels ... *C. plantagineum*

1. Craterostigma plantagineum *Hochst.*

Perennial rosette herb with acute, ovate to elliptic leaves, c 1–4cm long, the petiole often reddish, the margin entire or serrate, the undersurface pilose. Peduncle 3–5cm long, much longer than the pedicels which may be very short (thus making the inflorescence spicate) or up to 1.5cm long (thus giving the inflorescence a lax corymbose appearance).

This plant is rather polymorphic and I have included all long-peduncled forms under this name, though it is not certain they are all the same species. The genus is in need of revision.

Apparently a rare plant found between 1600 and 2500m in a number of scattered localities: Suq al Jum'ah, J. Ash Shirq (Wood 1952); above Ḥadīyah (Wood 3298); Ḥajarah, Ḥarāz (Wood 3329); Wadi Ḍahr (Radcliffe-Smith and Henchie 4630); Ḥūth (Miller 3182).

2. Craterostigma pumilum *Hochst.*

Differs from the previous species in the peduncle, which is absent or less than 1cm long and thus much shorter than the pedicels which are mostly 2–3cm long.

Widely distributed between 1900 and 2900m on the escarpment, high plateau and southeastern mountains, apparently scattered in occurrence but perhaps overlooked because of its short flowering season.

14. LIMOSELLA *L.*

Small glabrous rosette herbs spreading by stolons. Leaves entire. Flowers solitary in the leaf axils, very small; corolla with a short tube, pale pink. Fruiting pedicels strongly recurved.

Leaves linear. Ripe fruit almost enclosed by the calyx ... *L. macrantha*
Leaves elliptic or ovate. Calyx much shorter than the ripe fruit ... *L. africana*

1. Limosella africana *Gluck*

Small herb. Petioles 5–8cm long. Leaves very variable in shape and size, usually elliptic, rarely ovate but always with an expanded leaf blade, 1–2.5cm long. Corolla almost regular, the tube shorter than the calyx. Mature fruiting capsule longer than the calyx, the pedicels c 1cm long.

A rare escarpment plant which grows between 2000 and 3000m on mud at pond margins. The plants are covered by water during the rainy season but begin flowering as the water recedes at the onset of the dry season. Known only from J. Ḥubaysh; J. Sumārah; near Suq al Jum'ah, J. Ash Shirq; Kawkabān; J. Masār, Ḥarāz; Maḥwīt.

2. Limosella macrantha *R.E. Fries*

Minute herb. Leaves linear without a perceptible petiole, 3–6cm long. Corolla 2-lipped, the tube longer than the calyx. Mature fruiting capsule about as long as the calyx, the pedicels very short, 1–2mm long.

In similar habitats to *L. africana*, but much rarer and only known between 2600 and 3000m on J. Ḥubaysh and J. Manār.

15. SCOPARIA *L.*

1. Scoparia dulcis *L.*

Erect, much-branched, nearly glabrous, bushy perennial to c 80cm. Leaves opposite, usually in whorls of 3–5, petiolate, linear-oblanceolate, 1–4cm long, punctate, the larger leaves serrate. Inflorescence a leafy panicle, the flowers 1–3 in the leaf axils; pedicels thread-like, 5–8mm long; calyx 4-lobed; corolla white, rotate, 4-lobed, without a tube.

A very rare plant of the gravelly margins of ± permanently flowing wadis 300–1500m; recently seen only in Wadi Sukhnah and at Suq al Jammah, but recorded by Forsskal from the Wadi Mawr area.

16. CAMPYLANTHUS *Roth*

Low, wiry, almost leafless grey undershrubs. Leaves (where present) alternate. Flowers in small racemes; calyx lobes free to the base, triangular, gradually narrowed into a fine point; corolla rotate with a long narrow cylindrical tube; stamens 2. Capsule strongly laterally compressed.

The goegraphical distribution of *Campylanthus* is of considerable interest. Like *Aeonium* and some other plants it has two centres of distribution, one in the Canary Islands and the other in southern Arabia and northeast tropical Africa. There is no obvious explanation for such a disjunct distribution.

The following three species are superficially very similar.

1. Raceme (10–)15–25-flowered; corolla blue *C. junceus*
Raceme less than 5-flowered; corolla mauve **2**
2. Plant hirsute *C. yemenensis*
Plant glabrous *C. pungens*

1. Campylanthus junceus *Edgw.*

Decumbent or ascending, branched undershrub to 1m, the branches mostly straight, rod-like, ± glabrous, dull green. Leaves, if present, linear, 2–8mm long, ± adpressed to the stem. Racemes (10–)15–25-flowered; calyx lobes small, c 2–2.5mm long; corolla blue, glabrous, c 8mm long.

A rare plant of arid stony hills between 500 and 2000m mostly in the southeastern mountains: J. Ra's; Rada'; between Dhī Nā'im and Bayḍā'; between Naqīl Hajlah and Ḥarīb. This species is known from both sides of the Gulf of Aden.

2. Campylanthus pungens *O. Schwartz*

Similar to the previous species but of lower stature (usually less than 50cm), the branches glaucous, slightly flexuous, not rod-like. Racemes 2–4-flowered, often becoming black in dry conditions; calyx lobes 3.5–4mm long; corolla mauve.

A characteristic plant of dry rocky hills on the high plateau from Dhamār northwards, growing from 2000 to 2800m; also near Majz (Steinberg). This species is endemic to southwest Arabia.

3. Campylanthus yemenensis *A.G. Miller*

Low, intricately branched, spinescent undershrub reaching c 15cm in height only, otherwise similar to *C. pungens* except for its hirsute indumentum which is particularly marked on the outside of the corolla.

A characteristic species of stony steppe in the southern Tihama growing between 100 and 400m from Nukaybah southwards. Endemic to Yemen.

17. VERONICA *L.*

Speedwells. Herbs with opposite leaves. Flowers blue or white; corolla rotate, 4-lobed with a very short tube; stamens 2. Capsule laterally depressed.

1. Flowers in axillary racemes *V. anagallis-aquatica*
Flowers solitary in the leaf axils **2**
2. Capsule lobes divergent. Corolla c 8–12mm diameter *V. persica*
Capsule lobes not divergent. Corolla c 4–8mm diameter *V. polita*

1. Veronica anagallis-aquatica *L.*

V. anagallis L.

Erect or ascending annual herb to 40cm, usually glabrous below but glandular-hairy above. Leaves sessile, ovate or lanceolate, acute, up to 8cm long, the margin usually serrate. Flowers in axillary racemes; corolla pale blue, 5–6mm in diameter.

Widespread and frequent between 700 and 2900 by permanent streams on the escarpment and on the western part of the high plateau. This species is a cosmopolitan weed of wet places.

2. Veronica persica *Poir.*

V. tournefortii Gmel.

A decumbent pubescent annual, usually much branched at the base. Leaves shortly petiolate, ovate, 1–2cm long, serrate. Flowers solitary in the leaf axils, corolla 8–12mm diameter. Capsule ciliate with keeled, divergent lobes so that it is almost twice as broad as long.

A widespread but infrequent weed found between 1500 and 3100m on the escarpment and high plateau but either in high rainfall areas or in irrigated land.

3. Veronica polita *Fries*

Similar to *V. persica* with which it can be easily confused, but the corolla is deep blue and only 4–8mm diameter and the capsule is crisped-pubescent and only slightly broader than long, the lobes neither divergent nor keeled.

Found between 1100 and 3000m in similar places to *V. persica*, but slightly more common.

18. ALECTRA *Thunb.*

Erect parasitic herbs with alternate leaves. Flowers subsessile in terminal bracteate, spike-like racemes; calyx campanulate; corolla regular with a very short tube, yellow. Capsule globular.

Leaves scale-like, less than 0.5cm long *A. parasitica*
Leaves not scale-like, 2–5cm long *A. sessiliflora*

1. Alectra parasitica *A. Rich.*

Sparingly branched, pubescent, annual parasite 10–25cm high. Leaves scale-like, c 4mm long, adpressed to the stem. Calyx c 6mm long, slightly shorter than the bracts and the yellow corolla.

Widely distributed on the escarpment from Turbah north to Aḥjar, but inconspicuous and always in small quantity. It grows on scrubby banks and eroded stony slopes in areas of good rainfall between 1500 and 2700m.

2. Alectra sessiliflora (*Vahl*) *Kuntze*

A. arabica Deflers

Sparingly branched, thinly hairy, annual semi-parasite 20–40cm high. Leaves petiolate, ovate or lanceolate, 2–5cm long, the margin dentate. Calyx 8–9mm long, much shorter than the leaf-like bracts; corolla 12–15mm long, yellow with dark veins.

A local plant found between 2000 and 3200m on the escarpment from J. Ṣabir north to Maḥwīt and Aḥjar. It grows in montane grassland in high rainfall areas where it is often associated with *Cyanotis barbata* and *Striga yemenica*. It is much more common in the Ibb region than elsewhere.

19. HARVEYA *Hook.*

1. Harveya obtusifolia (*Benth.*) *Vatke*

Erect, glandular-pubescent, semi-parasitic herb to 50cm, stem usually unbranched. Leaves opposite, scale-like, 2–4cm long. Flowers sessile in a terminal leafy spike; bracts about as long as the regular, pale pink, funnel-shaped corolla.

A very rare plant only known from Salafīyah in J. Raymah where it was collected by Gillespie at about 1600m between Maḥāfilah and 'Abdullah.

20. BUCHNERA *L.*

1. Buchnera hispida *Buch.-Ham.*

Erect, scabrid-pubescent, branched, semi-parasitic annual herb to 40cm. Leaves opposite, very shortly petiolate, narrowly elliptic or oblong, 2–5cm long, entire. Flowers in terminal spikes; calyx tubular; corolla blue, purple or white, rotate, with a narrow cylindrical tube.

A rare plant of grassland on shallow soils in areas of high rainfall between 1200 and 2000m: Ibb; Bani Sa'id, Ja'farīyah, J. Raymah; near 'Ukaybir, J. Milḥān.

21. STRIGA *Lour.*

Erect, parasitic annual herbs, often scabrid. Leaves opposite or alternate. Flowers in terminal spikes; calyx with a cylindrical tube and 5 spine-like lobes; corolla with a long, sharply bent, cylindrical tube, 2-lipped, often red.

1. Leaves reduced to scales. Plant glabrous or hairy, but never scabrid — *S. gesnerioides*
   Leaves not scale-like. Plant scabrid — 2
2. Leaves 2–7cm long. Corolla tube more than 1.5cm long — *S. hermonthica*
   Leaves less than 2cm long. Corolla tube less than 1cm long — 3
3. Calyx 5-ribbed — *S. yemenica*
   Calyx with 10 or more ribs — *S. asiatica*

1. Striga asiatica (*L.*) *Kuntze*

S. hirsuta Benth

S. lutea sensu Deflers

Slender, sparingly branched, scabrid, semi-parasitic herb to 20cm though commonly less. Leaves mostly alternate, linear, up to 1.5cm long. Calyx with 10 or more ribs, 6–8mm long; corolla bright crimson or white, the tube c 1cm, long, bent just below the rather broad lobes.

There are two distinct forms of this species in Yemen. The scarlet-flowered plant is an uncommon plant of thin grassland on shallow rocky soils found between 1300 and 2400m in areas of good rainfall on the escarpment: Wadi Ḍabāb, Taiz; J. Ṣabir; Rawnah, Shar'ab; Ibb; Yarīm; Suq al Jum'ah, J. Ash Shirq; Bani Sa'id and Ḥadādah, J. Raymah; Maḥwīt.

The white flowered form in which the bracts are much longer than the calyx is also uncommon plant but grows as a weed of dry, gravelly fields where it may be parasitic on Sorghum. It is found below 1300m mainly in the south: Barakānī, Hujariyah; Misrākh and Wadi Dabāb near Taiz; between Taiz and Rawnah, Shar'ab; Al Khawkah (Miller); between Bājil and Khamīs Bani Sa'd.

2. Striga gesnerioides (*Willd.*) *Vatke* — Fijil Al Arḍ

S. orabanchoides (R. Br.) Benth.

Very variable parasitic herb, 5–30cm high, stem usually stout, often unbranched, glabrous or pubescent, but never scabrid. Leaves scale-like, up to 1cm long. Calyx 5-ribbed, c 5mm long, about equalling the bracts, corolla pink or white, the tube c 8mm long, bent above the middle.

Widely distributed between 300 and 2500m along the escarpment north to Ḥūth, but always in small quantity. In Yemen probably always parasitic on shrubby Euphorbias and usually found on barren stony soil in between the Euphorbia bushes.

3. Striga hermonthica (*Del.*) *Benth.* — 'Udār

Sparingly branched, scabrid parasite, often very robust and up to 1m high. Leaves opposite, linear-lanceolate, 2–7cm long. Calyx 5-ribbed, 9–10mm long, ± equalling the bracts; corolla dull red, the tube 1.5–2cm long, bent at or just below the middle.

An attractive, conspicuous plant when in flower in autumn, but also a pernicious weed which is parasitic on the roots of sorghum. It is widespread and often abundant from 200 to 1800m in sorghum fields on the escarpment and alluvial plains in the east of the Tihama.

4. Striga yemenica *Musselman & Hepper*

Slender, sparingly branched, dark green, semi-parasitic herb, 10–25cm high. Leaves mostly opposite, linear, up to 2cm long. Calyx 5-ribbed, 5–6mm long shorter than the lower bracts, but exceeding the upper ones; corolla dull red, the tube c 8mm long, bent above the middle.

A characteristic plant of high montane grassland from 2000 to 3200m growing on all the main Ibb mountains, extending northwards through Waṣāb al 'Ālī to J. Bārad, Kusmah in J. Raymah.

This species is related to *S. aspera* (Willd.) Benth. but differs in its less robust habit and much smaller flowers. It also occurs in Ethiopia.

22. CYCNIOPSIS *Engl.*

1. Cycniopsis humifusa (*Forssk.*) *Engl.*

Prostrate, strigose trailing herb. Leaves opposite, shortly petiolate, obovate or elliptic, 5–10mm long, entire. Flowers solitary in the leaf axils, sessile, all parts drying blue-black; calyx tubular; corolla white, ± rotate with a long, narrow, slightly bent tube 10–12mm long.

An uncommon plant of short grassland in high rainfall areas between 1400 and 2500m, growing in similar situations to *Dyschoriste radicans*, very inconspicuous except when in flower: tolerably frequent in suitable localities near Ibb; above 'Adūf near Turbah; Jabalayn, 'Udayn; Kusmah region, J. Raymah; Ḥajarah, Ḥarāz; Maḥwīt; Ḥajjah. Outside Yemen only known from Ethiopia.

23. BARTSIA *L.*

1. Bartsia longiflora *Benth.*

Dark green, shrubby perennial, much branched at the base with numerous, erect, sub-glabrous branches to 50cm. Leaves opposite, sessile, oblanceolate, 1–2cm long, serrate. Flowers shortly pedicellate in small, glandular-hairy, leafy terminal racemes; corolla yellow, 2-lipped with a straight cylindrical tube c 13mm long.

A very rare plant found between 2700 and 2900m on terrace walls and cliffs on J. Sumārah.

24. BELLARDIA *All.*

1. Bellardia trixago (*L.*) *All.*

Bartsia trixago L.

Erect, usually unbranched, yellow-green, glandular-hairy annual to 30cm. Leaves opposite, sessile, lanceolate or oblong, 2–3.5cm long, coarsely dentate. Flowers sessile in densely glandular, terminal, bracteate spikes; corolla yellow, 2-lipped, with a short straight tube.

Abundant between 2000 and 2900m in meadowland on the mountains between Ḥadīyah and Kusmah, J. Raymah; unknown elsewhere in Yemen.

The Snapdragon, **Antirrhinum majus** *L.*, an attractive garden plant with terminal racemes of reddish-purple flowers, has been cultivated in Sana'a since at least the 1930s when it was collected by Rathjens.

OROBANCHACEAE

Parasitic herbs, lacking chlorophyll and growing on the roots of other plants. Stems erect, usually simple. Leaves alternate, scale-like. Flowers in a terminal spike or raceme; sepals united, at least at the base; corolla 5-lobed, regular or 2-lipped. Fruit a capsule containing numerous small seeds.

All species are called 'Fijil Al Arḍ' or 'Zub Al Qa' '.

Calyx lobes obtuse. Robust plants with a corolla at least 2cm long — 1. *Cistanche*
Calyx lobes acute. Slender or moderately robust plants, the corolla not exceeding 22mm long — 2. *Orobanche*

1. CISTANCHE *Hoffmanns. & Link*

Stems stout, simple. Flowers in dense terminal spikes, numerous, large, subtended by a large bract and 2 smaller bracteoles; calyx with obtuse lobes. Corolla 5-lobed, nearly regular.

Corolla uniformly deep rose-red, c 3cm long — *C. rosea*
Corolla yellow, or partly suffused with purple, 4–5cm long — *C. phelypaea*

1. Cistanche phelypaea (*L.*) *Cout.*

C. tubulosa (Schenk) R. Wight

Conspicuous, robust, nearly glabrous parasite to 50cm. Spike relatively lax, c 5–25cm long; calyx shortly tubular or ± campanulate, 12–15mm long; corolla yellow, sometimes suffused with purple especially on the tube, funnel-shaped, 4–5cm long, becoming bent at maturity; anthers hairy.

This species is sometimes divided into two. *C. phelypaea* is said to have an entirely yellow corolla and acute anther lobes whereas *C. tubulosa* has an at least partly purple corolla and obtuse anther lobes. Corolla colour, though, is not well correlated with the shape of the anther lobes and it is difficult to assess the significance of these features in any case. Both plants with acute anther lobes (Hepper 6223) and obtuse anther lobes (Hepper 5898, Wood 1114) occur in Yemen.

Widespread but local, growing between 500 and 2400m on the escarpment and high plateau but rare in the north and at altitudes below 1200m; mostly found on field margins in heavily cultivated alluvial plains in medium rainfall areas notably the Taiz-Qā'idah region, the montane plains between Sana'a and Dhamār and the Mafḥaq region. It is parasitic on *Salsola bottae* and probably on *Commicarpus*.

2. Cistanche rosea *Baker*

Stout, erect, nearly glabrous parasite to 25cm, the whole plant deep rose-red except for the conspicuously white, hairy anthers. Spikes very dense, 10–15cm long; calyx 12–15mm long; corolla rose-red, tubular, c 3cm long, apparently remaining straight at maturity.

Only known from the margins of cultivated land at 1300m at Wadi Ḥidrān near Taiz. Outside Yemen it is a rare plant of a few scattered localities in tropical Arabia.

2. OROBANCHE *L.*

Broomrapes. Stems slender or moderately stout, simple or branched. Flowers in dense spikes, each flower subtended by a single broad bract and sometimes additionally by 2 bracteoles; calyx with narrow acute lobes, sometimes 4-lobed, the lobes united below so making the calyx ± campanulate; corolla 2-lipped.

1. Bracteoles present. Stem usually branched — *O. ramosa*
   Bracteoles absent. Stem simple — **2**
2. Corolla tube inflated, white, glossy (best seen in dried specimens) — *O. cernua*
   Corolla tube neither inflated nor glossy — *O. minor*

1. Orobanche cernua *Loefl.*

Robust parasite with stout, simple, sparsely glandular-pubescent stems to 40cm. Spike very dense, many-flowered, obtuse, up to 20cm long; calyx with 2 bifid segments; corolla blue, 12–20mm long, the tube inflated, white, glossy, ± constricted above the capsule.

The distinctive glossy corolla tube is not particularly obvious on fresh specimens but is very clear when the plant has dried.

Apparently a rare plant, usually parasitic on tomatoes, found in a few scattered localities between 1400 and 2500m on the escarpment and high plateau: Ṣayfarah, Taiz; 'Usil and Manākhah in the Ḥarāz (Schweinfurth); Qa' Bakīl (Hepper); Ḥūth; J. Minmār; the Sana'a area, where it appears to be tolerably frequent in tomato fields.

2. Orobanche minor *Sm.*

Slender or robust parasite with simple, glandular-pubescent stems to 70cm. Spike lax, few- to many-flowered, acute, up to c 20cm long; calyx with 2 entire or bifid segments; corolla cream, yellow or dull purple, 10–18mm long.

Rather variable in the indumentum of the corolla in Yemen. In some specimens (Hepper 5799) the corolla is almost glabrous whereas in others (Wood 521, 1318) the corolla is densely glandular-hairy. These latter forms approach *O. pubescens* D'Urv. but the corolla is never villous and the filaments are not inserted more than 3mm above the base of the corolla.

Widespread, frequent but never in abundance, growing from 1600 to 2900m on the escarpment, but only once recorded from the high plateau at Sana'a (Wissmann). It grows on a great variety of hosts including *Crepis*, *Scabiosa*, *Flaveria*, *Cichorium* and doubtless many others.

3. Orobanche ramosa *L.*

O. mutelii F.W. Schultz

Slender parasite with branched glandular-pubescent stems to 25cm. Spike lax, usually with less than 25 flowers, acute, up to 15cm long; bracteoles present; calyx united below with a campanulate tube and 4 equal teeth; corolla blue.

Two subspecies differing mainly in corolla size occur in Yemen but intermediate forms (e.g. Wood 1319) occur. Subspecies **ramosa** (e.g. Hepper 6090) has a corolla 10–13mm long wheras subspecies **mutelii** (*F.W. Schultz*) *Coutinho* (e.g. Hepper 6315) has a corolla 18–22mm long. The distribution of the two subspecies is not accurately known.

O. ramosa as a whole is widespread on the escarpment between 1500 and 2900m, growing in open habitats on a wide variety of hosts. It is, however, very scattered in occurrence and like *O. minor* never occurs in large quantity.

ACANTHACEAE

Herbs, characteristically woody below or undershrubs. Leaves without stipules, opposite, simple, often with minute white rod-like deposits known as cystoliths. Inflorescence various but often spike-like and with conspicuous bracts; flowers bisexual, irregular; sepals 4 or 5, free or joined; petals joined with a 5-lobed or 2-lipped corolla; stamens 2 or 4, the anthers with 1 or 2 large cells, sometimes tailed; style long. Fruit a capsule, often opening explosively when ripe, the seeds borne on small hook-like outgrowths.

A tropical family with many escarpment species. There are a number of endemics particularly in *Barleria*. Unfortunately the generic distinctions are rather technical and it is necessary to examine the flowers carefully before using the key. Among the useful characters are the number of stamens, the number, arrangement and shape of the anther cells, the arrangement and shape of the bracts and the appearance of the seed. Species from different genera often have a strikingly similar superficial appearance so care is needed.

1. Fertile stamens 4 — **2**
   Fertile stamens 2 — **10**
2. Corolla with 1 lip only — **3**
   Corolla conspicuously 2-lipped or ± regularly 5-lobed — **5**
3. Leaves viciously spiny. Shrub 0.5–3m high with large dark-red flowers — 2. *Acanthus*
   Leaves usually unarmed, but if spiny, plant less than 0.5m high and flowers blue — **4**
4. Calyx segments 4. Flowers blue or white — 1. *Blepharis*
   Calyx segments 5. Flowers pink — 3. *Crossandra*
5. Corolla 2-lipped — 9. *Lepidagathis*
   Corolla ± regularly 5-lobed — **6**
6. Corolla with a tube 5–12cm long — 6. *Ruellia*
   Corolla tube less than 3cm long — **7**
7. Inflorescence terminal, sometimes axillary as well. Flowers white — **8**
   Inflorescence all axillary. Flowers blue, sometimes very pale — **9**
8. Bracts broadly ovate to subreniform, rounded — 5. *Phaulopsis*
   Bracts lanceolate, acute — 8. *Asystasia*
9. Anthers with a short spur-like tail. Corolla small, less than 8mm wide at the mouth — 4. *Dyschoriste*
   Anthers without a short spur-like tail. Corolla at least 10mm wide at the mouth — 2. *Ruellia*
10. Corolla 5-lobed. Plants often spiny — 7. *Barleria*
    Corolla 2-lipped. Plants never spiny — **11**
11. Shrubs 1–2m high. Flowers red-brown — **12**
    Undershrubs or herbs less than 1m high. Flowers of various colours but never red-brown — **13**

12. Anthers with a short spur-like tail. Leaves less than 2cm long — 12. *Justicia*
Anthers without a spur-like tail. Leaves 2–5cm long — 14. *Anisotes*
13. Flowers enclosed in a pair of free or joined bracts. Flowers pinkish-mauve, white only in *Hypoestes forskalei* — **14**
Flowers with a single bract or ebractiate, never with an involucre of 2 bracts. Flowers of various colours — **16**
14. Anthers 1-celled. Flowers single between the bracts — 17. *Hypoestes*
Anthers 2-celled. Flowers paired between the bracts — **15**
15. Inflorescence paniculate. Bracts very unequal in size — 16. *Peristrophe*
Inflorescence of dense axillary clusters. Bracts equal or nearly equal in size — 15. *Dicliptera*
16. Anthers with a short spur-like tail — **17**
Anthers without a short spur-like tail — **18**
17. Capsule 2-seeded, seeds smooth. Bracts suborbicular, conspicuous, concealing the calyx — 13. *Monechma*
Capsule 4-seeded, seeds rough. Bracts inconspicuous, rarely concealing the calyx and not suborbicular — 12. *Justicia*
18. Anther cells at equal heights. Flowers green or violet — **19**
Anther cells arranged one above the other. Flowers white — 11. *Isoglossa*
19. Corolla tube longer than the lobes; upper lip notched — 10. *Ecbolium*
Corolla tube shorter than the lobes; Upper lib entire — 11. *Megalochlamys*

1. BLEPHARIS *Juss.*

Annual or perennial herbs with leaves in whorls of 4, two smaller than the others. Flowers many, in bracteate spikes or in a modified spike consisting of a solitary, terminal flower and several empty bracts below; calyx 4-lobed, 1 lobe much larger than the others; corolla with a short tube, and a 3-lobed lower lip, the upper lip being absent; stamens 4, anthers 1-celled. Capsule 2-seeded, the seeds discoid, hairy.

1. Spike 1-flowered; bracts with bristles but not spines; corolla white — **2**
Spike many-flowered; bracts spiny; corolla blue — *B. ciliaris*
2. Bracts with long retrorse-barbed, clinging bristles. Corolla 1–1.5cm long — *B. maderaspatensis*
Bracts with short bristles which are not retrorse-barbed. Corolla c 2cm long — *B. sp.A*

1. Blepharis ciliaris (*L.*) *B.L. Burtt.* — Zughāf

B. edulis (Forssk.) Pers.
B. persica (Burm. f.) Kuntze
Pubescent woody annual with prostrate or erect stems up to 60cm high. Leaves in 4s, often absent, usually narrowly oblong, 1–7cm long, spine-tipped, entire or sinuate-lobed, the lobes sometimes spine-tipped. Inflorescence a many-flowered axillary bracteate spike 1–25cm long; bracts 1–5cm long, ovate or lanceolate, ± folded, with an acuminate spine-tip and 3–5 spines on each side; corolla 1.5–2cm long, the lip bright blue.

A very variable plant not easily divided into discreet entities. In general, plants growing on sand in the Tihama are tall and upright with very long bracts and flower spikes (Wood 74/408, 3426) whereas plants from rocky ground both in the Tihama and on the mountains are prostrate with bracts up to 2cm long and a very short flower spike (Wood 74/104, 3409, 3439). Intermediates between the forms occur (e.g. Miller 492 from 'Usil) however and it seems best to include all blue-flowered *Blepharis* in Yemen under one species including plants from Hanish Island, named *B. linariifolia* Pers.

Widespread and often abundant on sand and rocky ground from sea level to 2500m throughout Yemen except for the Sands; absent only from high rainfall areas such as the Ibb region.

A plant of characteristically Sahara-Sindian distribution.

2. Blepharis maderaspatensis (*L.*) *Heyne ex Roth*

B. boerhaviifolia Juss.
Procumbent or ascending woody herb with pubescent, often purplish stems. Leaves in whorls of 4, elliptic, entire, 2–6cm long, scabrid above, hispid and grey below. Inflorescence of short, solitary or clustered, axillary flower spikes each with a single terminal flower; bracts 8–12mm long, variable in shape, obovate or oblanceolate, with clinging, retrorse-barbed bristles on the margins; corolla 1–1.5cm long, the lip dirty white.

A characteristic plant of scrubby banks in the valleys of the central escarpment between 500 and 1500m. Although unknown north of the Wadi Lā'ah it is frequent in similar habitats further south, notably in the Ibb-'Udayn region and also in Wadi Ḍabāb, Misrākh and Aḥkūm in Taiz province.

3. Blepharis sp.A

Prostrate woody herb with sparsely adpressed-hairy, purplish stems. Leaves in whorls of 4, elliptic to ± rhomboid, entire, 2–3cm long, dark green and slightly scabrid above, grey below. Inflorescence similar to that of *B. maderaspatensis* but the bracts 10–15mm long, oblanceolate to elliptic with very short marginal bristles which are not retrorse-barbed; corolla about 2cm long, the lip white with brown veins.

Only known from a steep grassy bank at 1600m below the school at Rawnah, Shar'ab (Wood 74/281).

2. ACANTHUS *L.*

1. Acanthus arboreus *Forssk.* — Sanaf

Pubescent spiny shrub up to 4m high, but usually much less. Leaves oblong, sinuate-lobed, 8–15cm long, the main veins prominent, yellowish (contrasting with the green leaf) and excurrent as sharp spines. Flowers in dense terminal spikes; bracts yellowish, lanceolate, spiny-margined; calyx 4-lobed, the lobes unequal; corolla c 5cm long, the upper lip absent, the lower lip 3-lobed, dark red; stamens 4, anthers 1-celled. Capsule 2-seeded, seeds glabrous.

Flowers very sporadically and mostly in winter.

An unmistakable species whose thistle-like leaves are a feature of large areas of the escarpment and high plateau between 1000 and 3000m. At low altitudes it is mostly a plant of scrubby banks along watercourses while in dry country, particularly on the high plateau, it is found on cliffs and in rocky gullies. It is at its most frequent on the escarpment between 2000 and 2800m where it is often dominant on steep eroded slopes which are consequently painful to climb. It is also found on terrace walls, field borders and in a variety of waste places and is likely to become more frequent if the terrace system breaks down. It is unpalatable to grazing animals and vain attempts are made to control its spread by burning.

3. CROSSANDRA *Salisb.*

Undershrubs with opposite or whorled, entire leaves. Flowers in dense, long pedunculate, bracteate spikes; bracteoles 2, linear; calyx with 5 unequal chaffy segments; corolla with a long linear tube and a single 3-lobed lip;stamens 4, anthers 1-celled. Capsule 2–4 seeded; seeds flat, scaly.

1. Crossandra johanninae *Fiori*

C. infundibulariformis sensu Schwartz
C. wissmannii Schwartz
Straggling, leggy, woody perennial, scrambling over scrub vegetation and reaching 1½m in height. Leaves opposite or in whorls, entire, narrowly elliptic, 3–7cm long, gradually narrowed into an obtuse apex and a long attenuate base. Flower spike 2.5–5cm long; bracts closely imbricate, oblanceolate-elliptic, sharply acute, gland-dotted and densely covered on the upper surface and margins with long, white, large-celled hairs; corolla with a broad, 3-lobed, pinkish-mauve lower lip and a tube 1.5–2cm long, stamens 4, anthers 1-celled. Capsule 2–4 seeded; seeds flat, scaly. The flower colour persists long after the flower is dried.

Locally frequent on the escarpment between 500 and 1500m, particularly in the valley scrub communities of the central escarpment but also occurring in quite dry areas in bushy places protected from grazing.

4. DYSCHORISTE *Nees*

Herbs or undershrubs. Leaves obscurely crenate. Flowers in subsessile, axillary cymes; bracts linear, very short; calyx tubular, the 5 lobes free for half their length, tapering to long subulate points; corolla funnel-shaped, 5-lobed, blue; stamens 4, anther cells 2, shortly tailed. Capsule 4-seeded; seeds discoid, pubescent.

Prostrate herb. Leaves 8–15mm long	*D. radicans*
Sprawling undershrub. Leaves 4–7cm long	*D. perrottetii*

1. Dyschoriste perrottetii (*Nees*) *Kuntze*

Untidy, sprawling, pubescent undershrub scrambling to 1m in scrub. Leaves elliptic, acute, 2–5cm long. Flowers in axillary clusters of 3–8; corolla c 12mm long, slightly exceeding the calyx, blue.

An uncommon plant of bushy overgrown fences along ditches at the edge of cultivated land found between 1300 and 2300m in the south: Wadi Ḥidrān and Wadi Ḍabāb, Taiz; Dhī Sufāl; Wadi Sūdān, Qā'idah; below Mashwarah towards 'Udayn; between Ibb and Suq As Sabt.

2. Dyschoriste radicans *Nees*

Trailing pubescent or glabrescent herb from a woody rootstock. Leaves obovate, mucronate, 8–15mm long. Flowers solitary or in axillary clusters of 2–3; corolla c 14mm long, much exceeding the calyx, blue.

A plant of bare patches in shallow, stony grassland between 1700 and 2500m; widespread in suitable habitats in the central escarpment but elsewhere known only from between Ibb and Suq As Sabt.

5. PHAULOPSIS *Willd.*

1. Phaulopsis imbricata (*Forssk.*) *Sweet*

Pubescent trailing herb; stems woody below. Leaves ovate or elliptic, entire or crenate, 2–5cm long. Inflorescence of dense terminal (sometimes also axillary) cylindrical spikes of small flower clusters borne in the axils of rounded, broadly ovate, bract-like leaves; calyx 5-lobed, the lobes unequal; 2 subulate, 2 linear, 1 large, ovate-elliptic; corolla funnel-shaped with 5 ± equal lobes, white, c 10mm long; stamens 4, the anthers untailed. Capsule flattened, 4-seeded; the seeds hairy on the margins.

Locally frequent on the escarpment between 1300 and 2500m from Turbah north to Maḥābishah but mostly in the high rainfall areas of the central escarpment and the Ibb region. It is a plant of damp, partially shaded places particularly gullies on cliffs and in coffee terraces but it also grows by ditches and in grassland in areas of plentiful mist.

6. RUELLIA *L.*

Woody herbs or undershrubs. Leaves entire. Flowers 1–3 together in subsessile axillary cymes or clusters; bracts leaf-like; bracteoles present, exceeding the deeply 5-lobed calyx; corolla with or without a long cylindrical tube, campanulate above, ± equally 5-lobed; stamens 4, anther cells 2, not tailed. Capsule many-seeded; seeds discoid, appressed hairy on the margin.

The flowers of all species open at sunset and fall by midday the next day.

1. Corolla white, with a cylindrical tube 5–10cm long *R. grandiflora*
   Corolla blue, cylindrical portion of the tube absent or less than 1cm long 2
2. Bracts, bracteoles and leaf petioles covered with numerous, long, spreading, white, large-celled hairs which may be 5mm long *R. praetermissa*
   Bracts, bracteoles and leaf petioles without long, spreading, large-celled hairs 3
3. Corolla very pale violet. Calyx 10–18mm long, the lobes subulate. Capsule shortly pubescent *R. prostrata*
   Corolla blue. Calyx 4–6mm long, the lobes linear-acute. Capsule glabrous *R. patula*

1. Ruellia grandiflora (*Forssk.*) *Blatter*

R. longiflora Vahl

Woody undershrub to 1m, young branchlets stellate-hairy. Leaves ovate, stellate-hairy on both surfaces, dark green above, grey below, 2–4cm long. Flowers solitary in the axils of the upper leaves; calyx c 15mm long, the lobes ovate-triangular, 5mm wide; corolla white with a cylindrical tube 5–10cm long. Capsule c 2cm long, pubescent.

Frequent on cliffs and on shallow stony soils in *Acacia* bushland on the escarpment from 500–1500m, but less common in the wetter areas of the central escarpment; also widespread between 1600 and 2000m in the southeastern mountains and the Mashriq being recorded from J. Sawraq, Damt, Dhī Nā'im, Wadi Khārid, the Fardah pass, Khaywān and the Wadi 'Azū near Majz (Steinberg).

It is endemic to southwest Arabia.

2. Ruellia patula *Jacq.* Quṣīf

R. pallida Vahl

Prostrate or erect woody perennial, much branched at the base and never exceeding 50cm high. Leaves ovate or elliptic, pubescent, sometimes with glandular hairs but without stellate hairs, cystoliths or long, large-celled white hairs, dark green, mostly 1–3cm long. Flowers 1–3 in axillary cymes; calyx 4–6mm long, the lobes linear-acute, c 1mm wide; corolla blue (very rarely white), the cylindrical tube c 5mm long and usually ± hidden by the bracts. Capsule 10–18mm long, glabrous.

A rather variable species with two distinct forms in Yemen. The commonest of these is a trailing plant (Wood 74/129, 2450) with eglandular stems and one-flowered axillary cymes. It grows on rocky hillsides. The other form is an erect plant (Wood 2294, 2720) with numerous stalked glands on the upper part of the stem and 2–3-flowered axillary cymes. It is a plant of the margins of irrigated fields in the escarpment foothills. These forms require further investigation but intermediates probably occur.

An abundant and conspicuous plant of stony ground in the scrub communities of the escarpment growing from the first rocky outcrops in the Tihama up to 1800m; it is also perhaps the most characteristic species of terrace walls at low altitudes. Like the previous species it occurs in the Mashriq but is rare and only known from Majz (Steinberg) and the Fardah pass.

3. Ruellia praetermissa *Lindau*

Rather similar to some forms of *R. patula* but the ascending woody stems with white bark and the leaves up to 8cm long and densely covered in cystoliths. The bracts, bracteoles, the lower part of the leaves and their petioles are covered in conspicuous long, white, spreading, large-celled hairs. Flowers 1–3 in axillary cymes; corolla blue; calyx 10–12mm long, the lobes linear-acuminate. Capsule 2–2.5cm long, sparsely adpressed-hairy.

An uncommon plant of *Acacia* bushland between 500 and 1000m, mostly in the rainshadow areas of the central escarpment; between Mishrāfah and Masbaḥ, Waṣāb As Sāfal; above Ḥillah, J. Bura' (Schweinfurth); Rajamah, Ḥajjaylah; Khamīs Banī Sa'd; between Baḥal and Mawsanah in the Ḥarāz (Wood 2108, 3177).

4. Ruellia prostrata sensu *T. Anders.*

Similar to the previous two species. Stems ascending or erect up to 1m high. Leaves very variable in size, 2–10cm long, strigose and with cystoliths on the upper surface, pubescent below but without the long, spreading white hairs of the last species. Flowers 1–3 in axillary cymes; corolla very pale violet (almost dirty white); calyx 10–18mm long, the lobes gradually narrowed into long, subulate points. Capsule 1.5–2.5cm long, shortly but densely pubescent.

Widespread on the escarpment from 500 to 1400m from Misrākh near Taiz north to Maḥābishah, but mainly in areas of good rainfall; it grows in scrubby areas protected from grazing, particularly thorn fences along wadis and by paths.

7. BARLERIA *L.*

Herbs or shrubs, often spiny. Leaves entire. Flowers axillary, solitary, in dense 1-sided racemes or in few-flowered cymes. Bracts leaf-like or modified into a simple or toothed spine; calyx of 4 unequal segments, 2 large and 2 small and linear; corolla 5-lobed, the tube either cylindrical throughout or cylindrical below and funnel-shaped above; stamens 2, anthers 2-celled, not tailed. Capsule 2–4-seeded, the seeds pilose.

***Barleria*, like *Aloe*, is remarkable for the complementary nature of the distribution of allied species with a similar ecology. *Barleria prionitis* and *B. trispinosa* never grow together and nor do the four small blue-flowered species *B. argentea*, *B. hochstetteri*, *B. orbicularis* and *B. parviflora*. Three of the following species are endemic to Yemen and many more are rare species restricted to the southern Red Sea area.**

1. Plant spiny **2**
Plant unarmed **9**
2. Corolla white (drying blue), equally lobed and with a long cylindrical tube. Spines with small lateral spiny teeth **3**
Corolla yellow or pale blue, funnel-shaped with one lobe free to about half way down the tube. Spines without lateral teeth **6**
3. Large calyx segments similar to the bracts and prominently white-villous on the margins *B. hillcoatiae*
Large calyx segments very unlike the bracts, broad, papery and without prominent white marginal hairs **4**
4. Large calyx segments ovate. Flowers in 1-sided racemes **5**
Large calyx segments narrowly elliptic. Flowers solitary *B. lanceata*
5. Large calyx segments with numerous small spines along the margin. Corolla tube less than 4.5cm long *B. mucronifolia*
Large calyx segments unarmed except for the terminal spines. Corolla tube more than 5cm long *B. acanthoides*
6. Corolla blue *B. bispinosa*
Corolla yellow or orange **7**
7. Bracts numerous, ovate, imbricate, unlike the leaves *B. trispinosa*
Bracts resembling the spines, not imbricate **8**
8. Corolla tube 1.5–2cm long. Leaves obovate-elliptic, 1–1.5cm long *B. proxima*
Corolla tube 3–4cm long. Leaves lanceolate 2–12cm, long *B. prionitis*
9. Large calyx segments linear-lanceolate, about 3 times as long as broad **10**
Large calyx lobes ovate or elliptic up to 1½ times as long as broad **11**
10. Leaves ± sessile. Flowers solitary, borne on peduncles 1–2cm long *B. parviflora*
Leaves petiolate. Flowers in shortly pedunculate cymes *B. argentea*
11. Flowers subsessile in the leaf axils. Corolla large *B. ventricosa*
Flowers in pedunculate cymes. Corolla small **12**
12. Large calyx lobes ovate-cordate *B. orbicularis*
Large calyx lobes elliptic with a cuneate base *B. hochstetteri*

1. Barleria acanthoides *Vahl*

Pubescent spiny undershrub to 1m high. Spines with short lateral spiny teeth. Leaves petiolate, oblong, elliptic or oblanceolate with a mucronate apex, 1.5–5cm long. Flowers in short, dense, 1-sided racemes; bracts similar to the spines; large calyx segments papery, ovate, entire apart from the spiny excurrent midrib and the occasional excurrent lateral nerve; corolla white (drying blue) with a cylindrical tube up to 9cm long and 5 equal lobes.

A night-flowering species which is rather variable in indumentum, some plants (e.g. Wood 2563) being sticky and very glandular.

A plant of open *Acacia-Commiphora* scrub on stony and rocky soils in dry areas; common in suitable habitats from 200 to 1500m in the Tihama and on the escarpment, being most common in the first foothills and in Taiz province; it also occurs in the east but very rarely and is only known there from Qa'ṭabah (Wissmann) and the Jawf (Zeller).

2. Barleria argentea *Balf. f.*

B. diffusa (Oliv.) Lindau

Unarmed undershrub with pubescent stems up to 50cm high. Leaves shortly petiolate, oblanceolate or oblong, acute, sparsely strigose, 1–3cm long. Flowers in small, shortly pedunculate, axillary cymes, sometimes reduced to a single flower; bracts linear-oblanceolate, large calyx segments linear-lanceolate, acuminate; corolla small, pale blue, funnel-shaped, 5-lobed, sometimes appearing 2-lipped, the tube c 1cm long.

Differs from other small blue-flowered species in the prominent bracts arranged immediately below the calyx.

A local plant of stony ground in *Acacia* scrub between 600 and 1800m: widespread in rainshadow areas of the central escarpment particularly around Ḥajjaylah; between Riḥāb and Waṣāb; Wadi Sukhnah; Al 'Ashshah; Suq al Jumli between Jifān and Rāziḥ; Wadi 'Azu near Majz (Steinberg).

3. Barleria bispinosa (*Forssk.*) *Vahl* Shakadh

B. spinicyma Nees

Pubescent spiny shrub up to 2½m high, though usually less. Spines without lateral teeth, similar to those of *B. prionitis* but shortly pubescent and sometimes much reduced in size in shade conditions. Leaves petiolate, obovate or elliptic, acute, 2–9cm long. Flowers in short, dense, 1-sided racemes, sometimes solitary; bracts a pair of spines; calyx segments oblong, obtuse, unarmed; corolla very pale blue, with a funnel-shaped tube 3–4cm long, 5-lobed with one lobe free to about half way down the tube.

Widespread and frequent on the escarpment between 500 and 2000m, growing on shallow rocky soils in open bushland. It is often the only shrub species of barren, crumbly, volcanic cliffs and is much more tolerant of dry conditions than *B. prionitis* with which it sometimes grows. It is endemic to southwest Arabia and has no obvious counterpart in Africa.

4. Barleria hillcoatiae *J.R.I. Wood*

Pubescent spiny undershrub to 50cm. Spines with lateral spinescent teeth. Leaves shortly petiolate, obovate with a mucronate apex and dense white marginal hairs, 1–2cm long. Flowers in short, dense, 1-sided racemes; bracts similar to the spines, densely white-villous; large calyx segments similar to the spines, lanceolate, acuminate, spine-tipped and with 1–2 marginal spines, white-villous; corolla white (drying blue) with cylindrical tube up to 12cm long and 5 equal lobes.

A night-flowering species.

Widespread and apparently frequent on open stony hills from 1800 to 2000m in the area bounded by Rada', Bayḍā' and Watadah. Endemic to southwest Arabia.

5. Barleria hochstetteri *Nees*

Unarmed pubescent undershrub to about 50cm. Leaves petiolate, oblong-elliptic, obtuse, sparsely strigose, 2–5cm long. Flowers in lax, long-pedunculate, 1–3-flowered, axillary cymes; bracts like minute leaves, remote from the flowers; large calyx segments elliptic with a cuneate base and a rounded, minutely mucronate apex; corolla small, pale blue, funnel-shaped, equally 5-lobed, the tube c 1cm long.

A rare plant of open stony ground in *Acacia-Commiphora* scrub in the first foothills of the escarpment between 300 and 500m: between Bājil and Khamīs Banī Sa'd; Khalīfah; Suq As Sabt, J. Bura'. There is also a record by Wissmann from J. Jinān.

6. Barleria lanceata (*Forssk.*) *C. Chr.*

B. triacantha Hochst. ex Nees
B. noctiflora sensu Schwartz

Pubescent spiny undershrub up to 30cm high rather like a diminutive version of *B. acanthoides* but differing in the following characters. Internodes very short so the plant has a dense bushy habit. Leaves obovate, up to 1cm long. Flowers solitary, axillary, subsessile; large calyx segments papery, narrowly elliptic, acuminate, the margin with short spreading spines; corolla with a tube up to 4cm long.

A very rare plant of gravel banks by seasonal watercourses in the east of the Tihama: Surdūd (Forsskal); Hays; Nukaybah.

Outside Yemen a rare plant of the Red Sea coast from Sudan to Somalia.

7. Barleria mucronifolia *Lindau*

Similar to *B. acanthoides* with which it has often been confused, but differing in a number of characters. The leaves are obovate to suborbicular, 1–1.5cm long. The inflorescence is similar to that of *B. acanthoides* but the bracts are lanceolate and much broader, commonly exceeding 3mm in width at their narrowest point, the outer calyx segments have a spiny margin and the corolla a tube 4–4.5cm long.

A rare but overlooked plant found in dry bushland at about 1000m in Taiz province: 20km west of Taiz (Kamal Ibrahim); Shūb, Maqātirah (Wood 958). There is also a specimen collected by Forsskal but no locality is known.

Outside Yemen it is a rare and ill-understood species known from a few localities in Ethiopia and Somalia.

8. Barleria orbicularis *Hochst. ex T. Anders.*

Identical to *B. hochstetteri*, except for the two large calyx segments which are ovate-cordate in shape.

An uncommon plant of dry stony hills between 200 and 1000m on the escarpment south of the Wadi Rima' and the adjacent stony Tihama steppe: Mishrāfah; between Mishrāfah and Al Masbaḥ, Waṣāb As Sāfal; between Nukaybah and Mafraq; Al Barḥ; Ḥajdah.

Outside Yemen it is a rare plant found in Somalia and Ethiopia.

9. Barleria parviflora *R. Br. ex T. Anders.*

B. scabriuscula O. Schwartz

Unarmed undershrub with decumbent or ascending strigose stems. Leaves sessile, oblong or elliptic, the apex acute or rounded, ± glabrous, 1–2.5cm long. Flowers solitary in the leaf axils, the peduncle 1–2cm long; bracts sepal-like, enclosing the calyx; large calyx segments broadly lanceolate, acute; corolla small, pale blue with a short tube c 7mm long and 5 ± equal lobes.

Locally abundant on open stony hillsides in dry country between 1000 and 2000m; rather local on the escarpment and mostly in the Taiz area, 'Anis and the Ḥarāz; widespread in the Mashriq from Damt and Bayḍā' north to Majz and often associated with *Barleria proxima*, *Chrysopogon plumulosus* and *Solanum cordatum*.

10. Barleria prionitis *L.* — Shakadh

B. induta sensu Schwartz
B. appressa Forssk. (Vahl)

Adpressed-pubescent, spiny undershrub up to 1m high. Spines without lateral teeth. Leaves petiolate, lanceolate (rarely oblong), acuminate, 2–12cm long. Flowers solitary, axillary, sessile; bracts a simple spine; calyx segments broadly lanceolate, long acuminate, unarmed; corolla peach-coloured with a funnel-shaped tube 3–4cm long, 5-lobed with one lobe free to about half way down the tube.

The Yemen plant is subspecies **appressa** (*Forssk.*) *Brummitt & J.R.I. Wood*.

Locally abundant on the escarpment from 600 to 2500m from the Ḥujarīyah north to Shahārah but completely absent from J. Sharafayn and Khawlan Ash Sham where it is replaced by *B. trispinosa*; it grows in open bushy places, particularly along wadis and around cultivation and is restricted to high rainfall areas, only descending below 1500m in the central escarpment.

B. prionitis is widespread in the Old World tropics but subspecies *appressa* is only known from Yemen.

11. Barleria proxima *Lindau*

B. prionitis sensu Schwartz p.p.
B. smithii Rendle

Sparsely pubescent spiny undershrub up to 30cm high rather like a diminutive version of *B. prionitis* except for the following characters. Leaves obovate-elliptic, mucronate, 1–1.5cm long. Flowers in the axils of the uppermost leaves only, forming a condensed head-like inflorescence; bracts a distinctly broadened spine; corolla orange with a tube 1.5–2cm long.

A locally abundant plant of open, rocky hills in dry country between 500 and 2300m; absent from the escarpment except for Taiz province where it is found from Maqbanah southwards, extending east to Taiz; widespread in the Mashriq from Bayḍā' northwards extending westwards to Rada', Ma'mar, Al 'Ashshah and Suq al Jumlī between Ṣa'dah and J. Rāziḥ.

12. Barleria trispinosa (*Forssk.*) *Vahl* — Shakadh

An adpressed pubescent spiny undershrub up to 1m high with ovate or elliptic, acuminate leaves. Similar to *B. prionitis* in general appearance but readily distinguished by the large ovate, acute, gland-dotted imbricate bracts.

A rare plant in the central escarpment where it is found in valley scrub at about 500m below Ḥadīyah, at Ḥillah on J. Bura' and above Walajah on J. Milḥān; further north in Khawlan Ash Sham and J. Sharafayn where *B. prionitis* is absent it extends up the mountains to 1500m and is much more common.

Outside southwest Arabia it is a rare plant of the Red Sea hills in Sudan, Ethiopia and Somalia.

13. **Barleria ventricosa** *Hochst. ex Nees*

Pubescent, decumbent or scrambling, unarmed woody herb. Leaves petiolate, ovate-elliptic, acute, 2–6cm long, sparsely to densely adpressed hairy. Flowers 1(–3) in subsessile axillary cymes, those in the upper leaves congested forming a dense terminal spike; bracts linear-oblanceolate; larger calyx segments broadly elliptic, often becoming scarious with age; corolla large, blue, funnel-shaped with a short cylindrical tube c 1cm long and 5 ± equal lobes.

In rough grassland and on bushy banks in high rainfall areas of the escarpment between 1800 and 2500m; found from Misrākh and J. Ḥabashī north to Shahārah but far more common in the Ibb region than elsewhere.

8. ASYSTASIA *Blume*

Annual or perennial, often woody herbs. Leaves entire. Flowers in terminal (rarely axillary) spikes or racemes; calyx with 5 nearly free segments; corolla white, funnel-shaped, with a wrinkled palate and 5 equal lobes; stamens 4, the anthers 2-celled. Capsule 4-seeded, the seeds rough, glabrous.

1.	Bracts minute. Flowers pedicellate	*A. gangetica*
	Bracts at least 5mm long, ± concealing the calyx. Flowers sessile	2
2.	Annual. Calyx c 5mm long	*A. schimperi*
	Woody perennial. Calyx c 10mm long	*A. guttata*

1. **Asystasia gangetica** (*L.*) *T. Anders.*

Sparsely pubescent perennial woody herb, decumbent or scrambling over scrub vegetation to 80cm. Leaves up to 10cm long. Flowers in lax axillary and terminal, 1-sided racemes, sometimes forming a panicle; bracts minute, subulate, distant from and much shorter than the 5mm long calyx; corolla white with violet markings on the palate.

Widespread on the escarpment between 500 and 1800m, growing in partially shaded scrub vegetation by ditches and wadis with at least a semi-permanent flow of water; often associated with *Commicarpus plumbagineus*.

2. **Asystasia guttata** (*Forssk.*) *Brummitt* Sumīrah

A. somalensis (Franch.) Lebrun & Toussaint
A. petalidioides Deflers
Ruellia guttata Forssk.

Sparsely pubescent decumbent perennial woody herb. Leaves obovate or elliptic, up to 5cm long. Flowers in short, dense, terminal bracteate spikes; bracts sparsely hairy, lanceolate, 1–1.5cm long, concealing the 10mm long calyx; corolla large, white, spotted with green.

Locally abundant on grassy banks around fields and by tracks in the wetter parts of Ibb and Taiz provinces between 1500 and 2500m. Unknown elsewhere in Arabia but occurring in Somalia.

3. **Asystasia schimperi** *T. Anders.*

Erect or decumbent, pubescent annual herb to 40cm. Leaves ovate to elliptic, up to 6cm long. Flowers in short, dense, terminal, bracteate spikes; bracts linear-elliptic 1–1.5cm long, concealing the 5mm long calyx and covered with many long, bristly white hairs; corolla white, spotted with green.

A rare weed of coffee plantations at 1500m in Wadi Sukhnah.

9. LEPIDAGATHIS *Willd.*

Undershrubs. Leaves entire. Inflorescence of congested, terminal spikes or heads; bracts leaf-like; calyx of 5 unequal, free sepals, all with spinescent points, the uppermost large, scarious, bract-like; corolla strongly 2-lipped; stamens 4, anthers 2-celled. Capsule 2- or 4-seeded, the seeds discoid, densely hairy.

Inflorescence ± globular; calyx pilose	*L. scariosa*
Inflorescence elongate: calyx sparsely hairy	*L. calycina*

1. **Lepidagathis calycina** *Nees*

Decumbent or ascending, sparsely pubescent, much-branched woody undershrub. Leaves lanceolate, up to 4cm long, minutely grey-hairy below, some of the hairs being stellate. Inflorescence terminal, elongate, 1-sided, up to 5cm long; uppermost sepal obovate with a spinescent, mucronate apex, pale green, quickly becoming scarious; corolla white.

A very rare plant found in open bushland on the limestone escarpment east of Al 'Ashshah at c 1300m.

Outside Yemen a rare plant known from Oman and Ethiopia.

2. **Lepidagathis scariosa** *Nees*

Ascending or erect, much-branched woody undershrub to 1m, the whole plant densely hairy with normal and stellate hairs. Leaves ovate-elliptic, up to 4cm long, dark green above, grey below. Inflorescence of terminal, globular heads up to 3cm in diameter; uppermost sepal oblong-obovate with a spinescent mucronate apex, scarious; corolla pale lilac.

Widespread on the escarpment between 400 and 1500m but rare in Taiz and Ibb provinces where it is only known from Kāhim (Botta) and Bīr Ghubāḥ, Shar'ab. It is a plant of scrub in areas of above average rainfall and is most frequent in the valley scrub communities of the central escarpment.

10. ECBOLIUM *Kurz*

Undershrubs. Leaves entire. Inflorescence of terminal and/ or axillary spikes, usually densely bracteate; calyx of 5 free, ± equal, linear-lanceolate lobes; corolla 2-lipped, the upper lip notched, the lower 3-lobed; stamens 2, anthers 2-celled, 1 cell above the other. Capsule 2-seeded, seeds flat, discoid, smooth or rough. Our species are night-flowering and the flowers fall in the morning.

Corolla green. Inflorescence with numerous ovate, imbricate bracts — *E. viride*
Corolla dull violet. Inflorescence without bracts — *E. gymnostachyum*

1. Ecbolium gymnostachyum (*Nees*) *Milne-Redhead*
E. parvibracteatum Rendle
Pubescent much-branched undershrub to 1m. Leaves ovate to elliptic, 3–8cm long. Flowers in terminal spikes 5–10cm long; bracts absent but minute bracteoles present; corolla dull violet with a narrow cylindrical tube 1.5–3cm long. Capsule c 2.5cm long, the seeds muricate on the margins.

A plant of disturbed shady places often near cultivation occurring on cliffs and steep slopes immediately above cultivated fields, on track sides and in waste places close to human activity; widespread on the escarpment from 500 to 2000m and most common in the coffee areas of the central escarpment and the middle-altitude wadis; also occurring rather unexpectedly in the southeast mountains where it has been recorded from J. Sawraq, Damt and Rada'.

2. Ecbolium viride (*Forssk.*) *Alston* — Maraydah
E. linneanum Kurz
Very similar to *E. gymnostachyum* but with a longer inflorescence (up to 15cm long), bearing numerous, large, ovate-mucronate, imbricate bracts and a remarkable green corolla, the colour being unique in the Yemen flora.

Locally abundant on field borders, waste ground and rough scrubby slopes near cultivation. Like *E. gymnostachyum* it is seldom found far from human activity and is most common on rock and gravel outcrops in the heavily cultivated eastern Tihama plain and along the wadis penetrating the escarpment. It is found from 200 to 1200m.

11. MEGALOCHLAMYS *Lindau*

1. Megalochlamys violacea (*Vahl*) Vollesen
Ecbolium anisacanthus (Schweinf.) C.B. Cl.
E. violaceum (Vahl) Hillcoat & J.R.I. Wood
Justicia violacea Vahl
Decumbent or ascending undershrub with densely puberulent stems. Leaves lanceolate-oblong, up to 4cm long, more richly green below than above. Flowers in axillary and terminal bracteate spikes 2–5cm long; bracts elliptic with an acuminate, spinescent apex and a ciliate margin; calyx with 5 equal lobes; corolla violet, the tube about 1cm long, shorter than the lobes, the upper lip entire; anthers 2-celled, 1 cell above the other. Capsule c 1cm long, 2-seeded, the seeds flat, smooth. Another night-flowering species.

Locally abundant on dry rocky plain and in open *Acacia* bushland from 400 to 1700m; widespread on the escarpment north to Al 'Ashshah but with a southern tendency and most common in the dry foothills from the Wadi Rima' southwards; also in the Mashriq on the Fardah pass.

12. ISOGLOSSA *Oerst.*

1. Isoglossa punctata (*Vahl*) *Brummitt & J.R.I. Wood*
I. oerstediana Lindau
Justicia punctata Vahl
Sparsely pubescent, trailing woody herb. Leaves ovate-acuminate, entire, up to 6cm long, petiolate except immediately below the inflorescence. Inflorescence of simple or branched, terminal, spike-like racemes; bracts lanceolate, very small; calyx with 5 equal, subulate segments; corolla 2-lipped, white with violet spots, c 12mm long; stamens 2, anthers 2-celled. Capsule c 15mm long, 4-seeded, the seeds discoid, hispid.

A very rare plant collected by Forsskal in the mountains of Ḥadīyah but only known to me from the damp, shaded gully, leading northwards from Ṣa'bah on J. Milḥān at c 1700m. Here it grows on steep stony banks in association with various uncommon ferns, *Desmodium repandum* and *Oplismenus compositus*.

13. JUSTICIA *L.*

Herbs or shrubs. Leaves entire. Flowers in axillary or terminal spikes or in axillary clusters formed from reduced cymes; bracts present; calyx with 5 linear-lanceolate segments; corolla 2-lipped, the upper lip 2-lobed, the lower lip usually 3-lobed, the mouth typically wrinkled; stamens 2, anthers 2-celled, one cell above the other and the lower one tailed. Capsule 4-seeded, the seeds ± muricate.

1. Shrub c 1½m high. Corolla dull red, the lower lip bent sharply back, obscurely 3-lobed — *J. anisotoides*
   Herbs or undershrubs, rarely reaching 1m. Corolla not as above — **2**
2. Flowers in 1-sided pedunculate axillary spikes; corolla white — *J. calyculata*
   Flowers in terminal spikes or sessile in the leaf axils; corolla never white — **3**
3. Flowers in terminal spikes — **4**
   Flowers in the leaf axils — **5**
4. Corolla blue — *J. caerulea*
   Corolla yellow — *J. flava*
5. Perennial undershrub. Corolla yellow — *J. odora*
   Herbaceous annual. Corolla pale purple — *J. heterocarpa*

1. Justicia anisotoides *J.R.I. Wood*
Small shrub with stiff, erect stems to 1½m, shortly but densely pubescent. Leaves obovate or oblong, rounded at the apex, 1–3cm long. Flowers in shortly pedunculate axillary cymes, often reduced to a single flower, the plant producing few flowers at any one time; bracts minute; corolla c 25mm long, brick red on the outside, pale green inside, the upper lip forming a minutely-notched continuation of the tube, the lower bent backwards and obscurely 3-lobed. Capsule 15mm long.

A rare plant occurring in small colonies on nondescript rocky slopes on the inner escarpment between 2300 and 2700m: on the east of the Qa' Bakīl, 'Anis; just below Khamīs Madyur, al Ḥaymatayn; by the waterfall, Maḥjur in Wadi Aḥjār.

Outside Yemen it is known only from Somalia.

2. Justicia caerulea *Forssk.*

J. nuda (Nees) O. Schwartz

Decumbent or ascending, pubescent woody herb. Leaves lanceolate or ovate, narrowed at both ends, 4–8cm long. Flowers in terminal interrupted spikes; bracts linear-oblanceolate, c 7mm long; corolla c 10mm long, blue. Capsule c 13mm long.

Widespread on the wetter parts of the escarpment from 1000 to 2400m, growing in partially shaded bushy places, particularly scrub-filled gullies between fields in the Ibb region and cliff gullies in the central escarpment.

3. Justicia calyculata *Deflers*

J. exigua S. Moore

J. leptostachya (Nees) O. Schwartz

J. anselliana sensu Schwartz

Slender, decumbent or erect, woody annual or perennial, the stem with two opposite lines of pubescence. Leaves narrowly ovate-elliptic, obtuse, 1–3.5cm long. Inflorescence of long pedunculate, 1-sided, 4-flowered axillary spikes; bracts minute; corolla c 6mm long, white, streaked with pale pink in the throat. Capsule c 5mm long.

Abundant on grassy banks around fields between 800 and 1800m in Taiz province, becoming steadily less frequent in similar habitats on the escarpment northwards but reaching Ḥajjah (Rathjens) and Ḥūth.

4. Justicia flava (*Vahl*) *Vahl*

J. sulcata (Vahl) Vahl

Pubescent woody herb with decumbent or erect, channelled stems. Leaves very variable, linear to ovate, obtuse, usually 1–3cm long, but sometimes much more. Inflorescence of terminal spikes, often interrupted below; bracts broadly lanceolate or oblong, c 7mm long; corolla 7–10mm long, yellow. Capsule 7–8mm long.

Very variable in leaf shape, some plants have linear leaves, particularly from the Taiz – Qā'idah area while others have very large ovate leaves 6cm or so long (the type of *J. sulcata*).

Widespread on the escarpment between 500 and 2500m, growing in grassland, on field borders and in open scrub but only common in areas of high rainfall or near irrigation ditches.

5. Justicia heterocarpa *T. Anders.*

Erect pubescent annual herb to 50cm. Leaves ovate or elliptic, acute, 2–5cm long. Flowers in axillary clusters; bracts elliptic, c 3mm long; corolla 3–4mm long, pale purple. Capsules 3–4mm long, the normal capsules often mixed with abnormal, fimbriate-margined, indehiscent, 1-seeded ones.

A local weed of coffee terraces and of disturbed grassy places, found between 1000 and 2000m on the escarpment from Misrākh north to Maḥābishah but only common in the coffee regions of the central escarpment.

6. Justicia odora (*Forssk.*) *Vahl*

Small, erect or ascending, woody undershrub with channelled, glabrous (rarely pubescent) stems, the whole plant pleasantly scented even when dry. Leaves oblong or elliptic, acute, usually glabrous, 0.5–1.5cm long. Flowers 1–3 together in the leaf axils; bracts minute; corolla yellow, c 10mm long; capsule c 10mm long.

Widespread between 600 and 2500m, growing in well-drained open scrub on the escarpment, high plateau and southeast mountains; although found in a great variety of dry rocky habitats it is most characteristic of rainshadow hills such as J. Sawraq and J. Minmār where it is associated with *Acacia etbaica*, *Psiadia punctulata*, *Clutia lanceolata*, *Lavandula* etc.

Justicia adhatoda *L.* is cultivated at Taiz and Ḥays and perhaps elsewhere in the Tihama. It is a shrub, 1–2m high, with large white flowers.

14. MONECHMA *Hochst.*

Similar to *Justicia* but the capsule 2-seeded with smooth, discoid seeds. Our species have conspicuous suborbicular bracts.

The two species below are usually regarded as conspecific but in Arabia and apparently also in northeast tropical Africa, they behave as quite distinct species. Final judgement on their status will have to wait until the genus is revised throughout Africa; it seems best to treat them as separate for the time being.

Erect, sparingly branched annual. Leaves longer than the internodes ... *M. bracteatum*

Decumbent, much-branched perennial. Leaves shorter than the internodes ... *M. debile*

1. Monechma bracteatum *Hochst.*

Erect, often unbranched, crisped-hairy annual to 50cm. Leaves lanceolate-oblong, acute, 4–6cm long, exceeding the internodes. Flowers in short dense axillary spikes; bracts petiolate, broadly ovate to sub-orbicular, often conspicuously white-ciliate on the margins; corolla dirty white, c 5mm long. Capsule c 5mm long.

A rare plant of gravelly volcanic rock slopes in open *Acacia* scrub between 900 and 1400m: Misrākh; 'Usil and Mawsanah in the Ḥarāz; Wadi Yūr.

2. Monechma debile (*Forssk.*) *Nees*

Justicia debilis (Forssk.) Vahl

Decumbent or ascending, crisped-hairy, woody perennial herb, much branched from the base. Leaves broadly lanceolate-oblong, acute or obtuse, 1–4cm long, shorter than the internodes. Flowers in short axillary spikes, though commonly reduced to 1–2 flowers; bracts sessile, broadly ovate to elliptic, the margin with white cottony hairs but scarcely ciliate; corolla purple c 10mm long; capsule 5–6mm long.

Locally abundant in stony grassland and on field borders between 1500 and 2000m from the Ḥujāriyah north to the Mudhaykhirah – Ta'kar range.

Apparently a local plant outside Yemen and perhaps restricted to hills on the African side of the Red Sea.

15. ANISOTES *Nees*

1. Anisotes trisulcus (*Forssk.*) *Nees* Mudhīdh, Mudh
Finely puberulent, erect shrub to 2m. Leaves ovate with a truncate base and obtuse apex, entire, yellow-green, 2–5cm long. Flowers in very shortly pedunculate, 2-flowered axillary clusters; bracts small, linear; calyx with 5 linear segments; corolla 3–4cm long, brick-red, divided nearly to the base into 2 lips, the upper lip notched, the lower much narrower and weakly 3-lobed; stamens 2, anthers 2-celled, one cell above the other. Capsule 4-seeded, the seeds rough.

Abundant on the stony hills from the first rocky outcrops in the Tihama at 100m up to about 1500m and perhaps the commonest shrub in degraded *Acacia-Commiphora* scrub, where it is often the dominant species. The popping of the ripe pods is one of the most characteristic sounds of the foothills of the Yemen mountains. Although abundant in tropical Arabia this species is known only from one area in Africa.

16. DICLIPTERA *Juss.*

Herbs or undershrubs. Leaves entire. Flowers in pairs between two large bracts which are appressed to each other, the flower pairs solitary in the leaf axils or in dense, shortly pedunculate axillary umbels; calyx of 5 equal, linear segments; corolla 2-lipped, mauve; stamens 2, the anthers 2-celled. Capsule ovoid, 4-seeded.

1. Flowers in subsessile axillary umbels. Herbs with leaves more than 2cm long — 2
   Flower pairs solitary in the leaf axils. Wiry undershrub with leaves c 1cm long — *D. suffruticosa*
2. Bracts linear-elliptic, less than 4mm wide — *D. verticillata*
   Bracts elliptic, up to 10mm wide — *D. foetida*

1. Dicliptera foetida (*Forssk.*) *Blatter*
D. bivalvis sensu Schwartz
D. zeylanica Nees
Herb. Leaves ovate, acuminate, sparsely hispid, 6–7cm long. Inflorescence of few-flowered axillary umbels (? always 1–3 flower pairs); bracts ovate or elliptic, sometimes acuminate, up to 15mm long and 10mm wide; corolla ? pink.

A mysterious plant found by Forsskal at Ḥadīyah and never re-collected in Yemen. It may yet be refound as the mountains of Ḥadīyah are very extensive.

2. Dicliptera suffruticosa *J.R.I. Wood*
Small, roughly pubescent undershrub with many ascending, deeply channelled, wiry branches arising from a woody rootstock. Leaves ovate to oblong, c 1cm long. Inflorescence of solitary, shortly pedunculate, axillary flower pairs enclosed in a pair of bracts; bracts elliptic, mucronate, c 10mm long, 6mm wide; corolla mauve, 8mm long.

Locally frequent on open rocky plain between 2000 and 2400m along the east of the high plateau from Damt north to Rada' and Sanabān, thence to Wa'lān, the Sana'a area and Khulāqah in Nihm. It is often associated with *Barleria parviflora* and *Crotalaria emarginella*.

Outside Yemen known only from the Somali highlands.

3. Dicliptera verticillata (*Forssk.*) *C. Chr.*
D. micranthes Nees
D. chinensis sensu Schwartz
Weedy decumbent herb with glabrous or sparsely pubescent 6-angled stems. Leaves ovate to elliptic, acuminate, 2–5cm long. Inflorescence of dense, many-flowered, subsessile, axillary umbels; bracts linear-elliptic, mucronate, 10–12mm long, 3–4mm wide, usually whitish-green in the lower half; corolla mauve, c 8mm long.

An uncommon plant of trampled wet places in areas of high rainfall on the escarpment between 500 and 1700m: Wadi Ḥidrān, Taiz; Wadi Sūdān, Qā'idah; between Ibb and Suq As Sabt; Wadi Zabīd (Forsskal); Waṣāb (Forsskal); Ḥadīyah; Ḥajjaylah and 'Usil in the Ḥarāz (Schweinfurth).

17. PERISTROPHE *Nees*

Similar to *Dicliptera*, from which it can be readily distinguished by the lax paniculate inflorescence formed of pedunculate axillary cymes. The 2 bracts of each pair are very unequal in size, the larger being linear-oblanceolate and nearly twice the size of the smaller one; each bract pair contains 2 flowers each with 2 linear bracteoles which resemble the calyx segments.

Longest bract and corolla both c 1cm long — *P. paniculata*
Longest bract 1.5–2cm long; corolla c 2.5cm long — *P. sp.A*

1. Peristrophe paniculata (*Forssk.*) *Brummitt*
P. bicalyculata (Retz.) Nees
Weedy, much-branched, probably annual herb to 1m, stems 6-angled, almost glabrous. Leaves broadly lanceolate, acuminate, up to 8cm long. Flowers in axillary cymes forming a diffuse panicle; longest bract linear, c 1cm long; calyx 3–4mm long; corolla mauve, strongly 2-lipped, c 1cm long.

Widespread and frequent between 200 and 2500m, growing in the east of the Tihama and on the escarpment in disturbed bushy places.

2. Peristrophe sp.A
Similar to the previous species but more robust in all parts and the axillary cymes fewer-flowered and more densely glandular. The longest bract is oblanceolate, rather than linear, 1.5–2cm long, the calyx c 6mm long and the corolla very large, reaching 2.5cm in length. Intermediates with the previous species have not been observed.

A rare plant known from rocky gullies from 2300–2500m but possibly overlooked: in 'Anis between Qa' Bakīl and Suq al Jum'ah (Wood 1959); Manākhah (Radcliffe-Smith and Henchie 4656).

18. HYPOESTES *Soland. ex R. Br.*

Herbs or undershrubs. Flowers variously arranged in the leaf axils but always borne singly between a pair of bracts; bracteoles present; calyx of 5 equal segments; corolla 2-lipped, the upper lip reflexed; stamens 2, 1-celled. Capsule 4-seeded.

Flowers in axillary spikes. Corolla white with violet markings *H. forskalei*
Flowers in small axillary clusters, often in 3s. Corolla pale pink *H. triflora*

1. Hypoestes forskalei (*Vahl*) *Soland. ex Roem. & Schult.* Sawrab
H. verticillaris auct.
Woody, yellow-green perennial herb, sometimes decumbent at the base and rooting at the nodes, the stems eventually erect, glabrescent, up to 80cm long. Flowers in short axillary spikes forming a terminal, ± paniculate inflorescence, the upper branches usually without their subtending leaves; bracts united in their lower halves, slightly exceeding the calyx; corolla white with violet markings.

Widespread and usually abundant between 800 and 3200m on the escarpment, high plateau and southeast mountains. It grows on all kinds of open rocky hillsides and is only less common at lower altitudes and in dry areas.

2. Hypoestes triflora (*Forssk.*) *Roem. & Schult.*
H. radicans Deflers
Decumbent, straggling dark green annual herb with sparsely pilose stems. Leaves elliptic, acuminate, up to 10cm long, the margins obscurely crenate. Flowers in axillary clusters of 1–5 (usually 3) on short lateral branches with reduced leaves; bracts free to the base, unequal in size, the largest c 1.5cm long, much exceeding the calyx; corolla pale pink.

A plant of damp shady places, particularly stream sides; widespread in suitable habitats on the escarpment from 1000 to 2900m; on the high plateau known only from Ḥaddah.

PEDALIACEAE

Annual or perennial herbs, usually easily recognized by the white dot-like glands, which become slimy when wetted and are found on the young leaves and stems. Leaves opposite, without stipules. Flowers bisexual with a tubular, often 2-lipped, corolla.

PEDALIUM *L.*

1. Pedalium murex *L.*
Erect, branched, slightly succulent annual herb to 50cm, the whole plant dotted with minute, white glands but otherwise glabrous. Leaves petiolate, obovate, or elliptic, coarsely toothed, 2–4cm long. Flowers solitary, axillary, shortly pedicellate; corolla yellow, 2–3cm long. Fruit a 4-angled, wrinkled capsule, armed at the base with 4 spreading spines, c 1.5cm long.

An uncommon but apparently increasing weed of roadsides, field borders and stony wadi beds found between 300 and 600m in the foothills of the northern escarpment: between Bājil and Khamīs Banī Sa'd; between Kadān and J. Milḥān; between Aṭ Ṭūr and Suq al Aman; Wadi 'Abs; Wadi Līyah.

Sesame, **Sesamum indicum** *L.*, know as 'Simsim' or 'Juljul' is commonly cultivated in the Tihama and in parts of the escarpment up to 1500m. It is an annual herb with a conspicuous 2-lipped, white corolla. The seeds are contained within a capsule and are commercially very valuable. They can be pressed to obtain sesame oil and the dried shells are mixed with sugar to make a sweet, sticky paste very popular throughout the Middle East.

BIGNONIACEAE

A family of trees and shrubs with opposite leaves without stipules. Sepals united, the calyx ± campanulate; corolla 5-lobed, large, showy, campanulate or funnel-shaped, sometimes 2-lipped. Fruit usually a capsule.

Tree. Leaves pinnate with 5–9 leaflets 2. *Stereospermum*
Shrub. Leaves ternate 1. *Rhigozom*

1. RHIGOZOM *Burch.*

1. Rhigozom somaliense *Hall. f.*
Glabrous shrub to 2½m with spine-tipped branchlets, resembling and often growing with *Commiphora spp.* Leaves in alternate clusters of 1–4, entire, ternate; the terminal leaflet obovate, usually emarginate, c 5mm long; the lateral leaflets smaller, ± oblong. Flowers solitary in the leaf axils; pedicel filiform; calyx, small, half as long as the corolla tube; corolla funnel-shaped, yellow c 2.5cm in diameter. Capsule grey-brown, flattened, c 2.5cm long.

Locally abundant between 100 and 1000m in *Commiphora* bushland in the stony steppe of the southern Tihama and the dry foothills of Waṣāb, J. Ra's and Maqbanah; not known north of Waṣāb.

Outside Yemen it is only known from northern Somalia.

2. STEREOSPERMUM *Cham.*

1. Stereospermum kunthianum *Cham.*
Tree to 9m with grey bark, flaking to expose pale patches. Leaves imparipinnate; leaflets 5–9, oblong-elliptic, c 5–10cm long, entire on adult trees but crenate-serrate on saplings. Flowers in large pendulous racemes, pale pink to lilac, fragrant, usually appearing when the tree is leafless. Fruit a persistent, dark brown cylindrical capsule c 40–60cm long.

A very rare plant restricted to high rainfall areas on the escarpment: a single tree at 600m above Suq As Sabt, J. Bura' (Wood 3286), several trees in Wadi Sukhnah at

1500m (Wood 3357). It has never been collected in flower in Yemen.

Two members of this family are commonly cultivated, especially in Taiz. **Tecoma stans** (*L.*) *H.B.K.* is a shrub with serrate, glabrous, pinnate leaves and racemes of large, yellow, funnel-shaped flowers. **Jacaranda mimosifolia** *D. Don* is a tree with bipinnate leaves and racemes of very attractive blue flowers.

LENTIBULARIACEAE

UTRICULARIA *L.*

1. Utricularia australis *R. Br.*
U. neglecta Lehm.
A free-floating insectivorous herb of still water. Stems slender, up to 25cm long, bearing green leaves pinnately divided into filiform segments and small bladders to trap water insects. Inflorescence a small leafless raceme of 2-lipped, spurred, yellow flowers exserted above the surface of the water but never observed in Yemen. Fruit a capsule.

Only known at 2700m from pools at the head of the Wadi Banā near Ribāṭ.

CAMPANULACEAE

Our species are herbs with alternate, simple leaves. Flowers bisexual, regular; sepals usually joined; petals joined, often forming a bell-shaped corolla. Fruit a capsule

Perennial. Flowers large; calyx with a reflexed appendage between the lobes — 2. *Campanula*
Annual. Flowers very small; calyx without a reflexed appendage between the lobes — 1. *Wahlenbergia*

1. WAHLENBERGIA *Schrad.*

1. Wahlenbergia hirsuta (*Edgw.*) *Thyn.*
Slender hispid annual with an erect, often zigzag stem to 10cm. Leaves shortly petiolate, oblanceolate to elliptic, crenate, 1–1.5cm long. Flowers in loose, axillary and terminal corymbs; corolla white, c 3mm long, about equalling the calyx.

Only known from thin, shallow grassland at 1800m on the lower slopes of J. Ba'dān, just north of Ibb.

2. CAMPANULA *L.*

1. Campanula edulis *Forssk.* 'Anwak
C. rigidipila Steud. & Hochst. ex A. Rich.
An attractive, variable perennial herb arising from a stout woody rootstock; stems decumbent, branched. Leaves sessile, oblong, dentate, 1–2cm long. Flowering stems much branched, the flowers solitary and terminal on the branches; calyx with a reflexed appendage between the lobes; corolla bell-shaped, blue or white, 1.5–2.5cm long. Capsule pendulous.

Several distinct races occur but though the extremes are very distinct all sorts of intermediate forms are found. The commonest form is a subglabrous plant with blue flowers about 1.5cm long (Wood 75/520). It is the prevalent form of the central escarpment and most areas below 2500m in areas of good rainfall. A densely hispid form (Wood 74/230) with a white corolla of similar size is common on the mountains in Taiz and Ibb province both at low and high altitudes but appears to be absent further north. At high altitudes of about 3000m and above large-flowered forms are found with a corolla 2–2.5cm long. These forms are usually moderately hispid and generally blue-flowered, particularly in the Ibb area (e.g. Wood 1118) but large white-flowered forms also occur (Wood 3070). In heavily grazed grassland dwarf forms are found with a minute, strikingly blue-purple corolla.

A plant terrace walls, rock outcrops, cliffs and disturbed grassland between 1500 and 3500m; widespread on the escarpment and high plateau and often abundant in high rainfall areas.

RUBIACEAE

Herbs, shrubs or trees with simple, entire, opposite or whorled leaves; stipules always present though sometimes leaf-like as in *Galium*. Flowers regular, usually bisexual; calyx tubular or with free teeth; petals united; fruit various but the calyx often persistent on the ovary as it enlarges.

1. Trees or shrubs, the stem woody — **2**
Herbs, sometimes with a woody base, but the stem always herbaceous — **7**
2. Undershrubs with white branchlets and linear leaves less than 2cm long — 11. *Neogaillonia*
Trees or shrubs. Branchlets not white, leaves not linear, usually exceeding 3cm long — **3**
3. Tree. Leaves often in whorls of 3. Flowers in a globose head — 1. *Breonadia*
Shrub. Leaves in opposite pairs. Flowers not in heads — **4**
4. Flowers all axillary — 8. *Canthium*
Flowers all terminal — **5**
5. Corolla tube at least 1cm long and usually much more — 9. *Pavetta*
Corolla tube less than 5mm long — **6**
6. Leaves glabrous. Flowers in a dense corymb — 7. *Tarenna*
Leaves pubescent below. Flowers in a panicle — 2. *Wendlandia*
7. Leaves in whorls of 4 or more — 13. *Galium*
Leaves in opposite pairs or whorls of 3 — **8**
8. Flowers sessile, axillary — **9**
Flowers pedunculate, axillary or terminal — **10**
9. Annual. Fruit a capsule — 12. *Spermacoce*
Perennial. Fruit of two mericarps — 10. *Anthospermum*
10. Inflorescence terminal — **11**
Inflorescence axillary — **12**
11. Flowers in a dense corymb. Corolla up to 3cm long — 3. *Pentas*
Flowers in a very lax terminal cyme. Corolla less than 5mm long — 6. *Kohautia*

12.	Flowers in pedunculate cymes	4. *Pentodon*
	Flowers 2–3 in axillary clusters	5. *Oldenlandia*

1. BREONADIA *Ridsdale*

1. Breonadia salicina (*Vahl*) *F.N. Hepper & J.R.I. Wood* Diraḥ, Dundul

Adina microcephala (Del.) Hiern.
Breonadia microcephala (Del.) Ridsdale

Tall, spreading tree to 20m. Leaves in opposite pairs or more commonly in 3s, dark green, glabrous, lanceolate to elliptic, up to 20 × 3cm, acute, very reminiscent of Oleander leaves. Flowers white, in dense axillary, globose heads on peduncles 5–6cm long, the peduncles with bracts at about the middle. Fruit a capsule.

Widespread and very characteristic as a riverine tree (almost always by permanent flowing water) along the whole escarpment from 500 to 1800m but less common in the south; also a characteristic tree of the coffee country on the west-facing slopes of the central escarpment where it is allowed to grow as tall scattered individuals among the coffee. It is very commonly pollarded as its new branches grow up very straight and are a valued source of good straight poles for house building.

2. WENDLANDIA *Bartl.*

1. Wendlandia arabica *Deflers* Sura'

Shrub to 4m. Leaves lanceolate, acuminate, up to 25cm long, pubescent on the main veins below, the undersurface much paler than the upper. Flowers very numerous in a large pubescent terminal panicle; corolla small, pale pink, the tube c 2mm long. The fruit has never been observed.

Flowers in spring.

A rare plant restricted to the Ḥujarīyah where it is found by ± permanent water in rocky gullies between 1000 and 2000m: frequent on all the cliff gullies in the Turbah region; Naqīl Jawjib between Aḥkūm and Qadas; Musalla, Aḥkūm; Wadi Rujaymah, 'Arūq; J. Ghayāmī (Deflers); J. Jāḥ.

Known only from a few other localities in South Arabia and from Somalia.

3. PENTAS *Benth.*

1. Pentas lanceolata (*Forssk.*) *Deflers*

Pubescent perennial herb or undershrub, woody and branched below with several erect stems to 60cm. Leaves lanceolate, up to 10cm long. Flowers in dense terminal corymbs; calyx lobes very unequal, some more than twice the size of others; corolla white, pink or mauve, the tube up to 3cm long. Fruit a capsule.

An abundant and very characteristic species of the escarpment from the Ḥujarīyah to Khawlan Ash Sham. It is most common in areas where winter mists are frequent.

4. PENTODON *Hochst.*

1. Pentodon pentandrus (*Schum.*) *Vatke*

Trailing, slightly fleshy, glabrous perennial herb. Leaves sessile, ovate, acuminate, up to 7cm long. Inflorescence a lax pedunculate axillary cyme; flowers pedicelled; corolla small, pale blue with a short tube c 1–2mm long. Fruit a capsule, the pedicel often recurved.

A local plant of streamsides between 500 and 1900m: quite common in suitable habitats on the west side of all the mountains of the central escarpment including J. Sharafayn; elsewhere only known from Wadi Ḥidrān near Taiz.

5. OLDENLANDIA *L.*

1. Oldenlandia corymbosa *L.*

Erect or decumbent, glabrous annual herb. Leaves linear, up to 2cm long. Flowers axillary, pedicellate, solitary or in pairs; corolla very small, white, c 1mm long. Fruit a capsule, the pedicel often slightly recurved.

Local but sometimes abundant, growing either as a field weed or in sandy or gravelly wadi beds from 500 to 1900m on the escarpment north to Maḥwīt. It is most frequent in high rainfall areas and is distinctly southern in its distribution.

6. KOHAUTIA *Cham. & Schlecht.*

Herbs with entire, linear leaves. Flowers in very lax terminal monochasial or dichasial cymes; corolla 4-lobed with a cylindrical tube. Fruit a globose or ellipsoid capsule crowned by the 4 persistent calyx lobes.

1.	Corolla tube up to 6mm long. Fruit scabrid. Annual weed	2
	Corolla tube 10–12mm long. Fruit glabrous. Plant usually perennial	*K. caespitosa*
2.	Corolla red. Fruit oblong	*K. coccinea*
	Corolla white. Fruit sub-globose	*K. aspera*

1. Kohautia aspera (*Roth.*) *Bremek.*

Oldenlandia strumosa (A. Rich.) Hiern.

Small, erect, scabrid annual to 30cm, usually much branched from the base. Leaves linear or linear-lanceolate, up to 4cm long. Corolla white with a tube about 5mm long, the lobes very narrow. Fruit sub-globose, scabrid.

A weed of stony fields found between 200 and 1600m mostly on the escarpment; quite frequent in Taiz province where it is often associated with *Cleome angustifolia* and *Striga asiatica*. Elsewhere rare and known only from Suq As Sabt near Ibb, between Riḥāb and Waṣāb, 'Usil (Schweinfurth), Wadi 'Ayn 'Alī near Ḥajjah, Al 'Ashshah and in the Tihama from the Wadi Rima' (Hepper) and from between Bayt al Faqīh and Sa'īd Miṣr.

2. Kohautia caespitosa *Schnizl.*

Oldenlandia schimperi (C. Presl.) T. Anders.

Erect perennial (rarely annual) herb, usually with a woody rootstock and many erect stems to 40cm, variable in indumentum, sometimes scabrid below (but never on the capsule or pedicels), sometimes shortly pubescent, sometimes ± glabrous. Leaves linear, up to 2cm long. Corolla white with a tube 10–12mm long, the lobes very narrow. Fruit sub-globose, glabrous.

Widely distributed in Yemen from sea level to 2400m; frequent in the Tihama on sand dunes and as a weed of sandy fields; rare on the escarpment, occurring most commonly in the sandstone country in the Maḥwit-Ṭawīlah region but also in the rainshadow areas between Riḥāb and Waṣāb and anomalously on J. Bura'; locally frequent in the Mashriq from Damt north to Majz but mostly in the sandstone areas around Rada' and Ṣa'dah.

3. **Kohautia coccinea** *Royle*

Very similar to *K. aspera* but usually unbranched, the corolla scarlet with a tube c 6mm long and rather broader lobes. Fruit oblong-ellipsoid, scabrid.

A common weed of sorghum fields between 1700 and 2500m in the Ibb area, unknown elsewhere in Yemen.

7. TARENNA *Gaertn.*

1. **Tarenna graveolens** (*S. Moore*) *Bremek.* Sar

Shrub to 3m with light brown bark. Leaves elliptic to oblanceolate, up to 10 × 3.5cm, leathery, glabrous, dull dark green, becoming blackish when dry. Flowers in dense terminal corymbs, fragrant; corolla cream, the tube c 5mm long, about equalling the lobes, glabrous outrside. Fruit a globose, 1-seeded drupe c 5mm in diameter, crowned with the persistent calyx lobes.

Dried specimens usually turn black.

With the exception of a fruiting specimen from above Ḥillah, J. Bura' (Schweinfurth 344), which has broad rather shiny leaves, all specimens from Yemen probably belong to subspecies **arabica** (*Cuf.*) *Bridson*, though the dimensions of some specimens (e.g. Wood 789) approximate to those of subspecies *graveolens*.

Widespread between 500 and 1500m on the escarpment north to J. Sharafayn but only common at low altitudes on the western slopes of the central escarpment where it is a characteristic constituent of the valley scrub communities. In dry country it seems to prefer limestone, for example occurring in the Ḥujarīyah only on the limestone of J. Jirdad, Bani 'Umar.

8. CANTHIUM *Lam.*

Shrubs. Stipules triangular. Inflorescence an axillary cyme reduced in our species to a shortly pedunculate umbel. Fruit a bilobed drupe.

Inflorescence 1–5-flowered. Pedicels 1–2mm long. Leaves dull green c 7–10mm wide ... *C. phyllanthoideum*
Inflorescence 5–50-flowered. Pedicels c 1cm long. Leaves shiny c 2.5–3.5cm wide ... *C. schimperianum*

1. **Canthium phyllanthoideum** *Baill.*

C. bogosensis (Mart.) Penzig

Small shrub to 1½m. Leaves dull green, rather leathery, narrowly elliptic, acute, mostly 3–4cm × 7–10mm. Flowers cream, usually 3–5 in an umbel; pedicels very short, 1–2mm long, glabrous, becoming recurved in fruit; corolla tube c 4mm long.

Flowers in March and April, but only for about a week. For the rest of the year it is extremely nondescript and inconspicuous and it is likely that it is more common than my field records suggest.

Known from a few localities between 1500 and 2000m on the escarpment, usually growing on open very rocky hillsides with rather sparse vegetation: several localities in the Ḥujarīyah (Bani Ali, 'Abūs; 'Arūq; Rubaysah near Turbah); J. Sawraq; near Rawnah and Najdayn, Shar'ab; 'Udayn; between Ibb and Suq As Sabt; between Dan and Suq Ar Rubū', Waṣāb Al'Ālī; Maḥwīt (Miller); Maḥābishah (Wood 75/91, 1031).

2. **Canthium schimperianum** *A. Rich.* Shaqas

Plectronia schimperiana (A. Rich.) Vatke

In Yemen a shrub to 3m, but in Africa often becoming a tree with dark grey bark. Leaves shiny green, ovate-elliptic, acute, c 5–7 × 2.5–3.5cm. Flowers yellow-green, fragrant, c 5–50, in a shortly pedunculate or sessile umbel, many failing to mature, pedicels c 1cm long, pubescent; corolla tube c 5mm long.

Flowers in spring.

Local on the outer escarpment from 900 to 2200m; frequent in the Ibb-'Udayn region and the central escarpment including Maḥwīt and Maḥābishah; also in 'Utmah, and on J. Razih and J. Marran in Khawlan Ash Sham. Although sometimes an important constituent of *Acacia* scrub on open hillsides, notably just west of Mirkiz 'Utmah, it is more usually found in small quantity in shaded scrub-filled gullies.

9. PAVETTA *L.*

Small shrubs up to 1½m high. Leaves dark green, leathery, glabrous or hairy below. Flowers in many-flowered corymbs which are terminal on short lateral branches, usually very fragrant; corolla cream with a tube 1–2.5cm long. Fruit a globose drupe, crowned with the persistent calyx lobes, 6–8mm in diameter.

Pavetta **is one of the most complex genera in the Yemen flora and the present treatment can only be regarded as provisional. All species are superficially very similar. In the following account much reliance is placed on the size and shape of the calyx lobes as this is a useful character in African Pavettas. Other characters used are indumentum, leaf shape and pedicel length but these seem less reliable. Nearly all Yemen plants fall into the very variable *P. longiflora*. All are known as 'Shuwāf' or 'Bunān'.**

Pavettas mostly flower in April and May following the spring rains, but also flower erratically at other seasons. They are widespread on the escarpment and the south-eastern mountains but rather rare on the high plateau. They become less frequent northwards. They grow from 500 to 2600m in cliff gullies, bushy places around cultivation, valley scrub and *Euphorbia* scrub but appear to be largely absent from *Acacia* bushland.

1. Calyx lobes linear, lanceolate or oblong, scarcely widened at the base, more than 2mm long 2
   Calyx lobes triangular, distinctly widened at the base, less than 1.5(–2)mm long *P. longiflora*
2. Calyx lobes linear, less than 0.5mm wide. Petiole less than 5mm long *P. villosa*
   Calyx lobes lanceolate-oblong, at least 0.75mm wide. Petiole more than 1cm long 3
3. Leaves pubescent below. Calyx lobes 4–5mm long *P. sp.A*
   Leaves glabrous below. Calyx lobes 2–2.5mm long *P. sp.B*

1. Pavetta longiflora *Vahl*

P. arabica Bremek.
P. deflersii Bremek.
Leaves mostly elliptic or oblong, up to 12 × 5cm, glabrous, pubescent or tomentellous but never pilose on the undersurface. Inflorescence usually dense, the pedicels up to 1cm long, but commonly much less; calyx with very short triangular teeth up to 1.5(–2)mm long; corolla tube glabrous or pubescent, 1.5–2.5mm long.

A very variable plant divisable into two subspecies:

Subspecies **longiflora** (*P. arabica* Bremek.) has narrow, glabrous leaves (up to 12 × 3.5cm), a relatively lax inflorescence with pedicels 4–11mm long and the corolla and calyx glabrous or nearly so. It is a rare plant known only from Taiz province and neighbouring parts of South Yemen and within our area is only known certainly from between 1300 and 1800m at Taiz (Forsskal; Deflers 675; Hepper and Wood 5825) and on the Turbah cliffs (Wood 3413), but it doubtless occurs elsewhere in that region.

Subspecies **A** (*P. deflersii* Bremek ?) is especially variable, but can be separated into two intergrading varieties. The first of these (Wood 2318, 3224, 3289) is very widespread on the escarpment growing from 500 to 2500m but is most common in high rainfall areas. It has broad, ± elliptic leaves (2.5–5cm wide) which are glabrous or pubescent below. The corolla is hairy with a tube 2–2.5cm long. The second variety (Wood 3218, Radcliffe and Henchie 4760) is a local plant found mainly around the montane plains. It has narrow, ± oblong leaves (0.5–2.5cm wide) which are shortly tomentose below. The corolla is glabrous with a tube less than 2cm long. Subspecies *A.* is not definitely known outside southwest Arabia but is similar to *P. venenata* Hutch. and E. A. Bruce from Somalia which differs only in the shape of its stipules.

2. Pavetta villosa *Vahl*

Leaves with short petioles less than 5mm long, elliptic, large, up to 14 × 4.5cm, densely and coarsely pilose on the lower surface. Inflorescence rather dense, the pedicels 3–6mm long; calyx with narrow linear lobes, 2–6mm long, coarsely pubescent; corolla tube glabrous, c 2cm long.

A rare plant found between 1600 and 2600m in high rainfall areas: Kusmah (Forsskal): below Jirgir above Ḥadīyah (Wood 3299); J. Ba'dān (Wood 3405). It is probably a relic of highland forest where it occurs in Somalia.

3. Pavetta sp.A

Leaves long petiolate, elliptic, up to 13 × 5.3cm, densely pubescent below. Inflorescence rather lax with pedicels 10–18mm long; calyx lobes oblong-lanceolate, 4–5 × 1–1.5mm, pubescent; corolla not seen.

A very rare plant known from a single collection (Radcliffe-Smith and Henchie 4902) made at c 600m in the wadi above Suq Ar Ribāṭ, J. Raymah.

4. Pavetta sp.B

Leaves long petiolate, elliptic, up to 10 × 5.5cm, glabrous. Inflorescence rather lax with pedicels 9–13mm long; calyx lobes oblong-lanceolate, 2–2.5 × 0.75–1mm sparsely pubescent; corolla with a sparsely pubescent tube c 11mm long and rather broad lobes 3mm wide.

Only known from sandstone cliff gullies to the south of Wadi Ḍahr at 2300m (Wood 3440).

10. ANTHOSPERMUM *L.*

1. Anthospermum herbaceum *L.f.*

A. muriculatum Hochst. ex A. Rich.
Low, perennial, dioecious or monoecious herb with many decumbent or ascending branches arising from a woody base; stems square or nearly so, scabrid-puberulent when young, the epidermis flaking when old. Leaves often whorled, subsessile, lanceolate, up to 2.5cm long. Flowers minute, greenish, in the leaf axis. Fruit of 2 mericarps.

A characteristic plant of montane grassland and cliffs in good rainfall areas on the escarpment; found on all the higher mountains from 2500 to 3200m, but descending to 1600m in the mist regions of the central escarpment where it is common in the meadow communities.

11. NEOGAILLONIA *Lincz.*

Undershrubs with white, scabrid stems. Leaves sessile, small. Flowers small, tubular. Fruit a nutlet.

Leaves up to 2cm long. Calyx enlarging in fruit *N. calycoptera*
Leaves up to 1cm long. Calyx not enlarging in fruit *N. sp.A*

1. Neogaillonia calycoptera (*Decne.*) *Puff*

Pterogaillonia calycoptera (Decne.) Lincz.
Gaillonia calycoptera (Decne.) Lincz.
Erect undershrub to 50cm. Leaves linear, up to 2cm long. Inflorescence of branched terminal spikes; flowers surrounded by an involucre formed of conspicuous papery bracts, sessile, rather distant, solitary or in small clusters; corolla white with a long tube, 5–6mm long; calyx with two prominent lobes, enlarging in fruit and reaching c 1cm in length.

A rare plant of exposed rocky hills known from a few scattered localities between 2000 and 2300m in the Mashriq and on the high plateau: J. Nuqum (Rathjens); J. Shamsān near Wadi Ḍahr (Lavranos); Majz (Steinberg).

2. Neogaillonia sp.A

Erect undershrub to 50cm. Leaves linear or oblanceolate, up to 1cm long. Flowers sessile in the leaf axils, solitary; corolla white with a tube c 2mm long; calyx with 4–5 lobes, not enlarging in fruit.

This little-known plant needs re-collecting. Its exact identity is not known but it is close to, if not identical with, the Socotran endemic species *Neogaillonia tinctoria* (Balf. f.) Lincz.

Only known from J. Ra's where there is a small colony on the south side at 1000m (Miller and Long 3502).

12. SPERMACOCE *L.*

1. Spermacoce sphaerostigma *(A. Rich.) Oliv.*

Arbulocarpus sphaerostigma (A. Rich.) Tennant

Erect, sparingly branched, scabrid-pubescent annual herb to 20cm. Leaves sessile, linear or lanceolate, up to 5cm long. Flowers sessile in dense axillary clusters in the enlarged stipular sheath; corolla small, white or mauve, the tube about 5mm long. Fruit a capsule crowned by the persistent calyx, dehiscing from below.

A local plant found in high rainfall areas on the escarpment between 1400 and 2500m growing as a field weed, on banks by fields and in open places in grassland. It is found from the Turbah region north to the Wadi Lā'ah but has a strong southern tendency, being particularly common in the Ibb-'Udayn region.

13. GALIUM *L.*

A distinctive genus of annual or perennial herbs usually armed with stiff retrorse hairs so that the plants cling to clothing. Leaves in whorls of 4–10. Flowers in small axillary or terminal pedunculate cymes; calyx minute; corolla small, 4-lobed, with a short tube. Fruit of 2 1-seeded mericarps joined together.

1. Perennial. Ovary and fruit long-white-pilose — *G. kahelianum*
   Annual. Fruit bristly or granulate, not pilose — 2
2. Flowers terminal on small lateral branches. Corolla purple-black — *G. aparinoides*
   Flowers axillary. Corolla white — 3
3. Fruiting pedicels recurved. Fruit granulate — *G. tricornutum*
   Fruiting pedicels straight. Fruit bristly — *G. spurium*

1. Galium aparinoides *Forssk.* Malasi'

G. aparine var. hamatum sensu Schwartz p.p.

Decumbent annual often drying black, stems long, retrorse-barbed, clinging. Leaves in whorls of 4 or 6, oblanceolate, mucronate, 9–14mm long. Flowers terminal on small lateral branches; corolla purple-black. Fruit blackish, covered in short, white, hooked bristles.

A local plant of stream sides and cliff gullies found between 2500 and 3200m on the higher escarpment mountains from J. Ṣabir north to J. Maswār and J. Mada'.

2. Galium kahelianum *Deflers*

G. yemense auct.

Prostrate perennial with a woody rootstock and short, scabrid (not retrorse-barbed) stems up to c 15cm long. Leaves in whorls of 5, linear, acute but not mucronate, 2–6mm long. Flowers in axillary cymes, forming a small leafy panicle; corolla white. Ovary and fruit densely white-pilose.

A plant of terrace walls and cliffs occurring in scattered localities near villages between 1500 and 3000m on the escarpment from Ribāṭ (near Yarīm) and J. Ḥiṣn Madnān in Waṣāb al 'Ālī north to Shahārah. Endemic to Yemen.

3. Galium spurium *L.*

G. aparine L. var. *hamatum* sensu Schwartz p.p.

Decumbent annual, stems long, retrorse-barbed, clinging. Leaves in whorls of 6(–8), linear-oblanceolate, mucronate, 6–10mm long. Flowers axillary, usually solitary, long pedicellate; corolla white. Fruit covered in short, white, hooked bristles.

In Yemen there is only subspecies **africanum** *Verdc.*

A weed of earth banks around cultivation and in open scrub, found between 1200 and 2700m in the central escarpment and in the area around Ibb.

4. Galium tricornutum *Dandy*

Decumbent annual, stems rather short, up to 35cm long, retrorse-barbed, clinging. Leaves in whorls of 6–8, linear-lanceolate, mucronate, 12–18mm long. Flowers in 1–3-flowered axillary cymes; corolla white. Fruit granulate, the fruiting pedicels strongly recurved.

A weed of wheat and barley fields on the higher escarpment mountains from J. Ṣabir north to J. Maswar where it reaches 3100m but descending as low as 2000m at Maḥwīt and elsewhere in the central escarpment. It also occurs rarely in fields by streams on the high plateau at Baw'ān, Wa'lān and Yarīm.

Two members of the *Rubiaceae* have been cultivated in Yemen.

Madder, **Rubia tinctorum** *L.*, 'Fuwah', was formerly cultivated for the red dye that can be obtained from its roots but there are no records of its cultivation this century.

Coffee, **Coffea arabica** *L.*, known in Yemen as 'Bun', is the best-known agricultural product of Yemen. The coffee tree is not indigenous but was introduced from Ethiopia, possibly in the fifteenth century as it is not mentioned in Arabic literature at an earlier date. In the sixteenth century Yemen became a province of the Turkish Empire and it was from Yemen that the habit of coffee drinking spread first to Constantinople and thence to the rest of the world. Coffee exports reached their peak at the beginning of the eighteenth century when coffee was exported through the port of Mukhā to the rest of the world. The town of Mukhā, where no coffee has ever in fact been grown, has in this way given its name to coffee in many languages of the world.

Since the beginning of the eighteenth century Yemen coffee production has steadily declined though it is still an important crop. Coffee is grown along the whole escarpment between 700 and 2600m, but large-scale production is only found in two areas. The most important of these is the central escarpment where coffee is grown on terraces on the steep west-facing, mist-covered slopes between 800 and 1800m. Three places in this area are outstanding for their coffee cultivation: the Jāfarīyah region of J. Raymah, the slopes above Ḥillah on J. Bura' and J. Sāfān in the Ḥarāz. The other area of significant coffee cultivation is in the middle-altitude wadis such as Ḍabāb near Taiz, the 'Udayn wadis, Wadi Suknah and Wadi Lā'ah. Here there is no mist but the coffee is irrigated for at least part of the year. 'Udayn is much the largest producer in this area. Nowadays the beans are mostly exported through Hodeida, but the husks ('Qishr') are sold within Yemen and made into a hot drink which is flavoured with ginger.

The coffee tree is about 3m high with shiny, dark green, ovate leaves and axillary clusters of fragrant white flowers which appear in the spring. The berry is red when ripe.

DIPSACACEAE

Annual or perennial herbs. Leaves opposite, without stipules. Flowers in dense heads subtended by involucral bracts. In our species the individual flowers are bisexual and consist of a small epicalyx of fused bracteoles, a small calyx which in fruit develops long awn-like hairs or bristles and a tubular corolla with 5 equal lobes. Fruit an achene enclosed in the epicalyx.

Flowers pink. Awns of fruiting calyx plumose 2. *Pterocephalus*
Flowers blue. Awns of fruiting calyx glabrous 1. *Scabiosa*

1. SCABIOSA *L.*

1. Scabiosa columbaria *L.*
Scabious. Sparsely pubescent perennial herb with a thick rootstock, much branched at the base with many sparingly branched flowering stems to 50cm. Leaves oblanceolate or obovate, deeply pinnatifid or (rarely) entire, up to 5cm long, becoming more finely divided upwards. Flower-heads 1.5–2.5cm diameter on long slender peduncles; calyx terminating in 5 black glabrous bristles, corolla pale blue, the outer flowers much larger than the inner.

A characteristic escarpment plant which is widespread and frequent on cliffs, stony slopes and in mountain grassland between 2000 and 3200m from Turbah north to J. Razih.

2. PTEROCEPHALUS *Adans.*

Undershrubs. Fruiting calyx with plumose awns. Flowers pink.

Trailing undershrub. Leaves up to 2cm long. Corolla lobes ± 5mm long *P. frutescens*
Erect undershrub. Leaves up to 5cm long. Corolla lobes ± 1mm long *P. pulverulentus*

1. Pterocephalus frutescens *Hochst.*
Prostrate trailing undershrub with finely puberulent stems. Leaves linear–oblanceolate, entire, mostly 1–1.5cm long. Flower-heads c 2.5cm diameter on peduncles 1–6cm long; calyx with c 12–25 dull pink plumose awns; corolla pink, the lobes elliptic, c 5mm long.

In East Africa this plant is very variable, particularly in habit and leaf toothing. The Yemen plant, variety **pumila** (*Deflers*) *O. Schwartz*, however is quite constant and is always prostrate and with entire leaves.

A characteristic plant of the mountains around the high plateau, growing from 2100 to 3300m on open rocky hillsides, often in rock crevices. It extends west and south to J. Sumārah, Maḥwit and J. Maswār and is noticeably more frequent on sandstone than on other rocks.

2. Pterocephalus pulverulentus *Boiss. & Bl.*
Undershrub, much branched below with numerous erect puberulent stems to 50cm. Leaves linear, entire or pinnatifid with small acute, distant segments, 2–5cm long. Flower-heads c 2cm broad on peduncles 12–22cm long; calyx with c 12–15 dull pink plumose awns; corolla pink, the lobes triangular, ± 1mm long.

A very rare plant collected twice on roadsides and stony hills at 2300m between Sana'a and Wa'lān (Wood 589, 2279). An Irano-Turanian species known from Syria south to Yemen.

COMPOSITAE

A large family of herbs and shrubs. Leaves alternate or opposite, without stipules. Inflorescence a head of numerous small sessile flowers (florets) surrounded by an involucre of bracts, the whole resembling an individual flower. The involucral bracts are arranged in one or more rows and are often imbricate. There are 2 distinctive types of floret: a regular floret with a tubular corolla and 5 equal apical lobes and an irregular floret with a spreading 3 or 5 toothed ligule at the top of the tube. The florets may be unisexual or bisexual. In the following account the regular florets are called tubular florets and the irregular florets ligulate florets, the ligules being referred to as rays. The florets are attached to a receptacle which may also bear scales and these are found in some species between the florets. The fruit is an achene which often bears hairs, scales or bristles at the apex; these are modified sepals and known as pappus. They are used to disperse the seeds and are of considerable taxonomic importance. The flower-heads are usually bisexual but are sometimes unisexual; they are massed together into various types of compound inflorescence but are typically corymbose.

The individual composite species in Yemen are mostly clearcut and present few taxonomic problems. Exceptions include *Conyza*, *Volutaria*, *Crepis* and *Pulicaria*, but they are not significant. In contrast the generic boundaries are far from settled. In particular there has been no universally accepted division of the genera around *Conyza* (Nos 8–11), *Helichrysum* (Nos 15–18) and *Senecio* (Nos 42-45) and almost every flora has the species in these groups differently arranged. Although the present arrangement is the most up-to-date available, I very much doubt it will be the final one. The difficulties of dividing the genera in these areas needs to be borne in mind when using the key and it may sometimes be necessary to follow both dichotomies.

All the usual geographical elements found in the Yemen flora are present in the Compositae but two are outstanding. The family is particularly rich in introduced weeds, most of them originating in America. *Xanthium spinosum*, *Tridax procumbens*, *Bidens biternata*, *Flaveria trinervia*, *Tagetes minuta* have all become locally abundant and in some areas dominant over considerable areas of disturbed ground. Other species are of more recent introduction and though at present rather rare may eventually become as common. The second outstanding feature is the large number of endemic species. These are scattered over many genera but *Vernonia bottae*, *Helichrysum arwae*, *Centaurothamnus maximus* and *Cichorium bottae* are not only endemic but extremely attractive in appearance.

1. Achenes entirely enclosed in a capsule-like casing which is covered in hooked spines **2**
Achenes not enclosed in a spiny capsule-like casing (but cf. *Sclerocarpus* where the casing is not spiny) **3**
2. Leaves alternate, palmately 3–5-lobed 25. *Xanthium*
Leaves opposite, unlobed 24. *Acanthospermum*
3. Flower-heads with ligulate florets only. Herbs with milky latex and yellow or blue (*Cichorium* and *Lactuca* only) flower-heads **4**
Flower-heads with tubular florets; ligulate florets present or absent. Herbs or shrubs without milky latex. Flower-heads variously coloured **15**
4. Pappus of small scales. Mat-forming plant with large blue flower-heads 59. *Cichorium*
Pappus of hairs. Flower-heads yellow or if blue, very small **5**
5. Involucral bracts 8, all similar and arranged in one row 68. *Urospermum*
Involucral bracts numerous, in at least two rows, the outer differing from the inner **6**
6. Flowers borne on a simple, leafless, hollow peduncle 63. *Taraxacum*
Flowering stem branched or with leaves or scales **7**
7. Pappus plumose **8**
Pappus simple **9**
8. Plant glabrous. Achenes beakless 69. *Scorzonera*
Plant ± bristly. Achenes beaked 61. *Picris*
9. Achenes beaked, the pappus distinctly stalked **10**
Achenes beakless, the pappus sessile on the achene **12**
10. Achenes flattened. Flower-heads blue or yellow 64. *Lactuca*
Achenes ± cylindrical. Flower-heads yellow **11**
11. Flower-heads in ± sessile clusters along the inflorescence branches. Involucre glabrous 67. *Launaea*
Flower-heads pedunculate in a corymbose inflorescence. Involucre hairy 62. *Crepis*
12. Involucre completely glabrous **13**
Involucre hairy or at least cottony towards the base **14**
13. Flower-heads mostly less than 1cm broad, usually with a number of scales immediately below the involucre 67. *Launaea*
Flower-heads 1–2cm broad, without scales below the involucre 65. *Reichardia*
14. Achenes flattened. Pappus hairs cottony, numerous 66. *Sonchus*
Achenes ± cylindrical. Pappus hairs 5–10, bristly 60. *Tolpis*
15. Leaves spiny **16**
Leaves unarmed **19**
16. Inflorescence a globular head formed of numerous 1-flowered flower-heads 50. *Echinops*
Inflorescence not a globular head; flower-heads with many florets **17**
17. Stems with spiny wings formed from the decurrent leaf bases. Flower-heads purple **18**
Stems without spiny wings. Flower-heads whitish-yellow 51. *Atractylis*
18. Pappus plumose. Involucral bracts all erect 52. *Cirsium*
Pappus simple. Lowermost involucral bracts reflexed 53. *Onopordum*
19. Involucral bracts terminating in a spine **20**
Involucral bracts not terminating in a spine **21**
20. Bracts with at least 1 pair of small lateral spines at the base of the central spine. Flower-heads all terminal 55. *Centaurea*
Bracts with a single spine. Flower-heads lateral as well as terminal 57. *Dicoma*
21. Leaves all in a basal rosette. Flowering stem entirely leafless, unbranched 58. *Piloselloides*
Leaves not all in a basal rosette. Flowering stem with one or more leaves, usually branched **22**
22. Leaves or some of them opposite **23**
Leaves all alternate **39**
23. Leaves pinnate **24**
Leaves simple though sometimes deeply incised **25**
24. Involucral bracts fused, forming a tubular sheath round the achene. Pappus not retrorse-barbed 39. *Tagetes*
Involucral bracts not fused to form a sheath round the achenes. Pappus retrorse-barbed 35. *Bidens*
25. Stems winged. Involucral bracts with spreading spinescent tips 23. *Geigeria*
Stems unwinged. Involucral bracts without spreading spinescent tips **26**
26. Achenes each enclosed in a woody capsule-like casing. Involucral bracts leaf-like, dentate 28. *Sclerocarpus*
Achenes not enclosed in a woody casing. Involucral bracts rarely leaf-like, always entire **27**
27. Outer involucral bracts spathulate, spreading, covered in large stalked glands 26. *Sigesbeckia*
Involucral bracts not as above **28**
28. Inflorescence a head of 1-flowered flower-heads, each enclosed in a tubular involucre formed of the fused bracts 22. *Lagascea*
Inflorescence not as above **29**
29. Ray florets present **30**
Ray florets absent **37**

30. Pappus of plumose hairs. Flowers solitary on peduncles 10–20cm long 32. *Tridax*
Pappus, if present, not of plumose hairs. Peduncles less than 10cm long **31**
31. Ray florets white. Leaves less than 5cm long **32**
Ray florets yellow or white, but if white, leaves more than 5cm long **33**
32. Pappus present. Erect herb with corymbose inflorescence 37. *Galinsoga*
Pappus usually absent. Prostrate herb with single pedunculate flower-heads from the leaf axils 27. *Eclipta*
33. Leaves densely white-hairy below. Upper leaves alternate 33. *Verbesina*
Leaves not white-hairy below. Upper leaves all opposite **34**
34. Pappus absent. Leaves lanceolate to oblong 34. *Guizotia*
Pappus present (though quickly deciduous). Leaves usually ovate, rarely lanceolate or oblong **35**
35. Achenes glabrous. Flower-heads large, more than 2cm in diameter 31. *Melanthera*
Achenes hairy. Flower-heads less than 2cm in diameter **36**
36. Achenes sharply 3-angled. Scales found between the florets deeply and sharply toothed 29. *Blainvillea*
Achenes flattened, weakly 2-angled, scales found between the florets shallowly and obscurely toothed 30. *Aspilia*
37. Inflorescence of dense clusters of yellow flower-heads. Pappus absent 38. *Flaveria*
Flower-heads not densely clustered, not yellow. Pappus present **38**
38. Florets white. Achenes glandular, sticky 2. *Adenostemma*
Florets blue. Achenes ± glabrous, not sticky 3. *Ageratum*
39. Dioecious shrub 1–4m high, the leaves strongly aromatic, white-tomentose below 7. *Tarconanthus*
Plant never dioecious and without the above combination of characters **40**
40. Inflorescence of small pedunculate globular heads of flower-heads, c 1cm in diameter. Shrub 1–3m high 12. *Blepharispermum*
Inflorescence of flower-heads variously arranged but never aggregated as above **41**
41. Pappus absent. Florets never purple **42**
Pappus present. Florets variously coloured **48**
42. Ray florets present (rather inconspicuously so in *Chrysanthellum*) **43**
Ray florets absent **47**
43. Leaves ± entire. Achenes curved or coiled 48. *Calendula*
Leaves toothed or pinnatifid **44**
44. Rays white 40. *Anthemis*
Rays yellow **45**
45. Leaves coarsely dentate to ± pinnatifid 49. *Osteospermum*
Leaves ± bipinnatifid with the primary lobes deeply toothed to pinnatifid **46**
46. Flower-heads less than 0.5cm broad 36. *Chrysanthellum*
Flower-heads more than 2cm broad 41. *Chrysanthemum*
47. Leaves bipinnatisect with linear segments. Inflorescence a racemose panicle 42. *Artemisia*
Leaves simple, or pinnatifid in the lower half only. Inflorescence a loose panicle 4. *Dichrocephala*
48. Ray florets white or blue, conspicuous **49**
Ray florets absent or if present, yellow **50**
49. Flower-heads long pedunculate 5. *Felicia*
Flower-heads sessile or nearly so 19. *Macowania*
50. Involucre cylindrical, formed of numerous parallel-sided, ± coherent, linear bracts with or without a second row of small, reduced bracts at the base of the head **51**
Involucre not as above, rarely ± cylindrical then with many rows of overlapping, inbricate bracts **54**
51. Achenes flattened. Lower leaves suborbicular, deeply incised 44. *Cineraria*
Achenes not flattened. Leaves not as above **52**
52. Involucral bracts in one row. Succulent plants 46. *Kleinia*
Involucral bracts in 2 rows; small reduced bracts present at the base of the cylindrical involucre. Plants succulent or not **53**
53. Florets mauve or purple 43. *Crassocephalum*
Florets yellow 45. *Senecio*
54. Marginal ligulate florets present. Plants usually prominently yellow-rayed **55**
Florets all tubular **57**
55. Involucral bracts in a single row. Branch tips with dense clusters of linear leaves 47. *Euryops*
Involucral bracts in several rows. Leaves not arranged in dense clusters at the branch tips **56**
56. Pappus of hairs surrounded by a ring of connate scales. Flower-heads usually prominently rayed 21. *Pulicaria*
Pappus of hairs only. Flower-heads not prominently rayed **71**
57. Pappus plumose 20. *Pegolettia*
Pappus sometimes scabrid but never plumose **58**
58. Leaves glabrous. Young growth covered in shiny gum 6. *Psiadia*
Leaves hairy. Young growth not covered in shiny gum **59**
59. Involucre hemispherical or campanulate, the bracts not scarious but with hardened tips which differ in colour from the rest of the bract **60**
Involucre of varying shape but, if hemispherical or campanulate, some or all of the bracts scarious, the tip not differing in colour or texture from the rest of the bract **62**
60. Florets yellow 55. *Centaurea*
Florets mauve **61**
61. Leaves simple, densely white-tomentose below 56. *Centaurothamnus*
Leaves pinnatifid, not white-tomentose below 54. *Volutaria*

62. Florets bluish. Florets and pappus usually much exceeding the involucre 1. *Vernonia*
Florets yellowish, pinkish or bluish; if bluish, the florets and the pappus ± equalling the involucre **63**

63. Flower-heads solitary or paired **64**
Inflorescence compound, the flower-heads in corymbs, panicles or heads **65**

64. Involucral bracts white with a pink base 17. *Helichrysum*
Involucral bracts yellow-green to brown 14. *Phagnalon*

65. Involucral bracts (or at least the inner ones) scarious. Leaves grey-tomentose on the lower surface **66**
Involucral bracts ± herbaceous, though sometimes with scarious margins. Leaves never grey-tomentose except in *Conyza incana* and *Laggera decurrens* **71**

66. Perennials. Involucral bracts yellow, white or white-tipped **67**
Annuals. Involucral bracts pallid or uniformly pale brown **69**

67. Involucral bracts yellow. Leaves sessile but not decurrent 17. *Helichrysum*
Involucral bracts white or yellow, but if yellow, leaves decurrent **68**

68. Involucral bracts yellow or pure white 18. *Achyrocline*
Involucral bracts brown with white tips 16. *Gnaphalium*

69. Inflorescence a raceme 16. *Gnaphalium*
Inflorescence a dense corymb or head **70**

70. Flower-heads arranged in globular heads subtended by leaves 13. *Filago*
Flower-heads in dense clusters which are not subtended by leaves 15. *Pseudognaphalium*

71. Stem winged from the decurrent leaves 10. *Laggera*
Stem not winged. Leaves sometimes auriculate but never decurrent **72**

72. Flower-heads bluish. Anthers tailed **73**
Flower-heads yellowish. Anthers not tailed 8. *Conyza*

73. Shrubs 11. *Pluchea*
Herbs 9. *Blumea*

1. VERNONIA *Schreb.*

Herbs or shrubs with alternate, simple leaves. Flower-heads usually in terminal corymbs; involucral bracts in many rows; florets all tubular, bisexual, purple or white. Achenes columnar, usually ridged and pubescent; pappus usually of two kinds, the outer consisting of short bristles, the inner of scabrid hairs.

1. Flower-heads solitary, very large, 2–3cm in diameter *V. bottae*
Inflorescence corymbose or paniculate, the flower-heads never solitary and always less than 1.5cm in diameter **2**

2. Annual herb *V. cinerea*
Perennial undershrubs, always woody below **3**

3. Leaves usually less than 1cm long, grey-tomentellous on both surfaces, often with prominent side lobes **4**
Leaves (1.5–)2–15cm long, never lobed nor grey-tomentellous on the upper surface **5**

4. Pappus pure white, clearly of 2 kinds. Leaves rhomboid in outline *V. arabica*
Pappus straw-coloured, the short outer bristles obscure, resembling the hairs on the achene. Leaves oblanceolate in outline *V. spatulata*

5. Leaves white-tomentellous on the undersurface *V. leopoldii*
Leaves green on both surfaces **6**

6. Leaves obovate, ± rounded at the apex *V. cinerascens*
Leaves lanceolate or narrowly elliptic, acute **7**

7. Involucre less than 6mm wide, the outer bracts obtuse *V. amgydalina*
Involucre 7–15mm wide, the outer bracts acute *V. abyssinica*

1. Vernonia abyssinica *Sch. Bip.*

Much-branched, erect undershrub to 1½m, stem brown, tomentose in young parts. Leaves petiolate, narrowly elliptic, acute, serrate, green on both surfaces but paler and pubescent below, 7–10cm long. Inflorescence corymbose; flower-heads 7–15mm wide, pale purple; involucral bracts all acute. Achenes pubescent; pappus straw-coloured, of two kinds.

A local plant of open stony banks, terrace walls and field borders found from 1000 to 2500m in areas of good rainfall on the escarpment from J. Ba'dān northwards; most common in the central escarpment.

Only known from Yemen and the Ethiopian highlands.

2. Vernonia amygdalina *Del.*

Large shrub or small tree up to 5m high, branchlets purple-brown, sparsely pubescent. Leaves petiolate, narrowly elliptic, acute, serrate, glabrescent, green on both surfaces but paler and gland-dotted below, 6–12cm long. Inflorescence a large corymbose panicle; flower-heads 4–6mm wide, white; outer involucral bracts obtuse. Achenes pubescent; pappus straw-coloured, the outer absent.

Flowers in winter.

A rare plant of scrubby streamsides at 1600 to 1700m in the Ibb area: between Ibb and Suq As Sabt; Wadi Sudān, Qā'idah.

3. Vernonia arabica *F.G. Davies*

An entirely grey-tomentellous undershrub similar to *V. spatulata* but differing in its leaves and pappus. Leaves usually less than 1cm long, ± rhomboid in outline, coarsely 1–6-lobed. Pappus of two distinct kinds, pure white.

On cliffs, at 1400m on the Naqīl Wusūf, Qabayṭah.

Widespread in southeast Arabia but known only from a single locality in Africa.

4. Vernonia bottae *Jaub. & Spach*

Undershrub with short, pubescent, purple-brown, prostrate stems from a woody base. Leaves shortly petiolate, oblanceolate, acute, green on both surfaces, glabrescent, gland-dotted, 1–2.5cm long, the margin with a few large teeth. Flower-heads solitary, terminal, 2–3cm broad, purple;

involucral bracts all acute. Achenes pubescent; pappus dirty white, of two kinds.

One of the most attractive endemic species which is restricted to cracks on flat, exposed rock masses between 1000 and 2100m: Wadi Tha'bat, J. Ṣabir; near Khalawah, J. Sawraq; abundant on the Dhubhān plateau around Turbah; on the western slopes of J. Bura'. Since it only occurs in any frequency in the last two localities, this species is well worth searching for elsewhere and might be expected to occur on J. Ḥushā', J. Sāmi' and Qa'alat Manṣūrah.

5. **Vernonia cinerascens** *Sch. Bip.*

Much-branched, erect undershrub to 60cm, the stems densely white adpressed-pubescent. Leaves sessile, obovate, rounded, subentire, green on both surfaces, glabrescent, gland-dotted, 2–4cm long. Inflorescence a corymb; flower-heads 3–10mm wide, purple; involucral bracts usually purple-tipped, the outer ones obtuse. Achenes pubescent; pappus white, of two kinds.

Widespread and frequent in open *Acacia* scrub from 500 to 1600m on the escarpment; more local in the Mashriq where it reaches 2000m and has been recorded from the Rada' area, Nihm, Ḥūth and around Majz (Steinberg). It is often associated with *Barleria proxima*, *Solanum cordatum*, *Psilotrichum gnaphalobryum* and other species of dry bushland.

6. **Vernonia cinerea** (*L.*) *Less.*

Chrysocoma ovata Forssk.

Erect, pubescent annual to c 60cm, the stems green, striate. Leaves shortly petiolate, lanceolate, ovate or obovate, rounded at the apex, subentire, green on both surfaces, pubescent, 2–4cm long. Inflorescence a terminal panicle; flower-heads, 4–6mm wide, purple; involucral bracts all acute. Achenes pubescent; pappus white, of two kinds.

A widespread and quite frequent field weed found on the escarpment from 500 to 1500m, most common in irrigated land in the foothills.

7. **Vernonia leopoldii** *Vatke*

Much-branched undershrub to 1m, the stems striate, white-tomentellous. Leaves petiolate, oblong-elliptic, acute, serrate, green on the upper surface, white-tomentellous below, 6–15cm long. Inflorescence a large terminal corymb; flower-heads 5–7mm wide, purple; involucral bracts often purple-tipped, the outer obtuse, mucronate. Achenes glabrous, strongly ridged; pappus white, of two kinds, the outer very short.

A conspicuous and very characteristic plant of the escarpment mountains between 2000 and 3000m; it is common in a variety of rocky habitats including cliffs, terrace walls, eroded hillsides and overgrown gullies between fields.

Outside Yemen known only from the Ethiopian mountains.

8. **Vernonia spatulata** (*Forssk.*) *Sch. Bip.*

V. atriplicifolia Jaub. & Spach

Much-branched woody undershrub with stiff, almost spinescent branches, the whole plant grey-tomentellous. Leaves sessile, linear-oblanceolate in outline, rounded at the apex, usually with an opposite pair of lyrate-lobes, c 1cm long. Inflorescence a lax corymb or ± cymose; flower-heads 4–6mm wide, purple; involucral bracts all acute. Achenes pubescent; pappus straw-coloured, the outer obscure and similar to the hairs on the achene.

A plant of rocky hillsides in sparse, dry, *Acacia* scrub between 1000 and 2000m. Widespread on the escarpment and southeast mountains north to Mafḥaq, Jihānah and Dhī Nā'im but scattered in occurrence, usually in small quantity and only at all common in Taiz province.

Outside Yemen it is a rare plant found in a few places in South Yemen and Somalia.

2. ADENOSTEMMA *J.R. & G. Forst.*

1. **Adenostemma caffrum** *DC.*

A. viscosum sensu Schwartz

Erect or ascending, sparsely pubescent glandular herb to 80cm. Leaves opposite, petiolate, ovate or rhomboid, coarsely serrate, 4–10cm long, Inflorescence laxly corymbose with long pedunculate, white flower-heads; involucre saucer-shaped, the bracts in 1 row, similar; florets all tubular, adherent to each other. Achenes flattened, glandular, sticky, the pappus of 4 spreading glandular bristles.

An uncommon plant found in ditches and by ponds in high rainfall areas of the escarpment between 1500 and 2800m: above Dhī Sufāl; J. Ba'dān; Wadi Dūr, 'Udayn; near Ribāṭ at the head of the Wadi Banā; 'Attārah and Manākhah, Ḥarāz; Bayt Bashārī, Maḥwīt; Ṭawīlah; Aḥjar.

3. AGERATUM *L.*

1. **Ageratum conyzoides** *L.*

Erect, branched, pubescent annual herb to 80cm. Leaves opposite, petiolate, ovate, serrate, 2–5cm long, the base almost truncate. Inflorescence of dense terminal corymbs of small blue-purple flower-heads; involucral bracts in several rows; florets all tubular. Achenes ± glabrous, black, columnar, angled; the pappus of about 5 awned scales.

A locally abundant weed found from 400 to 1500m in irrigated fields around Mishrāfah and on the western slopes of J. Raymah, particularly the coffee terraces of the Ja'farīyah. This species is an introduction from America and is likely to spread to other areas.

4. DICHROCEPHALA *DC.*

Herbs with alternate, dentate, often pinnatifid leaves. Flower-heads globular, sometimes solitary but usually in a loose panicle or raceme; involucre very small with 1 row of bracts; all florets tubular. Achenes small, flattened, obovate, glabrous, without pappus.

Leaves sessile. Flower-heads 5–8mm wide — *D. chrysanthemifolia*

Leaves petiolate. Flower-heads 3–4mm wide — *D. integrifolia*

1. **Dichrocephala chrysanthemifolia** (*Bl.*) *DC.*

Erect, densely pubescent, sparingly-branched annual to 50cm. Leaves narrowly obovate, coarsely dentate and sometimes pinnatifid towards the sessile, auriculate base,

3–8cm long. Flower-heads greeny-yellow (the central florets sometimes reddish-purple), few, solitary or in loose racemes or panicles, always with stout peduncles at least 1cm long.

Widespread on the escarpment and high plateau from 2000 to 3600m; frequent but usually in small quantity growing on terrace walls, cliffs, in orchards and in similar partially shaded, periodically disturbed, rocky places.

2. Dichrocephala integrifolia (*L.f.*) *O. Kuntze*
D. latifolia DC.
Ascending, pubescent, sparingly branched herb to 50cm. Leaves petiolate, ovate, coarsely dentate and usually pinnatifid in the lower half, 3–9cm long. Flower-heads whitish-red, rather small, arranged in dense, conical, terminal panicles, the peduncles slender and mostly 1–5cm long.

An uncommon plant of damp, shady, disturbed places between 1400 and 2400m in the central escarpment, mostly in the coffee country: Jirjir above Ḥadīyah; Ḥadādah, J. Raymah; Manākhah (Schweinfurth); Maḥwīt; near Suq Ath Thalūth, Wadi Lā'ah.

5. FELICIA *Cass.*

Woody-based perennial herbs. Leaves alternate, simple, entire. Flower-heads conspicuously rayed; involucral bracts in several rows; outer florets ligulate, blue or white; inner florets tubular, yellow. Achenes flat; pappus of bristles.

Leaves linear, c 1mm wide. Ray florets white — *F. abyssinica*
Leaves linear-oblanceolate, 3–6mm wide. Ray florets blue — *F. dentata*

1. Felicia abyssinica *A. Rich.*
Much-branched, woody-based perennial with erect, cottony-hairy stems 10–15cm high. Leaves numerous, linear, bristly-hairy, 1–1.5cm long, c 1mm wide. Flower-heads long-pedunculate, solitary; rays white (drying blue). Involucral bracts with fimbriate margins.

A plant of open stony places from 2000 to 3400m; widespread and abundant on the high plateau, in Khawlan Ash Sham and on the inner escarpment extending south and west to J. Ba'dān, Waṣāb, Salafīyah, the Ḥarāz, J. Ḥufāsh and Ḥajjah, thus absent from the outermost escarpment mountains. This is the conspicuous daisy-like flower seen in abundance shortly after heavy rain.

2. Felicia dentata (*A. Rich.*) *Dandy*
F. richardi Vatke
Much-branched, woody-based perennial with erect, densely pubescent stems. Leaves linear-oblanceolate, bristly-hairy, 1.5–3cm long, 3–6mm wide. Flower-heads usually solitary and long pedunculate but sometimes with 1–2 flower-heads developing in the upper leaf axils below the main flower-head; rays pale blue; involucral bracts entire.

Widespread and frequent between 2000 and 3000m on the escarpment and high plateau, growing on steep, often scrubby rocky slopes including cliffs and terrace walls, but unlike the previous species occurring in small quantity.

6. PSIADIA *Jacq.*

1. Psiadia punctulata (*DC.*) *Vatke* — Futaḥ
P. arabica Jaub. & Spach
An erect, much-branched, rounded, glabrous shrub to 2m. Leaves alternate, linear-lanceolate, narrowed into an indistinct petiole, acuminate, 3–7cm long, the margin entire or undulate and the whole leaf glistening with resin. Inflorescence of dense corymbs with numerous yellow flower-heads; involucre with several rows of pale bracts; tubular florets only present. Achenes c 1mm long, columnar, ribbed, ciliate; pappus of dirty white bristles.

Widespread and often abundant on open rocky hillsides between 1000 and 3000m on the escarpment, high plateau and southeast mountains. Its distribution is similar to that of *Dodonaea viscosa* and *Tarconanthus camphoratus* and like them is most abundant on rainshadow mountains such as J. Minmār, J. Sawraq and the Ḥujarīyah hills.

7. TARCONANTHUS *L.*

1. Tarconanthus camphoratus *L.* — Bayaḍ, Mukār
Untidy, strongly aromatic, dioecious shrub to 3m, the branchlets white-tomentose. Leaves alternate, shortly petiolate, lanceolate to narrowly elliptic, entire, grey-green above, white-tomentose below, 4–15cm long. Male flower-heads in terminal racemes, whitish-brown, the florets tubular; female flower-heads in terminal panicles, involucre campanulate with one row of 4 bracts, florets few, tubular, surrounded by a mass of white woolly hair. Achenes woolly, without pappus.

Widespread on open stony hillsides on the escarpment between 1200 and 2500m. It is often associated with *Dodonaea viscosa* and, like it, is most common on rather dry rainshadow hills in the Ḥujarīyah and Khawlan Ash Shan, but also occurs as scattered individuals in the wettest areas of the central escarpment.

8. CONYZA *Less.*

Herbs or shrubs with alternate, simple leaves. Inflorescence a terminal corymb; involucre ± campanulate with many rows of bracts; flower-heads yellow, the inner florets all tubular, the outer florets either ligulate or narrowly tubular. Achenes flattened, glabrous or hairy; pappus of minutely scabrid hairs.

A genus of rather weedy plants, mostly distinguished by small, somewhat imprecise characters. In the following key and descriptions the leaves referred to are the stem leaves. The basal leaves are usually ephemeral and of different shape and size to those on the stem.

1. Leaves grey-tomentose especially on the undersurface — *C. incana*
   Leaves green on both surfaces — 2
2. Leaves small (less than 2cm long), mostly 3-lobed, the lower half very narrow and petiole-like — *C. stricta*
   Leaves mostly more than 2cm long, simple, or if pinnatifid with more than 3 lobes — 3

3. Leaves dark green, linear, less than 5mm wide *C. bonariensis*
   Leaves green, ovate, lanceolate or oblong, more than 5mm wide **4**
4. Involucral bracts 2–4mm long. Shrubs 0.5–3m high **5**
   Involucral bracts more than 4mm long. Herbs less than 50cm high **6**
5. Leaves entire or obscurely toothed, usually petiolate. Corymbs very dense, some flower-heads sessile *C. pyrrhopappa*
   Leaves usually serrate, sessile. Corymbs not very dense, the flower-heads all distinctly stalked *Pluchea dioscoridis*
6. Stems and involucral bracts densely hairy; the bracts greenish or with purplish margins **7**
   Stems and involucral bracts sparsely hairy or ± glabrous; bracts greenish or at most purple-tipped **10**
7. Involucral bracts subulate, 0.5mm broad. Leaves usually pinnatifid *C. aegyptiaca*
   Involucral bracts at least 1mm broad. Leaves never pinnatifid **8**
8. Involucral bracts green with a white or yellowish margin *C. hochstetteri*
   Involucral bracts with purple margins **9**
9. Leaves densely hispid-pilose 2–4cm long. Involucral bracts up to 7mm long *C. variegata*
   Leaves sparsely hispid 4–8cm long. Involucral bracts up to 4mm long *C. tigrensis*
10. Stem leaves gradually narrowed to a broad sessile base, entire or serrate 11
    Stem leaves at first narrowed but then widened to a broad auriculate base, coarsely toothed with triangular, often apiculate teeth 2–3mm long *C. steudelii*
11. Involucral bracts 1–2mm broad. Leaf margin bristly *C. gouanii*
    Involucral bracts 2–2.5mm broad. Leaf margin glabrous or minutely echinate *C. sp.A*

1. **Conyza aegyptiaca** (*L.*) *Ait.*

Erect woody annual to 80cm, stems pilose. Leaves sessile or clasping, oblong to oblanceolate in outline, coarsely dentate to pinnatifid with forward-pointing lobes, coarsely hairy, 4–8cm long. Inflorescence racemose or weakly corymbose with few, rather large flower-heads borne on elongating peduncles; involucral bracts subulate, pilose. Achenes thinly pubescent, the hairs forked at the tips.

Schwartz records this species from Manākhah (Wissmann), Naqīl Shijah (Wissmann) and from the Sana'a area (Rathjens and Wissmann), but it has never been found by anyone else in Yemen or indeed in Arabia and it seems probable the records are erroneous. *C. aegyptiaca* is, however, a relatively distinct species readily recognised by its usually pinnatifid leaves and few, large, long-pedunculate flower-heads.

2. **Conyza bonariensis** (*L.*) *Cronq.*

Erigeron linifolius Willd.

Erect woody annual to 1m, stem very leafy, ridged, hairy with long and short hairs mixed. Leaves dark green, sessile, linear, entire, pubescent, up to 5cm long and 5mm wide. Inflorescence a panicle with numerous flower-heads; involucral bracts subulate, pilose.

Readily recognised by the numerous linear, dark green leaves.

A common weed found from 1100 to 2700m throughout the escarpment and high plateau; it grows generally in waste places but shows a preference for grassy banks by streams and irrigation ditches.

3. **Conyza gouanii** (*L.*) *Willd.*

Rather like a depauperate version of the much more common *C. hochstetteri* from which it can be distinguished by its sparser indumentum and fewer flower-heads. Stems sparsely pilose, with prominent longitudinal green lines. Leaves bristly on the margins and veins below but otherwise nearly glabrous. Inflorescence a dense terminal corymb with c 5–8 flower-heads; involucral bracts 1–2mm wide, with broad scarious margins, sometimes purple-tipped.

Apparently a rare plant of disturbed damp places, particularly irrigation ditches: Musalla, Aḥkūm; Wadi Qābah, south of Mudhaykhirah; between Suq Ar Rubū' and Dan', Waṣāb Al 'Ali; Dhamār (Wood 75/517, 1558, 2305, 2462).

4. **Conyza hochstetteri** *Sch. Bip. ex A. Rich.*

Perennial herb, branched at the base and with erect, usually simple, densely bristly-hairy stems 30–40cm high. Leaves sessile, clasping or auriculate, lanceolate or oblong, entire or dentate, bristly-hairy, 3–5cm long. Inflorescence of dense terminal clusters of flower-heads; involucral bracts 1–2mm wide, up to 5mm long, densely hairy, often obtuse, with scarious or yellow margins, sometimes purple-tipped.

Closely related to several other species but usually recognised by its very hairy indumentum, dense inflorescence and green involucral bracts.

Widespread and frequent between 1200 and 2900m on the escarpment; also extending to the high plateau where it is less common. Unlike all closely related species it is usually a plant of dry habitats, growing on cliffs, terrace walls, rocky slopes, field borders and in waste ground around villages.

5. **Conyza incana** (*Vahl*) *Willd.* Bayaḍ

A distinctive, much-branched, densely grey-tomentose undershrub to 50cm with very leafy stems. Leaves sessile, linear or oblanceolate, grey above, white below, entire to ± pinnatifid, terminating in a fine mucro, 2–5cm long. Inflorescence ± corymbose, formed of dense terminal clusters of many small flower-heads; involucral bracts subulate, less than 3mm long.

Widespread and frequent on steep, bare rocky hillsides from 1600 to 3100m on the escarpment and high plateau. This species requires little soil and is common in very eroded places such as the higher parts of J. Milḥān.

Outside Yemen known only from Ethiopia.

6. Conyza pyrrhopappa *Sch. Bip. ex A. Rich.*
Erect, much-branched, pubescent undershrub to 2m. Leaves sessile or shortly petiolate, oblanceolate to narrowly elliptic, ± entire, mucronate, 4–7cm long. Inflorescence a dense terminal corymb with many small flower-heads; involucral bracts subulate, pubescent, 2–3mm long.

Widespread and frequent on open rocky hillsides between 1500 and 3000m on the escarpment and high plateau. It has a very similar distribution and ecology to the previous species and they often grow together though *C. incana* is slightly less common.

7. Conyza steudelii *Sch. Bip. ex A. Rich.*
Decumbent or ascending, much-branched, sparsely-hairy perennial; stem with alternate yellow and green longitudinal lines, a few minute spines and thin scurfy indumentum. Leaves ± elliptic, widened to an auriculate base, glabrous except for the margins and veins below, 5–7cm long, the margin coarsely and deeply dentate. Inflorescence a dense terminal corymb; involucral bracts subulate, 4–5mm long, c 1mm wide.

A rare plant of damp scrub and rocky gullies found on the escarpment between 2100 and 2900m: between J. Manār and Suq As Sabt; Markah, J. Milḥān; Kawkabān (Wood 1325, 2385).

8. Conyza stricta *Willd.*
C. triloba Decne.
Woody herb to 40cm, much branched at the base; stem pubescent, with alternate yellow and green longitudinal lines. Leaves sessile, attenuate at the base, usually 3-lobed though sometimes simple above or 5-lobed below, mucronate, pubescent, 1–2cm long. Inflorescence a dense, many-flowered terminal corymb; involucral bracts subulate, often reddish-tipped.

Widespread and often abundant between 1500 and 3000m on the escarpment and high plateau, growing on cliffs, terrace walls, rocky hillsides, field borders, roadsides and in derelict fields.

9. Conyza tigrensis *Sch. Bip. ex Oliv. & Hiern*
Similar to *C. hochstetteri* from which it is distinguished by the following characters: stems usually taller, less densely hairy. Leaves with sparse, rough hairs, mostly restricted to the margins and veins below. Inflorescence more branched, with flower-heads not only in dense terminal clusters but also in pedunculate, axillary clusters; involucral bracts with distinctive dark-purple margins similar to those in *C. variegata* but smaller and not usually exceeding 4mm long.

A rare plant of streamsides between 2500 and 2900m on the high plateau: Yāzil; in the gorge above Tin'am (Wood 2190, 2342).

10. Conyza variegata *Sch. Bip. ex A. Rich.*
C. nana sensu Deflers
Perennial herb branched at the base with erect, densely white-pilose stems to 30cm, though usually much less. Leaves sessile, sometimes weakly auriculate, lanceolate to oblong, coarsely toothed, pilose, 1.5–3cm long. Inflorescence a terminal corymb usually reduced to 1–2 large flower-heads; involucral bracts rather large, up to 7mm long and 2mm broad, with prominent purple margins.

The long white hairs are a distinctive field character.

A local plant found on cliffs and steep rocky slopes from 2500 to 3500m on the higher mountains from J. Ṣabir north to J. Maswār and J. Lawz (Wood 75/249).

11. Conyza sp. A
Erect or ascending, almost glabrous, branched perennial herb, the stems slightly ridged, red-brown. Leaves sessile, sometimes auriculate, narrowly oblong to oblanceolate, coarsely serrate or sometimes entire, sparsely hairy on the veins below, 3–8cm long, the margin glabrous or minutely echinate. Inflorescence a dense terminal corymb with c 5–6 flower-heads; involucral bracts 2–2.5cm broad, greenish-brown with scarious margins, rounded at the apex, concave.

An undescribed species related to *C. gouanii* but easily recognised by its sparse indumentum and broad involucral bracts.

An uncommon, apparently endemic species found on the escarpment by streams between 1900 and 2900m: in the wadi above 'Adūf near Turbah; J. Ṣabir; above Dhī Sufāl; J. Ba'dān; J. Manār; at the head of the Wadi Banā near Ribāṭ; Baw'an area; J. Maswār (Wood 75/353, 1330; Hepper & Wood 5885).

9. BLUMEA *DC.*

Similar to *Conyza* but the flower-heads purplish and the anthers tailed.

1. Plant densely pubescent to pilose. Leaves oblong-obovate, more than 5mm wide — 2
   Plant glabrescent or with a few sparse hairs. Leaves linear, less than 5mm wide — *B. bovei*
2. Leaves sessile, often decurrent as interrupted wings on the stem. Inflorescence lax, the flower-heads c 1.5cm wide — *B. aurita*
   Larger leaves petiolate, never decurrent. Inflorescence very dense, the flower-heads c 0.75cm wide — *B. lacera*

1. Blumea aurita *(L.f.) DC.*
Laggera aurita (L.f.) Sch. Bip.
Much-branched, bushy, aromatic, glandular, pilose annual woody herb to 1m. Leaves sessile, often auriculate and decurrent with interrupted wings on the stem below, oblong-obovate, coarsely toothed to ± pinnatifid, 1–5cm long. Inflorescence of lax, terminal and axillary corymbose racemes with few, rather large, long-pedunculate purplish flower-heads c 1.5cm wide. Achenes sparsely pilose.

An awkward species taxonomically which can be placed as easily in *Laggera* as in *Blumea*.

An uncommon plant found by streams and irrigation ditches in the foothills of the escarpment between 500 and 1000m from Waṣāb As Sāfal northwards.

2. Blumea bovei (*DC.*) *Vatke*
B. abyssinica Sch. Bip. ex A. Rich.
Slender glabrous or very sparsely pilose perennial herb with ascending stems to c 20cm. Leaves sessile, linear, obscurely toothed, 2–5cm long. Inflorescence a lax terminal corymb with few, long-pedunculate, bluish flower-heads c 1cm wide. Achenes sparsely pilose.

A very widespread but rather uncommon plant found in scattered localities between 1000 and 2600m on the escarpment, high plateau and Mashriq. It usually grows on ± permanently wet rocks in rather dry country, particularly dripping cliffs where it is associated with *Xerophyta* and *Samolus*, but it is also tolerant of quite high salt levels and occurs in springs in the Mashriq at 'Ujlī, Hammām Damt and the Sirwāh area.

3. Blumea lacera (*Burm. f.*) *DC.*
Branched, aromatic, glandular, densely pubescent and pilose perennial herb to 50cm. Leaves petiolate (except when immature), narrowly oblanceolate, coarsely dentate, 3–8cm long. Inflorescence of dense, terminal and lateral panicles, the purple flower-heads, c 0.75cm wide, very numerous, becoming matted together. Achenes glabrous.

A rare plant found by ditches and streams between 1500 and 2000m in high rainfall areas of the escarpment: Wadi Sūdān, Qā'idah; Wadi Dūr, 'Udayn; between Ibb and Suq As Sabt; 'Attārah, Ḥarāz.

10. LAGGERA *Sch. Bip. ex Benth.*

Similar to *Conyza* but the stems always winged from the decurrent leaf bases. The flower-heads are white, purplish or pale yellow and the achenes are sparsely pilose.

1. Leaves entire. Whole plant grey-tomentose — *L. decurrens*
   Leaves toothed. Plant greenish — 2
2. Stem wings dentate or interrupted — 3
   Stem wings entire ± continuous — *L. alata*
3. Plant pilose. Stem wings short or ± absent — *Blumea aurita*
   Plant puberulent and shortly glandular but not pilose. Stem wings long, continuous, dentate — *L. crispata*

1. Laggera alata (*D. Don*) *Sch. Bip. ex Oliv.*
Erect, glandular, slightly aromatic, grey-green annual herb to 1½m, the stems tomentose, often unbranched, with long, entire wings decurrent from the leaf bases. Leaves sessile, linear or oblong, serrate, glabrescent above, grey-tomentose below, 5–9cm long. Inflorescence a large, cylindrical terminal panicle usually 30–40cm long; flower-heads numerous, pedunculate, purplish, 1–1.5cm wide.

L. alata is part of a complex of species in tropical Africa and the Yemen plant is not entirely uniform; in Wood 2159, a broad-leaved form, the outer involucral bracts have recurved tips and are more densely glandular than in narrow-leaved forms (Wood 2150). Such variation is probably worth investigating.

A widespread but uncommon species found between 1400 and 2700m on the escarpment and high plateau. It grows on roadsides and well-drained stony hillsides with a strong preference for volcanic soils; it is most common in the Dhamār area.

2. Laggera crispata (*Vahl*) *F.N. Hepper & J.R.I. Wood*
L. pterodonta (DC.) Sch. Bip. ex Oliv.
L. purpurascens (A. Rich.) Sch. Bip.
Erect, glandular, slightly sticky aromatic herb to 70cm, the stems sparsely pubescent, often unbranched, with dentate wings decurrent from the leaf bases. Leaves sessile, oblanceolate, glabrescent, 4–8cm long. Inflorescence loosely paniculate, the flower-heads pedunculate, purplish, 1–2cm wide.

A plant of damp, nitrogen-rich habitats usually growing by ditches in villages and showing a strong preference for sewage outlets, the margins of streams used for washing clothes and similar places. It is widespread and quite frequent between 1500 and 2900m on the escarpment but rare on the high plateau where it is only known from Wadi Ḍahr and Khamir.

3. Laggera decurrens (*Vahl*) *F.N. Hepper & J.R.I. Wood* — Sadaq
Conyza arabica Willd.
Blumea garieppina DC.
Erect, much-branched, grey, aromatic, woody perennial to 1m, the stems tomentellous and with entire wings decurrent from the leaf bases. Leaves sessile, linear-oblanceolate, entire, grey-tomentose. Inflorescence loosely paniculate, the flower-heads mostly solitary and terminal on short, leafy secondary branches; flower-heads pedunculate, dull yellow, c 1cm in diameter.

Widespread between 1000 and 2500m in the drier parts of the escarpment, on the high plateau and in the Mashriq, preferring alluvial soils and growing on roadsides, field borders and especially in derelict fields where it is often abundant.

This common, undemanding weedy species is rather curious since it appears to be far more abundant in Yemen than elsewhere in Arabia and north Africa, though it is common in the southern half of Africa.

11. PLUCHEA *Cass.*

Similar to *Conyza* but shrubby in habit and the flower-heads usually pale purple in colour.

Leaves softly tomentose, rounded at the apex, less than 3cm long. Involucre 5–6mm long — *P. sordida*
Leaves ± pubescent, acute 3–7cm. Involucre 3–4mm long — *P. dioscoridis*

1. Pluchea dioscoridis (*L.*) *DC.*
An erect, much-branched, aromatic, pubescent or rather scurfy shrub 1–3m high. Leaves sessile, obovate to oblong, acute, almost always coarsely serrate, grey-green, 3–7cm long. Inflorescence of terminal corymbs, the flower-heads

pink, pale purple or (rarely) dull yellow, stalked; involucre 3–4mm long.

A characteristic plant of wadis in dry country where ground water is close to the surface, often associated with *Tamarix*, *Calotropis* and *Typha domingensis*. Widespread and often abundant between 100 and 2000m, particularly in dry parts of the escarpment and in the Mashriq.

2. Pluchea sordida *Oliv. & Hiern*

Much-branched, weakly aromatic shrub to 1m, the young stems and leaves softly tomentose. Leaves very shortly petiolate, obovate, entire or obscurely toothed near the rounded apex, brownish-green, 2–3cm long. Inflorescence dense, terminal, subglobose, the flower-heads purple, shortly stalked or sessile; involucre 5–6mm long.

Only known from thorn fences by streams at 1300m in Wadi Hiḍrān near Taiz.

This species is unknown elsewhere in Arabia or in adjacent parts of Africa but is common on the Kenyan coast.

12. BLEPHARISPERMUM *Wight*

1. Blepharispermum yemense *Deflers* Naḥd

Erect, much-branched, glabrous shrub to 3m. Leaves alternate, oblanceolate, rounded at the apex, gradually narrowed into an indistinct petiole, entire, fragrant, 3–5cm long. Inflorescence a terminal white spherical head of flower-heads, solitary on a long peduncle; involucre of scarious bracts, inconspicuous, largely hidden by pubescent concave scales; florets tubular. Achenes of two kinds, the outer flattened with ciliate wings c 2.5mm long, the inner columnar; pappus of small scales.

A rare plant restricted to dry bushland between 1000 and 1300m in the eastern Ḥujarīyah: Ḥamīd region, southeast of Raḥidah; between Qābayn and Ḍabāb, Yūsafin; J. Masanah (Deflers).

Elsewhere known only from Somalia.

13. FILAGO *L.*

Grey-tomentose annual herbs with alternate leaves. Flower-heads small, in terminal or axillary globose clusters which in our species are subtended by an involucre of leaves; involucral bracts in several rows, the outer woolly, the inner scarious, yellowish-brown to red; florets all tubular, pale yellow. Achenes small; pappus of hairs.

Similar in general appearance to *Gnaphalium* and *Pseudognaphalium* but usually recognised by the cymose branching of the stem, each branch ending in a globular cluster of flower-heads.

***Filago* is a complex genus in which much contemporary research is being done. The names used here are tentative and may need revision.**

Plant with a distinct main stem. Involucral bracts usually reddish, not spreading at a wide angle in fruit *F. abyssinica*
Plant without a main stem, the primary cluster of flowers being at the base of the stem and surrounded by a small rosette of leaves. Involucral bracts rarely red, spreading at a wide angle in fruit *F. prolifera*

1. Filago abyssinica *Sch. Bip. ex A. Rich.*

Densely grey-tomentose annual herb with a distinct, simple or sparingly branched erect stem to 10cm. Leaves sessile, linear-oblanceolate, 8–15mm long. Inflorescence of globular clusters of 10–20 flower-heads which are slightly overtopped by the subtending leaves; involucral bracts brown, often tinged with red, remaining ± erect in fruit.

Known only from the summit ridge of J. Ta'kar at 3100m (Wood 1167).

Elsewhere known only from Ethiopia.

2. Filago prolifera *Pomel*

F. germanica sensu Schwartz

Prostrate annual herb of similar but less heavy indumentum to the last species, readily distinguished by the absence of a main stem, the primary flower cluster being borne at ground level, the subtending leaves forming a basal rosette. Leaves similar to those of *F. abyssinica*. Flower clusters with 1–8 flower-heads which ± equal the subtending leaves; involucral bracts pale brown, rarely tinged with red, the inner bracts widely spreading in fruit.

An uncommon plant of stony banks around fields and similar flat open places with little competition from other plants, found between 2200 and 3300m on the high plateau and the surrounding mountains. It is known with certainty from J. Shibām, Ḥarāz (Schweinfurth) J. Sumārah (Wood 1586), J. Nabī Shu'ayb (Wood 3244), Baw'an (Wood 733), Raydah (Hepper 6185) and Wadi Shalālah near Yarīm (Hepper 6258) but field records from 'Amrān, J. Manār and J. Lawz probably belong here.

14. PHAGNALON *Cass.*

Low undershrubs with simple, alternate leaves. Flower-heads c 1cm broad, solitary at the ends of the branches; involucre hemispherical, the bracts in many rows, sometimes scarious; florets yellowish, all tubular. Achenes cylindrical, pubescent; pappus of hairs.

Leaves less than 2cm long, the margin revolute, entire *P. harazianum*
Leaves more than 2cm long, the margin dentate, never revolute *P. scalarum*

1. Phagnalon harazianum *Deflers*

P. persicum sensu Schwartz

Conyza tomentosa Forssk. p.p.

Low, much-branched undershrub with very leafy, erect white-tomentose stems to 15cm. Leaves sessile, linear, grey and thinly tomentose above, densely white-tomentose below, the margin revolute and entire, 7–17mm long. Flower-heads borne on peduncles c 1cm long; involucral bracts brown with scarious tips, the outer reflexed at maturity.

Widespread on the escarpment and high plateau between 2100 and 3500m growing on terrace walls and rock faces but uncommon except at high altitudes on the escarpment.

This endemic species is closely related to *P. persicum* Boiss. and its taxonomic status needs investigation.

2. Phagnalon scalarum *Schweinf. ex Schwartz*
P. hypoleucum sensu Schwartz
Branched undershrub with ascending, white-tomentose stems to 40cm. Leaves shortly petiolate, oblanceolate, glabrescent above, white-tomentose below, 2–6cm long, the margin irregularly dentate, never recurved. Flower-heads borne on long slender peduncles 2–10cm long; involucral bracts yellow-green, scarious towards the tips, spreading or erect.

Widespread and frequent on terrace walls and rocky banks between 1400 and 3000m on the escarpment and high plateau, most plentiful in areas with frequent winter mists.

15. PSEUDOGNAPHALIUM *Kirp.*

1. Pseudognaphalium luteo-album (*L.*) *Hilliard & Burtt*
Gnaphalium luteo-album L.
Simple or branched, erect annual herb to 30cm, the stem white-tomentose. Leaves alternate, sessile, linear, ± entire, grey-tomentose on both surfaces, 2–5cm long. Flower-heads very small, in globose clusters arranged in dense terminal corymbs; involucral bracts in 3 rows, imbricate, scarious, pale brown, obtuse; florets all tubular, pale yellow. Achenes small, glabrous or hairy; pappus of hairs.

Widespread and fairly frequent on the escarpment and high plateau from 1200 to 2900m growing as a field weed and on seasonally flooded mud plain or on the dried beds of ponds and streams.

16. GNAPHALIUM *L.*

Very similar to *Helichrysum* from which it can only be consistently distinguished by very technical characters. Our species are low, uniformly grey-tomentose annual or perennial herbs with small flower-heads and pale brown involucral bracts which may or may not be white-tipped.

Perennial. Flower-heads in congested corymbs; involucral bracts white-tipped *G. unionis*
Annual. Flower-heads in terminal racemes; involucral bracts uniformly pale brown *G. polycaulon*

1. Gnaphalium polycaulon *Pers.*
G. indicum auct.
Grey-tomentose annual herb rather similar in habit to *Pseudognaphalium luteo-album*; the stem branched at the base, decumbent or erect to 25cm. Leaves sessile, oblanceolate, 4–7cm long. Flower-heads in dense terminal racemes; involucral bracts in 3 rows, uniformly pale brown.

Only known from 1500m at Ḥajjah (Rathjens); not seen recently.

2. Gnaphalium unionis *Sch. Bip. ex Oliv. & Hiern*
Grey-tomentose perennial, the stem at first prostrate and rooting at the nodes, finally ascending. Leaves sessile, linear 1–3cm long. Flower-heads in congested corymbs; involucral bracts in 4 rows, the outermost dark brown, the innermost pale brown with distinctive white tips.

An uncommon plant found between 2000 and 3100m, growing on mud at the margin of seasonal ponds drying up at the end of the rainy season: J. Ḥubaysh; on the ridge between J. Sumārah and J. Manār; between Sanabān and Dhamār; between the Qa' Bakīl and Suq al Jum'ah, 'Anis; Kawkabān; Maḥwīt; J. Maḥdad.

17. HELICHRYSUM *Mill.*

Everlasting flowers. Perennial herbs or undershrubs with alternate, usually entire, often tomentose leaves. Flower-heads solitary or in terminal heads or corymbs; involucre ± tubular with many rows of scarious, sometimes brightly coloured bracts; florets all tubular, yellow. Achenes glabrous or hairy; pappus of scabrid hairs.

1. Flower-heads solitary; involucral bracts white with a pinkish base *H. arwae*
   Flower-heads numerous in dense corymbose inflorescences; involucral bracts yellow **2**
2. Flower-heads 10–20mm broad *H. foetidum*
   Flower-heads less than 5mm broad **3**
3. Leaves less than 1cm long. Inflorescence with up to 10 flower-heads *H. pumilum*
   Leaves more than 1cm long. Inflorescence dense, usually with 20 or more flower-heads **4**
4. Leaves 1–3cm long, grey-tomentose on both surfaces. Leafy undershrub *H. forskahlii*
   Leaves more than 3cm long, green and glabrescent above. Rosette herb with a sparingly leafy flowering stem *H. nudifolium*

1. Helichrysum arwae *J.R.I. Wood*
Silvery mat-like undershrub with much-branched prostrate woody stems. Leaves in dense clusters at the tips of short side branches, sessile, linear, densely appressed white-tomentose on both surfaces, 5–6mm long, 1–1.5mm wide. Flower-heads borne on short peduncles arising from the tips of the leafy branches, solitary, c 2cm wide; involucral bracts white with a pinkish-maroon base, 7–9mm long.

Flowers in April and May.

A spectacular endemic species which carpets north-facing volcanic cliffs between 3000 and 3200m on J. Ta'kar and J. Manār.

2. Helichrysum foetidum (*L.*) *Cass.*
Erect, sparingly branched perennial to 1m, the stems densely grey-tomentose and very leafy to the bottom of the inflorescence. Leaves sessile, oblong-lanceolate, apiculate, dull green above but grey-tomentose on the midrib, margins and the undersurface, 4–8cm long, 7–15mm wide. Inflorescence a compact corymb or head of about 5–12 large flower-heads, each 10–20mm broad; involucral bracts yellowish, 6–8mm long.

Flowers mainly in autumn.

Widespread on the escarpment between 1900 and 3400m growing on cliffs and rocky slopes in areas of average to good rainfall; very rare on the high plateau but recorded from Al Zīl, Al Ḥadā'.

3. Helichrysum forskahlii *(J.F. Gmel.) Hilliard & Burtt*
H. cymosum auct.
H. fruticosum auct.
Erect, much-branched undershrub to 40cm, the stems densely white-tomentose and very leafy except immediately below the inflorescence. Leaves sessile, linear, grey-tomentose, 1–3cm long, 1–3mm wide. Inflorescence a dense, ± globular corymb or head of numerous small flower-heads, each 1–1.5mm broad; involucral bracts yellow, c 1mm long.

Flowers mainly in spring.

Locally abundant in montane grassland and on rocky hillsides between 2000 and 3500m on the escarpment; its distribution appears to be related to the presence of mist as it occurs at much lower altitudes in the grassland areas of the central escarpment than elsewhere.

4. Helichrysum nudifolium *(L.) Less.*
Gnaphalium globosum sensu Schwartz
Perennial herb with a basal rosette of petiolate leaves from which arises an erect, white tomentose stem to 60cm; stem leafy in the lower half, almost leafless above. Stem leaves sessile, oblong to narrowly elliptic, green and glabrescent above, grey-tomentose and strongly 3-veined below, 3–15cm long, 1–3.5cm wide. Inflorescence a dense terminal umbel of flower-heads, each c 4–5mm wide; involucral bracts yellow, 2–3mm long.

Flowers mostly in early summer.

An attractive and locally abundant plant of montane grassland in high rainfall, mist regions of the escarpment between 2000 and 3200m; particularly characteristic of the J. Raymah meadows and the orchid-rich high-altitude grasslands of the Ibb region.

5. Helichrysum pumilum *(Klatt.) Moes.*
Low, wiry, much-branched, grey-tomentose undershrub to 15cm. Leaves sessile, linear, 5–8mm long, 1–3mm wide. Inflorescence a small terminal cluster of 2–10 flower-heads, each 1–2mm wide; involucral bracts pale yellow, 2–3mm long.

Widespread and frequent on open rocky plain on the high plateau, particularly in sandstone country; also in similar situations near Ṣa'dah, on the Dhubhān plateau near Turbah, at Najd Dhamrān in the east Ḥujarīyah and in Bilad Aṭ Ṭā'am (Wissmann).

Outside southwest Arabia this species is only known from the Somali highlands.

Helichrysum abyssinicum Sch. Bip. was recorded by Deflers from J. Shibām, Ḥarāz. There is no specimen at Paris and it has not been refound in this exceptionally well-known locality so it was almost certainly an error, perhaps for *H. foetidum* which was not listed by Deflers.

18. ACHYROCLINE *(Less.) DC.*

Similar to *Helichrysum* but the flower-heads with few florets, the female outnumbering the hermaphrodite.

1. Involucral bracts yellow. Leaves decurrent onto the stem — *A. stenopterum*
   Involucral bracts white. Leaves not decurrent — 2
2. Leaves linear, sessile, grey-tomentose on both surfaces — *A. glumaceum*
   Leaves ovate or elliptic, petiolate, dark green and ± glabrescent above — *A. schimperi*

1. Achyrocline glumaceum *(DC.) Oliv. & Hiern*
Helichrysum glumaceum DC.
Low, grey-tomentose undershrub, much-branched at the base with many erect or ascending almost leafless branches to 30cm. Leaves sessile, linear to oblanceolate, apiculate, 2–4cm long, 2–12mm wide, Inflorescence a dense globose head of small flower-heads, each 1–3mm broad; involucral bracts pure white, 2–3mm long.

Widespread on the escarpment and high plateau between 1500 and 2500m but of very scattered occurrence; it is usually found either on sandstone plateau or on broken volcanic ridges where it is associated with *Becium serpyllifolium*, *Elionurus argenteus* and *Lotus quinatus*.

2. Achyrocline schimperi *Sch. Bip. ex A. Rich.*
Helichrysum schimperi (Sch. Bip. ex A. Rich.) Moes.
Vigorous, decumbent undershrub with grey-tomentose stems scrambling over scrub vegetation to 1½m. Leaves petiolate, ovate or elliptic, dark green and glabrescent above, grey-tomentose below, obscurely crenate on the margins, 3–5cm long. Inflorescence much branched with many dense terminal corymbs; flower-heads 1–2mm wide; involucral bracts white, 2–3mm long.

Mostly autumn-flowering.

A plant of disturbed bushy places such as cliff gullies and overgrown fences, often associated with *Acanthus arboreus* and *Vernonia leopoldii*; widespread and fairly frequent on the escarpment between 1800 and 3000m; on the high plateau only at Maḥalayn on J. Lawz.

3. Achyrocline stenopterum *(DC.) Hilliard & Burtt*
Helichrysum stenopterum DC.
Decumbent, thinly grey-tomentose herb with long stems which are sparingly leafy especially in the upper half. Leaves sessile, decurrent onto the stem as long entire wings, linear-oblanceolate, more thinly tomentose above than below, 3–9cm long, 0.5–1.5cm wide. Inflorescence a dense terminal corymb of small flower-heads, each 1–1.5mm wide; involucral bracts yellow, c 2mm long.

An uncommon plant of bushy grassland between 2000 and 2900m in high rainfall areas of the escarpment: J. Ṣabir; J. Ba'dān; J. Manār; J. Ḥiṣn Madnān, Waṣāb al 'Ālī; J. Kabūrah above Hadīyah; Markah, J. Milḥān.

19. MACOWANIA *Oliv.*

1. Macowania ericifolia (*Forssk.*) *Burtt & Grau*

Antithrixia abyssinica sensu Schwartz

Small wiry undershrub with branched decumbent or ascending stems, the lower parts of the branches covered with persistent, slightly cottony, old leaf bases. Leaves densely crowded towards the branch tips, alternate, sessile, linear, green above, white-tomentose below, 0.5–1.5cm long. Flower-heads solitary, terminal on the branches, the central florets yellow, tubular, the marginal florets ligulate with white rays; involucre of many rows of yellowish, glabrous, imbricate bracts. Achenes cylindrical, pilose; pappus of white hairs.

A local plant of exposed, rocky slopes found between 2500 and 3200m on the higher mountains of the escarpment and high plateau from J. Ṣabir north to Kawkabān.

Known elsewhere only in Ethiopia.

20. PEGOLETTIA *Cass.*

1. Pegolettia senegalensis *Cass.*

Erect, aromatic, slightly sticky, gland-dotted crisped-pubescent woody annual to 50cm, often unbranched in the lower half. Leaves alternate, linear-oblanceolate, subsessile, serrate, 1–2.5cm long. Inflorescence a lax corymb of small, pale yellow flower-heads; involucral bracts in several rows, subulate to linear; ligulate florets absent. Achenes ribbed; pappus of two kinds, the inner consisting of plumose bristles which are much longer than the fringed scales forming the outer pappus.

Widespread between 1000 and 2300m; locally abundant in dry country on the escarpment growing in disturbed bushland and as a weed of roadsides and stony fields particularly in Taiz province; elsewhere very rare but found in a few scattered localities on the high plateau and in the Mashriq.

21. PULICARIA *Gaertn.*

Perennial herbs or undershrubs, often glandular and aromatic. Leaves alternate. Inflorescence loosely corymbose; flower-heads yellow with tubular and ligulate florets, often prominently rayed; involucral bracts in many rows. Achenes cylindrical, ribbed, the pappus of two kinds, the outer consisting of a short scarious cup, the inner of hairs.

1. Leaves petiolate — 2
   Leaves sessile — 4
2. Leaves linear-oblanceolate, entire — *P. somaliensis*
   Leaves lanceolate to ovate, toothed — 3
3. Flower-heads ± 1cm broad; peduncles less than 5cm long — *P. petiolaris*
   Flower-heads ± 1.5cm broad; peduncles 5–8cm long — *P. grandidentata*
4. Flower-heads 1.5–3.5cm broad, prominently rayed — 5
   Flower-heads 0.5–1.5cm broad, the rays very short or apparently absent — 6
5. Stem and leaves white-woolly. Rays 5–10mm long — *P. schimperi*
   Stem and leaves tomentose, grey-green. Rays 3–5mm long — *P. jaubertii*
6. Young stems white-woolly. Achenes glabrous — *P. crispa*
   Young stems green or reddish, not woolly. Achenes pilose — 7
7. Leaves scabrid with a broad, ± auriculate base. Involucral bracts villous — *P. inuloides*
   Leaves adpressed-pilose, narrowed into an attenuate base. Involucral bracts glandular but otherwise ± glabrous — *P. somaliensis*

1. Pulicaria crispa (*Forssk.*) *Oliv.* — Jithjāth

Much-branched, bushy undershrub to 40cm, usually white-woolly when young but becoming green and less woolly when old. Leaves sessile, linear-oblanceolate with an undulate, toothed margin, 0.5–3cm long. Inflorescence very laxly corymbose, the flower-heads 5–10mm broad, often appearing solitary; rays very short; involucral bracts linear, glabrous. Achenes glabrous.

Widespread and locally abundant on the high plateau and in the Mashriq between 1000 and 2600m usually growing on alluvial plain with a particular preference for derelict fields. It is also found on alluvial banks by streams, on field borders and in similar habitats. It is widespread but local on the escarpment.

A plant with a characteristic Sahara-Sindian distribution.

2. Pulicaria grandidentata *Jaub. & Spach*

Similar to the much more common *P. petiolaris*, differing in its peduncles which are 5–8cm long and in its larger flower-heads which are c 1.5cm in diameter.

A plant of steep rocky hillsides between 2200 and 3000m along the inner escarpment with outlying records from J. Ba'dān, J. Ta'kar and J. Shibām in the Ḥarāz.

This endemic species is separated from *P. petiolaris* by rather unsatisfactory characters. The large flower-heads are readily recognised in the field but whether it is really anything more than a large-flowered variety of *P. petiolaris* is far from certain.

3. Pulicaria inuloides *DC.*

P. arabica sensu Deflers

Much-branched woody perennial herb with purplish stems to 40cm, scabrid when mature but densely pilose on young growth. Leaves sessile, ± auriculate, linear-oblanceolate, obscurely toothed, slightly sticky, 2–6cm long. Flower-heads c 1cm broad, the rays short, rarely exceeding 2mm long; involucral bracts linear, pilose, glandular. Achenes thinly hairy.

A common plant of streams, ponds and irrigation ditches between 1600 and 2900m on the high plateau and inner escarpment, extending southwards through the Ibb and Taiz mountains to the Ḥujarīyah; in similar places in Khawlan Ash Sham but apparently rare in the Mashriq and only recorded from Nihm.

4. Pulicaria jaubertii *Gamal Ed Din*

Khawah, Munas, 'Ansīf

P. orientalis sensu Schwartz

P. undulata sensu Schwartz

Much-branched, very aromatic, grey-villous woody herb to 40cm. Leaves sessile, oblanceolate, obscurely toothed, 2–4cm long. Inflorescence with few flower-heads, scarcely corymbose; flower-heads 1.5–2cm wide, the rays 3–5mm long; involucral bracts oblanceolate, villous. Achenes thinly hairy.

A very widespread species found throughout Yemen east to Mārib and from sea level to 2500m, being absent only from high rainfall areas of the escarpment. In the mountains it grows on open rocky ground but in the Tihama it is mainly found in irrigated sandy fields. It is perhaps more common in disturbed habitats. The whole plant is strongly aromatic and is worn by women together with Rue and other fragrant herbs and it is sometimes seen for sale in markets.

5. Pulicaria petiolaris *Jaub. & Spach*

Much-branched, erect, aromatic undershrub to 40cm, covered with varying amounts of long and short hairs and numerous sessile or stalked glands. Leaves petiolate, lanceolate to ovate, dentate, 1.5–5cm long. Inflorescence distinctly corymbose; flower-heads c 1cm broad, the rays short, 1–2mm long; involucral bracts linear, glandular but otherwise ± glabrous. Achenes sparsely hairy.

Widespread and frequent from 1200 to 3200m on the escarpment and high plateau, growing on steep rocky banks including cliffs and terrace walls.

6. Pulicaria schimperi *DC.*

Superficially similar to *P. jaubertii* but readily recognised by the white-woolly indumentum which covers the whole plant. In addition it is only weakly aromatic, the leaves are obovate and strongly dentate, the flower-heads 2–3.5cm wide with the rays especially prominent, being 5–10mm long.

An uncommon plant of derelict fields and open stony banks by roads and around cultivation, found between 1300 and 2300m on the northern escarpment: at the head of the Wadi Yūr; J. Ḥufāsh; Bayt Braz, Maḥwīt; between Bayt 'Adhāqah and Wadi Lā'ah; Ḥajjah; on the eastern slopes of Maḥābishah.

A plant from the cliffs at Akāḥilah near Turbah (Wood 75/35) has characters intermediate between *P. schimperi* and either *P. petiolaris* or *P. grandidentata*. Gamal Ed Din has named it as the hybrid *P. petiolaris x schimperi* but since the latter plant does not occur anywhere in the region, this seems unlikely. It may prove to be a distinct species.

P. schimperi is endemic to the western side of Arabia.

7. Pulicaria somaliensis *O. Hoffm.* Huzār

P. glutinosa sensu Schwartz

P. adenensis Schweinf.

Erect, much-branched, aromatic undershrub to 1½m high, the leaves and young stems glandular, sticky and ± adpressed pilose. Leaves gradually narrowed at the base into an indistinct petiole, linear-oblanceolate, entire, 2–5cm long. Inflorescence corymbose; flower-heads 0.5–1cm wide, the rays apparently absent; involucral bracts linear, glandular but otherwise ± glabrous. Achenes pubescent.

Locally abundant on stony hillsides in open *Acacia* scrub between 400 and 1500m in the Ḥujarīyah north to Al Barḥ and Taiz; also in Wadi Rakwah in Khawlan Ash Sham (Steinberg).

P. somaliensis is known only from tropical Arabia and Somalia.

22. LAGASCEA *Cav.*

1. Lagascea mollis *Cav.*

Much-branched, erect annual herb to 50cm, stem scabrid-pubescent. Leaves opposite, petiolate, ovate, serrate, tomentose below, 2–4cm long. Inflorescence a long-pedunculate, scabious-like head, subtended by a whorl of silky, obovate bracts and consisting of numerous flower-heads, each with a single white tubular floret enclosed in a tube formed from the fused involucral bracts. Achenes black, oblanceolate in outline, without pappus, enclosed in the involucral tube.

An introduction from America only known at present on a roadside at 1000m on the escarpment below Maḥābishah.

23. GEIGERIA *Griessel.*

1. Geigeria alata (*DC.*) *Oliv. & Hiern*

Much-branched, glabrous annual to 30cm, the stems prominently winged. Leaves opposite, sessile, oblong-elliptic, stiff, scabrid, reticulate, mucronate, 2–4cm long. Flower-heads small, subsessile, 1–3 together in the forks of the branches; involucre campanulate, the bracts imbricate with hardened, spreading, spinescent tips; ligulate and disc florets present, few, yellow. Achenes pubescent with a pappus of bristles and scales.

A rare plant known only from the lower Wadi Jawf at 1100m (Zeller).

24. ACANTHOSPERMUM *Schrank*

1. Acanthospermum hispidum *DC.*

Untidy, erect or sprawling, much-branched, coarsely hairy, woody annual. Leaves opposite, subsessile, elliptic, serrate, minutely gland-dotted, 1.5–3cm long. Flower-heads axillary, sessile, yellow, small, each head developing 4–5 yellowish spiny fruits formed from the enlarged bracts of the outer (female) florets, the two apical spines much larger than the other hooked spines. The achene is enclosed within the spiny capsule-like fruit.

A plant of roadsides and gravelly wadi banks between 1000 and 1900m in Taiz province: Maqātirah; above 'Adūf, near Turbah; Barakānī; Wadi Ḍabāb. Although rare at the moment this unpleasant weed is likely to increase in the future. It is an introduction from America.

25. XANTHIUM *L.*

Monoecious annual woody herbs. Leaves alternate, palmately lobed. Flower-heads axillary, solitary or in clusters, the male heads in the upper part of the plant and the female below; male heads with 1 row of involucral bracts and numerous tubular florets; female heads with 2 tubular florets which are enclosed by the 2 fused bracts which form a hard capsule-like casing covered in hooked spines round the achene. Achenes compressed, without pappus.

Plant with 1–2 trifid spines at the base of each leaf petiole ... *X. spinosum*
Plant unarmed except on the fruit ... *X. strumarium*

1. Xanthium spinosum *L.*

Erect, much-branched annual to 60cm, armed with 1–2 yellow trifid spines from the base of each leaf petiole. Leaves shortly petiolate, narrowly rhomboidal, 3–5-lobed, entire, dark green above, white-felted below, 1–4cm long. Fruit ellipsoid, pubescent, covered in hooked spines and terminating in a spine-tipped beak, 8–12mm long.

An introduction from America which is now an abundant weed of fields and waste places almost everywhere between 1200 and 2700m.

2. Xanthium strumarium *L.*

X. brasilicum Vell.

Erect, much-branched, unarmed annual to 1m, pubescent and slightly scabrid in all parts. Leaves long-petiolate, triangular-ovate, 3–5-lobed, dentate, green, 5–12cm long. Fruit ellipsoid, pubescent, covered in hooked spines and terminating in a short pointed beak, 12–18mm long.

Another species introduced from America but known in Yemen since Forsskal's time. It is a local plant of high rainfall areas of the escarpment between 500 and 2200m where it grows on roadsides and in waste places. It is only common around Taiz and Ibb.

26. SIGESBECKIA *L.*

1. Sigesbeckia orientalis *L.*

Erect, glandular-pubescent woody annual with a branched, purplish stem to 80cm. Leaves opposite, petiolate, lanceolate or elliptic, incised-dentate, scabrid, 5–10cm long. Inflorescence a lax panicle of small yellow, shortly-rayed flower-heads; involucre 2-rowed, sticky, covered in stalked glands, the outer bracts linear-spathulate, spreading, much larger than the florets, the inner bracts oblong, erect, equalling the florets. Achenes columnar, 4-angled, glabrous, without pappus.

A local plant of bushy places around fields and near streams in the central escarpment and the lush middle-altitude wadis between 900 and 1700m from Misrākh north to the Wadi Lā'ah.

27. ECLIPTA *L.*

1. Eclipta prostrata (*L.*) *L.*

E. alba (L.) Hassk.

Decumbent, scabrid annual herb. Leaves opposite, sessile, lanceolate or narrowly elliptic, obscurely serrate, 1–4cm long. Flower-heads small, white, inconspicuously rayed, solitary (rarely paired) on slender axillary peduncles; involucre hemispherical, the bracts in 1 row; ligulate and tubular florets present. Achenes flattened or 3-angled, tuberculate, with or without a pappus of 2 small bristles.

A plant of sandy and gravelly banks by permanent streams, which sometimes spreads into irrigated fields on sandy soil. It is widespread and frequent in suitable habitats in the Tihama and on the escarpment from 100 to 2000m, being particularly characteristic of the larger wadis flowing through the foothills where it is often associated with *Phyla nodiflora*.

28. SCLEROCARPUS *Jacq. ex Murr.*

1. Sclerocarpus africanus *Jacq. ex Murr.*

Erect, branched, scabrid-pubescent annual to 1m. Leaves opposite, petiolate, ovate to elliptic, serrate, 3–9cm long. Flower-heads large, yellow, solitary and terminal on the branches or 2–3 together in a lax corymb; involucral bracts in one row, leafy; florets few, all tubular, enclosed by scales which enlarge and harden to form a rugose, capsule-like casing round the achene which is without pappus.

An uncommon plant found along bushy wadi margins between 500 and 1500m in the central escarpment; also Wadi Dūr at 'Udayn, Bani Miftāḥ between Riḥāb and Waṣāb, Wadi Sukhnah.

29. BLAINVILLEA *Cass.*

1. Blainvillea gayana *Cass.*

Erect, sparingly branched, scabrid-pubescent annual herb to 1m. Leaves opposite, petiolate, ovate, acuminate, serrate, 5–14cm long. Inflorescence of terminal cymes; flower-heads small, with yellow tubular inner florets and inconspicuously white-rayed outer ligulate florets; involucre ± tubular with 1 row of green bracts. Achenes columnar, sharply 3-angled, blackish-green, shortly pubescent; pappus of a small cup from which arise 2 pubescent bristles.

Vegetatively almost identical to *Aspilia helianthoides* and best distinguished by the sharply 3-angled achenes and the deeply toothed scales which are folded around the florets.

A rare plant of dry open stony bushland in areas of seasonally heavy rainfall between 1000 and 1500m: between Riḥāb and Waṣāb; Wadi Khufīt near 'Usil in the Ḥarāz (Schweinfurth); Wadi Sukhnah.

30. ASPILIA *Thou.*

Scabrid herbs with opposite leaves. Inflorescence much branched, ± cymose; the flower-heads mostly solitary with tubular inner florets and shortly rayed, ligulate outer florets; involucre with 1–2 rows of bracts. Achenes oblanceolate, flattened, weakly 2-angled, silky; pappus of a small cup from which arise 2 bristles.

Flowers dark maroon to black	*A. kotschyi*
Flowers yellow	*A. helianthoides*

1. Aspilia helianthoides (*Schum. & Thonn.*) *Oliv. & Hiern*

Erect annual herb to 1m. Leaves shortly petiolate, lanceolate to elliptic, acuminate, 4–15cm long. Flower-heads yellow with yellow rays; involucral bracts in 2 rows, 3–5mm long.

The Yemen plant is subspecies **ciliata** (*Schum.*) *Adams*. It is easily confused with *Blainvillea gayana* but can be distinguished by the achenes and by the broadly truncate, obscurely toothed scales which are found among the florets.

A local field weed found between 500 and 2200m on the escarpment from J. Ḥabashī north to the Wadi Lā'ah; it grows in irrigated land or, more usually, in high rainfall areas; it is only common in the Ibb region.

2. Aspilia kotschyi (*Sch. Bip.*) *Oliv.*

Similar to *A. helianthoides* but with a purple-blotched stem and dark-maroon florets. The involucre consists of a single row of bracts 10–15mm long and the scales found between the florets are narrowed to an acuminate apex.

A local plant found in disturbed places in open bushland in areas of average to good rainfall on the escarpment between 1000 and 1700m: frequent at 'Udayn; between Mukhādir and Riḥāb; between Riḥāb and Waṣāb; Suq Ar Rubū', Waṣāb al 'Ālī; above Ḥillah, J. Bura' (Schweinfurth); J. Sāfān, Ḥarāz; between Ḥammām Shamsān and Maḥwīt.

31. MELANTHERA *Rohr.*

1. Melanthera abyssinica (*Sch. Bip. ex A. Rich.*) *Oliv. & Hiern*

Wurschmittia abyssinica Sch. Bip. ex A. Rich.

Wedelia abyssinica sensu Schwartz

Decumbent or weakly erect, scabrid perennial to 1½m. Leaves opposite, subsessile, broadly lanceolate, acuminate, serrate, 5–15cm long. Inflorescence sparingly branched with large, long-pedunculate, yellow-rayed flower-heads; involucral bracts in 1 row; tubular and ligulate florets present. Achenes oblanceolate, slightly flattened, 2-angled, glabrous; pappus with a cup from which arise 1–4 easily deciduous bristles.

A rare plant of bushy places in high rainfall areas of the escarpment between 1500 and 1900m: between Ibb and Suq As Sabt; north of the Suq, 'Udayn; above Suq Ath Thalūth, Wadi Lā'ah.

32. TRIDAX *L.*

1. Tridax procumbens *L.*

Trailing perennial herb with erect or ascending pubescent flowering stems to 40cm. Leaves opposite, petiolate, ovate, coarsely incised-dentate, 2–4cm long. Inflorescence a solitary, very long-pedunculate, terminal white-rayed flower-head; involucral bracts in 2 rows, numerous; outer florets with short, white rays, inner florets tubular, yellow, interspersed with numerous long, scarious scales. Achenes c 2mm long, ± cylindrical, silky-hairy, the pappus of many plumose bristles.

An introduction from America which is now locally abundant on the escarpment from 500 to 1500m though absent from Taiz and Ibb provinces apart from an isolated record from 'Udayn. It grows on tracksides and terrace walls, in wadi beds and in cultivated land extending onto derelict terraces and open rocky hills and is particularly common at low altitudes in the central escarpment.

33. VERBESINA *L.*

1. Verbesina encelioides (*Cav.*) *A. Gray*

Ascending or untidily sprawling, much-branched, appressed-hairy annual herb to 1m. Leaves opposite below, alternate above, petiolate, ovate, coarsely dentate, dark green above, densely white-hairy below, 3–12cm long; petiole winged, with an auriculate base. Flower-heads yellow, conspicuously rayed, long pedunculate, solitary or forming a very lax corymbose inflorescence; involucral bracts in 2 rows; ligulate and disc florets present. Achenes c 7mm long, flattened, obovate, winged, notched at the apex, pubescent; pappus of 2 bristles.

An introduction from America, now well established on roadsides and waste places in Taiz and spreading north to Qa'idah and south to Nashimah in the Ḥujarīyah; found from 1300 to 1800m.

34. GUIZOTIA *Cass.*

1. Guizotia scabra (*Vis.*) *Chiov.* Barah

G. schultzii Hochst. ex Sch. Bip.

Rather variable erect, glabrous, scabrid or pilose annual herb to 80cm. Leaves opposite, sessile, lanceolate to oblong, entire or dentate, 6–12cm long. Inflorescence a lax corymbose cyme; involucral bracts in 1 row, leaf-like; tubular and ligulate florets present. Achenes columnar, glabrous, without pappus.

Two subspecies whose distribution is not accurately known are recorded from Yemen. Subspecies **scabra** is a glabrous or scabrid perennial, usually with 8 involucral bracts and 8–16 ray florets. Subspecies **schimperi** (*Sch. Bip.*) *J. Baagoe* is a pilose annual usually with 5 involucral bracts and only 6–8 ray florets. Subspecies *scabra* appears to be the more common of the two.

Abundant between 2000 and 3000m on roadsides, field borders and disturbed mountain slopes in the Ibb area where it is one of the most conspicuous autumn flowers; uncommon elsewhere but extending south to the Taiz area where it descends to 1400m in flooded fields in Wadi Ḍabāb and north in scattered localities through the central escarpment to Maḥwīt.

35. BIDENS *L.*

Annual herbs. Leaves opposite, pinnately lobed. Flower-heads solitary, terminal, pedunculate; involucral bracts in 2 rows, the outer herbaceous, the inner scarious-margined; tubular florets yellow; ligulate florets yellow or white, sometimes absent. Achenes linear, glabrous; pappus of 2–5 retrorsely-scabrid bristles.

The 3 following species are all introductions from America and may be difficult to separate but in contrast the genus itself is all too familiar, the achenes sticking to all manner of clothing.

1.	Pinnae lobed almost to the midrib	*B. bipinnata*
	Pinnae unlobed	**2**
2.	Rays white. Achenes up to 12mm long	*B. pilosa*
	Rays yellow. Achenes up to 18mm long	*B. biternata*

1. Bidens bipinnata *L.*

Erect, almost glabrous, branched annual to 1m. Leaves petiolate, with up to 4 pairs of leaflets, the lower of which are lobed almost to the midrib, the lobes of varying shape but usually coarsely serrate. Flower-heads distinctly longer than wide, yellow-rayed or without rays. Achenes 8–18mm long, black, with a pappus of 2–4 yellowish bristles.

A local, perhaps overlooked, but sometimes abundant species of roadsides and eroded, stony mountainsides found between 1200 and 2500m on the escarpment north to the Ḥarāz; most common in the Ibb area where it is often associated with the next species (Wood 75/811, 1235).

2. Bidens biternata (*Lour.*) *Merr. & Sherff.*

Very similar to the previous species, differing only in the leaves which are simply pinnate, the leaflets all being serrate but unlobed.

The most common *Bidens* in Yemen, being found everywhere on the escarpment between 1200 and 2700m and occurring in great abundance in areas of high rainfall, notably around Ibb. It grows in a variety of disturbed stony places including eroded hillsides, field borders, wadi beds and roadsides.

3. Bidens pilosa *L.*

A plant with simply pinnate leaves similar in appearance to *B. biternata* from which it can usually be distinguished by the presence of white ray florets. Unfortunately these are not always present but the flower-heads are about as broad as long and the achenes do not exceed 12mm in length and are often less than this.

A local plant found in the wetter parts of the escarpment from 1500 to 2700m, but unlike the previous two species it seldom grows in large quantity. It grows in damp shaded places particularly stream sides and cliff gullies in the central escarpment where it is most frequent.

Related to *Bidens* is the garden flower, **Cosmos bipinnata** *Cav.* This is commonly cultivated at Sana'a and elsewhere in villages on the escarpment and high plateau.

36. CHRYSANTHELLUM *Rich.*

1. Chrysanthellum americanum (*L.*) *Vatke*

Small, much-branched, procumbent or ascending, nearly glabrous annual herb. Leaves alternate, ± bipinnately divided into small, linear segments, 1–3cm long. Inflorescence lax, subcorymbose, with small, pedunculate, inconspicuously rayed, yellow flower-heads; involucral bracts in 1–2 rows; outer florets ligulate, the inner tubular. Achenes 2–2.5mm long, flattened, oblong, black with brownish margins; pappus absent.

An uncommon weed of stony fields and field borders, found on the escarpment between 1000 and 1800m from Raḥidah and Misrākh north to the Al 'Ashshah area but most characteristic of the middle-altitude wadis with a good rainfall, including those at 'Udayn, Wadi Ḍabāb and Wadi Sukhnah. It is an autumn-flowering plant which appears to require a long, relatively uninterrupted period of summer rain.

37. GALINSOGA *Ruiz & Pav.*

1. Galinsoga parviflora *Cav.*

Erect, much-branched, glabrous or sparsely pilose, annual herb to 70cm. Leaves opposite, petiolate, ovate, obscurely crenate, 2–5cm long. Inflorescence of small, terminal corymbs of small, white-rayed flower-heads; involucral bracts in 1 row, ovate; ligulate florets usually 5, white rayed, tubular florets yellow. Achenes of 2 kinds, the central achenes 1mm long, ovoid, black, with a pappus of white bristles; marginal achenes without pappus.

A locally common weed of waste places and cultivated land, found from 1100 to 2900m on the escarpment and high plateau; it is noticeably more frequent in high rainfall areas, particularly Ibb province and the coffee region of the central escarpment.

It is an introduction from America.

38. FLAVERIA *Juss.*

1. Flaveria trinervia (*Spreng.*) *Mohr.*

F. repanda Lag.

Erect, glabrous, branched annual herb to 1m, the branches in opposite pairs arising at right angles to the reddish, main stem. Leaves opposite, petiolate, oblanceolate, prominently 3-nerved below, obscurely serrate, 1–5cm long. Inflorescence of dense terminal clusters of small, yellow flower-heads, each with a single rayed floret and 1–3 tubular florets; involucre a tube formed of the fused bracts. Achenes cylindrical, black, without pappus.

Probably introduced from America in the 1920s and now widespread and often abundant throughout Yemen apart from the Sands and the high mountains above 2300m. It grows on waste ground, field borders and along wadi beds, even in quite saline conditions. It is almost dominant on waste ground on the north side of Taiz.

39. TAGETES *L.*

1. Tagetes minuta *L.*

Erect, strong-smelling, glabrous, woody annual to 1½m. Leaves opposite, pinnate with linear to elliptic, serrate leaflets, 2–10cm long. Inflorescence of dense terminal corymbs with numerous small pale yellow flower-heads; involucre tubular, formed of fused bracts; florets all tubular, rather few in each flower-head. Achenes 6–8mm long, black, cylindrical, the pappus consisting of 4–5 bristles.

Locally abundant on roadsides, waste ground and field borders between 1400 and 2500m in Taiz province, extending northwards through the Mudhaykhirah region to 'Udayn. This is an introduced species from tropical America which is likely to become more widespread in the future.

The French Marigold, **Tagetes patula** *L.* and the African Marigold, **Tagetes erecta** *L.* are both commonly cultivated in the larger towns. The latter is also a common potherb on village rooftops.

40. ANTHEMIS *L.*

1. Anthemis tigrensis *A. Rich.*

A. yemensis Podl.

Prostrate or ascending, aromatic, pubescent annual or perennial herb. Leaves alternate, deeply pinnatisect with small linear segments, 1–4cm long. Flower-heads solitary, terminal, conspicuously white-rayed; involucral bracts numerous, in many rows, black-margined; outer florets ligulate with a white ray; inner florets tubular, yellow. Achenes 4-angled, ribbed, without pappus.

Rather a variable species. Plants from cultivated land at lower altitudes (Wood 75/170) being annual, ascending in habit and sparsely pubescent whereas plants from exposed mountain tops (Wood 3241) are perennial, prostrate and white-hairy.

A characteristic plant of the high plateau which grows on open alluvial plain, earth banks and in cultivated land, from 2300m to the summit of J. Nabī Shu'ayb at 3666m. It also occurs at Baraṭ (Fleurentin) and in similar habitats in Khawlan Ash Sham at Sāqayn and J. Marran.

Outside Yemen it is known only from the East African mountains.

41. CHRYSANTHEMUM *L.*

1. Chrysanthemum coronarium *L.*

Much-branched, glabrous, erect woody annual to 80cm. Leaves alternate, sessile, clasping, deeply bipinnatifid, dark green, 3–5cm long. Flower-heads solitary, terminal, large, 3–6cm diameter, with prominent yellow rays; disc florets yellow; involucral bracts in several rows, glaucous with a pale brown, scarious margin. Achenes without pappus, of 2 kinds, the outer winged, triangular, the inner columnar, without wings.

Cultivated in gardens in Sana'a and doubtless elsewhere; well naturalised on roadsides at Dhamār.

Two other species from related genera are cultivated in Yemen. Feverfew, **Tanacetum parthenium** (*L.*) *Sch. Bip.* (*Chrysanthemum parthenium* (L.) Bernh., *C. parthenifolium* Willd.) is grown as a potherb and used as a medicinal plant. It is a very aromatic plant with pinnate leaves and has flower-heads 1–2cm wide with white ray florets. **Dendranthema indicum** (*L.*) *Desmoul* (*Chrysanthemum indicum* L.) is recorded from Sana'a (Rathjens) and Salafīyah (Wissmann) but has not been seen recently. It resembles the florists' chrysanthemum and has flower-heads 3–4cm wide with white ray florets.

42. ARTEMISIA *L.*

1. Artemisia abyssinica *Sch. Bip. ex Oliv. & Hiern*

A. abrotanum sensu Schwartz

Erect, aromatic, woody annual or short-lived perennial to 2m, the stems sparingly branched, grooved, silky-hairy, especially above. Leaves alternate, grey-green, deeply bipinnatisect with linear segments, 4–10cm long. Inflorescence a long narrow racemose panicle with numerous, small dull yellow flower-heads; involucral bracts in several rows, ovate, scarious-margined; all florets tubular. Achenes small, cylindrical, without pappus.

Widespread on the high plateau from 2200 to 3600m and often abundant on roadsides, alluvial plain and abandoned fields, reaching nearly to the summit of J. Nabī Shu'ayb; rare in similar habitats on the escarpment and recorded from J. Ṣabir, J. Ta'kar, J. Ḥubaysh, Manākhah (Schweinfurth), Shahārah and Sāqayn.

Wormwood, **Artemisia arborescens** *L.* (*A. abyssinica* sensu Schwartz), known as 'Rand', is commonly cultivated as a potherb. It is an aromatic, grey, silky hairy plant with pale yellow flower-heads and is well known as a stimulant and an analgesic.

43. CRASSOCEPHALUM *Moench.*

1. Crassocephalum crepidioides (*Benth.*) *S. Moore*

Gynura crepidioides Benth.

Erect, branched, pubescent annual herb to 2m. Upper leaves sessile, pinnatifid; lower leaves petiolate, dentate, up to 20cm long, sometimes entire, ovate, more usually deeply pinnatifid with a large elliptic terminal lobe and 1–2 pairs of smaller lateral lobes. Inflorescence a lax terminal corymb; flower-heads few, long-pedunculate, purplish, unrayed, 1–2cm wide; involucre and achenes as in *Senecio*.

A rare plant of damp bushy places between 1300 and 1600m in high rainfall areas of the escarpment: Wadi Dūr, 'Udayn; between Ibb and Suq As Sabt; above Ḥillah, J. Bura' (Schweinfurth); between Ḥabash and Ṣa'bah, J. Milḥān.

44. CINERARIA *L.*

1. Cineraria abyssinica *Sch. Bip. ex A. Rich.*

Decumbent to erect annual herb, usually cottony-tomentose on young growth, particularly the undersurface of the leaves, but glabrescent with maturity. All leaves dark green on the upper surface, paler below, deeply incised-dentate;

lower leaves long-petiolate, suborbicular, 1.5–2cm in diameter; upper leaves sessile, sometimes auriculate, oblong-oblanceolate, deeply pinnatifid. Inflorescence a terminal corymb; flower-heads yellow, shortly rayed, 3–5mm wide; involucre 2-rowed, as in *Senecio*. Achenes flattened, glabrous; pappus of hairs.

A characteristic plant of the higher escarpment mountains from J. Ṣabir north to J. Maswār; it grows from 2400 to 3600m in disturbed rocky places, such as pathsides, terrace walls and rocky banks around villages.

45. SENECIO *L.*

Herbs or shrubs with alternate, sometimes succulent leaves and yellow, often rayed, flower-heads. Inflorescence usually corymbose; involucre usually tubular, 2-rowed, the inner row with parallel-sided, equal, ± coherent bracts, the outer with few, short, unequal bracts; inner florets tubular; ligulate outer florets usually present. Achenes cylindrical, ribbed; pappus of simple hairs.

1. Leaves deeply pinnatifid, cut to at least half their width — 2
   Leaves entire to coarsely dentate but never pinnatifid — 5
2. Decumbent perennial undershrubs. Terminal leaf-lobe clearly triangular in outline — 3
   Erect annual herbs. Terminal leaf-lobe cuneate to rounded in outline, never triangular — 4
3. Glabrous succulent. Ray florets absent — *S. bojeri*
   Thinly pubescent non-succulent. Ray florets present — *S. lyratus*
4. Ray florets present. Leaves with a long narrow petiole-like base, never auriculate — *S. glaucus*
   Ray florets usually absent. Upper stem leaves scarcely narrowed towards the broad, auriculate base — *S. vulgaris*
5. Plant densely covered in sticky glandular hairs — *S. sumarae*
   Plant neither sticky, nor glandular-hairy — 6
6. Succulent glabrous shrub — *S. hadiensis*
   Annual or perennial herbs, never succulent and always hairy at least on the peduncles and in the leaf axils — 7
7. Annual. Leaves ± glabrous, less than 1cm broad — *S. schimperi*
   Perennial. Leaves hairy, at least below, more than 1cm broad — 8
8. Leaves persistently white-tomentose below. Involucral bracts ± 8mm long — *S. asirensis*
   Leaves ± white-tomentose below when young but rapidly becoming dark green with maturity. Involucral bracts ± 6mm long — *S. harazianus*

1. Senecio asirensis *Boulos & J.R.I. Wood*

Rather similar to *S. harazianus* but distinguished by its persistent cottony-tomentose indumentum and larger dimensions. The leaves are mostly sessile, 5–12cm long, incised-dentate and persistently white-tomentose below. The inflorescence is laxer with large flower-heads 15–25mm wide.

Frequent at about 2000m on terrace walls on J. Marran and J. Rāziḥ in Khawlan Ash Sham (Wood 2933, 3032).

This species is known outside Yemen from the Asir mountains where it appears to be common.

2. Senecio bojeri *DC.** — Sila'

S. subscandens Hochst.
S. sonchifolius sensu Schwartz
Cacalia angulata Vahl

Glabrous, succulent undershrub with long decumbent or scrambling stems. Leaves sessile, usually auriculate, oblanceolate in outline, deeply pinnatifid with a large triangular to rhomboidal terminal lobe, the margin irregularly dentate, 5–15cm long. Inflorescence of dense, nearly globular umbels borne at the ends of long branches; flower-heads unrayed, 3–5mm wide. Achenes pilose.

Widespread in the wetter areas of the escarpment between 1000 and 2400m, usually growing on steep scrubby banks above streams and irrigation ditches.

* This species has been recently placed in a separate genus under the name *Solanecio angulatus* (Vahl) C. Jeffrey.

3. Senecio glaucus *L.*

S. desfontainei Druce

Erect, usually unbranched, glabrescent annual to 60cm, often with tufts of cottony hair in the axils of the young leaves. Leaves with a narrow petiole-like base, obovate in outline, deeply pinnatifid with linear-cuneate side lobes and a rounded-cuneate terminal lobe, the margin irregularly dentate, 4–10cm long. Inflorescence a terminal corymb; flower-heads rayed, 10–15mm wide. Achenes pilose.

An uncommon weed of damp disturbed places between 2300 and 2600m on the high plateau, probably introduced from the Mediterranean region in recent years: Dār al Ḥajjah, Wadi Ḍahr; above the gorge near Tin'am.

4. Senecio hadiensis *Forssk.* — Sila', Khidhrābah

Glabrous, succulent undershrub with numerous decumbent or ascending stems 1–2m long arising from a woody base. Leaves subsessile, broadly lanceolate to elliptic, entire or dentate but never pinnatifid, 8–18cm long. Inflorescence a dense, many-flowered terminal corymb with a strong, unpleasantly sweet smell; flower-heads rayed, 5–10mm wide. Achenes shortly pubescent.

Widespread and frequent on the escarpment between 800 and 2800m though mostly in areas of good rainfall. It grows in rocky places including cliffs, terrace walls, eroded gullies, open hillsides and stony wadi beds.

S. hadiensis is named after Ḥadīyah and still grows in the wadi just above the market. It is only known in southwest Arabia and Ethiopia.

5. Senecio harazianus *Deflers*

Erect, much-branched perennial herb to 60cm, the young growth covered in white cottony tomentum but this disappearing as the plant matures so old parts are ± glabrous. Leaves mostly shortly petiolate, lanceolate to elliptic, coarsely dentate but never pinnatifid, 5–10cm long. Inflorescence a loose terminal corymb; flower-heads rayed, 12–20mm wide. Achenes densely pubescent.

Widespread on the escarpment between 2000 and 2900m from Milḥān, Maswār and Kuḥlān southwards but becoming less common in Ibb province and completely absent from Taiz province. It is almost always found in villages, where it grows on waste ground, terrace walls and by sewage outlets, showing a marked preference for nitrogen-rich habitats.

S. harazianus was named after the Ḥarāz mountains and is endemic to Yemen.

6. Senecio lyratus *Forssk.*

S. lyratipartitus A. Rich.
S. auriculatus Vahl
Cineraria schimperi Sch. Bip.

Thinly pubescent perennial with wiry, prominently ridged, decumbent or scrambling stems. Leaves sessile, auriculate, pinnatifid with 1–2 pairs of small cuneate side lobes and a large ovate triangular terminal lobe, the margin regularly dentate, 5–9cm long. Inflorescence a loose terminal panicle; flower-heads rayed, 8–15mm wide. Achenes shortly pubescent.

A locally common plant found between 1500 and 2600m in high rainfall areas of the escarpment from Dhī Sufāl north to Maḥābishah but virtually restricted to the Ibb region and the central escarpment. It grows in steep, scrubby grassland and is associated with *Hybanthus* and *Pelargonium*.

7. Senecio schimperi *Sch. Bip. ex A. Rich.*

S. biflorus Vahl (probably)

Erect, sparingly branched, annual herb to 60cm, sparsely pubescent or glabrescent but usually with small tufts of cottony hair in the axils of the young leaves. Leaves narrowed to a petiole-like base, linear, irregularly dentate or ± entire, 5–10cm long. Inflorescence of lax terminal corymbs; flower-heads usually without rays, 5–10mm wide. Achenes shortly pubescent.

Widespread and frequent on the escarpment and high plateau between 1500 and 2900m. It grows in a variety of disturbed habitats including waste ground in villages, the margins of irrigation ditches, field borders, track sides and eroded, rocky hillsides.

8. Senecio sumarae *Deflers*

Erect, sticky glandular-hairy perennial to 30cm, branched at the base only. Leaves sessile, auriculate, oblong or oblanceolate, regularly dentate, never pinnatifid, 3–8cm long. Inflorescence a few-flowered terminal raceme or lax corymb; flower-heads rayed, 2–2.5cm wide. The only Yemen species with glabrous achenes.

An uncommon plant found on rocky slopes between 2600 and 3400m on the highest escarpment mountains: on all the Ibb mountains extending south to J. Ṣabir and north to J. Ḥiṣn Madnān in Waṣāb al 'Ālī; rare on the high mountains northeast of Sana'a but found on J. Nabī Shu'ayb, J. Ta'iz and J. Maswār.

S. sumarae is named after J. Sumārah where it still occurs. It is only found in Yemen and the Asir region of Saudi Arabia.

9. Senecio vulgaris *L.*

Groundsel. An erect glabrescent annual herb to 30cm, usually with tufts of cottony hairs on the peduncles and in the axils of young leaves. Leaves sessile, oblong or oblanceolate, pinnatifid and irregularly dentate, 4–8cm long. Inflorescence a dense terminal corymb, the peduncles elongating as the flowers open; flower-heads almost always unrayed, 3–5mm wide.

An uncommon weed of gardens and fields between 2300 and 2900m: J. Ṣabir; Yarīm; Sana'a; below the waterfall at Baw'an.

Senecio bipartitus Sch. Bip. is listed by Schwartz on the basis of a plant collected by Botta. The name was never in fact published and I was unable to find any specimen of Botta's at Paris under that name.

46. KLEINIA *Mill.*

Similar to *Senecio* but always succulent and lacking the outer row of involucral bracts. All our species have unrayed flower-heads.

1. Inflorescence umbellate with small pale yellow flower-heads. Plant with stiff, erect, mostly leafless stems — *K. odora*
   Flower-heads solitary, long-pedunculate, red or purple, at least 1.5cm wide. Stems decumbent or erect, leafy or leafless — **2**
2. Leaves reduced to scales. Flower-heads red — *K. pendula*
   Leaves fully developed, always present. Flower-heads purple — *K. semperviva*

1. Kleinia odora (*Forssk.*) *DC.* 'Udhkhur

Senecio odorus (Forssk.) Sch. Bip.

Erect, much-branched, glabrous undershrub with ridged, succulent, often leafless stems to 2m. Leaves at the tips of the branches, quickly deciduous to leave prominent scars, sessile, linear-lanceolate, entire, succulent, 1–2cm long. Inflorescence of terminal umbels; flower-heads pale yellow, 2–4mm wide. Achenes villous.

Widespread and often abundant on open, well-drained stony hillsides between 600 and 2500m; common within all suitable areas of Yemen except the east of the Mashriq.

2. Kleinia pendula (*Forssk.*) *DC.*

Senecio pendulus (Forssk.) Sch. Bip.

Glabrous leafless plant with thick, branched, cylindrical stems which are decumbent or pendulous on cliffs. Leaves reduced to subulate scales up to 3cm long, these arising from wart-like outgrowths on the stem. Inflorescence of solitary, terminal, long-pedunculate, red flower-heads 2–4cm wide. Achenes villous.

A curious and attractive-flowered succulent found on cliffs and steep rocky slopes between 800 and 1900m. It is tolerably frequent in the central escarpment and in the Hujarīyah but otherwise is a rare plant of scattered localities in the south of the country: J. Jinān (Wissmann); Al Ḥaradah, Damt; Qā'idah; between Ḥays and Taiz (Botta): Wadi Ḥidrān, Taiz. This last is the only locality I know of where it flowers at all regularly.

3. Kleinia semperviva (*Forssk.*) *DC.*

Senecio semperviva (Forssk.) Sch. Bip.

Clump-forming undershrub with short erect, simple or forked stems 10–20cm high. Leaves clustered at the branch tips, sessile, oblong-oblanceolate, entire, glabrous, succulent, farinose when young, 3–6cm long. Inflorescence of solitary, terminal, long-pedunculate purple flower-heads 2–3cm wide. Achenes villous.

A common plant of cliffs and rocky slopes between 2000 and 3000m all over the high plateau; rare on the escarpment and only known from J. Raymah, J. Ḥiṣn Madnān in Waṣāb al 'Ālī, J. Sawraq and J. Ḥabashī.

47. EURYOPS *Cass.*

1. Euryops arabicus *Steud.* — Shajarah Al Kans

E. pinifolius auct.

Small, much-branched, erect undershrub to 40cm, covered in the woody scale-like remains of the leaf bases on old wood. Leaves densely crowded towards the branch tips, alternate, sessile, linear, glabrous, 2–4cm long. Inflorescence a dense mass of rayed yellow flower-heads arising in the axils of the leaves towards the branch tips; tubular and ligulate florets present, both yellow; involucre of a single row of similar, yellowish bracts. Achenes cylindrical, ribbed, pubescent; pappus of short white hairs.

Locally abundant between 2000 and 3000m in open rocky places on the mountains surrounding the high plateau, in similar areas of Khawlan Ash Sham and on the Dhubhān plateau in the Ḥujarīyah. it is commonly associated with *Juniperus*, *Psiadia*, *Lavandula* etc.

48. CALENDULA *L.*

1. Calendula arvensis *L.*

C. aegyptiaca Desf.

Very variable, sparsely hispid, ascending annual herb, usually branched at the base. Leaves alternate, sessile, weakly clasping, linear-lanceolate, 2–4cm long. Flower-heads solitary, yellow, with both tubular and ray florets; involucral bracts in one row, equal. Achenes without pappus, boat-shaped, curved or ring-like, spiny to smooth on the outer surface, glabrous.

A frequent field weed on the high plateau between 2000 and 3000m though more common on irrigated land or in areas of high rainfall; rarer on the escarpment but known from J. Ṣabir, Manākhah, Shahārah and Saqayn.

49. OSTEOSPERMUM *L.*

Woody herbs with alternate, shallowly pinnately-lobed leaves. Inflorescence loosely racemose to subcorymbose; involucral bracts in one row, all alike, all with scarious margins; outer florets ligulate; inner florets tubular. Achenes obscurely 3-sided, glabrous, sometimes winged, without pappus.

Achenes with conspicuous scarious wings. Flower-heads large, exceeding 1cm diameter — *O. vaillantii*
Achenes not winged. Flower-heads small, 0.5–1cm diameter — *O. muricatum*

1. Osteospermum muricatum *E. Mey.*

Much-branched, bushy, grey-green woody herb to 50cm, gland-dotted and sticky but otherwise ± glabrous. Leaves sessile, linear, pinnately-lobed, 1.5–2cm long. Inflorescence a few-flowered leafy raceme of small, rayed, yellow flower-heads. Achenes dark brown, muricate on the back, without wings, 4–5mm long.

An uncommon plant found on alluvial soils near streams and irrigation ditches and on the margins of irrigated fields between 2200 and 3000m on the high plateau and in similar areas of Khawlan Ash Sham: Wa'lan (Wissmann); Sana'a (Rathjens); Thilā; Baw'an; between 'Amrān and Raydah; between Ḥūth and Khamir; Sāqayn; between J. Minmār and Ṣa'dah.

A plant with a most unusual distribution, being known outside Yemen only from Somalia and Namibia.

2. Osteospermum vaillantii (*Decne.*) *T. Norlindh.*

Tripteris vaillantii Decne.

Much-branched, decumbent or ascending woody herb to 70cm, gland-dotted and scabrid but otherwise nearly glabrous. Leaves shortly petiolate, oblong, pinnately-lobed or merely dentate, 2–4cm long. Inflorescence a lax, usually 3–5-flowered, weakly corymbose raceme with large rayed yellow flower-heads. Achenes pale brown with 2 large scarious wings, c 8mm long.

Widespread and often abundant on terrace walls and open rocky hillsides between 1500 and 3200m on the escarpment and high plateau. It is one of the commonest large yellow-flowered composites in Yemen, particularly in areas of reasonable rainfall.

50. ECHINOPS *L.*

Robust woody herbs. Leaves alternate, pinnately-divided, spiny. Inflorescence a large globose head, formed of numerous 1-flowered flower-heads; involucral bracts free, forming a tubular involucre; florets tubular. Achenes cylindrical, pilose, the pappus consisting of a ring of short scales.

A poorly understood genus which requires further collecting, particularly so that the very variable *E. spinosissimus* can be studied more carefully.

1. Stem and the involucral bracts densely covered with stalked and sessile glands. Plant 1½–3m high — *E. sp.A*
   Stem and involucral bracts eglandular or with a few glands only. Plant less than 2m high — 2
2. Leaves simply pinnatifid, usually more than 5mm wide between the lobes — *E. sp.B*
   Leaves bipinnatifid, always less than 5mm wide between the lobes — *E. spinosissimus*

1. Echinops spinosissimus *Turra* — Būrī, Jīrdab

E. spinosus L.

E. boranensis Lanza

Erect, sparingly branched perennial to 1½m; stem purplish and cobwebby in young parts, usually with a few scattered stalked glands below the inflorescence. Leaves 20–40cm long, sessile, green or glaucous above, white-cobwebby below, bipinnatifid with the veins terminating in hard yellowish spines. Inflorescence terminal, 4–8cm in diameter, the florets white or bluish; involucral bracts, pale glaucous, spinescent, the outermost developing into hard spines which often exceed the florets, ± glabrous, mostly 1.5–2cm long.

An extremely variable plant, almost every collection differing in some respect from the next. A lot of further material is needed to see whether the variation is continuous or whether there are in fact a number of reasonably well-defined species or subspecies. Two forms appear outstanding at the present time and will probably be worthy of some sort of taxonomic recognition in the future. The first represented by Wood 3160 is found on sand dunes at Ḥarīb and possibly also in the Jawf, growing at a lower altitude (c 1100m) than other *Echinops* species in Yemen. It is distinguished at a distance by its strikingly blue flowers and on close examination by the glandular leaves, the upper surface and the veins below being densely covered in stalked and sessile glands. Another very distinct form represented by Wood 2880 is found at high altitudes mainly on the inner escarpment. In this the involucre develops numerous long spines so the inflorescence appears to be a spiny ball and the flowers are almost completely hidden. The leaves are reduced to green-winged yellow spines.

Apart from the isolated records from Ḥarīb and the Jawf *E. spinosissimus* is found only on the escarpment and high plateau between 1500 and 3600m. It is widespread and often abundant on open rocky hillsides, terrace walls, derelict fields and cliffs.

2. Echinops sp.A

Similar to *E. spinosissimus* but readily recognised by its robust habit and glandular indumentum. Erect perennial 1½–3m high with dark purple stems which are never cobwebby but densely covered in numerous sessile and stalked glands. Leaves 20–80cm long, dark green and glabrous above apart from scattered glands, white-cobwebby below, bipinnatifid (some basal leaves ± 3-pinnatifid), the veins terminating in spines. Inflorescence up to 10cm in diameter, florets white; involucral bracts glandular, more than 2cm long.

Flowers mostly in July.

An uncommon plant found on cliffs between 2500 and 3000m: Shaḥir, J. Milḥān; J. Shibām, Ḥarāz; Kawkabān; Maḥjar in Wadi Aḥjar; Kuḥlān; J. Maswār. Particularly fine colonies can be seen on the cliffs on the west side of Kawkabān (Wood 2445, 3337).

3. Echinops sp.B

Slender, erect or ascending, completely eglandular plant similar to *E. spinosissimus* but distinguished by its leaves. These are simply pinnatifid and relatively shallowly lobed, usually being at least 5mm wide between the lobes. The inflorescence is 3–5cm in diameter, the florets bluish-white and the involucral bracts only 10–12mm long.

Apparently a rare plant of steep, partially shaded slopes between 1300 and 1900m in the wettest parts of the central escarpment: J. Ẓallamlam above Ḥadīyah; between Ḥabash and Ṣa'bah, J. Milḥān (Wood 2851).

51. ATRACTYLIS *L.*

Spiny thistle-like undershrubs. Leaves alternate, pinnatifid, the lobes spine-tipped. Flower-heads solitary, terminal, bell-shaped, whitish-yellow, subtended by a whorl of leaves; involucral bracts in many rows, spine-tipped; tubular florets only present. Achenes cylindrical, silky-hairy with plumose pappus.

Involucral bracts acute, densely and uniformly covered in grey, cobwebby hair — *A. carduus*
Involucral bracts emarginate with a spine arising from the sinus, somewhat cottony at the apex but not uniformly grey-cobwebby — *A. kentrophylloides*

1. Atractylis carduus (*Forssk.*) *C. Chr.*

A. flava Desf.

Much-branched villous undershrub with decumbent or ascending stems reaching 50cm high. Leaves sessile, but not decurrent onto the stem, rather widely spaced, the margins indistinctly white-cottony, 3–6cm long. Flower-heads large, tightly embraced by the erect subtending leaves; involucral bracts grey, cobwebby-hairy, acute.

An uncommon plant of gravelly plain between 1800 and 2300m in the Mashriq: near Sana'a airport; Bani Hārith; above the Fardah pass, Nihm; Majz (Steinberg).

2. Atractylis kentrophylloides (*Baker*) *F.G. Davies*

A. yemensis Podl.
Carduncellus kentrophylloides Baker

Very spiny undershrub c 30cm high with many short spreading or ascending branches. Leaves sessile, closely arranged along the stem with very short internodes, the margins white-cottony, 2–4cm long. Flower-heads loosely embraced by the ascending subtending leaves; involucral bracts green or purplish, often white-cottony near the emarginate apex, but otherwise glabrescent.

Widespread but uncommon on the north of the high plateau and in the Mashriq from Aḥjar, Sana'a and J. Lawz northwards; it grows on dry cliff slopes and open rocky plain, particularly on limestone from 1900 to 3000m.

Safflower, **Carthamnus tinctorius** *L.*, known as 'Qurtum', is occasionally cultivated, particularly in Arḥab and on the Dhubhān plateau where it is grown round the edges of fields. It is a spiny-leaved woody annual with orange flower-heads from which a yellow dye can be extracted.

52. CIRSIUM *Mill.*

1. Cirsium vulgare (*Savi*) *Ten.*

C. lanceolatum (L.) Scop.

Spear Thistle. Spiny, cottony-hairy biennial herb producing a basal rosette in the first year from which arises an erect stem to 1½m in the second year. Stem leaves alternate, sessile with the base decurrent as spiny wings on the stem, oblong, deeply pinnatifid with spiny lobes, 3–12cm long. Flower-heads in small clusters, sometimes solitary; involucre bell-shaped with many rows of spine-tipped bracts; florets all tubular, purple. Achenes c 3mm long, yellowish, with a pappus of long, plumose, silky hairs.

A plant of marsh and damp grassland found between 1600 and 2900m, with a similar distribution to several temperate plants such as *Apium nodiflorum* and *Lotus corniculatus*; widespread on the high plateau, extending south through the Ibb region to Wadi Sūdān at Qā'idah and Mudhaykhirah, but only known elsewhere from J. Shibām, Manākhah (Schweinfurth).

53. ONOPORDUM *L.*

1. Onopordum arabicum *Podl.*

O. sibthorpianum sensu Schwartz
O. ambiguum auct.
Scotch Thistle. Spiny, grey, cottony-hairy, biennial herb, producing a basal rosette in the first year from which arises an erect stem to 2m in the second year. Stem leaves alternate, sessile, with the base decurrent as spiny wings on the stem, oblong-elliptic, with small, irregular, triangular spine-tipped lobes 6–15cm long. Flower-heads solitary, large, terminal; involucre bell-shaped with many rows of spiny bracts, the outer ones recurved, florets all tubular, purple. Achenes 4-angled, glabrous, the pappus with simple hairs.

Widespread and frequent between 2200 and 2900m on the high plateau growing on roadsides and in abandoned fields; also in Khawlan Ash Sham at Sāqayn.

The Globe Artichoke, **Cynara scolymus** *L.*, is cultivated on a small scale at Sana'a for its edible involucral bracts. It is a tall thistle-like plant with unarmed, pinnatifid leaves.

54. VOLUTARIA *Cass.*

Similar to *Centaurea* but the leaves always pinnatifid, the involucral bracts unarmed and the achenes pubescent.

Leaf lobes dentate. Stem densely white-tomentose — *V. albicaulis*
Leaf lobes ± entire (rarely an odd tooth present). Stem greenish with thin cottony indumentum — *V. abyssinica*

1. Volutaria abyssinica (*Sch. Bip. ex A. Rich.*) *C. Jeffrey ex Cuf.*

Centaurea hochstetteri (Buch.) Oliv. & Hiern
Erect, branched perennial herb to 50cm, the whole plant covered in thin cottony indumentum. Leaves sessile, pinnatifid, 3–8cm long, the lobes oblong, entire or (rarely) with 1–2 teeth. Flower-heads mauve, solitary, terminal on long peduncles; involucre much shorter than the florets, the bracts with a dark hardened apex.

There is some doubt as to whether *V. abyssinica* is the correct name for this plant (Wood 75/159, 1598) as it is part of a very complex group of species which also includes *V. somaliensis* (Oliv. & Hiern) C. Jeffrey ex Cuf. I suspect the two African species are conspecific with the Yemen plant and have therefore used the oldest name available.

A plant of open stony hills between 1500 and 2500m, often growing on limestone; widespread and common on the high plateau, extending into the Mashriq and onto the escarpment in the limestone areas around Ḥajjah and Maḥābishah; also in the Ṣa'dah area.

2. Volutaria albicaulis (*Deflers*) *J.R.I. Wood* comb. nov.

Volutarella albicaulis Deflers in Bull. Soc. Bot. de France 43:109 (1896)
Centaurea albicaulis (Deflers) O. Schwartz
Similar to the previous species but less robust, apparently not exceeding 20cm and the stem densely white-tomentose. Leaves pinnatifid with oblong, dentate lobes. Flower-heads as in *V. abyssinica* but more shortly pedunculate and the involucre ± equalling the florets.

Known only from cliffs at 1600 to 2000m in the Ḥujarīyah: Rubaysah near Turbah; Najd Dhamrān. (Wood 75/36)

V. albicaulis is known only from southwest Arabia. Unfortunately the type specimen is missing but Wood 75/36 agrees quite well with plants from Dhofar and with Deflers' description so there is little doubt that this plant is correctly named.

55. CENTAUREA *L.*

Herbs or shrubs with alternate, simple or pinnatisect leaves. Flower-heads usually solitary, rarely 2–3 together, at the ends of the branches; involucre hemispherical to campanulate with many rows of closely appressed, often spine-tipped bracts; all florets tubular with long, often spreading lobes. Achenes glabrous, pappus of bristles.

1. Involucral bracts spine-tipped — 2
   Involucral bracts not spine-tipped — *C. schimperi*
2. Flower-heads reddish-purple. Involucral bracts with a single pair of lateral spines at the base of the central spine — *C. yemensis*
   Flower-heads yellow. Involucral bracts with at least two pairs of lateral spines at the base of the central spine — *C. pseudosinaica*

1. Centaurea pseudosinaica *Czerep.* — Laḥīb

C. pallescens sensu Schwartz
C. melitensis sensu Schwartz
Crisped-pubescent perennial herb to 80cm. Leaves sessile, decurrent onto the stem, lanceolate, coarsely dentate or pinnatifid especially below, up to 10cm long and 1cm wide. Flower-heads always solitary, 1–2.5cm long; involucral bracts with a central spine 1–3cm long and with 2 pairs of lateral spines at the base, the spines yellowish; florets yellow.

The Yemen plant is subspecies **niebuhrii** *Wagenitz.*

A plant of cultivated and waste land, particularly field borders and roadsides, which sometimes extends onto rocky hillsides; it is widespread between 400 and 3400m in all parts of Yemen except the Tihama and the Sands but is generally rare at low altitudes and on the escarpment; on the high plateau it is one of the most conspicuous plants in waste places.

2. Centaurea schimperi *DC.*

Woody perennial, much branched at the base with numerous erect, almost leafless flowering branches to 50cm, the whole plant thinly covered in cottony indumentum. Leaves sessile, linear, entire, dentate or obscurely pinnatifid, 1–3cm long, 1–3mm wide. Flower-heads solitary, long-pedunculate, c 2cm long; involucral bracts yellowish, without spine-tips, florets yellow.

A very local but sometimes abundant plant of bare stony hills between 1700 and 2400m with a very pronounced preference for sandstone: abundant on sandstone hills east of Rada'; at the head of the Wadi Ḍahr near Suq Bayt An Nā'im; frequent in the Ṣa'dah area, extending south to 'Ujlī and west to Sāqayn and Jifān.

3. Centaurea yemensis *Wagenitz*

Much-branched perennial herb to 30cm, ± cottony-tomentose when young but becoming crisped-pubescent with maturity. Leaves sessile, linear-oblanceolate, weakly dentate but never pinnatifid, 3–5cm long, 3–9mm wide. Flower-heads solitary, 2–2.5cm long; involucral bracts with a central spine 10–15mm long and with a single pair of small lateral spines arising from its base; florets reddish-purple.

A distinctive endemic species which is frequent on the Dhubhān plateau around Turbah between 1900 and 2100m, growing on grassy banks around fields and on stony hillsides.

56. CENTAUROTHAMNUS *Wagenitz & Dittrich*

1. Centaurothamnus maximus (*Forssk.*) *Wagenitz & Dittrich* Bayaḍ, Khambūb

Centaurea maxima Forssk.

Sparingly branched shrub to 2m, the stems densely white-tomentose. Leaves alternate, petiolate, lanceolate to narrowly elliptic, green above, densely white-tomentose below, 8–15cm long, 2–6cm wide. Flower-heads solitary at the ends of the branches, 3–4cm long; involucre campanulate with many rows of bracts; florets purple, all tubular. Achenes glabrous, pappus of bristles.

A spectacular and distinct species, found on cliffs between 1700 and 3200m; widespread and frequent on the escarpment; rare on the high plateau. The species and genus are both endemic to the mountains of southwest Arabia.

57. DICOMA *Cass.*

1. Dicoma tomentosa *Cass.*

Much-branched, erect, woody annual with purplish, white cottony stems to 60cm. Leaves alternate, linear, mucronate, serrate, dull green above, white-tomentose below, 2–5cm long. Flower-heads solitary, subsessile, axillary or on short lateral branches; florets all tubular, yellow-brown; involucre bell-shaped, the bracts in many rows, tipped with yellow spines. Achenes cylindrical, pilose; pappus of plumose hairs.

Only known from open *Acacia* scrub on a steep, stony hillside at 1200m in Wadi 'Ayn 'Alī near Ḥajjah.

58. PILOSELLOIDES (*Less.*) *C. Jeffrey*

1. Piloselloides hirsuta (*Forssk.*) *C. Jeffrey*

Gerbera piloselloides (L.) Cass.

Leaves all in a basal rosette, petiolate, obovate-oblong, 2–6cm long, the upper surface dark green, the lower surface grey-tomentose. Flower-heads large, solitary, terminal on an unbranched, purple, villous peduncle 5–25cm long; involucral bracts in many rows, grey, villous; outer florets ligulate with white rays which are pinkish on the under-surface; inner florets tubular, white. Achenes 7–8mm long, beaked, strigose, the pappus of numerous silky hairs.

A local plant found in grassland between 2000 and 3000m on the escarpment mountains from Akāhilah near Turbah north to J. Milhān. Its distribution is closely related to the frequency of winter mists and it occurs at lower altitudes in the central escarpment and Turbah than in the Ibb area.

59. CICHORIUM *L.*

1. Cichorium bottae *Deflers*

A very distinctive, stemless, sparsely hairy, mat-forming perennial from a woody rootstock. Leaves in small dense tufts, linear, entire or sinuate-dentate, 1.5–2.5cm long. Flower-heads blue, sessile or with short thick peduncles, solitary and terminal on the leafy tufts; involucral bracts in 2 distinct rows; ligulate florets only present. Achenes c 2mm long, brown, obscurely ridged, crowned with a very short, pale brown, scaly pappus.

A characteristic plant of the high plateau and surrounding mountains, found between 2300 and 3600m growing on grassy banks particularly around fields and often associated with *Dianthus uniflorus* and *Convolvulus thymoides*. It also occurs further north at Shahārah and on J. Jifān and at Madaq (Steinberg) in Khawlan Ash Sham. It is one of the most attractive and distinctive endemic species, being found from Yemen north to Taif in Saudi Arabia.

The endive, **Cichorium endivia** *L.*, is occasionally cultivated in Sana'a.

60. TOLPIS *Adans.*

1. Tolpis altissima (*Balb.*) *Pers.*

Erect, sparsely cottony-hairy, glabrescent perennial with an erect stem to 1m and both basal and stem leaves. Leaves narrowly oblanceolate, with few, large marginal teeth or (rarely) ± entire. Flower-heads corymbose, yellow, the florets all ligulate; involucral bracts 2-rowed, linear, greenish black. Achenes cylindrical, ribbed, glabrous, brown; pappus of 5–10 stiff white hairs.

An uncommon plant of grassy banks usually by streams, found on the northern escarpment between 2000 and 3000m: above Baw'an; Wadi Aḥjar; Bayt Bashārī near Maḥwīt; near Bayt 'Adhāqah.

61. PICRIS *L.*

Annual or perennial bristly-hairy herbs with a simple or branched flowering stem. Leaves mostly in a basal rosette but a few on the stem. Flower-heads solitary, yellow, the florets all ligulate; involucral bracts blackish-green with pale margins, ± 2-rowed, the outer much shorter than the inner. Achenes ± cylindrical, curved, ribbed, beaked; pappus of plumose dirty-white hairs.

Flower-heads mostly 10–20mm wide. Flowering stems densely and uniformly covered in spreading bristly hairs *P. scabra*
Flower-heads mostly 5–10mm wide. Flowering stems with few spreading bristly-hairs especially in the upper half *P. sp.A*

1. Picris scabra *Forssk.* Murār
P. abyssinica sensu Schwartz
P. sulphurea sensu Schwartz
Perennial herb densely covered in spreading bristly hairs, up to 30cm high. Flowering stems erect, simple or once branched, usually with 1–2 leaves and a few scattered scales. Basal leaves oblong-oblanceolate, obtuse, entire or (rarely) weakly pinnatifid with an undulate margin, 3–10cm long. Flower-heads solitary, yellow, 1–2cm wide. Achenes brown.

Widespread and frequent in montane grassland and on rocky slopes in areas of good rainfall between 2000 and 3200m on the escarpment and more rarely on the mountains surrounding the high plateau.

P. scabra is endemic to southwest Arabia but closely related to *P. abyssinica* Sch. Bip. from Ethiopia.

2. Picris sp.A
Rather similar to *P. scabra* but more slender, reaching only 20cm and only sparsely bristly on the leaves and stems. The leaves are clearly oblanceolate in outline and almost always pinnatifid, the margin otherwise entire, not undulate. Flower-heads only 5–10mm wide.

A plant of derelict stony fields and alluvial plains between 1500 and 3000m; it is widespread and often abundant on the high plateau, making a show of yellow flowers following heavy rain; more local on the northern escarpment and in Khawlan Ash Sham but recorded from Ḥajjah, Shahārah, the Wadi Badr and Majz (Steinberg). (Wood 75/599, 1609, 3234)

P. sp.A is also apparently endemic but probably extends into Saudi Arabia.

62. CREPIS *L.*

Herbs (ours perennial) with leaves in a basal rosette. Stems branched, leafless or leafy. Flower-heads corymbose; involucre ± tubular, the bracts in 2 distinct rows or overlapping so the rows are indistinct; florets all ligulate, yellow. Achenes cylindrical, ribbed, minutely scabrid, beaked; pappus of white hairs.

1. Stem leaves always present, hispid on the veins and margins. Involucre and peduncles very bristly *C. sp. B*
   Stem leaves usually absent, but if present glabrous. Involucre and peduncles with a few sparse bristles 2
2. Involucral bracts in 2 dissimilar rows; inner bracts less than 1.5mm wide at the base *C. sp. A*
   Involucral bracts overlapping, similar and not differentiated into rows at least at flowering time; inner bracts more than 2mm wide at the base in fruit *C. nabishaibica*

1. Crepis rueppellii *Sch. Bip.* Murār
C. forskalii Babc.
Perennial herb to 50cm, the flowering stems slender, branched, almost glabrous, leafless or with 1–2, often reduced leaves. Basal leaves linear-oblanceolate, gradually narrowed to a petiole-like base, dentate to pinnatifid, glabrous, 5–30cm long. Flower-heads medium sized, the outer florets red underneath; involucre cylindrical c 7–8mm long and 4–5mm wide, adpressed-hairy and with a few black bristles, the indumentum spreading to the peduncle below; involucral bracts in 2 rows, the outer very different from the inner which are linear and 1–2mm broad. Achenes with about 10 ribs.

A very variable species, apparently responding strongly to the influence of light and water. Plants from shaded wet places have leaves in excess of 30cm long (Wood 683) while plants from open, grazed habitats in dry country have short, strongly pinnatifid leaves and are of quite different appearance (e.g. Wood 2339 from Tin'am).

Widespread on the escarpment and high plateau between 1200 and 3000m; in dry country restricted to the vicinity of streams but on the wetter parts of the escarpment abundant in a wide variety of habitats including cliffs, terrace walls, montane grassland, wadi beds, irrigation ditches and field borders; it sometimes grows as a weed under qat.

2. Crepis sp. A
Similar to *C. rueppellii* but more compact in habit and not exceeding 15cm high. It is best distinguished by the involucral bracts which at flowering time are not arranged in distinct inner and outer rows but merge imperceptibly with each other. The inner bracts on fruiting flower-heads are linear-lanceolate, 8–10mm long, 2–3mm wide at the base.

Only known from the upper slopes of J. Nabi Shu'ayb (3200–3600m) but worth looking for on other high mountains northwest of Sana'a (Miller 160; Wood 2967, 3240).

3. Crepis sp. B
Perennial herb to 25cm, the flowering stem branched, leafy, bristly especially above. Leaves oblanceolate with a petiole-like base, dentate, sparsely bristly on the margins and the veins below (especially on the stem leaves), 4–15cm long. Flower-heads fairly large, the outer florets red underneath; involucre bell-shaped c 12–15mm long, 5–10mm wide, pubescent and bristly; involucral bracts irregularly arranged, the outer merging with the inner which are linear and up to 14mm long and 2.5–4mm wide. Achenes with 15–20 ribs.

Locally abundant on cliffs and steep rocky slopes between 2900 and 3200m on a few of the Ibb mountains: J. Ta'kar; J. Manār; J. Sumārah. (Wood 1149, 2481, 2877)

63. TARAXACUM *Weber*

Dandelions. Herbs with a rosette of entire or pinnatifid leaves and solitary yellow flower-heads borne on a leafless hollow, simple peduncle; florets all ligulate; involucral bracts in 2 rows, the inner much longer than the outer. Achenes cylindrical, ribbed, long-beaked; pappus of simple white hairs.

Flower-heads small, c 1cm wide. Outer involucral bracts erect and ± appressed to the inner bracts — *T. sp.A*
Flower-heads large, 2–4cm wide. Outer involucral bracts reflexed — *T. officinale*

1. Taraxacum officinale *Weber* s.l.
Almost glabrous perennial. Leaves oblong, pinnatifid (rarely entire), 5–18cm long. Flower-heads large, 2–4cm wide; involucral bracts green, the inner erect, c 15mm long, the outer reflexed to ± spreading. Achenes usually brown, 2–3 times shorter than the beak.

An introduction from Europe, naturalised in the lawns of the British Embassy in Sana'a at 2300m.

2. Taraxacum sp.A
Slender glabrous perennial. Leaves oblong or oblanceolate, entire or pinnatifid, 5–8cm long. Flower-heads small, c 1cm wide; involucral bracts green, the inner erect, c 10mm long, the outer erect and usually appressed to the inner bracts. Achenes olive green, about as long as the beak.

Although some populations apparently have only pinnatifid leaves (Wood 2321) and others have only entire leaves (Wood 2413), still others have leaves of both kinds (Wood 3193). Since the Yemen plants do not seem to differ in any other way, it seems best to regard all Yemen plants as conspecific until evidence can be found to demonstrate that the entire-leaved forms differ specifically from the pinnatifid-leaved forms. *Taraxacum* is in any case a very complex genus.

An uncommon endemic plant of seasonally flooded grassland between 2000 and 2600m on the high plateau, often associated with *Falkia oblonga*: near Ribāṭ at the head of the Wadi Banā; between Yarīm and Raḍmah; Rada'; Zīl, Al Ḥadā'; in the wadi north of Wa'lan.

64. LACTUCA *L.*

Annual or perennial herbs with both basal and stem leaves. Inflorescence a bracteate panicle of small, cylindrical flower-heads; involucral bracts in many rows, glabrous, with scarious margins; florets all ligulate. Achenes ribbed, flattened, beaked; pappus of simple whitish hairs.

Flowers blue or white. Leaves unarmed — *L. inermis*
Flowers yellow. Leaves with bristly spines on the margins and veins below — *L. serriola*

1. Lactuca inermis *Forssk.*
L. capensis Thunb.
L. yemensis Deflers
Variable, almost glabrous perennial to c 60cm, the stem decumbent to erect, usually branched only in the upper half. Basal leaves linear or oblong, entire or dentate, unlobed or pinnatifid, glabrous or with a few bristles on the margins and veins, 8–15cm long; stem leaves entire, sessile, often sagittate. Inflorescence diffuse with many small blue or whitish flower-heads. Achenes dark brown, obscurely ribbed, glabrous, finely beaked.

A plant of damp, partially shaded places which is only common in the mist areas of the escarpment. It grows in a wide variety of suitable habitats between 1200 and 3000m, including grassland, cliff gullies, terrace walls and rocky banks. On the high plateau it is rare and restricted to streamsides.

2. Lactuca serriola *L.*
Biennial, nearly glabrous woody herb with an erect, sparingly branched stem to 1½m. Leaves oblong, entire or deeply pinnatifid, often sagittate at the base, glaucous, glabrous but with bristly spines on the margins and veins below, 10–30cm long. Inflorescence pyramidal with rather few, small, untidy, yellow flower-heads. Achenes olive-grey, prominently ribbed, bristly at the apex, with a slender white beak.

A local, probably introduced species found on roadsides and in wadi beds between 1100 and 2400m: Qā'idah; Madīnat Al 'Abid; Nādirah in the Wadi Banā; 'Attārah in the Ḥarāz (Wissmann); Sana'a.

Lettuce, **Lactuca sativa** *L.*, 'Salātah', is an increasingly popular salad vegetable. It is commonly grown in market gardens at Sana'a and to a lesser extent at the other larger towns.

65. REICHARDIA *Roth*

1. Reichardia tingitana (*L.*) *Roth* — Murār
Erect glabrous herb to 40cm. Leaves mostly basal, obovate to oblanceolate, dentate and often ± pinnately lobed, 2–6cm long; stem leaves 0–2, cordate, clasping, with rounded auricles. Peduncle with a few scales near the top, simple or sparingly branched; flower-heads large, yellow, terminal; involucre with many rows of ovate, glabrous, white-margined bracts; florets all ligulate. Achenes c 2mm long, unbeaked with a pappus of silky white hairs.

A widespread and common plant found from 600 to 2900m on the escarpment and high plateau, growing in disturbed places including roadsides, terrace walls and cultivated land.

66. SONCHUS *L.*

Sow-thistle. Annual or perennial herbs with white latex. Leaves often glaucous, the stem leaves sessile, clasping. Inflorescence terminal, cymose; involucre with several rows of bracts; florets all ligulate, yellow. Achenes flattened, ribbed; pappus of cottony hairs.

1. Perennial. Involucre and pedicels with numerous black glandular hairs — *S. saudensis*
   Annual. Involucre and pedicels without black glandular hairs — **2**

2. Achenes smooth. Stem leaves with rounded auricles *S. asper*
   Achenes rugose particularly on the ribs. Stem leaves with pointed auricles *S. oleraceus*

1. Sonchus asper (*L.*) *Hill*

Erect, sparingly-branched, annual herb with glabrous, hollow stems to 1m. Stem leaves sessile with rounded auricles, lanceolate to obovate, pinnatifid or not, the margin spinous-dentate, usually 10–15cm long. Inflorescence an umbel-like cyme of small yellow flower-heads; involucral bracts glabrous when mature; achenes c 3mm long, smooth.

A widespread weed of disturbed damp places, particularly irrigation ditches and irrigated fields found from 1100 to 2700m on the escarpment and high plateau, but occurring in widely scattered localities and rather rare.

2. Sonchus oleraceus *L.*

Similar to *S. asper* but the leaves almost always pinnatifid, the auricles pointed and the achenes distinctly rugose, especially on the ribs.

A widespread weed of cultivated land found from 100 to 2900m extending from Mawza' in the Tihama east to Mārib but mostly in irrigated land or in high rainfall areas; much more common than *S. asper*.

3. Sonchus saudensis *L. Boulos*

S. melanolepis sensu Schwartz

Decumbent or ascending, nearly glabrous perennial herb with a tuft of basal leaves and an almost leafless stem. Stem leaves sessile always (?) without auricles, oblanceolate, deeply pinnatifid, the margin dentate, 5–10cm long. Inflorescence usually with 3 flower-heads, variously arranged in a terminal cluster or cyme or along a very long axis; involucral bracts with a tuft of wool at the base and numerous black, glandular hairs; achenes c 2.5mm long, smooth.

Widespread on the escarpment from 1500 to 3200m and common on terrace walls and cliffs but mostly restricted to areas of good rainfall. It is endemic to the mountains of southwest Arabia.

67. LAUNAEA *Cass.*

Herbs with a branched, leafy or leafless flowering stem. Leaves in a basal rosette, usually pinnatifid and dentate. Flower-heads yellow with ligulate florets only; involucral bracts in several rows, usually with scarious margins and merging into scales on the peduncle. Achenes cylindrical, beakless or with a very short beak; pappus of simple white hairs.

1. Flower-heads sessile or borne on peduncles less than 5mm long, usually clustered or in pairs 2
   Flower-heads pedunculate, the peduncles at least 10mm long, not clustered 5
2. Flower-heads mostly in clusters at the ends of the stems. Achenes broadly winged *L. capitata*
   Flower-heads mostly in rather distant clusters (often paired) along the branches of the inflorescence. Achenes not winged 3
3. Annual. Stem leaves numerous, 5–25cm long *L. intybacea*
   Perennial. Stem leaves absent or if present few, less than 6cm long **4**
4. Pappus quickly deciduous and easily detached from the achenes; outer achenes acuminate *L. procumbens*
   Pappus persistent on the achene and not easily detached; all achenes truncate *L. nudicaulis*
5. Flower-heads c 2mm wide. Leaves spiny-dentate *L. massauensis*
   Flower-heads 5–10mm wide. Leaves pinnatifid but otherwise entire *L. mucronata*

1. Launaea capitata (*Spreng.*) *Dandy*

L. glomerata (Cass.) Hook. f.

Short-lived glabrous perennial with ascending, usually leafless stems to 20cm. Leaves oblong, pinnatifid, glaucous, 4–8cm long, the margin dentate with small white teeth. Flower-heads subsessile, mostly in clusters at the ends of the branches, pale yellow, 5–9mm wide. Achenes broadly winged, pale yellow, 3–4mm long.

Only known from the Bani Ḥushaysh at c 2300m where it is common on sandy plain particularly where water accumulates in the rainy season.

2. Launaea intybacea (*Jacq.*) *Beauverd*

Lactuca arabica Jaub. & Spach

L. goraeensis (Lam.) Sch. Bip.

Glabrous annual herb with a basal rosette and an erect, leafy stem, which is much branched in the upper half and reaches 1m high. Leaves oblanceolate, pinnatifid, 5–25cm long, the margin finely spiny-dentate. Flower-heads subsessile or shortly pedunculate in rather distant clusters of (1–) 2 (–4) along the branches of the inflorescence, pale yellow, 3–7mm wide. Achenes brown, unwinged, very shortly beaked, 2.6–4mm long.

A widespread but not very frequent weed from sea level to 1500m in the Tihama and the lower slopes of the escarpment. It grows as a field weed and in fences and scrub around cultivation.

3. Launaea massauensis (*Fresen.*) *Chiov.*

Delicate glabrous annual herb with a basal rosette and a much-branched leafless stem to 40cm. Leaves lanceolate, pinnatifid, 5–10cm long, the margin spiny-dentate. Flower-heads long-pedunculate in a paniculate inflorescence, pale yellow, c 2mm wide. Achenes brown, unwinged, 2.6–4mm long.

A plant of sloping banks where there is little competition from other species; usually growing in partial shade on cliffs, in scrub or by shaded footpaths. Widespread between 500 and 1900m on the escarpment, but generally uncommon except in the central escarpment; in the Mashriq at Dhī Nā'im and on the Fardah pass.

4. Launaea mucronata (*Forssk.*) *Muschl.*

Glabrous, probably perennial, bushy, woody herb with a much-branched leafy stem to 30cm. Leaves oblong, pinnatifid but otherwise entire, 4–10cm long. Flower-heads long-pedunculate in a paniculate inflorescence, yellow, 5–10mm wide. Achenes brown, unwinged, 4.5–5mm long.

A local plant found on gravelly and sandy plain between 1000 and 2500m in the Mashriq and the high plateau: between Wa'lān and the Yislah pass; Sana'a; at the head of the Wadi Sirr, Bani Ḥushaysh; between Khamir and Ḥūth; Suq Bayt An Nā'im; Arḥab; Ḥarīb.

5. Launaea nudicaulis (*L.*) *Hook. f.* Sirbah
Glabrous perennial herb with a basal leaf rosette and ascending, branched, almost leafless stems to 30cm. Basal leaves oblong-oblanceolate, pinnatifid, with an obscurely toothed margin, 5–10cm long; stem leaves linear, serrate, not pinnatifid, 2–6cm long. Flower-heads subsessile or shortly pedunculate in rather distant pairs along the branches of the inflorescence, yellow, 3–6mm wide. Achenes brown, unwinged, truncate, 4.5–5.5mm long, the pappus persistent.

Widespread on the escarpment, high plateau and south-east mountains from 500 to 2600m, growing mainly in open areas with a reasonable soil depth including field borders, alluvial plains and roadsides; uncommon in high rainfall areas and most frequent on the high plateau and in Taiz province.

6. Launaea procumbens (*Roxb.*) *Dandy*
L. fallax Jaub. & Spach
Very similar to *L. nudicaulis* but the outer achenes acuminate and the pappus quickly deciduous and easily removed by hand.

Known only from the Wadi Jawf (Zeller) but possibly overlooked elsewhere.

68. UROSPERMUM *Scop.*

1. Urospermum picroides (*L.*) *Scop. ex F.W. Schmidt*
Robust, erect, much-branched, sparsely hispid annual herb to 60cm. Leaves sessile, auriculate, lanceolate, spiny-dentate on the margin, 5–15cm long. Flower-heads solitary, yellow, 2–3cm wide, the florets all ligulate; involucre consisting of 8 broadly lanceolate, dull green, accrescent bracts in 1 row. Achenes strongly curved, muricate, glabrous, brown-beaked; pappus of white plumose hairs.

An uncommon weed of fields and roadsides, found between 1800 and 2700m at 'Amrān and on the escarpment from the Ibb region north to Maḥābishah.

69. SCORZONERA *L.*

1. Scorzonera musilii *Velen.*
S. yemensis Podl.
Decumbent or ascending, much-branched, dull green, glabrescent perennial herb without a basal rosette. Leaves sessile, narrowly-linear, entire, 3–10cm long, of similar appearance to the stems. Inflorescence sparingly branched with solitary, yellow flower-heads 5–10mm wide; florets all ligulate; involucral bracts in 2 rows, the outer very short, the inner grey-green, linear. Achenes cylindrical, ribbed, glabrous, greeny-brown, unbeaked; pappus of brownish-white, plumose hairs.

An inconspicuous plant of sandy plain, the margins of watercourses, derelict fields and loose stony hillsides between 1000 and 3000m: widespread and frequent on the high plateau; in the Mashriq only known from Majz (Steinberg) and Ḥarīb but probably overlooked elsewhere; on the escarpment only at Ḥajjah.

HYDROCHARITACEAE

All our plants grow submerged in shallow sea water and are often associated with species of *Cymodoceaceae*. They have a creeping rhizone from which arise small tufts of leaves. The flowers have never been seen in Yemen but are dioecious, the 3–6 perianth segments being enclosed in a spathaceous bract. Fruit many-seeded, breaking up irregularly.

1. Leaves oblong or elliptic, up to 6cm long 3. *Halophila*
 Leaves linear, 5–60cm long **2**
2. Leaves 1–1.5cm wide, up to 60cm long 1. *Enhalus*
 Leaves c 0.5cm wide, up to 15cm long 2. *Thalassia*

1. ENHALUS *Rich.*

1. Enhalus acoroides (*L.f.*) *Rich. ex Steud.*
Marine plant with a long creeping rhizome from which arise clusters of 2–3 linear, strap-like leaves, 1–1.5mm wide and up to 60cm long, the leaf bases surrounded by the fibrous remains of old leaves. Fruit borne on a long spiral peduncle, reaching to the water surface.

Washed up on the shore at Hodeida (Schweinfurth) and Luḥayyah.

2. THALASSIA *Banks ex Konig*

1. Thalassia hemprichii (*Ehrenb.*) *Aschers.*
Submerged marine herb with a long creeping rhizome from which arise sessile leaf clusters, their base covered in scarious sheaths. Leaves linear, slightly falcate, 4–6mm wide, up to 20cm long, the apex rounded, entire. Flowers small, the male pedunculate, the female sessile.

Superficially very similar to *Cymodocea serrulata* and *Thalassodendron ciliatum* but differing in the scarred rhizome, sessile leaf clusters and entire leaf tips.

Found at Hodeida (Deflers) but not seen again.

3. HALOPHILA *Thou.*

Submerged marine herbs with petiolate, oblong or obovate leaves arising in pairs from a creeping rhizome. Flowers small, the male pedunculate, the female sessile.

Leaves 1.5–3cm long, long-petiolate, the petiole not hidden by sheaths *H. ovalis*
Leaves 3–5cm long, shortly petiolate, the petiole ± hidden by scarious sheaths *H. stipulacea*

1. Halophila ovalis (*R. Br.*) *Hook. f.*
Leaves oblong-obovate, glabrous, entire, 1.5–3cm long, 7–10mm wide, borne on a petiole 1–2cm long, the petiole naked and without scarious sheaths.

Washed up on the shore at Hodeida (Schweinfurth), Manẓār and Mukhā.

2. Halophila stipulacea *(Forssk.) Aschers.*
Leaves oblong, pubescent, minutely toothed, often puckered, 3–5cm long, 4–10mm wide, the petiole very short and ± hidden by the two prominent scarious sheaths.

A plant of relatively deep sea water which has not been refound in Yemen since Forsskal's original collection at Mukhā.

POTAMOGETONACEAE

POTAMOGETON *L.*

Perennial anchored waterweeds (pondweeds) with elongate, floating stems. Leaves alternate, thin, translucent and submerged or coriaceous and floating, usually with a sheathing stipule. Inflorescence a pedunculate spike usually raised above the surface of the water; flowers inconspicuous, bisexual, without petals but with 4, scale-like perianth segments. Fruit a drupe.

1. Floating leaves present — 2
   Floating leaves absent — 3
2. Submerged leaves linear. Floating leaves decurrent for a short distance along the petiole and appearing folded at the base — *P. natans*
   Submerged leaves with an expanded lanceolate blade. Floating leaves neither decurrent nor folded at the base — *P. nodosus*
3. Leaves linear, grass-like — *P. pusillus*
   Leaves elliptic, at least 15mm broad — *P. schweinfurthii*

1. Potamogeton natans *L.*
Floating leaves very variable in size but mostly 6–10 × 3–4cm, elliptic, coriaceous, opaque, rounded at the base, with 2 decurrent wings along the petiole which appear as folds in dried specimens; submerged leaves linear 15–30cm × 1–3mm. Fruiting spike dense, 3–8cm long.

Only known with certainty from a pond in Wadi Ahjar at 2500m (Wood 3345) but records from Sana'a (Deflers) and Ḥaddah (Rathjens) may be correct.

2. Potamogeton nodosus *Poir.*
Floating leaves mostly 5–12 × 2–3.5cm, oblong-elliptic, coriaceous, ± cuneate at the base and without decurrent wings; submerged leaves mostly 10–15 × 1.5–3cm, broadly lanceolate, translucent. Fruiting spikes similar to *P. natans*.

P. nodosus is difficult to distinguish from *P. natans* when submerged leaves are absent as is often the case. The floating leaves of *P. nodosus* have clearly visible transverse veins and lack the distinct leaf base of *P. natans*.

In ponds and slow-moving streams from 300–2900m; in suitable habitats on the escarpment and high plateau and even occurring in the Tihama near Sa'īd Misr. (Wood 1362, 2947, 2981)

3. Potamogeton pusillus *L.*
Floating leaves absent; submerged leaves 1–4cm × 0.5–1mm, narrowly linear, 1-nerved (2 lateral nerves obscure), translucent, the stipular sheaths open. Fruiting spikes 5–10mm long.

A plant of ponds, cisterns and slow-moving streams between 1900 and 3000m; widespread on the high plateau but very local on the escarpment and only known from Turbah, Manākhah and J. Maswār; usually dominant where it occurs.

4. Potamogeton schweinfurthii A. Benn.
Floating leaves absent; submerged leaves mostly 8–12 × 1.5–2.5cm, oblong or oblong-elliptic, many-nerved, translucent, the apex acute to cuspidate with the midrib exserted as a 0–2.5cm long point. Fruiting spikes 4–6cm long.

A rare plant of ponds between 2500 and 3000m: on the ridge between J. Sumārah and J. Manār; at the head of the Wadi Banā near Ribāṭ.

RUPPIACEAE

RUPPIA *L.*

1. Ruppia maritima *L.*
Submerged waterweed, the slender filiform stem rooted below, floating above. Leaves alternate or opposite, filiform, 3–5cm long, with a conspicuous dilated sheathing base. Flowers small, bisexual, borne in pairs on slender pedicels but the inflorescence rapidly developing as the plant matures so the fruiting inflorescence is umbel-like with a long flexuous peduncle, 1–5cm long. Fruit a small nut.

Only known at 1500m from a brackish pool near Taiz airport.

NAJADACEAE

NAJAS *L.*

1. Najas graminea *Del.*
Submerged aquatic herb with slender, very leafy, repeatedly forked stems. Leaves in lateral bunches, though appearing opposite or whorled, c 2cm long, linear, grass-like with a sheathing base and a minutely spiny margin. Flowers 1–3 in the leaf axils, minute, monoecious. Fruit a small ellipsoid nut, c 1.5mm long.

A rare plant of pools and slow-moving steams between 300 and 600m: between Sa'īd Miṣr and 'Alūjah; in the wadi below Ḥadīyah.

ZANNICHELLIACEAE

ZANNICHELLIA *L.*

1. Zannichellia palustris *L.*
Horned Pondweed. Submerged waterweed with much-branched, filiform leafy shoots, often creeping over the bottom of streams and ponds. Leaves opposite or whorled, capillary, translucent, 1.5–5cm long, the base sheathing. Flowers monoecious, minute, axillary. Fruit a crescent-shaped beaked achene, 2–3mm long, muricate on the back; achenes 2–6 together, sessile or on a short common peduncle;

Fig. 19. **A** *Potamogeton nodosus*; **B** *Ruppia maritima*; **C** *Cymodocea serrulata*.

only the pedunculate form, variety **pedicellata** *Wahl.* has been confirmed from Yemen. (Wood 950)

A local plant of shallow streams and ponds between 1800 and 3000m: in the wadi above 'Aduf near Turbah; frequent in mountain ponds and in the streams around the head of the Wadi Banā; Bani Maṭar (several places); Aḥjar; between Sāqayn and Haydān.

CYMODOCEACEAE

Dioecious marine waterweeds submerged in shallow water. Leaves in erect stalked clusters from a creeping stem and with prominent, pale sheathing bases. Flowers in cymes (*Syringodium*) or solitary in a short terminal branch, the male consisting of 2 stamens and the female of 2 carpels and a style. Fruit a 1-seeded nutlet.

These sea grasses are usually found washed up on the shore after gales but can also be found by careful searching in water about 1m deep. Flowers are rarely seen.

1.	Leaves cylindrical, grooved	4. *Syringodium*
	Leaves flat, not grooved	**2**
2.	Leaves 1-nerved	3. *Halodule*
	Leaves many-nerved	**3**
3.	Leaf cluster on a woody stalk up to 40cm long, the stalk covered in numerous ring-like scars	2. *Thalassodendron*
	Leaf cluster on a stalk 0–4cm long; scars if present, not forming a complete ring	1. *Cymodocea*

1. CYMODOCEA *Koen.*

Leaves 3–4mm broad, the apex entire	*C. rotundata*
Leaves 4–8mm broad, the apex serrate	*C. serrulata*

1. Cymodocea rotundata *Aschers. & Schweinf.*
Leaf clusters sessile or nearly so; leaves linear, straight, obscurely 9–13-nerved, 5–12cm long, 3–4m wide, the apex rounded, entire.

Known with certainty only from Hodeida (Schweinfurth) and Mukhā but doubtless all along the coast.

2. Cymodocea serrulata (*R. Br.*) *Aschers. & Magnus.*
Leaf clusters sessile or more commonly with stalks up to 4cm long, the stalks sometimes scarred but the scars not forming a complete ring; leaves linear, straight or nearly so, obscurely 12–18-nerved, 6–15cm long, 4–8mm broad, the apex rounded, serrate.

Known with certainty from Luḥayyah, Hodeida (Schweinfurth) and Mukhā, but doubtless elsewhere along the coast.

2. THALASSODENDRON *den Hartog*

1. Thalassodendron ciliatum (*Forssk.*) *den Hartog*
Cymodocea ciliata (Forssk.) Ehrenb. ex Aschers.
Leaf clusters with woody stalks up to 40cm long, the stalk covered in ring-like scars; leaves linear, falcate, obscurely 12–20-nerved, 6–15cm long, 7–10mm broad, the apex serrate.

Known from Hodeida (Deflers), Mukhā (Forsskal) and Ḥanish Island (Slade).

3. HALODULE *Endl.*

1. Halodule uninervis (*Forssk.*) *Aschers.*
Diplanthera uninervis (Forssk.) Aschers.
Leaf clusters sessile or nearly so; leaves linear, straight, 1-nerved, 4–15cm long, 1–2mm broad, the apex truncate, 2–3-toothed.

Known from Mukhā, Khawkhah, Manẓār and Hodeida, but doubtless elsewhere on the coast.

4. SYRINGODIUM *Kutz.*

1. Syringodium isoetifolium (*Aschers.*) *Dandy*
Cymodocea isoetifolia Aschers.
Leaf clusters with stalks up to 3cm long, rarely sessile; leaves linear, cylindrical, grooved but apparently without nerves, 3–12cm long, 1–2mm broad, the apex truncate but untoothed.

Known from Hodeida (Schweinfurth) and Mukhā.

PALMAE

Palms. A very distinctive family of trees and shrubs having an unbranched trunk (branched in *Hyphaene thebaica*) with a terminal crown of large, stiff leaves with sheathing bases. Inflorescence in our species axillary, paniculate; flowers small, unisexual, with spathaceous bracts enclosing 3 sepals and 3 petals. Fruit a drupe or berry.

The palms of Yemen are little known and need further study before exact names can be given to most species. It is important that habit photographs as well as leaves, male flowers (*Phoenix*) and fruit (*Hyphaene*) are collected; unfortunately because of the height of mature trees it is often difficult to collect good material, but it needs to be stressed that leaves alone are insufficient for accurate identification.

Leaves pinnate	1. *Phoenix*
Leaves fan-shaped	2. *Hyphaene*

1. PHOENIX *L.*

Trunks unbranched, but often clustering at the base. Leaves pinnate with folded, acuminate, often spine-tipped leaflets. Inflorescence with a single bract, the peduncle elongating and branching at the tip to produce a cluster of spike-like branches. Fruit a date or date-like berry.

1.	Mature trees in the colony with trunks at least 2m high and commonly much more	**2**
	Mature trees in the colony apparently without a trunk or with a short trunk up to 2m high	*P. caespitosa*
2.	Leaves ± glaucous, often 4-ranked when young. Mature fruit 2–5cm long, with thick, edible flesh (date)	*P. dactylifera*
	Leaves green, 2-ranked. Mature fruit 1–2cm long, the flesh thin, scarcely edible	*P. reclinata*

1. Phoenix caespitosa *Chiov.*
P. arabica Burrett
P. reclinata sensu Schwartz
Gregarious plant forming large colonies, the trunk absent or up to 2m high and with a dense crown of green leaves 1–2m long. Fruit 1–2cm long, resembling a small, withered date, dry and tasteless when eaten.

Widespread between 400 and 2000m in scattered localities along the escarpment and in the Mashriq, but characteristic of two habitats in particular. It is locally abundant in the valley scrub communities of the central escarpment, notably on J. Bura' where it is subdominant at about 700m in the valley forest above Suq As Sabt. It also forms large communities in marsh and along streams in dry country where there is a moderately high salt level and good examples can be seen at Suq al Jammah, between 'Ashshah and Shahārah, at Damt and in the Ḥujarīyah valleys around Barakānī and Bani 'Umar.

2. Phoenix dactylifera *L.* — Nakhl
Date palm. Tall straight tree up to c 20m high, often suckering. Leaves glaucous, often 4-ranked, stiffer and more spinescent than in related species, the partially developed leaflets at the top of the petiole being particularly hard and spiny. Fruit 2–5cm long, fleshy, sweet, usually yellow in Yemen.

Although odd date palms can be found throughout Yemen, wild trees are not a particularly characteristic part of the scenery as in other Middle Eastern countries. They are tolerably frequent throughout the Tihama but only occur as scattered individuals elsewhere, reaching about 2500m. Immature plants are often found by roads and tracks where they have presumably grown up from discarded stones.

In a few places in the southwest of the country and to a much lesser extent in the Mashriq (Harīb, Khabb) there are extensive date palm plantations but production is on a modest scale. There are plantations by permanent streams in the dry southwest escarpment, notably by the Wadi Rusyān but all the really large plantations are found at the outflow of the major wadis into the Red Sea. Despite the proximity of the sea, fresh ground water is close to the surface and the palms flourish. Huge areas are found north of Yakhtul and Khawkah and between Ṭa'if and Manẓār.

3. Phoenix ? reclinata *Jacq.*
Straight tree growing in small colonies and reaching about 8m in height. Leaves green, 2-ranked, not particularly stiff and spinescent and so resembling those of *P. caespitosa*. Fruit 1–2cm long, resembling a small withered date.

Only known at c 1500m on the steep, scrub-covered slopes between Ḥadīyah and Bani Mukhtār in J. Raymah, but possibly overlooked elsewhere in that region.

2. HYPHAENE *Gaertn.*

Trunk branched or unbranched. Leaves fan-shaped and usually deeply lobed. Flowers borne in pits on long catkin-like inflorescence branches. Fruit a large globose-quadrangular drupe.

1.	Leaves shallowly lobed to less than a quarter of their length	*H. sp.B*
	Leaves deeply lobed almost to the base	**2**
2.	Trunk repeatedly forked	*H. thebaica*
	Trunk unbranched	*H. sp.A*

1. Hyphaene thebaica (*L.*) *Mart.* — Dawm, Dūm
A very distinctive tree on account of its tall, repeatedly forked trunk which may reach 20m high. Leaves fan-shaped, deeply lobed into linear-lanceolate segments forming a crown at the end of each branch; petiole with hooked spines. Fruit a large, dark-brown, globose-quadrangular, punctate drupe 6–8cm long.

The fleshy fruit is sweet and edible but dry and rather unpalatable. Although sold in Tihama markets it is mostly eaten by small children. The leaves are used to make mats and baskets.

Widely distributed from 0–500m in the Tihama but probably always cultivated and virtually restricted to villages and towns. It is the most characteristic tree of several of the older Tihama towns including Zabīd where it is grown in most of the town gardens.

2. Hyphaene sp.A
Gregarious palm growing in large colonies. Trunk unbranched, up to c 5m high, almost entirely covered in the dead bases of old leaves. Leaves fan-shaped, deeply lobed into linear-lanceolate segments and forming a crown at the top of the stem. Flowers and fruit not known.

The exact distribution is not known but it is locally abundant near the Red Sea where ground water is close to the surface. It has been noted north of Khawkah, near Al Fāzah, north of Hodeida near the outflow of the Wadi Surdūd and at Barīdah near Ibn 'Abbās (K. Brockie).

3. Hyphaene sp.B
A little-known species readily distinguished by its entire or very shallowly lobed fan-shaped leaves. The habit, flowers and fruit are not known.

Recorded at Zabīd and at the mouth of the Wadi Rima' (Hepper 5726).

The Coconut, **Cocus nucifera** *L.*, Narjīl, was cultivated at Mukhā according to Forsskal, but it is not there today and was perhaps washed away in the storms that destroyed most of the Mukhā date palms earlier this century.

PANDANACEAE

PANDANUS *L.*

Pandanus odoriferus (*Forssk.*) *Chiov.* Kadhī
P. odoratissimus L.f.
P. tectorius Sol.
Screw pine. Small branched palm-like dioecious tree to 4m, often sending down aerial roots. Leaves in clusters at the branch tips, stiff, sword-shaped, glaucous, 40–70cm long, the margin with whitish spine-like teeth 5–6mm long. Flowers densely crowded on racemose spadices which are subtended by whitish spathes; male flowers of numerous, whitish stamens, fragrant; female flowers developing into a cone-like fruit, similar to a small pineapple.

Widespread by flowing streams in the escarpment foothills from 500 to 1500m, most common in high rainfall areas. The fragrant male flowers are wrapped in leaves and sold on roadsides and in markets.

Only male plants seem to occur in Yemen and the tree is propagated vegetatively. Forsskal was almost certainly right to suggest it was introduced from India where it is used to make perfumes.

ARACEAE

Herbs. Leaves basal or spirally arranged on the stem. Flowers minute, bisexual or unisexual, crowded onto a dense spike (spadix) which is enclosed in a brightly coloured spathe. Fruit a berry.

1. Leaves simple, peltate — 1. *Colocasia*
   Leaves compound, not peltate — **2**
2. Leaves and flowers produced together. Leaves more than 1, leaflets nearly equal in size — 3. *Arisaema*
   Flowers appearing when the plant is leafless. Leaves 1, the leaflets unequal in size, the terminal much larger than the basal — 2. *Sauromatum*

1. COLOCASIA *Schott.*

1. Colocasia esculenta (*L.*) *Schott.* Kurkūm, Kurn, Kirkir
C. antiquorum Schott.
Herb with a thickened underground corm and a tuft of long-petiolate, simple, peltate, ovate-cordate, glabrous, dark green leaves, usually 20–50cm long. Spathe 15–40cm long, yellow, much longer than the flower spike. Berries enclosed in the persistent tube of the spathe.

Rarely flowers anywhere in the world and is spread by the breaking off of parts of the corm which readily root in damp conditions.

Originally introduced but appearing quite wild, growing by the sides of small slow-moving streams in the wetter parts of the escarpment from 500 to 1500m. The roots were formerly eaten in times of famine but do not seem to be used for any purpose nowadays.

2. SAUROMATUM *Schott.*

1. Sauromatum venosum (*Ait.*) *Kunth.*
Herb to c 60cm with a small disc-shaped corm and solitary leaves appearing separately from the flowers. Leaves pedately divided into up to 7 sessile, broadly lanceolate leaflets which are very unequal in size, the terminal c 25 × 9cm, the basal only 6 × 3cm. Spathe elongate, tubular, purple-spotted, up to 30cm long, slightly exceeding the purple-brown flower spike. Fruit an ellipsoid head of berries c 5cm long.

The flowers have never been collected in Yemen.

A rare, shy-flowering plant of shaded places in high rainfall areas of the escarpment between 900 and 1500m: on scrub-covered river cliffs, Wadi Dūr, 'Udayn; quite frequent in damp, shaded coffee terraces above Ḥadīyah.

3. ARISAEMA *Mart.*

Herbs with erect, palmately-divided leaves.

Leaflets all arising from the same place. Spathe green or purple-brown — *A. bottae*
Leaflets pedate (i.e. borne on branching lateral petioles). Spathe yellow — *A. flavum*

1. Arisaema bottae *Schott.* Dakhāf
Arum pentaphyllum sensu Forsskal
Erect herb to 1m. Leaves palmately divided into 5–7 sessile, obovate or elliptic, dark green leaflets up to 20 × 10cm, all arising from the same place. Spathe 12–20cm long, green to purple-brown with paler longitudinal lines, flower spike ± enclosed in the rather long spathe tube. Fruit a pedunculate head of berries.

Flowers April to August.

A gregarious plant of damp mountain gullies at high altitudes, often dominant in suitable situations and associated with *Anthriscus sylvestris* and *Crinum album*; common on J. Ṣabir and the Ibb mountains but rarer further north and only recorded elsewhere from J. Al 'Amal in 'Utmah, J. Kabūrah above Ḥadīyah, J. Milḥān and J. Maswar (Shirnamah).

Elsewhere only in Ethiopia where it is rare.

2. Arisaema flavum (*Forssk.*) *Schott.* Dakhāf
A. enneaphyllum sensu Schwartz
Erect herb to 80cm arising from a corm. Leaves usually 2 or 3, pedately divided into 5–9 oblong-acute, equal or slightly unequal leaflets, up to 10 × 2.5cm, some of which are distinctly pedicelled (cf. *Sauromatum*). Spathe 5–8cm long, yellow, flower spike slightly exserted from the short spathe tube. Berries in a head on a recurved peduncle.

Flowers April to July.

Widespread on the escarpment between 1700 and 2900m, growing as scattered individuals on rocky slopes, path sides and grassy banks.

It is widely distributed from Ethiopia to the Himalayas.

LEMNACEAE

LEMNA *L.*

Our species are free-floating aquatic herbs sometimes carpeting the surface of still water. They consist of tiny leaf-like fronds each with a single root. They mainly reproduce by producing small lateral daughter fronds which soon separate from the parent. Flowers which are rarely produced are monoecious, enclosed in a small sheath and produced on the side of a thallus.

All species are known as duckweed in English and 'Blaysnān' in Arabic.

1. Frond ± hemispherical with large inflated cells below — *L. gibba*
   Frond thin — 2
2. Root cap acute. Root with a distinct winged sheath at the junction with the frond — *L. perpusilla*
   Root cap obtuse. Root without a distinct winged sheath at the junction with the frond — *L. minor*

1. Lemna gibba *L.*

Distinguished by its hemispherical fronds, whose undersurface has large spongy cells. The root cap is obtuse.

The commonest duckweed in Yemen, found in cisterns and on ponds between 1300 and 3000m on the escarpment and high plateau.

2. Lemna minor *L.*

Fronds thin, never spongy below. Root cap obtuse, the root joined to the base of the frond without a distinct sheath.

The only confirmed records are from the Wadi Shalālah near Yarīm (Hepper 6280) and from Madīnat Al 'Abid (Radcliffe-Smith and Henchie) but it is probably widespread in ponds and cisterns on the high plateau and possibly also on the escarpment, growing from 1200 to 2600m.

3. Lemna perpusilla *Torrey*

Very similar to *L. minor* differing only in the key characters.

Only known with certainty from 600 to 1500m on J. Raymah (below Ḥadīyah; between 'Alūjah and Kusmah) but quite possibly elsewhere on ponds and slow-moving streams on the escarpment. It is the only species likely to occur at low altitudes. (Wood 75/847, 2248)

COMMELINACEAE

Succulent herbs. Leaves alternate, simple, entire, the base (petiole) forming a sheath around the stem. Inflorescence of simple or compound, terminal or axillary cymes, often partially enclosed in a spathaceous bract; flowers regular or irregular, sepals and petals 3, free or united; stamens 6, all fertile or some sterile or reduced to staminodes. Fruit a capsule.

Although the family is readily recognised, the individual species of *Commelinaceae* can look confusingly similar until one is familiar with the plants and has learnt to look for differences between the species, most of which are in fact quite clear-cut.

1. Inflorescence enclosed in or closely subtended by a spathaceous bract — 2
   Inflorescence neither enclosed in nor subtended by a spathaceous bract — 3. *Aneilema*
2. Filaments glabrous. Petals free, 2 large and one small — 1. *Commelina*
   Filaments hairy. Petals joined to form a tube, equal — 2. *Cyanotis*

1. COMMELINA *L.*

Annual or perennial herbs with erect or trailing stems and fibrous roots (in our species). Inflorescence terminal and leaf-opposed, consisting of 1–2 cymes enclosed in a folded spathe; sepals free; petals 3, unequal, the upper 2 long-clawed, larger than the lower one; stamens 6 but only the lower 3 fertile; filaments glabrous; capsule usually 2-valved, 1–5-seeded.

Some species develop subterranean cleistogamous flowers at the end of the season. They are particularly characteristic of *C. forskalaei*.

All species are called 'Wa'lān' or 'Jilīf' in Arabic.

1. Spathes free to the base (i.e. simply folded with 2 sides free) — 2
   Spathes united for part of their length (i.e. only the upper side completely free) — 4
2. Flowers blue — *C. diffusa*
   Flowers yellow or pinkish-brown — 3
3. Flowers yellow. Leaves ovate, more than 1cm broad — *C. africana*
   Flowers pinkish-brown. Leaves linear, less than 0.5cm broad — *C. subulata*
4. Spathes on peduncles less than 1.5cm long. Leaf base never auriculate or cordate — 5
   Spathes on peduncles 1.5–8cm long, but if less the leaf base auriculate or cordate — 8
5. Angle between the fused and folded side of the spathe (i.e. the angle to which the peduncle is joined) more than 90°. Leaves with conspicuous undulate margins — *C. forskalaei*
   Angle between the fused and folded side of the spathe less than 90°. Leaf margin not undulate — 6

6. Leaves usually less than 3 times as long as broad. Leaves with white or coloured hairs at least 3mm long at the apex of the sheath *C. benghalensis*
Leaves usually more than 5 times as long as broad. Leaves without long, white or coloured hairs at the apex of the sheath **7**
7. Folded edge of the spathe straight or slightly curved. Flowers blue *C. erecta*
Folded edge of the spathe strongly curved. Flowers often mauve, rarely truly blue *C. albescens*
8. Capsule square. Seeds spherical *C. latifolia*
Capsule longer than broad. Seeds elongate *C. imberbis*

1. Commelina africana *L.*

C. edulis A. Rich.
C. divaricata Vahl

Trailing herb with prostrate stems from a stout perennial base. Leaves ovate to oblong, c 4–6 × 1.5–2cm, often yellowish-green, usually with long, white or coloured hairs at the mouth of the sheath. Spathes on peduncles 1–3cm long, scattered along the stem, usually solitary, 1.5–4cm wide, the folded edge slightly curved, the short edge free; petals yellow. Capsule usually 1-seeded, the seed slightly elongate, obscurely reticulate.

Very variable in tropical Africa but the Yemen plant always has broad, sparsely hairy leaves and is the typical variety **africana**.

Widespread and frequent on the escarpment between 1500 and 3000m, thus growing higher than any other *Commelina*. It is a plant of shaded rocky places including stream beds, terrace walls, rocky hillsides and cliff gullies.

2. Commelina albescens *Hassk.*

Perennial herb with a thickened stem base and ascending or erect stems, sometimes reaching 50cm in the protection of bushes. Leaves linear or narrowly elliptic c 8–12 × 0.5–1.3cm, the lower surface usually whitish. Spathes subsessile in dense clusters at the branch tips, 1–2.2cm wide, strongly falcate, the folded edge curved, the short edge fused for most of its length; petals mauve or lilac, very rarely (never?) blue. Capsule usually 3-celled, one cell indehiscent, the seeds elongate, smooth.

The most drought-resistant *Commelina*, usually immediately recognised by its narrow whitish leaves and strongly falcate spathes. It is widespread from 200 to 2300m, growing in open, well-drained rocky places; it occurs on stony outcrops in the Tihama and in the drier areas of the escarpment, particularly below 1000m, reappearing on the high plateau and extending east to Ṣa'dah, Ḥūth and Rada'. It is often chewed down by goats and only attains a reasonable height in the protection of thorn fences and spiny shrubs.

3. Commelina benghalensis *L.*

C. canescens Vahl

Short-lived (? annual) much-branched herb with ascending or erect branches to c 40cm; cleistogamous flowers sometimes present. Leaves ovate or elliptic, 4–5 × 1.5–2cm, unequal-based, narrowed into a distinct false petiole with long, white or coloured hairs at the mouth of the sheath. Spathes on short peduncles less than 0.5cm long, 1 or few together at the tips of the branches, 1–1.5cm wide, the folded edge nearly straight, the short edge fused; petals blue. Capsule up to 5-seeded, the seeds kidney-shaped, reticulate.

C. benghalensis can be difficult to recognise but the small, ± triangular spathes, petiolate leaves and the presence of long hairs at the top of the sheath are useful characters.

Widespread and frequent on the escarpment, growing from 500 to 2500m but most commonly in areas of good rainfall below 3000m. It grows in scrub, on field borders and as a field weed.

4. Commelina diffusa *Burm. f.*

Slender decumbent herb, the stems rooting at the nodes. Leaves ovate or oblong, mostly c 5 × 1.5cm, narrowed into a short false petiole, with long white hairs at the mouth of the sheath. Spathes shortly pedunculate, usually solitary towards the tips of the branches, 2–2.5cm wide, the folded edge slightly curved, the short edge free; cymes well exserted from the spathes; petals blue. Capsule 3-celled, one cell indehiscent, the seeds black, reticulate.

A local plant of slow-moving streams, ditches or marsh where there is semi-permanent water; it is found between 800 and 2500m on the escarpment from Wadi Ḥidrān north to Rujum, but is mostly in the high rainfall areas of Ibb and the central escarpment.

5. Commelina erecta *L.*

Tufted herb with erect or ascending stems to 1m from a perennial base. Leaves usually lanceolate (rarely ovate or oblong), c 6–13 × 1–3cm, sometimes narrowed into a false petiole, green or glaucous, the lower surface often pale. Spathes subsessile, solitary or in small clusters at the tips of the branches, 1–2.5cm wide, the folded edge straight or slightly curved, the short edge fused; petals blue. Capsule 1–3-seeded, the seeds smooth.

A very variable plant in Yemen, probably divisible into several varieties or subspecies. Plants from heavy shade (Wood 2174, 2529) tend to be subglabrous with broader, ovate or oblong leaves, whereas plants from open habitats are hairy, with narrower leaves. Some plants have 2-celled capsules (Wood 2119, 2586) whereas others are 3-celled (Wood 2629). The seeds are globose in Wood 2119 but distinctly elongate in 1890. It is difficult to assess the significance of such variation without cultivation experiments and further collections.

A local plant found between 500 and 1800m on the escarpment from Wadi 'Annah, 'Udayn north to the Wadi 'Ayn 'Alī, Ḥajjah but virtually restricted to the central escarpment. Robust scrambling plants are frequent in the valley scrub communities whereas smaller erect plants are found on grassy field borders at higher altitudes.

6. Commelina forskalaei *Vahl*

Short-lived (? annual) herb with trailing stems which root at the nodes; cleistogamous flowers often present at the end of the season. Leaves oblong, c 3–6 × 0.5–1cm with distinct undulate margins and long white hairs at the mouth of the sheath. Spathes on short peduncles 3–8mm long, scattered along the stem, solitary, 1–1.5cm long, the folded edge straight, the short edge fused and lying at an angle of more than 90° to the folded edge; petals blue. Capsule 3-celled, one cell indehiscent, the other with 2 smooth globose seeds.

Widespread and often abundant as a field weed, also growing on disturbed field borders and open rocky ground where there is little competition from grasses. It is found from 200 to 2200m in the east of the Tihama, on the escarpment and on the southeast mountains east to Damt.

7. Commelina imberbis *Ehrenb. ex Hassk.*

C. petersii sensu Schweinf.

Vigorous herb from a perennial base, the stems either decumbent or scrambling high over bushes. Leaves lanceolate or ovate, c 6–12 × 1.5–3cm, the base cordate or auriculate and usually clasping the stem, long white hairs present at the mouth of the sheath. Spathes on peduncles 1–4cm long, scattered along the stem, but mostly at the branch tips, solitary, 2–3.5cm long, the folded edge straight, the short edge fused for part of its length; petals blue. Capsule elongate with 4 seeds, the seeds elongate, nearly smooth or with a few obscure transverse ridges on the back.

Widespread between 600 and 2300m on the escarpment from Turbah north to Maḥwīt but restricted to areas of good rainfall where it is found in scrub, thorn fences around cultivation, on cliff gullies and on terrace walls.

Schweinfurth's *C. petersii* should almost certainly be included here. I have been unable to trace a specimen but his description of the seed as having 3–5 folds on the back indicates *C. imberbis*, which grows today near 'Usil where Schweinfurth collected his plant. *C. petersii* Hassk. is similar to *C. imberbis* but has leaves which are not auriculate or cordate and seeds with conspicuous transverse ridges.

8. Commelina latifolia *Hochst. ex A. Rich.*

Virtually identical to *C. imberbis* except for the capsule which is almost square with globular seeds.

In similar situations to *C. imberbis*, with which it often grows but much less common and recorded only from Madāhit near Turbah, Dhī Sufāl, Ibb, Suq al Jum'ah on J. Ash Shirq, above Ḥadādah in J. Raymah and at Bayt Mughlad on J. Shibām in the Ḥarāz.

9. Commelina subulata *Roth.*

Tufted annual herb with erect stems to c 30cm. Leaves linear, 5–10 × 0.2–0.5cm with long white hairs at the mouth of the sheath. Spathes sessile, scattered along the stem, solitary, 0.5–1cm wide, often reddish-veined, the folded edge slightly curved, the short edge free; petals pinkish-brown. Capsule with up to 5 seeds, the seeds globular, reticulate.

A rare plant found on seasonally wet, black soils between 1800 and 2300m on the escarpment and usually associated with *Craterostigma* spp., *Kyllinga chlorotropis* and *Ophioglossum polyphyllum*: Ẓulmah, J. Ḥubaysh; Suq al Jum'ah, J. Ash Shirq; Rujum.

Two species of *Commelina* described from Yemen have not been included in the above account as their exact identity is not known. The specimen of *C. commelinoides* Forssk. has been lost and it is impossible to know from the description what species Forsskal had found. It may not even have been a *Commelina* at all. I have also been unable to trace any specimens of *C. ussilensis* Schweinf. which was collected at 'Usil in the Ḥarāz. The description fits *C. albescens* well but Schweinfurth knew that plant and collected it himself from Yemen. Although the description does not fit very well it seems possible that it was a form of *C. erecta* since it grows near 'Usil and is the only *Commelina* found in the area which Schweinfurth did not otherwise collect.

2. CYANOTIS *D. Don*

Herbs with erect or decumbent stems. Roots very variable and of great taxonomic significance. Leaves oblong-lanceolate. Cymes sometimes subtended by spathaceous bracts, sessile or pedunculate; flowers regular; sepals 3, joined; petals 3, joined to form a tube; stamens 6, all fertile, the filaments hairy. Capsule 5–6-seeded.

1. Roots tuberous. Sterile rosettes present at flowering time — *C. nyctitropa*
   Roots fibrous, corms present or absent. Sterile rosettes absent at flowering time — 2
2. Perennial. Corms present. Spathe-like bracts reduced in size towards the branch tips, pubescent but not woolly underneath — *C. vaga*
   Annual. Corms absent. Spathe-like bracts not reduced in size, towards the branch tips, often paired, woolly underneath — *C. lanata*

1. Cyanotis lanata *Benth.*

Much-branched annual with erect, woolly stems to 25cm, the roots fibrous, basal rosette absent. Leaves up to 6cm long, woolly below. Flowers in the upper 3 leaf nodes, sessile, the spathaceous bracts often paired, not diminishing in size towards the branch tips, woolly below; corolla white or more rarely pale blue.

A very local plant growing on rock outcrops at 1500m in the lush middle-altitude wadis: rare in Wadi Sukhnah; in great abundance near the suq, 'Udayn.

2. Cyanotis nyctitropa *Deflers*

C. hirsuta sensu Schwartz

C. foecunda sensu Schwartz

Tillandsia decumbens Forssk.

Tufted perennial with thinly tuberous roots, non-flowering leaf-rosettes and pilose, decumbent or ascending flowering stems to c 25cm. Leaves up to c 15cm long, pilose on the margins but otherwise ± glabrous. Flowers in the upper 3–7 leaf nodes, sessile, the spathaceous bracts, solitary, diminishing in size towards the branch tips, glabrous below; corolla mauve.

Grows up rapidly and is in flower in about a month after heavy rain but dies back quickly if the rains stop. During the winter dry season the leaves die away completely and the plant cannot be seen.

Widespread on the escarpment and on the southeast mountains east to Damt growing from 900 to 2500m on shallow stony soil and on terrace walls. It occurs locally in great abundance in well-drained open scrub in areas of good rainfall notably the *Euphorbia* scrub around Qā'idah and under *Acacia hockii* north of Ibb.

C. nyctitropa is endemic to southwest Arabia but closely related to the Ethiopian *C. polyrihiza* which differs in always having pedunculate spathes.

3. Cyanotis vaga (*Lour.*) *Schultes & Schultes f.*

C. barbata D. Don.
C. parasitica Hassk.
C. hirsuta C. A. Mey.

Tufted perennial, growing from a corm with erect pilose stems to 40cm, roots fibrous, basal rosette absent. Leaves up to 6cm long, pilose below but not woolly. Flowers clustered into dense sessile heads, the spathaceous bracts solitary, diminishing in size towards the branch tips; corolla purple.

Begins to grow after the onset of the spring rains, matures relatively slowly and flowers in August and September, dying away completely in the winter dry season. Its abundance in the Ibb area and its slow maturation suggests that it requires a relatively continuous summer rainfall.

Locally abundant in high montane grassland between 2300 and 3200m in Ibb province, extending north to J. Ḥisn Madnān in Waṣāb al 'Āli, J. Al 'Amal in 'Utmah and Suq al Jum'ah in J. Ash Shirq.

3. ANEILEMA *R. Br.*

Herbs with erect or trailing stems. Cymes not enclosed in spathaceous bracts but forming a lax terminal panicle-like inflorescence; sepals 3, free; petals 3, free, the upper 2 large and clawed, the lower smaller; stamens 6, the upper 3 reduced to staminodes, filaments glabrous. Capsule with 2 or more seeds.

Annual. Bracteoles spaced 5–8mm apart. Sepals and pedicels sparsely pubescent — *A. forskalei*
Perennial. Bracteoles spaced 1–3mm apart. Sepals and pedicels densely pubescent — *A. woodii*

1. Aneilema forskalei *Kunth.*

A. tacazzeanum sensu Schwartz p.p.

Annual with pubescent stems, decumbent and rooting at the nodes or ascending and scrambling over scrub vegetation. Leaves ovate, narrowed into a short petiole-like base, up to 15cm long, sparsely pubescent below. Inflorescence 3–9cm long, bracteoles 5–8mm apart, pedicels and sepals sparsely pubescent; petals pale lilac, almost white. Capsule shiny, c 5mm long, the pedicels recurved, 7–9mm long.

A local plant of valley scrub in areas of high rainfall between 500 and 1500m; widely distributed in the central escarpment, but elsewhere only known from Wadi Ḍabāb near Taiz (Miller), 'Udayn and Wadi Sukhnah.

2. Aneilema woodii *R.B. Faden*

A. tacazzeanum sensu Schwartz p.p.

Similar to the previous species but perennial, more vigorous, the leaves definitely pubescent below, the bracteoles close together (the gap 1–3mm), the pedicels and sepals strongly pubescent and fruiting pedicel sharply decurved.

This plant was collected by Schweinfurth at Rajamah near Ḥajjaylah and at Ḥillah, J. Bura'. I have failed to refind it at Rajamah but there is a considerable quantity in the valley above Ḥillah. Cultivation of material from this valley by Dr Faden has shown that whereas some plants, notably from the cliffs just above the suq, are quite fertile others are sterile and with defective pollen and are presumably the hybrid *A. forskalei x woodii*. This very rare plant is endemic to Yemen.

JUNCACEAE

JUNCUS *L.*

Rushes. Glabrous, annual or perennial herbs, usually rhizomatous. Leaves with an open sheath, long and narrow, entire, flat, cylindrical or grooved, sometimes reduced to scales. Inflorescence a monochasial cyme, occasionally reduced to a head, formed of solitary or clustered flowers; flowers small, bisexual, regular; perianth with 6 green or brown segments. Fruit a many-seeded capsule.

A predominantly temperate genus of marshy places which, in Yemen, is largely restricted to the high plateau. The general word for a Rush in Yemen is 'Numās'.

1. Small, readily uprooted annual; stems less than 25cm high. Leaves deeply channelled, filiform — *J. bufonius*
   Perennials, usually difficult to uproot; stems mostly 30–150cm long. Leaves terete, never filiform — 2
2. Stem leaves present, conspicuously transversely partitioned. Flowers in globular heads — 3
   Leaves all basal or absent, not transversely partitioned. Flowers not in globular heads — 4
3. Heads lax, with 6–12 greenish flowers; perianth segments 3–4mm long — *J. fontanesii*
   Heads dense, with 10–100 brown flowers; perianth segments 2.5–3mm long — *J. punctorius*
4. Inflorescence an apparently lateral cluster, the stem-like bract much longer than the inflorescence — *J. inflexus*
   Inflorescence an obviously terminal cyme, the stem-like bract shorter than the inflorescence — 5
5. Capsule yellow-green, about equalling the perianth — *J. rigidus*
   Capsule chestnut brown, much exceeding the perianth — *J. socotranus*

1. Juncus bufonius *L.*

Tufted annual herb to 25cm. Leaves soft, filiform, channelled, 1–5cm long. Inflorescence much branched, leafy, occupying the greater part of the plant; flowers mostly solitary along the upper side of the inflorescence branches; perianth segments 3–6mm long, green with a hyaline margin, slightly longer than the blunt, oblong capsule.

A plant of damp sand and mud at the edge of streams and pools between 2000 and 3000m, often associated with *Isolepis setacea*; widespread in suitable habitats on the high plateau but known elsewhere only from Majz (Steinberg) and Sāqayn in Khawlan Ash Sham.

Fig. 20. **A** *Juncus punctorius*; **B** *Juncus bufonius*; **C** *Juncus inflexus*.

2. Juncus fontanesii *Gay*

Stoloniferous perennial herb with weak, decumbent or ascending stems, 30–50cm long, each with 2–4 leaves. Leaves soft, terete, transversely partitioned, 8–15cm long. Inflorescence much branched with numerous stalked flower clusters, each with 6–12 flowers; perianth segments 3–4mm long, green becoming red-brown when old, slightly shorter than the distinctly beaked capsule.

The Yemen plants are all subspecies **pyramidatus** (*Lah.*) *Snog.*, which lacks the rhizome usually found in *Juncus* and has a capsule with a beak 1–2mm long.

A plant of marsh and stream sides from 1800 to 2900m; widespread and common on the high plateau but elsewhere rare and only known from the Wadi Rakwah in Khawlan Ash Sham (Steinberg), Manākhah (Schweinfurth), Dhī Sufāl and the Wadi Sūdān near Qā'idah.

3. Juncus inflexus *L.*

Densely tufted perennial with erect, dull-glaucous, ridged stems up to 60cm high. Leaves reduced to glossy, dark brown sheaths. Inflorescence with many short branches, forming an apparently lateral cluster, the bract, stem-like, spine-tipped and much exceeding the inflorescence; perianth segments 2.5–4mm long, brown, equalling the glossy, brown, ovoid, mucronate capsule.

An uncommon rush of marsh between 2300 and 2900m on the high plateau: Yarīm; Dhamār; between Dhamār and the Shirqabī pass; Wa'lān; Al Zīl, Al Ḥadā'; Baw'an.

4. Juncus punctorius *L.f.*

Perennial with a long rhizome and erect stems to 1m. Basal leaves reduced to pale brown sheaths; stem leaf solitary, transversely partitioned, often spine-tipped, up to 25cm long. Inflorescence much branched, with many stalked flower clusters, each very dense and with 10–100 flowers; perianth segments 2.5–3mm long, pale brown, shorter than the glossy, brown, ovoid, mucronate capsule.

A plant of springs and marsh between 1900 and 3200m; rare on the outer escarpment and only known from the wadi above 'Adūf near Turbah; widespread and frequent on the high plateau; widespread in the Mashriq but apparently rare and known only from Majz (Steinberg), between Ṣa'dah and J. Minmār, 'Ujlī south of Ṣa'dah and Khulāqah in Nihm.

5. Juncus rigidus *Desf.*

J. arabicus (Aschers. & Buch. ex Boiss.) Adams.

Densely tufted perennial with erect stems to 1m. Leaves few, basal, terete, spine-tipped, up to 35cm long. Inflorescence subtended by a pungent bract, much branched, the flowers in numerous clusters of 1–3; perianth segments c 3mm long, whitish-green, slightly shorter than the narrowly ellipsoid, pale green, acuminate capsule.

Widespread in the Mashriq from Dhī Nā'im north to the Ṣa'dah area, growing from 1200 to 2000m in marsh and by springs and streams. It is often associated with *Scirpoides holoschoenus* and is tolerant of moderately high salt levels.

6. Juncus socotranus (*Buch.*) *Snog.*

Densely tufted perennial with erect stems to 1m. Leaves few, basal, terete, spine-tipped, up to 30 cm long. Inflorescence subtended by a pungent bract, often elongated to 20cm or more, much branched, consisting of several distinct lax or contracted cymes; flowers mostly solitary or paired, perianth segments 2mm long, pale brown, much shorter than the subglobose, glossy chestnut brown capsule.

Abundant in marsh and along wadis with a high salt level north of Rada' but elsewhere a rare plant of marsh between 1900 and 2400m: Al Ḥaradah, Hammām Damt; Dhamār; in the wadi north of Wa'lān.

According to Schwartz, *Juncus sparganiifolius* Boiss. & Kotschy was collected by Rathjens at Ghaymān, but the occurrence of this species in Yemen is very improbable since it is otherwise restricted to Turkey and Syria. It is a very distinct species with flat, grass-like leaves but it has been confused by some collectors with *J. fontanesii* and it may have been that species which Rathjens found.

CYPERACEAE

A family of grass-like, usually rhizomatous herbs. Stems commonly solid, 3-angled, unbranched and leafless. Leaves usually in a basal tuft, having a distinct closed sheath as well as a blade, but often lacking the ligule found in grasses. Flowers in spikelets, bisexual or unisexual, each in the axil of a small floral bract (glume); perianth of scales or bristles, often absent. Fruit a nut, trigonous or biconvex, depending on the number of style branches.

The *Cyperaceae* are a neglected family partly because of their relative scarcity and partly because they do not have the economic importance of grasses. They are also difficult taxonomically and it is of great importance that specimens are collected which have ripe fruit and show the rootstock clearly. Most species grow in wet places and many are rare.

1. Inflorescence a long interrupted spike, the spikelets in 5–14 distant clusters along the rhachis — 13. *Scleria*
   Inflorescence not as above — **2**
2. Nut enclosed in a flask-shaped perianth (utricle). Flowers unisexual — **3**
   Nut not enclosed in a utricle. Flowers bisexual — **4**
3. Utricles containing female flower parts only — 15. *Carex*
   Utricles containing both female and male flower parts — 14. *Schoenoxiphium*
4. Spikelets solitary, terminal, not subtended by a conspicuous bract — **5**
   Inflorescence compound, the spikelets very rarely solitary, but if so, the spikelet subtended by a conspicuous bract — **6**
5. Spikelet ovate, greenish, at least 5mm long. Style base not persistent on the nut — 6. *Fimbristylis*
   Spikelet cylindrical, brownish, rarely appearing ovate, but if so less than 5mm long. Style base persistent on the nut — 5. *Eleocharis*
6. Spikelets compressed, the glumes in 2 opposite ranks — **7**
   Spikelets not compressed, the glumes spirally arranged, not 2-ranked — **10**

7. Inflorescence head-like, enclosed by 2–3 bracts, the lowermost with a very broad sheathing base. Lower leaf-sheaths blackish — 12. *Schoenus*
   Lowermost bract without a broad sheathing base. Lower leaf sheaths not blackish — **8**
8. Style branches 2. Nutlet, 2-angled — 3. *Pycreus*
   Style branches 3. Nutlet 3-angled — **9**
9. Rhachilla persistent after the glumes have fallen — 2. *Cyperus*
   Rhachilla and glumes falling together — 4. *Mariscus*
10. Glumes grey-green, pubescent, terminating in a conspicuous bristle — 11. *Fuirena*
   Glumes not as above, though pubescent in *Fimbristylis sieberiana* — **11**
11. Inflorescence apparently lateral, one bract upright and appearing to be a continuation of the stem — **14**
   Inflorescence obviously terminal — **12**
12. Inflorescence head-like, made up of 1–many sessile spikelets — **13**
   Inflorescence rayed, the spikelets (or some of them) clearly pedicelled — 6. *Fimbristylis*
13. Leaves filiform with a tuft of hairs at the mouth of the sheath. Bracts scarcely exceeding the inflorescence. Nuts not compressed — 7. *Bulbostylis*
   Leaves neither filiform nor with a tuft of hairs at the mouth of the leaf sheath. Bracts much exceeding the inflorescence. Nuts compressed — 1. *Kyllinga*
14. Spikelets crowded into globular heads — 8. *Scirpoides*
   Spikelets variously arranged, but never in globular heads — **15**
15. Inflorescence of 1–3 spikelets in a dense head. Bract less than 1cm long — 10. *Isolepis*
   Inflorescence usually with more than 3 spikelets but if spikelets 1–3, then the bract more than 1cm long — 9. *Schoenoplectus*

1. KYLLINGA *Rottb.*

Tufted annuals or perennials. Inflorescence a sessile head of one or more dense globose or shortly cylindrical spikes subtended by one or more long bracts which in our species much exceed the inflorescence; styles bifid, the nuts laterally compressed, smooth, ellipsoid.

Glumes dark brown. Stem bases swollen, bulbous — *K. chlorotropis*
Glumes whitish. Stem bases not swollen — *K. odorata*

1. Kyllinga chlorotropis *Steud.*

Slender perennial, 5–25cm high, the base bulbous and covered with the remains of old leaf sheaths. Leaves all basal, sparsely pilose, 5–20cm long, 0.5–1mm wide. Inflorescence of 1–3 crowded, dark brown spikes, the central one largest, 5–7mm long; bracts usually 3, the longest 2–9cm long; glumes ovate with an excurrent mucro.

A small, easily overlooked plant of bare patches in grassland where water accumulates in the rainy season, often associated with *Craterostigma* spp., *Microchloa kunthii*, *Commelina subulata* etc. It is found between 2000 and 3200m in scattered localities on the Ibb mountains, the central escarpment and J. Ash Shirq. Outside Yemen it is only known from the Ethiopian highlands.

2. Kyllinga odorata *Vahl*

Rhizomatous perennial with erect stems 5–30cm high. Leaves all basal, prominently keeled, glabrous, 5–20cm long, 3–4mm wide. Inflorescence of 1–3 crowded, whitish spikes, the central one largest, 7–12mm long; bracts usually 3, the longest 3–12cm long; glumes with a prominent green keel, ovate, acute, but not mucronate.

Only known from damp grassland at c 2000m between Shaḥir and 'Ukaybir, J. Milḥān.

2. CYPERUS *L.*

Mostly perennial, leafy plants of wet places, very variable in the nature of their rootstock which is consequently of considerable taxonomic significance. Inflorescence usually umbellate but quite often reduced to a dense head subtended by 1–many bracts which vary considerably in size and number; spikelets compressed with the glumes in 2 opposite ranks; style 3-fid (except in *C. laevigatus*). Nuts trigonous, smooth in our species.

A large, rather difficult genus in which vegetative characters are valuable in distinguishing species. It is, therefore, important to collect the underground parts when making specimens.

1. Stems transversely partitioned. Plant leafless — *C. articulatus*
   Stems not transversely partitioned. Plant usually leafy — **2**
2. Leaves inrolled, stiff, strongly curved, with hard pungent tips — *C. effusus*
   Leaves not as above — **3**
3. Inflorescence a solitary, dense head — **4**
   Inflorescence rayed and ± umbellate or a simple spike — **10**
4. Inflorescence apparently lateral being subtended by a conspicuous bract which appears to be a continuation of the stem — *C. laevigatus*
   Inflorescence clearly terminal — **5**
5. Spikelets whitish — **6**
   Spikelets brown — **7**
6. Glumes 3–5mm long, conspicuously veined. Leaves much shorter than the inflorescence — *C. obtusiflorus*
   Glumes 5–6mm long, lateral veins obscure. Leaves about as long as the inflorescence — *C. conglomeratus*
7. Glumes obtuse with a green keel but no mucro. Stooniferous perennial of high mountains — *C. fissus*
   Glumes mucronate, without a green keel. Tufted perennials or annuals without stolons — **8**
8. Spikelets 2–3mm wide. Leaves as long as the stem — *C. jeminicus*
   Spikelets 3–5mm wide. Leaves shorter than the stem — **9**

9. Annual. Spikelets light brown. Glumes with the mucro bent backwards so having a spreading appearance *C. rubicundus*
Perennial. Spikelets dark brown. Glumes with an erect mucro so appearing to be appressed to the rhachilla *C. wissmannii*
10. Inflorescence of one or more spikes with a distinct elongate axis **11**
Inflorescence of spikelet clusters without a distinct axis **19**
11. Spikes dense, ± cylindrical, 1–4cm long. Vigorous plants with greeny brown spikelets **12**
Spikes ± ovate, never cylindrical, usually with few spikelets, mostly less than 1.5cm long. Spikelets brown or orange, never greenish **13**
12. Glumes strongly compressed, spreading away from the axis of the spikelet at maturity. Spikes 1–2cm long *C. nutans*
Glumes weakly compressed, remaining erect against the axis of the spikelet at maturity. Spikes 2–4cm long *C. alopecuroides*
13. Stems arising from bulbs. Leaves numerous, equalling or overtopping the inflorescence which is a simple or branched spike **14**
Bulbs absent. Leaves few to many, shorter than the rayed inflorescence **15**
14. Bracts few, short, inconspicuous. Inflorescence a simple spike *C. blysmoides*
Bracts 3–6, at least as long as the inflorescence. Inflorescence branched with several spikes *C. bulbosus*
15. Glumes mucronate. Vigorous plants usually 40–70cm high *C. nubicus*
Glumes usually obtuse, rarely acute but never mucronate. Plants slender or vigorous **16**
16. Spikelets dark brown or blackish in colour (common above 2000m) *C. rigidifolius*
Spikelets orange, pale brown or reddish-brown in colour (rare above 1500m) **17**
17. Stolons stout, at least 3mm thick, never bearing tubers *C. longus*
Stolons more slender, up to 2mm thick, usually bearing tubers **18**
18. Spikelets red-brown. Glumes with indistinct lateral nerves *C. rotundus*
Spikelets golden. Glumes with prominent lateral nerves *C. esculentus*
19. Bracts more than 10, forming a whorl around the inflorescence. Tall leafless plant usually exceeding 1m high *C. alterniflorus*
Bracts 1–4. Slender, often annual species less than 50cm tall **20**
20. Spikelets in globular heads, greeny brown *C. difformis*
Spikelets in lax, untidy heads, dark brown *C. fuscus*

1. Cyperus alopecuroides *Rottb.*

Leafy perennial with stout, erect, triangular stems, 1–1½m high. Leaves about as long as the stem, very broad, 1–2cm wide. Inflorescence large, but much exceeded by the very long bracts, umbellate, with rays up to 15cm long; spikelets arranged in cylindrical spikes which are 2–4cm long and clustered at the ends of the rays; spikelets lanceolate, straw-coloured, 4–8mm long, 1.5–2mm wide; glumes 2–2.5mm long, rounded (not strongly compressed), mucronate, pale brown with a greenish central band.

A rare plant of streamsides and marsh between 300 and 1200m: Wadi Ḥabab above Madīnat al 'Abīd; between Sa'īd Miṣr and 'Alūjah; below Ḥillah, J. Bura'.

2. Cyperus alternifolius *L.*

Tufted perennial; stems stout, erect, triangular or terete, channelled, slightly scabrid, up to 1½m high. Leaves absent. Inflorescence umbellate, consisting of numerous rayed clusters of 3–7 spikelets, all partially concealed by the numerous long bright green leaf-like bracts; spikelets linear-lanceolate, greeny brown, 3–8mm long, 1.5–2mm wide; glumes 1.5–2mm long, sharply keeled, acute, pale brown with hyaline margins.

The Yemen plant is subspecies **flabelliformis** (*Rottb.*) *Kuek*

Widespread on the escarpment between 500 and 2000m from the Wadi Ḥidrān, near Taiz north to the Wadi Līyah, but only frequent in the central escarpment; it grows by small permanent streams but is particularly characteristic of stagnant rock pools in wadis that otherwise dry up after the rainy season.

3. Cyperus articulatus *L.*

Stoloniferous perennial with erect, soft, cylindrical, transversely partitioned stems 1–2m high. Leaves absent, reduced to large, basal leaf sheaths. Inflorescence bracts very short, less than 1cm long; inflorescence umbellate, the spikelets in clusters at the ends of the rays; spikelets greeny brown or straw-coloured, linear, 8–30mm long, 1mm wide; glumes c 2mm long, obtuse, pallid with a green keel.

An uncommon plant of ditches and streamsides between 200 and 1300m: Wadi Ḥidrān, Taiz; Suq Ar Rubū', Waṣāb As Sāfal; between Sa'īd Miṣr and 'Alūjah; Wadi Ḥār (Deflers) and 'Ajārah (Schweinfurth) near Ḥajjaylah; below Ḥillah, J. Bura'; near Al Kadān.

4. Cyperus blysmoides *Hochst. ex C.B.Cl.*

C. bulbosus Vahl var. *spicatus* Boeck.

Tufted perennial to 30cm high arising from a small black bulb. Leaves slender, c 1mm wide, about as long as the stem. Inflorescence a single spike of 3–6 erect spikelets, subtended by a minute, inconspicuous bract 3–5mm long; spikelets linear, black, 8–12mm long, 1–2mm wide; glumes c 4mm long, reddish-black, with a green keel exserted as a short mucro.

A rare plant only known definitely as a field weed at about 2400m on the high plateau at Kitāb (Wood 587) and near Risābah (Radcliffe-Smith and Henchie 4747). Records from J. Nuqum (Rathjens), J. Ta'kar, J. Ḥubaysh and J. Maḥdad may be referable to this species.

5. Cyprus bulbosus *Vahl*

Rather similar to the last species but more robust, reaching 50cm and with very numerous leaves which are up to 3mm wide. Inflorescence a lax, simple or branched spike with numerous spreading spikelets, subtended by 3–6 bracts, the largest of which exceeds the inflorescence; spikelets linear, dark red-brown, 10–20mm long, 1–2mm wide; glumes c 3mm long, reddish-brown with the keel exserted as a short mucro.

Known definitely only from 1500m on a roadside at Mafḥaq (Wood 716) but records from Bani 'Umar in the Ḥujarīyah, J. Milḥān and Madīnat Al 'Abīd (Wissmann) probably refer to this species.

6. Cyperus conglomeratus *Rottb.* Qassīs

Tufted grey-green perennial with erect stems to 40cm, the base surrounded by soft, white, papery leaf sheaths. Leaves as long as the stem, sinuous and slightly wiry but neither rigid nor pungent. Inflorescence a dense terminal cluster subtended by 3 ascending bracts which reach 8cm in length; spikelets broadly-lanceolate, white, 5–20mm long, c 5mm wide; glumes 5–6mm long, acuminate, weakly keeled, whitish-green with a pale brown central area.

A plant of sand dunes and sandy wadi beds at c 1000m at Ḥarīb and Mārib (Wood 75/328, Maxwell Darling 202).

The Yemen plant cannot be separated from specimens collected on sand dunes in Palestine and Egypt which are presumably representative of the type. Whether the specimen collected by Mordtmann on Kamaran Island and cited as *C. conglomeratus* by Kükenthal in his monograph is the same plant is not certain but there are no modern records of this species in the Tihama.

7. Cyperus difformis *L.*

Tufted annual or short-lived perennial with erect, triangular, slightly winged stems to 50cm. Leaves nearly as long as the stem. Inflorescence umbellate, consisting of shortly rayed, dense, globular heads of spikelets, sometimes reduced to a single head; bracts much longer than the inflorescence; spikelets linear, greeny brown, 2–8mm long, 1–1.5mm wide; glumes 0.5–1mm long, rounded on the back, acute, the green keel with brown lines on either side, margin hyaline.

Only known from marsh at 1000m near Suq Ar Rubū', Waṣāb As Sāfal.

8. Cyperus effusus *Rottb.*

C. conglomeratus var effusus (Rottb.) Kuek.

Stiff, tufted perennial to 40cm, though usually much less, the base surrounded with rigid, brown sheaths which only become fibrous very tardily. Leaves convolute, rigid, glaucous, commonly recurved to form almost a complete circle, shorter than the stem, the apex pungent and perhaps brittle as it is often missing. Inflorescence either umbellate with shortly rayed spikelet clusters or reduced to a solitary cluster subtended by stiff bracts which rarely exceed 3cm in length; spikelets linear (lanceolate when young), usually straight, pale brown, 4–18mm long, c 3mm wide; glumes c 4mm long, pale brown with a green keel in the upper half extended as a short mucro.

Widespread on sand in the Tihama, mostly near the sea but extending inland to Bayt al Faqīh and Suq al Jammah where it reaches 300m; often growing at the edge of brackish marsh or in areas where the water table is close to the surface (Wood 75/227, 3057). It has also been found on Ḥanish Island (Slade 8).

9. Cyperus esculentus *L.*

Very similar to *C. rotundus*, differing principally in its small, golden spikelets which do not exceed 12mm long. The glumes are conspicuously green-nerved and are c 3.5mm long.

The record of this species by Rathjens at Shūb near Sana'a at 2300m is unconfirmed but the occurrence of this species in Yemen is quite probable.

10. Cyperus fissus *Steud.*

Stoloniferous perennial with erect stems to 20cm, the stem bases surrounded with brown fibrous leaf remains. Leaves shorter than the stem. Inflorescence a dense terminal head subtended by 3 bracts which reach 4cm in length; spikelets lanceolate, dark brown, 6–10mm long, 2–3.5mm wide; glumes c 5mm long, dark brown, with a green keel and prominent lateral veins, obtuse.

C. fissus is closely related to *C. rigidifolius* but is easily distinguished by the head-like inflorescence and the large, prominently veined glumes.

A rare plant of high-altitude grassland between 2900 and 3200m on J. Ṣabir and J. Ta'kar (Wood 75/212, 1708).

Outside Yemen it is a rare species known only from a few mountains in Ethiopia.

11. Cyperus fuscus *L.*

Tufted annual with erect triangular stems to 30cm. Leaves about half as long as the stem, withering early. Inflorescence similar to that of *C. difformis* but the spikelets dark brown in rather lax, untidy heads; glumes c 1mm long, acute, scarcely keeled, dark brown with a green central area.

Only known from swampy fields by the Wadi Sūdān near Qā'idah at 1700m.

12. Cyperus jeminicus *Rottb.*

C. conglomeratus var. multiculmis (Boeck.) Kuek.

Tufted perennial with erect stems to 40cm, the base surrounded with soft, pale brown, fibrous leaf sheaths. Leaves as long as the stem, sinuous and wiry but not pungent. Inflorescence a dense terminal cluster subtended by 3 ascending bracts up to 8cm in length; spikelets linear, often curved, brown, 10–20mm long, 2.5–3mm wide; glumes 3–4mm long, brown with broad hyaline margins, obtuse, with a short mucro which is bent backwards so the glumes appear spreading.

Known only from Ḥadīyah (Forsskal) and Majz (Steinberg 239). Its ecology is not known.

13. Cyperus laevigatus *L.*

Juncellus laevigatus (L.) C.B.Cl.

Rhizomatous mat-forming perennial with erect subcylindrical stems to 60cm. Leaves all basal, terete, grooved, much shorter than the stem. Inflorescence an apparently lateral

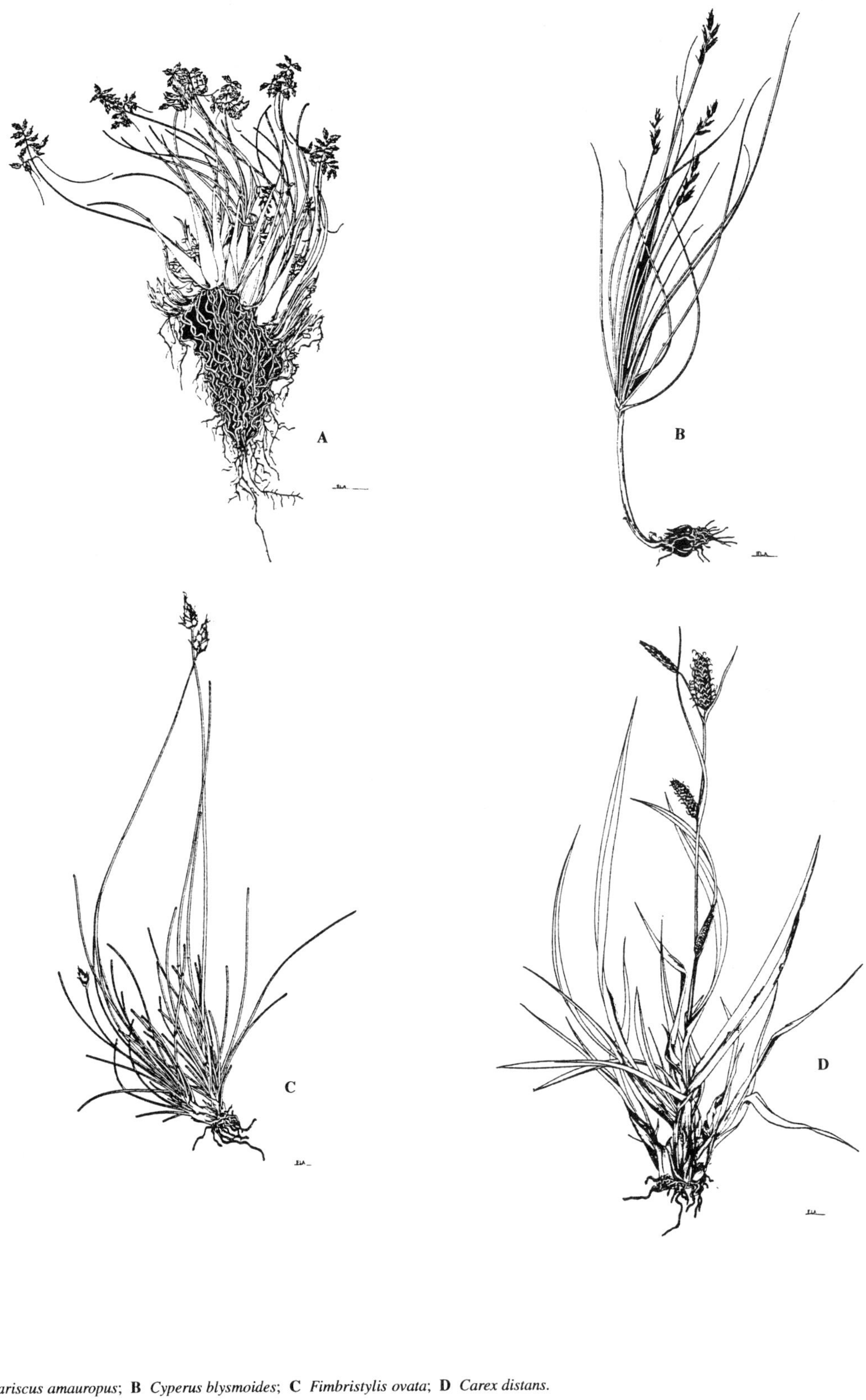

Fig. 21. **A** *Mariscus amauropus*; **B** *Cyperus blysmoides*; **C** *Fimbristylis ovata*; **D** *Carex distans*.

cluster, overtopped by a bract which appears to be an extension of the stem; spikelets linear, greenish-yellow, 5–12(–20)mm long and 2–2.5mm wide; glumes 2–3mm long, rounded on the back, acute, uniformly dull yellow-green; style bifid, the nut biconvex.

Vegetatively this species has a strong resemblance to the Rushes (*Juncus*), and the inflorescence is similar to that of a *Schoenoplectus* but the spikelets are compressed.

Very widely distributed in wet places from the Red Sea coast east to the Mārib area and reaching 2400m in the marshes of the high plateau. In freshwater marshes it is a frequent but not dominant species but in brackish marsh it is often dominant, giving a lawn-like appearance when heavily grazed. Such communities are found on the Red Sea coast at Al Fāzah and elsewhere and inland around Hawbān near Taiz and particularly at the hot springs of Hammām 'Alī and Hammām Damt.

14. Cyperus longus *L.*

Tall plant with a leafy stem to 1m similar to *C. rotundus* subspecies *tuberosus* and *C. nubicus* but distinguished from both by its stout stolons which never bear tubers and are at least 3mm thick. The glumes are obtuse and usually pale brown with a broad pallid margin.

Recorded from Madīnat Al 'Abīd (Wissmann) but it has never been refound and it seems probable that the record was erroneous, though the occurrence of *C. longus* is not impossible.

15. Cyperus nubicus *C.B.Cl.*

A robust plant, 40–70cm high with long red-brown spikelets, closely related to *C. rotundus* but readily distinguished from it and all related species by the shortly mucronate glumes. It is very similar in habit to *C. longus* with which it can easily be confused but it has relatively slender stolons, only c 2mm thick, as well as the glume difference.

Known with certainty only at c 1300m from a wadi between Taiz and Rawnah in Shar'ab (Wood 1544) and from the Wadi Ḥabab between Madīnat Al 'Abīd and Ḥammām 'Alī (Wood 3003) but field records from 'Adūf near Turbah, Wadi Ḍabāb and J. Milḥān probably refer to this species.

16. Cyperus nutans *Vahl*

C. eleusinoides Kunth

Perennial with a woody rhizome and stout, erect, triangular stems to 1m. Inflorescence similar to that of *C. alopecuroides* but the rays slightly shorter and the spikes of spikelets only 1–2cm long; spikelets linear, dull brown, 4–8mm long; glumes 2mm long, strongly compressed, pale brown with a green keel exserted as a short mucro, spreading at a wide angle in fruit.

The Yemen plant is variety **eleusinoides** (*Kunth*) *Haines.*

An uncommon plant of small streams in areas of high rainfall between 1100 and 1900m on the escarpment: on the west side of J. Ba'dān, north of Ibb; Wadi Dūr, 'Udayn; Wadi Ḥabab above Madīnat Al 'Abīd; between 'Alūjah and Kusmah, J. Raymah; Wadi Ḥijān, Ḥarāz: Wadi Sukhnah.

17. Cyperus obtusiflorus *Vahl*

Erect tufted perennial to 30cm, the stem base thickened and covered in dark brown sheaths. Leaves shorter than the stem, narrow and rather sinuous. Inflorescence a crowded terminal cluster, subtended by 2–4 spreading or deflexed bracts which are dilated at the base; spikelets strongly compressed, ovate to oblong, white, 5–10mm long, 4–8mm wide; glumes 3–5mm long, white with a pale green keel which is extended as a short mucro, conspicuously veined.

Widespread and locally frequent on shallow stony soils mainly on the plains of the high plateau and in sparse stony grassland on the wetter parts of the escarpment from 1500 to 2900m.

18. Cyperus rigidifolius *Steud.*

Perennial to 30cm spreading by relatively stout stolons c 2mm thick and with a strong superficial resemblance to *C. rotundus* from which it differs in its blackish spikelets which are only 6–10mm long. The glumes are 3–3.5mm long, obtuse, uniformly blackish apart from the green midrib.

Widespread from 2100 to 2800m growing by streams and in marsh on the escarpment and high plateau from the Ibb area north to Wadi Rakwah in Khawlan Ash Sham.

19. Cyperus rotundus *L.* Bisalah, Su'ad

Tufted perennial to 40cm spreading by slender stolons which are mostly 1–2mm thick and bear occasional ellipsoid tubers. Leaves few, about half as long as the stem. Inflorescence usually overtopped by the longest of the 2–4 subtending bracts, umbellate, the spikelets in short ovate spikes at the ends of the rays; spikelets linear, red-brown, 10–20mm long, 1–2mm wide; glumes c 3mm long, red-brown with a green keel and hyaline margins, obtuse or acute.

As well as subspecies **rotundus** described above, subspecies **tuberosus** (*Rottb.*) *C.B.Cl* also occurs. This is a more robust plant, usually 40–60cm high, with a thicker stem and paler spikelets which are 2–2.5mm wide (Wood 773).

Subspecies *rotundus* grows from 100 to 1500m in the eastern Tihama and on the escarpment and is also found at 2300m in irrigated land in the Bani Ḥushaysh and at Sana'a on the high plateau. It can be found throughout the year by irrigation ditches and in seasonally flooded land, particularly irrigated fields where it is one of the commonest weeds at low altitudes but after the rains it appears in all types of cultivated and waste land. Subspecies *tuberosus* is only known from the Wadi 'Ayn 'Alī near Ḥajjah at 1200m but has probably been overlooked elsewhere.

20. Cyperus rubicundus *Poir.*

C. teneriffae Poir.

Tufted annual; stems slightly thickened at the base, erect, up to 30cm. Leaves with membranous sheaths, shorter than the stem. Inflorescence a terminal cluster, subtended by 2 spreading or deflexed bracts; spikelets linear or lanceolate, brown, 5–20mm long, 3–5mm wide; glumes 3–5mm long, with a dark keel in the upper half extended as a short mucro which is distinctly bent backwards.

Widespread and common on the escarpment from 600 to 2900m, growing in shallow stony ground in open bushland

and rocky mountain slopes where there is little competition from taller plants.

21. Cyperus wissmannii *O. Schwartz*

Similar in general appearance to *C. rubicundus*, but perennial with stout, swollen stem bases which are surrounded by the dark brown remains of old leaf sheaths, seldom exceeding 20cm in height. Leaves filiform, about half as long as the stem. Inflorescence similar to that of *C. rubicundus* but the spikelets dark brown and the glumes with a shorter mucro which is not bent backwards so the glumes appear to be appressed to the rhachilla.

A rare plant of exposed stony cliff-tops and mountain ridges in a few localities between 2000 and 2500m: Wa'lān (Wissmann): J. Ḍawrān (Wissmann); Suq al Jum'ah; J. Ash Shirq (Wood 1950); Turbah (Wood 512). It may have a preference for sandstone but this cannot be confirmed.

Outside Yemen it is a rare plant known only from a few localities in the Horn of Africa.

3. PYCREUS *P. Beauv.*

Similar to *Cyperus* in which it is sometimes included but the style bifid and the nut obovoid or ellipsoid and laterally compressed. All our species are tufted plants of wet places. The inflorescence is of umbelled spikelet clusters (sometimes reduced to a single cluster), subtended by several prominent bracts.

1. Glumes dark brown or blackish — 2
   Glumes yellow or yellow-brown — 3
2. Glumes with 2 deep whitish furrows on either side of the green keel, becoming progressively darker brown towards the margins. Stem leafy for about half its length — *P. sanguinolentus*
   Glumes not furrowed, black except for the green keel. Stem leafy at the base only — *P. elegantulus*
3. Inflorescence apparently lateral, being subtended by a conspicuous bract which appears to be a continuation of the stem — *Cyperus laevigatus*
   Inflorescence clearly terminal — 4
4. Spikelets acuminate, light brown, c 1mm wide. Glumes with a green keel — *P. polystachyos*
   Spikelets abruptly narrowed, dull yellow, c 2mm wide. Glumes uniformly dull yellow — *P. flavescens*

1. Pycreus elegantulus (*Steud.*) *C.B.Cl.*

Cyperus elegantulus Steud.

Rhizomatous tufted perennial, 20–70cm high. Leaves slightly shorter than the stem. Inflorescence subtended by 3 (–5) long bracts, rayed or more commonly reduced to a single dense head of black spikelets, each spikelet c 1.5mm wide; glumes black with a green (eventually brown) keel. Nuts smooth.

A plant of marsh and streamsides from 2300 to 3100m; common in suitable habitats on the escarpment and high plateau from J. Ṣabir north to Bayt 'Adhāqah.

2. Pycreus flavescens (*L.*) *P. Beauv. ex Reichenb.*

Tufted annual, 15–30cm high. Leaves about half as long as the stem. Inflorescence subtended by 2–4 long bracts, usually umbellate with stalked spikelet clusters but sometimes reduced to a solitary cluster; spikelets dull yellow, linear-oblong, rather abruptly narrowed into an acute apex; glumes uniformly dull yellow. Nuts wrinkled (not easily seen with a hand lens).

A rather uncommon plant of pondsides and stream margins occurring in scattered localities between 1000 and 2100m on the escarpment from the wadi above 'Adūf near Turbah, north to Wadi 'Ayn 'Alī, Ḥajjah; often growing with *Fimbristylis sieberiana*

3. Pycreus polystachyos (*Rottb.*) *P. Beauv.*

Cyperus polystachyos Rottb.

Tufted perennial, 50–100cm high. Leaves long, about equalling the stem. Inflorescence subtended by 2–4 long bracts, usually umbellate with dense stalked spikelet clusters but sometimes reduced to a single dense head; spikelets yellow-brown, gradually tapering almost from the base into a long acuminate tip, 1mm broad; glumes yellow-brown with a green keel. Nuts smooth.

Recorded by Deflers from 'Attārah in the Ḥarāz and at Taiz, but not found by anyone else. The specimen is not at Paris and the record may well be an error of identification.

4. Pycreus sanguinolentus (*Vahl*) *Nees*

Densely tufted annual, 15–50cm high, the stems more leafy than in our other species. Leaves often longer than the stems. Inflorescence subtended by 3 bracts, one of which is very long, usually very dense and head-like but sometimes with a few stalked spikelet clusters; spikelets dark brown, 2mm broad; glumes dark reddish-brown with a green (eventually brown) keel and 2 pale furrows on either side of the keel. Nuts smooth.

A rare sedge of wet places on the escarpment from 1300 to 2500m: in the wadi above 'Adūf near Turbah; Wadi Ḍabāb, Taiz; by a cistern Ḥajarah, Ḥarāz; by a stream a few kilometres east of Rujum.

4. MARISCUS *Gaertn,.*

Similar to *Cyperus*, but the rhachilla and glumes falling together. The nuts are trigonous.

1. Annual. Glumes with a recurved awn — *M. squarrosus*
   Perennial. Glumes without a recurved awn — 2
2. Spikelets less than 1mm wide. Stems arising from a hard woody rhizome, not much swollen below — *M. cyperinus*
   Spikelets more than 1mm wide. Stem base swollen, often slightly succulent — 3
3. Inflorescence rayed (rarely contracted) with distinct spikes — 4
   Inflorescence a dense, subglobose head, the spikes not distinct — 5

4. Glumes usually widely spaced so the rhachilla is clearly visible. Spikelets pale reddish brown. Leaf sheaths pale *M. amauropus*
Glumes always closely imbricate so the rhachilla is hidden. Spikelets dark red-brown or greenish. Leaf sheaths dark brown *M. schimperi*
5. Spikelets variegated whitish-green and blackish-brown *M. plateilema*
Spikelets uniformly greenish-white *M. dubius*

1. Mariscus amauropus (*Steud.*) *Cuf.*

Cyperus leptophyllus Hochst.
Tufted perennial to 40cm, the stem bases swollen, slightly succulent and covered in the pale remains of old leaf sheaths. Leaves shorter than the stem, up to 2mm wide. Inflorescence bracteate, usually rayed, with 3–5 spikes borne on slender, filiform peduncles 0–3cm long; spikelets up to 10mm long, 2–3mm broad, pale reddish-brown; glumes widely spaced so the rhachilla is usually visible, very pale except for the keel, which is extended as a very short mucro.

Locally abundant between 900 and 1500m on loose stony ground in dry *Acacia* scrub, mostly in rainshadow areas of the central escarpment: above Ḥillah, J. Bura' (Schweinfurth); 'Usil, Ḥarāz (Deflers, Schweinfurth); Mawsanah, Ḥarāz; Mafḥaq; Wadi Sukhnah. This distribution parallels that of several species including *Euphorbia arabica* and *Svensonia laeta*.

2. Mariscus cyperinus *Vahl*

Tufted perennial to 40cm, the stem not thickened below but arising from a hard woody nodule on a short horizontal rhizome. Leaves slightly shorter than the stem, up to 5mm wide. Inflorescence a cluster of up to 12 sessile spikes, subtended by leafy bracts; spikelets up to 3mm long, 0.5–1mm broad, green, often 1-flowered; glumes ovate-oblong, obtuse, not keeled.

A very rare plant found between 2000 and 2500m in grassland in the central escarpment: near Bayt Mughlad, J. Shibām, Ḥarāz; J. Kabūrah above Ḥadīyah.

3. Mariscus dubius (*Rottb.*) *Hutch.*

Tufted perennial to 30cm, the stem bases swollen and covered with the brown remains of old leaf sheaths. Leaves almost as long as the stem, up to 4mm wide. Inflorescence a dense, bracteate, ovoid, greenish-white head; spikelets up to 8mm long, 2–3mm wide; glumes prominently nerved, obtuse, scarcely keeled.

Recorded in Schwartz from J. Ḍawrān, 'Usil in the Ḥarāz, Salafīyah and Bilad Aṭ Ṭa'am (all Wissmann), but it has not been found by other collectors and the records may be errors for *M. schimperi* var. *viridis* which is very similar.

4. Mariscus plateilema *Steud.*

Tufted perennial to 20cm, the stem bases swollen and densely covered in the brown, fibrous remains of old leaf sheaths. Leaves equalling or slightly shorter than the stem, up to 2mm wide. Inflorescence a dense ovoid, bracteate head, variegated whitish-green and blackish-brown in colour; bracts with distinct dilated bases; spikelets up to 8mm long, 1–2mm wide; glumes acuminate, weakly keeled.

Only known in montane grassland at 3100m on J. Ta'kar.

5. Mariscus schimperi *Hochst.*

Cyperus neo-schimperi Kuek.
Tufted perennial to 40cm, the stem bases swollen, slightly succulent, covered in the dark brown remains of old leaf sheaths. Leaves nearly as long as the stem, up to 3mm broad. Inflorescence bracteate, usually rayed, with 3–6 spikes borne on rather stout peduncles up to 2cm long, sometimes reduced to a head, but the spikes remaining ± distinct from each other. Spikelets up to 10mm long, 1–2mm broad, red-brown; glumes ovate, prominently nerved, the keel extended as a very short mucro.

As well as the type described above, variety **viridis** (*Schweinf.*) *C.B.Cl.* also occurs. It is distinguished by its pale greenish-brown spikelets. Specimens with a contracted inflorescence are very similar to *C. dubius* but the spikes are ± distinct and the glumes shortly mucronate.

The typical plant is found rarely in montane grassland between 1900 and 2900m on the central escarpment: Kusmah, J. Raymah; J. Shibām, Ḥarāz; Markah and 'Ukaybir, J. Milḥān; below Shamsān, Maḥābishah (Wood 835, 2374). Variety *viridis* is locally abundant on rocky hillsides, terrace walls and cliffs between 600 and 2100m in the wetter parts of the escarpment. It is particularly common in the coffee regions of the central escarpment (Wood 75/290; Radcliffe-Smith and Henchie 4419; Miller 18).

6. Mariscus squarrosus (*L.*) *C.B.Cl.*

Small tufted annual, rarely exceeding 10cm and with a distinct peppery smell even when dry, stem bases neither swollen nor covered in old leaf sheaths. Leaves about as long as the stem, 1–2.5cm broad. Inflorescence bracteate with several sessile and shortly pedunculate ginger-coloured spikes; spikelets up to 12mm long, 1–2mm broad; glumes prominently nerved, brown except for the green keel which is extended into a recurved awn.

A rare plant found between 1100 and 1500m on sand banks by small streams: between Riḥāb and Waṣāb; Wadi Sukhnah.

5. ELEOCHARIS *R.Br.*

Glabrous annuals or perennials with terete stems. Leaves reduced to sheaths. Inflorescence of solitary, terminal spikelets, not subtended by bracts, glumes spirally arranged, the lowermost empty and often of different shape to the others; style base enlarged and persistent on the top of the nut. Both our species have 2-branched styles and biconvex nuts.

Glumes strongly keeled. Spikelets 2–5mm long *E. geniculata*
Glumes scarcely keeled. Spikelets 6–30mm long *E. palustris*

Fig. 22. **A** *Kyllingia chlorotropis*; **B** *Fimbristylis cymosa*; **C** *Carex brunnea*; **D** *Mariscus squarrosus.*

1. Eleocharis geniculata (*L.*) *Roem. & Schult.*
E. capitata R. Br.
E. caribaea (Rottb.) Blake
Rhizomatous or tufted, annual or perennial up to 20cm high. Spikelets ovoid, 2–5mm long; glumes ovate, obtuse with a prominent green keel. Nut smooth, black; style base conical.

A plant of wet places, typically silt banks in and beside slow-moving streams, found in scattered localities between 500 and 1900m; widespread on the escarpment from 'Adūf, near Turbah north to Maḥābishah; rare in the Mashriq and known only from Ḥūth and Wadi Rakwah near Majz (Steinberg).

2. Eleocharis palustris (*L.*) *Roem & Schult.* Khazaj
Rhizomatous perennial to 60cm. Spikelets cylindrical, 6–30mm long; glumes ovate, acute, not keeled. Nut smooth, brown; the style base squat, about as broad as long.

The Yemen population of *E. palustris* is rather isolated and most collections made (e.g. Wood 2286, 2985) are of a form with a rather large lower glume which partly encircles the spikelet, thus showing some approach to the European *E. uniglumis* (Link) Schult. One collection (Wood 2414) from between Yarīm and Raḍmah is remarkable for its very large spikelets but the nuts are similar to the more common form. The Yemen plant needs careful investigation and it is likely that one or more distinct varieties can be described.

Widespread and common between 2200 and 3000m in marsh and by streams and ponds on and around the high plateau extending west to Ṭawīlah and J. Ḥubaysh but otherwise absent from the outer escarpment: in Khawlan Ash Sham known from between Ḥaydān and Sāqayn.

6. FIMBRISTYLIS *Vahl*

Tufted perennials with flat, linear leaves. Except in *F. ovata* the inflorescence is of stalked spikelets forming a terminal compound umbel; bracts 2–3, usually shorter than the inflorescence; glumes spirally arranged, the lowermost empty, the middle ones bisexual, the uppermost male or sterile. Nuts biconvex or trigonous, the style base not persistent on the nut.

1.	Spikelets solitary	*F. ovata*
	Spikelets many. Inflorescence umbellate	2
2.	Spikelets in small clusters. Glumes obtuse	*F. cymosa*
	Spikelets single, never clustered. Glumes acute	3
3.	Glumes pubescent	*F. sieberiana*
	Glumes glabrous	4
4.	Stem compressed and winged below the inflorescence. Style 3-fid	*F. complanata*
	Stem compressed but not winged below the inflorescence. Style 2-fid	*F. dichotoma*

Fimbristylis complanata (*Retz.*) *Link*
Tufted perennial, 15–40cm high, the stem strongly flattened and winged below the inflorescence. Leaves 10–30cm long, 2–5mm wide, glabrous, often shiny grey on the undersurface. Inflorescence lax with numerous, single, glabrous, chestnut brown, very narrowly ovoid spikelets, each 3–6mm long. Nuts trigonous, pale, papillose or with a few wavy ridges.

Apparently a rare plant of grassy marsh at c 2300m in Ibb province: above Dhī Sufāl; between Saddah and Nādirah in the Wadi Banā.

2. Fimbristylis cymosa *R. Br.*
F. obtusifolia (Lam.) Kunth
F. spathacea Roth
Tufted perennial, 10–40cm high. Leaves 5–20cm long 0.5–2mm wide, glabrous, glaucous, rather stiff (particularly in salty conditions). Inflorescence dense to somewhat lax, consisting of shortly pedicellate clusters of glabrous, pale brown spikelets, each 3–5mm long; glumes obtuse with broad hyaline margins. Nuts trigonous (sometimes biconvex), smooth, black.

A plant of sandy places in marsh and by streams growing in scattered localities in the Tihama and on the escarpment from sea level (Al Khawkah) to 1800m (above 'Adūf, near Turbah). It tolerates brackish conditions but is not restricted to places with a high salt level.

3. Fimbristylis dichotoma (*L.*) *Vahl*
Tufted perennial, 20–60cm high, the stem flattened but not winged below the inflorescence. Leaves 10–30cm long, 0.5–1.5mm wide, glabrous or pubescent, greyish. Inflorescence lax with numerous single, glabrous, chestnut brown, ovoid spikelets, each 5–10mm long. Nuts pale, biconvex, with longitudinal ridges.

A rare plant known from marsh between 1800 and 2300m in the Ibb region: on the lower slopes of J. Ba'dān; above Dhī Sufāl. Very probably overlooked elsewhere.

4. Fimbristylis ovata (*Burm. f.*) *Kern*
F. monostachyos Vahl
Tufted perennial, 5–30cm high. Leaves 3–10cm long, c 1mm wide, often slightly inrolled, glabrous. Inflorescence a solitary, terminal, whitish-green, glabrous, ovoid spikelet, 8–18mm long; glumes distichous below, spirally arranged above. Nuts pale brown, trigonous, tuberculate.

A plant of scattered localities on the escarpment from 1800 to 2800m from Turbah north to Rujum, usually growing in grassland in areas of prolonged winter mists; most characteristic of the grassland areas of the central escarpment.

5. Fimbristylis sieberiana *Kunth*
F. ferruginea sensu Schwartz (probably)
Tufted perennial, 20–100cm high, rather like a robust version of *F. dichotoma*. Leaves 5–30cm long, 1–2mm broad, glabrous or pubescent, greyish, the basal sheaths herbaceous. Inflorescence overtopped by the lowest bract, lax with numerous, single, pubescent, brown, ovoid, obtuse spikelets, each 7–15mm long; glumes greyish-pubescent. Nuts pale, biconvex, smooth.

A plant of marsh and the gravelly margins of streams between 400 and 2200m, quite tolerant of saline conditions; very rare in the Tihama and only known from Suq al Jammah; widespread and frequent on the escarpment:

elsewhere known only from 'Ujlī, south of Ṣa'dah and from between Yarīm and Raḏmah.

F. sieberiana has until recently been included in *F. ferruginea* (L.) Vahl but is best treated as a separate species. No authentic material of *F. ferruginea* has been traced from Yemen but it might occur. It can be distinguished by its coriaceous lower leaf sheaths, smaller leaves (3–10cm long, 0.5–1.5mm wide), shorter bracts and acute spikelets.

7. BULBOSTYLIS *Kunth ex C.B.Cl.*

Tufted annuals or perennials. Leaves filiform with a tuft of long hair at the mouth of the sheath. Inflorescence (in our species) a dense terminal head of 1–20 sessile spikelets with 2–3 short bracts; glumes spirally arranged. Nutlets trigonous or biconvex, obovoid, crowned by the persistent style base.

Inflorescence of 1–2 spikelets. Styles bifid, the nut biconvex — *B. humilis*
Inflorescence of 2–15 spikelets. Styles 3-fid, the nut trigonous (though not strongly so) — *B. atrosanguinea*

1. Bulbostylis atrosanguinea (*Boeck.*) *C.B.Cl.*

Densely tufted perennial with swollen bulb-like stem bases to 40cm. Leaves nearly as long as the inflorescence, filiform, sparsely pilose, grass-green. Inflorescence of 2–15 blackish spikelets, each c 6mm long; glumes ovate, acute, keeled, uniformly blackish; nuts trigonous, grey, almost smooth.

Only known at 2000m in the grassland on the northwest slope of J. Ẓallamlam above Ḥadīyah.

2. Bulbostylis humilis *Kunth*

B. striatella C.B.Cl.

Small tufted annual, 4–20cm high. Leaves 4–10cm long, glabrous. Inflorescence of 1–2 greenish-brown spikelets, each 5–8mm long; glumes ovate, mucronate, pale brown with a prominent green keel; nuts biconvex, pale green, finely longitudinally lined.

A rare plant of bare patches in montane grassland found between 2800 and 3200 on J. Ta'kar and Al Manār on J. Ḥubaysh.

8. SCIRPOIDES *Seguier*

1. Scirpoides holoschoenus (*L.*) *Sojak*

Scirpus holoschoenus L.
Holoschoenus vulgaris Link

Densely tufted perennial, ½ – 1m high. Leaves up to 60cm long, stiff and slightly rough on the margins, hemispherical in cross-section with a deep groove on the flat side. Inflorescence with 2 bracts, one stem-like, much larger than the inflorescence and up to 20cm long; inflorescence umbellate with sessile and pedunculate, dense, globular spikelet clusters; glumes brown, the green keel extended into a short mucro; style 3-fid. Nut trigonous, smooth.

Superficially very similar to *Juncus punctorius*.

A characteristic plant of springs and streams in the Mashriq; tolerant of brackish conditions and often associated with *Juncus rigidus*, it is found from 1200 to 2000m: Dhī Nā'im; between the 'Azum and Watadah passes; between Mārib and the Fardah pass; near Khulāqah in Nihm; between Ṣa'dah and J. Minmār.

9. SCHOENOPLECTUS (*Rchb.*) *Palla*

Annuals or perennials, leaves reduced to sheaths or if present basal only. Inflorescence terminal (though appearing lateral), subtended by a single bract which appears to be a continuation of the stem, umbelliform or contracted into a head of 1–many spikelets; glumes spirally arranged; perianth of bristles often present. Nuts biconvex or trigonous.

1. Inflorescence a solitary dense head, sometimes reduced to a single spikelet. Bract much longer than the inflorescence — **2**
   Inflorescence umbelliform with at least one spikelet cluster rayed. Bract shorter than or equalling the inflorescence — **4**
2. Stem triangular in cross-section — *S. mucronatus*
   Stem rounded in cross-section — **3**
3. Low annual, usually less than 20cm high. Spikelets 1–7 in a cluster — *S. proximus*
   Vigorous perennial 30–100cm high. Spikelets 10 or more in a cluster — *S. inclinatus*
4. Some spikelets more than 10mm long. Style 2-branched. Coastal plant — *S. littoralis*
   All spikelets less than 8mm long. Style 3-branched. Inland plants — **5**
5. Glumes emarginate, mucronate, the margins fringed. Spikelets dark red-brown — *S. lacustris*
   Glumes acute, shortly mucronate, glabrous. Spikelets of various colours but usually blackish when young — *S. inclinatus*

1. Schoenoplectus inclinatus (*Del.*) *Lye* — Ḥalal

Scirpus inclinatus (Del.) Aschers. & Schweinf.
S. brachyceras A. Rich.
S. corymbosus auct.

Robust leafless perennial, ½–2m high. Leaves reduced to the basal sheaths. Inflorescence bract stem-like, usually 1–2cm long, shorter than the umbellate inflorescence which consists of sessile and pedunculate spikelet clusters; spikelets 4–7mm long, pale brown to black, lanceolate; glumes acute, mucronate, glabrous; stigmas 3. Nut trigonous, smooth.

Rather atypical inflorescences are sometimes seen. In Wood 1002 from J. Sumārah the inflorescence is reduced to a single cluster and is subtended by a bract 5–6cm long and is thus of very different appearance.

A plant of ponds and slow-moving streams between 1600 and 3000m; common in suitable habitats on the high plateau but rare on the escarpment where it is known only from Suq As Sabt, near Ibb, J. Shibām in the Ḥarāz (Schweinfurth) and on the Maḥwīt-Ṭawīlah ridge.

2. Schoenoplectus lacustris (*L.*) *Palla*

Scirpus lacustris L.

Similar in general appearance to *S. inclinatus* but more robust, reaching 3m high. Best distinguished by its fatter, ovoid, red-brown spikelets and its emarginate mucronate glumes whose upper margins are distinctly fringed.

A local plant of ponds and very wet swamp between 2300 and 2700m on the high plateau: in a number of places around Yarīm; Wadi Wa'lān; Al Zīl, Ḥadā'; Ẓula' (Deflers); Wadi Aḥjar.

3. Schoenoplectus littoralis (*Schrad.*) *Palla*
Scirpus littoralis Schrad
Similar to the previous two species but the stem often triangular below the inflorescence. The spikelets are brown, subcylindrical, mostly 8–12mm long; the bracts obtuse, mucronate, but not fringed, the style bifid and the nut consequently biconvex.

A rare plant of coastal lagoons where the main wadis flow into the sea: at the mouth of the Wadi Zabīd at Fāzah, at the mouth of the Wadi Rima' (Hepper) and at the mouth of the Wadi Sihām between Ṭayf and Manẓār.

4. Schoenoplectus mucronatus (*L.*) *Palla*
Tufted perennial, 50–75cm high, the stem triangular with winged angles. Leaves reduced to basal sheaths. Inflorescence bract stem-like, 4–8cm long, longer than the inflorescence which is a lateral cluster of greeny-brown spikelets; glumes ovate, acute, green along the midrib becoming brown towards the margin; style 3-fid. Nut trigonous, smooth.

Only known from a stagnant pond at c 2000m between Rujum and Ṭawīlah.

5. Schoenoplectus proximus (*Steud.*) *Raynal*
Slender tufted annual, 3–10cm high. Leaves filiform, furrowed, short. Inflorescence bracts usually 2, the longest stem-like with a sheathing base and more than half the height of the plant; inflorescence a lateral cluster of 1–2 greenish spikelets; glumes hyaline with a green midrib, c 1.5mm long; style 3-fid. Nut trigonous, transversely wrinkled.

On drying mud by a pond c 10km west of Dhamār at 2300m.

10. ISOLEPIS *R. Br.*

1. Isolepis setaceus (*L.*) *R. Br.*
Scirpus setaceus L.
Slender tufted annual, 3–15cm high. Leaves filiform, furrowed, slightly shorter than the stems. Inflorescence bracts 3–9mm long, longer than the inflorescence which consists of 1–3 ovoid greeny-brown spikelets in an apparently lateral cluster; glumes with a green midrib, purple-brown above becoming paler below; style 3-fid. Nut trigonous, vertically ribbed.

A local plant of pond margins and streamsides found in scattered localities between 2200 and 3000m on the escarpment and high plateau.

11. FUIRENA *Rottb.*

Perennials with noded, triangular, leafy stems. Leaves with a short membranous ligule. Inflorescence essentially a panicle but contracted in our species into an apical cluster which is usually overtopped by the bract; glumes spirally arranged, grey-green, pubescent, terminating in a long, weakly recurved bristle. Nuts trigonous, obovoid.

Spikelets narrowly ellipsoid, 9–15mm long. Nut with brown, reticulate ridges *F. felicis*
Spikelets broadly ellipsoid, 6–10mm long. Nut smooth *F. pubescens*

1. Fuirena felicis *Hooper*
Rhizomatous perennial with erect stems to c 80cm. Leaves mostly on the stem, up to 30cm long. Inflorescence a terminal cluster of 1–10 grey-green narrowly ellipsoid spikelets each 9–15mm long and 2–2.5mm wide; glumes distinctly hooded, terminating in a mucro c 1mm long. Nut pale green with brown, reticulate ridges.

Only known from the wadi above 'Adūf near Turbah, where it grows at 1900m by small springs emerging from sandstone rocks (Wood 953). It is endemic to Yemen.

2. Fuirena pubescens (*Lam.*) *Kunth*
Similar in general appearance to the last species but the spikelets broadly ellipsoid, 6–10mm long and c 5mm broad, the glumes not hooded and the nuts smooth.

Only known from marsh at c 2500m in Wadi Aḥjar (Wood 3343).

12. SCHOENUS *L.*

1. Schoenus nigricans *L.*
Densely tufted glabrous perennial with wiry stems 30–80cm high, leafy at the base only. Leaves 15–60cm long, involute, the basal sheaths glossy, black, stiff. Inflorescence of 5 or more flattened, brown spikelets crowded into a dense terminal head and enclosed by 2–3 conspicuous bracts, the lowest up to five times as long as the inflorescence and with a broad sheathing base; glumes distichous, scabrid-hairy on the keels; nut trigonous or nearly globular.

Restricted to dripping sandstone cliff ledges between 1800 and 2900m; very local but abundant where it occurs: all over the Turbah cliffs; Aḥjar; J. Maswar near Bayt 'Adhāqah.

13. SCLERIA *Berg.*

1. Scleria bulbifera *Hochst. ex A. Rich.*
Rhizomatous perennial to 80cm with swollen, bulbous stem bases. Stems leafy, triangular. Leaves strongly keeled 10–25cm long, the mouth of the sheath hairy. Inflorescence a simple interrupted spike with the spikelets arranged in 5–14 distant clusters along a rhachis 5–20cm long; spikelets reddish-brown, 4–6mm long with female flowers below and male flowers above. Nuts white, glabrous, pitted.

Only known at 1900m in grassland on the northwest slopes of J. Ẓallamlam above Ḥadīyah.

14. SCHOENOXIPHIUM *Nees*

1. Schoenoxiphium sparteum (*Wahl.*) *Kuek.*
Slender, erect, tufted perennial with rather leafy stems to 50cm. Inflorescence long and narrow with many lateral, pedunculate spikes; glumes acute, chestnut brown with a

green midrib and hyaline margins, about as long as the nut; styles 3, nut 2.5–3mm long, trigonous, longitudinally ridged, green (becoming brown), glabrous, with a long entire beak.

Distinguished from *Carex* by having both male and female flower parts within each utricle.

Only known from montane grassland at c 3100m on J. Ta'kar (Wood 1150).

15. CAREX *L.*

Rhizomatous tufted perennials with triangular, usually leafy stems. Leaves linear, keeled. Flowers unisexual in 1-flowered spikelets which are arranged into spikes which may be bisexual or unisexual, clustered into a head or scattered along the inflorescence rhachis. The female flowers are surrounded by a beaked perianth (utricle) which persists round the trigonous or biconvex nut.

1. Spikes bisexual, similar in apperance, crowded into a dense head. *C. stenophylla*
   Spikes mostly unisexual, dissimilar in appearance, variously arranged on the axis 2
2. Nuts biconvex. Spikes all ± sessile forming a rather dense terminal inflorescence *C. brunnea*
   Nuts trigonous. At least the lowermost spike, strongly pedunculate. Inflorescence usually lax 3
3. Spikes very dense with numerous flowers. Utricle glabrous *C. distans*
   Spikes rather lax with few flowers. Utricle pubescent *C. negrii*

1. Carex brunnea *Thunb.*

Slender, tufted perennial to c 30cm; stem almost leafless. Inflorescence short, dense, with about 5 sessile spikes forming a terminal inflorescence; glumes mucronate, greyish-brown with a green midrib; styles 2, nut 3–3.5mm long, biconvex, flattened, the utricle longitudinally ridged, shortly pubescent, with a long bifid beak.

The Yemen plant is the endemic variety **arabica** *Hooper*.

A rare plant of rocky grassland between 2000 and 2700m on the escarpment mountains: Madāḥit near Tūrbah; J. Ṣabir (Wood 74/378); above Dhī Sufāl (Hepper & Wood 5864): J. Shibām, Ḥarāz (Miller 357); J. Kabūrah, Ḥadīyah.

2. Carex distans *L.*

C. burchelliana sensu Schweinf.

Densely tufted perennial, 10–30cm high. Leaves slightly glaucous, strongly keeled, mostly basal or towards the bottom of the stem, 5–20cm long. Inflorescence of a single terminal male spike and 3–4 female spikes variously arranged on the axis below, the lower spike(s) pedunculate, the peduncle partially concealed by the leaf-like sheathing bracts; spikes 1.5–2.5cm long; female glumes broadly ovate, brown with a prominent green keel which is extended into a scabrid mucro; styles 3. Fruit trigonous with a notched beak.

The Yemen (and Arabian) plant is variety **sinaica** (*Nees*) *Boeck.* which is in fact intermediate between the mainly European *C. distans* and the central Asian *C. diluta* M. Bieb. In *C. distans* the gap between the male spike and the uppermost female spike is at least 1cm long and the beak of the fruit is very shallowly notched. In *C. diluta* the uppermost female spike is contiguous with the male spike and the beak of the fruit is deeply notched. Some Yemen plants (Wood 1246) approach *C. distans*, some approach *C. diluta* (Wood 2282) while others are hard to place (Wood 74/35).

A local plant growing by springs and in marsh from 2300 and 2900m on the high plateau, particularly the Yarīm-Dhamār area; on the escarpment it is only known on J. Shibām, Ḥarāz (Schweinfurth) and on J. Ba'dān.

3. Carex negrii *Chiov.*

Slender tufted perennial with leafy stem, to 50cm. Inflorescence long, with many lateral pedunculate spikes; glumes acute, chestnut brown with a green midrib and hyaline margins; styles 3, nut 2.5–3mm long, trigonous, longitudinally ridged; utricle green, pubescent with a deeply bifid beak.

Only known from a steep grassy slope at c 2700m on J. Sumārah (Wood 75/182).

4. Carex stenophylla *Wahl.*

Slender tufted perennial, 10–25cm high. Leaves green, 5–15cm long. Inflorescence a dark brown, ovate, terminal head, 1–1.5cm long, subtended by a short bract and consisting of several crowded bisexual spikes; glumes ovate, brown with scarious margins and a green keel extended into a short mucro; styles 2. Fruit biconvex.

Widespread in marsh between 2300 and 3200m on the high plateau but usually in small quantity.

GRAMINEAE

The grasses comprise one of the most easily recognised families. Typically they have fibrous roots and are branched at ground level forming a rosette or tuft. The stems are cylindrical and hollow. The leaves are borne along the stem, arising from a node, the lower part or sheath tightly enclosing the stem, the upper part or blade (referred to as the leaf in the following account) usually linear and flat though quite frequently inrolled or folded. At the junction of the blade and sheath there is a hairy or membranous rim (ligule).

The leafless inflorescence develops from the upper leaf sheaths. Its basic unit analagous to the flower is the spikelet which may be arranged in spikes, racemes or panicles. It is usually bisexual. The spikelet consists of various scales arranged along an axis called a rhachilla. The two lowermost scales, which are empty, are called glumes. Above the glumes are one or more scales, known as lemmas, each of which enclose the parts of an individual flower including the anthers and a delicate scale known as the palea. The base of the lemma, where it is joined to the rhachilla may be of distinctive shape and is known as the callus. The lemma and, less commonly, the glumes may be produced into one or more long bristles known as awns. The fruit grain is dry, hard and achene-like, sometimes with the palea or some other part of the spikelet adhering to it.

The grasses are the largest plant family in Yemen and economically much the most important. The main cereal crops, including Sorghum, Bulrush Millet, Maize, Wheat, and Barley, are all grasses. They also provide most of the food for grazing, though well-developed grassland is rare in Yemen and mostly restricted to certain valleys, the margins of cultivated land, the high mountains of the Ibb region and certain areas of the central escarpment where there are extensive meadowlands on shallow soils.

While there are considerable problems in grass taxonomy, the species are generally not so difficult to identify as is popularly supposed. The following key is based on inflorescence type and, while admittedly it is unnatural and does not deal well with certain anomalous species, it is definitely less off-putting to a beginner than a natural one based on grass tribes, as technical characters can be left to a later stage in the key and may not be necessary at all. Anyone, though, who wishes to identify their grasses reliably will sooner or later have to examine the grass spikelet. All characters used in this account can be seen with a hand lens (though with difficulty in a few genera with small spikelets like *Agrostis*) and someone unfamiliar with grasses would be wise to examine species with large spikelets first (e.g. *Bromus*, *Panicum*).

Although tribal divisions are not used in the following key, certain tribes are easily recognised. The 3-branched awn of the *Aristideae* (Genera 26 and 27), the paired spikelets (one sessile, one pedicelled) of the *Andropogoneae* (Genera 78–93) and the 2-flowered spikelets with the awnless bony upper lemma of the *Paniceae* (Genera 61–77) are usually distinctive.

1. Inflorescence a dense head of closely packed spikelets, never more than 5cm long, usually much less — Group A **7**
Inflorescence not as above — **2**

2. Inflorescence of spikes or racemes variously arranged: solitary, paired, digitate, sessile along an axis or panicled — **3**
Inflorescence paniculate, the spikelets not arranged in racemes or spikes. (A few specimens may be hard to assign to one or other group, in which case both dichotomies may have to be followed) — **6**

3. Inflorescence a solitary terminal spike or raceme — Group B **12**
Spikes or racemes not solitary — **4**

4. Inflorescence a digitate cluster of spikes or racemes at the top of the stem — Group C **39**
Inflorescence not digitate — **5**

5. Spikes and racemes ± sessile along a short or long axis, thus forming 'a spike of spikes' — Group D **52**
Spikes and racemes clearly pedunculate, often in pairs and/or partially enclosed in the spathe-like sheaths of the upper leaves, forming a 'panicle of spikes' — Group E **76**

6. Spikelets awned — Group F **86**
Spikelets not awned — Group G **104**

GROUP A

Inflorescence a dense globose, ovoid or ellipsoid head of spikelets, 1–4 (–5)cm long.

7. Spikelets with awns at least 2mm long — **8**
Spikelets awnless though sometimes the lemmas or glumes ending in a short point — **9**

8. Annual not exceeding 10cm high. Leaf sheaths with bulbous-based hairs — 25. *Triraphis*
Perennial, exceeding 15cm high. Leaf sheaths glabrous except at the mouth — 53. *Fingerhuthia*

9. Inflorescence glabrous or nearly so. Leaves never pungent — **10**
Inflorescence densely hairy. Leaves usually pungent — **11**

10. Inflorescence partially enclosed by the upper leaf sheaths, up to 2cm long. Spikelets 1-flowered — 54. *Crypsis*
Inflorescence not enclosed by the upper leaf sheaths, mostly more than 3cm long. Spikelets 3-flowered — 13. *Phalaris*

11. Erect grass to 2m. Leaf sheaths glabrous. Lemmas 3-nerved — 42. *Odyssea*
Prostrate or ascending grass to 15cm. Leaf sheaths hairy at least below. Lemmas 9–11-nerved — 29. *Aeluropus*

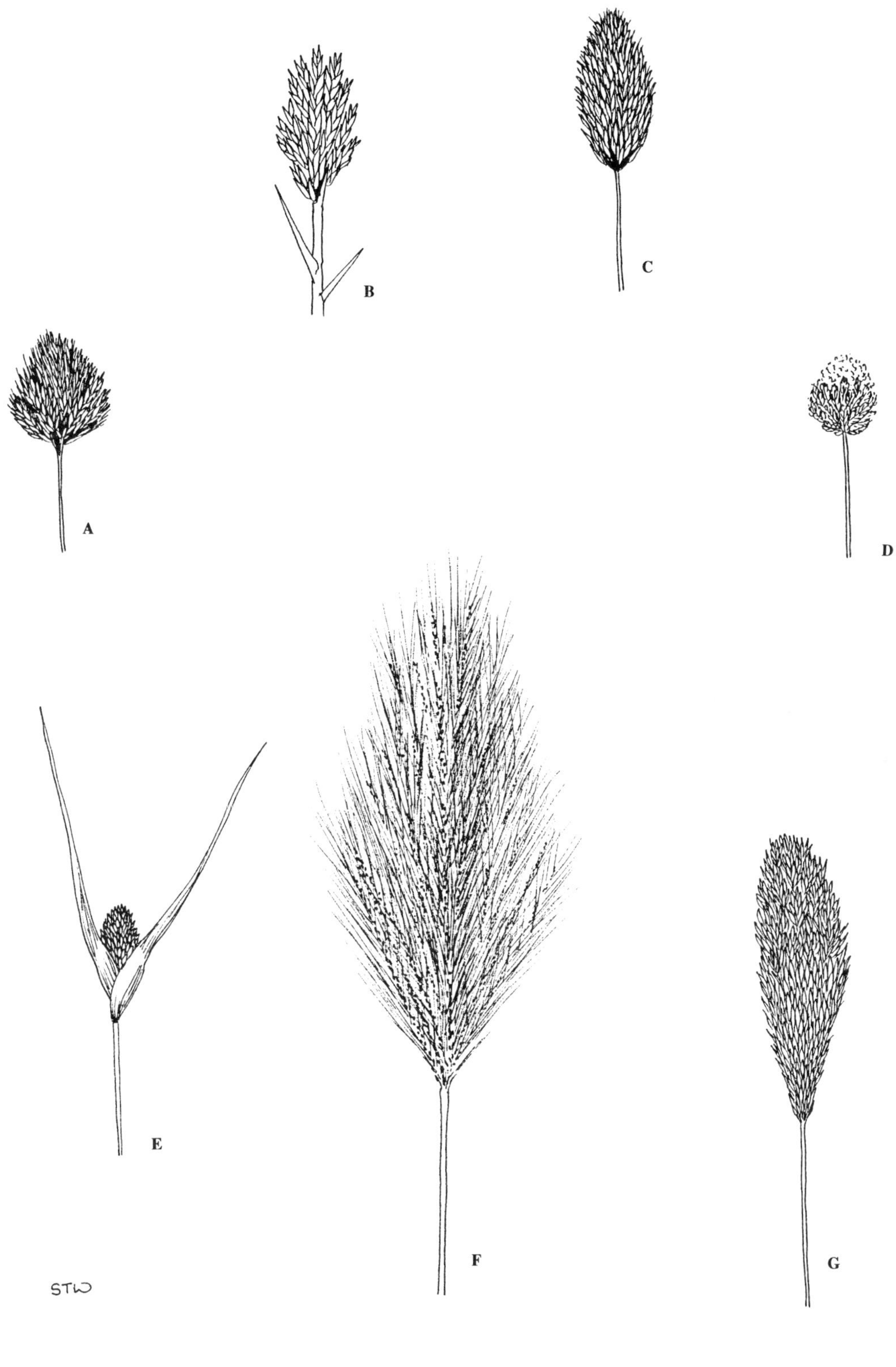

Fig. 23. Group A Grasses. **A** *Triraphis pumilio*; **B** *Odyssea mucronata*; **C** *Fingerhuthia africana*; **D** *Aeluropus lagopoides*; **E** *Crypsis schoenoides*; **F** *Chloris roxburghiana*; **G** *Phalaris paradoxa*.

GROUP B

Inflorescence a solitary terminal spike or raceme, always linear in shape, the spikelets sessile or shortly pedicelled.

12. Spikelets surrounded by an involucre of bristles or scales which may or may not be joined to each other. (Sometimes the involucre is of a solitary bristle) **13**
Spikelets not surrounded by an involucre of scales or bristles **17**
13. Involucre herbaceous, 4–6-lobed, one lobe very long and awnlike, up to 25mm long. Aquatic plant with inflated leaf sheaths 74. *Odontelytrum*
Involucre not as above, usually of bristles, never herbaceous. Terrestial plants **14**
14. Involucre of bristles and/or scales which, if present, are narrowly triangular in shape **15**
Involucre of broad stiff scales which are widest in the middle. Bristles absent 73. *Anthephora*
15. Bristles falling with the spikelet **16**
Bristles persistent after the spikelet has fallen 64. *Setaria*
16. All bristles free to the base, never flattened 76. *Pennisetum*
Bristles joined at the base or higher up, some or all distinctly flattened 77. *Cenchrus*
17. Spikelets awned **18**
Spikelets awnless **27**
18. Leaves stiff, pungent, terminating in a hard, spine-like point 20. *Rhytidosperma*
Leaves not stiff nor terminating in a hard, spine-like point **19**
19. Awns bent, the lower part (column) twisted, more than 2cm long. Spikelets paired, one sessile, one pedicelled **20**
Awns straight, less than 2cm long. Spikelets not in pairs **21**
20. Upper lemma acute, but without an awn-like bristle. Awns commonly twisted together 88. *Heteropogon*
Upper lemma terminating in an awn-like bristle, awns not becoming twisted together 84. *Sehima*
21. Inflorescence a spike-like panicle; spikelets irregularly arranged along the rhachis, some clearly pedicelled **22**
Inflorescence a spike, either one-sided or with the spikelets arranged in opposite ranks on either side of the rhachis **23**
22. Annual. Spikelets 1-flowered 14. *Gastridium*
Perennial. Spikelets 2–5-flowered 5. *Festuca*
23. Spikes 1-sided **24**
Spikelets arranged in opposite ranks on either side of the rhachis **25**
24. Spikes strongly curved or coiled. Spikelets 1-flowered, inserted into the rhachis of the spike 45. *Oropetium*
Spikes strongly curved. Spikelets 3-flowered, not inserted into the rhachis of the spike 49. *Enteropogon*
25. Lateral spikelets with a single glume; terminal spikelet with 2 glumes 4. *Lolium*
All spikelets with 2 glumes **26**
26. Spikelets less than 1cm long *43. Tripogon*
Spikelets 1–3cm long 9. *Brachypodium*
27. Spikelets with short hooked spines or hooked tubercles **28**
Spikelets without hooked spines or tubercles **29**
28. Pedicels broad, flattened. Inflorescence lax, the spikelets mostly more than 5mm apart 59. *Leptothrium*
Pedicels absent or, if present, short, terete. Inflorescence very dense 58. *Tragus*
29. Upper glume terminating in a stiff awn-like point 60. *Dignathia*
Upper glume not terminating in a stiff awn-like point **30**
30. Inflorescence with long, woolly hairs. Spikelets usually paired with 1 (–2) sessile and 1 pedicelled spikelet **31**
Inflorescence glabrous or pubescent but never woolly. Spikelets not paired **32**
31. Glumes uniformly woolly to the bifid tips. Leaves coiled 90. *Elionurus*
Glumes mostly woolly but the bifid tip ± glabrous. Leaves never coiled 91. *Lasiurus*
32. Spikelets 2- or more flowered **33**
Spikelets 1-flowered **36**
33. Spikelets sessile **34**
Spikelets, or the lower ones, clearly pedicelled **35**
34. Spikelets pubescent, less than 3mm long 29. *Aeluropus*
Spikelets glabrous, 5–20mm long 4. *Lolium*
35. Pedicels pubescent, spreading or deflexed 38. *Harpachne*
Pedicels glabrous, erect 32. *Eragrostis*
36. Spikelets sunk into hollows in the rhachis 45. *Oropetium*
Spikelets not sunk into hollows in the rhachis **37**
37. Spikes 1-sided. Leaves not pungent 47. *Microchloa*
Spikes ± cylindrical, not 1-sided. Leaves pungent **38**
38. Spikelets glabrous 56. *Sporobolus*
Spikelets hairy 55. *Urochondra*

Fig. 24. Group B Grasses. **A** *Tragus berteronianus*; **B** *Enteropogon macrostachyus*; **C** *Lolium multiflorum*; **D** *Pennisetum thunbergii*; **E** *Cenchrus ciliaris*; **F** *Tripogon multiflorus*; **G** *Brachypodium distachyon*.

Fig. 25. Group C Grasses. **A** *Ochthochloa compressa*; **B** *Dactyloctenium aegyptium*; **C** *Cynodon dactylon*; **D** *Digitaria ternata*; **E** *Chloris pycnothrix*; **F** *Andropogon distachyos*; **G** *Eleusine indica*; **H** *Dichanthium annulatum*; **J** *Eustachys paspaloides*.

GROUP C

Inflorescence of spikes or spike-like racemes in a digitate cluster at the top of the stem. Included are some species in which the spikes are subdigitate with an axis of up to about 1cm in length.

39. Spikelets awned **40**
Spikelets without awns **47**
40. Spikes 1–2 together **41**
Spikes 3 or more together, rarely some inflorescences reduced to 2 spikes **42**
41. Spikelets 3–10-flowered. Awns straight 52. *Tetrapogon*
Spikelets 2-flowered. Awns bent 85. *Andropogon*
42. Spikelets usually in pairs, one sessile and one on a short pedicel. Awns sharply bent **43**
Spikelets all sessile, not arranged in pairs. Awns straight or wavy **45**
43. Leaves with a broad, ± amplexicaul base, less than 6cm long 83. *Arthraxon*
Leaves without a broad, amplexicaul base, mostly more than 6cm long **44**
44. Pedicels and rhachis internodes solid 81. *Dichanthium*
Pedicels and rhachis internodes hollow (thus appearing striped when seen with a hand lens) 80. *Bothriochloa*
45. Spikelets golden brown. Awn less than 1.5mm long 50. *Eustachys*
Spikelets green, purplish or white. Awn more than 1.5mm long **46**
46. Spikelets 1-flowered. Awns 2–3cm long, wavy 46. *Schoenefeldia*
Spikelets 2–4-flowered. Awns less than 2cm long, straight 51. *Chloris*
47. Spikes 2 71. *Paspalum*
Spikes 3 or more together **48**
48. Spikelets 1–2-flowered; spike narrow, less than 2mm broad **49**
Spikelets with 3 or more flowers; spike stout, at least 3mm broad **50**
49. Spikelets 2-flowered, falling entire, some pedicellate 72. *Digitaria*
Spikelets 1-flowered, the glumes persistent after the rest of the spikelet has fallen, sessile 48. *Cynodon*
50. Rhachis of the spike terminating in a naked point 34. *Dactyloctenium*
Rhachis of the spike terminating in a spikelet, not a naked point **51**
51. Inflorescence strictly digitate. Lemma villous. Stoloniferous plant of sand dunes 36. *Ochthochloa*
Inflorescence of digitate spikes, usually with one or more spikes on the axis below. Lemma glabrous. Tufted plants not growing on sand dunes 37. *Eleusine*

GROUP D

Inflorescence of 2 or more spikes or spike-like racemes arranged along a short or long axis, the spikes and racemes sessile or nearly so or only the lowest with a peduncle up to about 1cm long.

52. Racemes numerous, forming a dense feathery inflorescence **53**
Racemes or spikes few to many, not forming a dense feathery inflorescence **54**
53. Annual. Leaves very short, less than 10cm long 44. *Leptocarydion*
Perennial. Leaves more than 10cm long 51. *Chloris*
54. Coarse robust grass to 2m, the inflorescence consisting of more than 30 short pale brown spikes along an axis 25–40cm long 31. *Desmostachya*
Grasses usually less than 1m high, the inflorescence with less than 30 spikes along an axis usually c 5–15cm long **55**
55. Spikelets surrounded by an involucre of bristles which persists after the spikelets have fallen 64. *Setaria*
Spikelets without an involucre of bristles **56**
56. Glumes ending in a short awn **57**
Spikelets awnless or with awns arising from the lemma **58**
57. Spikelets contiguous. Spikes up to 2cm long 39. *Dinebra*
Spikelets distant. Spikes 1–5cm long 65. *Oplismenus*
58. Spikelets 1–2-flowered **59**
Spikes 3- or more flowered **70**
59. Lemmas awned **60**
Lemmas awnless **61**
60. Spikelets paired, one sessile, one pedicelled. Awns strongly bent about the middle 80. *Bothriochloa*
Spikelets not in pairs. Awns straight 67. *Echinochloa*
61. Glumes longer than the spikelet, persistent after the rest of the spikelet has fallen 40. *Leptochloa*
Glumes (or at least the lower) much shorter than the spikelet **62**
62. Spikelets in 2–4 rows **63**
Spikelets irregularly arranged or in 1 row **68**
63. Lower glume absent **64**
Lower glume present **65**
64. Spikelets with a bead-like structure at the top of the pedicel. Spikes 4–20 68. *Eriochloa*
Spikelets without a bead-like structure at the top of the pedicel. Spikes 2–5 71. *Paspalum*
65. Spikes rather distant, shorter or equalling the internodes, arranged along a long axis. Spikelets ± glabrous 66. *Paspalidium*
Spikes usually much longer than the internodes, arranged along a short axis at the top of the stem. Spikelets usually hairy **66**
66. Upper lemma mucronate. Lower glume on the side opposite the rhachis 70. *Urochloa*
Upper lemma acute or mucronate. Lower glume facing the rhachis **67**
67. Glumes cuspidate 67. *Echinochloa*
Glumes obtuse 69. *Brachiaria*

Fig. 26. Group D Grasses. **A** *Paspalidium geminatum*; **B** *Dinebra retroflexa*; **C** *Leptochloa rupestris*; **D** *Brachiaria eruciformis*; **E** *Oplismenus hirtellus*; **F** *Digitaria velutina*; **G** *Urochloa trichopus*; **H** *Coelachyrum yemenicum*.

68. Spikelets 5–6mm long — 69. *Brachiaria*
Spikelets 1–3mm long — **69**
69. Spikes less than 2mm broad — 72. *Digitaria*
Spikes more than 3mm broad — 67. *Echinochloa*
70. Spikelets very large, at least 5mm broad. (Plant of coastal sand dunes) — 30. *Halopyrum*
Spikelets less than 5mm broad. (Plants not growing on coastal sand dunes) — **71**
71. Awn more than 1cm long, twisted and bent — 24. *Phaenanthoecium*
Awn, if present, less than 5mm long, neither twisted nor bent — **72**
72. Lemmas emarginate or shortly bifid — **73**
Lemmas acute or rounded — **74**
73. Lemma with a mucro from the apical cleft. Ligule 3–6mm long — 41. *Diplachne*
Lemma without a mucro from the apical cleft. Ligule 1.5mm long — 40. *Leptochloa*
74. Lemma extended into a short awn-like mucro. Spikes 5–11cm long — 35. *Acrachne*
Lemma acute or rounded. Spikes 1–7cm long — **75**
75. Lemmas glabrous — 37. *Eleusine*
Lemmas hairy — 33. *Coelachyrum*

GROUP E

Inflorescence of pedunculate racemes forming a panicle of racemes. Racemes often paired and/or partially enclosed in the spathe-like sheath of the upper leaves.

76. Spikelets without bristles or awns, of greatly modified appearance — **77**
Spikelets with bristles or awns — **79**
77. Spikelets densely hairy. Leaves wine red — 90. *Elionurus*
Spikelets glabrous. Leaves green — **78**
78. Sessile spikelets globose. Lower glume with pits and tubercles — 92. *Hackelochloa*
Sessile spikelets flattened. Lower glume smooth — 93. *Rottboellia*
79. Spikelets with an involucre of 1–many bristles, awnless — 76. *Pennisetum*
Spikelets without an involucre of bristles, awned — **80**
80. Racemes in pairs — **81**
Racemes solitary — **83**
81. Column of awn hairy — 87. *Hyparrhenia*
Column of awn glabrous — **82**
82. Racemes deflexed at maturity, seldom exceeding the spatheole in length. Panicle often compound, dense. Leaves aromatic in smell and taste — 86. *Cymbopogon*
Racemes not deflexed at maturity, usually longer than the spatheole, 1–2 together. Leaves not aromatic — 85. *Andropogon*
83. Awn up to 5mm long, straight — **84**
Awn more than 1cm long, bent — **85**
84. Spikelets solitary with a single awn — 75. *Snowdenia*
Spikelets paired, each with 2 awns — 57. *Tetrachaete*
85. Awns up to 5cm long. Pedicelled spikelet much longer than the sessile spikelet — 85. *Andropogon*
Awns up to 2.5cm long. Pedicelled spikelet equalling or slightly smaller than the sessile spikelet — 81. *Dichanthium*

Fig. 27. Group E Grasses. **A** *Dicanthium foveolatum*; **B** *Hyparrhenia quarrei*; **C** *Snowdenia petitiana*; **D** *Tetrachaete elionuroides*; **E** *Cymbopogon schoenanthus*.

GROUP F

Inflorescence an open or contracted panicle, spikelets awned.

86. Tall aquatic grass to 3m with a densely plumose panicle — 79. *Saccharum*
Grasses of varying size but if panicle plumose then less than 1m high — **87**
87. Awns 3-branched — **88**
Awn simple — **89**
88. Awn branches all glabrous — 26. *Aristida*
Central awn branch plumose — 27. *Stipagrostis*
89. Lemmas with 9 awns; inflorescence often feathery in appearance — 28. *Enneapogon*
Lemmas with 1–3 awns — **90**
90. Awns 4–8cm long — **91**
Awns less than 3cm long — **92**
91. Spikelets in red-brown heads, partially enclosed by a spatheole, the heads consisting of 1–2 fertile spikelets surrounded by an involucre of 4 sterile spikelets — 89. *Themeda*
Inflorescence a whitish elongate panicle, the spikelets not arranged in heads as above — 17. *Stipa*
92. Awns (or some of them) bent sharply at about the middle, the lower half (column) spirally twisted — **93**
Awns ± straight, the lower half not spirally twisted — **97**
93. Annual. Spikelets large, at least 15mm long — 11. *Avena*
Perennial. Spikelets up to 10 (–14)mm long — **94**
94. Panicle branches terminating in a whorl of golden hair below the spikelets — 82. *Chrysopogon*
Panicle branches without a whorl of hair below the spikelets — **95**
95. Ligule membranous — 12. *Helictotrichon*
Ligule a ring of hairs — **96**
96. Spikelets in groups of 1–3, c 10mm long — 21. *Danthoniopsis*
Spikelets solitary, c 4mm long — 23. *Pentaschistis*
97. Spikelets 1–3-flowered — **98**
Spikelets 5- or more flowered — **101**
98. Spikelets 1-flowered. Leaf sheaths glabrous. Ligule membranous — **99**
Spikelets 2–3-flowered. Leaf sheaths pubescent or pilose. Ligule a ring of hairs — **100**
99. Annual. Panicle dense, linear. Grain neither large nor shiny — 14. *Gastridium*
Perennial. Panicle usually lax and open. Grain large, shiny black — 16. *Oryzopsis*
100. Inflorescence silky-hairy (rarely ± glabrous). Spikelets gaping, 2-flowered. Annual or short-lived perennial — 63. *Rhynchelytrum*
Inflorescence without silky hairs. Spikelets not gaping, 2–3-flowered. Perennial — 22. *Centropodia*
101. Spikelets 1.5–2.5cm long — 8. *Bromus*
Spikelets less than 1.5cm long — **102**
102. Pedicels densely pubescent — 2. *Lamarckia*
Pedicels glabrous — **103**
103. Perennial, more than 20cm high — 5. *Festuca*
Annual, up to 20cm high — 7. *Vulpia*

GROUP G

Inflorescence an open or contracted panicle, spikelets without awns.

104. Inflorescence silky-hairy — **105**
Inflorescence without long silky hairs — **109**
105. Glumes and lemmas silky-hairy. Plant of dry rocky places up to 60cm high — 62. *Tricholaena*
Glumes glabrous. Tall stout plants of wet places, 50cm to 5m high — **106**
106. Spikelets 2-flowered. Panicle branches hairy — **107**
Spikelets 2–6-flowered. Panicle branches glabrous — **108**
107. Panicle up to 20cm long. Rhachis not breaking up readily at maturity — 78. *Imperata*
Panicle usually 30–40cm long. Rhachis breaking up readily at maturity and freely shedding spikelets — 79. *Saccharum*
108. Lemmas hairy. Rhachilla glabrous — 18. *Arundo*
Lemmas glabrous. Rhachilla villous — 19. *Phragmites*
109. Spikelets with an involucre of one or more bristles which persists after the spikelet has fallen — 64. *Setaria*
Spikelets without an involucre of bristles — **110**
110. Spikelets 1-flowered — **111**
Spikelets 2- or more flowered (lemmas very dissimilar in appearance in some of the genera) — **113**
111. Spikelets 5–6mm long. (Spikelets are in fact awned, but the awn falls early; careful search will usually reveal an awn on a young spikelet) — 16. *Oryzopsis*
Spikelets 1–3mm long — **112**
112. Lower glume longer than the spikelet. Ligule membranous — 15. *Agrostis*
Lower glume shorter than the spikelet. Ligule a ring of hairs — 56. *Sporobolus*
113. Spikelets 2-flowered, the lemmas very dissimilar, the lower like the upper glume, the upper usually polished and bony — **114**
Spikelets 2- or more flowered, the lemmas all similar in appearance — **116**
114. Spikelets with a bead-like structure at the top of the pedicel — 68. *Eriochloa*
Spikelets without a bead-like structure at the top of the pedicel — **115**
115. Panicle diffuse, much-branched, the secondary branches usually long — 61. *Panicum*
Panicle sparingly branched, the secondary branches, if present, very short and less than 2cm long — 69. *Brachiaria*
116. Panicle narrow with a hairy axis and shiny, hyaline lemmas; glumes longer than the lowest lemma — 10. *Koeleria*
Panicle not as above, the axis glabrous and the glumes shorter than the lowest lemma — **117**
117. Spikelets 3-flowered, 1 fertile and 2 sterile, 5–6mm long — 1. *Ehrharta*
Spikelets 2- or more flowered, but if 2–3- flowered, less than 5mm long — **118**
118. Panicle flattened, rigid. Small annual — **6.** *Catapodium*
Panicle neither flattened nor rigid. Annuals or perennials — **119**

Fig. 28. Group F Grasses. **A** *Bromus pectinatus*; **B** *Danthoniopsis barbata*; **C** *Enneapogon schimperianus*; **D** *Aristida congesta*; **E** *Rhynchelytrum repens*; **F** *Festuca obturbans*.

Fig. 29. Group G Grasses. **A** *Brachiaria comata*; **B** *Panicum atrosanguineum*; **C** *Poa schimperiana*; **D** *Agrostis viridis*; **E** *Catapodium rigidum*; **F** *Eragrostis cilianensis*; **G** *Sporobolus coromandelianus*.

119. Lemmas 1–3-nerved. Ligule a ring of hairs (except *E. japonica*) 32. *Eragrostis*
Lemmas 5–7-nerved. Ligule membranous 3. *Poa*

1. EHRHARTA *Thunb.*

1. Ehrharta erecta *Lam.*

E. abyssinica Hochst.

Loosely tufted perennial with ascending or erect stems to 80cm. Leaves flat, up to 15cm long; ligule membranous, c 2mm long. Inflorescence a narrow panicle (rarely much reduced and almost racemose), with short erect branches and few spikelets, 3–10cm long; spikelets 3-flowered, the lowest 2 reduced to lemmas, glabrous, awnless, whitish-green, c 5.5mm long; glumes unequal, ovate, much shorter than the lemmas.

The Yemen plant is variety **abyssinica** (*Hochst.*) *Pilg.* which has larger spikelets than the type.

Widespread between 2000 and 3100m on the main escarpment mountains from J. Ṣabir northwards; also near Ḥaydān in Khawlan Ash Sham. It is a nondescript plant of rocky, shaded places such as terrace walls, cliffs and mountain gullies and appears to be frequent in suitable habitats.

2. LAMARCKIA *Moench.*

1. Lamarckia aurea (*L.*) *Moench.*

Erect tufted annual to 20cm. Leaves flat, 10–20cm long, often overtopping the inflorescence; ligule a long white membrane, 4–8mm long. Inflorescence a crowded, narrow, one-sided panicle, 2–5cm long; spikelets arranged in spreading or pendulous, glabrous, whitish-green, shiny clusters on short, villous peduncles; spikelet clusters with a single small fertile spikelet, several larger sterile spikelets and a reduced spikelet similar to the fertile one; fertile spikelets, 1-flowered, c 2mm long with 2 straight awns; sterile spikelets awnless with 2–8 empty lemmas, 5–7mm long; reduced small spikelet awned, borne on the pedicel of one of the sterile spikelets.

A local plant of mountain gullies, terrace walls, cliff ledges and similar stony places protected from grazing and competition from vigorous perennials; it is found from 1800 to 3100m on the wetter escarpment mountains: present on nearly all the high mountains around Ibb; Kusmah and Al Jabin, J. Raymah; near 'Ukaybir, J. Milḥān; J. Shibām, Ḥarāz; Baw'ān; Ṭawīlah (Rathjens); Shibām-Kawkabān.

3. POA *L.*

Annuals or perennials. Leaves flat; ligule membranous. Inflorescence an open or contracted panicle; spikelets 2–10-flowered, awnless; glumes slightly unequal, 1–3-nerved, glabrous; lemmas concave, keeled, 5–7-nerved, often with a tuft of hairs at the base.

A large temperate genus, notoriously complex taxonomically but with few species in Yemen. Unfortunately, these are far from straightforward and there are few reliable characters that can be used to separate them.

1. Annual, usually less than 15cm tall, very rarely slightly more *P. annua*
Perennials, almost always more than 25cm tall, very rarely less **2**
2. Panicle contracted, linear, the branches erect, appressed to the rhachis *P. leptoclada*
Panicle open, some branches spreading or deflexed **3**
3. Lower panicle branches flexuous, usually deflexed. Stems weak and straggling. Anthers 0.5–1mm long *P. schimperiana*
Lower panicle branches spreading or ascending, not flexuous. Stems erect. Anthers ± 1.5mm long **4**
4. Ligule very short, less than 1mm long. Basal leaf sheaths papery *P. pratensis*
Ligule 2–4mm long. Basal leaf sheaths becoming fibrous when old *P. simensis*

1. Poa annua *L.*

Tufted annual, sometimes briefly perennial, with spreading or ascending stems to 15(–30)cm. Leaves up to 14cm long; ligule 2–5mm long. Panicle open, ovate or triangular, usually 1–4cm long; spikelets 3–10-flowered, whitish-green; upper glume ± elliptic, definitely broadest in the middle; lemma obtuse, glabrous or hairy on the nerves; anthers 0.7–1.2mm long.

A cosmopolitan weed which in Yemen is an uncommon plant found between 2000 and 3300m on the escarpment and high plateau growing by ditches and as a field weed.

2. Poa leptoclada *Hochst. ex A. Rich.*

Tufted perennial with weak, straggling ascending stems to 75cm. Leaves up to 12cm long; ligule 0.6–4.5mm long. Panicle contracted, linear, 5–19cm long, the branches erect, appressed to the axis; spikelets 2–5-flowered, whitish-green, or purplish; upper glume narrowly elliptic to lanceolate, broadest below the middle; lemma acute, glabrous or pubescent; anthers 0.5–1mm long.

Readily distinguished from other species by the linear panicle.

A rare plant of high-altitude grassland, cliffs and wet, stony places, found between 2200 and 3500m on the escarpment from J. Nu'mān in J. Ḥubaysh north to J. Maswar.

A puzzling collection from Bayt 'Alama on J. Maḥdad (Wood 2514) is probably best placed here until further material of the same plant becomes available. It is a small tufted perennial with a very short narrow inflorescence, smaller and more wiry than typical *P. leptoclada*.

3. **Poa pratensis** *L.*

P. menachensis Schweinf.

Tufted perennial spreading by rhizomes and stolons with stems to 80cm, the basal leaf sheaths papery. Leaves up to 30cm long, the ligule always very short, less than 1mm long. Panicle usually open, ovate, 2–10cm long, the lowest branches usually spreading (never deflexed) and in whorls of 5; spikelets 2–5-flowered, whitish-green; upper glume ovate to elliptic, broadest just below the middle; lemma acute, always pubescent on the nerves and with a large tuft of hair at the base; anthers ± 1.5mm long.

Only known from the northwest side of J. Shibām near Manākhah where it was collected by Schweinfurth. It has not been refound.

4. **Poa schimperiana** *Hochst. ex A. Rich.*

Very similar in habit and spikelet measurements to *P. leptoclada* but differing in its panicle which is open, very lax and 8–35cm long, thus much longer than in other *Poa* species in Yemen; the lower branches are typically deflexed and long, sometimes as much as 6cm in length.

Like *P. leptoclada*, found between 2000 and 3200m on the escarpment from the Ibb area north to J. Maswar, but more frequent in the central escarpment where it occurs in the meadow communities; elsewhere it is a plant of damp, shaded cliffs and rocky streamsides.

5. **Poa simensis** *Hochst. ex A. Rich.*

A little-known species very similar to *P. pratensis* but apparently without stolons and rhizomes and with the basal leaf sheaths becoming fibrous when old. The spikelet dimensions are similar to *P. pratensis* and it appears always to have the same pubescent lemma; it differs most strikingly in the longer ligule which is usually about 4mm long.

Only definitely known at about 2600m above Dimarayn on J. Ṣabir (Wood 75/202) but field records from J. Ta'kar, Ḥiṣn Ḥab (J. Ba'dān) and J. Nu'mān (J. Ḥubaysh) in the Ibb area may be this species. Outside Yemen it is a rare grass of one or two places in northern Ethiopia.

4. LOLIUM *L.*

Annuals or perennials. Leaves flat or inrolled, ligule membranous. Inflorescence a stiff, 2-sided spike, the spikelets alternate in opposite rows; spikelets large, 3–15-flowered, often more than 1cm long; lower glume absent except on the terminal spikelets; upper glume large, many-nerved; lemmas 5–9-nerved, awned or awnless.

All species are probably introduced into Yemen and their taxonomy is difficult since the species are ill defined and intergrade with each other.

1.	Upper glume equalling or exceeding the spikelet which becomes swollen at maturity	*L. temulentum*
	Upper glume shorter than the spikelet. Spikelets not becoming swollen at maturity	2
2.	Perennial. Lemmas awnless	*L. perenne*
	Annual. Lemmas awned	*L. multiflorum*

1. **Lolium multiflorum** *Lam.*

Italian Rye Grass. Annual or biennial with erect stems to 50cm. Leaves up to 30cm long. Spike 8–20cm long; spikelets 8–15-flowered, 1–2cm long; glume usually about half as long as the spikelet; lemmas with a straight awn, 5–8mm long.

Typical vigorous European strains of this plant have been introduced in Sana'a in recent years (Wood 2052), but an older population also exists as a weed of wheat and barley fields between 2700 and 3100m on J. Ba'dān, J. Manār and J. Ta'kar. This older population (Wood 1145, 1858) is a slender, weedy annual with 5–9-flowered spikelets and a relatively long lemma. It approaches forms of *Lolium persicum* Boiss. & Hoh. which is common in central Asia.

2. **Lolium perenne** *L.*

Perennial Rye Grass. Perennial grass similar to the previous species but often less vigorous. Spike 8–15cm long; spikelets 8–11-flowered, 5–15mm long; glume half as long as the spikelet; lemmas unawned.

A casual of field borders, ditches and waste ground found between 2000 and 2600m mainly on the high plateau; Saddah in the Wadi Banā; Yarīm (Hepper): Sana'a and the surrounding villages; 'Amrān; Khamr; Yāzil; J. Ẓallamlam above Ḥadīyah.

One collection (Miller 3384) from the Marib road, Sana'a, is apparently annual but is best referred to *L. perenne* since the spikelets are not conspicuously rigid nor embedded into concavities in the rhachis as is typical of the most closely related annual species, *Lolium rigidum* Gaud.

3. **Lolium temulentum** *L.*

Darnel. Annual with stiff erect stems to 50cm. Leaves dark green, up to 40cm long. Spikes rigid, 10–30cm long; spikelets 4–10-flowered, becoming stiff and swollen at maturity, 10–25mm long; glume equalling or exceeding the spikelet; lemmas with or without awns.

The seeds are well known for their poisonous properties.

A rare casual found in Sana'a by Rathjens in the 1930s and not seen since.

5. FESTUCA *L.*

Tufted perennials. Leaves flat or convolute; ligule membranous. Inflorescence an open or contracted panicle; spikelets several-flowered; glumes ± equal, the lower 1-nerved, the upper 3-nerved; lemmas 5-nerved with an awn-point from the tip.

Glumes longer than the spikelet	*F. cryptantha*
Glumes about ½ as long as the spikelet	*F. obturbans*

1. **Festuca cryptantha** T. Cope

Densely tufted perennial with erect stems to 30cm. Leaves convolute, filiform, often curved or even coiled, smooth, up to 15cm long. Panicle linear, spike-like, 5–10cm long, the upper branches absent and the spikelets sessile on the axis, the lower branches short, erect, adpressed to the axis; spikelets whitish-green or purplish, glabrous, usually 2-flowered; glumes distinctly longer than the rest of the spikelet, becoming purplish when old, the upper lan-

Fig. 30 **A** *Festuca cryptantha*; ***B*** *Stipa tigrensis*; **C** *Phragmites australis*; **D** *Koeleria capensis*.

ceolate, 6–7mm long; lowest lemma 4–5mm long, terminating in a short awn.

An uncommon endemic plant of cliff ledges, exposed grassy ridges and mountain tops growing from 2700 to over 3600m on the highest escarpment mountains: J. Ta'kar; J. Khudrā' on J. Ḥubaysh; J. Sumārah; J. Nabī Shu'ayb (abundant at high altitudes); J. Maḥdad; J. Maswār.

2. **Festuca obturbans** *St. Yves*

Densely tufted perennial with erect stems to 60cm. Leaves inrolled, filiform, very scabrid, up to 30cm long. Panicle linear, 6–20cm long, the short branches erect, adpressed to the axis; spikelets yellow-green, glabrous, 4–10-flowered; glumes about ½ as long as the spikelet, the upper lanceolate, 3.5–6mm long; lowest lemma 5–7mm long with a short awn.

A rare plant of cliff ledges and steep grassy banks between 2500 and 3300m on the high escarpment mountains west of Sana'a: J. Ḍawrān; J. Nabī Shu'ayb; near Bayt Yarīsh, Ḥaymat ad Dakhilīyah; J. Shibām, Ḥarāz; between Shibām and Kawkabān.

Outside Yemen this is a rare plant restricted to a few high mountains in Ethiopia, Kenya and Tanzania.

Red Fescue, **Festuca rubra** *L.*, has been introduced as a lawn grass in Sana'a but is not yet naturalised.

6. CATAPODIUM *Link.*

1. **Catapodium rigidum** (*L.*) *C.E. Hubbard*

Annual, much-branched at the base with erect or ascending branches to 10 (–20)cm. Leaves flat, up to 10cm long; ligule membranous, up to 3cm long. Inflorescence a flattened, green panicle with short, rigid, ascending branches, 4–10cm long; spikelets 3–6-flowered, glabrous, awnless, 2–4mm long; glumes shorter than the lowest lemma; lemmas obtuse.

A very rare grass of bare patches on steep banks at 2200m in the grassland areas of J., Ẓallamlam above Ḥadīyah. This grass is widely distributed in Europe and around the Mediterranean and is naturalised in temperate regions throughout the world. Whether the Yemen population is a recent introduction or an isolated southern extension of its native range is not clear.

7. VULPIA *C.C. Gmel.*

1. **Vulpia bromoides** (*L.*) *S.F. Gray*

Loosely tufted annual with slender, erect stems to 30cm. Leaves very narrow, flat, rarely more than 10cm long; ligule a short, truncate membrane. Inflorescence a narrow panicle, 3–10cm long; spikelets 3–10-flowered, whitish-green or purplish, glabrous, 7–10mm long; glumes unequal, the lower 1-nerved, the upper 1–3-nerved; lemmas scabrid, awn straight, up to 12mm long.

A common weed in temperate countries known in Yemen only from a steep, bare slope below the summit of J. Ta'kar at 3100m.

8. BROMUS *L.*

Annuals or perennials. Leaves flat; ligule membranous. Inflorescence a loose panicle; spikelets large, compressed, many-flowered; glumes unequal, shorter than the lemmas, often with scarious margins; lemmas 5–9-nerved, bifid, awned.

Annual. Lower glume 3-nerved (requires careful examination in a good light)	*B. pectinatus*
Perennial. Lower glume 1-nerved	*B. leptoclados*

1. **Bromus leptoclados** *Nees*

B. cognatus Steud.

Tufted perennial with erect or ascending stems to c 1m. Leaves up to 30cm long, sheaths hispid or glabrous. Panicles large and loose, often partly nodding; spikelets 5–10-flowered, whitish-green, glabrous, 1–3cm long; glumes terminating in a short awn point, the lower 1-nerved, the upper 3-nerved; lemmas with a short straight awn, 2–12mm long.

Widely distributed between 2200 and 3200m on the higher escarpment mountains from J. Ṣabir north to J. Maswār; it is a plant of ungrazed, shaded places such as cliffs and montane scrub and is restricted to high rainfall areas or the margins of streams and ditches.

2. **Bromus pectinatus** *Thunb.*

B. tectorum sensu Hepper

B. adoensis Steud.

Tufted annual with several erect stems to 30cm. Leaves up to 20cm long, the sheaths pubescent. Panicles loose, usually small; spikelets 4–7-flowered, whitish-green becoming purplish when old, usually ± puberulent, 1–3cm long; glumes lanceolate, acute, but without an awn point, the lower 3-nerved, the upper (obscurely) 7-nerved; lemma with an awn 9–17mm long.

A common field weed on the higher escarpment mountains growing between 2200 and 3000m from J. Ṣabir north to J. Maswār and extending as far east as Dhamār (Hepper) on the high plateau. It is sometimes abundant in high-altitude wheat and barley fields but usually in areas of good rainfall or near small streams and ditches.

9. BRACHYPODIUM *P. Beauv.*

Annuals or perennials. Leaves flat or inrolled, thin; ligule membranous. Inflorescence a spike-like raceme composed of several subsessile spikelets, usually arranged on opposite sides of the axis; spikelets ± cylindrical, many flowered, large; glumes unequal, 3–7 nerved; lemmas awned from the acuminate apex.

Annual. Leaves sparsely pilose	*B. distachyon*
Densely tufted perennial. Leaves glabrous	*B. retusum*

1. **Brachypodium distachyon** (*L.*) *P. Beauv.*

Trachynia distachya (*L.*) Link.

Slender, erect annual to 50cm. Leaves flat, sparsely pilose, up to 10cm long. Inflorescence of 1–3 (–7) spikelets, crowded at the top of the stem; spikelets 6–15-flowered, laterally compressed, glabrous or hairy, 2–3cm long; lemma gradually narrowed into an awn of about equal length, the whole c 25mm long.

A rare grass of cliff slopes between 2300 and 3000m on two northern mountains: J. Maḥdad; J. Shahārah near the southern gate to the city. This species is widely distributed around the Mediterranean and east to India and reaches its southernmost limits in Yemen and the Ethiopian region.

2. Brachypodium retusum (*Pers.*) *P. Beauv.*

B. ramosum Schult.
B. flexum sensu Schwartz

Densely tufted, erect or ascending perennial to 70cm. Leaves minutely scabrid, but otherwise glabrous, inrolled or flat, up to 12cm long, 1–2 (–3)mm broad, often appearing distichous in dried specimens. Inflorescence of 1–4 spikelets, laxly arranged at the top of the stem; spikelets 5–15-flowered, ± terete, glabrous, 1–3cm long; lemma rather longer than the awn into which it is gradually narrowed, the whole 10–18mm long.

Rather variable in Yemen, plants from exposed mountain ridges being of low stature with 1–2 spikelets (Wood 1143) while those from protected cliffs are taller with more spikelets (Wood 1687). The two forms freely intergrade.

Widespread on cliffs, overgrown terrace walls and in rocky grassland between 2000 and 3000m on the outer escarpment from 'Akāhilah near Turbah north to J. Milḥān and J. Maswār; very rare on the inner escarpment and known only from Baw'ān, J. Maḥdad and J. Zāfin.

Brachypodium is a difficult genus with ill-defined species. Yemen plants have been identified as *B. flexum* Nees which is widespread in tropical Africa. That species though is of quite different appearance with consistently broader leaves (3–6mm wide) and 5–8 spikelets. The Yemen plant agrees well with *B. retusum* from southern Europe and I agree with Deflers' identification made when the plant was originally found in Yemen nearly a hundred years ago.

10. KOELERIA *Pers.*

1. Koeleria capensis (*Steud.*) *Nees*

Tufted perennial with erect stems to 40cm. Leaves convolute or flat, hairy or glabrous, up to 15cm long; ligule membranous, up to 2mm long. Inflorescence a narrow, glistening, silvery-purplish panicle, 4–10cm long, with short, erect branches and a pubescent axis; spikelets 1–3-flowered, glabrous or minutely puberulent, awnless; glumes with a hyaline margin, unequal, the upper 3–6mm long, exceeding the lower lemma; lemmas hyaline or nearly so 3–6mm long.

K. capensis is unusual in that it comes into flower with the spring rains but apparently does not flower in the autumn as well.

A very local plant of rocky hillsides from 2300 to 3100m: near Dimarayn, J. Ṣabir; J. Kabūrah, Ḥadīyah; near Shaḥir, J. Milḥān; J. Shibām, Ḥarāz; J. Maswār; J. Madhbur in J. Lawz.

11. AVENA *L.*

Oats. Annuals. Leaves flat; ligule a scarious membrane. Inflorescence an open panicle with large pendulous awned spikelets; spikelets up to 6-flowered; glumes equal, lanceolate, usually becoming scarious towards the acuminate apex, many-nerved; lemmas herbaceous, 5–9-nerved, with a bent awn.

Only wild species of oat occur in the Yemen where they grow mainly as field weeds. The differences between the species are often obscure and the present account must be regarded as tentative.

Rhachilla breaking up above the glumes. Lemma entire or shortly bifid with teeth less than 0.5mm long ... *A. fatua*
Rhachilla breaking up irregularly. Lemmas terminating in 2 fine bristles 1–3mm long ... *A. abyssinica*

1. Avena abyssinica *A. Rich.*

A. sativa L. var. *abyssinica* (A. Rich.) Engl.

Annual with erect stems to 1½m. Leaves up to 40cm long. Panicle usually large with spreading or pendulous branches. Spikelets whitish-green, c 2cm long, the rhachilla breaking up tardily and irregularly; glumes glabrous; lemmas usually glabrous, bifid with 2 short bristles 1–3mm long; awn bent, 3–4cm long.

A. abyssinica is said to be a cultivated derivative of *Avena vaviloviana* (Malzev.) Mordv. which is virtually restricted to Ethiopia. Like most wild species the rhachilla breaks up readily in *A. vaviloviana* and it can be further distinguished according to Baum's monograph by the heart-shaped scars left when the flowers fall. No plants with this last feature have been found in Yemen. Both species are closely related to *A. barbata* Pott. ex Link from the Mediterranean region.

A. abyssinica is an uncommon weed of disturbed roadsides and wheat and barley fields found between 2000 and 2800m at 'Akāhilah near Turbah (Wood 1431), on J. Ṣabir (Hepper 6144), at Ribāṭ near Yarīm (Wood 1342), at Yarīm (Hepper 6144), on J. Ba'dān (Wood 818), at Manākhah (Schweinfurth 1769) and Sana'a (Rathjens 462).

2. Avena fatua *L.*

Very similar in general appearance to the previous species, differing in the key characters. In addition the lemmas are usually densely bearded from the callus at their base. (It is usually possible to see the point of fracture on the rhachilla by carefully removing the glumes and hairs before examination with a lens.)

Known with certainty only from near the summit of J. Ta'kar at 3100m (Wood 1714) and from Dhamār at 2400m (Wood 2304). It grows in disturbed ground and as a field weed.

12. HELICTOTRICHON *Schult.*

Tufted perennials. Leaves flat, ligule membranous. Inflorescence a narrow panicle; spikelets 2–6-flowered, shiny; glumes unequal, hyaline; lemmas with hyaline margins and tips, bifid with a bent awn.

Lower glume 1-nerved. Leaves up to 2mm wide, ± filiform *H. angustum*
Lower glume 2–3-nerved. Leaves up to 5mm wide *H. elongatum*

1. Helictotrichon angustum *C. E. Hubbard*

Erect tufted perennial to 30 (–90)cm. Leaves up to 10 (–25)cm long and 2mm wide. Panicle linear, 5–9 (–20)cm long, interrupted, the branches erect and ± adpressed to the axis; spikelets 3–5-flowered; lower glume 1-nerved, 4–6mm long, upper glume 3-nerved, 6–8mm long; awn 9–16mm long.

The Yemen plant differs from the type, which is known from a few localities in Kenya, in its smaller size and smooth lemmas. How significant these differences are is difficult to say without further material.

On rocky slopes at 2800m on the east side of J. Ta'kar. (Wood 114)

2. Helictotrichon elongatum *(A. Rich.) C.E. Hubbard*

Erect tufted perennial to 1½m. Leaves up to 45cm long and 5mm wide. Panicle often nodding, usually linear, 10–45cm long, the branches ascending; spikelets 2–3-flowered; lower glume 2–3-nerved, 5–10mm long; upper glume 3–5-nerved, 7–11mm long; awn 15–20mm long.

A local plant of montane grassland between 2200 and 2900m in high rainfall areas of the escarpment from J. Ṣabir north to J. Ẓallamlam above Ḥadīyah; also occurring by ditches and in marsh at Aḥjar and in the southern part of the high plateau near Dhamār and round the headwaters of the streams feeding the Wadi Banā in the Yarīm area.

13. PHALARIS *L.*

1. Phalaris paradoxa *L.*

Tufted annual, the stems at first weak and decumbent, finally erect, up to c 60cm high. Leaves flat, up to 15cm long; ligule a white membrane, 2–3 (–5)mm long. Inflorescence a dense obovoid, whitish-green head-like panicle, 2–5cm long; spikelets in groups with one large, sessile, fertile spikelet surrounded by 4–6 reduced sterile ones; fertile spikelet flattened, 3-flowered, the glumes 6–8mm long, with a tooth-like wing and a long attenuate, bristle-like apex, much exceeding the lemmas.

The only Yemen collection (Wood 2288) is variety **praemorsa** *(Lam.) Coss. & Dur.* in which some of the sterile spikelets are reduced to club-like structures. This is most easily seen in the lower part of the inflorescence.

Only known from a roadside ditch at 2400m in Ḥaddah near Sana'a.

14. GASTRIDIUM *P. Beauv.*

1. Gastridium phleoides *(Nees & Meyen) C.E. Hubbard*

G. lendigerum sensu Schwartz?

Annual with slender erect stems to 30cm. Leaves flat, up to 10cm long; ligule a membrane 2–7mm long. Inflorescence a narrow, whitish-green, spike-like panicle with short, erect branches, 2–8 (–18)cm long; spikelets 1-flowered, linear-oblong, ± glabrous, 5–6 (–7.5)mm long; glumes lanceolate, unequal; lemma pubescent, with a broad, toothed apex, c 1.5mm long; awn bent, 4–7mm long.

Gastridium lendigerum (L.) Desv. is recorded in Schwartz on the basis of a record by Rathjens from Ṭawīlah. At that time no distinction was made between the essentially African *G. phleoides* and the Mediterranean species *G. ventricosum* (Gouan) Schinz & Thell. (*G. lendigerum*) and without seeing the original specimen it is impossible to be certain which species Rathjens found. *G. ventricosum* differs from *G. phleoides* in its smaller spikelets (± 4mm long), subglabrous lemma and shorter awn (3–4mm long).

A rare plant of shaded terrace walls and bare earth banks on the higher escarpment mountains from 2300 to 3000m: J. Ṣabir; J. Ta'kar; J. Ba'dān; J. Manār; J. Huzār and J. Bārad near Kusmah; Shaḥir, J. Milḥān; J. Mahḍad; Ṭawīlah (Rathjens).

15. AGROSTIS *L.*

Annuals or perennials. Leaves linear, flat; ligule membranous. Inflorescence an open or contracted panicle; spikelets small, linear-lanceolate, 1-flowered; glumes ± equal, usually 1-nerved; lemmas shorter than the glumes, 3–5-nerved.

Our three species are similar-looking perennials of wet places with awnless spikelets and scabrid glumes. Unfortunately the spikelets are very small and it is difficult to see the nerves on the lemma without a microscope.

1. Spikelets falling entire. Panicle branches usually covered in spikelets for all their length *A. viridis*
   Glumes persistent long after the flower has fallen. Panicle branches often bare of spikelets near their base 2
2. Lemmas 5-nerved, glabrous *A. schimperiana*
   Lemma 3-nerved, pilose (very rarely glabrous) *A. lachnantha*

1. Agrostis lachnantha *Nees*

Loosely tufted perennial, sometimes stoloniferous or rhizomatous, stems slender, ascending to 80cm. Leaves up to 20cm long. Panicle narrow, weakly contracted, up to 20cm long, the branches usually several together on one side of the axis, bare of spikelets towards the base; spikelets light green or purplish, 2–2.5mm long; glumes persistent long after the flower has fallen; lemma 3-nerved, pilose (rarely glabrous).

A rare plant of marsh and mountain springs found between 2500 and 3000m; known definitely from J. Ba'dān, J. Sumārah, Shibām and between Baw'ān and Suq As Silf, but probably widespread on the higher mountains. (Wood 75/401, 2365, 2405; Radcliffe-Smith and Henchie 4962)

2. Agrostis schimperiana *Steud.*
A. hirtella Steud.
Virtually identical to the previous species, differing in its 5-nerved, glabrous lemma, which sometimes has a short awn-like point up to 1.5mm long.

A rare, but probably overlooked plant found in the same situations as the last species; Manākhah (Schweinfurth 1413); Dhamār (Wood 2281); Baw'ān (Wood 2913); Wadi Rakwah, Khawlan Ash Sham (Steinberg).

3. Agrostis viridis *Gouan*
A. verticillata Vill.
A. semiverticillata (Forssk.) C. Chr.
Polypogon semiverticillatus (Forssk.) Hyl.
Loosely tufted, stoloniferous perennial with erect stems to 80cm. Leaves up to 10cm long. Panicle contracted, up to 12cm long, the branches short, whorled, densely covered in spikelets, the whorls not contiguous with each other below, so the panicle is ± interrupted; spikelets whitish-green, 2–2.5mm long; glumes falling with the rest of the spikelet; lemma 5-nerved, puberulent.

Widespread and common between 1300 and 3000m on the escarpment and high plateau, growing on open ground by streams, ditches and ponds often in association with *Veronica anagallis-aquatica*.

Unlike the previous two species which are African in their distribution, this is a plant of the Mediterranean region which extends as far south as Ethiopia and Yemen.

16. ORYZOPSIS *Michx.*

1. Oryzopsis holciformis (*M. Bieb.*) *Hack.*
Tufted perennial with erect stems to 80cm. Leaves flat, up to 40cm long; ligule membranous, up to 7mm long. Inflorescence an open or contracted panicle, 10–30cm long; spikelets 1-flowered, 6–8mm long, green, glabrous, awned, though often appearing awnless as the awn is quickly deciduous; glumes equal, concave, 3–5-veined, acuminate, the tip often scarious and becoming purplish; lemma shorter than the glumes, pubescent; awn deciduous, c 10mm long; grain black, shining.

A local grass of cliffs and ungrazed rocky slopes between 2500 and 3200m on the higher escarpment mountains from J. Ta'kar and J. 'Ayfū north to J. Milḥān and J. Maswār.

O. holciformis is a plant of the East Mediterranean and central Asia and has its most southerly outposts in the Yemen mountains. The Yemen population (e.g. Wood 75/256, 1241) has consistently smaller spikelets than more northern populations and may be worthy of varietal status.

17. STIPA *L.*

1. Stipa tigrensis *Chiov.*
Densely tufted, erect perennial to 70cm. Leaves involute, filiform, up to 40cm long; ligule a short membrane. Inflorescence a contracted linear panicle, 5–15cm long; spikelets 1-flowered, shiny-green; glumes hyaline, unequal, acuminate, the lower c 10mm long, the upper c 15mm long; lemma pilose, c 6mm long including the prominent silky callus; awn 5–7mm long, bent, with a twisted, villous column.

A characteristic grass of cliffs and rocky slopes on the higher mountains around the high plateau growing between 2500 and 3500m from J. Sumārah north to Kuhlān and east to J. Lawz. On the outer escarpment it is rare and known only from Al Manār on J. Ḥubaysh, J. Bārad above Kusmah and J. Shibām in the Ḥarāz.

This species is far more common in Yemen than anywhere else in the world and is only known from two isolated localities elsewhere, one in Ethiopia and one in Uganda.

18. ARUNDO *L.*

1. Arundo donax *L.* Qaṣab, 'Alāl
Tall, woody, rhizomatous grass with erect, usually simple, bamboo-like, hollow stems to 4m, usually growing in large clumps. Leaves stiff with an auriculate base, up to 50cm long and 5cm wide; ligule a very short membrane. Inflorescence a large, compound, silvery, plume-like panicle with glabrous branches, up to 50cm long; spikelets 2–3-flowered; rhachilla glabrous; glumes glabrous, 12–13mm long, ± equalling the spikelet; lemmas long-pilose, 3-nerved, the nerves sometimes exserted to form short aristate points.

A. donax flowers in summer but it is a shy flowerer and the flowers rarely appear except beside flowing water in hot country.

Abundant along the margins of permanent flowing wadis on the escarpment from 500 to 1500m, but it is commonly planted elsewhere and forlorn isolated clumps are seen near villages as high as 3000m.

The stems are cut and used as 'bamboo' poles and the leaves are made into matting, baskets and similar items. It is an extremely useful plant in the traditional economy.

19. PHRAGMITES *Adans.*

Reeds. Vigorous aquatic perennials with stout, tall stems. Leaves linear; ligule a line of hairs. Inflorescence a large plumose panicle; spikelets 3–11-flowered; rhachilla silky-hairy; glumes unequal, 3–5-nerved, shorter than the spikelets; lemmas awnless.

Upper glume 5–9mm long. Rhachilla hairs 6–10mm long ... *P. australis*
Upper glume 4–6mm long. Rhachilla hairs 4–7mm long ... *P. karka*

1. Phragmites australis (*Cav.*) *Steud.*
P. communis Trin.
P. mauritianus sensu Schwartz.
Perennial, spreading by rhizomes to form large colonies. Stems erect, up to 3m. Leaves smooth beneath and with long filiform tips, up to 60cm long. Panicle up to 25cm long, the lowest node usually few-branched; spikelets 12–18mm long, reddish-purple, the rhachilla hairs 6–10mm long; glumes glabrous, the upper 5–9mm long, usually apiculate; lemmas glabrous.

Although said by Forsskal to be frequent by streams, the only certain records of this grass are from saline marsh at 2000m in the Mashriq at Rada' and in the reed-beds at the head of the Wadi Khārid in Arḥab. (Wood 2692, 3079)

2. Phragmites karka (*Retz.*) *Steud.*
Similar to the last species but more robust, the stems up to 5m or more in height. Leaves usually rough below, the tips rather stiff. Panicle up to 50cm long, the lowest node with many branches in a whorl; spikelets 9–12mm long, the rhachilla hairs 4–7mm long; upper glume 4–6mm long, acute.

A very rare grass known only from swampy scrub at 1600m between Ibb and Suq As Sabt. (Wood 2980)

20. RHYTIDOSPERMA *Steud.*

1. Rhytidosperma subulata (*A. Rich.*) *T. Cope*
Danthonia subulata A. Rich
Densely tufted perennial to 40cm. Leaves erect, tightly inrolled, setaceous and rigid with hard spinescent tips, up to 25cm long; ligule a ciliate rim. Inflorescence a linear spike-like panicle, 2–4cm long, most of the spikelets subsessile on the axis but the lower on short branches appressed to the axis; spikelets whitish-green, 3–4-flowered, 10–12mm long; glumes ± equal, hyaline, c 8mm long, about as long as the lowest lemma; lemmas sparsely pilose, bifid; the awn with a twisted column, bent about the middle, c 7mm long.

Locally frequent in grassy gullies on and between cliffs from 2800 to 3100m on the Sumārah-Manār ridge and on J. Maḥdad. Outside Yemen this distinctive grass with spine-tipped leaves is only known from the Ethiopian highlands.

21. DANTHONIOPSIS *Stapf.*

Tufted perennials with long stem bases. Leaves flat, often glaucous; ligule a ring of hairs. Inflorescence a loose, narrow panicle; spikelets 2-flowered; glumes unequal, 3-nerved; lower lemma as long as the spikelet; upper lemma with a prominent bearded callus, 2-lobed, each lobe terminating in a short bristle, the main awn bent about the middle with a conspicuous spirally twisted column.

Spikelets in groups of 3 (rarely solitary or 2 together).
Lower half of upper lemma pilose ... *D. barbata*
Spikelets solitary. Lower half of the upper lemma glabrous ... *D. sp.A*

1. Danthoniopsis barbata (*Nees*) *C.E. Hubbard*
Densely tufted perennial with erect stems to 1m. Leaves glaucous with a conspicuous pale margin, pilose or glabrous, up to 10cm long and 5mm wide. Panicle up to 15cm long; spikelets (1–) 3 together, becoming purple-brown with maturity, ± 10mm long; glumes glabrous, the lower 5mm long, the upper 9mm long; lower lemma similar in appearance to the upper glume; upper lemma c 7mm long, with a prominent transverse line of hairs below the lobes, pilose in the lower half below this line; awn c 2cm long, the column much shorter than the rest of the awn.

Frequent at 1500m on stony hills in open *Acacia-Commiphora* bushland in the Mafḥaq-Banī Manṣūr region.

2. Danthoniopsis sp.A
Densely tufted perennial with erect stems to 40cm, bearing a strong superficial resemblance to *Chrysopogon plumulosus*. Leaves mainly just above the base, numerous, appearing distichous, glabrous, up to 5cm long and 2mm wide. Panicle 5–10cm long; spikelets solitary, whitish, the dimensions similar to those of *D. barbata*, but differing in the upper lemma which is completely glabrous in the lower half and in the awn whose column is about half its length.

Only known from open stony hillsides near Suq al Kawlah at 500m in the Wadi Zabīd. (Wood 1420)

It is difficult to assess the significance of the characters by which this grass differs from *D. barbata* but I have not seen any intermediate specimens at all and it seems probable this is a good new species.

22. CENTROPODIA (*R. Br.*) *Reichb.*

1. Centropodia forsskalii (*Vahl*) *T. Cope*
Asthenatherum forsskalii (Vahl) Nevski
Danthonia forsskalii (Vahl) R. Br.
Tufted perennial with erect stems to 50cm. Leaves up to 12cm long, linear, glaucous, flat, usually pubescent on both surfaces, the sheaths pubescent; ligule a ring of hairs. Panicle dense, contracted, linear, up to 12cm long, partly enclosed by the spathe-like upper leaf sheath; spikelets 2–3-flowered, whitish-green or purplish tinged, 8–10mm long; glumes equal, 5–7-nerved, glabrous; lemma shorter than the glumes, 9-nerved, pilose, 2-lobed, with an awn c 10mm long from the gap between the lobes.

An uncommon grass of sandy plain found between 1000 and 2400m in the Mashriq: Wadi As Sirr, Bani Ḥushaysh; Mārib (Miller); Jūbah.

23. PENTASCHISTIS *Stapf.*

1. Pentaschistis pictigluma (*Steud.*) *Pilger*
Densely tufted perennial with erect or ascending stems to 50cm, though usually less than 10cm in Yemen. Leaves narrowly linear, up to c 5cm long; ligule a ring of hairs. Inflorescence a linear spike-like panicle, 1–6cm long; spikelets 2-flowered, shiny, 4–5mm long; glumes equal, hyaline, 1-nerved, glabrous; lemmas pubescent with 2 bristle-tipped lobes and an awn from the gap between them; awn 3–5mm long, bent about the middle, the column twisted.

A rare plant of exposed mountain rocks between 3100 and 3300m on a few high mountains northeast of Sana'a; J. Nabī Shu'ayb (Lavranos); J. Ta'iz, J. Mada'; J. Maswār. On the last three mountains it is a plant of sandstone rocks on the summits.

Outside Yemen it is a plant of the Ethiopian highlands.

24. PHAENANTHOECIUM *C.E. Hubbard*

1. Phaenanthoecium koestlini (*Hochst.*) *C.E. Hubbard*
Slender trailing perennial to 15cm, the stems rooting at the nodes but eventually weakly ascending. Leaves very narrow, slightly inrolled, stiff and spreading, 1–4cm long; ligule absent. Inflorescence of 1–4 ascending shortly pedunculate spike-like racemes on a slender axis about 1cm long;

racemes 5–15mm long; spikelets 3–6-flowered, whitish-green, c 3mm long; glumes ± equal, scarious; lemmas with a ciliate callus and a bent, twisted awn 1–2cm long.

An inconspicuous and easily overlooked grass found from 2600 to 3200m on the higher escarpment mountains from J. Ta'kar north through Raymah and the Ḥarāz to J. Maswār; it grows on earth banks where there is little competititon from other plants and has been found on anthills, between large boulders and on broken cliff slopes; it is often associated with *Gastridium phleoides*.

25. TRIRAPHIS *R. Br.*

1. Triraphis pumilio *R. Br.*

Small tufted annual herb, much branched at the base with erect stems to c 10cm. Leaves flat, up to 4cm long, the sheaths with long, white, bulbous-based hairs; ligule a very short, ciliate membrane. Inflorescence a dense, ovoid, silvery head-like panicle 1–2.5cm long; spikelets many-flowered, 4–5mm long; glumes unequal, terminating in a short mucro; lemmas hyaline, bifid, densely ciliate, 3-awned, 1.5–2mm long, awns thread-like, straight, the longest c 2mm long.

A very rare desert plant known only at 1100m from the sandy bed of the Wadi Saba' at Mārib.

26. ARISTIDA *L.*

Tufted annuals or perennials. Leaves linear, mostly basal; ligule a fringe of hairs. Inflorescence an open or contracted panicle; spikelets 1-flowered; glumes long, scarious, equal or unequal; lemma glabrous or scabrid, hardened, produced into a 3-branched awn; awn with or without a column, jointed or not, glabrous or scabrid but never plumose.

1. Lemma passing directly into the awn without a joint and without an intervening twisted column — 2
   Lemma prolonged into a twisted column which may be jointed at the base or just below the branching point of the awns — 4
2. Glumes ± equal, slightly exceeding the branching point of the awns. Lemma smooth or scabrid on the keels only — *A. ferrilateris*
   Glumes unequal, the shortest always shorter than the branching point of the awns. Lemmas scabrid, often coarsely so — 3
3. Upper glume more than 1cm long. Perennial — *A. pennei*
   Both glumes less than 1cm long. Annual or short-lived perennial — *A. adscensionis*
4. Column jointed at the base. Awns at least 3.5cm long and commonly much more — 5
   Column jointed at the summit. Awns less than 3cm long — 6
5. Annual. Column of the awn more than 2cm long — *A. funiculata*
   Perennial. Column of the awn less than 1cm long — *A. migiurtina*
6. Annual. Panicle whitish-green, usually open — *A. mutabilis*
   Perennial. Panicle purplish, always contracted and spike-like — *A. congesta*

1. Aristida adscensionis *L.* Thunayb

A. caerulescens Desf.

Tufted annual or short-lived perennial with erect or ascending stems to 50cm. Leaves flat, up to 15cm long. Panicle usually contracted, more rarely open, up to 25cm long, the primary branches mostly ascending or erect; glumes usually unequal, purplish or green, acute but not awned, the upper 5–10mm long, the lower slightly shorter, both usually shorter than the lemma; lemma 5–13mm long, scabrid, passing into the awns without a joint or distinct column; awns very variable, equal, or the laterals shorter, 7–25mm long.

A very variable species often divided up into several varieties or even species. *A. caerulescens* is often regarded as a separate species but differs only in its perennial habit, and since this character is unsupported by any secondary distinctions and is itself unreliable in many dry-country plants, the two species are best regarded as one. Perennial forms of *A. adscensionis* are quite common in Yemen and are found mainly in years of good rainfall.

A. adscensionis is a pantropical grass abundant throughout Yemen up to 3000m and absent only from the eastern parts of the Mashriq, though less common in the Tihama where it is often replaced by *A. mutabilis*. It grows in all manner of open stony ground and is a frequent field weed. It is often subdominant on open, barren plains on the high plateau, coming up after the rains in great abundance. The hardened fruiting awns are an uncomfortable nuisance to stockinged feet.

2. Aristida congesta *Roem. & Schult.*

A. elytrophoroides Chiov.

Densely tufted erect perennial to 60cm. Leaves folded or convolute, up to 20cm long. Panicle dense, contracted, spike-like, 3–12cm long, the primary branches erect; glumes unequal, purplish, acuminate, terminating in a short awn point, the upper 7–10mm long, the lower slightly shorter; lemma 4–5mm long, scabrid, connected to the awns by a 2–4mm long twisted column, which is jointed just below the branching point of the awns; awns ± equal, 12–30mm long.

Widespread from 1800 to 2300m on the high plateau from Rada' and Ma'bar north to Arḥab and extending westwards to Bayt 'Adhāqah and Rujum on the escarpment, but only common in sandstone country. It is particularly abundant on the open sandy plain northwest of the sandstone hills near Ghirās where it is almost the only species present.

3. Aristida ferrilateris *S.P. Phillips*

A. adoensis auct.

Densely tufted erect perennial to 60cm. Leaves flat or involute, up to 25cm long. Panicle contracted, 5–20cm long, the primary branches mostly erect, but the lowest often spreading or deflexed, sometimes pilose in the axils; glumes ± equal, purplish, acuminate, terminating in a short awn point, 8–12mm long, much longer than the lemma; lemma 6–8mm long, smooth or slightly scabrid on the keel, passing directly into the awns without a joint or column; awns ± equal, 1.5–2cm long.

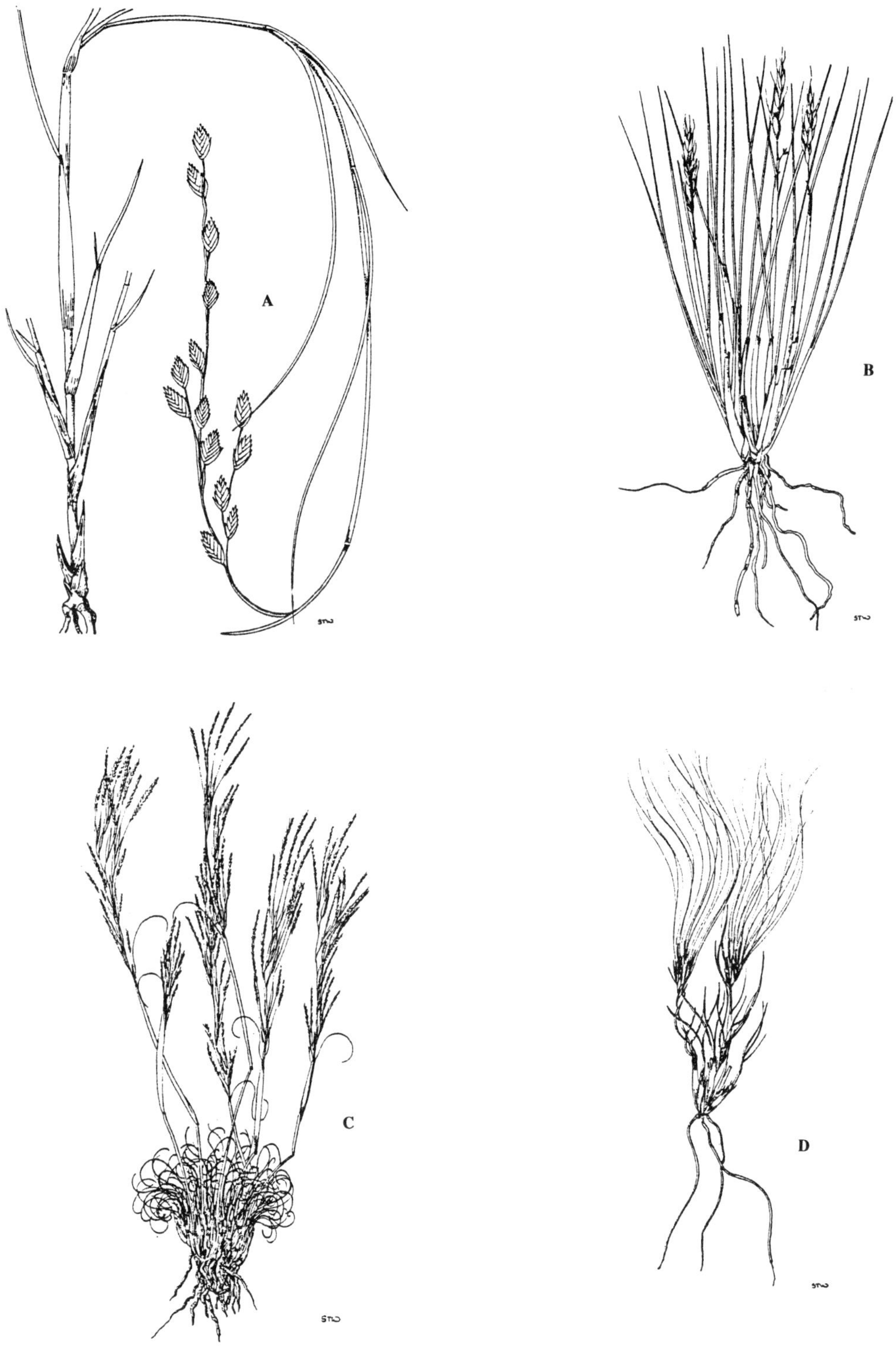

Fig. 31 **A** *Halopyrum mucronatum*; **B** *Rhytidosperma subulata*; **C** *Stipagrostis obtusa*; **D** *Aristida migiurtina*.

Widespread and frequent between 1400 and 2600m in areas of average or good rainfall on the escarpment; it grows in shallow grassland among rocks and on open grassy cliff ledges.

4. Aristida funiculata *Trin. & Rupr.*
Tufted annual with numerous erect stems to 25cm. Leaves flat or folded up to 10cm long. Panicle few-flowered, contracted, up to 10cm long, the primary branches erect; glumes unequal, whitish-green, drawn out into a fine point, the upper 20–30mm long, the lower about 5mm shorter; lemma 4–5mm long, scabrid, connected to the awns by a 3–4.5cm long, twisted column which is jointed at the base; awns ± equal, 3.5–7cm long.

A rare species known only from stony plain between Ḥaraḍ and Suq al Jammah at c 300m.

5. Aristida migiurtina *Chiov.*
Dwarf tufted perennial with decumbent to erect stems up to 10cm high. Leaves involute, up to 5cm long. Panicle subcapitate, c 2cm long (excluding the awns); glumes very unequal, purple, acuminate with a short awn point, the upper c 20mm long, the lower about 8mm shorter; lemma c 6mm long, smooth, connected to the awns by a 4mm long, twisted column which is jointed at the base; awns ± equal 3.5–5.5cm long.

A rare plant found on barren stony slopes between 3300 and 3400m on the highest mountains; J. Nabī Shu'ayb; J. Ta'iz.

Outside Yemen known only from the northern Somalia mountain ranges.

6. Aristida mutabilis *Trin. & Rupr.*
A. meccana Trin. & Rupr.
A. cassanellii Terr.
Loosely tufted annual with erect or ascending stems to 50cm. Leaves usually convolute, up to 12cm long. Panicle variable, contracted or spreading, 6–15cm long, the primary branches ascending; glumes ± equal, whitish-green, acute or with a short awn point, 4–7mm long; lemma 4–5mm long, scabrid, connected to the awns by a 2–5mm long, twisted column, which is jointed just below the branching point of the awns; awns ± equal, 10–30mm long.

Widespread and often abundant on sand dunes and stony plain throughout the Tihama up to c 500m.

Outside Yemen it is a plant with a characteristic Sahara-Sindian distribution.

7. Aristida pennei *Chiov.*
A. jemensis Henrard ex Schwartz
Erect tufted perennial to 25cm. Leaves involute, up to 8cm long. Panicle few-flowered, contracted, up to 10cm long, the primary branches erect; glumes unequal, reddish-purple, acuminate, terminating in a short awn point, the upper 12–13mm long, the lower about 5mm shorter; lemma c 10mm long, scabrid, passing directly into the awns without a joint or column; awns ± equal, 1.5–2.5cm long, becoming noticeably stiff and spreading at right angles to each other at maturity.

Widespread but uncommon on stony and sandy plain and more rarely on mountain slopes between 2000 and 3000m on and around the high plateau from Dhamār north to Ma'mar and from Banī Maṭar east to J. Lawz. Like *A. congesta* with which it often grows, it shows a strong preference for sandstone country.

Until recently this plant was thought to be endemic to Yemen, but it has turned up in a number of places in Arabia and may prove to be quite common.

27. STIPAGROSTIS *Nees*

Similar to *Aristida* but the central awn plumose. All five species are characteristic of arid stony situations.

1. Stem nodes villous. Base of the column of the awn hollow, attached to the upper part of the lemma — *S. ciliata*
   Stem nodes glabrous. Base of the column of the awn solid, attached to the tip of the lemma — **2**
2. Glumes shortly pilose. Central awn plumose for all its length — *S. hirtigluma*
   Glumes glabrous. Lower ⅓ of the central awn not plumose — **3**
3. Stems with a single node. Central awn without an exserted naked tip — *S. obtusa*
   Stems (1–) 2–3 noded. Central awn with a naked tip — **4**
4. Central awn 3–5 (–6)cm long, pale brown. Basal leaf sheaths usually woolly — *S. plumosa*
   Central awn 6–7.5cm long, dark violet. Basal leaf sheaths always glabrous — *S. paradisea*

1. Stipagrostis ciliata (*Desf.*) *De Winter*
Aristida ciliata Desf.
Densely tufted perennial with erect stems to 60cm; basal leaf sheaths pale yellow, papery, glabrous; stem nodes 1–3, densely villous. Leaves convolute, often wavy, smooth, up to 20cm long. Panicle 10–20cm long, lax; spikelets pallid with a distinctive dark violet base; glumes glabrous, the upper ± 12mm long; awns with a hollow column joined to the upper part of the lemma, the central awn plumose in the upper ⅔ except for the tip, 5–6cm long, the lateral awns 1.5–2cm long.

Widespread on the high plateau from Rada' and Ma'bar north to 'Amrān, growing from 1900 to 2600m on open hillsides, usually on steep, loose, stony slopes. It is less common than *S. obtusa*.

2. Stipagrostis hirtigluma (*Trin. & Rupr.*) *De Winter*
Aristida hirtigluma Trin. & Rupr.
Densely tufted annual or short-lived perennial up to 80cm; basal leaf sheaths papery, glabrous; stem nodes usually 3, glabrous. Leaves convolute, or ± flat, scabrid, up to 20cm long. Panicle 10–15cm long, contracted or open; spikelets pallid or purplish; glumes shortly pilose, the upper 10–13mm long; awns with a solid column joined to the tip of the lemma, the central awn plumose for all its length except for the tip, 3.5–5.5cm long, the lateral awns c 1cm long.

Rather variable; plants from the escarpment foothills (e.g. Wood 75/437, 1406) being obviously annual with longer central awns ± 5cm long whereas plants from the high plateau (Wood 75/164) are perennial with shorter central awns ± 4cm long.

Locally abundant in dry stony *Acacia-Commiphora* bushland between 300 and 1500m in the east of the Tihama and on the escarpment from Bani 'Umar in the Ḥujarīyah north to Suq al Jammah; on the high plateau it is widespread but infrequent between 2000 and 2400m from Damt north to 'Amrān; it is also near Majz (Steinberg) and doubtless elsewhere in the Ṣa'dah area.

3. Stipagrostis obtusa (*Del.*) *Nees*

S. bifida (J. Karl) De Winter
Aristida obtusa Del.
Very densely tufted perennial to 30cm; basal leaf sheaths papery, glabrous or pilose; stem nodes always 1, glabrous. Leaves numerous, densely crowded at the base, convolute, filiform, strongly curled and sometimes coiled, glabrous or pilose, up to 5cm long. Panicle 5–10cm long, very narrow; spikelets pallid; glumes glabrous, the upper 9–11mm long; awns with a solid column joined to the tip of the lemma, the central awn plumose in the upper ⅔ and lacking the naked tip found in all our other species, 1.5–3cm long, the lateral awns 1–1.5cm long.

This species is divisible into two distinct varieties in southwest Arabia, one with glabrous leaves (Wood 75/124) and one with pilose leaves (Wood 75/163). Both varieties are common but intermediates have not been observed.

Widespread and frequent on open stony plain between 1500 and 2600m throughout the high plateau and the western part of the Mashriq. This is a plant of northern distribution which reaches its southernmost point in Arabia in the Rada' area. It is noticeably more common in the Ṣa'dah area than elsewhere and extends westward into the mountains of Khawlan Ash Sham, reaching as far as Sāqayn and Suq al Jumlī; further south it never penetrates the western mountains.

4. Stipagrostis paradisea (*Edgw.*) *De Winter*

Aristida paradisea Edgw.
Densely tufted perennial to 45cm; basal leaf sheaths horny, glabrous; stem nodes usually 2–3, glabrous. Leaves convolute, strongly curved and rigid, glabrous, up to 10cm long. Panicle 5–15cm long, contracted; spikelets pallid; glumes glabrous, the upper 14–16mm long; awns with a solid column joined to the tip of the lemma, the central awn plumose in the upper ⅔ except for the naked tip, 6–7.5cm long, the lateral awns 2–3cm long.

Only known definitely from 1100m on the barren rocks of the Naqīl Ḥajlah between Jūbah and Ḥarīb, but Botta also recorded this plant, probably from Yemen, but without locality.

5. Stipagrostis plumosa (*L.*) *Munro ex T. Anders.*

Aristida plumosa L.
Densely tufted perennial to 40cm; basal leaf sheaths horny, woolly, the wool deciduous as the plant matures (sometimes glabrous in other countries); stem nodes 2–3, glabrous. Leaves convolute, strongly curved and rigid, glabrous or hairy, up to 5cm long. Panicle 5–15cm long, contracted; spikelets pallid; glumes glabrous, the upper 15–18cm long; awns with a solid column joined to the tip of the lemma, the central awn plumose in the upper ⅔ except for the naked tip, 3–5 (–6)cm long, the lateral awns 1–2cm long.

The sole Yemen collection (Wood 75/934) is abnormal in that the glumes are very unequal, the lower being much longer than the upper.

Only known from sandy plain in the Tihama, growing at 100m between Mukhā and Mafraq.

28. ENNEAPOGON *Desv. ex P. Beauv.*

Tufted annuals or perennials with stiff, wiry stems. Leaves flat or involute; ligule a fringe of hairs. Inflorescence a contracted panicle; spikelets 3–flowered, pubescent; glumes slightly unequal, longer than the lemmas; lemmas villous with the nerves excurrent as 9 ciliate awns.

The following four grasses can mimic each other quite effectively in their superficial appearance and it is wise to rely on technical characters to separate them. They are confused in Schwartz and frequently misidentified by contemporary botanists as well.

1. Uppermost flower 0.5–2mm long — 2
   Uppermost flower 4–7mm long — *E. schimperianus*
2. Anthers 0.3–0.5mm long. Basal leaf sheaths fibrous — *E. desvauxii*
   Anthers 1–1.8mm long. Basal leaf sheaths not fibrous — 3
3. Stem tufted without a bulbous base. Leaves not wiry — *E. cenchroides*
   Stem usually branched above the sub-bulbous base. Leaves wiry — *E. scoparius*

1. Enneapogon cenchroides (*Roem. & Schult.*) *C.E. Hubbard*

Pappophorum molle (Lehm.) Kunth
Robust, tufted annual or short-lived perennial, usually 20–80cm high but sometimes less, basal leaf sheaths not fibrous. Leaves glabrous or softly hairy, up to 25cm long. Panicle 2–16cm long, usually whitish-green; upper glume 3.5–5.5mm long; lowest lemma 1.5–2mm long with ciliate awns 2.5–5mm long; upper lemma 0.7–2mm long including the awns; anthers 1–1.8mm long.

Known with certainty only from dry *Acacia-Commiphora* bushland in the escarpment foothills from 300 to 1000m. (Hepper 5771)

2. Enneapogon desvauxii *P. Beauv.*

Pappophorum brachystachyum Jaub. & Spach
Tufted perennial, 5–30cm high, basal leaf sheaths persistent, fibrous. Leaves usually involute, rarely flat, pubescent, up to 10cm long. Panicle spike-like, 2–8cm long, greyish-white; spikelet dimensions as for *E. cenchroides* except for the anthers which are 0.3–0.5mm long.

Widespread and common between 1800 and 3000m on open stony plain on the high plateau with a marked preference for sandstone country; elsewhere its distribution is not accurately known but it definitely occurs on sandstone on the Dhubhān plateau and probably elsewhere in the Ḥujarīyah and in Khawlan Ash Sham.

3. Enneapogon schimperianus (*A. Rich.*) *Renv.*
E. elegans (Steud.) Stapf
Pappophorum molle sensu Schwartz p.p.
Tufted perennial, 5–70cm high, basal leaf sheaths not fibrous. Leaves usually involute, rarely flat, pubescent, up to 12cm long. Panicle spike-like, 2–10cm long, usually blackish-grey; upper glume 5.5–8.7mm long; lowest lemma 1.5–2.4mm long with ciliate awns 5–7mm long; upper lemma 4–7mm long including the awns; anthers 0.6–1.3mm long.

Widespread between 600 and 2500m on the escarpment and high plateau, growing in stony bushland and on open rocky plain, often in association with one or other of the two previous species.

4. Enneapogon scoparius *Stapf*
Differs from *E. cenchroides* only vegetatively, being less obviously tufted, the stem often branched above the sub-bulbous base and with fine, wiry leaves. The panicle is usually blackish-grey.

Distribution not accurately known but found on well-drained stony hills on J. Shibām, Manākhah, above Wadi Dhār (Wood 75/123) and in Salafiyah (Gillespie 79) at 2000–2500m.

29. AELUROPUS *Trin.*

1. Aeluropus lagopoides (*L.*) *Trin. ex Thw.* Kirshah
A. repens (Desf.) Parl.
A. brevifolius Nees ex Steud.
A. massauensis (Fres.) Mattei
Variable perennial, sometimes densely tufted, sometimes with long prostrate stems rooting at the nodes, but always low and less than 15cm high. Leaves glaucous, flat or inrolled, pungent and distichous or soft, glabrous or hairy, up to 3cm long; ligule a ring of hairs. Inflorescence a dense globular head made up of hairy one-sided spikes, sometimes reduced to a single one-sided spike, 5–15mm long; spikelets 4–18-flowered, unawned; glumes unequal; lemmas 9–11-nerved.

In Yemen divisible into two quite distinct forms which are sometimes treated as separate species, though they intergrade in other countries.

The first form is a tufted plant with glabrous, pungent, spine-tipped, distichous leaves and a dense head-like inflorescence (Wood 76/1065). This is abundant along the whole Red Sea littoral, growing in salt marsh and subkhah where it is within range of salt spray. It is almost always associated with *Suaeda fruticosa*.

The second form has creeping prostrate stems which root at the nodes, soft, densely hairy leaves and an inflorescence reduced to a single one-sided spike (Wood 3212). This is a rare plant only known at 2000m from seasonally flooded salt plains north of Rada' where it is associated with *Diplachne fusca*, *Marsilea aegyptiaca* and *Polygonum corrigioloides*.

30. HALOPYRUM *Stapf*

1. Halopyrum mucronatum (*L.*) *Stapf*
Tussocky perennial with stiff, woody, erect stems to 1m, spreading by stolons to form colonies. Leaves stiff, linear, involute, up to 45cm long, ligule a fringe of hairs. Inflorescence very lax, consisting of several short racemes scattered along a long central axis; racemes with 3–7 rather distant, pedicellate spikelets, spikelets 8–25-flowered, unawned, glabrous, straw-coloured, mostly 15–20 × 5–6mm; lemmas shortly mucronate.

A distinctive grass which is locally abundant on sand dunes by the sea; between Hodeida and Manẓār (Deflers); between Khawkhah and Mawshij.

31. DESMOSTACHYA *Stapf*

1. Desmostachya bipinnata (*L.*) *Stapf.* Ḥalafī
Coarse, robust, tufted perennial to 1½m, spreading by rhizomes to form dense colonies, the dead leaf sheaths long-persistent around the base of the stem. Leaves stiff, inrolled, tapering to a long fine point, up to 50cm long; ligule a dense ring of hairs. Inflorescence narrow, cylindrical, consisting of numerous, crowded, short, sessile spikes grouped in clusters along an axis 25–40cm long; spikes 1–4cm long; spikelets in two rows along one side of the spike, 8–18-flowered, strongly compressed, awnless, pale brown, glabrous, 4–7mm long.

A very distinctive, local but sometimes abundant grass of sandy wadis where the water table is not far below the surface, found from sea level to 1100m in the Tihama and the Mashriq: near the sea at the mouth of the Wadi Rima' (Hepper) and the Wadi Sihām; by the Wadi Mawr, Wadi 'Abs and Wadi Ḥaraḍ where they enter the Tihama; by the Wadi Dhanah at the Mārib dam; western Wadi Jawf (J. King).

32. ERAGROSTIS *Wolf*

Annuals or perennials often with sessile glands on the leaf margins and the pedicels and axils of the inflorescence. Leaves linear, usually flat; ligule replaced by a tuft of hairs (except in *E. japonica*). Inflorescence an open or contracted or spike-like panicle; spikelets many-flowered, compressed, awnless; glumes usually equal.

***Eragrostis* is a difficult genus taxonomically and the species are best identified when beginning to fruit so that the manner in which the spikelets break up can be observed. Although spikelet size is usually significant it is very variable within species and it is difficult to quantify and little use has been made of it in the key. The form of the inflorescence is likewise characteristic and this and other keys place much reliance on it though aberrant forms may not key out satisfactorily. Despite the problems, most species are readily recognised and in Yemen identification is made easier by the fact that only five species are at all common: *E. barrelieri*, *E. cilianensis*, *E. ciliaris*, *E. papposa* and *E. pilosa*. Of these *E. papposa* is the only perennial and the only one usually found in natural habitats.**

1. Spikelets with a ciliate palea **2**
Spikelets glabrous **3**
2. Panicle spike-like, the branches very short. Lemmas with a few hairs *E. ciliaris*
Panicle spreading. Lemmas glabrous *E. tenella*
3. Ligule membranous, the mouth of the leaf sheath glabrous *E. japonica*
Ligule a line of hairs **4**
4. Small wart-like glands present on the leaf margin and pedicels **5**
Leaf margin and pedicels eglandular **6**
5. Lemmas 2–2.8mm long. Spikelets 2–4mm broad *E. cilianensis*
Lemmas 1.5–2mm long. Spikelets 1.5–2mm broad *E. minor*
6. Stem and inflorescence rhachis pilose. Annual with lower spikelets in whorls *E. porosa*
Stem and inflorescence rhachis glabrous. Spikelets whorled or not **7**
7. Annuals with a lax spreading panicle and small spikelets which are always shorter than their pedicels **8**
Annuals or perennials. Panicle spike-like or spreading, usually with short stiff branches but, if lax and the spikelets with long pedicels, then plant perennial **12**
8. Spikelets breaking up from the top downwards. Leaf sheaths usually pilose *E. aspera*
Spikelets breaking up from the bottom upwards (very tardily in *E. tef*). Leaf sheaths ± glabrous **9**
9. Palea persistent after the lemma has fallen. Glumes ± equal *E. macilenta*
Palea falling at the same time as the lemma. Glumes unequal, the lower smaller than the upper **10**
10. Lemmas 2–2.5mm long; spikelets pale brown or reddish *E. tef*
Lemmas less than 1.6mm long; spikelets dark or whitish-green **11**
11. Axils of the panicle branches with a tuft of hairs. Lemma 1–1.6mm long *E. pilosa*
Axils of the panicle branches glabrous. Lemma 0.7–1mm long *E. aethiopica*
12. Spikelets few, large, 1.5–4mm broad, ovate, elliptic or oblong; glumes boat-shaped, ± equal; lemmas cartiliginous with indistinct veins. Leaves and leaf sheaths usually pilose **13**
Spikelets usually numerous, linear, 1–2mm broad, glumes unequal, the lemma membranous with ± distinct veins. Leaves and leaf sheaths usually glabrous **15**
13. Panicle contracted, spike-like, the branches very short and appressed to the axis **14**
Panicle spreading, the branches mostly more than 1cm long *E. schweinfurthii*
14. Leaves mostly more than 30cm long, overtopping the inflorescence. Lemmas c 3mm long *E. longifolia*
Leaves less than 10cm long, much shorter than the inflorescence. Lemmas c 1.5mm long *E. schweinfurthii*
15. Inflorescence narrow, spike-like, the branches ± appressed to the axis. Tall, densely tufted perennial *E. braunii*
Inflorescence a spreading panicle but, if ± spike-like then plant annual **16**
16. Annual. Inflorescence dense, usually with very short branches and many shortly pedicellate purplish spikelets *E. barrelieri*
Perennial. Inflorescence lax, spikelets usually about as long or longer than the pedicels, dark olive green **17**
17. Basal leaf sheaths strongly nerved, adpressed pilose *E. curvula*
Basal leaf sheaths weakly nerved, glabrous **18**
18. Spikelets 1–3mm wide, dark green, margins serrate *E. tenuifolia*
Spikelets 1mm wide, grey-green, margins entire *E. papposa*

1. Eragrostis aethiopica *Chiov.*

Slender tufted annual to 50cm. Leaves up to 20cm long, the sheaths glabrous. Panicle diffuse and spreading with delicate, sometimes whorled branches which are glabrous in the axils; spikelets linear, very small, grey-green, breaking up from the base, the palea falling at the same time as the glumes; glumes unequal; lemma 0.7–1mm long.

Only found once as a weed of a sandy field between Nukaybah and Mafraq at 200m.

2. Eragrostis aspera (*Jacq.*) *Nees*

Vigorous tufted annual to 80cm. Leaves up to 30cm long, the sheaths pilose. Panicle large, diffuse, with long delicate branches which have tufts of hair in the axils; spikelets linear, rather small, often purplish, breaking up from the top downwards; lemmas 1.2–1.7mm long.

The large diffuse panicle is very distinctive and it is only likely to be confused with *E. macilenta*, in which species however the spikelets break up from the bottom upwards.

A rare plant of disturbed stony places found between 1200 and 1500m in areas of good rainfall on the escarpment: by a small wadi, Misrākh; near the suq, 'Udayn; Wadi Sukhnah; near Suq Ath Thalūth, Wadi Lā'ah.

3. Eragrostis barrelieri *Daveau*

Tufted annual to 30cm. Leaves up to 10cm long, the sheaths glabrous. Panicle condensed, sometimes almost spicate with short branches and many shortly pedicelled spikelets, axils glabrous but with a prominent gland; spikelets linear, commonly more than 1cm long, purplish, breaking up from the bottom upwards, the palea persistent after the lemma has fallen; glumes very unequal, the lower small and hyaline; lemmas 1.7–2.5mm.

Plants with a more spreading inflorescence are very similar to *E. papposa*, but differ in their annual habit (not always obvious), purplish spikelets and longer lemmas.

A very widespread weed of cultivated land occurring wherever there is cultivation from the Tihama east to Mārib and from 200 to 2500m; generally abundant except in the Tihama.

Fig. 32 **A** *Eragrosis barrelieri* ; **B** *Coelachyrum poiflorum*; **C** *Tripogon leptophyllus*; **D** *Acrachne racemosa*.

4. **Eragrostis braunii** *Schweinf.*

Vigorous tufted perennial to 90cm. Leaves stiff, up to 13cm long, the sheaths glabrous or sparsely pilose towards the top. Panicle long and narrow, spike-like, the branches erect and ± appressed to the axis, the axils glabrous; spikelets linear, usually more than 1cm long, grey-black, breaking up from the bottom upwards, the palea persistent after the lemma has fallen; glumes unequal; lemmas 1.7–2.2mm long.

One of a group of grasses restricted to the high escarpment mountains northwest of Sana'a where it is found from 2500 to 2900m on cliffs and steep grassy slopes: J. Shibām, Ḥarāz; Shibām; J. Maswār.

5. **Eragrostis cilianensis** (*All.*) *Lut.*

E. multiflora sensu Schwartz

Tufted annual to c 60cm. Leaves usually with marginal wart-like glands, glabrous or sparsely pilose, up to 15cm long, the sheaths usually pilose above. Panicle open with rather short, stiff branches, the axils often hairy; pedicels glandular; spikelets large, narrowly ovate or lanceolate, usually leaden-grey, breaking up from the bottom upwards, the palea persistent after the lemma has fallen; lemmas glandular on the keel, 2–2.8mm long.

E. cilianensis is a very variable species and the distinctive glands are not always present on the leaf margins and pedicels. Such forms can readily be confused with *E. schweinfurthii*, but that species is often perennial. One curious variant of *E. cilianensis* was collected on an old volcano near Shamlān (Wood 2018); in this specimen the spikelets break up from the top downwards, the palea falling with the lemma.

Like *E. barrelieri*, a very widespread weed of cultivated and waste places, found almost everywhere from 200 to 2500m and common wherever there are suitable habitats, including the Tihama.

6. **Eragrostis ciliaris** (*L.*) *R. Br.*

Small tufted annual to 30cm. Leaves up to 12cm long, leaf sheaths pilose. Panicle spike-like with short erect branches appressed to the axis and densely clustered spikelets; spikelets small, ovate, pinkish, densely pilose from the palea, breaking up from the top downwards; lemmas 0.9–1.5mm long, usually with a few hairs.

Widespread in the Tihama and on the escarpment foothills from sea level to 800m growing in disturbed sandy places such as fields and wadi beds, but inconspicuous, rather local and seldom in large quantity.

7. **Eragrostis curvula** (*Schrad.*) *Nees*

Densely tufted perennial to c 30cm in Yemen, though much more elsewhere. Leaves up to 20cm long, the lower leaf sheaths prominently ridged, adpressed pilose. Panicle open or contracted, the lowest branches often whorled and with a tuft of hairs in the axils; spikelets linear, grey-green, breaking up from the bottom upwards, the palea persistent after the lemma has fallen; glumes unequal; lemma 1.8–2.6mm long.

Most likely to be confused with *E. papposa* but differing in its hairy, ribbed lower leaf sheaths.

A very rare species of stony mountains around the high plateau found from 2700 to 3100m between Shibām and Kawkabān (Hepper 5810) and above Maḥalayn on J. Madhbur in J. Lawz (Wood 3399).

The distribution of this species is of some interest. It is a South African species introduced and naturalised north to southern Kenya growing at altitudes below 1800m. In addition there are isolated colonies at altitudes above 2600m in Yemen and Eritrea. Since the plant was collected by Schweinfurth in Eritrea as long ago as the 1890s it is difficult to believe that these isolated high-altitude colonies are casual introductions and it is not easy to give a satisfactory explanation for their existence.

8. **Eragrostis japonica** *Thunb.*

E. namaquensis Schrad.

E. diplachnoides Steud.

Tufted annual to 40cm (more in other countries). Leaves up to 20cm long, the sheaths glabrous. Panicle narrow, with short ascending branches and numerous shortly pedicellate spikelets ± appressed to the secondary branchlets, the panicle axils glabrous; spikelets ovate, whitish-green, very short and not exceeding 2.5mm long, breaking up from the top downwards; lemmas 0.7mm long.

Readily distinguished from all other species by the truncate, fibriate, membranous ligule.

Only known from a sandy irrigated field at 100m near Ḥaraḍ.

9. **Eragrostis longifolia** *Hochst. ex Steud.*

Tufted perennial to 35cm. Leaves up to 35cm long, the sheaths and the leaf blades pilose. Panicle narrow, spike-like, very short, the axis 2–4cm long and with less than 10 spikelets; spikelets oblong, olive-green, mode of breaking up not known but presumably from the bottom; lemmas ± 3mm long, cartiliginous with indistinct nerves.

A very distinctive species known only at 2200m in shallow grassland at Banī Mukhtār above Ḥadīyah. Outside Yemen it is a rare plant known only from a few places in Ethiopia.

10. **Eragrostis macilenta** (*A. Rich.*) *Steud.*

E. decidua Hochst.

Tufted annual to 60cm. Leaves up to 20cm long, the leaf sheaths glabrous or with a few hairs above. Panicle spreading, diffuse, with long, delicate branches and long pedicellate spikelets, the lower branches sometimes whorled and with tufts of hair in the axils; spikelets oblong, small (up to 6mm long), green or dark purple, breaking up from the bottom upwards, the palea persistent after the lemma has fallen; glumes unequal, acuminate; lemmas 1.3–1.6mm long.

An uncommon plant found in high rainfall areas of the escarpment between 1000 and 2200m growing as a weed of shaded cultivated land, particularly coffee terraces in the central escarpment: 1km north of Ibb; Wadi Sukhnah; Manākhah (Schweinfurth); Bani Sa'īd, Ja'farīyah; around Mirzin, J. Bura'; Ḥabash, J. Milḥān.

11. Eragrostis minor *Host*

Tufted annual to 60cm, though usually much less. Leaves usually glabrous, up to 12cm long, the margins with warty glands. Panicle narrow or open, rather small; pedicels less than 3mm long, usually glandular; spikelets narrowly oblong, usually less than 1cm long and 2mm broad, variously coloured, breaking up from the bottom upwards, the palea persistent after the lemma has fallen; lemmas usually glandular on the keel, 1.5–2mm long.

This species looks like a glandular form of *E. barrelieri*.

A rare plant known only from a wadi bed below Aṭ Ṭūr (Muller-Hohenstein and Deil).

12. Eragrostis papposa (*Roem. & Schult.*) *Steud.*

Tufted perennial to 40cm. Leaves flat or involute, up to 15cm long, the leaf sheaths glabrous. Panicle lax and spreading, the pedicels usually much longer than the spikelets, the axils of the branches glabrous; spikelets linear, grey-green, breaking up from the bottom upwards, the palea persistent after the lemma has fallen; glumes unequal; lemmas 1.1–1.7mm long.

The most abundant *Eragrostis* in Yemen, growing on open stony hillsides and dry alluvial plains almost everywhere on the escarpment, high plateau and the Mashriq from 500 to 3200m, though less common at both ends of its altitude range.

13. Eragrostis pilosa (*L.*) *P. Beauv.*

Tufted annual to 50cm. Leaves up to 20cm long, the leaf sheaths glabrous. Panicle lax, open, the lowest branches usually whorled and with a few long white hairs in the axils; spikelets linear, very small (less than 7mm long), grey-green, breaking up from the bottom upwards, the palea and the lemma falling together; glumes unequal, the lower very small; lemmas 1–1.6m long.

Very similar to *E. aethiopica* and *E. tef* but distinguished from both by the hairs in the axils of the lower panicle branches.

A field weed found between 400 and 2800m on the escarpment and high plateau just entering the Tihama near Bājil and the Mashriq at Majz (Steinberg) and Ḥūth. Although very widespread and usually abundant where it occurs it is surprisingly local and is probably a recent introduction. It is significant that Schwartz has no records of this plant from tropical Arabia. (Deflers' records from South Yemen are of *E. tef*) and this suggests that the plant has arrived within the last 30 years.

14. Eragrostis porosa *Nees*

Tufted annual to 60cm. Leaves up to 15cm long, the sheaths pilose. Panicle open, with a pilose axis and ascending branches, the lowest of which are whorled; spikelets linear, very small (up to 5mm long), grey, breaking up from the bottom upwards, the palea persisting after the lemma has fallen; lemmas 1.1–1.5mm long.

A very local species of roadsides and field borders in dry scrubland: Bani Manṣūr; frequent between Riḥāb and Waṣāb. (Wood 1626)

15. Eragrostis schweinfurthii *Chiov.*

E. multiflora auct.

E. kiwuensis Jedw.

Annual or short-lived perennial to 40cm. Leaves up to 10cm long, the sheaths and the leaf blades usually pilose. Panicle open, with short stiff spreading branches, each bearing rather few spikelets, the axils glabrous or (variety **kiwuensis** (*Jedw.*) *S.M. Phillips*) contracted and spikelike; spikelets usually narrowly ovate, large, dark olive-green, breaking up from the bottom upwards, the palea persisting after the lemma has fallen; lemmas 1.5–1.7mm long, cartiliginous, the nerves indistinct.

A rare plant of montane grassland found from 2000 to 3200m on a few escarpment mountains in high rainfall areas: below 'Arūs, J. Ṣabir; J. Ta'kar; J. Manār; J. Ḥiṣn Madnān, Waṣāb al 'Ālī; J. Al 'Amal, Utmah; Al Markah, J. Milḥān. Variety **kiwuensis** occurs in the same habitats but is less common: 'Arūs, J. Sabir; J. Sumārah; J. Ḥiṣn Madnān, Wasāb al 'Alī; Bughah to Kusmah (Forsskal) and J. Kabūrah in Raymah.

16. Eragrostis tef (*Zucc.*) *Trotter* Taḥaf

Very similar to *E. pilosa*, differing in the absence of hairs in the axils of the panicle branches, the pale brown or reddish spikelets which break up very slowly at maturity, the longer lower glume (1.2–2.5mm long) and the longer lemma (2–2.7mm long). Although the grain is large in Ethiopian specimens, this seems not to be the case in Yemen where it is used not for its grain but for forage.

Only known from flood channels beside the Wadi Saba' at Mārib at 1100m.

In Ethiopia *E. tef* is cultivated for its grain and this seems also to have been the practice in medieval Yemen though its cultivation seems to have ceased in modern times. It is impossible to say whether *E. tef* is indigenous to Yemen or an ancient introduction from Ethiopia, though the name 'Taḥaf', from which 'Tef' is doubtless derived, seems to be of impeccable Semitic origin.

17. Eragrostis tenella (*L.*) *Roem. & Schult.*

Small tufted annual to 50cm. Leaves up to 10cm long, the sheaths glabrous. Panicle open with short spreading branches, the axils glabrous or hairy; spikelets oblong, very small (up to 2.5mm), green or purplish, sparsely ciliate from the palea, breaking up from the top downwards; lemmas 0.7–1mm long, glabrous.

On sand banks at 500m by the wadi at Suq As Sabt, J. Bura'.

18. Eragrostis tenuifolia (*A. Rich.*) *Steud.*

Tufted perennial to 70cm with compressed basal shoots. Leaves flat, up to 30cm long, the leaf sheaths glabrous. Panicle lax and open, the pedicels usually longer than the spikelets; spikelets linear, 1–3mm wide, dark green, the florets soon spreading and so making the spikelet margin serrate, breaking up from the bottom upwards; glumes unequal; lemmas 1.7–2.5mm long.

E. tenuifolia is similar to vigorous examples of *E. papposa* but can be distinguished by its wider, serrate spikelets.

Known only from the wadi by the Agricultural Institute, just west of Ibb (R. Spellenberg).

33. COELACHYRUM *Hochst. & Nees*

Perennials. Leaves flat; ligule membranous, lacerate. Inflorescence raceme-like, consisting of several spike-like racemes spaced along a short or long axis at the top of the stem; spikelets borne in two rows along one side of the raceme only, sessile, many flowered; glumes ± equal, lemmas 3-nerved without awns.

Inflorescence rhachis 1–1.5cm long. Racemes up to 1.5cm long *C. poiflorum*
Inflorescence rhachis 5–15cm long. Racemes 2–3 (–6)cm long *C. yemenicum*

1. Coelachyrum poiflorum *Chiov.*

C. induratum Pilger
Erect tufted perennial to 30cm, spreading by stolons to form small colonies. Leaves dark green, 2–3cm long. Inflorescence of 2–4 ascending spike-like racemes on a short axis 1–1.5cm long; racemes 1–1.5cm long; spikelets 6–10-flowered, blackish, pilose; lemmas shortly mucronate, pilose below.

Widespread but rather local, growing on rocky or sandy plain from 1400 to 3000m; it is most common on the high plateau but occurs sporadically on the escarpment (Taiz, Raḥidah, 'Udayn, Wadi Badr) and in the Mashriq (Ḥūth, Nihm, Majz (Steinberg)).

2. Coelachyrum yemenicum (*Schweinf.*) *S.M. Phillips*

Cypholepis yemenica (Schweinf.) Chiov.
Eragrostis yemenica Schweinf.
Erect tufted perennial to 1m. Leaves narrow, up to 20cm long. Inflorescence narrow, consisting of 2–8 erect spike-like racemes on an axis 5–15cm long; racemes 2–4 (–6)cm long; spikelets 7–12-flowered, whitish-green, glabrous outside; lemmas obtuse, sparsely adpressed clavate-pilose below.

Widespread but infrequent in open stony ground, usually in *Acacia* scrub, between 1000 and 2300m on the escarpment and high plateau from Qa'idah north to Sāqayn and extending to the east of Rada'. It appears to be most frequent in the Ḥarāz and the 'Anis – J. Ash Sharq region.

It is widespread in east and southern Africa and the name *yemenicum* was only used because it was first discovered by Schweinfurth in Yemen.

34. DACTYLOCTENIUM *Willd.*

Annual or perennial grasses similar to *Eleusine*. Inflorescence of short, thick, digitate, sessile, 1-sided spikes; spikelets in 2 rows, many-flowered; upper glume with a short awn-like point; lemmas acute or mucronate, awnless.

***Dactyloctenium* can be distinguished from all similar genera by the spikes which have the rhachis tip bare of spikelets.**

Annual. Spikes 1–5cm long, the rhachis straight *D. aegyptium*
Stoloniferous perennial. Spikes less than 1.5cm long, the rhachis curved *D. scindicum*

1. Dactyloctenium aegyptium (*L.*) *Willd.*

Robust annual spreading to form small patches, the stems at first decumbent and rooting at the nodes, finally ascending to 60cm. Leaves flat, pilose or ± glabrous, often with purple markings, up to 20cm long. Spikes 3–9, spreading or ascending, 1–5cm long, the rhachis straight; spikelets 3–4-flowered, whitish-green, glabrous.

A common weed of fields and waste places growing from sea level to 1500m in the Tihama and on the escarpment.

2. Dactyloctenium scindicum *Boiss.*

D. glaucophyllum Courb.
Stoloniferous perennial forming extensive mats; stems swollen below, erect, up to 30cm. Leaves flat, sparsely pilose often glaucous, up to 10cm long, but usually much less. Inflorescence more head-like than in *D. aegyptium*; spikes 3–4, 1–1.5cm long, the rhachis strongly curved upwards; spikelets 3–9-flowered, whitish-green, glabrous.

Widespread and often abundant in the alluvial plains in the east of the Tihama and along the main valleys, penetrating the escarpment up to about 1500m; elsewhere very rare but known from Sana'a (Rathjens) and from the southeastern mountains at Qa'ṭabah (Wissmann) and Dhī Nā'im. It grows on stony plain and in open *Acacia* scrub but shows a particular preference for earth banks around cultivation and the margins of irrigation ditches.

35. ACRACHNE *Chiov.*

1. Acrachne racemosa (*Roem. & Schult.*) *Ohwi*

Eleusine verticillata Roxb.
Erect tufted annual to 80cm. Leaves flat, light green, up to 22cm long; ligule a fringed membrane, c 1.25mm long. Inflorescence of 1–5, slender, ascending, sessile spikes arranged digitately at the top of the stem and with whorls of decreasing size on the axis below; spikes up to 11cm long, bearing the spikelets along one side only; spikelets sessile, 6–25-flowered, usually whitish-green, glabrous; glumes unequal, the upper shortly awned; lemmas 3-nerved, the central nerve ending in a short awn, the lateral shortly excurrent.

An uncommon plant found between 500 and 800m in disturbed places in the valley scrub of the central escarpment, usually growing in gravelly wadi beds: 'Alūjah, J. Raymah; above Ḥillah and below Nashām, J. Bura'; Walajah, J. Milḥān (Schweinfurth); Rajamah near Ḥajjaylah and J. Sāfān in the Ḥarāz.

36. OCHTHOCHLOA *Edgw.*

1. Ochtochloa compressa (*Forssk.*) *Hilu*

Eleusine compressa (Forssk.) Aschers. & Schweinf. ex C. Chr.
E. flagellifera Nees
Wiry perennial grass with decumbent stolon-like stems which root at the leafy nodes. Leaves flat or slightly inrolled, grey-green, up to 10cm long; ligule a lacerate membrane, 0.5mm long. Inflorescence of 3–6 digitate sessile spikes; spikes 2–3.5cm long, bearing the spikelets in 2 rows along one side only; spikelets 3–6-flowered, whitish-green; lemma 3-nerved, acute, villous, awnless.

A characteristic grass of sand dunes in the Tihama, growing in abundance in suitable habitats from sea level to 500m.

37. ELEUSINE *Gaertn.*

Tufted annuals or perennials. Leaves folded or flat; ligule membranous, short. Inflorescence of sessile spikes arranged digitately at the top of the stem, usually with 1 or 2 spikes below, rarely shortly racemose; spikelets borne in 2 rows on one side of the spike only, several-flowered; lemma 3-nerved, awnless.

1. Perennial. Leaves with tufts of white hairs along the margins *E. floccifolia*
Annual. Leaves without tufts of hairs along the margins **2**
2. Spikes 1–3cm long. Lemmas mucronate or cuspidate *E. multiflora*
Spikes more than 3cm long. Lemmas acute, but not mucronate *E. indica*

1. Eleusine floccifolia (*Forssk.*) *Spreng.*
Coarse tufted perennial to c 80cm. Leaves usually folded, up to 30cm long, the margins with tufts of whitish hairs. Inflorescence ± digitate, usually with a single spike below the apical cluster; spikes mostly 3–7cm long, ascending to weakly reflexed; spikelets 4–7-flowered, grey-green, glabrous; lemmas acute.

Widespread and frequent between 1500 and 3000m on the escarpment and high plateau but unrecorded north of the Wadi Mawr. In high rainfall areas it grows on grassy field borders and roadsides but in other areas is restricted to ditch sides, pond margins and similar wet places.

2. Eleusine indica (*L.*) *Gaertn.*
Tufted annual (often appearing to be perennial) to 1m. Leaves usually folded, dark green, up to c 20cm long. Inflorescence ± digitate, usually with a single spike below the apical cluster; spikes mostly 3–7cm long, ascending; spikelets 3–9-flowered, whitish-green, glabrous; lemmas acute.

Most if not all *E. indica* from Yemen belongs to subspecies **africana** (*Kennedy – O'Byrne*) *S.M. Phillips* which is more robust than the type and has a distinctly ciliate ligule (e.g. Wood 74/65, 1788).

Locally frequent by streams and ditches between 1100 and 1800m in the high rainfall areas of Ibb province south to J. Ḥabashī and Misrākh; elsewhere very rare and only known in the central escarpment from a number of places on J. Bura' and near Bani Daw'ān on J. Raymah.

3. Eleusine multiflora *A. Rich.*
Low tufted annual with numerous stems which rarely exceed 10cm. Leaves usually flat, sparsely pilose, dark green, less than 10cm long. Inflorescence racemose with 3–5 spikes borne on a short axis; spikes 1–3cm long, ascending; spikelets 5–15-flowered, whitish-green, glabrous; lemmas cuspidate or mucronate.

Widespread on the higher escarpment mountains from J. Ṣabir north to Shahārah growing from 2200 to 3000m in disturbed, often well-trodden, gravelly places such as tracksides, cistern edges and village streets; sometimes occurring as a field weed.

The Finger Millet, **Eleusine coracana** (L.) Gaertn. ('Kanab') is recorded in medieval agricultural treatises but although it has been cultivated this century in South Yemen and Saudi Arabia, there is no evidence of its cultivation in North Yemen in recent times.

38. HARPACHNE *Hochst. ex A. Rich.*

1. Harpachne schimperi *Hochst. ex A. Rich.*
Slender tufted perennial with erect stems to 30cm. Leaves flat, pilose, up to 10cm long, ligule a fringe of hairs. Inflorescence a solitary terminal raceme, the spikelets hanging on slender, villous pedicels from a central rhachis 3–6cm long; spikelets 6–13-flowered, wedge-shaped, awnless, glabrous, purplish, 5–15mm long; glumes unequal; lemmas long-acuminate, 4–7mm long.

A distinctive species widely distributed between 1800 and 2800m in the central escarpment where it is frequent on open stony ground; elsewhere rare and known only from Ibb, the Shibām-Aḥjar region and J. Rāziḥ.

39. DINEBRA *Jacq.*

1. Dinebra retroflexa (*Vahl*) *Panzer*
Loosely tufted annual with slender, straggling, ascending stems to 50cm (–1m). Leaves flat, dark green, 5–15cm long; the ligule a short lacerate membrane. Inflorescence narrow, consisting of 10–40 shortly pedunculate, one-sided spikes arranged irregularly along a central axis 5–20cm long; spikes mostly 0.5–2cm long, at first ascending but becoming pendulous with maturity; spikelets in 2 rows, 1–3-flowered, greenish-white, glabrous; glumes ± equal, membranous, acuminate, ending in a fine aristate point, 6–8mm long, much longer than the spikelets; lemmas without awns.

A rare plant of damp, partially shaded irrigated fields and streamsides from 600 to 1700m: Wadi Ḥidrān, Taiz; Wadi Sūdān, Qā'idah; in a banana plantation below Ḥadīyah.

40. LEPTOCHLOA *P. Beauv.*

Annuals or perennials. Leaves flat; ligule short, membranous, lacerate. Inflorescence raceme-like, consisting of several slender, sessile spike-like racemes arranged along a central axis at least 10cm long, the spikelets borne in 2 rows along one side of the raceme only; glumes ± equal, lemmas 3-nerved, awnless.

Spikelets 1-flowered. Leaves up to 10cm long *L. rupestris*
Spikelets 6–12-flowered. Leaves normally more than 10cm long *L. obtusiflora*

1. Leptochloa obtusiflora *Hochst.*

Eleusine obtusiflora (Hochst.) Schweinf.

Vigorous tufted perennial to 1½m. Leaves glaucous, c 10–30cm long; ligule c 1.5mm long. Racemes ascending, up to 10cm long; spikelets nearly sessile, 6–12-flowered, whitish-green, glabrous, 4–7mm long; glumes shorter than the spikelets; lemma shortly bifid.

A local grass found between 600 and 1800m on the escarpment, growing on shallow stony soil usually where protected from grazing by thorn fences, spiny *Euphorbia* and *Acacia* bushes; nearly all records are from the central escarpment or Taiz province but it may have been overlooked elsewhere.

2. Leptochloa rupestris *C.E. Hubbard*

? *L. uniflora* sensu Schweinfurth

Slender perennial spreading by rhizomes and sending up small clusters of weak ascending stems which reach c 1m in length. Leaves light green, up to 10cm long and 5mm wide; ligule c 1mm long. Racemes ascending, mostly 2–4cm long; spikelets nearly sessile, 1-flowered, whitish-green (becoming purplish with age), glabrous, 1.5–2.5mm long; glumes longer than the spikelets; lemma acute, not bifid.

Widespread between 600 and 1500m on the escarpment from Wadi Ḍabāb north to Khawlan Ash Sham, but only frequent in areas of good rainfall. It grows in rough scrub and in the protection of small shrubs which give support to its weak stems. In drier country it is usually found in thorn fences by irrigation ditches.

The very closely related *Leptochloa uniflora* A. Rich. was recorded by Schweinfurth but probably in error for *L. rupestris* which was not described at that time. All specimens I have seen can be assigned to *L. rupestris*, though *L. uniflora* is likely to occur. It is a usually annual, tufted grass with broader leaves (6–18mm wide) and a longer inflorescence.

41. DIPLACHNE *P. Beauv.*

1. Diplachne fusca (*L.*) *P. Beauv.*

Erect tufted perennial to 1m, though usually much less in Yemen. Leaves dark green, inrolled, up to 20cm long; ligule an acute membrane, 3–8mm long. Inflorescence an open panicle composed of nearly sessile, lax, spike-like racemes; racemes at first erect, finally spreading, 3–7cm long; spikelets rather distant, 3–7-flowered, awnless, olive-green, 6–10mm long, glumes unequal; lemma pilose below, the tip bidentate with a short mucro from the cleft.

Only known from the salt flats just north of Rada' at 2000m.

42. ODYSSEA *Stapf*

1. Odyssea mucronata (*Forssk.*) *Stapf* Shawkham

Stiff, woody, spiny, much-branched perennial to 2m, spreading by rhizomes through sand to form dense colonies. Leaves distichous, glaucous, stiff, spine-tipped, 2–5cm long; ligule a fringe of hairs. Inflorescence dense, head-like, c 2 × 1.5cm, composed of about 10 short crowded racemes, each with few spikelets; spikelets pedicelled, 5–9-flowered, villous, 5–10mm long; glumes ± equal, scarious; lemmas purplish.

Abundant on sand dunes near the sea in the Tihama but decreasing in frequency inland and not found above 250m or east of Bayt al Faqīh and Zaydīyah.

This very distinctive, unpleasantly spiny grass is restricted to the Red Sea basin.

43. TRIPOGON *Roem. & Schult.*

Small densely tufted perennials with slender erect stems. Leaves narrow, filiform; ligule a ciliate membrane. Inflorescence a solitary terminal spike with the spikelets arranged alternately in 2 ranks on opposite sides of the rhachis; spikelets 2–many flowered; glumes unequal, 1-nerved; lemmas 3-nerved, bidentate, awned.

A poorly understood genus neglected by most collectors since the species are inconspicuous and of a dried-out yellow-green colour which blends into a stony or earthy background.

1. Spikelets 8–19-flowered (spikelets in the middle of the spike ± 11-flowered). Upper glume obtuse, up to half as long as the spikelet — *T. multiflorus*
   Spikelets usually 3–6-flowered (rarely a few spikelets with up to 10 flowers in *T. leptophyllus*). Upper glume acuminate, more than half as long as the spikelet — **2**
2. Basal leaf sheaths breaking up to form fibrous cushions. Lateral nerves of the lemma not excurrent. Low plants with yellowish or purplish spikelets — **3**
   Basal leaf sheaths not becoming fibrous. Lateral nerves of the lemma excurrent as short awns. Mostly taller plants, more than 20cm high with blackish-green spikelets — *T. montanus*
3. Spikelets 3 (–6)-flowered. Callus ciliate. Awn c 1mm long — *T. purpurascens*
   Spikelets 3–10-flowered. Callus shortly appressed hairy. Awn 2–4mm long — *T. leptophyllus*

1. Tripogon leptophyllus (*A. Rich.*) *Cuf.*

Densely tufted perennial to 20cm, the basal leaf sheaths very persistent, breaking up, becoming fibrous and forming a cushiony base to the plant. Leaves, sparsely pilose, slightly shorter than the inflorescence, up to 18cm long. Spike 2–10cm long, rather lax, the rhachilla internodes relatively long so the separate spikelets can be clearly seen without a lens; spikelets 3–10-flowered, yellow-green to purplish; upper glume acuminate, slightly shorter than or as long as the spikelet; lemmas 2.2–3.3mm long, bidentate, with a single awn 2–4mm long; callus shortly appressed hairy (the lemma thus lacking the tuft of hairs from the base found in all our other species).

A rare grass of steep, seasonally wet, grassy slopes between 1800 and 2300m in the central escarpment: J. Kabūrah above Ḥadīyah (Wood 75/553); Bayt Braz, Maḥwīt (Wood 72/24); near Rujum (Wood 2547); a field record from between Markah and Ṣa'bah, J. Milḥān is almost certainly this species.

Outside Yemen only known from Ethiopia and J. Marra in the Sudan.

2. Tripogon montanus *Chiov.*

Densely tufted perennial to 35cm, the basal leaf sheaths papery, never fibrous. Leaves sparsely pilose, up to 25cm long, usually much shorter than the inflorescence. Spike 3–8cm long, the spikelets tightly appressed to the rhachis; spikelets 2–5-flowered, blackish-green; upper glume acuminate, about ¾ as long as the spikelet; lemmas 2.5–5mm long, the lateral nerves exserted as short awns 0.5–1.5mm long, the central awn very variable in length 1.5–13mm long; callus pilose.

A very variable plant, each collection (from Yemen, at least) differing significantly from the others but all with the distinctive papery basal sheaths, blackish spikelets and excurrent lateral nerves on the lemma. The two extremes are represented by Wood 672 and 1857. The former has 2–3-flowered spikelets, a short lemma 2.5–3mm long and a very short central awn c 1.5mm long. The latter has 3–5-flowered spikelets, a lemma c 4mm long, and central awns up to 13mm long. Since these two extremes are connected by other specimens it seems best to regard them all as one species.

An uncommon plant of grassy cliffs, usually near streams, found between 2500 and 2900m on a few escarpment mountains: J. Ba'dān (Wood 1857); by the waterfall, Baw'ān (Wood 2912); between Shibām and Kawkabān (Wood 2367); J. Shibām, Ḥarāz (Wood 672). Field records from the gorge near Ribāṭ at the head of the Wadi Banā and from Ḍawrān are almost certainly this species.

Known elsewhere from one locality in Uganda, J. Marra in the Sudan and Ethiopia.

3. Tripogon multiflorus *de Miré & Gillet*

Densely tufted perennial to 30cm (more outside Yemen), the basal leaf sheaths breaking up tardily and becoming somewhat fibrous. Leaves flat or convolute, pilose or glabrous, up to 20cm long, much shorter than the inflorescence. Spike 5–20cm long, oblanceolate, the upper spikelets conspicuously longer than the lower; spikelets 8–19-flowered, whitish-green or olive-green; upper glume obtuse, less than half as long as the spikelet; lemmas 3–4.5mm long, bidentate or entire, with a central awn 1–2mm long; callus pilose.

Readily recognised by the many-flowered spikelets and relatively short, obtuse upper glume, but variable. Wood 2020 approximates the type and has convolute, glabrous leaves and erect, olive-green spikelets. Other plants, mainly from the escarpment (Wood 75/288, 2825) have some of the leaves flat and pilose and the inflorescence whitish-green with loosely ascending spikelets. There is also a tendency for the basal leaf sheaths to be more fibrous; similar variation though is seen in African plants.

The most drought-resistant *Tripogon*, found on a few well-drained rocky slopes between 1200 and 2300m on the escarpment and high plateau: north of the suq, 'Udayn (Wood 75/288); between Ḥadīyah and Na'mah, J. Raymah (Wood 2825); Wadi Sukhnah (Wood 2962); on an old volcanic cone near Shamlān (Wood 2020); on limestone cliffs at the head of the Wadi Khārid, Arḥab (Wood 2775).

Outside Yemen it is known from two places in Kenya, one in Eritrea and from most of the mountains of the central Sahara.

4. Tripogon purpurascens *Duthie*

Densely tufted perennial to 20cm, similar in appearance to *T. leptophyllus* with which it has been confused but usually becoming purplish when old, basal leaf sheaths breaking up and becoming fibrous. Leaves pilose, up to 15cm long, shorter than the inflorescence. Spike 2–10cm long, the spikelets erect and appressed to the rhachis; spikelets 3 (–6)-flowered, purplish; upper glume acuminate, nearly as long as the spikelet; lemmas 2.5–3mm long, bidentate with a single awn c 1mm long; callus pilose.

Only known from a steep earthy bank on a river cliff at 1500m at the bottom of the Sumārah pass near Dalīl. (Wood 2902, 3333)

A grass of northwest India extending to Afghanistan and through Oman to Yemen.

44. LEPTOCARYDION *Stapf.*

1. Leptocarydion vulpiastrum (*De Not.*) *Stapf*

Loosely tufted annual with erect or ascending stems to 1m. Leaves ovate or lanceolate, abruptly narrowed and ± cordate at the base, up to 5 (–10)cm long, ligule a short membrane. Inflorescence reddish-purple, feathery, spike-like, 5–20cm long, composed of numerous, crowded, slender racemes up to 4cm long; spikelets subsessile, 6–12-flowered, 5–11mm long; glumes unequal; lemmas pubescent with ciliate margins, c 3mm long, the apex minutely toothed with the midrib extended as a straight awn 2.5–5mm long.

A very rare grass known only from the protection of a thorn fence on a gravelly bank by a small wadi at c 1300m near Misrākh.

45. OROPETIUM *Trin.*

Tiny, inconspicuous tufted perennials. Leaves flat or folded, glabrous or hairy; ligule a line of hairs. Inflorescence a solitary, curved or coiled, terminal spike; spikelets embedded into hollows in the rhachis, 1-flowered; lower glume minute, hyaline; upper glume appressed to the rhachis, sometimes awned, exceeding the lemma.

Spikes curved. Upper glume 2–3mm long, unawned — *O. capense*

Spikes curved or coiled. Upper glume 6–20mm long including the awn — *O. minimum*

1. Oropetium capense *Stapf*

Densely tufted perennial to 12cm. Leaves usually pilose, obtuse or acute, up to 4cm long. Spike curved, up to ,0cm long; spikelets usually in opposite ranks; upper glume 2–3mm long, acute but not extended into an awn.

Only known at 1800m on the Fardah pass (Miller). It is easily overlooked and may occur elsewhere.

2. Oropetium minimum (*Hochst.*) *Pilg.*

Densely tufted perennial to 12cm. Leaves densely pilose, acute, up to 5mm long. Spike curved or coiled, up to 8cm long; spikelets usually in adjacent ranks thus making the spike appear one-sided; upper glume c 3mm long, drawn out into an awn 3–17mm long.

An inconspicuous plant, growing on open rocky ground between 200 and 1900m in a number of widely scattered localities in the Tihama, on the escarpment and in the Mashriq: near al Mafraq; between Riḥāb and Waṣāb; Wadi Sukhnah; Shahārah; Majz (Steinberg). It is usually found soon after the onset of heavy rains which bring it into flower.

46. SCHOENEFELDIA *Kunth*

1. Schoenefeldia gracilis *Kunth.*

Erect tufted annual to 1m. Leaves flat, narrow, up to 20cm long; ligule a hairy rim. Inflorescence of 2–4 slender, digitate, 1-sided spikes, c 8–15cm long; spikelets in 2 rows, 1-flowered, whitish-green, c 2mm long; glumes unequal, the upper terminating in a short awn; lemmas shorter than the glumes, ciliate, shortly bifid; awn wavy, 2–3cm long.

A characteristic weed of sandy fields in the Tihama from the Wadi Zabīd northwards; growing from 100 to 500m, sometimes in abundance and often associated with *Cassia nigricans* and *Lasiurus scindicus*.

47. MICROCHLOA *R. Br.*

1. Microchloa kunthii *Desv.*

M. indica sensu Schwartz ?

Tufted perennial with erect stems to 40cm, the base surrounded by the fibrous remains of old leaf sheaths. Leaves filiform, up to 8cm long, ligule a short membranous rim. Inflorescence a solitary, terminal, slender, one-sided curved spike, 2–20cm long; spikelets sessile in 1 row, 1-flowered, purplish, 2.5–5mm long; glumes as long as the spikelet, nearly equal, glabrous; lemma enclosed by the glumes, ciliate, awnless.

Widespread on the escarpment and high plateau from 1200 to 3200m, but rather local or perhaps overlooked since it is usually inconspicuous. Although growing in montane grassland at high altitudes and in the meadow communities above Ḥadīyah it is most characteristic of small hollows in rocky plain where water accumulates in the rainy season. Here it is associated with *Ophioglossum* and *Craterostigma* spp and like them comes into flower very soon after the rains.

Schwartz only records *Microchloa indica* (L.f.) P. Beauv. from Yemen but probably in error for *M. kunthii*. *M. indica*, though, is quite likely to occur; it differs only in its annual habit.

48. CYNODON *Rich. in Pers.*

1. Cynodon dactylon (*L.*) *Pers.* Wubal, zīl

Stoloniferous perennial spreading mainly by rhizomes to form extensive swards; stems erect, slender, up to 40cm. Leaves flat, often glaucous, up to 10cm long; ligule a membranous ciliate rim. Inflorescence of 4–6 slender, digitate 1-sided spikes, 1.5–5cm long; spikelets in 2 rows, 1-flowered, purplish, c 2cm long; glumes ± equal; lemma silky on the margin.

One of the most widespread and abundant plants in Yemen, occurring almost everywhere up to 2800m, apart from the driest areas; in areas of good rainfall it occurs as a field weed and in disturbed grassy places such as tracks, villages, roadsides and field borders; in drier areas it is virtually restricted to irrigated fields, permanently wet places and alluvial banks by seasonal watercourses.

This pantropical grass is known as Star Grass or Bermuda Grass and is widely used for making lawns in Yemen and elsewhere in the tropics, but can become a persistent weed of cultivated land if not properly controlled. Its value for grazing is disputed.

49. ENTEROPOGON *Nees*

1. Enteropogon macrostachyus (*A. Rich.*) *Benth.*

Tufted perennial with erect stems to 1m. Leaves flat, up to c 30cm long; ligule a ciliate rim. Inflorescence a solitary, terminal, one-sided, curved spike 8–20cm long; spikelets sessile in 2 rows, 3-flowered, glabrous, green or straw-coloured, 7–10mm long; glumes unequal, terminating in a bristle-like point; lowest lemma scabrid, with an awn 10–18mm long from the minutely bidentate tip.

Frequent from 500 to 1500m on scrubby west-facing hillsides in the central escarpment including Maḥwīt and J. Sharafayn; elsewhere rare in similar habitats at Wadi Ḥidrān near Taiz, Mafḥaq, Wadi Sukhnah and Wadi Qu'sa' near Madīnat Al 'Abīd.

50. EUSTACHYS *Desv.*

1. Eustachys paspaloides (*Vahl*) *Lanza & Mattei*

Chloris petraea Thunb.

Tufted perennial, often conspicuously flattened at the base and the lower leaf nodes; stems at first geniculate and rooting at the nodes, finally erect, up to 70cm in height. Leaves flat, 2–10cm long, ligule a fringe of hairs. Inflorescence of 2–6 digitate, 1-sided spikes, 3–6cm long; spikelets 2-flowered, brown, c 1.5mm long; glumes ± equal, the upper with a short awn; lemmas ciliate along the nerves, emarginate with white membranous lobes.

Locally abundant on the escarpment in open scrub and in grassland on shallow rocky soils from 1000 to 2600m; widespread in the central escarpment from J. Raymah to the Wadi Lā'ah and along the inner escarpment from Ibb to Aḥjar, but unknown elsewhere.

51. CHLORIS *Sw.*

Perennials or annuals. Leaves flat; ligule membranous, ciliate. Inflorescence of slender, 1-sided, digitate, sessile spikes, less commonly racemose with the spikes on a short axis; spikelets sessile, 2–4-flowered, in two, sometimes obscure rows; glumes unequal, the upper sometimes awned; lemmas awned.

1. Spikes numerous, densely clustered along a long axis and forming a feathery head-like inflorescence — *C. roxburghiana*
   Spikes 5–15, arranged digitately or nearly so, not forming a feathery head-like inflorescence — 2
2. Perennial. Spikes 10–25cm long — *C. mensensis*
   Annual. Spikes 2–6 (–10)cm long — 3
3. Leaves 2–5 (–10)cm long, the apex obtuse. Lemmas glabrous — *C. pycnothrix*
   Leaves 5–20cm long, tapering to an acute apex. Lemmas with a crown of hairs at the apex — *C. virgata*

1. Chloris mensensis (*Schweinf.*) *Cuf.*

Tufted perennial with long straggling stems reaching 2m with the support of surrounding bushes. Leaves up to 50cm long, tapering to an acute apex. Inflorescence sub-digitate; spikes 3–5, 10–25cm long; spikelets 2-flowered, 2-awned, whitish-green, c 5mm long; lemmas glabrous, the longest awn c 10mm long.

A very rare grass known only from the valley forest above Suq As Sabt, J. Bura' at 600m. Outside Yemen it is only known from Ethiopia and Dhofar.

2. Chloris pycnothrix *Trin.*

C. radiata sensu Schwartz
C. leptostachya Hochst.

Stoloniferous annual with distinctive flattened leafy shoots spreading to form mats and with short, often geniculate flowering stems to 30cm. Leaves yellow-green, very short with an obtuse apex, 2–5 (–10)cm long. Inflorescence digitate; spikes 2–7, 2–6cm long; spikelets 2-flowered, 1–2-awned, purplish, c 2.5mm long; lemmas glabrous, the longest awn c 20mm long.

Widespread and frequent between 800 and 2400m on the escarpment and high plateau, growing in disturbed places where competition from taller grasses is slight; in high rainfall areas it is common on tracksides and field borders but elsewhere is restricted to irrigated grassland including lawns and the margins of irrigation ditches.

3. Chloris roxburghiana *Schult.*

C. myriostachya Hochst.

Tufted perennial to 1m or more. Leaves 10–30cm long, tapering to an acute apex. Inflorescence of numerous spikes, arranged along an axis 4–10cm long forming a dense elongated feathery head; spikes 3–8cm long; spikelets 3–4-flowered, 3–4-awned, whitish, 1–2mm long; lemmas sparsely ciliate, the longest awn 10–15mm long.

A rare plant of stony slopes in open scrub between 600 and 1200m in the valleys of the central escarpment; Ḥadīyah, J. Raymah; Ḥillah, J. Bura' (Schweinfurth); Wadi Ḥijān, Ḥarāz (Deflers); Wadi Yūr.

4. Chloris virgata *Sw.*

Weakly tufted erect annual to 80cm, sometimes rooting at the lower nodes. Leaves 5–20cm long, tapering to an acute apex. Inflorescence digitate; spikes 4–15, 2–7 (–10)cm long; spikelets 3-flowered, 2–3-awned, purplish, 2–3mm long, quickly deciduous; lemmas ciliate and with a crown of long hairs at the apex, the longest awn 5–10mm long.

Widespread and common from sea level to 1500m in Taiz province, becoming rarer further north and not found north of the Wadi Sihām in the Tihama or Waṣāb on the escarpment; there are also isolated records from Al 'Ashshah and Mārib.

52. TETRAPOGON *Desf.*

Tufted annuals or perennials. Leaves usually flat; ligule a short ciliate membrane. Inflorescence of solitary or paired terminal spikes; spikelets cuneate, 4–9-flowered (the upper sterile, and reduced), sessile, in 2 rows along the rhachis; glumes similar, glabrous, membranous; lemma coriaceous, usually ciliate, entire or bifid, awned, the awn in our species c 1cm long.

1. Lowest lemma glabrous or with short hairs on the nerves — *T. tenellus*
   Lowest lemma villous — 2
2. Spikes ± enclosed by the inflated sheath of the upper leaf. Basal leafy shoots at most slightly flattened — *T. cenchriformis*
   Spikes not enclosed by the upper leaf sheath. Basal leafy shoots strongly flattened — *T. villosus*

1. Tetrapogon cenchriformis (*A. Rich.*) *W.D. Clayton*

Lepidopironia cenchriformis A. Rich.

Tufted annual (or short-lived perennial) with erect stems to 60cm, the basal leafy shoots slightly flattened. Leaves slightly glaucous, 3–10cm long. Spikes 3–8cm long, usually solitary, ± enclosed by the inflated sheaths of the upper leaves; spikelets 5–9-flowered, each flower with a reddish-purple awn; upper glume 10–12mm long, reddish-purple, slightly exceeding the densely white-villous lemmas.

Widespread on the escarpment but infrequent, growing from 500 to 1500m on well-drained stony ground in open *Acacia* scrub.

2. Tetrapogon tenellus (*Roxb.*) *Chiov.*

T. macroanthus (Jaub. & Spach) Benth.
Chloris tenella Roxb.

Loosely tufted annual (or short-lived perennial) with ascending, often reddish stems to 40cm. Leaves dark green, 3–15cm long. Spikes 3–6cm long, solitary or paired, not enclosed by the upper leaf sheaths; spikelets 5–6-flowered, the lower 3–5 fertile and with a reddish-purple awn; upper glume 5–6mm long, reddish-purple, slightly shorter than the whitish, glabrous or sparsely ciliate lower lemma.

Widespread and frequent on the escarpment from the first foothills in the Tihama at 300m up to 1500m, growing like *T. cenchriformis* on well-drained stony hillsides in open *Acacia* scrub.

Fig. 33 **A** *Tetrapogon villosus*; **B** *Sporobolus africanus*; **C** *Setaria pumila*; **D** *Sporobolus ioclados*.

3. **Tetrapogon villosus** *Desf.*

Densely tufted perennial with erect stems to 50cm, the basal leafy shoots strongly and conspicuously flattened. Leaves glaucous 2–4 (–12)cm long. Spikes 2–9cm long, usually paired but often stuck together, not enclosed by the upper leaf sheaths; spikelets 4–6-powered, all with a white or purple-red awn, the lowest only fertile; upper glume mucronate, 3–4mm long, slightly exceeding the white-villous lemmas.

Widespread and often abundant on dry rocky hills from 600 to 2700m on the escarpment, high plateau and western parts of the Mashriq, though absent from high rainfall areas and avoiding the loose stony slopes favoured by the two previous species. It is particularly characteristic of the dry limestone and sandstone country north and east of Sana'a.

53. FINGERHUTHIA *Nees*

1. **Fingerhuthia africana** *Lehm.*

Densely tufted perennial with erect stems to 30cm. Leaves flat, narrow, up to 6cm long, ligule a densely ciliate rim. Inflorescence a dense, narrowly ovoid, greenish-purple head, 1–3cm long, consisting of numerous subsessile 1-flowered spikelets; glumes ± equal, hyaline, 4–5cm long, with a ciliate keel and prominent dark purple midrib which is extended into a short awn, 2–3mm long; lemma green, terminating in a short point, c 6mm long.

A very distinctive grass found in a number of scattered localities on dry stony hills between 2000 and 2600m on the high plateau and in Khawlan Ash Sham: Rada'; between the Yislah pass and Wa'lān (Wissmann); J. Nuqum (Rathjens); on cliffs south of Wadi Ḍahr; J. Minmār; Wadi 'Azu' (Steinberg).

This is one of a small group of plants found in South Africa and then from the Horn of Africa to Baluchistan.

54. CRYPSIS *Ait.*

1. **Crypsis schoenoides** (*L.*) *Lam.*

Tufted annual, much branched from the base with many prostrate or ascending branches, up to c 20cm high. Leaves flat, up to 10cm long; ligule a ring of hairs. Inflorescence a dense, ovoid or ellipsoid, whitish-green head, 0.5–3cm long, formed of numerous subsessile spikelets and usually partially enclosed by a pair of spathe-like upper leaf sheaths; spikelets 1-flowered, awnless, c 3.5mm long; glumes ± equal, shortly ciliate on the keels, slightly shorter than the lemma.

A rare plant of seasonally flooded mud known only from between 1800 and 2000m by the main cistern in Ḥūth and on the salty plains north of Rada'.

55. UROCHONDRA *C.E. Hubbard*

1. **Urochondra setulosa** (*Trin.*) *C.E. Hubbard*

Densely tufted perennial with short rhizomes and stiff erect stems to 1m. Leaves glaucous, inrolled, up to 30cm long, the apex sharp, pungent, the nerves shortly pilose, the sheaths persistent; ligule a line of hairs. Inflorescence a dense cylindrical spike-like panicle 4–13cm long, the persistent yellow anthers contrasting with the grey spikelets; spikelets 1-flowered, flattened, 2.5–4mm long; glumes ± equal, scabrid, acuminate, about as long as the spikelet; lemma ending in a short mucro.

Distinguished from *Sporobolus* by the beaked grain.

A plant of coastal sand dunes recorded by Bové from Yemen without locality but most probably in the Hodeida area.

56. SPOROBOLUS *R. Br.*

Annuals or perennials with erect stems. Leaves flat or inrolled; ligule a ring of hairs. Inflorescence an open or contracted panicle, ± spicate in a few species; spikelets small, 1-flowered, glabrous, awnless; glumes nerveless or 1-nerved, unequal, shorter than the lemma; lemma usually 1-nerved, shiny; palea similar to the lemma. Grain relatively large, round.

Although *Sporobolus* is easily recognised as a genus, the individual species are often hard to separate and useful taxonomic characters are few. Most species, however, have distinct ecological requirements.

1. Panicle branches whorled. Panicle always open **2**
Panicle branches not whorled. Panicle open or spike-like **5**
2. Annuals **3**
Perennials **4**
3. Spikelets 2–2.7mm long. Lowest panicle branches with 0–2 spikelets only *S. panicoides*
Spikelets 1–1.4mm long. Lowest panicle branches with 2–8 spikelets *S. coromandelianus*
4. Upper glume as long as the lemma *S. ioclados*
Upper glume distinctly shorter than the lemma *S. kentrophyllus*
5. Primary panicle branches bare of spikelets at the base. Panicle open **6**
Primary panicle branches with spikelets at the base. Panicle often spike-like or contracted, rarely open **9**
6. Rhizomatous perennial, basal leaf sheaths neither horny nor fibrous. Leaves flat 2–6mm broad *S. confinis*
Densely tufted perennials, basal leaf sheaths horny or fibrous. Leaves usually inrolled 1–3mm broad **7**
7. Spikelets 1–1.5mm long. Leaf sheaths very fibrous *S. festivus*
Spikelets 1.7–3mm long. Leaf sheaths horny or moderately fibrous **8**
8. Basal leaf sheaths fibrous. Spikelets 2–3mm long. Upper glume slightly more than half as long as the spikelet *S. angustifolius*
Basal leaf sheaths horny. Spikelets 1.7–2.1mm long. Upper glume nearly as long as the spikelet *S. nervosus*
9. Panicle spike-like, usually short, the primary branches all less than 7mm long. Low plants with stiff leaves **10**
Inflorescence clearly paniculate, though sometimes narrow. Some primary branches at least 1cm long. Often vigorous plants commonly exceeding 30cm high **11**

10. Stoloniferous plant with pungent leaf tips. Lower glume about a quarter as long as the whitish spikelet *S. spicatus*
Tufted plant, leaf tips not pungent. Lower glume about half as long as the purplish spikelet *S. tourneuxii*
11. Lower glume about as long as the spikelet. Spikelets usually pale **12**
Lower glume up to three-quarters as long as the spikelet, though often much less. Spikelets usually dark green **13**
12. Stems stout, 3.5–8mm diameter at the base *S. consimilis*
Stems slender, wiry, 1–2mm diameter at the base *S. helvolus*
13. Basal leaf sheaths fibrous. Leaves convolute. Upper glume half to three-quarters as long as the spikelet *S. pellucidus*
Basal leaf sheaths papery or horny, never fibrons. Leaves flat. Upper glume one quarter to one half as long as the spikelet **14**
14. Upper glume obtuse. Panicle branches spreading *S. pyramidalis*
Upper glume acute. Panicle branches erect to spreading **15**
15. Panicle branches erect, up to 2cm long *S. africanus*
Panicle branches ascending to ± spreading, up to 7cm long *S. natalensis*

1. **Sporobolus africanus** (*Poir.*) *Robyns & Tournay*

S. indicus sensu Schwartz

Densely tufted perennial to 1m, the basal leaf sheaths broad and papery. Leaves usually flat, up to 20 (–40)cm long and 4mm broad. Panicle contracted, linear, usually 8–20cm long in Yemen; the primary branches very short, all less than 2cm long, erect, appressed to the main axis, densely covered in spikelets; spikelets dark green, 2–2.8mm long; glumes hyaline, the lower obtuse, a quarter as long as the spikelet, the upper acute, half as long as the spikelet.

A plant of montane grassland found between 2000 and 3200m on the wetter escarpment mountains from Turbah north to J. Maswār, most common on the high Ibb mountains; it also occurs rarely on the high plateau in grassland near ponds at Dhamār and Yarīm.

2. **Sporobolus angustifolius** *A. Rich.*

Densely tufted perennial to 70cm, the basal leaf sheaths becoming fibrous when old. Leaves usually convolute, up to 25cm long and 2mm broad. Panicle open, narrowly ovate, 5–20cm long, the primary branches up to 5cm long, spreading but not whorled, bare of spikelets towards their base; spikelets mostly arranged on one side of the panicle branches, grey-green, 2–3mm long; glumes grey, acute, the lower about one-third and the upper slightly more than half as long as the spikelet.

A local grass of shallow grassland on steep mountain slopes between 900 and 2000m in high rainfall areas of the escarpment, often associated with *Eustachys paspaloides* and *Tripogon* spp: Ibb; near Dalīl on the lower slopes of J. Sumārah; J. Kabūrah, Hadīyah; above Ḥillah, J. Bura'; around Markah, J. Milḥān; below Shamsān, Maḥābishah.

3. **Sporobolus confinis** (*Steud.*) *Chiov.*

Perennial grass to 70cm, not tufted but spreading by long rhizomes, the basal leaf sheaths few, papery. Leaves flat, up to 15cm long and 6mm broad. Panicle open, narrowly ovate, 6–15cm long; the primary branches up to 4cm long, ascending, nor whorled, bare of spikelets towards their base; spikelets mostly on one side of the panicle branches, dark green, 1.6–2.1mm long; glumes grey, acute, the lower half, the upper three-quarters as long as the spikelet.

Abundant between 1300 and 2000m in the Wadi Badr and on J. Rāziḥ, growing on field borders, tracksides and open stony slopes. Unknown elsewhere in Arabia and in Africa a local plant of Ethiopia, Kenya and Tanzania.

4. **Sporobolus consimilis** *Fresen.*

S. robustus sensu Schwartz & Deflers

Robust tussocky grass to 3m, spreading by rhizomes. Leaves flat, scabrid, rather rigid, up to 60cm long and 8mm wide. Panicle narrow, lanceolate, 25–30cm long; the primary branches simple, ascending, spike-like, densely covered in spikelets to the base, 3–8cm long; spikelets appressed to the rhachis of the panicle branches, yellow-green, 1.7–2.3mm long; glumes ± equal, acute, yellow-green, almost as long as the spikelet.

An uncommon plant of widely scattered localities between 500 and 1900m, but always growing in wet places with a high salt level: in saline swamp just north of Suq al Jammah; Wadi Ḥār near Ḥajjaylah (Deflers); Ḥammām Damt; between Bayḍā' and Dhī Nā'im; 20kms north of Rada'; in swamp at 'Ujlī, south of Ṣa'dah.

5. **Sporobolus coromandelianus** (*Retz.*) *Kunth*

Tufted annual to 20cm. Leaves flat, up to 10cm long and 4mm wide. Panicle open, ovate, 2–8cm long, the primary branches spreading, whorled, bare of spikelets for half their length, 1–2cm long; spikelets grey-green, 1–1.4mm long; lower glume a very small hyaline scale, upper glume acute, grey, as long as the spikelet.

Only known from waste ground and tracksides in Ḥajjaylah at 600m. It is probably an introduction from India since Ḥajjaylah was once a major caravansarai and it is odd that it was not found by the earlier botanists who collected in this area.

6. **Sporobolus festivus** *Hochst. ex A. Rich.*

Densely tufted perennial to 50cm, the basal leaf sheaths becoming fibrous when old. Leaves usually convolute, up to 5 (–7)cm long and 2mm wide. Panicle very delicate, open, narrowly ovate, 3–15cm long, the primary branches capillary, spreading but not whorled, bare of spikelets towards the base, up to 3cm long; spikelets grey-green, 1–1.5mm long; glumes greyish with hyaline margins, acute, the lower about one-third and the upper about half as long as the spikelet.

Locally abundant on steep, well-drained but seasonally wet, stony hillsides between 800 and 2300m where it is often associated with *Xerophyta arabica* and *Tripogon* spp. It is found on the escarpment from Shar'ab north to the Wadi Lā'ah but mostly in 'Udayn and the central escarpment where it is common; elsewhere it is only known from Ẓulmah on J. Ḥubaysh, Suq al Jum'ah in J. Ash Shirq and on the escarpment west of Dhī Nā'im.

7. Sporobolus helvolus (*Trin.*) *Dur. & Schinz*
S. glaucifolius (Steud.) Hochst. ex Dur. & Schinz
Tufted perennial with long stolons and a wiry, ascending stem to 1m (not erect as in other species), basal leaf sheaths few, ± papery. Leaves flat, up to 15cm long and 4 (–10)mm broad. Panicle narrow, linear or spicate, 4–12cm long, the primary branches erect, densely covered in spikelets to the base, up to 2cm long; spikelets whitish-green, 1.4–2mm long; glumes ± equal, hyaline, acuminate, about as long as the spikelet.

Widespread in the Tihama from sea level to 600m, growing as a field weed or in open scrub on alluvial plain where protected from grazing by thorn bushes; it appears to be absent from the lighter sandy soils.

8. Sporobolus ioclados (*Trin.*) *Nees*
S. marginatus A. Rich.
S. jemenicus Pilger
Tufted perennial to 80cm, the basal leaf sheaths papery. Leaves flat, up to 30cm long and 5mm wide. Panicle open, ovate, 3–20cm long; the primary branches spreading, whorled, bare of spikelets towards the base, up to 8cm long; spikelets grey-green or pale whitish-green, 1.5–2.2mm long; lower glume grey with hyaline margins, obtuse, about a quarter as long as the spikelet, upper lemmas grey, acute, as long as the spikelet.

Rather variable in general appearance; plants from the Tihama (e.g. Wood 776) having a larger, laxer panicle, often with more pallid spikelets, than plants from the high plateau. It seems very probable that *S. jemenicus* was based on the neat, tidy version of this plant found near Sana'a (e.g. Wood 1971) though I have been unable to trace the type specimens. Certainly the description fits *S. ioclados* well and no other suitable *Sporobolus* has been found in the area of the type collections.

A widespread plant found between 300 and 2400m, growing on heavy alluvial plains and hard banks round fields, sometimes in areas of high salinity. It is of rather local occurrence being mainly found in the alluvial plains in the east of the Tihama, on the high plateau and in dry alluvial basins on the escarpment such as the areas round Al 'Ashshah and Taiz airport.

9. Sporobolus kentrophyllus (*K. Schum.*) *W.D. Clayton*
Very similar to the last species differing mainly in the upper glume which is distinctly shorter than the spikelet. The spikelets are usually ± appressed to the panicle branches and of a distinctive pale grey colour.

Only known from the saline marsh at 'Ujlī south of Ṣa'dah at 1800m.

10. Sporobolus natalensis (*Steud.*) *Dur. & Schinz*
Tufted perennial virtually identical to *S. africanus* except for the open panicle with ascending or spreading primary branches up to 7cm long.

S. natalensis has a similar panicle to *S. pyramidalis* from which it differs in its acute upper glume. Indeed it may be a hybrid between *S. africanus* and *S. pyramidalis*.

Only known from grassy slopes below sandstone cliffs near Shibām at 2500m. (Wood 75/310)

11. Sporobolus nervosus *Hochst.*
Tufted perennial to 30cm, the basal leaf sheaths stiff and horny. Leaves flat or convolute, up to 10cm long and 3mm broad. Panicle open, ovate, 6–18cm long, the primary branches up to 6cm long, ascending, not whorled, bare of spikelets towards the base; spikelets mostly clustered at the tips of the secondary branchlets, pale grey-green, 1.7–2.1mm long; glumes pale grey, acute, the lower about half and the upper about three-quarters as long as the spikelet.

A very rare plant known only from the summit of an old volcanic cone near Shamlān northwest of Sana'a at 2300m.

12. Sporobolus panicoides *A. Rich.*
Loosely tufted annual to 60cm. Leaves flat, up to 30cm long and 6mm broad. Panicle open, narrowly elliptic, 4–15cm long, the primary branches up to 3cm long, simple, capillary, ascending, whorled, each branch bare for most of its length but with 0–4 spikelets at the tip, the lowest branches often entirely barren; spikelets greeny-brown, 2–2.7mm long; glumes acute, olive-green, the lower nearly half as long and the upper as long as the spikelet.

A local plant of stony banks and field borders between 1400 and 2000m in areas of high rainfall on the escarpment but restricted to the south of the country: Rubaysah near Turbah; Qā'idah – Dhī Sufāl; Ibb; 'Udayn; Riḥāb – Mukhādir; between Suq Ar Rūbū' and Dan, Waṣāb al 'Ālī; Suq al Jum'ah, J. Ash Shirq.

13. Sporobolus pellucidus *Hochst.*
Tufted perennial to 70cm, the basal leaf sheaths fibrous. Leaves convolute, up to 15cm long and 2mm wide. Panicle open or contracted, linear to lanceolate, 6–20cm long, the primary branches up to 2cm long, ascending, not whorled, covered in spikelets to the base; spikelets light to dark green, 1.7–2mm long; lower glume hyaline, about a quarter as long as the spikelet; upper glume acute, about three-quarters as long as and similar in colour to the spikelet.

A very local plant restricted to sandy run-offs in sandstone country between 1400 and 2300m: about 30kms east of Rada'; in several places between Shamlān and Wadi Ḍahr; east of Rujum; Al 'Ashshah.

14. Sporobolus pyramidalis *P. Beauv.*
Vigorous tufted perennial to 2m, though often much smaller, the basal leaf sheaths horny to slightly papery. Leaves flat, up to 50cm long and 8mm broad. Panicle open, narrowly pyramidal to ± oblong, 15–50cm long, the primary branches up to 8cm long, spreading, not whorled, covered in spikelets to the base; spikelets dark green, 1.7–2mm long; glumes hyaline, obtuse, the lower a quarter and the upper one-third as long as the spikelet.

Widespread between 1000 and 2500m on the escarpment but mainly in areas of good rainfall. It grows on field borders, banks by streams, grassy slopes, amongst rocks and similar disturbed grassy places where there is a good depth of soil; it is particularly characteristic of abandoned terraces in high rainfall areas where it is one of the most successful colonisers.

15. Sporobolus spicatus (*Vahl*) *Kunth*

Tufted perennial to 20 (–70)cm, spreading by means of looping stolons and sometimes forming mats, basal leaf sheaths papery. Leaves flat or convolute, stiff with pungent spiny tips, up to 10 (–30)cm long and 4mm wide. Panicle cylindrical, spicate, slender, 1.5–10cm, the primary branches up to 3mm long, closely appressed to the axis; spikelets whitish, 1.4–2.2mm long; glumes hyaline, obtuse, the lower a quarter and the upper two-thirds as long as the spikelet.

A local but sometimes abundant grass found from sea level to 2000m on hard clay and alluvial plain made periodically or permanently moist with salty water: in subkhah by the Red Sea between Hodeida and Ṭa'if and between Khawbah and Ṣalīf; in salt swamp north of Suq al Jammah; by streams between Ḥajdah and Ḥays; in the salt lands around Taiz; Ḥammām Damt; on the salt plains around Rada'.

16. Sporobolus tourneuxii *Cosson*

Densely tufted perennial to 20cm, the leaf sheaths hard and very horny. Leaves flat, narrowly triangular in shape, rather stiff, the margin whitish, up to 3cm long and 4mm wide. Panicle cylindrical, spicate, rarely narrowly conical, 1–3cm long, the primary branches up to 7mm long, appressed to the axis; spikelets purple-brown, 1.2–1.8mm long; glumes hyaline, ± equal, acute, about half as long as the spikelet.

An uncommon but easily overlooked plant of stony and sandy plain between 1800 and 2300m in the Mashriq: between Milāḥ and Sanabān; Bukta, 40kms northeast of Dhamār (Hepper); Bani Ḥushaysh; on a limestone hill northwest of Majz (Steinberg).

57. TETRACHAETE *Chiov.*

1. Tetrachaete elionuroides *Chiov.*

Loosely tufted annual with spreading or weakly ascending wiry stems to 20cm. Leaves flat, up to 3cm long; ligule a ring of hairs. Inflorescence sparingly branched with few, long-stalked, whitish-green, plumose spikes; the spikes 1–3.5cm long, partially enclosed by the spathe-like upper leaf sheaths; spikelets 1-flowered, sessile, in pairs which fall together at maturity; glumes bristle-like, ± equal, plumose at the base, 10–17mm long; lemma with an awn 3–4mm long.

An uncommon plant of dry stony plain from 200 to 1300m, mostly in the foothills of the southwestern escarpment: Bani 'Umar, Ḥujarīyah; near the old airport, Taiz; Al Mafraq; near Suq al Kawlah, Wadi Zabid; Ḥajjaylah.

Elsewhere only known from Dhofar and a few places in East Africa.

58. TRAGUS *Hall.*

Annual grasses (our species). Leaves flat, the margin ciliate; ligule a ring of hairs. Inflorescence a stiff, usually dense, spike-like raceme with the spikelets in small, shortly pedunculate clusters along an axis 2–12cm long; spikelets 1-flowered, awnless; glumes dissimilar, the lower reduced to a hyaline scale or absent, the upper as long as the spikelet, densely covered in hooked spines; lemma membranous.

Upper glume 2.5–3mm long. Spikelet clusters with 2 spikelets only — *T. berteronianus*
Upper glume 4mm long. Spikelet clusters with 2–4 fertile spikelets and 1–2 sterile ones — *T. racemosus*

1. Tragus berteronianus *Schult.*

T. alienus sensu Schwartz

Loosely tufted, erect annual to 30cm. Leaves up to c 3cm long. Inflorescence purplish, narrowly cylindrical, 2–5cm long, 3–4mm wide, usually partially enclosed by the sheath of the uppermost leaf; spikelet clusters with 2 fertile spikelets, the rhachis indistinct; upper glume 2.5–3mm long, 5-nerved.

Widely distributed between 500 and 2100m, growing both on the escarpment and in the Mashriq east to Mārib but apparently absent from the high plateau. It grows in scattered localities in stony fields, on roadsides and in bare trodden places.

2. Tragus racemosus (*L.*) *All.*

Tufted annual, the stems often decumbent and rooting at the nodes, finally ascending to 40cm. Leaves up to 4cm long. Inflorescence usually purplish, broadly cylindrical and more lax than in *T. berteronianus*, 5–12cm long, 7–12mm wide, often partially enclosed by the uppermost leaf sheath; spikelet clusters with 2–4 fertile spikelets and 1–2 sterile ones borne on a distinct rhachis, upper glume 4mm long, 7-nerved.

Growing in similar places to *T. berteronianus* but more common, reaching 3000m and occurring on the high plateau as well as on the escarpment and in the Mashriq.

59. LEPTOTHRIUM *Kunth*

1. Leptothrium senegalense (*Kunth*) *W.D. Clayton*

Latipes senegalensis Kunth

Tufted annual or short-lived perennial with erect, wiry stems to 50cm. Leaves flat, narrow, up to 10cm long; ligule a ring of hairs. Inflorescence a lax terminal spike-like raceme with the spikelets solitary or in pairs on short flattened peduncles along a rhachis 2–15cm long; spikelets 1-flowered, dark purple (contrasting strongly with the light green of the peduncle), awnless, lower glume modified into a flat recurved tail, the margins with hooked hairs, 2.5–8mm long, exceeding the rest of the spikelet; upper glume densely covered in spiny tubercles, tightly enclosing the flower.

Widely distributed in a variety of open well-drained habitats from sea level to 2300m: frequent on sand dunes and stony plain in the Tihama; locally frequent in open bushland on the escarpment but mainly in the foothills and in the rainshadow areas of the central escarpment, often associated with *Tetrapogon tenellus* and *Chrysopogon* spp; more local on the high plateau and in the Mashriq, growing on open stony plain, but with a preference for sandstone country.

60. DIGNATHIA *Stapf*

1. Dignathia hirtella *Stapf*

Loosely tufted annual with erect or decumbent stems to 20cm. Leaves flat, narrow, up to 6cm long, ligule a ciliate rim. Inflorescence a terminal spike-like raceme with the spikelets arranged in shortly pedunculate clusters of 2–3 along a rhachis 2–4cm long; spikelets 1-flowered, pubescent, papillose, yellow-green, suborbicular, 5–7mm long; lower glume crescent-shaped; upper glume swollen below, drawn out above into a stiff pointed beak; lemma shorter than the glumes.

A distinctive, but rare species found on stony slopes in open scrub at 600m below Ḥadīyah.

61. PANICUM *L.*

Annuals or perennials. Leaves linear, flat; ligule a short, ciliate membrane. Inflorescence a lax, open panicle. Spikelets pedicelled or sessile, awnless, 2-flowered, the upper flower only fertile; glumes membranous, unequal; lower lemma similar to the upper glume, both ± equalling the spikelet; upper lemma coriaceous.

Although there are few formal characteristics to separate *Panicum* from related genera, it is almost always recognisable by the lax paniculate inflorescence with well-developed secondary branches.

1. Annual. Leaves hairy — *P. atrosanguineum*
   Perennials. Leaves usually glabrous — 2
2. Lower glume nearly as long as the spikelet. Stiff woody desert grass — *P. turgidum*
   Lower glume about a quarter as long as the spikelet or less. Grasses neither stiff nor woody — 3
3. Lower panicle branches whorled — *P. maximum*
   Lower panicle branches not whorled — 4
4. Tufted plant. Lower glume membranous, coloured — *P. coloratum*
   Plant with long rhizomes. Lower glume hyaline — *P. repens*

1. Panicum atrosanguineum *Hochst. ex A. Rich.*

Tufted annual with erect or ascending stems to 30cm. Leaves pilose, up to 10cm long, the sheaths also pilose. Panicle loose and open, the branches not whorled; spikelets all clearly pedicellate, dark purple or greenish, 1.5–2mm long; lower glume herbaceous, about one third as long as the spikelet, upper lemma shiny, brown or blackish.

A local plant of open, periodically disturbed grassy places like field borders and tracksides found in scattered localities on the escarpment from 600 to 1800m north to Shahārah, but more frequent in the south.

2. Panicum coloratum *L.*

Very variable, tufted, shortly rhizomatous or shortly stoloniferous perennial with erect stems to 2m. Leaves glabrous, up to 30cm long, the sheaths glabrous or pilose. Panicle loose and open, often large, 4–30cm long, the branches not whorled; spikelets sessile or pedicelled, green, becoming dark purplish, 2–3mm long; lower glume less than a quarter as long as the spikelet, herbaceous, usually dark purple; upper lemma pale, shiny.

At least two very distinct varieties occur in Yemen and careful investigation may reveal others. Variety **minus** *Stapf ex Chiov.* is a small plant with short stolons from which arise a dense mass of short erect stems to 30cm in height. The dimensions are generally small with leaves up to 10cm long and 5mm wide, the panicle 4–7cm long and the spikelets 2–2.3mm long. This appears to be the common variety on the margins of sandy watercourses in the Tihama and the Mashriq (Wood 1426). Variety **A**, as yet apparently unnamed, occurs in swampy scrub and overgrown thorn fences, often growing with *Dicanthium aristatum* in the Wadi Hiḍrān near Taiz, Wadi Sūdān near Qā'idah and in the valleys north of Ibb (Wood 995, 1789). This is a very vigorous plant up to 2m high, the leaves more than 20cm long and 5–10mm broad with pilose leaf sheaths. The panicle is large with spikelets ± 2.5mm long.

Widely distributed between 300 and 1900m in the east of the Tihama, on the escarpment and in the Mashriq, but rather local; it is a plant of the margins of sandy watercourses in dry country, occurring more rarely on field borders or, in the case of variety *A*, in ± permanently wet scrub on the margins of cultivated land. It is a valuable fodder grass, readily grazed by animals and so usually growing in places at least partially protected from grazing.

3. Panicum maximum *Jacq.*

Vigorous, tufted, rhizomatous perennial with erect stems to 2m. Leaves glabrous or pilose (usually only above the sheaths) up to 40cm long, the sheaths usually ciliate at the mouth and at the nodes. Panicle loose and open, up to 30cm long, the lower branches in whorls; spikelets mostly pedicelled, dark purple, 3–3.5mm long; lower glume less than a quarter as long as the spikelet, herbaceous, purplish; upper lemma pale, shiny, slightly rugose.

Widespread between 600 and 2000m on the escarpment, growing in places protected from grazing such as thorn fences, *Acacia* and *Euphorbia* scrub, rock ledges and the margins of cultivated land. *P. maximum* is a much valued fodder grass which has been planted in many countries and an isolated occurrence of this plant on a roadside at Ma'bar on the high plateau may be a relic of an agricultural experiment.

4. Panicum repens *L.*

P. hygrocharis Steud.

Perennial with long rhizomes and surface stolons; stems erect, up to 60cm high. Leaves rigid, often distichous, glabrous except around the ligule and along the margins of the sheath, up to 15cm long. Inflorescence loose, open or slightly contracted, up to 20cm long, the branches not whorled; spikelets mostly pedicelled, pale, c 2.5mm long; lower glume less than a quarter as long as the spikelet, hyaline; upper lemma pale, smooth, shiny.

Known only from ditches in the coffee plantations at Rajamah near Ḥajjaylah at 850m (Schweinfurth) and at Jahālī near Madīnat Al 'Abīd (Wissmann). There are no modern records.

5. Panicum turgidum *Forssk.* Bukār

Untidily branched perennial with numerous woody, glaucous erect, ascending or spreading stems forming an untidy bush c 1m high. Leaves glabrous, usually less than 5cm long, the sheaths persistent, yellowish, glabrous. Inflorescence a small panicle with a few, ascending (never whorled) branches and few spikelets; spikelets shortly pedicelled, swollen, glaucous, c 4mm long; lower glume hyaline but with 7 distinct green veins, nearly as long as the spikelet; upper lemma smooth, pale, shiny.

Widespread and often abundant on sand dunes in the Tihama, growing from sea level to 300m; also on sand dunes around Mārib at 1100m.

Millet, **Panicum miliaceum** *L.*, was cultivated in the Wadi Surdūd region and elsewhere in the Tihama in the eighteenth century (*fide* Forsskal) but has not been recorded subsequently though it has been grown near Aden much more recently. It is a medium-sized annual cereal grass with roughly hairy leaf sheaths and large panicles.

62. TRICHOLAENA *Schrad.*

1. Tricholaena teneriffae (*L.f.*) *Parl.*

T. leucantha (A. Rich.) Hochst. ex Aschers. & Schweinf.

Densely tufted perennial to 60cm with erect, often brittle stems from a gnarled rootstock. Leaves flat or inrolled, glaucous, up to 15cm long, ligule a ciliate rim. Inflorescence a loose or weakly contracted panicle 4–12cm long; spikelets on delicate, filiform pedicels, silky-hairy with hairs much longer than the spikelet, 2-flowered, awnless, c 3mm long; lower glume hyaline, minute, sometimes absent, separated from the upper glume by a distinct gap; upper glume and lower lemma similar, as long as the spikelet, silky-hairy

Widespread between 600 and 2400m on the escarpment and high plateau, but local, occurring in scattered places on dry open stony hillsides showing a preference for volcanic rocks.

63. RHYNCHELYTRUM *Nees*

Annuals or perennials. Leaves flat, ligule a ciliate rim. Inflorescence an open panicle with capillary branches; spikelets 2-flowered, strongly compressed, gaping, often silky-hairy; lower glume small, separated from the upper glume by a distinct gap; upper glume similar to the lower lemma, as long as the spikelet, swollen in the lower half, 5-nerved, with a straight awn; upper lemma smooth, cartiliginous.

Spikelets glabrous or shortly hairy, 6–11mm long — *R. grandiflorum*
Spikelets villous, 2.5–8mm long — *R. repens*

1. Rhynchelytrum grandiflorum *Hochst.*

R. roseum sensu Schwartz

Tufted annual with erect, usually pubescent stems to 1m. Leaves up to 15cm long, the sheaths pilose. Panicle 5–10 (–15)cm long; spikelets 6–11mm long, yellow-brown, glabrous or pubescent, the pedicels glabrous; lower glume very small, less than 3mm long; upper glume with a long narrow beak, ciliate on the margins with an awn 2–10mm long; lower lemma similar.

Recorded by Schweinfurth at 900m on J. Bura' and by Wissmann at Jahālī near Madīnat Al 'Abīd. Apart from these two old records it is only known from a single collection by L. Gillespie in the Salafīyah region of J. Raymah but it seems probable that it has been confused with the next species. It is worth looking out for on open stony hillsides on the escarpment and may turn out to be quite frequent.

2. Rhynchelytrum repens (*Willd.*) *C.E. Hubbard*

R. villosum (Parl.) Chiov.

Similar to the previous species but sometimes perennial and with longer leaves up to 30cm in length. The panicle is distinctly fluffy with densely villous spikelets which are smaller than in *R. grandiflorum* being 2.5–8mm long.

Widespread and common on the escarpment, growing from 800 to 2000 (–2400)m on shallow rocky soil in poor grassland, in open scrubland and on field borders and cliffs. The fluffy, red-brown flowers are a characteristic feature of open rocky ground at lower altitudes for a short period after the rains.

64. SETARIA *P. Beauv.*

Annuals or perennials. Leaves flat, sometimes broad and plicate; ligule a ciliate rim. Inflorescences an open panicle or a cylindrical spike-like panicle; spikelets oblong or ovate, 2-flowered, glabrous, awnless, solitary or in clusters, surrounded by an involucre of ± scabrid bristles which persist after the spikelets have fallen; glumes unequal, the lower often clasping the base of the spikelet; lower lemma membranous; upper lemma coriaceous, convex with a strongly inrolled margin.

1.	Inflorescence a cylindrical spike	2
	Inflorescence an open panicle	4
2.	Bristles retrorse-barbed, 'clinging', usually dull violet	*S. verticillata*
	Bristles with forward pointing hairs, not 'clinging', usually orange	3
3.	Annual	*S. pumila*
	Perennial	*S. sphacelata*
4.	Leaves with a sagittate base	*S. sagittifolia*
	Leaves narrowed to an entire base	5
5.	Annual. Upper lemma rugose	*S. barbata*
	Perennial. Upper lemma smooth	*S. plicatilis*

1. Setaria barbata (*Lam.*) *Kunth*

Annual with weakly erect or ascending stems to 1m. Leaves light green, sparsely pilose, conspicuously plicate near the base, up to 20cm long and commonly more than 1cm broad. Inflorescence an open panicle up to 15cm long, with numerous short primary branches mostly 2–3cm long and few, very short secondary branchlets; spikelets 2–2.5mm long, subsessile, whitish-green, subtended by a single whitish-green bristle which rarely exceeds 1cm long; upper lemma rugose, shiny.

A local plant of open scrub in high rainfall areas between 500 and 1500m from Wadi Ḍabāb near Taiz north to the Wadi Lā'ah. It is very typical of the lush middle-altitude wadis where it grows in shade in the protection of thorn fences and in rough scrub around irrigated coffee plantations and sorghum fields.

2. **Setaria plicatilis** (*Hochst.*) *Hack.*

Perennial with erect or ascending stems to 1½m. Leaves dark green, scabrid, conspicuously plicate, up to 40cm long and often more than 1cm broad, especially below. Inflorescence a narrow panicle up to 20cm long, the short primary branches erect and lying against the axis; spikelets 3–3.5mm long, subsessile, whitish-green or purplish where exposed to the sun, subtended by a single whitish-green or purplish bristle which rarely exceeds 1cm long; upper lemma smooth, shiny.

A local plant of shaded streamsides, irrigation ditches and damp mountain gullies between 900 and 1900m on the escarpment from Musalla in Aḥkūm north to Wadi 'Ayn 'Alī near Ḥajjah; frequent in Taiz province and the central escarpment but absent from the Ibb region, its distribution perhaps dependent on the presence of winter mist at lowish altitudes.

3. **Setaria pumila** (*Poir.*) *Roem. & Schult.*

S. glauca auct.
S. pallide-fusca (Schumach.) Stapf & Hubbard
Slender, tufted annual with weak, ascending stems to c 60cm. Leaves sparsely pilose or glabrous, up to c 20cm long. Inflorescence a cylindrical spike-like panicle, 1–6cm long; spikelets 1.5–3.5mm long, subsessile, whitish-green, subtended by 5–10 orange bristles, ± 5mm long; upper lemma rugose.

Nearly all Yemen examples of this species are the African form, previously separated as *S. pallide-fusca* which has spikelets 1.5–2.5mm long (e.g. Wood 75/473), but the Mediterranean form with spikelets 3–3.5mm long has been recorded in the experimental farm at Ṣayfarah near Taiz where it may have been introduced with imported seed (Wood 75/9).

Widespread but uncommon, growing in scattered localities between 1100 and 2600m on the escarpment from Taiz north to J. Rāziḥ mainly in high rainfall areas, especially Ibb province; it grows as a field weed and as a colonist of open stony hillsides where there is little competition from other plants.

4. **Setaria sagittifolia** (*A. Rich.*) *Walp.*

Tufted annual with erect stems to 60cm. Leaves with a conspicuously sagittate base, up to 20cm long and commonly more than 1cm broad. Inflorescence an open panicle 5–10cm long with numerous, short, ascending primary branches 1–2.5cm long; secondary branchlets usually absent; spikelets 1–1.5mm long, subsessile, suborbicular, whitish, arranged along one side of the panicle branches, subtended by a single pale brown bristle, ± 5mm long; upper lemma shiny, rugose, gibbous, conspicuously keeled.

An uncommon plant of shaded, stony and gravelly banks by wadis found between 500 and 1500m on the escarpment, usually appearing after heavy summer rains: Bani Miftāḥ, between Riḥāb and Waṣāb; Madīnat Al 'Abīd; below Suq Ar Ribāṭ, J. Raymah; Rajamah near Ḥajjaylah, Ḥarāz; Wadi Sukhnah; between Ḥammām Shamsān and Suq Ath Thalūth, Wadi Lā'ah.

5. **Setaria sphacelata** (*Schumach.*) *Stapf & Hubbard*

S. aurea Hochst ex A.Br.
Tufted perennial with erect stems to 1½m. Leaves narrow, glaucous, up to 40cm long. Inflorescence a dense cylindrical spike-like panicle, 4–12cm long; spikelets 2–3mm long, subsessile, usually purplish, subtended by 6–10 rigid orange bristles, 3–5mm long; upper lemma rugose.

In Yemen there only occurs variety **aurea** (Hochst. ex A. Br.) W.D. Clayton, a tall plant in which the flattened basal leaf sheaths break up and become fibrous.

Abundant on grassy banks and field borders from 1700 to 2100m on the limestone at Maḥwīt, but unknown elsewhere in Yemen.

6. **Setaria verticillata** (*L.*) *P. Beauv.*

S. adhaerens (Forssk.) Chiov.
Weakly tufted annual with decumbent or ascending stems to c 30cm. Leaves usually sparsely pilose, yellow-green, up to 20cm long. Inflorescence a cylindrical spike-like panicle, 2–6cm long, becoming twisted and untidily stuck together at maturity; spikelets c 2mm long, subsessile, whitish-green, subtended by 1–5 dull violet, retrorse-barbed bristles 4–7mm long; upper lemma finely rugose.

Immediately distinguished from all other Setarias and most other grasses by the distinctive 'clinging' bristles.

A widespread weed found in suitable habitats from 600 to 2400m throughout Yemen; it grows on roadsides and banks by irrigation ditches, in wadi beds, in fields and in gardens but does not readily tolerate prolonged drought and is usually found in places which are seasonally wet or irrigated.

65. OPLISMENUS *P. Beauv.*

Weak, trailing perennials of shaded places with short, broad leaves. Inflorescence of slender, one-sided, spike-like racemes arranged along a central axis; spikelets mostly in pairs, 2-flowered; glumes unequal, both terminating in an awn.

The two following species are very closely related and superficially similar.

Spikelets distant, adjacent pairs on the lower racemes, 4–12mm apart. Lowest raceme 0.5–10cm long — *O. compositus*

Spikelets ± contiguous, adjacent pairs on the lower racemes 0.5–4mm apart. Lowest raceme 0.5–3cm long — *O. hirtellus*

1. **Oplismenus compositus** (*L.*) *P. Beauv.*

Perennial with long trailing stems which root at the nodes and weak ascending inflorescences. Leaves broadly lanceolate, up to 12cm long and 2.5cm broad, the sheath strongly striate, the ligule a short, ciliate membrane.

Inflorescence lax; the racemes 5–10, widely spaced along an axis 10–20cm long, the lowest 2.5–10cm long with adjacent spikelet pairs 4–12mm apart; spikelets green, shortly adpressed pubescent and also with a few long ciliate hairs from the base, 3–4mm long; lower glume with an awn up to 3mm long; upper glume with an awn up to 10mm long.

An uncommon plant of shaded woodland and coffee terraces between 600 and 1700m in the central escarpment: in the coffee groves above Ḥadīyah; in dense scrub above Suq Ar Ribāt, J. Raymah; in the valley forest above Suq As Sabt, J. Bura'; in a gully below Ṣa'bah, J. Milḥān. (Wood 576, 1926, 3461)

2. **Oplismenus hirtellus** (*L.*) *P. Beauv.*

Almost identical to the previous species, differing only in the key characters.

Known only from deep shade in coffee plantations at 1500m by the Wadi Dūr, 'Udayn (Hepper and Wood 5988, Wood 1999).

66. PASPALIDIUM *Stapf*

Similar to *Brachiaria* (69) but the inflorescence consisting of many short, sessile, 1-sided, 2-rowed, spike-like racemes which are widely spaced along a long axis, the racemes sometimes shorter than the internodes; the spikelets are glabrous and pressed tightly into hollows in the raceme rhachis which is consequently wavy; the upper lemma is acute. All our species are stoloniferous perennials.

The three following species are very close and superficially similar so care is needed in separating them.

1. Spikelets contiguous, narrowly ovate. Racemes about equalling or longer than the internodes **2**
   Spikelets widely spaced with a distinct gap of 1–2mm between them, broadly ovate. Racemes shorter than the internodes *P. flavidum*
2. Raceme rhachis winged. Stolons spongy *P. geminatum*
   Raceme rhachis triangular in cross-section. Stolons hard, not spongy *P. desertorum*

1. **Paspalidium desertorum** (*A. Rich.*) *Stapf*

Tufted perennial spreading by rhizomes and hard stiff stolons, stems ascending to 40cm. Leaves flat, glaucous, glabrous, up to 20cm long. Racemes solitary (rarely paired) up to 3cm long, the rhachis triangular in cross-section; spikelets contiguous, usually glaucous, narrowly ovate, 2–3mm long.

A local but sometimes abundant plant growing in sandy fields and on banks around cultivation between 100 and 800m on the alluvial plains in the east of the Tihama from the Wadi Zabīd northwards; it also penetrates the escarpment along the broader wadis; apparently most common around Bājil. (Wood 1630)

2. **Paspalidium flavidum** (*Retz.*) *A. Camus*

Tufted perennial with yellow-green leaves similar in general appearance to the last species but readily distinguished by the inflorescence: racemes short, up to 1.5cm long, widely spaced and much shorter than the internodes; spikelets broadly ovate, shiny and swollen in appearance, widely spaced with a gap of 1–2mm, not ± contiguous as in both our other species.

Only known from swampy pools at 300m between Sa'īd Miṣr and 'Alūjah (Wood 1363). This is an Indian species with its most westerly location in Yemen.

3. **Paspalidium geminatum** (*Forssk.*) *Stapf*

Very similar to *P. desertorum* but the stolons and rhizomes spongy in texture, the stem usually creeping, or floating on the surface of water and the leaves green (intermediate in this respect between the other two species). The inflorescence is almost identical to that of *P. desertorum*, but the raceme rhachis is distinctly winged and the spikelets are slightly smaller, rarely exceeding 2mm.

The most common *Paspalidium* in Yemen, growing on mud in or beside permanent water, occurring wherever there are suitable habitats on the escarpment between 400 and 1900m but becoming less common northwards; there is also an isolated record from the Wadi Khārid in the Mashriq at c 2000m. (Wood 75/29, 2950)

67. ECHINOCHLOA *P. Beauv.*

Similar to *Brachiaria* (69) and related genera but distinguished by its cuspidate, almost mucronate glumes. In addition, the spikelets in our species are not arranged in neat rows and form an untidy raceme.

Spikelets awned or acuminate, up to 4mm long. Racemes up to 8cm long *E. crus-galli*
Spikelets acute, up to 3mm long. Racemes up to 3cm long *E. colona*

1. **Echinochloa colona** (*L.*) *Link*

Tufted annual, the stems at first decumbent and rooting at the nodes, finally ascending to 40cm. Leaves flat, usually yellow-green, 5–30cm long. Inflorescence of 4–12, sessile, ascending, spike-like racemes along an axis 3–12cm long; racemes usually solitary, 1–3cm long; spikelets on 1 side of the raceme in 4 rather obscure rows, sessile, greenish-purple, acute, pubescent, 2–3mm long; glumes and lemmas pubescent.

Widespread and often abundant as a field weed, though in dry country restricted to irrigated land; also occurring in wetland habitats such as swamp, irrigation ditches and the margins of pools and streams. It grows throughout the Tihama and on the escarpment up to 1800m, but elsewhere is only known from irrigated fields at Mārib.

2. **Echinochloa crus-galli** (*L.*) *P. Beauv.*

Erect tufted annual to 1m. Leaves flat, up to 30cm long. Inflorescence of 10–20 sessile spike-like racemes along an axis 6–20cm long; racemes 1–3 together, 1–8cm long; spikelets in two untidy rows, shortly pedicellate, usually awned from the lower lemma, hispid, 3–4mm long.

Only known from Wadi Ḥawayat near Ḥajjaylah at 700m in the Ḥarāz where it was found by Deflers; not seen since.

68. ERIOCHLOA *Kunth*

Similar to *Brachiaria* (69), but readily distinguished by the absence of the lower glume and the presence of a thickened bead-like structure immediately below the spikelet.

Annual. Spikelets silky-pilose, terminating in a bristle *E. fatmensis*
Perennial. Spikelets glabrous, acute, but not terminating in a bristle *E. meyeriana*

1. Eriochloa fatmensis (*Hochst. & Steud.*) *W.D. Clayton*

E. nubica (Steud.) Hack. & Stapf ex Thell.
E. acrotricha (Steud.) Hack.
Tuifted annual with numerous ascending or erect stems to 60cm. Leaves flat, up to 15cm long; ligule a dense fringe of hairs. Inflorescence of 4–20 spike-like racemes arranged along an axis 2–8cm long; racemes solitary, nearly sessile, rather lax, 1–3cm long, not obviously 1-sided; spikelets in 2 rows, shortly pedicellate, whitish-green, c 3mm long; upper glume and lower lemma both silky-pilose and terminating in a bristle; upper lemma mucronate.

Very variable in indumentum; the common Yemen plant (Wood 894) is pilose on the spikelets only, whereas other plants, notably Wood 2512 from Al 'Ashshah, are more robust with pubescent leaves and the whole inflorescence densely pilose.

Locally frequent from 300 to 1500m on the escarpment and the alluvial plains in the east of the Tihama, but with a strong southern tendency in its distribution becoming rare north of the Wadi Rima' with only isolated records further north from Khalīfah, Ḥajjaylah (Schweinfurth), Wadi Sukhnah and Al 'Ashshah. It is a plant of wadi beds, sandy depressions where water accumulates in the rainy season, irrigated fields and other seasonally flooded disturbed places.

2. Eriochloa meyeriana (*Nees*) *Pilger*

Panicum meyerianum Nees
Perennial with a decumbent rooting base and erect stems to 2m. Leaves glaucous, flat, up to 20cm long, ligule a ciliate fringe. Inflorescence a rather narrow panicle, formed from 10–20 lax, pedunculate racemes arranged along an axis 4–18cm long; spikelets not obviously on one side of the rhachis nor in rows, pedicellate, whitish-green or purplish, glabrous, c 3mm long; upper glume and lower lemma acute but without a bristle; upper lemma shortly mucronate.

An uncommon plant often associated with *Dicanthium aristatum* and found growing in scrubby vegetation by permanent streams on the escarpment but mostly in the south: Wadi al Ḥammām, Barakānī, Ḥujarīyah; Wadi Ḥidrān, Taiz; Wadi Sūdān, Qā'idah; near Suq As Sabt, Ibb; 'Ajārah near Ḥajjaylah (Schweinfurth).

69. BRACHIARIA (*Trin.*) *Griseb.*

Annuals or perennials. Leaves flat, usually truncate or abruptly rounded at the base, the ligule reduced to a ring of hairs. Inflorescence either of spike-like racemes arranged along a central axis or more obviously paniculate with the spikelets loosely arranged along the panicle branches and sometimes also on small secondary branchlets; spikelets solitary or paired, sometimes arranged along the rhachis in 1 or 2 rows, 2-flowered, ovate or oblong; glumes unequal, the lower shorter than the spikelet and facing the rhachis, the upper similar to the lower lemma; upper lemma obtuse, acute or mucronate.

1. Spikelets in 1 or 2 rows, sessile. Upper lemma quite smooth **2**
   Spikelets irregularly arranged and not forming distinct rows, usually pedicellate. Upper lemma granulate or rugose **3**
2. Perennial. Spikelets in 1 row, ± 5mm long *B. brizantha*
   Annual. Spikelets in 2 rows, ± 2mm long *B. eruciformis*
3. Perennial, up to 2m in height *B. chusqueoides*
   Annual, rarely exceeding 50cm **4**
4. Glumes separated by a distinct internode. Lowest raceme commonly deflexed *B. leersioides*
   Glumes adjacent. Lowest raceme not deflexed **5**
5. Racemes compound with small secondary branches **6**
   Racemes simple, secondary branches absent **7**
6. Spikelets 1.5–2.5mm long, narrowly elliptic, purple *B. comata*
   Spikelets 2.5–3.5mm long, broadly elliptic, green *B. deflexa*
7. Spikelets 1.5–2.2mm long, mostly solitary **8**
   Spikelets 2.5–5mm long, mostly paired **9**
8. Lower glume very small, less than a quarter the length of the spikelet *B. reptans*
   Lower glume one-third – half the length of the spikelet *B. scalaris*
9. Pedicels (or some of them) longer than the spikelets *B. deflexa*
   Pedicels shorter than the spikelets **10**
10. Upper glume and lower lemma thinly cartiliginous, obtuse *B. ovalis*
   Upper glume and lower lemma membranous, acute **11**
11. Pedicels 0.3–0.8mm long *B. lata*
   Pedicels 1–2mm long *B. ramosa*

1. Brachiaria brizantha (*Hochst.*) *Stapf*

Stout, shortly rhizomatous, erect perennial to 1½m. Leaves yellow-green, up to 30cm long and 1.5cm broad. Inflorescence of 2–5 (–8) 1-sided, sessile, spike-like racemes arranged along an axis 5–10cm long; racemes 3–12cm long, the rhachis typically arching, pilose; spikelets solitary, sessile, regularly spaced, ± touching, in a single neat row, usually glabrous, light green with a purple apex, 4–6mm long; lower glume one-third – half as long as the spikelet; upper glume smooth, acute.

A very distinctive but rare species known only from grassy field borders at 1800m between Ibb and Suq As Sabt.

2. **Brachiaria chusqueoides** (*Hack.*) *W.D. Clayton*

Panicum chusqueoides Hack.

Rhizomatous perennial with long leggy stems reaching 2m with the support of dense scrub vegetation. Leaves yellow-green, up to 18cm long and 1cm broad. Inflorescence paniculate, very lax, with about 5 primary branches (each up to 10cm long) along an axis about 20cm long; spikelets solitary or in pairs, usually pedicelled, scattered irregularly along the primary branches or on short secondary branchlets, glabrous, light green, 3–4mm long; lower glume about half as long as the spikelet; upper lemma rugose, obtuse.

Readily recognized by its long leggy growth.

A rare plant restricted to dense scrub around 500–600m in the lush valleys of the central escarpment where it grows with *Chloris mensensis* and *Leptochloa rupestris*; above Suq As Sabt and Ḥillah, J. Bura'; below Ḥadīyah.

3. **Brachiaria comata** (*Hochst.*) *Stapf*

B. kotschyana (Hochst.) Stapf

Tufted annual with ascending or decumbent stems, the latter rooting at the nodes, rarely exceeding 40cm high. Leaves green, up to 9cm long and 1.5cm wide. Inflorescence paniculate with 6–15 short primary branches (up to c 5cm long) along an axis 5–15cm long; spikelets solitary, paired or in small clusters, usually pedicelled, arranged irregularly but rather densely along the primary branches and short secondary branchlets, glabrous or hairy, dark purple-brown, 1.5–2.5mm long; lower glume about one-third as long as the spikelet; upper lemma striate or granulose, acute.

Very variable in spikelet indumentum, some plants being glabrous (e.g. Wood 1783) whereas others are densely pubescent (e.g. Wood 1173), the latter forms being previously separated as *B. kotschyana*.

Easily confused with *B. reptans* and *B. scalaris*; all 3 species have spikelets of similar size, shape and colour though they differ in the size of the lower glume. *B. comata* can be distinguished from both by the presence of short secondary branchlets.

A rare, but probably overlooked, species growing as a field weed and on scrubby field borders in areas of high rainfall between 1300 and 2000m: Wadi Ḍabāb and Wadi Tha'bat near Taiz; J. Ḥubaysh; Wadi Sukhnah.

4. **Brachiaria deflexa** (*Schumach.*) *C.E. Hubbard ex Robyns*

B. regularis (Nees) Stapf

Tufted annual to 50cm with erect stems. Leaves green or glaucous, up to 15cm long and 1cm wide. Inflorescence paniculate with 6–15 primary branches (2–8cm long) along an axis 5–20cm long; spikelets solitary, paired or in small clusters, usually pedicelled, scattered along the primary branches and on small secondary branchlets, usually distant, glabrous or pubescent, green or glaucous, 2.5–3.5mm long; lower glume about one-third as long as the spikelet; upper glume and lower lemma thinly cartiliginous; upper lemma rugose, acute.

Easily confused with *B. lata*, *B. ovalis* and *B. ramosa* but differing from all in the distant spikelets whose pedicels are often longer than the spikelet and in the presence of secondary branchlets.

Widespread and probably common in the alluvial plains in the east of the Tihama and on the escarpment, growing from 200 to 1600m on stony hillsides in open bushland and less commonly in and around cultivated land. (Wood 75/429, 1637, 1732)

5. **Brachiaria eruciformis** (*Sm.*) *Griseb.*

B. isachne (Roth) Stapf

Loosely tufted annual, spreading by stolons, the stem at first decumbent and rooting at the nodes, finally ascending to about 30cm. Leaves grey-green, up to 6cm long and 0.5cm broad. Inflorescence of 5–10 sessile, 1-sided, spike-like racemes arranged along an axis 3–6cm long; racemes 1–2cm long, erect and lying against the axis; spikelets sessile, closely packed and ± imbricate in 2 rows, pubescent, grey-green, 1.5–2.5mm long; lower glume about a quarter as long as the spikelet; upper lemma smooth, obtuse.

A very distinctive species on account of its short, neat, 2-rowed, spike-like racemes.

A local weed of cultivated land on the escarpment and high plateau, usually betweem 1600 and 2500m, but rarely descending to 700m. It seems to be most common on alluvial plains and is thus most typical of places like the Turbah plateau, the Taiz-Qā'idah plain, the Ibb plains and the rich valleys around Sana'a.

6. **Brachiaria lata** (*Schumach.*) *C.E. Hubbard*

Urochloa insculpta (Steud.) Stapf

Vigorous tufted annual to 70cm. Leaves green, up to 15cm long and 2cm wide. Inflorescence of 5–15 sessile, 1-sided, spike-like racemes arranged along an axis 5–15cm long; racemes ascending, 2–5cm long; spikelets shortly pedicellate (pedicels less than 0.8mm long) arranged irregularly but densely along one side of the raceme, solitary or in pairs, usually glabrous, whitish-green, 2.5–4mm long, lower glume about one-third as long as the spikelet; upper glume and lower lemma membranous; upper lemma mucronate, finely granulate.

Virtually identical to *B. ramosa* apart from the key characters.

A rare but probably overlooked plant known only from a few localities on the escarpment between 500 and 1200m where it grows on stony field borders and in open bushland: Ḥajdah; Wadi Rusyān between Ḥajdah and Madrūbah; 'Ajārah near Ḥajjaylah (Schweinfurth); east of Bājil (Miller), Madīnat Al 'Abīd; below Suq Ar Ribāṭ, J. Raymah. (Wood 1843, 1931, 1956, 2074)

7. **Brachiaria leersioides** (*Hochst.*) *Stapf*

Tufted annual to 50cm. Leaves glaucous, up to 15cm long and 5cm wide. Inflorescence of 4–10 remote, shortly pedunculate, 1-sided, spike-like racemes arranged along an axis 5–20cm long; racemes 2–5cm long, mostly spreading but the lower 1 or 2 usually deflexed; spikelets shortly pedicellate, solitary or in pairs, irregularly arranged and rather distant from each other, glabrous, almost white, c 3mm long; lower glume about one-third as long as the spikelet from the rest of which it is separated by a distinct internode; upper lemma acute, finely granulate.

Fig. 34 **A** *Bracharia reptans*; **B** *Digitaria abyssinica*; **C** *Odontelytrum abyssinicum*; **D** *Chrysopogon plumulosus*.

A distinct species on account of its glaucous colour, deflexed lower racemes and the internode above the lower glume.

Widespread and often abundant on shallow stony soils in open scrub on the escarpment from 500 to 1500m.

8. **Brachiaria ovalis** *Stapf*

Very similar to *B. deflexa*, differing principally in the absence of secondary branchlets, the spikelets being arranged mostly in pairs along the primary panicle branches.

A rare plant found on dry stony hills in open *Acacia* scrub in the southwest foothills of the escarpment: between Hajdah and Madrūbah in Maqbanah; between Mishrāfah and Masbaḥ in Waṣāb As Ṣafal. (Wood 2073, 3175.)

9. **Brachiaria ramosa** (*L.*) *Stapf*

Very similar to *B. lata*, differing in the longer spikelet pedicels which are 1–2mm long and the consequent laxer appearance of the inflorescence.

A rare but almost certainly overlooked species of sand dunes, field borders and stony ground in open *Acacia* scrub in the Tihama and the escarpment foothills from 200 to 600m: Mishrāfah in the Wadi Rima'; below Ḥadīyah; Bayt al Faqīh; Bājil. (Wood 75/534, 1631, 2095; Hepper 5745)

10. **Brachiaria reptans** (*L.*) *Gardn. & C.E. Hubbard*

Urochloa reptans (L.) Stapf

Tufted annual, the stems often at first decumbent and rooting at the nodes but eventually ascending to 40cm. Leaves green, up to 6cm long and 1cm wide. Inflorescence of 2–12, sessile, 1-sided, spike-like racemes arranged along a short central axis 1–6cm long; racemes ascending, 1–3cm long; spikelets shortly pedicellate, densely but irregularly arranged, glabrous, dark purple-brown, 1.5–2.5mm long; lower glume a quarter as long as the spikelet; upper lemma granulose, mucronate.

Widespread and often abundant on the alluvial plains in the east of the Tihama and on the lower slopes of the escarpment, growing from 100 to 1500m in a great variety of habitats including sand dunes, cultivated land, roadsides and irrigation ditches but only appearing after heavy rain or in the vicinity of semi-permanent water. It is particularly common in the rich Tihama sorghum fields below the central escarpment around Al Kadān, Khalīfah and Sa'īd Miṣr. (Wood 74/409, 1388, 1442)

11. **Brachiaria scalaris** (*Mez.*) *Pilger*

B. distichophylla sensu Hepper

Very similar to the previous species, differing principally in the lower glume which is one-third – half as long as the spikelet. It is also similar to *B. comata* but there are no secondary branchlets.

Only known from the Wadi Ḥadabah, 'Udayn at 1500m. (Wood 904; Hepper and Wood 5976)

70. UROCHLOA *P. Beauv.*

Very similar to *Brachiaria* (69) but differing in the lower glume which is on the side of the spikelet facing away from the rhachis. In addition, the upper lemma is always mucronate and in our species the spikelets are always in 2 neat rows on one side of the raceme rhachis.

Lower glume up to half as long as the spikelet *U. panicoides*
Lower glume about three-quarters as long as the spikelet *U. trichopus*

1. **Urochloa panicoides** *P. Beauv.*

U. helopus (Trin.) Stapf

Loosely tufted annual herb, spreading by stolons, the stems at first decumbent and rooting at the nodes, finally ± erect, reaching 60cm. Leaves broad, flat, up to 15cm long. Inflorescence of 2–7 spike-like racemes arranged along an axis 1–9cm long; racemes solitary, shortly pedunculate, 1–4cm long, spikelets in 2 rows on one side of the raceme only, sessile, crowded, flattened, whitish-green, glabrous, c 3mm long; lower glume obtuse, less than half as long as the spikelet.

A local but probably overlooked plant of field borders, thorn fences and scrub in areas of fairly good rainfall from 500 to 1700m on the escarpment: Ḥajdah; between Taiz and Rawnah, Shar'ab; 'Udayn; Mukhādir; Bani Miftāḥ between Waṣāb and Riḥāb; below Ḥadīyah; Ḥajjaylah (Schweinfurth) and Mawsanah in the Ḥarāz. (Wood 1274)

2. **Urochloa trichopus** (*Hochst.*) *Stapf*

Erect, tufted, coarse annual to 80cm. Leaves broad, flat, up to 15cm long. Inflorescence of 3–20 spike-like racemes arranged along an axis 4–20cm long; racemes solitary, sessile, 1–5cm long; spikelets in 2 rows on one side of the raceme only, crowded, flattened, greenish-brown, 3–4mm long; lower glume acute, nearly as long as the spikelet; upper glume and lower lemma villous.

A characteristic weed of fields in the alluvial plains in the east of the Tihama, which in dry areas extends into the escarpment to reach the Wadi Badr, Al 'Ashshah, Madīnat Al 'Abīd and the Taiz area at about 1300m.

71. PASPALUM *L.*

Perennial grasses of wet places. Leaves flat; ligule short, membranous, truncate. Inflorescence of 1–20 spike-like racemes arranged along a short axis, often in pairs and appearing digitate; spikelets in 2 rows along one side of the rhachis only, similar in structure to *Brachiaria* (69), but the lower glume absent and the upper lemma always acute.

1. Racemes 1–5 (–20), 3–6cm long *P. scrobiculatum*
   Racemes usually 2, less than 3cm long 2
2. Racemes mostly 2–3cm long. Upper glume pubescent *P. paspalodes*
   Racemes mostly 1–1.5cm long. Upper glume glabrous *P. vaginatum*

1. Paspalum paspalodes (*Michx.*) *Scribn.*
P. distichum auct.
Rhizomatous and stoloniferous perennial, the stem at first decumbent, finally erect, reaching 30cm. Leaves dark green, 3–5 (–10)cm long. Inflorescence a sub-digitate pair of ascending racemes, each 1–3cm long; spikelets acute, whitish-green with a minutely pubescent upper glume and prominent black stigmas.

An introduction from America now thoroughly naturalised in swampy ditches between 1500 and 1700m in the Wadi Ḥadabah, 'Udayn and near Suq As Sabt, Ibb.

2. Paspalum scrobiculatum *L.*
P. commersonii Lam.
P. orbiculare Forst.
Tufted perennial with ascending stems to c 20cm in Yemen though much more in other countries. Leaves light green, often becoming bronze when old, 3–10cm long. Inflorescence of 1–5 (–20) ascending racemes arranged along an axis up to 5cm long; racemes 3–6cm long; spikelets rounded, glabrous, the upper lemma brown.

A rare plant found by streams and irrigation ditches between 600 and 1900m: in the wadi above 'Adūf near Turbah; Suq Ar Rūbū', Waṣāb As Sāfal; Khamīs Banī Sa'd (Miller).

3. Paspalum vaginatum *Sw.*
Similar to *P. paspalodes*, but smaller in all parts and rarely higher than 10cm. Leaves yellow-green, 1–3 (–5)cm long. Inflorescence a sub-digitate pair of spreading racemes, each 1–1.5cm long; spikelets acute, whitish-green, completely glabrous.

A very local, but sometimes abundant sward-forming grass of salt marsh found from sea level to 2000m: in coastal salt marsh at Khawkah and north to Hodeida; in salt marsh near Taiz airport; salt plains north of Rada'.

72. DIGITARIA *Heist. ex Fabr.*

Annual or perennial grasses. Leaves flat; ligule short, membranous or scarious. Inflorescence of long, slender, spike-like racemes, arranged digitately at the top of the stem often with one or two racemes below or, alternatively the racemes solitary, paired or in small whorls along a short axis; spikelets sessile or shortly pedicellate, usually elliptic, c 2mm long, 2-flowered, awnless, in groups of 1–5 which lie against the often wavy and/or winged rhachis; glumes very unequal, the lower sometimes absent, the upper usually hairy.

A very complex genus in tropical Africa, but in Yemen with rather few, clear-cut species.

1. Spikelet groups remote, mostly more than 1cm apart, the lower half of the rhachis plumose, without spikelets — *D. pennata*
   Spikelets lying close to each other, extending to the base of the rhachis — 2
2. Inflorescence a digitate whorl of 3–6 (–10) racemes (rarely a second whorl also formed on an extension of the inflorescence axis above the first whorl or a single raceme present on the axis below). Plant annual — 3
   Inflorescence of 3–20 racemes arranged along a short axis; racemes mostly solitary but some often in whorls of 2–4, never in a single digitate cluster. Annuals or perennials — 4
3. Spikelets in groups of 3 (–5). Plant tufted, not rooting at the nodes — *D. ternata*
   Spikelets in groups of 2. Plant creeping and rooting at the nodes, not tufted — *D. ciliaris*
4. Annual with decumbent stems rooting at the nodes. Leaves pilose — *D. velutina*
   Erect perennials. Leaves ± glabrous — 5
5. Inflorescence completely glabrous. Racemes few, c 5, the spikelets dark purple — *D. abyssinica*
   Inflorescence hairy. Racemes usually more than 8; the spikelets not purplish — 6
6. Pedicels with long white hairs which overtop the glabrous, brown spikelets — *D. diagonalis*
   Pedicels lacking long white hairs; spikelets pubescent, whitish-green — *D. nodosa*

1. Digitaria abyssinica (*Hochst. ex A. Rich.*) *Stapf*
D. vestita Fig. & De Not.
D. scalarum (Schweinf.) Chiov.
Perennial, spreading by rhizomes to form small patches and with short erect stems to c 30cm. Leaves glaucous, usually glabrous, 5–10cm long. Inflorescence of 2–6 (–9) ascending, solitary or whorled racemes on a short axis 2–5cm long; racemes sessile, 2–6cm long; spikelets in pairs, dark purple, glabrous; lower glume present.

Widespread and frequent from 1800 to 2800m on the escarpment and high plateau but with a strong preference for rich soils and thus usually found in and around cultivated land, often where water accumulates in the rainy season; it is also found by streams and sewage outlets in villages.

2. Digitaria ciliaris (*Retz.*) *Koel.*
D. marginata Link
Annual, at first decumbent and rooting at the nodes, finally ascending to c 40cm. Leaves dark green, glabrous or sparsely pilose, 5–15cm long. Inflorescence of 3–6 (–10) racemes in a digitate cluster (sometimes a second whorl formed on an extension of the axis above the first whorl); racemes sessile, 5–15cm long, spikelets solitary or in pairs, whitish-green, adpressed-villous when young, but prominently spreadingly villous when mature; lower glume absent.

A common weed of fields in the alluvial plains in the east of the Tihama and on the lower slopes of the escarpment, becoming progressively less common with increased altitude but reaching 1500m.

3. Digitaria diagonalis (*Nees*) *Stapf*
Erect, tufted perennial to 1m, the stem base slightly swollen and surrounded by the remains of dead leaves. Leaves green, glabrous, 10–30cm long. Inflorescence of 10–25 ascending solitary or whorled racemes on an axis 5–12cm

long; racemes sessile, 1–7cm long; spikelets overtopped by the long white hairs on their pedicels, paired, brown, glabrous; lower glume absent.

An uncommon but very characteristic species of the grassland areas of the central escarpment found between 1500 and 2200m: J. Kabūrah and J. Ẓallamlam above Ḥadīyah; between Bani Dawān and al Jabin, J. Raymah; near 'Ukaybir, J. Milḥān; Maḥwīt.

4. Digitaria nodosa *Parl.*

Erect tufted perennial to 1m, the stem base swollen and bulb-like. Leaves slightly glaucous, usually glabrous, 5–25cm long. Inflorescence of 5–15 ascending, whorled or solitary racemes on an axis 2–10cm long; racemes sessile, 2–15cm long; spikelets in pairs, whitish-green, silky-pubescent; lower glume present.

Widespread and common on dry stony hills from 600 to 2400m on the escarpment and high plateau extending eastwards into the Mashriq to the Ṣa'dah-Nihm-Bayḍā' line, but decreasing in frequency southwards and unknown south of the Wadi Rima' apart from isolated records from Al Masbaḥ in Waṣāb As Sāfal and Bīr Ghubāḥ in Shar'ab.

5. Digitaria pennata (*Hochst.*) *Cooke*

Tufted rhizomatous perennial with a swollen bulb-like stem base and long, weak ascending stems reaching 1½m with the support of surrounding vegetation. Leaves glaucous, glabrous, 2–10cm long. Inflorescence of 6–10 ascending racemes arranged in a whorl, sometimes with one or two additional racemes from the axis above the whorl; racemes without spikelets in the lower half, plumose, 5–15cm long; spikelets solitary or in pairs, remote and usually more than 1cm apart, whitish-green, glabrous; lower glume absent.

Widespread on the escarpment from 500 to 1400m but rather local and usually in small quantity growing in the protection of thorn bushes in dry *Acacia-Commiphora* scrub where it is associated with *Vernonia cinerascens*, *Corbichonia decumbens* and *Commicarpus helenae*.

6. Digitaria ternata (*A. Rich.*) *Stapf.*

Tufted annual herb with decumbent or erect stems to 50cm. Leaves green, glabrous, 2–18cm long. Inflorescence of 3–5 (–7) racemes in a digitate cluster, sometimes with a single raceme on the axis below; racemes sessile, 3–20cm long; spikelets usually in threes, whitish-green but with a prominent black upper lemma, pubescent; lower glume absent.

Local on the escarpment between 1200 and 2300m, growing in damp, shaded, often disturbed places such as tracksides, field borders and scrubby irrigation ditches where it is often associated with *Pennisetum unisetum*; frequent in the high rainfall areas of Taiz and Ibb provinces, becoming rarer northwards and not recorded north of the Wadi Lā'ah.

7. Digitaria velutina (*Forssk.*) *P. Beauv.*

D. horizontalis sensu Schwartz

Low annual with decumbent stems rooting at the nodes and ascending inflorescences. Leaves light green, pilose, 2–6cm long. Inflorescence of 8–15 spreading, solitary or paired racemes on an axis 3–6cm long; racemes sessile, 3–8cm long; spikelets in pairs, whitish-green, glabrous or with a few hairs; lower glume absent.

Widespread on the escarpment from 600 to 2200m but only common in high rainfall areas. It is usually found in partial shade where there is little competition from other plants and is thus found under bushes, in *Acacia* and *Euphorbia* scrub and as a weed of coffee and qat terraces.

73. ANTHEPHORA *Schreb.*

Tufted perennials with flat leaves; ligule membranous. Inflorescence a terminal cylindrical, spike-like raceme with a wavy rhachis; spikelets in clusters of 3–11, each cluster enclosed by an involucre formed by the scale-like lower glumes which are united at the base; spikelets 2-flowered, one sterile, one fertile.

Inflorescence densely hairy. Lower glume broadest below the middle — *A. pubescens*
Inflorescence sparsely hairy. Lowest glume broadest at the middle — *A. nigritana*

1. Anthephora nigritana *Stapf & Hubbard*

Tufted perennial to 1m. Leaves glaucous, up to 30cm long; ligule 2–3mm long. Inflorescence narrowly cylindrical, spike-like, shiny, straw-coloured, sparsely pilose, 8–16cm long; spikelets in clusters of 3–10, 4–6mm long, the spikelet cluster on a short pubescent peduncle; lower glumes (forming the involucre) elliptic (broadest in the middle), acute.

Frequent on open sandstone hills from 1800 to 2200m between Rujum and Ṭawīlah.

2. Anthephora pubescens *Nees*

A. hochstetteri Nees ex Hochst.

Very similar in habit to the last species, differing in the densely hairy inflorescence which frequently becomes blackish and in the lower glumes which are broadest below the middle and end in a long acuminate point.

Collected by Rathjens on J. Ḥadīdah and not seen again. Unfortunately I have been unable to trace the whereabouts of J. Hadīdah and so have been unable to search for the plant. There must remain some doubt as to whether *A. pubescens* really occurs in Yemen until the original specimen is seen. *A. nigritana* and *A. pubescens* were not regarded as separate species until the 1930s and it is uncertain whether Pilger, who identified Rathjens' grasses, regarded the two as distinct at that time.

74. ODONTELYTRUM *Hack.*

1. Odontelytrum abyssinicum *Hack.*

Perennial, growing in mud at the margins of pools, stems decumbent, purplish, rooting at the nodes and often spreading onto the surface of the water. Leaves flat, dark-green with a pale midrib, up to 20cm long, the sheaths purplish, inflated; ligule a very short, truncate membrane. Inflorescence a solitary terminal spike-like raceme, 4–12cm long, partially enclosed by the uppermost leaf sheath; spikelets shortly pedicelled, 2-flowered, glabrous, 10–14mm long, tightly enclosed by a 4–6-lobed herbaceous involucre, one of whose lobes is awn-like and up to 25mm long; involucre purplish, about two-thirds as long as the spikelet with which it falls; glumes very unequal, the lower scale-like or absent; lemmas whitish-green.

Abundant in and around pools between 2500 and 3100m on the Sumārah-Manār ridge and around the headwaters of the Wadi Banā near Riḥāt and Yarīm; unknown elsewhere apart from an isolated occurrence in a pond in 'Anis between Ḍawrān and Suq al Jum'ah.

Outside Yemen it is a rare plant known only from Ethiopia and a single locality in Tanzania.

75. SNOWDENIA *C.E. Hubbard*

Annuals. Leaves flat; ligule membranous. Inflorescence of cylindrical spike-like racemes borne on slender peduncles in the axils of the upper leaves; spikelets lanceolate, 2-flowered, the upper fertile, the lower sterile; glumes reduced to small hyaline scales; lower lemma awned; upper lemma membranous.

Leaf sheaths smooth. Lower lemma 2–3.3mm long, acute, obtuse or bidentate *S. polystachya*
Leaf sheaths scabrid. Lower lemma 3.2–4mm long, acuminate *S. petitiana*

1. Snowdenia petitiana (*A. Rich.*) *C.E. Hubbard*

Weak, straggling annual with ascending stems to 50cm. Leaves dark green, up to 13cm long, the leaf sheaths rather loose, slightly scabrid. Racemes 3–5cm long; spikelets green; lower lemma 3.2–4mm long, acuminate with a short awn 2–5mm long; upper lemma acute, with a short mucronate point.

Rather similar to *Pennisetum unisetum* but annual and lacking the involucral bristle found in that species.

A rare plant of loose rocky slopes in partial shade, found between 1900 and 2300m at Ḥaddah (Miller) and Shahārah.

2. Snowdenia polystachya (*Fresen.*) *Pilg.*

Beckera polystachya Fresen.

Densely tufted annual, the stems numerous, at first decumbent, finally erect to 50cm. Leaves dark green, up to 30cm long, the leaf sheaths smooth, rather loose. Racemes 2–5cm long; spikelets whitish-green; lower lemma 2–3.3mm long, the apex acute, obtuse or mucronate, with a short awn 1–7mm long; upper lemma obtuse.

Apart from two collections in the 1930s by Rathjens in Sana'a, this is entirely restricted to the escarpment from J. Ṣabir and J. Ḥabashī north to the Ḥarāz and Banī Maṭar. It is a local plant of roadsides, ditches, waste ground and field borders, found from 2200 to 3000m in areas of high rainfall.

It is a well-reputed fodder grass which has been introduced into a number of countries but is probably native in Yemen.

76. PENNISETUM *L. Rich.*

Perennials or annuals, mostly tufted. Leaves flat or convolute; ligule a ring of hairs or more rarely membranous. Inflorescence of terminal and axillary, cylindrical, spike-like panicles. Spikelets lanceolate or oblong, glabrous, 2-flowered, awnless, solitary or in small clusters, the spikelet clusters surrounded by an involucre of 1–many, slender, free bristles; glumes unequal; lower lemma membranous, upper lemma membranous or coriaceous.

1. Panicles terminal and axillary 2
   Panicles terminal only 3
2. Involucre of 15–30 bristles. Spikelets 4–8mm long *P. glaucifolium*
   Involucre of a single bristle. Spikelets 2–3mm long *P. unisetum*
3. Involucral bristles ciliate or plumose 4
   Involucral bristles glabrous 5
4. Inflorescence 2–5cm long, white. Lower glume rounded, scale-like, less than a quarter the length of the spikelet. Lower lemma 7–11-nerved *P. villosum*
   Inflorescence (5–) 10–20cm long, purplish. Lower glume acute, half as long as the spikelet. Lower lemma 3-nerved *P. setaceum*
5. Lower glume more than half as long as the spikelet. Clump-forming desert grass with stiff, woody stems and persistent inflated leaf sheaths *P. divisum*
   Lower glume less than half as long as the spikelet. Stems not woody except below; persistent leaf sheaths absent 6
6. Ligule membranous. Plant of dry places *P. stramineum*
   Ligule a ring of hairs. Plant of wet places 7
7. Lower lemma mucronate, less than three-quarters the length of the spikelet. Spikelets usually green *P. thunbergii*
   Lower lemma acuminate, more than three-quarters the length of the spikelet. Spikelets usually purplish *P. macruorum*

1. Pennisetum divisum (*J.F. Gmel.*) *Henr.* Thamām

P. dichotomum (Forssk.) Del.

Much-branched, tufted perennial growing in clumps with stiff, woody, glaucous, erect or geniculately ascending stems to 1½m. Leaves stiff, glaucous, inrolled or flat, 3–10cm long; ligule a shortly-hairy rim. Inflorescence a solitary, terminal, whitish, spike-like panicle, 4–10cm long; spikelets 6–7mm long, usually solitary, surrounded by an involucre of numerous unequal scabrid bristles up to 10mm long; lower glume about three-quarters as long as the spikelet.

Readily distinguished from other *Pennisetum* species by its stiff, woody stems, numerous swollen straw-coloured empty leaf sheaths and desert habitat but superficially very similar to *Lasiurus scindicus* from which it can be distinguished by the presence of involucral bristles.

Widely distributed by sandy watercourses between 1000 and 1900m in the Mashriq from the Wadi Nuwāsh and Ḥarīb northwards; I have no records north of the Jawf but it probably occurs there. There are a number of records from the Tihama but I have seen no specimens from that region and they are probably errors for *Lasiurus scindicus*.

P. divisum is a plant with a characteristic Sahara-Sindian distribution.

2. Pennisetum glaucophyllum *Hochst. ex A. Rich.*
Tufted perennial to 1m. Leaves flat, up to 35cm long; ligule membranous up to 3mm long. Inflorescence of a terminal spike-like panicle with 1–2 pedunculate axillary panicles below; panicles 4–10cm long, whitish; spikelets 4–7mm long, usually solitary, surrounded by an involucre of numerous bristles; lower glume ovate, obtuse, less than a quarter as long as the spikelet.

Known only at 2700m in the Wadi Shurfah on the north side of J. Shibām, Manākhah, where it was collected by Schweinfurth (No. 1949). I have not seen the plant collected by Schweinfurth, but suspect it is *P. stramineum* which differs from *P. glaucophyllum* mainly in having a solitary terminal panicle and slightly smaller spikelets. The inflorescence of both species is very similar in appearance, particularly when that of *P. glaucophyllum* is reduced to a single spike and, since *P. stramineum* was not described in Schweinfurth's days, he would not have tried to distinguish the two species. *P. stramineum* is frequent at Manākhah. *P. glaucophyllum* is a rare and little-known species of a few localities in Ethiopia.

3. Pennisetum macruorum *Trin.*
P. stenorrhachis Stapf & Hubbard
P. quartinianum A. Rich.
Rhizomatous perennial forming colonies by water, stems rather soft, decumbent or erect to 2m. Leaves flat or convolute, up to 40cm long; ligule a line of hairs. Inflorescence a solitary, terminal, purplish, spike-like panicle 6–15cm long; spikelets 2–5mm long, solitary, surrounded by an involucre of numerous unequal scabrid bristles 5–20mm long; lower glume about a quarter as long as the spikelet; lower lemma three-quarters as long as the spikelet.

Rather variable in size; the most vigorous forms with tall stems, long panicles and bristles 2cm long were formerly separated as *P. quartinianum* which is recorded from Manākhah (Schweinfurth 1699).

Widespread by ponds and streams from 2000 to 3000m from Turbah north to Khawlan Ash Sham but local and mostly along the inner escarpment.

4. Pennisetum setaceum (*Forssk.*) *Chiov.*
P. rueppellii Steud.
Densely tufted perennial with stiff, erect stems to 1m. Leaves flat, narrow, up to 30cm long; ligule a ciliate rim. Inflorescence a solitary, terminal, purplish, spike-like panicle, 4–18cm long, the rhachis pilose; spikelets 6–8mm long, solitary or in clusters of 2–3 on a short hairy peduncle, surrounded by an involucre of numerous, unequal, plumose bristles up to 40mm long; glumes acuminate, the lower half as long as the spikelet.

Plants from damp habitats (e.g. Wood 1826 from Ibb) develop rhizomes and have almost glabrous bristles, though in other respects they seem identical to *P. setaceum*. They are probably adaptations to a different environment but deserve investigation.

Abundant on open stony hillsides from 600 to 3000m on the escarpment, high plateau and southeastern mountains; it is one of the first plants to colonise roadside banks, rockfalls and stone debris by quarries and is also abundant on heavily eroded hillsides.

5. Pennisetum stramineum *Peter*
P. laxum sensu Schwartz?
Erect, much-branched perennial to 1m, the stems stiff and woody below. Leaves flat, glaucous, up to 12cm long; ligule a short lacerate membrane, c 1.5mm long. Inflorescence a solitary, terminal, whitish, spike-like panicle, 3–8cm long; spikelets c 5mm long, solitary, surrounded by an involucre of numerous unequal scabrid bristles up to 10mm long; lower glume less than a quarter as long as the spikelet, ovate, acute.

Widespread and frequent between 1800 and 2800m on cliffs and terrace walls on the high plateau extending eastwards into the Mashriq in Arḥab, Nihm and east of Rada' and westwards onto the escarpment at Shahārah, Ṭawīlah and Manākhah.

6. Pennisetum thunbergii *Kunth*
P. glabrum Steud.
Loosely tufted, rhizomatous perennial with soft, ascending stems, to 1m. Leaves flat, linear, up to 40cm long, ligule a ciliate rim. Inflorescence a solitary terminal, whitish (or dark violet), spike-like panicle 2–15cm long; spikelets 2.5–5mm long, solitary, surrounded by an involucre of numerous, unequal, glabrous, bristles up to c 10mm long; lower glume a quarter as long as the spikelet; lower lemma ± half as long as the spikelet, distinctly mucronate.

Widespread in marsh and by streams and ponds from 2000 to 3100m, often growing with *P. macruorum*, from which it can be distinguished by its small mucronate lower lemma. It is, however, much more characteristic of the outer escarpment and is particularly common in the Ibb region; it is rare on the high plateau where it is mainly found around Yarīm; unknown north of Bayt 'Adhāqah.

7. Pennisetum unisetum (*Nees*) *Benth.*
Beckeropsis uniseta (Nees) K. Schum
B. nubica sensu Hepper
Rhizomatous perennial with long weak stems scrambling amongst scrub vegetation to 3m. Leaves flat, up to 40cm, the lower leaves narrowed into a long petiole-like base; ligule a ciliate rim. Inflorescence of slender, purplish, terminal and axillary, pedunculate spike-like panicles 1–4cm long; spikelets 2–3mm long, solitary with an involucre of a single bristle, 5–10mm long (though more outside Yemen); lower glume a short rounded scale.

Might be confused with *Snowdenia petitiana* but that species has awned spikelets and no involucral bristle.

Widespread on the escarpment from 1800 to 2600m but restricted to areas of good rainfall. It grows in damp, scrubby places such as streamsides, damp gullies, and overgrown field borders and is often associated with *Senecio hadiensis*, *Maesa lanceolata* and *Acanthus arboreus*. It is most common in the south, particularly in the Ibb region, and thins out northwards, being surprisingly rare in the central escarpment.

8. Pennisetum villosum *R. Br. ex Fresen.*
P. yemense Deflers
Tufted perennial spreading by rhizomes and with short ascending stems to 30cm. Leaves flat or inrolled, up to 15cm long, very hairy at the mouth of the sheaths; ligule a ciliate rim. Inflorescence a short, solitary, terminal, whitish, spike-like panicle, 2–5cm long; spikelets 8–10mm long, solitary or in clusters of 2–4, subsessile, surrounded by an involucre of numerous long plumose bristles up to 50mm long; lower glume scale-like, rounded, less than quarter as long as the spikelet; lower lemma 7–11-nerved.

Widespread and often abundant between 1600 and 3000m on the escarpment and high plateau. The fluffy white inflorescences are a characteristic sight on banks around fields and along irrigation channels; it has a marked preference for places with deep soil and is rare on rocky eroded slopes.

Bulrush Millet, **Pennisetum glaucum** (*L.*) *R. Br.* (*P. typhoides* (Burm. f.) Stapf & Hubbard, *P. americanum* (L.) K. Schum.), known as 'Dukhn', is commonly cultivated in the Tihama and more rarely on the escarpment up to 1500m. It is a tall cereal grass, branched at the base with several stiff, erect stems to 2½m and an inflorescence about 30–50cm long. The seeds are used mostly to make a heavy dark brown bread which stays fresh longer than most other breads produced in Yemen. It is the staple food of many of the poorer Tihama people.

Bulrush Millet is planted in irrigated fields and areas with good rainfall but Sorghum or Maize are preferred in such places. It is, however, the most drought-resistant of all cereals in Yemen (and perhaps in the world) and is the favoured crop for planting in marginal land in the Tihama since it will thrive on a small rainfall insufficient for other cereals. In years of reasonable rainfall in the Tihama, huge areas of sand dunes around Bayt al Faqīh, Qutay, Qanāwis and other places are covered with Bulrush Millet and a large crop is harvested. In years of poor rainfall, the dunes are barren. The seeds are planted to coincide with the onset of the rains in the Tihama in late summer and the crop is harvested in winter.

Pennisetum clandestinum *Hochst. ex Chiov.*, a drought-resistant, stoloniferous grass with 1–4 spikelets, has been introduced recently as a lawn grass in Sana'a. It is of high grazing value and may become more frequent but is not easily propagated by seed.

77. CENCHRUS *L.*

Annual or perennial grasses with erect or ascending stems. Leaves flat; ligule a line of hairs. Inflorescence a solitary, terminal, spike-like panicle; spikelets 2-flowered, sessile, awnless, solitary or in small clusters, surrounded by an involucre of fused bristles, the involucre falling with the spikelet cluster; spikelets whitish, c 3mm long in our species; glumes unequal.

Very similar to *Pennisetum* but the often rigid bristles are fused, though only at the base in *C. ciliaris*.

1. Inner bristles rigid, the tip spinescent — 2
   Inner bristles flexuous, the tip not spinescent — 3
2. Bristles retrose-barbed, the inner erect, the outer spreading or reflexed, becoming straw-coloured when mature — *C. biflorus*
   Bristles not retrose-barbed, all erect, becoming green or purplish when mature — *C. setigerus*
3. Inner bristles joined at the base only — *C. ciliaris*
   Inner bristles united for 1–2.5mm above the base — *C. pennisetiformis*

1. Cenchrus biflorus *Roxb.* Kurbays
Annual with erect stems to 1m. Leaves up to 15cm long. Inflorescence straw-coloured, 3–10cm long; involucral bristles rigid, spinescent, retrorse-barbed, united below, the inner erect, c 5mm long, the outer spreading or reflexed, slightly shorter.

A characteristic plant of Tihama sand dunes, growing from sea level to 300m from Bayt al Faqīh north to Qanāwis; it is associated with *Panicum turgidum*, *Blepharis ciliaris* and *Indigofera argentea*.

2. Cenchrus ciliaris *L.*
Pennisetum ciliare (L.) Link
Tufted perennial with ascending or erect stems up to 1m. Leaves glabrous or pubescent, up to 25cm long. Inflorescence purplish, usually 2–5cm long, rarely more; involucral bristles flexuous, soft, united at the base only; the inner flattened and ciliate below, becoming slender and flexuous above, c 8–15mm long; the outer slender, scabrid, shorter.

Widespread and abundant everywhere from sea level to 3000m, except the more arid eastern part of the Mashriq and the Sands. It grows in a great variety of open, well-drained habitats including sand dunes, field borders, roadsides, open rocky plain, terrace walls, dry bushland and arid, stony hillsides. It has the reputation of being a valuable fodder grass.

3. Cenchrus pennisetiformis *Steud. & Hochst.*
Superficially identical to *C. ciliaris*, differing only in the inner involucral bristles which are united for 1–2.5mm of their length. This species is sometimes thought to have originated as a hybrid between *C. ciliaris* and *C. setigerus* and grows where the ranges of the two species overlap.

Only known definitely from the Wadi Rima', presumably in the Tihama (Pratt 68) and from 'Ajārah near Ḥajjaylah (Schweinfurth 894). It is probable that it has been overlooked and is in fact common in the Tihama and the escarpment foothills.

4. Cenchrus setigerus *Vahl.*

Annual or perennial with erect stems to 1m. Leaves glabrous or pilose, up to 15cm long. Inflorescence dull green or dark bluish-purple, 5–9cm long; involucral bristles rigid, spinescent, the inner erect, conspicuously flattened, united for c 2mm of their length, 5–6mm long, the outer absent or very few, not flattened, ± erect, much shorter than the inner ones.

Locally abundant from 100 to 1500m in the Tihama and on the lower slopes of the escarpment. It grows most frequently on the margins of irrigated fields in rich agricultural areas in the east of the Tihama and along the large wadis of the escarpment foothills, but also spreads onto roadsides, sand dunes and stony hills near cultivation but is usually much smaller in stature in poorer situations like these.

78. IMPERATA *Cyr.*

1. Imperata cylindrica (*L.*) *P. Beauv.*

Coarse, densely tufted perennial with erect, dark purple stems to 1m, spreading by rhizomes to form dense clumps. Leaves stiff and wiry, flat or inrolled, up to 25cm long in Yemen but much more elsewhere. Inflorescence a dense terminal, silvery spike-like panicle, 10–20cm long, the branches with long white hairs. Spikelets in pairs, 2-flowered, enveloped in long white hairs; glumes ± equal, long-pilose, 4–5mm long, exceeding the pilose lemmas.

Only known from marsh just north of Rada' at 2000m. Although very rare in Yemen this grass is a widespread and pernicious weed of eroded land in high rainfall areas of the Old World tropics.

79. SACCHARUM *L.*

Tall, woody, perennial grasses with solid stems to 3m. Leaves flat or concave; ligule a short, ciliate membrane. Inflorescence a large, terminal, plume-like panicle, made up of numerous simple or branched, fragile racemes which break up quickly shedding their spikelets; panicle branches covered in long white hairs; spikelets paired, one sessile, one pedicelled, 2-flowered, enveloped in long white hairs; glumes equal; lemmas hyaline, slightly shorter than the glumes.

Spikelets awned	*S. ravennae*
Spikelets unawned	*S. spontaneum*

1. Saccharum ravennae (*L.*) *Murr.*

Densely tufted grass to 3m. Leaves stiff, deeply concave with a prominent midrib, up to 1m long, the sheaths appressed pilose. Panicle silvery, 20–50cm long; the racemes branched; spikelets c 3.5–5mm long, about as long as the hairs from the callus; awns c 5mm long.

A very local grass of springs in dry country, restricted to a few places between 1300 and 1900m in the Mashriq: 'Ujlī, south of Ṣa'dah; Khulāqah, Nihm; between Mārib and the Fardah pass.

2. Saccharum spontaneum *L.*

Densely tufted grass to 2½m. Leaves stiff, deeply concave with a prominent midrib, up to 1m long, the sheaths glabrous. Panicle silvery, 20–45cm long; racemes apparently unbranched; spikelets c 5mm long, overtopped by the long hairs from the callus, awnless.

Widespread by springs and permanent streams from 200 to 2500m, growing in scattered colonies wherever there are suitable habitats in all regions; it is, however, local in occurrence, generally avoiding areas of high rainfall and showing a preference for damp sandy places.

Sugar cane, **Saccharum officinarum** *L.*, known as 'Qaṣab', is occasionally cultivated on a small scale in hot valleys where there is sufficient water for irrigation. There is no sugar industry in Yemen and it is grown merely to satisfy the small demand for fresh sections of sugar cane in the local markets.

80. BOTHRIOCHLOA *O. Kuntze*

Perennials. Leaves flat, rather narrow; ligule a short, ciliate membrane. Inflorescence usually of ± digitately arranged, villous, spike-like racemes with a fragile rhachis; racemes sometimes irregularly arranged along a short axis at the top of the stem; spikelets paired, one sessile, one pedicelled, 2-flowered, the lower flower reduced to a shiny hyaline lemma; glumes equal, usually pitted; awn glabrous, bent about the middle, the column spirally twisted.

Very similar to *Dichanthium* from which it is distinguished by the translucent centre of the pedicels and rhachis internodes which can be clearly seen when they are held up to the light and viewed with a hand lens. In addition, the racemes are never solitary in our species and some of the glumes are always pitted.

1.	Racemes sub-digitate, arranged along a short axis 0–4cm long	2
	Racemes arranged along an axis 4–20cm long	*B. bladhii*
2.	Lower glume of sessile spikelet pitted	*B. insculpta*
	Lower glume of sessile spikelet without pits	*B. radicans*

1. Bothriochloa bladhii (*Retz.*) *S.T. Blake*

Very vigorous, loosely tufted, erect perennial to 1½m. Leaves glaucous, up to 35cm long. Inflorescence of 20–30 slender, branched racemes arranged along an axis 4–20cm long; racemes 4–8cm long; spikelets shiny, purplish, c 3.5mm long; the lower glume of both spikelets pitted; awn up to 2cm long, bent just above the middle, the column dark brown.

Only known from rough grassland at the margins of fields about 2km north of Ibb (Wood 1976). *B. bladhii* is notorious for its readiness to hybridise with the next species and the single collection shows considerable variation in the length of the axis of the inflorescence (4–9cm). This suggests that the isolated Yemen population is in the active process of hybridisation and may cease to exist as a separate entity in the near future.

2. Bothriochloa insculpta (*A. Rich.*) *A. Camus*

B. pertusa auct.
Amphilophis pertusa sensu Schwartz
Tufted, shortly rhizomatous perennial with erect stems to c 50cm. Leaves glaucous, up to 15cm long. Inflorescence of 6–12 slender racemes arranged subdigitately or on a short axis 0–4cm long; racemes 2.8cm long; spikelets shiny, purplish, 3–4mm long; the lower glume of both spikelets pitted; awn 1–1.5cm long, bent just above the middle, the column dark brown.

Widespread and often abundant on the escarpment, growing from 600 to 2700m on grassy banks around fields and on stony hillsides, often where protectected from grazing by thorn bushes and fences. There are no confirmed records from the Tihama and on the high plateau it is very rare and apparently only known in the Rada' area and on the borders of irrigated fields between Sana'a airport and Arḥab; in the Mashriq it is known from Majz (Steinberg).

3. Bothriochloa radicans (*Lehm.*) *A. Camus*

Amphilophis radicans (Lehm.) Stapf
Identical to *B. insculpta* except for the lower glume of the sessile spikelet which does not have a prominent pit.

Only known from the collection made by Rathjens on J. Ḥadīdah. The plant has not been found again and the exact location of J. Ḥadīdah is unknown so it has not been possible to try and refind the plant.

81. DICHANTHIUM *Willem.*

Very similar to *Bothriochloa* from which it can be distinguished by the pedicels and rhachis internodes which are solid. In addition our species either have solitary, pedunculate racemes or, if the racemes are ± digitately arranged, the glumes are all without pits. As in *Bothriochloa*, the spikelets are in pairs in a slender, villous, spike-like raceme whose rhachis is fragile, the glumes are ± equal and the glabrous awn is bent about the middle.

1. Racemes in a terminal digitate cluster; the peduncle glabrous — *D. annulatum*
   Racemes axillary as well as terminal, usually solitary; the long, slender peduncle pubescent — **2**
2. Lower glume of the sessile spikelet usually with a single pit and an orange tip. Racemes densely villous — *D. foveolatum*
   All glumes without pits or orange tips. Racemes sparsely villous — *D. aristatum*

1. Dichanthium annulatum (*Forssk.*) *Stapf*

Tufted perennial with numerous erect stems to 1m. Leaves up to 15cm long, the nodes villous; ligule a glabrous membrane. Inflorescence a terminal, digitate cluster of 2–9 erect racemes on a glabrous peduncle; racemes purplish, densely villous, 2–6cm long; spikelets c 3mm long; glumes pitless, the lower rounded, pilose near the keels; awn 1.5–2.5cm long.

Superficially very similar to *Bothriochloa insculpta* and *B. radicans* but distinguished by the glumes which have no pits at all.

Frequent on stony and sandy plain between 200 and 600m in the east of the Tihama from the Wadi Rima' northwards. Also locally frequent on stony plain from 2000 to 2400m on the high plateau from Rada' and Ma'bar north to Ma'mar and Arḥab, showing a preference for sandstone areas.

2. Dichanthium aristatum (*Poir.*) *C.E. Hubbard*

Weakly tufted stoloniferous grass with erect or ascending stems to 2m. Leaves up to 15cm long, the nodes usually glabrous; ligule a short truncate membrane. Inflorescence terminal and axillary, the axillary branches arising from low down on the stem; racemes purplish, sparsely villous, erect, 4–8cm long, in long pedunculate digitate clusters, in Yemen always (?) reduced to a single raceme, the peduncle pubescent below the raceme; spikelets 4–5mm long; glumes pitless, the lower rounded, pilose near the keels; awn 1–2.5cm long.

A local plant of grassy ditch sides between 1000 and 1700m in Taiz and Ibb provinces: Musalla, Aḥkūm; Wadi Ḥidrān and Wadi Ḍabāb, Taiz; Wadi Sūdān, Qā'idah; between Ibb and Suq As Sabt.

This is a plant of India and southeast Asia found in one or two places in Africa where it may have been introduced. The status of the Yemen population is unclear but it appears to be native.

3. Dichanthium foveolatum (*Del.*) *Roberty*

Eremopogon foveolatus (Del.) Stapf
Tufted perennial with slender, erect or ascending stems to 50cm. Leaves often becoming reddish-brown when old, up to 10cm long, nodes villous; ligule a lacerate membrane. Inflorescence of solitary pedunculate, terminal and axillary racemes, often partially enclosed in a leaf sheath, peduncles slender, pubescent below the racemes; racemes pale brown, densely villous, 3–4cm long; spikelets c 4mm long; lower glume of the sessile spikelet with a single pit, ± glabrous, with an orange tip, awn 1.5–2cm long.

There is a good deal of variation as regards the presence and absence of pits on the glumes. The lower glume of the pedicelled spikelet may or may not have a pit but the presence of a pit on the lower glume of the sessile spikelet is usually taken as a distinguishing character of the species. However, one collection (Wood 1503) from subkhah between Ibn Abbās and Al Khawbah is unusual in that the lower glumes are mostly without pits but it is not clear what the significance of this variation is.

Widespread but infrequent, growing from sea level to 200m in the western part of the Tihama; it is almost always found on sand dunes, usually near the sea and is often associated with *Odyssea mucronata*.

82. CHRYSOPOGON *Trin.*

Tufted perennial herbs with elongated, decumbent stem bases and slender, erect stems. Leaves short; ligule a ciliate rim. Inflorescence an open panicle with whorled branches which are densely bearded just below the spikelet group; spikelets in groups of 3, one sessile and two pedicelled, 2-flowered, the lower flower reduced to a lemma, the upper flower bisexual in the sessile spikelet but male in the pedicelled spikelet; upper glume and upper lemma of all spikelets awned, but the awn of the upper lemma of the sessile spikelet bent in the middle and much longer than the others.

Awns plumose *C. plumulosus*
Awns pubescent but not plumose *C. aucheri*

1. Chrysopogon aucheri (*Boiss.*) *Stapf*

Densely tufted perennial to 60cm. Leaves glaucous, up to 10cm long. Spikelets c 6mm long, white or purplish (contrasting with the often golden tuft of hairs at the end of the panicle branches), usually conspicuously pilose at the top; awns pubescent, the awn of the upper lemma of the sessile spikelet 2–3cm long, brown, bent in the middle and with a twisted column; the other awns straight, 5–10mm long.

A rare but possibly overlooked plant of rocky hills at around 2500m on the high plateau: J. Nuqum (Deflers); near Suq Bayt An Nā'im. (Wood 75/316.)

This is a species of eastern Arabia, Iran and Pakistan.

2. Chrysopogon plumulosus *Hochst.*

C. aucheri variety *quinqueplumis* (Hochst. ex A. Rich.) Stapf

Identical to the previous species but differing in that the 5 shorter awns of each spikelet are plumose; the longest awn, from the upper lemma of the sessile spikelet, though, is pubescent as in the previous species and not plumose.

Widespread on dry, open, stony hills including *Acacia* bushland, extending from the first foothills in the Tihama at 200m through the escarpment and high plateau into the Mashriq and reaching an altitude of about 2500m. It is absent from areas of high rainfall and, more surprisingly, from the rocky foothills of the southern Tihama; indeed its distribution is complementary to that of *Cymbopogon schoenanthus* and, although growing in similar conditions, the two species rarely occur together.

C. plumulosus is a plant of northeast Africa and western Arabia.

83. ARTHRAXON *P. Beauv.*

Annuals or perennials with short, ovate-lanceolate leaves abruptly narrowed to form a ± auriculate base; ligule a short ciliate membrane. Inflorescence of ± digitate spike-like racemes; spikelets paired, one sessile, one pedicelled, the latter sometimes absent, 2-flowered; glumes equal; lemma acute, entire; awn bent, 8–10mm long.

Annual. Racemes 1–2cm long. Lower glume of sessile spikelet not spiny *A. lancifolius*
Perennial. Racemes 2–5cm long. Lower glume of sessile spikelet spiny-pectinate *A. prionodes*

1. Arthraxon lancifolius (*Trin.*) *Hochst.*

Slender annual herb, the stem at first decumbent and rooting at the nodes, finally ± ascending to reach 20cm, thus forming small dense colonies. Leaves mostly 2–3cm long, narrowed into a fine acuminate point, the sheaths somewhat inflated. Racemes ciliate, 3–9, 1–2cm long, borne on filiform peduncles; spikelets paired or with the pedicelled spikelet absent, whitish-purple, c 3mm long; lower glume of sessile spikelet not spiny-pectinate.

A rare but possibly overlooked grass found between 1500 and 1900m on the escarpment, growing on steep stony slopes or cliff faces where there is dripping water in the rainy season: in the wadi above 'Adūf near Turbah; in a gully at the foot of J. Ba'dān north of Ibb; Wadi Sukhnah.

2. Arthraxon prionodes (*Steud.*) *Dandy*

A. lanceolatus sensu Schwartz

Perennial herb with short rhizomes from which arise scattered ascending stems to 70cm. Leaves 3–6cm long, narrowed into a fine acuminate point, the sheaths tight. Racemes ciliate, 2–5, 2–5cm long, borne on relatively stout peduncules; spikelets always paired, whitish-purple, 5–6mm long; lower glume of sessile spikelet pectinate-spiny on the margin and the nerves.

Widespread and frequent from 800 to 2700m on the escarpment and high plateau, growing on cliffs, steep rocky slopes, eroded hillsides and terrace walls, often in partial shade; in dry country it is virtually restricted to cliff gullies which become small streams in the rainy season; in the central escarpment it is one of the most characteristic plants of the terrace wall flora, being found over a wider altitudinal range than any other species in this habitat.

84. SEHIMA *Forssk.*

Annuals or perennials. Leaves linear, flat; ligule a line of hairs. Inflorescence a solitary terminal spike-like raceme; spikelets paired, one sessile and fertile, one pedicelled and sterile; sessile spikelet with the lower glume coriaceous, prominently nerved, often with a median groove, the upper glume boat-shaped, terminating in a bristle and the upper lemma with a bent awn, the column spirally twisted and puberulous.

The genus was named by Forsskal after the Arabic word 'Sahīm' which was applied to this grass in the Ḥadīyah area.

Perennial. Lower glume of the sessile spikelet with a membranous tip up to a quarter its length *S. nervosum*
Annual. Lower glume of the sessile spikelet with a membranous tip up to half its length *S. ischaemoides*

1. Sehima ischaemoides *Forssk.*

Erect annual to 60cm. Leaves glaucous, up to 30cm long. Racemes white, villous, 3–15cm long; sessile spikelets 9–15mm long, the lower glume membranous in its upper half, the upper lemma with an awn 4–7cm long.

Known only from Forsskal's type collection at Ḥadīyah and two collections by Wissmann from Bilad Aṭ Ṭa'am in J. Raymah which may not be correctly named.

It is unfortunate that Forsskal's type collection has disappeared since it seems probable that the plant he actually found was the next species, which is common at Ḥadīyah. Since Forsskal's description does not mention any feature that can be used to distinguish the two species, it is unlikely that the matter can ever be settled with certainty.

2. Sehima nervosum (*Rottl.*) *Stapf*

Tufted perennial with erect stems to 1m. Leaves glaucous, long-attenuate, up to 40cm long. Racemes whitish-green, villous, curved, 3–12cm long; sessile spikelets 6–10mm long, the lower glume membranous near the tip only, the upper lemma with an awn 2–4cm long.

Widespread between 500 and 2000m on the escarpment from Turbah north to the Wadi Lā'ah, growing in riverine scrub and *Acacia* bushland with a preference for areas of good rainfall; it is particularly common in the central escarpment and the Ibb-'Udayn region.

85. ANDROPOGON *L.*

Annuals or perennials. Leaves linear; ligule membranous or a ciliate rim. Inflorescence usually of paired (rarely digitate) terminal spike-like racemes, sometimes the racemes solitary or axillary as well as terminal; spikelets paired, one sessile, one pedicelled, the sessile spikelet fertile, 2-flowered, awned; glumes as long as the spikelet; awn bent about the middle, the column twisted, glabrous; pedicelled spikelet usually smaller than the sessile spikelet.

Most species are readily distinguished from related genera by their large racemes in terminal pairs. Some species can be confused with *Cymbopogon* but *Andropogon* leaves are never aromatic, the racemes remain erect at maturity and are much longer than the spatheole.

1. Racemes solitary *A. fastigiatus*
   Racemes paired **2**
2. Racemes in terminal pairs only **3**
   Racemes in axillary and terminal pairs *A. crossotos*
3. Sessile spikelet 9–12mm long. Base of plant silky-hairy *A. distachyos*
   Sessile spikelet 6–8mm long. Base of plant glabrous
4. Pedicelled spikelet awned. Sessile spikelet without a green membranous tip. Densely tufted grasses, not stoloniferous, rarely exceeding 50cm high **5**
   Pedicelled spikelet awnless. Sessile spikelet with a green membranous tip. Stoloniferous grass, usually more than 70cm high *A. greenwayi*
5. Pedicels and rhachis internodes swollen upwards and strongly clavate. Awn of pedicelled spikelet 4–10mm long *A. chinensis*
   Pedicels and rhachis internodes linear or slightly swollen upwards. Awn of pedicelled spikelet 1–3mm long *A. amethystinus*

1. Andropogon amethystinus *Steud.*

A. pratensis Hack.

Densely tufted, erect perennial to 60cm though commonly much less in Yemen, the basal leaf sheaths glabrous. Leaves green or glaucous, flat, up to 35cm long; ligule a ciliate membrane. Racemes terminal, paired, villous, whitish-green, 4–9cm long; pedicels and rhachis internodes linear or somewhat swollen upwards; sessile spikelet 6–8mm long, its lower glume with a mucro 3mm long and the fertile lemma with an awn 8–15mm long; pedicelled spikelet with a short awn 1–3mm long.

A very variable plant in appearance, the most common forms in Yemen being small and compact (e.g. Wood 1154) whereas some plants are very robust with long, very villous racemes and broader, longer, glaucous leaves (Wood 1691). The more robust forms are usually found on cliffs and similar places protected from grazing.

Locally abundant in high-altitude grassland and on cliffs and rocky mountain slopes found between 2700m and 3300m on the higher escarpment mountains from J. Ṣabir north to J. Maswār, but most common in the Ibb area.

2. Andropogon chinensis (*Nees*) *Merr.*

A. ascinoides C.B. Cl.

Very similar to robust forms of the previous species, the leaves flat, glaucous, 15–35cm long. The inflorescence is, however, more obviously spatheate and the raceme pairs often partially enclosed in the reddish sheaths of the upper leaves; the pedicels and rhachis internodes are clavate, being strongly swollen upwards and the awns are distinctly longer, that of the sessile spikelet being 15–25mm long and the awn of the pedicelled spikelet 4–10mm long.

Known only from rocky hillsides at 1900m on the ridge northeast of Markah, J. Milḥān.

3. Andropogon crossotos T. Cope

Densely tufted, erect perennial to 40cm; basal leaf sheaths glabrous, often reddish. Leaves flat, slightly glaucous, up to 10cm long; ligule a truncate membrane. Inflorescence of pedunculate, terminal and lateral raceme pairs, the racemes slender, hairy, usually purplish, 2–3.5cm long;

pedicels and rhachis internodes flattened, villous along the margins; sessile spikelet 3–4mm long, its lower glume acute but not mucronate, the fertile lemma with an awn 8–11mm long; pedicelled spikelet awnless.

Widespread and frequent between 1700 and 2400m on open rocky hills on the high plateau from Rada' and Risābah north to Ḥūth, extending onto the escarpment in the sandstone country near Rujum. Endemic to Yemen.

4. Andropogon distachyos *L.*

Densely tufted, erect perennial to 80cm, the basal leaf sheaths usually pilose. Leaves flat, glaucous, up to c 20cm long, though commonly less in Yemen; ligule a ciliate rim. Racemes terminal, paired, sparsely pubescent, becoming purplish when old, 4–8cm long; pedicels and rhachis internodes linear; sessile spikelet 9–12mm long, its lower glume with a short mucro 1–2mm long and the fertile lemma with an awn 20–30mm long; pedicelled spikelet very pale, with an awn 5–8mm long.

Widespread and frequent between 1800 and 3400m on the escarpment and high plateau growing on field borders, rocky hillsides and in disturbed places in montane grassland.

5. Andropogon fastigiatus *Sw.*

Diectomis fastigiata (Sw.) P. Beauv.

Tufted annual with ascending stems, to 1m, glabrous below. Leaves flat, up to 30cm long; ligule a glabrous membrane c 5mm long. Inflorescence of solitary, terminal and lateral, long-pedunculate villous racemes 4–6cm long, often partially enclosed by the reddish upper leaf sheaths; peduncles very slender, filiform; pedicels and rhachis internodes swollen upwards, clavate; sessile spikelet c 4mm long, its lower glume acute but not mucronate and the fertile lemma with an awn 30–50mm long; pedicelled spikelet much larger than the sessile spikelet, purplish c 10mm long, the awn 9mm long.

Only known on rocky hillsides just north of the suq at 'Udayn at 1500m.

6. Andropogon greenwayi *Napper*

Weakly tufted, slender, erect perennial to 1m, spreading by stolons and often growing in large colonies; basal leaf sheaths glabrous. Leaves flat, slightly glaucous, often wavy, up to 15cm long; ligule a ciliate membrane. Racemes terminal, paired, villous, whitish-green, 4–8cm long; pedicels and rhachis internodes linear; sessile spikelet 7–8mm long, with a light-green membranous tip, its lower glume not mucronate, the fertile lemma with an awn 8–12mm long; pedicelled spikelet awnless.

Locally abundant on field borders and open mountain slopes between 2000 and 3000m on the escarpment north to J. Rāziḥ, but apparently completely absent south of the Ta'kar-Mudhaykhirah range, including the whole of Taiz province, though it extends onto the high plateau in a few places including the Yislah pass (Hepper) and Khamr. It is particularly common around Ibb where it is often the dominant species on grassy field borders. It is only found in flower in autumn.

86. CYMBOPOGON *Spreng.*

Densely tufted perennials. Leaves linear or flat; ligule membranous. Inflorescence consisting of short, paired, unequal-based racemes, often partially enclosed by a reddish spatheole, and massed together into a compound panicle; spikelets in pairs; the sessile spikelet 2-flowered, the upper flower fertile, the lower glume deeply grooved, the upper glume boat-shaped, the lemma with a bent awn, the column twisted, glabrous; pedicelled spikelet sterile, awnless.

***Cymbopogon* is similar to *Hyparrhenia* and *Andropogon*. It differs from the former in the glabrous column of the awn and from the latter in its compound inflorescence with the racemes about as long as the spatheole. It differs from both in its racemes which are deflexed at maturity and in the aromatic oil which is always present in the leaves. This is sometimes readily apparent due to the pleasant smell of the plant but can be confirmed even on dried specimens by chewing the leaves which always have a distinct taste, some of liquorice, others of lemon.**

1. Lower glume of sessile spikelet with a *V*-shaped median groove in the lower third. Basal leaf sheaths not persistent. Leaves flat ... *C. caesius*
   Lower glume of sessile spikelet with a *U*-shaped median groove. Basal leaf sheaths persistent. Leaves flat or filiform ... 2
2. Lower glume of sessile spikelet with rounded keels. Pedicels and rhachis internodes densely pilose along the margins, glabrous or sparsely pilose on the back ... *C, commutatus*
   Lower glume of sessile spikelet with sharp keels. Pedicels and rhachis internodes densely pilose all over ... 3
3. Lowermost pedicel of the sessile raceme hard, shiny and swollen ... *C. schoenanthus*
   Lowermost pedicel of the sessile raceme not swollen ... *C. jwarancusa*

1. Cymbopogon caesius (*Nees*) *Stapf*

Densely tufted perennial with erect stems to 3m, basal leaf sheaths glabrous, not persistent. Leaves flat, sometimes wavy, glaucous, up to 40cm long. Panicle narrow, up to 40cm long; racemes whitish-green, 1–2cm long; pedicels and rhachis internodes densely pilose along the margins, glabrous or sparsely pilose on the back, the lowest pedicel of the sessile raceme swollen; sessile spikelets c 4mm long, the lower glume with a deep *V*-shaped groove in the lower third; awn strongly bent, 10–12mm long.

Widespread on the escarpment from Aḥkūm north to J. Marran, but local and with a strong southern tendency being mostly found in Ibb and Taiz province. It is also found in the alluvial plains in the east of the Tihama below the central escarpment around Bājil and Khalīfah where it grows to 3m in height (it is usually much less in the mountains). It is found from 300 to 2400m, growing in the Tihama in clumps on field borders and open grassy plain whereas on the escarpment it is mainly a plant of steep rocky hillsides and broken cliffs.

Fig. 35 **A** *Andropogon crossotos*; **B** *Hyparrhenia papillipes*; **C** *Hackelochloa granularis*; **D** *Elionurus muticus*.

2. Cymbopogon commutatus (*Steud.*) *Stapf*

C. parkeri Stapf

Densely tufted, erect perennial to 1m, basal leaf sheaths glabrous or hairy, persistent. Leaves flat or filiform, up to 20cm long. Panicle narrow, usually short but sometimes up to 40cm long; racemes green and brown, 1–2.5cm long; pedicels and rhachis internodes densely pilose along the margins, glabrous or sparsely pilose on the back, the lowermost pedicel of the sessile raceme swollen; sessile spikelets c 5mm long, the lower glume with a *U*-shaped groove for most of its length and distinctly rounded keels; awn strongly bent, c 15cm long.

C. commutatus is a variable plant and the sole Yemen collection (Wood 3013) is similar to forms found in Africa and has flat leaves with glabrous sheaths. Forms from the Gulf which were previously separated as *C. parkeri* look very different and have inrolled, filiform leaves with hairy sheaths, but have an identical inflorescence.

On open sandstone hillsides in the wadi above 'Adūf near Turbah (1900m).

3. Cymbopogon jwarancusa (*Jones*) *Schult.*

C. olivieri (Boiss.) Bor

Densely tufted perennial to 60cm with persistent, glabrous, basal leaf sheaths and narrow, wiry, filiform, flexuous leaves. Inflorescence similar to that of *C. commutatus*, but the racemes densely villous, the lowermost pedicel of the sessile spikelet not swollen, the lower glume of the sessile spikelet with sharp keels and the awn weakly bent or even straight.

The Yemen plant is subspecies **olivieri** (*Boiss.*) *Soenarko*.

A plant of Iran and northwest India which has been found twice in Yemen on arid stony plain between 700 and 1100m: between Ḥajdah and Baiḥ (Hepper and Wood 6032) and near Mārib (Wood 3165).

4. Cymbopogon schoenanthus (*L.*) *Spreng.*

C. circinnatus Hochst.

Differs from the previous species in that the lowermost pedicel of the sessile raceme is hard, swollen and shiny.

Abundant and sometimes dominant over huge areas of stony steppe and arid foothills from 100 to 1300m in the southern Tihama and southwestern foothills north to the Wadi Zabīd and inland as far east as the Wadi Ḥidrān near Taiz. In this area it provides much of the ground cover under *Acacia* and *Commiphora* bushes. It is also frequent on arid stony plain and hillsides between 1000 and 2300m in the Mashriq, extending from the Mārib area northwest to the Ṣa'dah area, but is seldom dominant here. It has also been found near Sana'a on the high plateau. (Wood 944, 2946)

C. schoenanthus is a plant of North Africa and the Arabian peninsula, differing from *C. jwarancusa* in the solitary key character. The two species grow together in Yemen and it seems doubtful to me whether the two can be maintained as separate species on the grounds of a single character of this nature, unsupported by any secondary distinctions.

87. HYPARRHENIA *Anderss. ex Fourn.*

Vigorous annuals or perennials. Leaves flat; ligule a short membrane. Inflorescence a lax spatheate panicle composed of paired spike-like racemes, each pair on a slender peduncle partially enclosed by a spathaceous sheath (spatheole); spikelets arranged in pairs along a fragile rhachis, one sessile, fertile, 2-flowered and awned, the other pedicellate, sterile and unawned; glumes equal; upper lemma of the lower spikelet bifid, the bent, hairy awn arising from the sinus.

Distinguished from all similar genera by the awn which is hairy on the lower part.

***Hyparrhenia* is a very difficult genus, but in Yemen the situation is simplified by the small number of species. *H. hirta* is very abundant whereas all other species are rare. *H. coleotricha* and *H. papillipes* are relatively distinct but the remaining four large perennial species are superficially very similar. Apart from *H. hirta*, the species are only normally in flower in late autumn and it is possible that their occurrence is related not so much to the quantity of rain as to the duration of the rainy season. It is perhaps significant that most species are virtually restricted to the Ibb region where a prolonged rainy season is the rule rather than the exception.**

1. Annual. Awns 4–6cm long. Pedicelled spikelet terminating in a bristle 6–8mm long — *H. coleotricha*
   Perennial. Awns less than 3.5mm long. Pedicelled spikelet terminating in a bristle up to 2mm long — 2
2. Raceme bases unequal (i.e. one raceme of each pair apparently sessile, the other on a stalk more than 1.5mm long) — 3
   Raceme bases equal (i.e. both racemes of each pair apparently sessile or nearly so) — 4
3. Tall perennial to 2½m with at least 10 raceme pairs in each panicle; racemes commonly deflexed, 1–1.5 (–2)cm long — *H. quarrei*
   Perennial to 1m with less than 10 raceme pairs in each panicle; racemes usually erect, 2–4cm long — *H. hirta*
4. Slender perennial to c 50cm. Racemes densely villous, mostly 3–4cm long — *H. papillipes*
   Stout perennials to 2½m. Racemes glabrous to villous, 1–2.5 (–3)cm long — 5
5. Awns 10–25 per raceme pair. Spatheoles 2.5–5cm long — *H. dregeana*
   Awns 3–8 per raceme pair. Spatheoles 1.4–2.6cm long — 6
6. Awns 3–5, each 1.8–3.2cm long. Spatheoles deeply concave, boat-shaped — *H. variabilis*
   Awns 6–8, each 0.8–1.8cm long. Spatheoles shallowly concave, scarcely boat-shaped — *H. formosa*

1. Hyparrhenia coleotricha (*Steud.*) *Anderss. ex W. D. Clayton*

Vigorous tufted annual with ascending or decumbent stems to about 1m in Yemen, though much more elsewhere. Leaves green, up to 30cm long. Panicle very loose with rather few nearly equal-based raceme pairs; peduncles 8–15cm long, usually pilose at the top; spatheoles 5–7cm long, lanceolate, green or slightly purplish; racemes 2–2.5cm long; pedicelled spikelet, villous, ending in a bristle 6–8mm long; awns 4–8 per raceme pair, 4–6cm long, golden-brown.

A very local plant of seasonally damp gullies in middle-altitude wadis at c 1500m: Wadi Sukhnah (rare); in several places around 'Udayn.

2. Hyparrhenia dregeana (*Nees*) *Stapf ex Stent*
Vigorous tufted perennial with erect stems to 2½m. Leaves often glaucous, up to 60cm long. Panicle narrow, moderately dense, with more than 20 equal-based raceme pairs; peduncles 1.5–5cm long, usually pilose at the top; spatheoles 2.5–5cm long, lanceolate, wine-red; racemes 2–2.5 (–3)cm long; pedicelled spikelet glabrous, ending in a small bristle-like point; awns 10–25 per raceme pair, 1.5–2cm long, golden-brown.

A rare plant of grassy field borders in high rainfall areas between 2100 and 2600m: Ibb; Mashwarah between Ibb and 'Udayn; near Bayt Mughlad, J. Shibām, Ḥarāz.

3. Hyparrhenia formosa *Stapf*
Robust perennial, similar in appearance to the last species. Panicle narrow, rather dense, with 30 or more equal-based raceme pairs; peduncles 0.2–1cm long, pilose at the top; spatheoles 1.8–2.6cm long, lanceolate, wine-red; racemes 1–1.5cm long; pedicelled spikelet glabrous except on the margins, ending in a small bristle-like point; awns 6–8 per raceme pair, 0.8–1.8cm long, brown.

Only known from montane grassland at 2100m on J. Kabūrah above Ḥadīyah where it grows in the protection of a fence round a qat plantation.

4. Hyparrhenia hirta (*L.*) *Stapf.*
Tufted perennial with erect or ascending stems to 1m. Leaves often glaucous, up to 15cm long. Panicle loose, narrow, with up to 10 raceme pairs only, the bases unequal; peduncles 3–10cm long, usually pilose at the top; spatheoles 3–8cm long, lanceolate, green or wine-red when old; racemes 2–3 (–4)cm long; pedicelled spikelet villous or pubescent, not terminating in a bristle; awns 8–16 per raceme pair, 1.5–2.5cm long, brown.

One of the most abundant and drought-resistant grasses in Yemen, growing from 500 to 3000m on terrace walls, stony hillsides and in montane grassland, occurring almost everywhere on the escarpment, southeastern mountains, high plateau and the western part of the Mashriq.

5. Hyparrhenia papillipes (*Hochst. ex A. Rich.*) *Anderss. ex Stapf*
Slender tufted perennial with erect stems to 50cm. Leaves green, up to 15cm long. Panicle loose and lax with up to 10 equal-based raceme pairs; peduncles 3–8cm long, pilose above, usually strongly flexuous so the raceme pairs are deflexed; spatheoles 4–7cm long, lanceolate, brown; racemes 2–4cm long; pedicelled spikelet villous, terminating in a short bristle-like point; awns 9–19 per raceme pair, 2–3cm long, brown.

Only known from the open stony ridge on the summit of J. Sumārah at 3000m.

6. Hyparrhenia quarrei *Robyns*
Robust perennial, similar in appearance to *H. dregeana*. Panicle narrow, rather dense with at least 30 raceme pairs, the bases unequal; peduncles 2–7cm long, pilose above, becoming flexuous so the raceme pairs are deflexed; spatheoles 3–5cm long, lanceolate, wine-red; racemes 1–1.5 (–2)cm long; pedicelled spikelet villous, not terminating in a bristle; awns 6–10 per raceme pair, 2–3.5cm long, brown.

Only known from grassy banks around fields on the Ibb-'Udayn pass below Mashwarah at c 2400m.

7. Hyparrhenia variabilis *Stapf*
Robust perennial, similar in appearance to *H. dregeana*. Panicle long, narrow, dense, with numerous equal-based raceme pairs; peduncles 0.3–0.9mm long, pilose at the top; spatheoles 1.4–2.4cm long, deeply concave and boat-shaped, wine-red; racemes 0.8–1.3cm long; pedicelled spikelet glabrous, terminating in a short bristle-like point; awns 3–5 per raceme pair, 1.8–3cm long, golden brown.

Only known in open swampy scrub at 1600m between Ibb and Suq As Sabt.

88. HETEROPOGON *Pers.*

1. Heteropogon contortus (*L.*) *P. Beauv. ex Roem. & Schult.*
Tufted perennial with erect stems to 1m. Leaves flat, glaucous, glabrous or hairy, up to 10 (–25)cm long, the sheaths conspicuously flattened; ligule a short, finely ciliate membrane. Inflorescence a solitary, terminal, spike-like raceme, sometimes partially enclosed by the uppermost leaf sheath, 4–7cm long; spikelets whitish-green, paired, one sessile, one pedicelled, the sessile one 2-flowered, 7–8mm long; glumes ± equal, pilose; awns golden-brown, bent about the middle with a long twisted column, up to 9cm long; all the awns of the raceme frequently becoming twisted together.

Widespread from 500 to 2900m, being common on the escarpment but local and mainly around Sana'a on the high plateau. It does not appear to tolerate heavy grazing and it is usually found where there is some measure of protection from grazing animals and is thus typical of *Acacia* scrub, thorn fences, cliff slopes, field margins and similar ungrazed situations.

89. THEMEDA *Forssk.*

1. Themeda triandra *Forssk.*
Tufted perennial with erect stems to 3m, the whole plant becoming reddish-brown when old. Leaves flat, hairy or glabrous, up to 30cm long; ligule a short ciliate membrane. Inflorescence a large lax panicle composed of clusters of 2–8 racemes enfolded by reddish spatheoles, the spatheoles 1.5–3.5cm long, glabrous or with bulbous-based hairs; each raceme composed of an involucre of four sterile sessile spikelets around one spikelet pair whose sessile spikelet is fertile; fertile spikelets 6–11mm long, brown, the lemma with a bent pubescent awn 4–6cm long.

Widespread and abundant in rough grassland, on field borders and on open stony hillsides between 1500 and

3000m on the escarpment and the high plateau. It is a valuable fodder grass.

The name *Themeda* is based on the local name 'Thamīd', cited to Forsskal by the people at Ḥadīyah.

90. ELIONURUS *Willd.*

Annual or perennial tufted grasses. Leaves various; ligule short, membranous, ciliate. Inflorescence of 1 or more villous, spike-like racemes; spikelets in pairs, one sessile and one pedicelled, the sessile fertile, 2-flowered; glumes equal, the lower with 2 conspicuous teeth or tails, the upper acute; lemmas partly or wholly hyaline, much shorter than the glumes, awnless.

Annual. Inflorescence of several racemes along a short axis, each raceme enclosed by the sheath of the upper leaf *E. royleanus*
Perennial. Inflorescence of a single terminal raceme not enclosed by the sheath of the upper leaf *E. muticus*

1. Elionurus muticus (*Spreng.*) *Kuntze*

E. argenteus Nees

Densely tufted, erect perennial to c 40cm. Leaves narrow, inrolled, usually coiled towards the tips, pilose for a short distance above the ligule, glaucous, 5–20cm long. Inflorescence a densely villous, solitary, terminal spike-like raceme, 4–8cm long; glumes of the sessile spikelets purple, pilose, the lower deeply bifid, 8–9mm long.

The densely tufted habit and the thin, wiry coiled leaves make this species one of the most easily identified Yemen grasses even when sterile. The woolly inflorescence is also very distinctive and appears very soon after the onset of the rains.

Widely distributed on the escarpment and high plateau, growing from 1500 to 2700m, rather local, though sometimes abundant where it occurs. It is the characteristic grass of a distinctive plant community found on bare, well-drained mountain ridges, cliff tops and similar situations, usually on crumbly volcanic rocks or broken sandstone where it is associated with *Becium serpyllifolium*, *Indigofera arabica* and *Lotus quinatus*.

2. Elionurus royleanus *Nees ex A. Rich.*

Tufted annual with erect or ascending geniculate stems to 20cm, the whole plant turning wine-red as it matures. Leaves flat, glabrous, 5–10cm long. Inflorescence branched, consisting of several pedunculate, villous, spike-like racemes; racemes 1–5cm long, enclosed by the inflated sheaths of the upper leaves, the rhachis very brittle; glumes of the sessile spikelet wine-red, the lower extended at the tip into 2 flat tails, fringed with a row of warts, each wart with a tuft of long white hairs, the whole glume c 12–15cm long.

Another very distinctive species only known from a stony trackside at 1500m on the lower slopes of J. Shahārah. It has a characteristic Sahara-Sindian distribution with its southernmost limits in Yemen and Somalia.

91. LASIURUS *Boiss.*

1. Lasiurus scindicus *Henrard* Thamām

L. hirsutus auct.

Rottboellia hirsuta sensu Schwartz

Clump-forming perennial, much-branched and sometimes woody at the base with many stiff, glaucous, ascending or erect stems to c 80cm. Leaves crowded below, stiff, glaucous, flat or involute, up to c 20cm long; ligule a fringe of hairs. Inflorescence a solitary, terminal, villous, spike-like raceme, 5–10cm long, the rhachis fragile. Spikelets paired or in 3s (one pedicelled and one–two sessile), 2-flowered, whitish-green; glumes of the sessile spikelets villous, unequal, the lower with an almost glabrous bifid tip, 6–7mm long, the upper shorter, acute; lemmas hyaline, awnless.

Rather easily confused with *Pennisetum divisum* but differing in the absence of an involucre of bristles round the spikelets and in the villous (not scabrid), fragile raceme rhachis.

Abundant on sand and stony steppe throughout the Tihama from sea level to 600m.

92. HACKELOCHLOA *Kuntze*

1. Hackelochloa granularis (*L.*) *Kuntze*

Erect branched annual to c 80cm, the whole plant thinly covered in long, white, bulbous-based hairs. Leaves lanceolate with a broad ± amplexicaul base, up to 10cm long; ligule a very short, ciliate membrane. Inflorescence of spike-like racemes arranged along an axis 1–10cm long; racemes 1–2cm long, partially enclosed by the spathe-like leaf sheaths; spikelets green, paired, the lower much modified, globose, glabrous, tuberculate, awnless, 1–1.5mm long, the upper reduced to a pair of empty glumes 2–2.5mm long.

A very distinctive species found on disturbed grassy slopes between 1500 and 1800m in high rainfall areas: between Ibb and Suq As Sabt; Wadi Sukhnah.

93. ROTTBOELLIA *L.f.*

1. Rottboellia cochinchinensis (*Lour.*) *Clayton*

R. exaltata L.f.

Erect, branched, fast-growing annual to 2m, sometimes sending down stilt-roots from the lower nodes. Leaves slightly glaucous, at least 1cm broad and up to 40cm long, the sheaths densely covered in white, bulbous-based hairs; ligule a short brown membrane. Inflorescence of stiff, erect, pedunculate, spike-like racemes arranged along a long axis; racemes 6–10cm long, whitish-green, partially enclosed by the spathe-like leaf sheaths; spikelets deeply sunk into the rhachis, glabrous, 4–6mm long, paired, the sessile spikelet paler and slightly larger than the green, pedicelled spikelet.

Widespread from 200 to 1800m in the Tihama and on the escarpment, growing as a field weed and on field borders from Hajdah north to the Wadi Badr but local and scattered in occurrence, being mainly found in irrigated land or in high rainfall areas.

It is only likely to be found in late autumn at the end of the growing season.

Apart from Bulrush Millet (*Pennisetum glaucum*), Finger Millet (*Eleusine coracana*) and Common Millet (*Panicum miliaceum*), all cultivated cereal grasses belong to genera which are not indigenous in Yemen.

Much the most important crop in Yemen is Sorghum, **Sorghum bicolor** (*L.*) *Moench.* which is generally known as 'Dhurah' but there are numerous cultivars all of which have local names mostly indicating the spikelet colour, such as 'Dhurah ḥamrā' but some with no obvious meaning such as 'Gharib'. These cultivars have been treated as distinct varieties or species (Schwartz lists three species, *S. durra* Stapf, *S. cernuum* Host. and *S. subglabrescens* Aschers. & Schweinf., and nine distinct varieties) but it is generally accepted today that there is one cultivated species with many interfertile varieties. *S. bicolor* is in the active process of producing more varieties and the distinctions between old established varieties is being broken down with improved communications and the spread of previously localised varieties over a much wider area.

Sorghum is readily recognised when in flower or fruit. It is a stout perennial, 1–3m high with a dense, erect or nodding panicle, the spikelets yellow, orange, red or even black. It is very variable in size, being only about 1m in height on the high plateau but usually about 3m in height in the valleys around Ibb. It is cultivated throughout Yemen from sea level to about 2500m but is uncommon in dry country where it is replaced by Bulrush Millet.

Sorghum is almost always rain-fed in Yemen, though it is grown in irrigated land in the Tihama, along the escarpment wadis and in parts of the Mashriq. In most areas it is planted in March or April to coincide with the onset of the spring rains. It has usually grown sufficiently to be able to withstand the partial drought of June and July and comes into flower with the onset of the summer rains at the end of July. The crop is ripe in October. Usually harvesting takes place in three stages: the leaves are first removed for fodder, the head of grain is cut off shortly afterwards and finally the stalks are harvested and stored for animal fodder during the winter.

Although Sorghum is of great value for animal feed, its prime importance is as a grain for human consumption. It is mainly eaten as bread or as a sort of porridge known as ''Asīda'. Until the end of the 1960s, Sorghum was the staple diet of most Yemenis but is now of lesser importance since the influx of large quantities of cheap imported wheat. The decline of Sorghum cultivation is quite marked at the present time and huge areas of marginal land have been abandoned, particularly on the lower slopes of the escarpment and in the Ḥujarīyah. This decline may be of enormous ecological significance as it means that the elaborate terrace system built up over the years may be allowed to decay, thus leading to rapid soil erosion on the steep hillsides.

Maize, **Zea mays** *L.*, known as 'Dhurah Ṣafrah', 'Dhurah ḥindi', or simply as 'Rūmī' or 'Ḥind', is a crop of increasing importance. Vegetatively it is similar to Sorghum but, unlike other grasses, is monoecious, the female flowers developing into the familiar cob of yellow corn. It is cultivated in irrigated land in the Tihama and along the main escarpment wadis. Some areas are able to produce three crops a year on the same land. Maize is cooked over open fires and eaten off the cob or it is ground into flour and made into dry, long-lasting cakes. It is also of great value as animal fodder and it is likely that its increasing cultivation is mostly to satisfy the considerable demand for animal feed.

Both the Common Wheat, **Triticum aestivum** *L.* (*T. vulgare* Vill.), known as 'Būrr' or, more rarely, 'Hanta' or 'Qamḥ', and the Macaroni wheat, **Triticum durum** *Desf.*, known as 'Samrah', are cultivated today. Both are very variable plants, growing to about 1m high in Yemen and are superficially similar. The Common Wheat has a relatively slender, glabrous, often awnless inflorescence, the glumes keeled only in the upper half, whereas the Macaroni Wheat has a relatively stout, often hairy, almost always awned inflorescence with the glumes strongly keeled from the bottom to the top. Both are widely cultivated on the escarpment and the high plateau from 2000 to 3200m but the Macaroni Wheat is much the less common of the two.

Unlike Sorghum, the Wheats seem to have no specific seed time in Yemen. Crops are frequently planted in the winter when little rain can be expected, though other sowings take place to coincide with the main rains. Local strains appear to ripen quickly but their yield is low. The whole plant is plucked from the ground at harvest time and the grain is removed on a threshing floor, the straw being used for animal fodder. Wheat flour is, of course, used primarily for bread making but nearly all the flour on sale in Yemen is now imported and the local grain is mostly consumed in the villages where it is grown.

Although only two species of Wheat are cultivated in Yemen today, a third, Emmer, **Triticum dicoccum** (*Schrank.*) *Schuebl.*, known as ''Alas', was cultivated in the Middle Ages and was found in Manākhah by Schweinfurth at the end of the last century. Emmer is the oldest known Wheat and, unlike the more modern cultivated species, the grain does not come away freely from the palea and lemma. Threshing is therefore difficult and cultivation is unusual anywhere nowadays.

Barley, **Hordeum vulgare** *L.* (*H. deficiens* Steud.), known as 'Sha'īr', is commonly cultivated in the same areas as the Wheats to which it bears a strong superficial resemblance. It differs from them, though, in having 3 (not 1) spikelets at each inflorescence node. Several varieties of Barley are known from Yemen but most, if not all, that is cultivated today is the 2-rowed form in which the spikelets are arranged in two ranks on opposite sides of the rhachis. 4-rowed and 6-rowed varieties have been observed in the past.

Cultivation methods for barley are similar to those of wheat but it is grown mainly for animal fodder. I have no knowledge of barley being used for human consumption.

Rice, **Oryza sativa** *L.*, known as 'Rūz', has been cultivated occasionally in Yemen since medieval times. It is very rare today but there is good evidence that it is cultivated in the areas between 'Abs and Muḥarraq and rumours that it is grown elsewhere. It needs large amounts of water for successful cultivation and so is unlikely ever to be an important crop in Yemen.

As well as in the production of cereal grains, grasses are also of enormous importance as animal feed. Sorghum and Maize are mainly grown for their grain but the dried stalks and leaves are stored on roof tops and in trees and are fed to animals through the dry season. They are not yet used to make silage, though some experimental projects in silage production are taking place in Yemen. Other cereal grasses are also used for fodder but they are of much less importance even though some, notably *Eragrostis tef*, are grown entirely as animal feed.

Despite the importance of the dried cereal grasses during the winter season, the main source of animal fodder is undoubtedly the indigenous grass flora. Almost all grasses except a few species with hardened spinescent leaves such as *Odyssea*, are of some grazing value when young. Indeed, it is very probable that a major proportion of the grass actually consumed by animals is provided by species which are only palatable at the beginning of the season. Such abundant species as *Aristida adscensionis*, *Stipagrostis obtusa* and *Heteropogon contortus* are readily eaten when they first come up but develop hardened awns when old and these can penetrate the mouths of grazing animals. Other species such as *Hyparrhenia hirta* and *Andropogon distachyos* become hardened and unpalatable when old. The following, though, is a short list of grasses which are common in some regions of Yemen and of undoubted grazing value and perhaps worthy of cultivation as animal fodder: *Brachiaria* spp. especially *B. deflexa* and *B. ramosa*, *Cenchrus ciliaris*, *Digitaria abyssinica*, *Echinochloa colona*, *Lasiurus scindicus*, *Panicum coloratum*, *P. maximum*, *Snowdenia polystachya*, *Tetrapogon* spp. There are, of course, a considerable number of species which are rare in Yemen but often cultivated for fodder in other countries and include *Brachiaria brizantha*, *Lolium multiflorum* and *L. perenne*. *Leptothrium senegalense* may be of value on sand dunes since its seed is easily collected.

In recent years grasses have been cultivated on an experimental scale for fodder. No accurate information is available as to what species have been grown but there are reports that species of fodder *Sorghum* and *Brachiaria* have been trialled. None of these fodder grasses has yet been cultivated on a wide scale nor is there any sign of their becoming naturalised, though isolated records of *Lolium perenne* and *Panicum maximum* on the high plateau may, in fact, be relics of cultivation.

Grasses have also been introduced for lawns in recent years and there are a variety of introduced species, not always very suitable, which are found in Sana'a. They include *Dactylis glomerata*, *Festuca rubra*, *Lolium multiflorum* and *Pennisetum clandestinum*.

TYPHACEAE

TYPHA *L.*

Reedmace, Bulrush. Gregarious, rhizomatous plants of wet places, usually growing in shallow water. Stems erect, simple, 1½–4m high. Leaves with broad sheathing bases, long and linear, mostly basal. Inflorescence a long, dense, terminal spike, usually 30–75cm long; the upper part male, yellowish, quickly deciduous; the lower part female, brown, persistent for months after the male flowers have fallen. Fruit a small nut.

Leaves hemispherical in cross-section with a rounded back, 5–10mm wide ... *T. domingensis*
Leaves strongly keeled on the lower surface, 20–30mm wide ... *T. elephantina*

1. Typha domingensis *Pers.* — 'Alāl

T. angustata Bory & Chaub.

Tall reed to 2m high. Leaves 5–10mm wide, flat on the upper surface, rounded below, hemispherical in cross-section. Inflorescence 30–40cm long, the male and female parts separated by a distinct gap.

Widely distributed throughout Yemen from 300 to 2400m but virtually restricted to the margins of slow-moving streams and thus of rather scattered, infrequent occurrence. It occasionally forms reed beds, notably by the Wadi Ḥidrān near Taiz.

2. Typha elephantina *Roxb.*

More robust than the last species, reaching 4m in height. Leaves 20–30mm wide, flat on the upper surface, strongly keeled below. Inflorescence 50–75cm long, the male and female contiguous.

In similar situations to the last species but much rarer: between 'Ubal and Madīnat Al 'Abīd (Ahmed Khattab); Wadi Sukhnah; at the head of the Wadi Khārid where it forms a massive reed bed.

MUSACEAE

Sterile cultivated hybrids of *Musa* (usually recorded under the name *M. paradisiaca* L.) are grown as Bananas (Mawz) below about 2000m on the escarpment and in the Tihama but they only flourish in the rich, irrigated valleys of the escarpment foothills. Several varieties are grown and banana trees with their huge, torn leaves and long pendulous flowering stems are a familiar sight in the low-lying valleys. Banana production though has fallen on hard times; as late as 1975 all bananas were locally produced but since then disease and an influx of imported fruit have driven nearly all the local produce off the market.

ZINGIBERACEAE

Two species of this family are cultivated in the central escarpment and perhaps elsewhere.

Turmeric, *Curcuma longa* L. (Ḥurūd) is a rhizomatous plant about 1m high with large, ovate leaves. The rhizomes are dried to obtain a yellow powder well known for flavouring and dyeing food but more commonly used by Yemeni women for face colouring. On J. Milḥān where this plant is grown, nearly all women have yellow-dyed faces. At present little of this plant is grown in Yemen and most turmeric is imported.

Ginger, *Zingiber officinale* L. (Zinjabīl) is another rhizomatous plant with erect stems about 1m high but with narrowly lanceolate leaves. It is cultivated in a few places below 1500m in the central escarpment, particularly on J. Bura' and the Bani Wākit district of the Ja'farīyah region of J. Raymah. The rhizomes produce the well-known ginger root which is mainly used in Yemen for flavouring Qishr, the drink made from the hulls of the coffee bean. Most ginger used in Yemen is imported.

CANNACEAE

CANNA *L.*

1. Canna indica *L.*

Erect herb to 1½m from a tuberous underground rhizome. Leaves borne on the stem, 40–50cm long, broadly lanceolate with a distinct midrib and a petiole forming a sheath round the stem. Inflorescence a large terminal raceme of red flowers, each formed of 3 brightly coloured linear-lanceolate perianth segments. Fruit a large, tubercled capsule.

Commonly cultivated in gardens in Sana'a and Taiz but also well naturalised on grassy banks by permanent wadis between 500 and 1500m in high rainfall areas: Wadi Ḍabāb, Taiz; Wadi Hadabah and Jabalayn, 'Udayn; above 'Alūjah, J. Raymah; Wadi Sukhnah; Wadi 'Ayn 'Ali, Ḥajjah.

LILIACEAE

Herbs or shrubs, usually with underground storage organs such as bulbs, rhizomes or thickened tuberous roots. Leaves usually narrow, often basal, withering in the dry season in most genera but in some (*Aloe*, *Kniphofia*) evergreen; rarely (*Asparagus*) reduced to scales. Flowers regular, bisexual, usually in racemes (umbels in *Allium*), the perianth segments free or united into a tube. Fruit a capsule or berry.

Many of the species are very attractive and there are many endemics, though only in the genera *Aloe* and *Kniphofia*.

1. Woody plants, usually climbing, sometimes spiny, always with leaves reduced to scales and replaced by needle-like cladodes. Fruit a berry — 1. *Asparagus*
Leaves normally developed. Plants never climbing or spiny, rarely woody. Fruit a capsule — **2**
2. Leaves fleshy, 2–10mm thick, usually with prominent, widely spaced marginal teeth. Inflorescence usually branched — 3. *Aloe*
Leaves at most slightly fleshy, 1mm or less thick, usually entire. Inflorescence simple (except in *Asphodelus*) — **3**
3. Leaves with scabrid-dentate margins. Perianth tubular with very short free lobes — 2. *Kniphofia*
Leaf margins entire. Perianth segments free or united near the base only — **4**
4. Leaves basal — **5**
Leaves in whorls on the upper part of the stem, the lower part of the stem leafless — 13. *Littonia*
5. Bulbous plants with purple flowers — **6**
Rhizomatous or bulbous plants with white, yellow or brown flowers — **7**
6. Plant stemless when flowering. Leaves linear, unspotted. Perianth segments 30–40mm long — 12. *Merendera*
Inflorescence borne on a distinct peduncle. Leaves lanceolate, usually spotted. Perianth segments c 6mm long — 11. *Scilla*
7. Flowers yellow or greenish-brown. Plant bulbous or rhizomatous — **8**
Flowers white or pinkish-tinged. Plant never bulbous — **10**
8. Filaments covered in yellow hairs. Plant rhizomatous — 5. *Bulbine*
Filaments glabrous. Plant bulbous — **9**
9. Flowers yellow. Bracts ± equalling the perianth — 10. *Albuca*
Flowers greeny-brown. Bracts much shorter than the perianth — 9. *Dipcadi*
10. Inflorescence of branched racemes. Leaves terete, scabrid — 4. *Asphodelus*
Inflorescence a simple raceme or umbellate. Leaves neither terete nor scabrid — **11**
11. Seeds grey. Peduncle pubescent below — 7. *Trachyandra*
Seeds black. Peduncle glabrous — **12**
12. Seeds pyramidal, angular. Flowers single on the raceme rhachis — 6. *Anthericum*
Seeds discoid, rounded. Flowers in pairs on the raceme rhachis or in umbels — 8. *Chlorophytum*

1. ASPARAGUS *L.*

Perennial woody herbs with branched, often climbing stems. Roots tuberous or fibrous. True leaves reduced to scarious scales or spines and replaced by green, leaf-like linear cladodes which are solitary or clustered. Flowers solitary, clustered or in racemes, axillary or terminal; perianth segments free, white, 2–3mm long. Fruit a small orange-red berry, usually 1-seeded.

1. Glabrous spiny plants. Flowers in racemes — **2**
Unarmed plants or, if spiny, scabrid on young growth. Flowers solitary or in clusters — **3**
2. Cladodes flattened — *A. falcatus*
Cladodes square or triangular in cross-section — *A. racemosus*

3. Flowers and cladodes usually solitary. Plant glabrous *A. virgatus*
Flowers and cladodes in clusters. Young growth scabrid *A. africanus*

1. Asparagus africanus *Lam.*

A. asiaticus auct.
A. mitis A. Rich.
Branched woody plant climbing to 3 or 4m over scrub vegetation. Stems smooth or slightly grooved, usually green, scabrid when young. Spines present on old wood only, straight. Cladodes in dense clusters of up to 15, terete, rigid, needle-like, 5–10mm long. Flowers in axillary clusters of 1–5 (–20), the pedicels c 5mm long.

Widespread between 1000 and 3400m on the escarpment, high plateau and Mashriq, growing in places protected from grazing, particularly cliffs, bushes along watercourses in dry country and in scrub on the escarpment.

2. Asparagus falcatus *L.*

Branched glabrous climber. Stems usually tortuous, smooth, grey. Spines straight or slightly falcate, numerous. Cladodes in small clusters of 1–5, flattened, linear, 2–3cm long, 1–3mm broad. Flowers in lax racemes 2–3cm long.

An uncommon plant of cliffrs and steep rocky slopes in the wetter parts of the escarpment between 1000 and 2200m: Naqīl Jūbirān near Turbah; Wadi Dūr, 'Udayn; Bani Mukhtār above Ḥadīyah; between Suq Ar Ribāṭ and Bani Dawān, J. Raymah; above Ḥillah, J. Bura'; Walajah-'Ukaybir, J. Milḥān; J. Jinān (Wissmann).

3. Asparagus racemosus *Willd.*

Branched glabrous climber. Stems grooved, yellowish when young. Spines straight, numerous. Cladodes in dense clusters, numerous, square or triangular in cross-section, 1–3cm long and up to 1mm broad. Flowers in solitary or paired lax racemes.

Only known from 1000–1200m above Ḥillah on J. Bura' (Schweinfurth). The plant has not been refound and it is possible that it is an error for *A. falcatus*, which grows in the same locality today.

4. Asparagus virgatus *Baker*

Erect, branched, glabrous undershrub to 50cm. Stems grooved, dark green. Spines absent. Cladodes usually solitary, rarely in clusters of up to 3, square in cross-section, up to 1cm long. Flowers solitary on pedicels c 1cm long.

Only known from the western slopes of J. Milḥān where it is frequent on cliffs and terrace walls between 1800 and 2400m. (Wood 1212, 2380, 2865)

The Yemen plant does not seem to differ from African examples of this species which is unknown north of southern Tanzania.

Asparagus officinalis *L.*, which is cultivated for its edible young shoots, was recorded by Schweinfurth in the late nineteenth century but there is no evidence of its cultivation in recent years.

2. KNIPHOFIA *Moench.*

1. Kniphofia sumarae *Deflers* Būnīt

Rhizomatous herb with an erect stem to 1m. Leaves in a basal tuft, linear, keeled, 15–40cm long, 5–10mm wide, the margin scabrid-dentate. Inflorescence a dense, many-flowered, simple raceme 6–18cm long; bracts short, scarious; perianths with a cylindrical tube and 6 linear lobes, deep yellow, 8–9mm long. Fruit an ovate capsule with black seeds.

Flowers May–July.

A spectacular endemic species restricted to steep grassy slopes and cliff gullies between 2700 and 3200m on the main Ibb mountains: J. Ṣabir, Ta'kar, Qarad, Ba'dān, Ḥubaysh, Manār and Sumārah. It is named after J. Sumārah.

3. ALOE *L.*

A distinctive genus of succulent herbs or shrubs. Stems erect, trailing or absent. Leaves fleshy with toothed (rarely entire) margins, usually clustered though sometimes ± distichous, the clusters either at the apex of the stem or, in stemless species, forming a basal rosette. Inflorescence of simple or branched racemes; bracts papery; perianth a cylindrical tube with free lobes, yellowish or reddish. Fruit a capsule with black seeds.

A large and complex genus in which all our species, with the possible exceptions of *A. inermis* and *A. vera* variety *officinalis*, are endemic to southwest Arabia. The species are not easily separated from each other but are much more distinct in the field than in the herbarium since the habit of the plant is of great importance. It is also of value to note whether the plant suckers or not, the leaf colour, the direction of the leaf tip, the flowering season and the flower colour, though most species have two colour forms.

The species belong to seven groups: *A. vera*; *A. sabaea*; *A. inermis* and *A. fleurentinorum*; *A. rubroviolacea* and *A. sp.A*; *A. pendens*, *A. yemenica* and *A. rivierei*; *A. vacillans* and *A. castellorum*; *A. tomentosa*, *A. doei*, *A. menachensis*, *A. niebuhriana*, *A. sp.B A. sp.C*. Although the seven groups are relatively well defined, the species within each group clearly have a common and fairly recent origin. Within a group the distribution of species is complementary and they do not in general occur together but, where the range of two species overlaps, intermediates may be expected to occur and such forms will not key out satisfactorily. Hybrids between species in different groups also occur but are very rare.

All Aloes contain a latex which is of considerable medicinal value. It is particularly efficacious in treating sores and infected wounds and is widely used in traditional medicine in Yemen and many other countries. Aloes are often called 'Sabir' but sometimes ' 'Asaf', 'Khīr' or other names.

1. Leaves entire 2
Leaves with toothed margins 3
2. Leaves green. Pedicels ± 7mm long *A. inermis*
Leaves blackish. Pedicels ± 15mm long *A. fleurentinorum*

3. Leaves 60–90cm long, pendulous from the apex of an erect stem 2–3m high. Perianths c 1cm broad *A. sabaea*
   Leaves mostly less than 60cm long, not pendulous from the apex of an erect stem. Perianths less than 7mm broad **4**
4. Bracts 20–30mm long **5**
   Bracts up to 20mm long **6**
5. Flowers red. Latex staining dark red *A. rubroviolacea*
   Flowers yellow. Latex not staining *A. sp.A*
6. Plants with a distinct erect or trailing stem. Pedicels slender, filiform **7**
   Plants stemless, rarely a short erect stem present on old plants, but always the majority of plants in a colony stemless. Pedicels stout, never filiform **9**
7. Leaves distichous, not clustered, about as thick as broad. Inflorescence simple. Stem trailing or pendulous *A. pendens*
   Leaves usually clustered at the top of the stem except in young plants, broader than thick. Inflorescence branched. Stem erect, trailing or pendulous **8**
8. Stem erect. Leaves up to 55cm long *A. rivierei*
   Stem erect or trailing but always becoming pendulous in the end. Leaves up to 40cm long *A. yemenica*
9. Leaf teeth c 7–8mm apart **10**
   Leaf teeth c 12–15mm apart **11**
10. Leaves yellow-green. Flowers yellow *A. castellorum*
    Leaves glaucous. Flowers yellow or red *A. vacillans*
11. Tihama plant suckering to form large clumps. Perianths glabrous. Leaves yellow-green with large, mostly spreading, teeth *A. vera*
    Plant usually solitary, rarely suckering. Perianths puberulent to coarsely hairy (very rarely glabrous). Leaves grey or grey-green with falcate teeth **12**
12. Perianths very shortly pubescent or subglabrous, the hairs not visible to the naked eye **13**
    Perianths clearly hairy to the naked eye **14**
13. Perianths red or yellow *A. sp.B*
    Perianths red with yellow-green tips. Leaves grey *A. niebuhriana*
14. Bracts c 5–7mm long **15**
    Bracts more than 10mm long **16**
15. Perianths with long hairs at the tips only, red or yellow *A. doei*
    Perianths uniformly hairy, whitish-green *A. tomentosa*
16. Perianths (? always) red; bracts 10–15mm long *A. menachensis*
    Perianths yellow (rarely red); bracts 15–20mm long *A. sp.C*

1. Aloe castellorum *J.R.I. Wood*

Similar to *A. vacillans* from which it is immediately distinguished by its yellow-green leaves with numerous small brown teeth. The flowers which are always yellow are found from July to November.

Locally abundant between 1300 and 2200m on west-facing mist-covered slopes on the northern escarpment: Ḥajjah; below Kuḥlān; Maḥābishah (abundant); Shahārah; J. Marran; J. Rāziḥ.

2. Aloe doei *Lavranos*

Stemless aloe with large rosettes of grey-green leaves similar to *A. tomentosa* but less robust, differing in the perianths which are coarsely hairy at the apex only and are equally commonly red or yellow in colour. It flowers from March to May.

Locally abundant on dry stony hills between 1000 and 2000m in the southeastern mountains extending from Taiz and Raḥida northeast to Damt, Rada', Watadah and Dhī Nā'im.

3. Aloe fleurentinorum *Lavranos & Newton*

Solitary stemless aloe with entire, often distichous leaves, similar to *A. inermis* but never suckering, the leaves always blackish, the pedicels much longer (c 15mm long) and the perianths always red. It flowers throughout the year but mainly after heavy rain.

Widespread but infrequent between 1300 and 2300m on dry stony, usually limestone, hills in the Mashriq from Wadi Ḍahr (Lavranos) and Nihm northwards; there are also isolated occurrences on the escarpment at Mafḥaq and on J. Sharafayn.

4. Aloe inermis *Forssk.* 'Iblī

Readily recognised by its entire, light green leaves but otherwise rather variable. Solitary or suckering to form small colonies, the stem short and decumbent or completely absent. Leaves distichous particularly in young plants or forming a basal rosette. Inflorescence branched; bracts short; pedicels c 7mm long; perianths yellow or more commonly red, glabrous, strongly constricted above the ovary. It flowers throughout the year following any considerable rainfall.

Widespread and locally abundant between 1000 and 2000m on dry stony hills in the southeast mountains, extending from the Ḥujarīyah, where it is common, and Taiz, northeast to Dhī Nā'im and the Wadi Nuwāsh north of Rada'.

A hybrid, *A. inermis x vacillans*, is known from Taiz where both parents grow together (Wood 1978).

5. Aloe menachensis (*Schweinf.*) *Blatter*

A. trichosantha Berg. var. *menachensis* Schweinf.

Very similar to *A. tomentosa*, differing in its perianths which are red and pubescent (not coarsely hairy) and in its bracts which are 10–15mm long. It flowers from March to May.

Known with certainty only from the villages of Muthmūr (Schweinfurth, 1685) and Bayt Mughlad (Wood 3233) on J. Shibām in the Ḥarāz and on J. Sumārah (Lavranos). It seems probable that this species is of hybrid origin.

6. Aloe niebuhriana *Lavranos*

Stemless aloe with large rosettes of grey leaves, usually suckering to form clumps. Leaves spreading or ascending, the marginal teeth 12–15mm apart. Inflorescence sparingly-branched; the bracts c 8mm long; pedicels c 5mm long; perianths brick-red with yellow-green tips, puberulent. Flowers following the first rains of the year, thus usually in spring.

Widespread and frequent from 200 to 1500m from the Wadi Zabīd southwards-growing in the stony steppe of the southern Tihama and on the lower escarpment slopes west of the watershed, particularly common near al Mafraq, in the Barakānī-Nashimah area and northwest of Taiz.

A. niebuhriana intergrades with *A. doei* around Taiz and with *A. sp.B* in the Wadi Zabīd area.

7. **Aloe pendens** *Forssk.*

A. arabica Lam.

A small distinctive suckering species with a short decumbent stem which is often pendulous on cliffs. Leaves widely spaced, distichous, c 20cm long, very fleshy and about as thick as wide, usually white-spotted or blotched when young and bronze when older, the marginal teeth minute, c 4–7mm apart. Inflorescence unbranched, recurved; bracts c 10mm long; pedicels filiform c 6mm long; perianths yellow or (rarely) red-brown, glabrous. Flowers throughout the year but mainly in summer.

Abundant between 1500 and 2500m on cliffs and steep rocky slopes on the mountains between Ḥadīyah and Kusmah, extending northwards on the western slopes of J. Raymah to Al Jabīn where it is less common.

8. **Aloe rivierei** *Lavranos & Newton*

A robust suckering species with an erect leafless stem c 1m high. Leaves in a dense cluster at the top of the stem, erect, pale green, up to 55cm long, the marginal teeth 4–7mm apart. Inflorescence erect, branched; bracts c 10mm long; pedicels filiform, c 10mm long; perianths red-brown (yellow only in Wadi Ḍabāb, Taiz), glabrous. Flowers from October to December. Might be confused with *A. sabaea* but easily distinguished even when sterile by the erect leaves.

Local on cliffs and steep rocky slopes between 1300 and 1900m in Taiz and Ibb provinces, usually in small quantity; it is most common in the Ḥujarīyah but the biggest colony is at Wadi Ḍabāb. There is an outlying colony at a suq below Shahārah.

9. **Aloe rubroviolacea** *Schweinf.*

A very distinctive and attractive aloe with a trailing stem which is commonly pendulous on cliffs. Leaves in a dense cluster at the apex of the stem, reddish-violet, the tips strongly deflexed. Inflorescence branched, erect, or on pendulous plants, recurved; bracts 25–35mm long; pedicels very short, c 2mm long; perianths red, minutely puberulent or subglabrous.

Flowers from October to December.

A characteristic plant of cliffs in areas of winter mist on the escarpment mountains; thus restricted to 2500–3300m in most areas but descending to 2000m in the central escarpment, at Turbah and on J. Rāziḥ, all areas of low-level winter mists. It often grows in inaccessible places and is associated with *Aeonium*.

10. **Aloe sabaea** *Schweinf.* Jishb

The largest aloe in Yemen, having an erect stem 2–3m high. Leaves clustered at the top of the stem, pendulous, green, 60–90cm long, the marginal teeth white. Inflorescence erect, branched; bracts c 15mm long; perianths red-brown, glabrous, c 1cm broad, much wider than in other species. Flowers from November to January.

Widespread on steep rocky hills, rarely on true cliffs, between 600 and 1800m along the whole escarpment; usually in small quantity, but sometimes abundant, notably around Hammām 'Alī and in the Wadi Ḥijān.

11. **Aloe tomentosa** *Deflers*

A. tricasantha var. *menachensis* sensu Schwartz

Solitary aloe with massive basal rosettes of large, spreading or ascending grey-green leaves with marginal teeth 12–15mm apart. Inflorescence erect, branched, the racemes often short and conical; bracts 5–7mm long; perianths whitish-green, coarsely and uniformly hairy, the indumentum spreading to the pedicels, bracts and peduncle. Flowers from April to September.

Although usually quite distinct, *A. tomentosa* intergrades with other species. Plants from the wadi above 'Adūf near Turbah and from al Qabayn are probably intermediates with *A. doei*. Plants from the gypsum mines near Ghirās are intermediate between *A. tomentosa* and *A. sp.C*. *A. menachensis* may be a hybrid between *A. tomentosa* and *A. sp.B*. All such intermediates require further investigation.

Frequent on rocky hillsides between 2400 and 3100m on the mountains surrounding the high plateau. It is absent from the outer escarpment apart from isolated records from J. Ṣabir, Mirkiz 'Utmah and J. Shibām in the Ḥarāz and generally shuns the damp misty slopes favoured by *A. rubroviolacea*.

12. **Aloe vacillans** *Forssk.*

A. dhalensis Lavranos

A. audhalica Lavranos & Hardy

Stemless aloe, usually solitary but sometimes suckering. Leaves in a basal rosette, rather small, erect, glaucous with small, brown marginal teeth c 7–8mm apart. Inflorescence branched; bracts 10–12mm long; pedicels 5–12mm long; perianths red-brown or yellow, glabrous. Flowers from August to November.

Very distinctive at any season because of the small, erect, glaucous leaves. Yellow-flowered forms predominate in Taiz province, both colours are about equally frequent in Ibb province, but elsewhere red-flowered forms are the rule.

The commonest aloe in Yemen, being found on open rocky hills between 1300 and 3000m everywhere on the escarpment, high plateau and southeast mountains.

13. **Aloe vera** (*L.*) *Burm. f.* Ṣabir

A. barbadensis Miller

Freely suckering aloe, spreading to form large colonies, either stemless with the leaves in a basal rosette or with a short erect stem less than 50cm long and the leaves in a cluster at the top of the stem. Leaves erect or ascending, narrow, yellow-green (sometimes white-spotted when young), with rather large, spreading marginal teeth 12–15mm apart. Inflorescence branched; bracts c 10mm long; pedicels stout, c 8mm long; perianths red or yellow, glabrous. Flowers mostly from October to January.

Can be difficult to separate from *A. sp.B* which also has both colour forms in equal frequency. *A.vera*, though, always grows in large clumps and its narrow, yellow-green leaves with spreading teeth are very distinctive.

The Yemen plant is variety **officinalis** (*Forssk.*) *Baker* (var. *angustifolia* Schweinf.) which differs from the aloe of commerce and medicine (var. *vera*) in its yellow-green leaves and rather slender peduncle. How significant these features are is uncertain and it may be that the cultivated Aloe in fact originated in Yemen.

Widespread between sea level and 800m from the Wadi Zabīd northwards with a single outlying station further south between Ḥays and Madrūbah. Although found on the western sandy parts of the Tihama it is characteristic of stony outcrops in the alluvial plains in the east and in the escarpment foothills where it sometimes occurs in great abundance, particularly below J. Raymah.

14. Aloe yemenica *J.R.I. Wood*

A. pendens sensu Schweinf.

Small aloe suckering and growing in clumps, the stems at first erect, finally bending over and becoming pendulous. Leaves loosely clustered at the top of the stem, sometimes ± distichous, spreading, with deflexed tips, yellow-green (sometimes white-spotted when young), c 30–40cm long, the marginal teeth 8–10mm apart. Inflorescence similar to that of *A. rivierei*, except that the peduncle is recurved on pendulous plants; pedicels filiform, 8–12mm long; perianths red (yellow only at Bayt Bashārī, Maḥwīt), glabrous.

Flowers from September to December.

Distinguished from *A. rivierei* by its more slender habit, pendulous stems and spreading leaves. Some plants approach *A. pendens* but the leaves are never as thick and most plants in the colony will have branched inflorescences.

A widespread but rather local aloe found from 800 to 2000m on the escarpment from Suq Ar Rūbū' in Waṣāb al 'Ālī north to Maḥābishah, but largely absent from J. Raymah where it is replaced by *A. pendens*. In the central escarpment, particularly on J. Milḥān and the Ḥarāz, it is common generally on cliffs but in drier areas (Wadi Sukhnah, Wadi 'Ayn 'Alī near Ḥajjah, Wadi Qūsa' near Madīnat Al 'Abid) it is restricted to cliffs above permanent flowing streams.

15. Aloe sp.A

A little-known plant probably similar in habit to *A. rubroviolacea* but with light green leaves lacking the dark red latex of that species. The perianths are yellow and the bracts are rather narrower than in *A. rubroviolacea*.

Only known from cliff ledges at 1800m below Turbah on the path to Maqāṭirah (Wood 961). It has not been refound despite repeated search.

16. Aloe sp.B

Solitary stemless aloe with large rosettes of spreading or ascending grey-green leaves, the marginal teeth 12–15mm apart. Inflorescence erect, branched; bracts 10–12mm long; perianths red or yellow, puberulent or subglabrous.

Flowers from April to June.

This is the common spring-flowering aloe found between 500 and 1800m on the escarpment from the Wadi Zabīd north at least to Maḥābishah. It is particularly fine in the Mafḥaq region. (Wood 3275)

17. Aloe sp.C

Similar to *A. tomentosa* from which it is readily distinguished by its much longer bracts, which are 15–20mm long. The perianths are yellow-green, not whitish. Flowers from April to June.

A. sp.C is a relatively widespread species but is little known and apparently rather variable. Some plants, notably Wood 1601 from Ḥūth, have rather glaucous leaves covered in a whitish bloom, though dull grey-green is the usual colour. Specimens intermediate with *A. tomentosa* occur at Ghirās. On the eastern slopes of J. Sharafayn plants intermediate with *A. sp.B* occur; they can be recognised by their shorter indumentum and may be red- or yellow-flowered.

A. sp.C is widely distributed in the Mashriq between 1000 and 2100m from Al 'Ashshah and Nihm north across the Saudi Arabian border and west to Maḥābishah and the mountains of Khawlan Ash Sham but is local and appears to be absent from considerable tracts of country. (Wood 3249, 3264) Its distribution closely parallels that of *A. fleurentinorum*.

4. ASPHODELUS *L.*

1. Asphodelus fistulosus *L.*

A. tenuifolius Cav.

Annual tufted herb with fibrous roots and an erect leafless stem to 60cm. Leaves basal, fleshy, terete, filiform, minutely scabrid, 10–20cm long. Inflorescence a panicle formed of sparingly branched, lax, erect racemes; flowers solitary or in small clusters; perianth segments, free, equal, white with a reddish midrib, 4–5mm long, deciduous before the fruit ripens. Capsule globose.

The Yemen plant is variety **tenuifolius** (*Cav.*) *Baker* and flowers throughout the year.

An uncommon plant found on stony roadsides and mountain slopes between 2400 and 2900m mainly in the high plateau: J. Ḍawrān (Wissmann); Yislah pass; between Yāzil and Baw'ān; Khamr; Shahārah.

5. BULBINE *Willd.*

1. Bulbine abyssinica *A. Rich.*

B. asphodeloides (*L.*) J.A. & J.H. Schult.

Shortly rhizomatous herb with fleshy roots and an erect stem to 40cm. Leaves in a basal tuft, erect, narrowly linear, soft, fleshy and ± terete, 5–20cm long. Inflorescence a simple raceme of bright yellow flowers which is conspicuously denser upwards, 5–10cm long; perianth segments free, c 8mm long, persistent round the globose capsule; filaments pilose.

Flowers in June and July.

A plant of stony hillsides from 2400 to 3000m on the higher mountains around Ibb, J. Shibām in the Ḥarāz, Kawkabān, and J. Maḥdad; it is usually in small quantity but sometimes grows in abundance, notably on J. Nu'mān.

6. ANTHERICUM *L.*

Rhizomatous perennials with fibrous or thickened roots, sometimes with tubers near the tips (*A. inconspicuum*). Leaves in a basal tuft. Inflorescence a raceme with solitary flowers in our species; perianth segments white with a dark central line, persistent. Capsule trigonous, shallowly lobed. Seeds black, sharply angled, irregularly compressed.

Perianth segments c 3mm long; pedicels jointed in the lower half — *A. inconspicuum*
Perianth segments c 10mm long; pedicels not jointed — *A. corymbosum*

1. Anthericum corymbosum *Baker*

Small herb with thickened, slightly tuberous roots and folded leaves up to 16cm long and 7mm wide, glabrous except for the short ciliate hairs on the margins. Inflorescence a 5–12-flowered, corymbose raceme; the peduncle ascending, glabrous, 3–6cm long; pedicels not jointed, erect, elongating in fruit; perianths c 10mm long.

Only known from grassland in the graveyard of the village on the east of the summit of J. Ta'kar at 2800m. (Wood 75/244)

Outside Yemen it is a rare plant of Ethiopia and Somalia.

2. Anthericum inconspicuum *Baker*

Small herb with distinctly swollen tubers near the root tips. Leaves linear, up to 27cm long but only 1–4mm wide, entirely glabrous. Inflorescence a lax raceme with 4–10 distant flowers; peduncle ascending, glabrous, c 5cm long; pedicels jointed in the lower half, ascending; perianths c 3mm long.

Only known from a rocky gully at 1900m on the escarpment west of Dhī Nā'im. (Wood 2748)

This aptly named plant is only known elsewhere from a few localities in Somalia, but is very easily overlooked.

7. TRACHYANDRA *Kunth*

1. Trachyandra saltii (*Bak.*) *Oberm.*

Rhizomatous perennial with narrowly linear, usually pilose leaves, 10–20cm long, 1–2.5mm wide, arranged in a basal tuft. Inflorescence a 5–20-flowered raceme, the peduncle 3–5 (–10)cm long, pubescent, usually strongly curved so the raceme itself lies outside the tuft of leaves; flowers single, the pedicels jointed at the top; perianth segments 7–9mm long, white with a brown central line. Capsule trigonous, shallowly lobed; seeds grey, angular but not irregularly compressed.

A rare plant of shallow grassland between 2200 and 2400m in Ibb province: in the forestry plantation south of Ibb; Suq Shīrqī, Mudhaykhirah. (Wood 75/267)

8. CHLOROPHYTUM *Kunth*

Closely related to both *Anthericum* and *Trachyandra* from both of which it can be reliably distinguished only by the fruit. The trigonous capsule is deeply lobed, the lobes ± oblong in shape; the black seeds are circular in outline and ± discoid in shape. Our plants are more robust than those in the two related genera and the inflorescence is either umbellate or, if racemose, the flowers are often paired, unlike those of *Anthericum* and *Trachyandra*.

Flowers in racemes. Leaves linear — *C. tordense*
Flowers in umbels. Leaves ovate — *C. tetraphyllum*

1. Chlorophytum tetraphyllum (*L.*) *Baker*

Low herb with swollen tubers near the root tips. Leaves flat, ovate or sword-shaped, up to 15cm long and 2.5cm wide, the margin lined with short ciliate hairs. Inflorescence umbellate, very short, the peduncle glabrous, flattened, less than 2cm long or apparently absent; umbel 5–20-flowered, subtended by long, scarious, linear bracts; pedicels 5–30mm long, not jointed; perianth white, 6–8mm long.

Abundant on open stony hills between 1600 and 2100m for a few kilometres north and west of Ibb. It is one of the most characteristic of a number of bulbous or tuberous plants which come up rapidly after the spring rains and are at their best in June. (Wood 75/390)

Outside Yemen known only from Ethiopia.

2. Chlorophytum tordense *Chiov.*

Herb to 40cm high, the roots stringy but apparently without tubers. Leaves flat or folded, up to 30cm long and 5mm wide, the margin glabrous. Inflorescence a raceme up to 15cm long, borne on a glabrous flattened peduncle 5–25cm long; flowers single or paired; pedicels 3–7mm long, jointed just below the middle; perianth white, 9–10mm long.

Scattered on the escarpment between 1200 and 1900m from Turbah north to the Wadi Lā'ah, but often found as isolated individuals. It is most typical of shallow grassland on steep slopes in high rainfall areas where it is associated with *Hypoxis angustifolia* and *Sporobolus festivus*.

9. DIPCADI *Medik.*

Bulbous herbs. Leaves basal, entire. Inflorescence a simple raceme; bracts small with scarious margins. Flowers greeny-brown, the perianth segments persistent, forming a tube, dissimilar, the 3 outer with long, acuminate, spreading tips, the 3 inner shorter, not acuminate. Fruit a capsule.

Leaves flat or folded, 5–15mm wide — *D. viride*
Leaves usually inrolled, less than 3mm wide — *D. filifolium*

1. Dipcadi filifolium *Baker*

Leaves inrolled (rarely flat) 10–30cm long, 1–3mm wide. Stem erect, 30–40cm high with 10–15 flowers along a rhachis 6–12cm long; perianth 12–14mm long in Yemen plants.

A very rare plant known from montane grassland at 2100m on J. Ẓallamlam above Ḥadīyah (Wood 2811) and from grassy cliff ledges at 2700m between Shibām and Kawkabān (Wood 2757).

2. Dipcadi viride (*L.*) *Moench.*
D. tacazzeanum (*Hochst.*) *Baker*
Leaves flat or folded, 4–20cm long, 5–15mm wide. Stem erect, 5–100cm high with 5–30 flowers along a rhachis 4–25cm long; perianth 15–20mm long in Yemen plants.

A variable species divisible into two forms which, though intergrading in Africa, appear to be distinct in Yemen. *Form 1* (Wood 75/200, 1578) is a very vigorous plant with leaves 10–20cm long, and an inflorescence 50–100cm high with 15–30 flowers along a long rhachis. It is apparently a rare plant found in a few localities between 1700 and 2500m in areas of high rainfall on the escarpment: above Dimarayn, J. Ṣabir; between Qā'idah and Sayyāni; Jabin, J. Raymah; J. Shibām, Manākhah; below Shamsān, Maḥābishah. *Form 2* (Wood 692, 2764; Steinberg 978, 1172) is a small plant with leaves 5–10cm long and an inflorescence 5–15cm high with 5–10 flowers along a short rhachis. It is widely distributed on dry stony plain in the Mashriq up to 2300m but also occurs rarely on the escarpment in similar places and has been found at Ḥajjaylah, Mafḥaq, Maḥābishah and in the Wadi Līyah in Khawlan Ash Sham, where it descends to 600m.

10. ALBUCA *L.*

1. Albuca abyssinica *Jacq.*
Bulbous herb with an erect stem to 80cm. Leaves usually 6–8, basal or arising from the lower part of the stem, erect, linear, entire, flat or folded, 30–60cm long, 0.5–2cm broad. Inflorescence a simple raceme 10–25cm long with 10–30 flowers; bracts papery, equalling the perianths; perianths persistent, 12–20mm long, the segments free to the base, yellow with a greeny-brown midrib, the inner 3 hooded and ± enclosing the filaments. Fruit a capsule with black seeds.

Widespread on cliffs and steep rocky slopes in the wetter parts of the escarpment from 1000 to 3100m north to Shahārah; it usually grows either in areas of winter mist or on river cliffs where it is at least partially moistened by spray.

11. SCILLA *L.*

1. Scilla hyacinthina (*Roth*) *Alston*
S. yemensis Deflers
S. sickenbergeri Deflers
Bulbous herb with several erect flowering stems to 20cm. Leaves all basal, ascending, lanceolate, 5–15cm long, 1–2.5cm wide, usually dark-spotted. Inflorescence a simple, many-flowered raceme, 3–8cm long; the rhachis, small bracts, pedicels and perianths all suffused with mauve; perianth segments, free, equal, c 6mm long, persistent around the capsule.

A plant of shallow stony soils in areas of average or good rainfall between 1200 and 2800m, very commonly associated with *Dorstenia barnimiana*: widespread but infrequent throughout Taiz province; abundant throughout Ibb province except the east; very rare elsewhere and only known from Dan in Waṣāb al 'Ālī, Suq al Jum'ah in J. Ash Shirq and Bani Mukhtār above Ḥadīyah.

12. MERENDERA *Ram.*

1. Merendera abyssinica *A. Rich.*
Bulbous herb. Flowers and leaves appearing together in a small tuft in early spring. Perianth segments free, oblanceolate, purple, 3–4cm long, nearly as long as the leaves. As the flowers die, a short stem, 1–3cm high, appears; leaves 3 per stem, linear, elongating in fruit and finally reaching 15cm in length and forming a perianth round the subsessile capsule. The only Yemen *Liliaceae* with 3 styles.

An attractive crocus-like plant which is locally abundant on grassy banks between 2100 and 3200m on the wetter escarpment mountains from Akāḥilah near Turbah north to J. Maswār. It is particularly abundant on J. Ta'kar. Outside southwest Arabia it is only known from Ethiopia and Somalia.

13. LITTONIA *Hook. f.*

1. Littonia minor *Deflers*
Bulbous herb with an erect stem to 20cm, leafless at the base and in the lower half. Leaves in whorls of 4–5 in the upper part of the stem, linear, 4–9cm long, 2–5mm wide. Inflorescence a 1–3-flowered terminal raceme; perianth segments free, linear-oblanceolate, white, persistent. Fruit an ellipsoid capsule, c 2cm long.

A plant of sandy plain known from the Wadi' Antirīyah, below Ḥayfān (Deflers). Outside southwest Arabia it is only known from Somalia.

Although many species of *Liliaceae* are cultivated as ornamental plants in temperate countries, the only species grown as an ornamental in Yemen is the white-flowered **Lilium candidum** *L.*, known as 'Rinjās', which is still cultivated in villages above Ḥadīyah just as it was in Forsskal's day.

One genus, *Allium*, is of considerable importance as the vegetables, onion, garlic and leek all belong to it.

The onion, **Allium cepa** *L.*, 'Baṣal', is widely cultivated in small irrigated fields and as a roof-top potherb. Many different cultivars are grown but in the villages the preferred varieties are of the *aggregatum* group which freely produces new bulbils. Onions are one of the most popular vegetables in Yemen; they are served in salads, fried with liver and small meat, mixed in omelettes and bean dishes and used as a garnish for fish.

Garlic, **Allium sativum** *L.*, 'Thūm' or 'Tūm', is a common potherb and the cloves are used to flavour all kinds of meat, egg and bean dishes.

The large European leek, **Allium porrum** *L.*, has been introduced in recent years and is cultivated around Sana'a for its edible elongated bulb but far more important is the local variety 'Karath' sometimes called Kurrat in English. This is a small plant usually less than 25cm high, grown in small irrigated beds in and around towns and villages. The flat leaves are cut or plucked and sold in bundles in the markets. They are used to flavour soups and most traditional meat dishes.

HYPOXIDACEAE

HYPOXIS *L.*

Small herbs from a perennial corm. Leaves in a basal rosette, linear, pilose. Inflorescence axillary, leafless, racemose or cymose; perianth segments 6, green and hairy on the outside, yellow on the inside. Fruit an ellipsoid capsule.

***Hypoxis* is in great need of revision and the following account must be treated as temporary. Our three plants are very similar but appear to be sufficiently distinct to be treated as separate species.**

1. Peduncle flattened and winged, 1–4mm broad. Flowers 2–4 in an umbel *H. villosa*
   Peduncle flattened, but not winged, less than 1mm broad. Inflorescence 1–4-flowered, clearly cymose or racemose if more than 2-flowered 2
2. Flowers 2–4 in a raceme; peduncle c 15cm long. Fruiting pedicels recurved *H. textilis*
   Flowers usually 2 (rarely 1 or 4) in a cyme; peduncle 2–10cm long. Fruiting pedicels not recurved *H. angustifolia*

1. Hypoxis angustifolia *Lam.*

Leaves up to 25cm long, 2–7mm broad, sparsely to densely pilose. Inflorescence (1–)2 flowered (4-flowered cymose inflorescences are known from Africa); peduncle 2–10cm long, flat but not winged; perianth 4–8mm long; fruiting pedicels erect.

A rare plant of grassland in high rainfall areas of the escarpment between 1600 and 2500m: J. Ḥiṣn Dafdaf, 'Udayn; on the lower slopes of J. Ba'dān, north of Ibb; J. Ḥuzār and J. Kabūrah above Ḥadīyah; between Rujum and Ṭawīlah. (Wood 485, 1835, 1856, 1740, 2800)

2. Hypoxis ? textilis *Nel.*

Leaves up to 30cm long, 5–10mm broad, densely pilose. Inflorescence a 2–4-flowered raceme borne on a flat but unwinged peduncle about 15cm long; perianth 7–9mm long; fruiting pedicels recurved.

There is some doubt as to the identity of this plant since the type has not been seen.

Only known from steep grassy slopes at c 1800m between Markah and Ṣa'bah, J. Milḥān. (Wood 2383, 2859)

3. Hypoxis villosa *L.f.*

Leaves up to 20cm long, 5–10mm broad, often falcate, mostly pilose on the margins and midvein only. Inflorescence a 2–4-flowered umbel borne on a flat, winged peduncle 4–6cm long, 1–4mm wide; perianth c 10mm long, densely villous on the outside; fruiting pedicels erect.

Only known in grassland on the west side of J. Nu'mān at 2600m. (Wood 75/382)

Hypoxis violacea Schult. f. was collected by Forsskal at Taiz but the specimen has been lost. It is impossible to know what the plant was but the description and locality suggest that it may have been *Scilla hyacinthina* and not a *Hypoxis* at all.

AMARYLLIDACEAE

Bulbous herbs with large, attractive flowers. Leaves basal. Inflorescence an umbel (sometimes reduced to a single flower) borne on a long, leafless peduncle and subtended by an often scarious spathe; flowers bisexual, regular, the perianth segments free or united; corona sometimes present. Fruit a capsule or a berry.

1. Inflorescence a globular umbel formed of 50–150 small red flowers 2. *Scadoxus*
   Inflorescence of 1–20 large white flowers, not globular **2**
2. Flowers solitary. Corona present 3. *Pancratium*
   Flowers in umbels of up to 20 flowers. Corona absent 1. *Crinum*

1. CRINUM *L.*

1. Crinum album (*Forssk.*) *Herb.* Tuḥāf

C. yemense Deflers ex Schweinf.

Vigorous perennial from a large bulb. Leaves 10–15, linear-lanceolate, up to 1¼m long, weakly ascending, the margins often undulate. Inflorescence an umbel with up to 20 flowers, subtended by 2 large, scarious, spathaceous bracts 9–10cm long and borne on a thick, unbranched peduncle, 50–100cm high; flowers pendulous, bell-shaped, 20–25cm long with conspicuous white lobes and a long narrow tube. Fruit a globose capsule about 5cm in diameter.

This very attractive endemic lily is widely distributed between 1200 and 3300m on the escarpment north to the Wadi Sharas and also occurs in a few places on the high plateau such as Wadi Ḍahr, Kumaym in Al Ḥadā' and Khamr. It is undoubtedly most common in Taiz and Ibb provinces where it occurs in a wide range of open rocky places including *Euphorbia* scrub. It usually grows in small quantity but on damp, high-altitude mountain slopes it sometimes occurs in great abundance, notably around the summits of J. Ta'kar and J. Nu'mān.

2. SCADOXUS *Raf.*

1. Scadoxus multiflorus (*Martyn*) *Raf.*

Baṣal al Ḥanash, Zirā'ah al Ḥanash

Haemanthus multiflorus Martyn

H. coccineus Forssk.

Fireball. Bulbous herb sending up a fleshy, spotted peduncle, 10–30cm high which bears a brilliant red globular inflorescence which is subtended by a scarious bract and consists of numerous small, red, tubular flowers, their long red filaments and their yellow anthers. Leaves borne on a short stem and appearing after the inflorescence, 3–5 per stem, elliptic, 10–20cm long, dark green with conspicuous purple, spotted sheaths which enclose the stem. Fruit a berry.

The conspicuous red inflorescences appear in January and February in the driest part of the winter, contrasting vividly with the dried-up brown vegetation and acting as a harbinger of spring.

Widespread between 1300 and 2900m on the escarpment, growing on stony hillsides, cliffs, terrace walls and particularly on scrubby banks around cultivation; it is particularly common in the Ibb region.

3. PANCRATIUM *L.*

1. Pancratium maximum *Forssk.* Baṣal Ar Rubāḥ
P. trianthum sensu Schwartz
Spider Lily. Bulbous herb with 4–5 linear leaves, 10–20cm long, 0.5–2cm wide, appearing at the same time as the flowers. Flowers solitary, resembling a white daffodil, consisting of a funnel-shaped perianth with 6 narrowly lanceolate lobes containing a conspicuous cup-like corona at the mouth of the slender tube which is 10–15cm long, the whole borne on a short peduncle and subtended by a large sheathing scarious spathe. Fruit a capsule.

The attractive flowers last less than 24 hours. They open in the late afternoon and wither before midday the following day.

Widespread on the escarpment from 500 to 1700m, but local and easily overlooked except when in flower. Although found in various scrub habitats including the Qā'idah *Euphorbia* scrub, it is most characteristic of cliffs and stony slopes in the valley scrub communities of the central escarpment where it is quite common. The Arabic name, meaning 'Baboon's Onion' is particularly apt since the bulbs are one of the main foods of the baboons that frequent these valleys.

P. maximum is closely related to the African *P. trianthum* Herb. and is in my opinion doubtfully distinct.

IRIDACEAE

Herbs with corms or rhizomes. Leaves flat, linear, usually 2-ranked. Inflorescence variable but usually terminal; flowers bisexual with 6 perianth segments regularly or irregularly arranged, often in two whorls of 3. Fruit a capsule.

1. Flowers solitary. Leaves less than 2mm wide — 2. *Romulea*
   Flowers in spikes or cymes. Leaves 2–30mm wide — **2**
2. Flowers in spikes. Perianth segments subequal, glabrous, red — 3. *Gladiolus*
   Flowers in cymes. The 3 outer perianth segments bearded, larger than the 3 inner ones, white or blue — 1. *Iris*

1. IRIS *L.*

1. Iris albicans *Lange* Binafsaq, Zaraynbākh
I. florentina sensu Schwartz
Wild Iris. Gregarious perennial with a creeping rhizome and short, erect flowering stems 10–30cm high. Leaves basal, distichous, sword-shaped, grey-green, 10–25cm long. Inflorescence a sparingly branched terminal cyme, subtended by leaf-like spathaceous bracts; flowers white (rarely blue) consisting of 3 larger, deflexed outer perianth segments which are yellow-bearded on the upper surface and 3 smaller, erect inner segments and 3 style branches.

Locally abundant between 2200 and 3000m on grassy and stony mountain slopes; widespread on the southern mountains from J. Ṣabir through the Ibb region and Waṣāb to the Kusmah region of J. Raymah; further north very rare and only known from the Manākhah area, Yāzil, Ṭawīlah (Rathjens) and Shibām. The blue form is much rarer than the white and is only known from Shibām, J. Ba'dān and Suq Shirqī in Mudhaykhirah.

Despite the seemingly native distribution pattern this species is probably an ancient introduction from the Mediterranean region. Although in great abundance on the upper part of J. Ṣabir where it appears quite native, it is strangely absent from other apparently suitable mountains, J. Ta'kar for example, and in many places grows in small quantity around villages and particularly on graveyards where it is undoubtedly planted. Indeed it is only abundant on mountains like J. Ṣabir and J. Ba'dan which are remarkable for their high population density and the two facts are probably related.

2. ROMULEA *Maratti*

1. Romulea fischeri *Pax*
Small crocus-like herb from a bulbous corm. Leaves all basal, subulate, c 5cm long at flowering time but growing on to 30cm. Peduncles simple, several per corm, 3–6cm long at flowering time but elongating to twice this later; flowers solitary, erect, violet, bell-shaped, c 15mm long, subtended by 2 lanceolate, acute, scarious bracts half the length of the flower.

An attractive little plant found rarely in moist, high-altitude grassland between 2800 and 3200m: J. Ṣabir; J. Ta'kar; J. Bārad, Kusmah.

3. GLADIOLUS *L.*

1. Gladiolus dalenii *Van Geel*
G. natalensis (Eckl.) Lodd.
Erect herb to 1m, arising from a corm and with both basal and stem leaves. Leaves distichous, linear, up to 40cm long. Inflorescence a simple terminal spike with up to 9 large, red or orange flowers each subtended by a pair of scarious spathaceous bracts; perianth with a strongly curved tube and unequal segments, the uppermost forming a hood over the stamens.

A local plant found on cliffs and steep grassy slopes between 2000 and 2700m from Turbah north to J. Milḥān but virtually restricted to the central escarpment and the Ibb region, where it is widespread and easily found in July and August. The Turbah locality is somewhat anomalous since the plant grows at a lower altitude than usual and in local abundance as a field weed.

VELLOZIACEAE

XEROPHYTA *Juss.*

Small perennials having a short, dichotomously branched woody stem with leaves clustered at the apex, the fibrous bases of old leaves remaining attached to the stem. Inflorescence a solitary flower on a leafless peduncle 5–20cm long; flowers bisexual with 6 equal, free, linear-lanceolate perianth segments. Fruit a hard, dry capsule.

Perianth 2–3cm long — *X. arabica*
Perianth c 5cm long — *X. sp.A*

1. Xerophyta arabica (*Baker*) *N. Menezes*
Vellozia arabica Baker
Barbacenia arabica (Baker) O. Schwartz
Stem very short, often appearing to be absent with the leaves in a basal tuft. Leaves 5–10cm long, 2–5mm wide, keeled, often folded, entire or minutely serrate particularly on larger leaves. Peduncle glandular upwards; perianth mauve-tinged, never (?) pure white, 2–3cm long.

Restricted to cliff ledges and often associated with *Tripogon* spp, apparently requiring periodic prolonged moistening from mists or dripping water but also tolerating long periods of drought. It is widely distributed between 800 and 2900m on the escarpment, favouring areas with winter mists, and also occurs in the Mashriq where it is probably more frequent than the following few records suggest: Hūth; Fardah pass; on the escarpment west of Dhī Nā'im; Hammām Damt. *X. arabica* is endemic to southwest Arabia.

2. Xerophyta sp.A
Similar to *X. arabica* but more robust; the leaves 10–20cm long, serrate with black marginal teeth, the pure white perianth c 5cm long.

Only known from 900m above Suq Ar Ribāṭ, J. Raymah (Radcliffe-Smith and Henchie 4858).

Although the specimens of this plant appear quite distinct, they were made from plants cultivated at Kew and it is possible that the larger dimensions are simply the result of better growing conditions.

AGAVACEAE

A family of rhizomatous herbs, shrubs or trees characterised by the presence of fibres in the stiff, sharp-pointed leaves. Our species have bisexual flowers in racemes or panicles, the perianth united to form a tube. Fruit a capsule or berry.

Tree. Leaves in clusters at the ends of the branches — 1. *Dracaena*
Woody herbs. Leaves arising from an underground rhizome or ± distichous along a short stem — 2. *Sansevieria*

1. DRACAENA *L.*

1. Dracaena ? serrulata *Baker* — 'Uzāf
Dragon's Blood Tree. Small tree to 6m with a repeatedly branched trunk, grey bark and a spreading crown. Leaves crowded into dense clusters at the ends of the branches, sword-shaped, rigid, dark green, keeled, 25–60cm long, 1–2.5cm wide, the margin rough. Inflorescence a large terminal panicle; perianths with a cylindrical tube and spreading lobes, probably white. Fruit an orange berry.

It probably flowers in spring but no flowers have ever been observed on *Dracaena* in Arabia and for this reason the identity of our plant is uncertain. It may be *D. ombet* Kotschy & Perr., found on the Red Sea hills in Sudan or *D. serrulata* from Dhofar, if these two species are indeed distinct.

An uncommon plant found in scattered localities between 1000 and 2600m on open stony hills, most common in Khawlan Ash Sham: Bathayjah, Ḥujarīyah; above Ḥillah, J. Bura'; around Ḥajjah; on limestone hills east of Maḥābishah; Shahārah; J. Marran; J. Minmār; Majz (Steinberg).

2. SANSEVIERIA *Thunb.*

Large woody herbs with thick, fibrous leaves. Inflorescence of terminal racemes, sometimes branched and paniculate, the flowers in clusters along the rhachis; perianths tubular with spreading lobes. Fruit a berry.

Leaves flat, much wider than thick — *S. forskaliana*
Leaves ± D-shaped in cross-section, as thick as wide — *S. ehrenbergii*

1. Sansevieria ehrenbergii *Schweinf. ex Baker* — Salaf
Shy-flowering rhizomatous plant growing in colonies. Leaves arranged in 3–6 opposite pairs along a short stem up to 20cm high, distichous, crowded, D-shaped in cross-section with several prominent longitudinal channels, dark green, rigid, terminating in a sharp point, 30–100cm long. Inflorescence a terminal panicle up to 1¼m long; perianths greenish, c 1cm long.

Widespread and often abundant on dry stony ground, being found in suitable habitats between 200 and 1900m in the east of the Tihama, on the escarpment and in the southeast mountains as far east as Qa'ṭabah. Although occurring in *Acacia* bushland it is most abundant in *Euphorbia* scrub, both in the escarpment foothills and in the relatively high altitudes of the Qā'idah area.

2. Sansevieria forskaliana (*Schult. f.*) *F.N. Hepper & J.R.I. Wood* — Danaq
S. abyssinica N.E. Br.
S. guineensis Willd.s.l.
Shy-flowering plant growing in colonies. Leaves 1–2 from an underground rhizome, erect, flattened, stiff, lanceolate, dark green with a reddish margin, terminating in a sharp point, c 30–50cm long, 4–6cm broad. Inflorescence a simple raceme about as long as the leaves; perianths cream, 3–4cm long.

Widely distributed on the escarpment between 500 and 1900m but less common than *S. ehrenbergii* and preferring deeper soils in areas of higher rainfall. It is most common in the Ibb-'Udayn region and the central escarpment, being found on steep banks in open scrub.

Plants similar to this are cultivated as house plants in Europe and are known as Mother-in-law's Tongue in Britain.

Sisal, **Agave sisalana** *Perr.*, known as 'Salaf', has been introduced but does not appear to have been grown commercially. It is easily recognised by its massive rosettes of leaves which are entire apart from the viciously sharp terminal spine. The inflorescence appears after many years and grows to a height of about 4m; at maturity a crop of bulbils is shed by the inflorescence and then the whole plant dies. It is found locally in the south, particularly by the road from Taiz to Ibb.

Polianthes tuberosa *L.*, a fragrant lily-like plant known as 'Sambaq' or 'Rinjās', is cultivated in the central escarpment at Ḥillah, J. Bura' (Schweinfurth) and at Suq al Qita' between Maḥwīt and Ḥufāsh.

ORCHIDACEAE

Our species are perennial terrestial herbs often with subterranean tubers. The stem is leafy or leafless and in *Eulophia* has swollen bases (pseudobulbs). Leaves alternate, entire, often distichous and with sheathing bases. Inflorescence a spike, raceme or panicle; flowers bisexual, irregular, bracteate; perianth of 3 petaloid sepals and 3 petals, the middle petal (lip) of strikingly different appearance to the other perianth segments and often with a backward-pointing spur; the sexual organs are united to form a column which lies opposite the lip. Fruit a capsule with numerous minute seeds.

This huge and fascinating family is poorly represented in Yemen. Nearly all the nineteen species recorded are rare and restricted to the wettest parts of the Ibb region and the central escarpment. They mostly flower for a short season and die down rapidly after flowering.

1. Stem leaves 0. Basal leaves 1–2, present or absent at flowering time 2
   Stem leaves present, 2 or more in number, always present at flowering time 3
2. Flowers small, less than 5mm long. Leaves 2, present or absent at flowering time 1. *Holothrix*
   Flowers large, more than 2cm long. Leaves 1, absent at flowering time 6. *Nervilia*
3. Stem base swollen, forming pseudobulbs. Very robust plants, usually more than 50cm high 8. *Eulophia*
   Stem base without pseudobulbs. Plants usually less than ½m high 4
4. Plants with rhizomes. Leaves linear or linear-lanceolate, with prominent raised nerves 5
   Plants with tubers. Leaves lanceolate, ovate or elliptic, without raised nerves 6
5. Flowers yellowish, pubescent, without spurs 7. *Epipactis*
   Flowers predominantly pink, glabrous, spurred 8. *Eulophia*
6. Spurs 2. Flowers brown 5. *Satyrium*
   Spur 1. Flowers pink, white or greenish 7
7. Flowers pink. Spur arising from the middle sepal, upward pointing 4. *Disa*
   Flowers white or greenish. Spur arising from the lip, not upward pointing 8
8. Spur more than 10cm long 3. *Bonatea*
   Spur less than 4cm long 2. *Habenaria*

1. HOLOTHRIX *Lindl.*

Small orchids with underground tubers and a pair of suborbicular, sessile basal leaves which lie flat on the ground. Inflorescence erect, unbranched, leafless. Flowers small, less than 5mm long; sepals free; petals longer than the sepals, the lip 3-lobed and spurred.

Plant glabrous. Leaves absent at flowering time *H. aphylla*
Plant pubescent. Leaves present at flowering time *H. arachnoidea*

1. Holothrix aphylla (*Forssk.*) *Reichb. f.*

Glabrous plant with 2 orbicular basal leaves, which are absent at flowering time. Stem usually 10–20cm high, densely covered in ± scarious sheaths. Flowers white, tinged with purple, bearing a prominent curved spur 4–9mm long.

Flowers in March and April before the onset of the rains.

Known with certainty from Kusmah, J. Raymah and Waṣāb (Forsskal), from J. Ba'dān and in some abundance on the hills between Mashwarah and Ibb, but probably widespread in shallow grassland on the wetter parts of the escarpment between 2200 and 2800m since the distinctive glabrous leaves have been seen as far north as J. Maswār.

2. Holothrix arachnoidea (*A. Rich.*) *Reichb. f.*

H. vatkeana Reichb. f.

Hairy plant with 2 orbicular basal leaves which are present at flowering time. Stem usually 10–20cm high, pubescent but without scales. Flowers green, tinged with purple, bearing a short conical spur, usually 1–2mm long but absent in Wood 751 from J. Ṣabir.

Flowers mainly in February to March, but occasionally summer-flowering plants are seen.

A shy-flowering plant known with certainty from J. Ṣabir, J. Shibām in the Ḥarāz (Schweinfurth) and Bayt Braz near Maḥwīt but probably widespread in the Ibb region and the central escarpment, growing on grassy slopes in areas with plentiful winter mist between 1800 and 3000m. The large paired orbicular hairy leaves are appressed to the ground and can be recognised at any season.

2. HABENARIA *Willd.*

Erect herbs with underground tubers. Leaves arranged along the stem. Inflorescence a terminal, unbranched raceme; flowers green and/or white; sepals free, the laterals spreading and the middle one erect and often forming a hood with the lateral petals; lip entire or 3-lobed, spurred.

All species flower after prolonged rain, mainly in July and August.

1. Lip 3–4cm long, the side lobes with 6–10 narrow threads on the outer margins — *H. macrantha*
   Lip less than 1cm long, without threads on the side lobes — 2
2. Lateral petals entire — 3
   Lateral petals deeply 2-lobed — 5
3. Sepals pubescent. Spur more than 8mm long — *H. attenuata*
   Sepals glabrous. Spur less than 3mm long — 4
4. Flowers very small, the spur less than 1mm long and the bracts not exceeding 8mm long — *H. lefebureana*
   Flowers larger, the spur 1.5–3mm long and the bracts up to 20mm in length — *H. peristyloides*
5. Middle sepal reflexed, leaving the column exposed. Pedicel and ovary 20–35mm long, the inflorescence thus very broad being 5–6cm in diameter — *H. schimperiana*
   Middle sepal forming a hood over the column. Pedicels and ovary 10–20mm long, the inflorescence less than 4.5cm in diameter — 6
6. Spur twisted in the middle. Bracts eglandular. Flowers usually cleistogamous — *H. cultrata*
   Spur untwisted, though sometimes bent in the middle. Bracts glandular-pubescent (at least sparsely). Flowers not cleistogamous — 7
7. Spur less than 10mm long. Petals ciliate — *H. antennifera*
   Spur 16–20mm long. Petals glabrous or papillose — *H. cultriformis*

1. Habenaria antennifera *A. Rich.*

Stem usually 20–40cm high. Leaves broadly lanceolate. Inflorescence 2.5–3.5cm in diameter; bracts slightly shorter than the flowers, glandular-pubescent; flowers yellow-green; middle sepal c 4mm long, forming a hood over the column; lateral petals ciliate-margined, deeply divided to the base, the lower segment linear-lanceolate, 8–12mm long; lip 3-lobed, 6–9mm long, the spur 5–10mm long, not twisted.

Known with certainty from near the summit of J. Ta'kar at 3100m (Wood 1718) and from Bani Sa'īd in the Ja'farīyah at 1400m (Wood 2713); in both localities it grows on steep grassy slopes in areas of high rainfall and it almost certainly occurs elsewhere in the central escarpment and the Ibb area.

Outside Yemen known only from Ethiopia and Sudan.

2. Habenaria attenuata *Hook. f.*

Stem usually 20–40cm high. Leaves oblong-lanceolate. Inflorescence usually long, lax and slender, 0.5–1.5cm in diameter; bracts longer than the flowers, glabrous; flowers green; sepals all ciliate, the middle one c 3mm long, equalling the unlobed lateral petals with which it forms a hood; lip deeply 3-lobed, 3–4.5mm long, the spur slender, pendent, 10–16mm long.

Only known from a grassy slope on the west side of J. Ba'dān at 2700m.

3. Habenaria cultrata *A. Rich.*

Similar to *H. antennifera* from which it is distinguished by its glabrous bracts, and the 12-15mm long spur which is distinctly twisted in the middle. The flowers are usually cleistogamous and seed is ripened without cross-pollination.

Known only at c 1800m on the grassy lower slopes of J. Ba'dān just north of Ibb but very probably elsewhere in the Ibb region.

Outside Yemen only known from Ethiopia.

4. Habenaria cultriformis *Kraenzl. ex Engl.*

Closely related to both *H. antennifera* and *H. cultrata* but more robust than either with the inflorescence 4–4.5cm in diameter and the untwisted spur 16–20mm long. The lateral petals are papillate, not ciliate, and the lower petal lobe is broadly ovate and abruptly narrowed just above the base, not gently tapering from the base as in the other two species. The bracts are sparsely glandular.

Known only from a shaded grassy bank at 1800m on J. Ḥiṣn Dafdaf, 'Udayn. Another species known elsewhere only from Ethiopia.

5. Habenaria lefebureana (*A. Rich.*) *Dur. & Schinz*

Stem usually 10–20cm high. Leaves elliptic. Inflorescence 0.8–1.3cm diameter, densely many-flowered, glabrous; bracts lanceolate, shorter than the flowers; flowers white, small, fragrant; sepals all similar, 3–4mm long; lateral petals entire; lip 2.5–3mm long, 3-lobed at the apex and with an obscure spur 0.5mm long.

Locally abundant in montane grassland between 2500 and 3100m: J. Ṣabir; J. Ta'kar; J. Nu'mān; J. Ba'dān; J. Manār; J. Ḥiṣn Madnān, Waṣāb al 'Ālī; on the mountains between Kusmah and Ḥadīyah.

Elsewhere only in Ethiopia.

6. Habenaria macrantha *Hochst. ex A. Rich.*

Stem 20–50cm high. Leaves ovate or broadly lanceolate. Inflorescence 7–9cm in diameter, lax; bracts lanceolate, glabrous, ± equalling the flowers; flowers large, green and white, smelling of plum; sepals green, papillose, the middle sepal erect and forming a hood with the 2 entire lateral petals, each 20–25mm long; lip 3–4cm long, deeply 3-lobed, the lateral lobes with 6–10 narrow threads on the outer margin; spur 2–3.5cm long, bent but not twisted.

Certainly the most conspicuous and probably the commonest *Habenaria* in Yemen, found from the Ibb region north to the Ḥarāz, Baw'ān (Carden) and J. Ashmūr (Christmann), from 2000 to 2900m. Unlike most other species, which are restricted to montane grassland, this orchid also grows on stony hillsides wherever there is a reasonable soil depth.

7. Habenaria peristyloides *A. Rich.*

Stem 10–40cm high. Leaves lanceolate. Inflorescence 1.5–3cm in diameter, many-flowered, glabrous; bracts equalling or exceeding the flowers; flowers yellow-green, sepals similar, c 3mm long, slightly longer than the entire lateral petals; lip c 5mm long, with an auriculate base, deeply 3-lobed, the lateral lobes spreading almost at right angles to the central lobe, spur 1.5–3mm long.

Only known at 2600m from the summit of J. Ḥiṣn Madnān in Waṣāb al 'Ālī.

8. Habenaria schimperiana *Hochst. ex A. Rich.*
Stem mostly 25–50cm in Yemen, though often more in Africa. Leaves lanceolate. Inflorescence 5–6cm in diameter, lax; bracts lanceolate, acuminate, much shorter than the pedicels; flowers brownish; middle sepal 5–8mm long, reflexed to expose the column, smaller than the laterals; lateral petals ciliate-margined, deeply lobed, the lower lobe c 15mm long and much larger than the upper lobe; lip 3-lobed, ± 15mm long; spur 10–16mm long, twisted.

Found from 2400 to 2900m in small quantity in grassland on most of the Ibb mountains, but elsewhere known only from J. Kabūrah above Ḥadīyah.

3. BONATEA *Willd.*

1. Bonatea steudneri (*Reichb. f.*) *Dur. & Schinz*
Habenaria arabica (Defl.) Kraenzlin
Herb with elongated tubers and an unbranched very leafy stem, usually 30–60cm high. Inflorescence usually subcorymbose; flowers greeny-white, smelling of cloves; pedicels and ovary together 4–7cm long, much exceeding the bracts; middle sepal erect, forming a hood, 2–2.5cm long; petals deeply divided to the base; lip 3-lobed to about the middle, the central lobe c 3cm long; the very long spur (10–15cm) distinguishes this species from all other Arabian orchids.

Flowers June to August.

Widespread on the escarpment, growing on rocky banks, cliffs and grassy field borders from 1800 to 2900m but apparently absent south of Taiz and in Khawlan Ash Sham. It is one of the commonest orchids in Yemen but seldom occurs in quantity.

4. DISA *Berg.*

1. Disa pulchella *Hochst. ex A. Rich.*
Herb with underground tubers and an erect leafy stem 15–30cm high. Leaves lanceolate. Inflorescence cylindrical, densely flowered, c 2cm diameter; bracts shorter than the pinkish-purple, fragrant flowers; middle sepal erect, c 5mm long, bearing an upward-pointing cylindrical spur 4–6mm long; lateral petals whitish, erect inside the middle sepal, obscurely 2-lobed; lip narrow, 5.5mm long, without lobes or spur.

Flowers in summer.

An attractive species known only from grassland at 3200m near the summit of J. Ta'kar.

Outside Yemen found only in Ethiopia.

5. SATYRIUM *Sw.*

1. Satyrium brachypetalum *A. Rich.*
Herb with underground tubers and an erect leafy stem, 20–35cm high. Leaves elliptic to lanceolate with sheathing bases. Inflorescence narrowly cylindrical, c 2cm diameter; bracts longer than the flowers, reflexed; flowers yellow-green, tinged with red-brown; sepals united to the petals and lip; lip uppermost, erect, hooded, with 2 spurs, each 11–14mm long.

Flowers in summer.

A rare orchid found in montane grassland between 2300 and 3200m: J. Ta'kar; J. Ba'dān; Al Manār on J. Ḥubaysh; J. Kabūrah and J. Bārad in J. Raymah.

Known only from Yemen and Ethiopia.

6. NERVILIA *Commers. ex Gaud.*

1. Nervilia umbrosa (*Reichb. f.*) *Schultr.*
Tuber nearly spherical. Leaves appearing after the flowers have died away, solitary, broadly ovate-cordate, prominently nerved, c 16cm long, borne on an erect petiole 10–20cm long. Inflorescence an erect, glabrous, few-flowered raceme, the peduncle with a few sheaths near the base; flowers large, pendulous; the sepals and petals similar, yellow-green, c 3cm long; lip white with purple veins, obscurely 3-lobed, without a spur.

Probably flowering in May–June.

A rare plant, not yet seen in flower in Yemen, known only from shaded banks between 1500 and 1600m in the wettest parts of Ibb province: Wadi Dūr, 'Udayn; between Ibb and Suq As Sabt.

7. EPIPACTIS *Sw.*

1. Epipactis veratrifolia *Boiss.*
Rhizomatous herb with an erect leafy stem 25–50 (–100)cm high. Leaves linear–lanceolate, acuminate, prominently veined, with sheathing bases. Inflorescence very lax, usually 5–10-flowered, the rhachis, pedicels and sepals all pubescent; bracts leaf-like, shorter than the flowers; sepals and lateral petals similar, yellow-green with purple margins, 1–2cm long; lip with erect sides, deeply constricted in the middle into two halves, yellow-green, without a spur.

A rare plant of wet places between 1900 and 2700m; characteristic of dripping sandstone rock faces in relatively dry country and often associated with *Schoenus nigricans*: on the Turbah cliffs; in marsh above Dhī Sufāl; frequent in marsh and on cliffs in Wadi Aḥjar; Majz (Steinberg).

Widely distributed from Turkey to the Himalayas, extending south to Yemen and Somalia.

8. EULOPHIA *R. Br. ex Lindl.*

Herbs with creeping underground stems either with underground tubers or with aerial pseudobulbs developing at the base of each year's growth. Inflorescence lateral, a raceme or panicle; bracts small, shorter than the flowers; sepals similar; petals often brightly coloured; lip with a swollen sack-like spurred base.

1. Plant more than 40cm high; pseudobulbs present. Flowers usually more than 10 — 2
   Plant 30–40cm high from a subterranean corm. Flowers about 6 — *E. clavicornis*
2. Spur more than 10mm long. Lip broadly obovate, apparently unlobed — *E. guineensis*
   Spur less than 4mm long. Lip 3-lobed — 3
3. Leaves thick, rigid, serrate near the apex. Lip white and pink — *E. petersii*
   Leaves not thick and rigid, entire. Lip yellow with red-brown striped side lobes — *E. streptopetala*

1. Eulophia clavicornis *Lindl.*

Herb with an underground rhizome and an erect stem, 30–45cm high; stem without pseudobulbs at the base. Leaves thin, prominently veined, linear, entire, up to 16cm long. Inflorescence a simple raceme with about 6 flowers; sepals green, apiculate, c 1cm long; petals dull pink: lip c 1cm long, shallowly 3-lobed, white, papillose on the mid-lobe with a spur c 2mm long.

The Yemen plant is variety **nutans** (*Sond.*) *Hall.* It flowers in June.

Only known from grassland at 1900m on the northwest slopes of J. Ẓallamlam above Ḥadīyah. (Wood 2818)

2. Eulophia guineensis *Lindl.*

E. quartiniana A. Rich.

Erect herb 40–100cm high with ovoid pseudobulbs and distinctive white, flattened, fleshy roots. Leaves thin, lanceolate or elliptic, plicate, entire, 20–50cm long. Inflorescence a lax, simple raceme with 7–20 very attractive flowers; sepals and lateral petals similar, lanceolate, acute, green, flushed with purple; lip broadly obovate or orbicular, unlobed, pink, 2–4cm long, with a spur 15–30mm long.

The Yemen plant is variety **purpurata** *Reichb. f. ex Kotschy* in which the leaves appear after the inflorescence and the lip is rounded. It flowers in May and June.

Only known from a steep bank in shaded scrub at 1500m near 'Udayn suq.

3. Eulophia petersii (*Reichb. f.*) *Reichb. f.*

E. schimperiana A. Rich.

Vigorous woody herb up to 3m high, with large ovoid or ellipsoid pseudobulbs. Leaves 2 or 3, spreading from the top of the pseudobulbs, thick, rigid, deeply channelled above, serrate towards the apex, 20–50cm long. Inflorescence a many-flowered panicle; sepals apiculate, greenish with purple veins; lateral petals similar in colour to the sepals; lip 3-lobed, white tinged with pink, 1.5–3cm long, with a spur 2–8mm long.

Flowers throughout the year.

The most drought-resistant and much the most common Yemen orchid. Widespread and frequent from 500 to 2000m in open stony *Acacia* and *Euphorbia* bushland on the escarpment and southeast mountains east to Damt.

It has superficially rather similar leaves to *Sansevieria ehrenbergii* with which it often grows.

4. Eulophia streptopetala *Lindl.*

Lissochilus arabica Lindl.

Large herb up to c 2m high with conical pseudobulbs. Leaves thin, lanceolate, plicate, entire, 30–60cm long. Inflorescence a simple raceme (rarely branched) produced before the leaves are fully developed; sepals obovate, mucronate, green mottled with purple-brown, lateral petals suborbicular, yellow; lip 3-lobed, yellow with red-brown veined lobes, 1–1.5cm long, with a conical spur 1.5–4mm long.

The Yemen plant is variety ***rueppellii*** (*Reichb. f.*) *Cribb* which has 5 raised crests on the lip. It flowers from March to May and is restricted to Yemen and Ethiopia.

An attractive but very local plant of steep scrubby slopes between 1500 and 2500m in the central escarpment: in several places on the mountains between Kusmah and Ḥadīyah; above Bani Daw'ān, J. Raymah; below Markah, J. Milḥān. Pseudobulbs probably belonging to this species have also been seen above Hiliyal on J. Sumārah and near Ẓulmah on J. Ḥubaysh but the flowers have not been seen in this area.

APPENDIX 1: Arabic Plant Names

This appendix gives the local names of plants found in Yemen. It is not comprehensive since I have only included names which have been given by more than one informant or more than one independent literary source. Hopefully these relatively stringent requirements have eliminated most names of doubtful application and extremely local usage.

It needs to be stressed that the names given only apply in Yemen and may be used for different plants in other countries or in classical Arabic. Thus, for example, Zaytūn is the guava in Yemen and not the olive. Quite frequently one name is applied to more than one plant. In some cases the usage is geographical: thus 'Ubāb refers to *Jatropha pelargoniifolia* in the Tihama but to *Withania somnifera* elsewhere. In other cases the Arabic name embraces a concept which does not coincide with a species: thus Ṭalḥ refers to more than one acacia and Ruqmah to any prostrate, pink-flowered herb. Wherever possible, I have indicated the concept of a particular name and the area in which it is used. In many cases, however, as with Juljul, no easy explanation is apparent.

Names are transliterated according to the system on page 52 and are usually given in the collective form rather than the singular. Comparatively few of the names are found in classical Arabic writings available to me and most have been supplied directly or indirectly from oral sources. This has inevitably resulted in problems in transcription and the following points should be borne in mind when searching for a particular name in the list. Short vowels are of very indeterminate quality so that it is often an arbitrary decision as to whether a particular word is transcribed with an a, i or u; 'Alb, 'Ilb and 'Ulb all sound very similar. The sound of g, as in the English 'go', is used for the letter j in the south of the country but for q in the north. The way in which this sound is transcribed will depend therefore on where the name was collected. There are also tendencies to interchange the following consonant pairs in some dialects: t/th; r/l; ḏ/ẓ. Both Thūm and Tūm, for example, are commonly used for the same plant in Yemen. Many other local variants in plant names are found and only a few of the most important, such as Baqām and Buqīm, are listed.

Names are listed in strict alphabetical order and the letter 'Ayn is ignored for this purpose. Words beginning with Th, will be found between Ta and Ti and the same principle applies to letters beginning with Dh, Kh and Sh.

Abal Calligonium crinitum
'Abīya' Hypericum revolutum (south only)
'Adan 1. Adenium obesum
2. Adenia venenata
'Adas Lens culinaris
'Afār Buddleja polystachya
'Alāl Any tall, reed-like grass, particularly Typha spp., Arundo donax
'Alas Triticum dicoccum
'Alb Zizyphus spina-christi
'Alka Dipterygium glaucum
'Amas Boscia angustifolia ('Udayn)
'Amb 1. Carica papaya
2. Mangifera indica
'Ambarūt Pyrus communis
'Amq Euphorbia ammak (sometimes misapplied to other succulent Euphorbias)
'Anamnamah Solanum villosum
'Ansīf Pulicaria jaubertii
'Antikār Atriplex farinosa
'Anwak Campanula edulis
'Ar'ar Juniperus procera (north)
'Arj Zizyphus spina-christi (Tihama)
'Arkis Climbing plants such as Maerua oblongifolia, Pentatropis nivalis, Zehneria scabra etc.
'Arm Carissa edulis (south)
'Asāf Aloe spp. (north)
'Asal Suaeda monoica & S. vermiculata
'Asam Combretum molle
'Asaq Acacia asak
'Asar Ehretia cymosa
'Ashaq Rosa abyssinica
'Ashraq Cassia italica
'Atar Pisum sativum
Athalamah Syzygium guineense
Athbīr Acokanthera schimperi (Ḥarāz)
Athl Tamarix spp.
Athrub Rumex nervosus
'Awsaj Lycium shawii
Azawa'rān Alkanna orientalis

Bādhinjān Solanum melanogena
Bajdah Plectranthus hadiensis
Baka Jatropha spinosa & J. pelargoniifolia
Balas Ficus carica & F. palmata
Bān Moringa aptera
Bāmiyah Abelmoschus esculentus
Baqām Solanum coagulans
Baqdūnis Petroselinum crispum
Baql Raphanus sativus
Barah Guizotia scabra
Barmūsh Opuntia ficus-indica
Barqūq Prunus armenaica
Baṣal Allium cepa
Baṣal Al Ḥanash Scadoxus multiflorus
Baṣal Ar Rubāḥ Pancratium maximum
Batātā Solanum tuberosum
Bayāḍ Any shrub with a white undersurface to its leaf, particularly Centaurothamnus maxima, Conyza incana, Debregeasia saenab and Tarconanthus camphoratus
Baybah Capsicum annuum
Bidaḥ Ficus sycomorus, F. glumosa F. ingens
Bilsim Lens culinaris
Binat Ochna inermis
Binafsaq Iris albicans

Bisa Euclea racemosa
Bisalah Cyperus rotundus
Biṣaṣ Diospyros mespeliformis
Bisbās Capsicum frutescens
Bishām Commiphora gileadensis
Blaysnān Lemna spp.
Bu'ar Zizyphus spina-christi (fruit)
Bukār Panicum turgidum
Bun Coffea arabica
Bunān Pavetta spp.
Būnīt Kniphofia sumarae
Buqīm Spiny Solanum spp. especially S. coagulans
Burar Portulaca quadrifida
Būrī Echinops spinosissimus
Burr Triticum aestivum
Burṭūm Grewia schweinfurthii
Burtaqal Citrus sinensis
Buyād See Bayād

Dabr Dobera glabra
Da' a' Amaranthus graecizans
Dahan Euphorbia schimperi
Daḥī Acacia asak
Dakhāf Arisaema bottae & A. flava
Dalūq Suaeda vermiculata
Danaq Sansevieria forskaliana
Darīf Ficus ingens
Ḍarīm Teclea nobilis
Dawm Hyphaene thebaica
Dawmir Ocimum forskolei
Ḍay'an Acacia mellifera (Mashriq)
Daymān Pithocellocium dulce
Dhajayj Oncoba spinosa (Milḥān)
Dhalīf Ficus glumosa (Ḥujarīyah)
Dharīm Teclea nobilis
Dhurah Sorghum bicolor
Dhurah ḥindī Zea mays
Dhurūb Combretum molle (south)
Difrān Acalypha fruticosa
Difrī Sterculia africana
Dijr 1. Digera muricata
2. Vigna unguiculata
Dima' See Duma'
Diraḥ Breonadia salicina
Draydah Indigofera argentea
Dubā Lagenaria siceraria
Ḍubar Dobera glabra
Dubbah See Dubā
Dukhn Pennisetum glaucum
Dūm Hyphaene thebaica
Duma' Jatropha spinosa (Mashriq)
Dundul 1. Jatropha variegata ('Udayn)
2. Breonadia salicina (Ḥujarīyah)

Fa' Euphorbia sp. aff. fruticosa (north)
Faja'ah 1. Crotalaria microphylla (Tihama)
2. Convolvulus arvensis
Fayjil Schowia purpurea
Fijil Raphanus sativus
Fijil Al Arḍ Any terrestial parasite such as Orobanche and Striga spp.
Filfil Schinus molle
Firsikh Prunus persica
Firūwak Nepeta deflersiana
Firūwish Juniperus procera
Ful Jasminium sambac
Fūl Vicia faba
Futaḥ Psiadia punctulata
Fuwah Rubia tinctoria

Gharāf Cordia sinensis (sometimes C. ovalis)
Gharīb Sorghum bicolor
Ghubār Applied to various Acanthaceae such as Ruellia patula, Lepidagathis scariosa etc.
Ghulāf (Ghulāq, Ghulāth) Caralluma acutangula C. penicillata
Ghurmil Commicarpus plumbagineus

Ḥab Al 'Azīz Arachis hypogaea
Ḥabaq Ocimum basilicum
Ḥabaqbaq Ocimum forskolei
Ḥabbah Sawda Nigella sativa
Ḥabḥab Citrullus lanatus
Ḥadaq Solanum glabratum & S. cordatum
Ḥadas Myrtus communis
Ḥadhḥadh Berberis holstii
Ḥakl Stachys yemenensis
Ḥalafī Desmostachya bipinnata
Ḥalaj Balanites aegyptiaca
Ḥalāl 1. Reed-like plants like Typha, Scirpus inclinatus, Arundo donax
2. Clematis spp. See Ḥalqa
Ḥalaq See Ḥalqa
Ḥalaṣ Cissus rotundifolia
Ḥalf Lepidium sativum
Ḥalka Dipterygium glaucum
Ḥalqa Cyphostemma digitatum
Ḥalqab Diplotaxis erucoides
Ḥamaḍmaḍ Oxalis corniculata & O. radicosa
Hamḍ Salsola spinescens
Ḥamshad Any plant with clinging hairs or fruit such as Achyranthes, Priva, Forsskalea, Triumfetta, Pupalia etc.
Ḥanqās Rubus spp.
Ḥanta Triticum aestivum
Ḥantafīs Rubus spp.
Ḥanẓil Citrullus colocynthis
Ḥaraqraq See Ḥariq
Ḥaras Acacia tortilis
Ḥarf Lepidium sativum
Ḥariq Spiny or stinging plants such as Tragia pungens, Xanthium spinosum etc.
Ḥarīr Ceiba pentandra
Ḥarm 1. Suaeda aegyptiaca
2. Salsola spinescens
Ḥarmil 1. Peganum harmala (high plateau)
2. Rhazya stricta (Mashriq)
Ḥaṣār Indigofera oblongifolia
Ḥashish 1. Any species of grass
2. Cannabis sativa
Ḥawjam Any spiny shrub, particularly Rosa abyssinica, Pterolobium stellatum, Oncoba spinosa etc.

Ḥawmir Any pinnate-leaved undershrubs such as Cadia purpurea, Sesbania leptocarpa, Cassia spp., Phyllanthus spp.
Ḥil Any small spiny pink-flowered plant especially Fagonia spp. and Indigofera spinosa
Ḥilwah Fagonia spp.
Ḥilwaq Climbing plants such as Clematis spp., Cocculus spp.
Ḥinah Lawsonia inermis
Ḥinah Rubāḥ Apparently applied to a number of small-leaved, spinescent shrubs including Flueggia virosa and Segeretia thea
Ḥind Zea mays
Ḥishrif Ficus exasperata
Ḥulbah Trigonella foenum-graecum
Ḥumar 1. Tamarindus indica
2. Cadia purpurea
3. Other Leguminosae shrubs with pinnate leaves
Ḥumaym Zygophyllum album
Ḥumays Rubus spp.
Ḥumayt Tragia pungens
Ḥummaṣ Cicer arietinum
Huqa (Huqil) Selaginella yemensis
Ḥuraysh Solanum glabratum
Ḥurūd Curcuma longa
Ḥurūmil Peganum harmala (Mashriq)
Ḥuwir (Ḥuwirwir) Indigofera tinctoria
Ḥuzār Pulicaria somaliensis

Ibiz Hypericum revolutum
Ibkī 1. Jatropha variegata (Ḥujarīyah)
2. Jatropha spinosa
'Ibli Aloe inermis (and other species?)
Ikrith Euphorbia cactus
'Imrīkh Acokanthera schimperi ('Udayn)
'Inab 1. Vitis vinifera
2. Cordia sinensis
Injāṣ Prunus domestica
'Inshāṭ Acalypha fruticosa (south)
'Irdhib Plectranthus hyemalis
'Irsim Solanum incanum
Istaqūtas Teucrium yemense
Ithāb Ficus salicifolia
Ithl Tamarix spp.
Ithrār Maytenus senegalensis

Jadmal Caralluma spp.
Jawz Juglans regia
Jazar 1. Daucus carota
2. Ipomoea batatas
Jidb Rhamnus staddo
Jilīf Commelina spp.
Jīr Ricinus communis (Mawza')
Jirdab Echinops spinosissimus
Jirjīr Eruca sativa
Jirūm Spiniluma oxyacantha
Jishb Aloe sabaea
Jithjāth 1. Pulicaria crispa
2. Salsola bottae
Jizīr Dichrostachys cinerea
Juljul 1. Sesamum indicum
2. Cassia occidentalis
3. Crotalaria retusa
Jumays 1. Anastatica hierochuntica
2. Euphorbia sp. aff. fruticosa
Junayd Polygala erioptera

Kabsarah Coriandrum sativum
Kadhi 1. Pandanus odorifera
2. Cadia purpurea
Kaf Anastatica hierochuntica
Kāfūr Eucalyptus spp.
Kanab Eleusine corocana
Kanāḥ Kanahia laniflora
Karath Allium porrum
Kash Cucurbita spp.
Khab Jasminum grandiflorum
Khabsh Tapinanthus rufescens
Khadar Grewia tenax & G. tembensis
Khadash Commiphora schimperi, C. sp. A and C. habessinica
Khalah Antiaris toxicaria
Khambūb Centaurothamnus maximus
Khanas Ficus sycomorus
Khardal 1. Brassica napus
2. Commicarpus plumbagineus
Khawah 1. Pulicaria jaubertii
2. Lavandula pubescens
Khawkh Prunus persica
Khawnab Cussonia holstii (Ḥarāz)
Khazaj Eleocharis palustris
Khidhrābah Senecio hadiensis
Khīr Aloe spp.
Khirmūsh Annona squamosa
Khisāl Ficus vasta (Ḥujarīyah)
Khishkhāsh Papaver somniferum
Khiyah Cucumis sativus
Khubayz Malva parviflora & M. verticillata
Khudām Cissus quadrangularis
Khudar Succulent Crassulaceae such as Aeonium, Cotyledon and Kalanchoe
Khulīyah Kedrostis foetidissima
Khurūb 1. Kalanchoe spp.
2. Cotyledon barbeyi
3. Plectranthus cylindraceus
4. Ceratonia siliqua
Kilāb Any spiny shrub such as Caesalpinia bonduc, Pterolobium stellatum or Rosa abyssinica
Killaḥ Euphorbia cactus
Kimb Barbeya oleoides
Kiranīt Ceratonia siliqua
Kirath Euphorbia cactus
Kirkir Colocasia esculenta
Kirmil Zygophyllum simplex
Kirshah Aeluropus lagopoides
Kisht Dolichos purpureus
Kubaj Brassica campestris
Kurad Phragmanthera regularis
Kurbays Cenchrus biflorus
Kurkūm Colocasia esculenta
Kurn Colocasia esculenta
Kūsah Cucurbita spp.
Kusmah 1. Huernia spp.
2. Any member of the Caralluma ango group

Lahas Carissa edulis
Laḥīb Centaurea pseudosinaica

Laṣaf Capparis cartiligenea
Lawz Prunus dulcis
Libākh Mimusops laurifolia
Libanah Small species of Euphorbia such as E. granulata, E. inaequilatera
Līm Citrus spp.
Līm ḥālī Citrus sinensis
Līm ḥāmidh Citrus aurantifolia
Līmum Segeretia thea
Līmun Citrus limon
Lisān Al Kabsh Crassula alba
Lisān Ath Thūr Echium rauwalfii
Līyah Aristolochia bracteata
Lizab Origanum marjorana
Lū' Applied to climbing plants such as Cocculus hirsutus and tree parasites like Plicosepalus curviflorus
Lūḍ Carissa edulis
Lūfah Luffa cylindrica

Malasi' Galium aparinoides
Manj Datura spp. especially D. stramonium
Manja Mangifera indica
Maraydah Ecbolium viride
Markh Leptodenia pyrotechnica
Ma'ṣ Maesa lanceolata
Mawmir Linum usitatissimum
Mawz Musa hybrids
Midan 1. Plectranthus spp.
2. Pulicaria petiolaris
3. Any oily aromatic shrub
Milaḥ Suaeda aegyptiaca
Mirū 1. Rhoicissus revoillii
2. Maerua crassifolia
Miru'ad Sarcostemma viminale
Mishmish Prunus armenaica
Miskhil Cordia sinensis
Misūq Ehretia obtusifolia
Mudaḥ 1. Ficus populifolia
2. F. glumosa (central escarpment)
Mudh (Mudhīdh) Anisotes trisulcus
Muḥadirah Heliotropium spp.
Muḥaraq Girardenia condensata, Tragia pungens and other stinging herbs
Muḥaṭ Achyranthes aspera
Mukār Tarconanthus camphoratus
Mulukhīyah Corchorus spp. especially C. olitorius
Munas Pulicaria jaubertii
Munīsh Ferula communis
Murār Any small yellow-flowered Dandelion-like Compositae especially Reichardia tingitana and Crepis rueppellii
Muraymirah Melia azederach
Mushaṭ (Mushayt) Any fern

Nafas Rubus spp.
Naḥd Blepharispermum yemense
Naḥaf Flueggia virosa
Najirān Maytenus spp.
Nakhl Phoenix dactylifera
Na'na' Cultivated Mentha hybrids
Narjil Cocos nucifera
Nasham Grewia bicolor, G. mollis, G. velutina and related species
Na'uḍ Mentha longifolia
Niba' Grewia tembensis
Nīl Indigofera spp.
Niṣaf Capparis cartiligenea
Nuqum Solanum incanum (north)
Numās Juncus spp.
Numnum Segeretia thea

Qa' Terminalia brownii
Qaḍab Cadaba rotundifolia
Qaḍb Medicago sativa
Qafal Commiphora kafal
Qa'in Jasminum grandiflorum
Qamḥ Triticum aestivum
Qaraḍ 1. Acacia etbaica
2. A. nilotica subsp. kraussiana
Qaranful 1. Dianthus caryophyllus
2. Ocimum tenuifolium and related species smelling of cloves
Qaṣab 1. Saccharum officinarum
2. Arundo donax
3. Any cane-like grass
Qaṣār Indigofera spinosa
Qaṣaṣ Euphorbia cactus and E. inarticulata
Qassīs Cyperus conglomeratus
Qāt Catha edulis
Qāt Ar Ra'ayān Maytenus spp.
Qaṭaf Commiphora kataf
Qatam Myrsine africana
Qaṭīn Nuxia congesta
Qawqa' Caucanthus edulis
Qīn Periploca visciformis
Qitat Acacia hamulosa
Qumbar Cardaria draba
Qurḍī Ochradenus baccatus
Qurtum Carthamnus tinctorius
Qusar Dorstenia foetida
Qushār Calotropis procera
Qusharī Vigna radiata
Quṣīf 1. Boerhavia diffusa
2. Ruellia patula
Quṭayb Spiny-fruited Tribulus spp. especially T. terrestris
Quṭīn Ceiba pentandra
Quṭn 1. Gossypium spp.
2. Vigna aconitifolia

Ra' Aerva javanica
Ra'd Ozoroa insignis
Raḍiḥ Sarcostemma viminalis
Raḥdh Ormocarpum yemenense
Ra'in 1. Aerva javanica
2. Abutilon pannosum
Rāk Salvadora persica
Ranayd Lavandula pubescens
Rand Artemisia arborescens
Ranf Delonix elata
Rīd Sarcostemma viminalis
Rifrāf Commicarpus spp.
Riḥān Ocimum basilicum
Riḥān al Baqr Salvia merjamie
Rijil Portulaca oleracea
Rinjās Lilies, particularly Lilium candidum and Polianthes tuberosa

Risha' Commiphora gileadensis
Riya' Syzygium guineense (Aḥjar)
Ruā Aerva javanica
Rukāb Piliostigma thonningii
Rumah Nuxia oppositifolia
Rumān Punica granatum
Rumayd Euphorbia schimperiana
Rūmī Zea mays
Ruqa' Trichelia emetica
Ruqmah Procumbent pink-flowered plants such as Boerhavia diffusa, Commicarpus helenae and Gisekia pharnaceoides
Rūz Oryza sativa
Sabānakh 1. Beta vulgaris
2. Spinacia oleracea
Sabaysab Euphorbia schimperiana
Ṣabir Aloe spp. especially A. vera
Sadaq Laggera decurrens
Safarjil 1. Cydonia vulgaris
2. Annona squamosa
Saḥīm Various grasses
Sakhil Cordia sinensis
Salaf 1. Sansevieria ehrenbergii
2. Agave sisalana
Salātah Lactuca sativa
Sambaq Polianthes tuberosa
Samrah Triticum durum
Samur Acacia tortilis & A. campoptila
Sana Cassia senna
Sanaf Acanthus arboreus
Sanafa (Sanafī) 1. Tephrosia purpurea
2. Digera muricata
San'am Saltia papposa
Sans Rhigazom somaliense
Sar Tarenna graveolens
Sa'tar Thymus laevigatus
Sawrab Hypoestes forskalii
Sawsab Euphorbia schimperiana
Say'am Saltia papposa
Saysībān Parkinsonia aculeata
Shadhab Ruta chalepensis
Shaḥ (Shahth) Dodonaea viscosa
Sha'īr Hordeum vulgare
Shajar The general term for a plant in Yemen
Shajarah al Kans Euryops arabicus
Shakadh Barleria bispinosa, B. prionitis & B. trispinosa
Shamar Foeniculum vulgare
Shambār Cassia fistulosa
Shammām Cucumis melo
Shandhaf Erica arborea
Shaqab 1. Premna resinosa
2. Otostegia fruticosa
3. Leucas alba
Shaqas Canthium schimperianum
Sha'r Plectranthus spp. especially P. asirensis and P. barbatus
Sharas Segeretia thea
Sharib Any plant with purgative properties, particularly Jatropha curcas
Sharijah Clematis spp. and sometimes other climbers
Sharjab Celtis africana
Shawḥat Grewia tembensis
Shawk A general term for spiny plants
Shawk al Kilab Pterolobium stellatum (and perhaps other spiny shrubs)
Shawkham Odyssea mucronata
Shibrad Halleria lucida
Shinzib Euphorbia cuneata
Shirī (Shirū) A general term for Cucurbitaceae such as Cucumis melo, Kedrostis gijef and Momordica balsamina
Shūrah Avicennia marina
Shurūr Succulents such as Caralluma spp. and Euphorbia fruticosa
Shuwāf Pavetta longiflora
Shuwayk Fagonia spp.
Sidad Capparis decidua
Sidr Zizyphus spina-christi
Siḥah Fagonia spp.
Sihām Various grasses
Sila' 1. Cissus quadrangularis
2. Succulent Compositae such as Kleinia spp., Senecio bojeri, S. hadiensis
Silām Acacia ehrenbergiana
Simsim Sesamum indicum
Sinab Debregeasia saenab
Sinayb Lavandula pubescens
Siraḥ Maerua crassifolia. Sometimes used for related species such as Boscia angustifolia and Cadaba farinosa
Ṣirak Grewia arborea
Sirbah Launea nudicaulis
Su'ad Cyperus spp. especially C. rotundifolius
Sulām Celtis toka
Sumīrah Asystasia guttata
Suqām Ficus sycomorus (Ḥujarīyah)
Sūr 1. Ficus sur
2. Balanites aegyptiaca (Ḥujarīyah)
Sura' Wendlandia arabica
Suraysah Crotalaria squamigera
Suyal Acacia nilotica subsp. indica and sometimes other Acacia spp. such as A. abyssinica
Suyayb 1. Euphorbia inarticulata (north)
2. E. triaculeata (Mawza')

Taḥaf Eragrostis tef
Ṭakhūb Sterculia africana (Waṣāb)
Ṭālab Rhus retinorrhaea & R. natalensis
Ṭalḥ Acacia gerrardii & A. origena but frequently misapplied to Acacia spp. generally
Talūq Ficus vasta
Tānib Cordia africana
Tankah (Tanqah) Dombeya schimperiana
Ṭawlab Rhus retinorrhaea & R. natalensis
Tawlaq Ficus vasta
Thāb Ficus salicifolia
Thamām 1. Lasiurus scindicus (Tihama)
2. Pennisetum divisum (Mashriq)
Thamīd Various grasses

Thū' Boscia angustifolia
Thū'ab Terminalia brownii (south)
Thū'am Myrica salicifolia
Thūm Allium sativum
Thunayb Aristida spp.
Tibḥī Adenium obesum ('Udayn)
Ṭibil Rumex steudelii
Tīn 1. Opuntia spp.
2. Ficus carica
Tīn Shawkī (Tīn Tūrkī) Opuntia spp.
Tinwāb Abrus bottae
Titūn Nicotiana spp.
Trūnj Citrus medica
Tubshah Ricinus communis
Ṭufāḥ Malus sylvestris
Tuflūq Cotyledon barbeyi
Tuḥāf Crinum album
Tūm Allium sativum
Tumāq Acokanthera schimperi
Tumār Berchemia discolor
Tumatah (Tumatim) Lycopersicum esculentum
Tumbac Nicotiana spp.
Tunah Acanthaceae spp. such as Ecbolium gymnostachyum & Hypoestes triflora
Tuqāḥ Celtis toka
Tushām Cassytha filiformis (Milḥān)
Tūt Morus niger

'Uām Cussonia holstii ('Udayn)
'Ubāb 1. Withania somnifera
2. Jatropha pelargoniifolia (Tihama)
'Udār Striga hermonthica
'Udhkhur Kleinia odora
'Ujal Convolvulus arvensis
'Ujām Cussonia holstii (J. Ṣabir)
'Uknī Cadaba heterotricha (south)
'Ulāb Zizyphus mucronata
'Ulb Zizyphus spina-christi
'Unīs Gynandropis gynandra
'Unqab Oncoba spinosa
'Untūr Carissa edulis
'Uraymbah Portulaca quadrifida
'Uraz Gramineae spp.
'Urfūṭ Acacia orfota
'Ushār Calotropis procera
'Ushriq Cassia italica
'Uthrub Rumex nervosus
'Utum Olea europaea
'Utur Carissa edulis
'Uzab Aromatic labiates such as Origanum marjorana and Ocimum tenuifolium
'Uzāf Dracaena serrulata
'Uzmuq Euphorbia uzmuk

Wakad Euphorbia sp. aff. fruticosa
Wālah Plectranthus amboinicus
Wa'lān Commelina and Aneilema spp.
Warad Rosa x bifera
Warad al Layl Mirabilis jalapa
Waraf Ehretia cymosa
Waras Flemingia sp.
Waykī Corchorus depressus
Wubal Cynodon dactylon

Yasar Ficus sur
Yasmin Jasminum fluminense
Yūsafī Citrus deliciosa

Zahar Tribulus arabicus & T. omanense
Zaraynbākh Iris albicans
Zarb Any spiny shrub particularly when cut
Za'tar Thymus laevigatus
Zaytūn Psidium guajava
Zīl Cynodon dactylon and other grasses
Zinjabīl Zingiber officinalis
Zirā'ah al Ḥanash Scadoxus multiflorus
Zub al Qa' 1. Orobanche spp.
2. Cistanche spp.
Ẓubah Acacia mellifera
Zughāf Blepharis ciliaris
Ẓulām Monotheca buxifolia

APPENDIX 2: Poisonous Plants

It is often important to know which plants are poisonous, partly because they may be eaten inadvertently by people (*Ricinus* is particularly dangerous) and partly because they may cause poisoning in livestock. The vast majority of plants listed below are seldom dangerous since they are not attractive to either people or livestock. In some cases only some parts of the plant are poisonous while in others they are only poisonous under certain conditions. Potatoes are one of the most important food crops but when the tubers are exposed to sunlight they become green and develop the alkaloid solanin. Similarly Sorghum is the most important animal fodder in Yemen but the dangers of eating the young growth were recognised by the Bughyat al Fallāḥin as long ago as the fourteenth century. Nearly all the poisonous plants have uses in folk medicine (see Fleurentin and Pelt: *Repertory of Drugs and Medicinal Plants of Yemen* (1982)) while others such as *Cannabis sativa, Catha edulis* and *Nicotiana* spp. are widely used for their narcotic effect.

The exact nature of the toxic element in each plant is not always known and some plants, indicated by an asterisk, are only suspected of being poisonous. Further information on the toxic nature and effects of different plants can be found in *Common Poisonous Plants of East Africa* by B. Verdcourt and E. C. Trump.

PTERIDOPHYTES
- Equisetum ramosissimum
- Pteridium aquilinum

ARISTOLOCHIACEAE
- Aristolochia bracteata

RANUNCULACEAE
- *Clematis spp.
- Nigella sativa
- Ranunculus multifidus

PAPAVERACEAE
- Argemone mexicana
- Papaver somniferum

CANNABACEAE
- Cannabis sativa

CHENOPODIACEAE
- Chenopodium spp. especially C. ambrosioides and C. schraderianum

CARYOPHYLLACEAE
- Small white-flowered species are suspect, especially
- *Arenaria leptodados
- *Minuartia filifolia
- *Stellaria spp.

GUTTIFERAE
- *Hypericum revolutum

MALVACEAE
- Gossypium spp.

PASSIFLORACEAE
- *Adenia venenata

CUCURBITACEAE
- Cucumis spp.
- *Cucurbita maxima

CAPPARACEAE
- *Cadaba farinosa
- *Maerua spp.
- Capparis tomentosa (very poisonous)

CRUCIFERAE
- Brassica napus

MYRSINACEAE
- Maesa lanceolata
- Myrsine africana

PRIMULACEAE
- Anagallis arvensis

CRASSULACEAE
- *Cotyledon barbeyi

LEGUMINOSAE
- Abrus precatorius (seeds very poisonous)
- Cassia spp. especially:
- C. didymobotrya
- C. occidentalis
- C. senna
- Crotalaria spp. especially:
- C. incana
- C. retusa
- C. spinosa
- Indigofera spicata
- Tephrosia spp.
- *Vigna spp.

THYMELIACEAE
- *Gnidia somaliensis

CELASTRACEAE
- Catha edulis

EUPHORBIACEAE
- Euphorbia spp. (The latex is caustic especially in the eyės)
- *Jatropha spp., especially J. curcas
- Phyllanthus maderaspatensis
- Ricinus communis (very poisonous)

VITACEAE
- Rhoicissus tridentata

LINACEAE
- *Linum usitatissimum

MELIANTHACEAE
- Bersama abyssinica

MELIACEAE
- Melia azederach (seeds)
- Trichelia emetica (seeds)
- *Turraea spp.

BALANITACEAE
- Balanites aegyptiaca

ZYGOPHYLLACEAE
- *Fagonia spp.
- Peganum harmala
- Tribulus spp.

OXALIDACEAE
- Oxalis spp.

UMBELLIFERAE
- Berula erecta
- Conium maculatum
- Ferula communis

APOCYNACEAE
- Acokanthera schimperi
- Adenium obesum
- Nerium oleander
- Rhazya stricta
- Thevetia peruviana

ASCLEPIADACEAE
- Calotropis procera
- *Gomphocarpus spp.
- Sarcostemma spp.

SOLANACEAE
- Datura spp.
- Lycium shawei
- Nicotiana spp.
- Solanum spp. especially S. incanum
- *Withania somnifera

BORAGINACEAE
- *Heliotropium spp.

VERBENACEAE
- Lantana spp.

ACANTHACEAE
- Many species such as Anisotes trisulcus and Hypoestes forskalei are bitter and avoided by animals but probably not poisonous

RUBIACEAE
- Pavetta spp.

COMPOSITAE
- Ageratum conyzoides
- Dichrocephala chrysanthemifolia
- Senecio spp.
- Xanthium spp.

GRAMINEAE
- Many species can build up hydrocyanic acid in their tissues under certain conditions (particularly when wilting following rapid growth). Sorghum bicolor, Cynodon dactylon and Themeda triandra are among many species known to do this

LILIACEAE
- *Aloe spp.

AMARYLLIDACEAE
- *Crinum album
- *Scadoxus multiflorus

IRIDACEAE
- *Iris albicans

APPENDIX 3: Gazetteer

The following gazetteer gives the coordinates (latitude and longitude) of all place names occurring in the text with the exception of a few which are always defined in relationship to a second locality (e.g. Minām, J. Ḥabashī). I have also included most of the localities at which earlier botanists made collections. With the exception of the three major towns, Sana'a, Taiz and Hodeida which are given their traditional spellings, all place names are spelt according to the system of transliteration given on page 00. In order to assist readers trying to find some of the old botanists' collecting localities, some variant spellings of the more important localities are given. It is impossible to include all variants because of limits of space but they can be very numerous; Ḥajjaylah for example may be spelt Hujjaylah, Hojeilah, Hodjeilah, Hodjela and Hadjeilah. The correct spelling can usually be found by searching for the word under a different initial vowel since this is often unstressed and of very indeterminate value. Words transliterated with *Dj* and *G* can be found under *J* and words beginning with *Ch* can be found under *Kh*.

Place names are listed alphabetically and the letter 'Ayn is ignored. Where the place name includes two or more words the first of these is almost always either the definite article (*Al, Ad, Adh, An, Ar, As, Ash, At, Ath, Az*) or one of the following descriptive terms: *'Aqaba(h)* (pass), *Bāb* (gateway, entry point), *Banī* (a tribal area, literally sons), *Bayt* (village, lit. a house), *Bīr* (well), *Dār* (village, lit. a house), *Dayr* (village, lit. houses), *Hammām* (hot springs), *Ḥiṣn* (fortress), *Jabal* (mountain), *Najd* (mountain), *Naqīl* (pass), *Qa'* (plain). *Sūq* (market), *Wādī* (valley). The article and the descriptive term are ignored for the purpose of alphabetical arrangement, thus when looking up Jabal Ash Shirq, the reader should ignore Jabal (the descriptive term) and Ash (the article) and refer to Shirq.

'Abs 16°01'N, 43°11'E
___, Wadi A small wadi flowing just south of 'Abs market
'Abūs A district centred on 13°17'N, 44°18'E
'Adhāqah, Bayt 15°37'N, 43°42'E
'Adūf 13°14'N, 44°08'E
Agara See 'Ajārah
Aḥad, Suq Al (Waṣāb As Sāfal) 14°16'N, 43°36'E
Aḥjar, Wadi 15°29'N, 43°53'E
Aḥkūm A district centred on 13°15'N, 44°15'E
Aḥmar, Najd Al The pass between Ibb and Taiz 13°52'N, 44°11'E
'Ajārah A district near Ḥajjaylah (14°59'N, 43°37'E), the location not exactly known
'Ajlub, Wadi A tributary of the Wadi Banā joining at Damt (14°07'N, 44°38'E)
'Akāhilah, Al 13°14'N, 44°10'E
'Akwa, Wadi See Rakwah
'Alab, J 17°33'N, 43°23'E
'Alama, Bayt 15°37'N, 43°46'E
'Ali, Banī A district in 'Abūs, 13°18'N, 44°19'E
'Ali, Ḥammām 16°32'N, 43°35'E
'Alūjah 14°35'N, 43°33'E
'Amal, J. Al 14°29'N, 43°54'E
Aman, Suq Al 1. (Hajjah) 15°40'N, 43°31'E
2. (Bani Matar) 15°10'N, 43°58'E
'Amrān 15°40'N, 43°56'E
'Anam, J. 17°01'N, 43°29'E
'Annah, Wadi One of the great wadis of 'Udayn, a tributary of the Wadi Zabīd, 13°59'N, 43°55'E
'Anis An important district centred on 14°46'N, 44°07'E
Arḥab A large district northeast of Sana'a, c. 15°50'N, 44°15'E
'Arūs, Ḥiṣn The summit of J. Ṣabir, 13°31'N, 44°03'E
'Arūq A district centred on 13°19'N, 44°15'E
'Ashshah, Al 16°16'N, 43°51'E
'Attārah 16°03'N, 43°40'E
Aybān, J. 15°18'N, 44°07'E
'Ayfū J. 13°52'N, 43°50'E
'Aymān, J. See Ayban, J.
'Ayn 'Ali, Wadi 15°43'N, 43°38'E
'Azū, Wadi 17°01'N, 43°33'E
'Azūm, Naqīl 15°11'N, 44°37'E

Ba'dān, J. 14°00'N, 44°15'E, the mountain mass immediately east of Ibb
Badī, Bayt 15°40'N, 43°52'E
Badr, Wadi 17°00'N, 43°22'E, one of the main wadis of Khawlan Ash Sham and the principal feeder to the Wadi Jīzān.
Baḥal, Bāb 15°07'N, 43°37'E, a pass in the Ḥarāz mountains
Bājil 15°03'N, 43°17'E
Bakīl, Qa' 14°47'N, 44°12'E
Balaq, J. The limestone mountains lying astride the Mārib dam at 15°24'N, 45°15'E
Balasān, Qa' 14°35'N, 44°23'E
Banā, Wadi The greatest of the eastward-flowing wadis beginning at 14°11'N, 44°20'E and flowing southeast through Saddah, Nādirah and Damt to South Yemen
Barad, J. 14°34'N, 43°39'E
Barah, J. See Barad, J.
Barakānī, Wadi 13°17'N, 43°58'E
Barāqish 16°00'N, 44°48'E

Baraṭ, J. A large ridge centred on 16°44′N, 44°17′E
Barḥ, Suq Al 13°27′N, 43°43′E
Bāshā, Bīr 13°36′N, 43°58′E
Bathayjah 13°08′N, 43°59′E
Bawn, Qa' Al 15°46′N, 44°02′E
Baw'ān 15°12′N, 43°59′E (sometimes referred to as Suq Al Khamīs)
Baws, Bayt 15°16′N, 44°12′E
Bayḍā, Al A town situated at 13°59′N, 45°34′E, which is also the name of a province
Bayḥān, J. 13°12′N, 44°07′E
Bishārī, Bayt Al 15°29′N, 43°35′E
Boka Forsskal's spelling of Bughah
Bosan, Qa' See Busan, Qa'
Braz, Bayt 15°29′N, 43°31′E
Bughah An area on the northwest slopes of J. Ḥuzār at 14°34′N, 43°37′E, where Forsskal collected many plants
Bukhta 14°45′N, 44°30′E (approx)
Bulgose One of Forsskal's collecting localities in the mountains of Ḥadīyah, perhaps an old name for Jirjir
Bura', J. 14°50′N, 43°28′E
Busān, Qa' 14°48′N, 44°32′E

Cahim One of Botta's localities – see Kāhim
Chuviet, Wadi One of Schweinfurth's localities – see Khufīt

Ḍabāb, Wadi 13°32′N, 43°47′E
Ḍabī 13°17′N, 44°18′E
Dafdaf, J. Ḥiṣn 13°59′N, 43°59′E
Ḍaḥī, Aḍ 15°13′N, 43°04′E
Ḍahr, Wadi 15°26′N, 44°07′E
Dalīl, Ad 14°07′N, 44°11′E
Damīm, Wadi A wadi south of Ḥays – 13°53′N, 43°30′E
Damt 14°06′N, 44°42′E
Dan, Ad 14°20′N, 43°51′E
Dawd, 'Aqabah Banī 15°28′N, 44°31′E
Dawān, Banī 14°42′N, 43°36′E
Ḍawrān, J. 14°44′N, 44°12′E
Dhamār 14°33′N, 44°24′E major town and provincial headquarters
Dhamrān, Najd 13°18′N, 44°20′E
Dhanah, Wadi The name of the Mārib wadi in its upper reaches above 15°09′N, 45°14′E
Dhī Nā' im 14°06′N, 45°26′E
Dhī Sufāl 13°50′N, 44°07′E
Dhubhān The plateau on which Turbah is situated, at c. 13°13′N, 44°08′E
Dimarayn 13°31′N, 44°01′E
Dorebat One of Forsskal's collecting localities, apparently near Taiz and possibly Wadi Dabāb
Dūr, Wadi 13°58′N, 44°01′E

Faqīh, Bayt Al 14°31′N, 43°19′E
Fardah, Naqil Al 15°45′N, 44°35′E
Fāzah, Al 14°07′N, 43°06′E

Ghalaybah 13°18′N, 44°18′E
Ghayāmī, J. 13°18′N, 44°18′E
Ghaylān, Naqīl 15°36′N, 44°24′E
Ghaymān 15°16′N, 44°21′E
Ghirās, Al 15°31′N, 44°20′E
Ghubāḥ, Bīr 13°43′N, 43°49′E
Ghuraz 16°54′N, 43°45′E

Ḥabab, Wadi 1. Madīnat Al 'Abīd – Ḥ ammām Ali 15°38′N, 44°00′E 2. Sirwah. 15°25′N, 44°54′E
Ḥabashī, J. 13°30′N, 43°52′E
Ḥadā, Al A large district northeast of Dhamār centered on 14°50′N, 44°30′E
Ḥadabah, Wadi One of the 'Udayn wadis, 13°57′N, 44°04′E
Ḥadādah 14°42′N, 43°36′E
Ḥaddah 15°18′N, 44°10′E
Hadīdah, J. One of Rathjen's collecting localities, its exact position unknown.
Hadie See Ḥadīyah
Ḥadīyah, Al Famous collecting locality of Forsskal, 14°31′N, 43°34′E
Ḥadnān 13°32′N, 44°00′E
Haguef ? Hajjaf. 13°32′N, 44°00′E
Ḥajarah 15°04′N, 43°37′E
Ḥajdah 13°35′N, 43°49′E
Ḥajjah Provincial headquarters, 15°41′N, 43°36′E
Ḥajjaylah, Al 14°59′N, 43°38′E
Ḥajlah, Naqīl 15°06′N, 45°30′E
Ḥamdān A district north of Sana'a, mostly covered in lava flows, 15°25′N, 44°05′E
Ḥamīd 13°20′N, 44°24′E
Ḥanish Island 13°44′N, 42°44′E
Ḥār, Wadi 1. 15°01′N, 43°35′E 2. 14°24′N, 44°20′E
Ḥaraḍ 16°25′N, 43°04′E
___, Wadi North of the town at c 16°28′N, 43°07′E
Ḥaradah, Al 14°05′N, 44°40′E
Ḥarāz A mountain range embracing the area round Manākhah and (as understood here) Ḥajjaylah, 15°05′N, 43°40′E
Ḥarf, Al 16°21′N, 44°05′E
Ḥarf Sufyān A large district lying mostly to the north of Al Ḥarf
Ḥarīb 14°55′N, 45°30′E
Ḥarith, Banī Al A district northeast of Sana'a, centred on 15°33′N, 44°18′E
Ḥasabah Asfāl 13°30′N, 44°01′E
Ḥaṣaṣ, Naqīl 13°24′N, 43°58′E
Ḥawah, J. approx. 14°40′N, 43°49′E
Ḥawāyat, Wadi One of Deflers' collecting localities near Ḥajjaylah, approx. 14°59′N, 43°38′E
Ḥawbān 13°35′N, 44°05′E
Ḥaydān 16°47′N, 43°25′E
___, Wadi One of Botta's collecting localities between Ḥays and Shar'ab, possibly Wadi Ḥayrān, 14°01′N, 43°48′E
Ḥayfān 13°18′N, 44°18′E
Ḥaymah Two districts lying between Sana'a and the Ḥarāz. Ḥaymat Ad Dakhilīyah at c 15°15′N, 43°50′E being north of Ḥaymat Al Kharijīyah at c 15°00′N, 43°52′E

Ḥays 13°55′N, 43°29′E
Ḥazm, Al 16°09′N, 44°46′E
Ḥidrān, Wadi The lower course of the Wadi Ḍabāb at c 13°36′N, 43°55′E
Ḥijān, Wadi The wadi dividing J. Masār from J. Sāfān in the Ḥarāz, c 15°03′N, 43°39′E
Ḥiliyal 14°10′N, 44°15′E
Ḥimyar, J. 13°47′N, 43°59′E
Ḥiṣn, J. 13°18′N, 44°17′E
Hodeida 14°48′N, 42°56′E
Hodjelah See Ḥajjaylah
Ḥubaysh, J. Mountain and district near Ibb, centred on 14°06′N, 44°05′E
Ḥufāsh, J. A mountain of the central escarpment lying in the rainshadow of J. Milḥān, centred on 15°16′N, 43°25′E
Ḥujarīyah As understood here, this region includes all the mountain areas south of Taiz excluding J. Ḥabashī and J. Ṣabir, thus extending from 13°25′N southwards and 43°50′E eastwards. The chief town is Turbah
Ḥusaynīyah 14°21′N, 43°19′E
Ḥushā, J. 13°46′N, 44°26′E
Ḥushaysh Banī A district around 15°24′N, 44°25′E
Ḥūth 16°14′N, 43°58′E
Ḥuzār, J. 14°33′N, 43°37′E

Ibb 13°58′N, 44°11′E, major town and provincial headquarters
Ibl, Bāb Al 15°08′N, 43°22′E
Ibl, Naqīl Al 13°34′N, 44°07′E
Ibn Abbās 15°21′N, 42°47′E
Idjz One of Wissmann's collecting localities, probably near Bani Manṣūr, 15°06′N, 43°52′E
Iriyān 14°16′N, 44°16′E
Isbil, J. 14°35′N, 44°40′E

Jabalayn, Al 13°56′N, 44°03′E
Jabīn, Al 1. (J. Raymah) 14°37′N, 43°36′E
2. (J. Ba'dān) 13°58′N, 44°12′E
Ja'farīyah District in the southwest of J. Raymah around 14°33′N, 43°34′E
Jāḥ, J. Al 13°18′N, 44°19′E
Jaḥālī One of Wissmann's collecting localities, its exact location not known but near Madīnat Al 'Abīd (14°38′N, 43°57′E)
Jalal, J. 14°47′N, 44°16′E
Jammah, Suq Al 16°31′N, 43°13′E
Janad, Al 13°40′N, 44°10′E
Jarraḥī, Al 14°08′N, 43°23′E
Jarūbah 14°22′N, 43°23′E
Jawf, Wadi Al A huge wadi descending from c 16°15′N, 44°30′E southeast to c 15°50′N, 45°25′E
Jawjib, Naqīl Pass between Aḥkūm and Qadas, 13°17′N, 44°10′E (? = Mawjib)
Jiblah 13°55′N, 44°08′E
Jifān 16°58′N, 43°30′E
Jiḥānah 15°13′N, 44°25′E
Jinān, J. One of Wissmann's collecting localities, the exact location is not known
Jirābī 15°15′N, 43°11′E
Jirdad, J. 13°10′N, 43°55′E
Jirjir, Al The pass between J. Ẓallamlam and J. Kabūrah, 14°33′N, 43°36′E
Jūbah, Al 15°03′N, 45°17′E
Juban 14°00′N, 44°52′E
Jubirān, Naqīl 13°13′N, 44°11′E
Jum'ah, Suq Al (Friday market) Applied to a village on J. Ash Shirq at 14°45′N, 44°02′E
Jumlī, Suq Al 17°00′N, 43°28′E

Kabūrah, J. 14°33′N, 43°36′E
Kadān, Al 15°15′N, 43°15′E
Kāḥil, J. 15°03′N, 43°45′E
Kāhim One of Botta's collecting localities lying north of Rawnah, Shar'ab (13°45′N, 43°48′E)
Kamaran Island 15°22′N, 42°35′E
Kanin, J. 15°04′N, 44°22′E
Karāwan One of Schweinfurth's collecting localities at (approx) 15°03′N, 43°43′E
Ka'taba See Qa'ṭabah
Kawkabān 15°30′N, 43°53′E
Kawlah, Suq Al (= Makawl) 14°06′N, 43°35′E
Khadrā, J. 1. 13°51′N, 44°11′E
2. 14°03′N, 44°04′E
Khalāwah 13°41′N, 44°18′E
Khalīfah, Al 14°53′N, 43°19′E
Khamir (Khamr) 15°59′N, 43°51′E
Khamīs, Suq Al Any place with a Thursday market particularly
1. 15°46′N, 43°12′E
2. Bani Sa'd 15°11′N, 43°30′E
3. Baw'ān 15°12′N, 43°59′E
4. Madyūr 15°09′N, 43°56′E
Khārid, Wadi A major tributary of Wadi Jawf rising at about 15°40′N, 44°25′E
Khatwah, Al 13°19′N, 44°16′E
Khawbah, Al 16°32′N, 43°44′E
Khawkhah, Al 13°48′N, 43°14′E
Khawlan Ash Sham As defined in this book, the whole mountainous district west of the Ṣa'dah-Majz main road and north of about 16°35′N
Khawlan At Tīyal A large and not very well-defined area of mountains between Sana'a and Sirwāḥ centred on Jiḥānah, 15°13′N, 44°25′E
Khaywan 16°16′N, 44°04′E
Khidār 14°59′N, 44°16′E
Khufīt, Wadi 15°02′N, 43°39′E
Khulāqah 15°38′N, 44°25′E
Khushm, Dayr 15°36′N, 43°11′E
Kitāb 14°11′N, 44°18′E
Kudmīyah One of Forsskal's collecting localities in the Wadi Mawr, c 15°42′N, 42°54′E, but the exact location is not known as several villages in this area have this name
Kuḥlān 15°44′N, 43°43′E

Kulabah, Wadi 15°46′N, 43°24′E
Kumaym 15°01′N, 44°22′E
Kurmah One of Forsskal's localities, see Kusmah
Kusmah 14°34′N, 43°38′E

Lā'ah, Wadi A major tributary of the Wadi Mawr separating Ḥajjah and Maḥwīt and flowing through Aṭ Ṭūr, c 15°35′N, 43°23′E
Lahanet One of Wissmann's collecting localities, probably near Madīnat Al 'Abid
Līyah, Wadi 16°47′N, 43°19′E
Lohaya See Luḥayyah
Luḥayyah, Al 15°42′N, 42°41′E

Maamara, J. See Māmarah, J.
Ma'bar 14°47′N, 44°17′E
Mada', J. 15°36′N, 43°47′E
Madān, Al 16°13′N, 43°37′E
Madaq (? Madak) 17°07′N, 43°28′E
Madfār, Wadi A wadi near Ḥajjaylah, 14°59′N, 43°38′E
Madhbūr, J. 15°22′N, 44°29′E
Madīnat Al 'Abīd 14°38′N, 43°57′E
Madnān, J. Ḥiṣn 14°19′N, 43°53′E
Madrūbah, Al 13°51′N, 43°35′E
Mafḥaq 15°06′N, 43°49′E
Mafraq (lit. a junction) 13°22′N, 43°36′E
Maghrabah (lit. a pass from which the setting sun is visible)
1. (Ḥarāz) 15°05°N, 43°44′E
2. (J. Bura') 14°52′N, 43°28′E
3. (at the head of the Wadi Yūr) 15°23′N, 43°23′E
Maghrib 'Ans A district centred on 14°28′N, 44°10′E
Maḥābishah 15°57′N, 43°26′E
Maḥalayn 15°22′N, 44°29′E
Maḥdad, J. 15°36′N, 43°46′E
Maḥjar 15°29′N, 43°53′E
Maḥwīt, Al 15°28′N, 43°32′E
Majz 17°04′N, 43°35′E
Makhadir 14°08′N, 44°12′E
Malāḥiṭ 16°46′N, 43°17′E
Māmarah (Maamara), J. Probably in Shar'ab ? c 13°45′N, 43°46′E (one of Botta's collecting localities)
Ma'mar 15°31′N, 44°06′E
Manākhah 15°04′N, 43°44′E
Manār, Al 14°03′N, 44°03′E
——, J. 14°02′N, 44°16′E
Mandab, Bāb al 12°40′N, 43°27′E
Manṣūr, Bani 15°06′N, 43°54′E
Manṣūrah, Qa'alat 13°19′N, 44°13′E
Manṣūrīyah 14°41′N, 43°18′E
Mantar See Manẓar
Manẓar 14°43′N, 42°58′E
Maqbanah A large district west of Taiz, centred on 13°36′N, 43°38′E
Maqsalah 17°07′N, 43°34′E
Marāshī 16°33′N, 44°17′E
Marāwi'ah 14°50′N, 43°09′E
Mārib 15°25′N, 45°20′E
Markah 15°24′N, 43°20′E
Marran, J. 16°48′N, 43°23′E
Masājid 15°18′N, 44°04′E
Masanah, J. One of Deflers' collecting localities, probably near Ḥayfān
Masār, J. 15°05′N, 43°41′E
Mashwarah 13°58′N, 44°06′E
Masna'a Ṭa'am, J. 14°50′N, 43°32′E
Maswār, J. 15°36′N, 43°36′E
——, Wadi 15°12′N, 44°32′E
Maṭar, Banī A district west of Sana'a, centred on 15°15′N, 44°03′E
Matnah 15°15′N, 44°01′E
Mawīyah 13°35′N, 44°21′E
Mawr, Wadi The longest wadi in Yemen, the main course separating Ḥajjah from J. Sharafayn and flowing through Zuhrah to Luḥayyah (15°41′N, 42°43′E)
Mawsanah 15°07′N, 43°42′E
Mawshij 13°43′N, 43°16′E
Mawza' 13°16′N, 43°30′E
Midaḥiṭ 13°14′N, 44°10′E
Mīdī 16°19′N, 42°48′E
Miftāḥ, Banī A district between Riḥāb and Waṣāb centred on 14°15′N, 44°07′E
Mighlāf, Al 15°17′N, 43°11′E
Mikhlāf A district centred on 13°43′N, 43°47′E
Milāḥ, Al 14°25′N, 44°47′E
Milḥān, J. 15°22′N, 43°21′E
Minmār, J. 16°54′N, 43°32′E
Mirbish, J. A mountain in Salafīyah, the exact location not known but c 14°39′N, 43°51′E
Mirkiz, Suq Al 13°13′N, 44°01′E
Mirzin, Al 14°54′N, 43°26′E
Mishrāfah 14°22′N, 43°32′E
Misrākh 13°27′N, 44°01′E
Mocha See Mukhā
Moghlaf See Mighlāf
Mudhaykhirah A district comprising the mountain range between 'Udayn and Taiz from 13°54′N, 43°51′E to 13°54′N, 44°03′E
Mughlād, Bayt 15°03′N, 43°43′E
Muharraq, Suq Al 16°01′N, 43°22′E
Muḥarras The old name for Najd Al Aḥmar 13°52′N, 44°11′E
Mujwarah 15°05′N, 43°39′E
Mukhā, Al 13°19′N, 43°15′E
Mukhadir See Makhadir
Mukhtār, Banī 14°32′N, 43°37′E
Munīf, J. 13°09′N, 44°07′E
Musalla 13°15′N, 44°15′E

Nabī Shu'ayb, J. 15°17′N, 43°57′E
Nādirah, An 14°06′N, 44°29′E
Nā'im, Suq Bayt An 15°25′N, 44°04′E
Najdayn, An 13°47′N, 43°55′E (As Salam)
Najrān, Wadi A wadi rising near Ṣa'dah and flowing northeast into Saudi Arabia at c 17°30′N, 44°15′E
Nakhlah, Wadi A wadi descending from Shar'ab and passing north of Ḥays at 13°56′N, 43°29′E
Na'mah 14°33′N, 43°35′E (J. Raymah)
Nasham 14°51′N, 43°29′E
Nashimah 13°23′N, 43°58′E

Nasr, Dār An 13°33′N, 44°01′E (J. Ṣabir)
Nihm A huge district northeast of Sana'a, centred on 15°45′N, 44°30′E
Nukaybah (? Nuqaybah) 13°33′N, 43°29′E
Nu'mān, J. 14°00′N, 44°06′E
Nuqum, J. 15°21′N, 44°15′E
Nuwāsh, Wadi 14°42′N, 44°54′E

Qābah, Wadi 13°47′N, 43°58′E
Qabayn, J. 13°18′N, 44°18′E
Qabayṭah, Al Applied in general to the eastern Ḥujarīyah and specifically to the southeast extremity around 13°15′N, 44°27′E
Qadas A district of the Hujariyah around 13°17′N, 44°07′E
Qā'idah, Al 13°45′N, 44°07′E
Qanāwis 15°29′N, 43°08′E
Qarad, J. See Himyar, J,
Qaṣab, Wadi Al 15°09′N, 43°34′E
Qa'ṭabah 13°51′N, 44°42′E
Qidam, Al 15°07′N, 43°37′E
Qimah, Al 14°54′N, 43°27′E
Qita', Suq Al 15°26′N, 43°27′E
Qubayṭah, Al See Qabayṭah
Qū sa', Wadi 14°34′N, 44°00′E
Qutay 14°54′N, 43°12′E

Rada' 14°25′N, 44°50′E
Raḏmah, Ar 14°12′N, 44°36′E
Rahbān 15°29′N, 43°35′E
Raḥidah, Ar 13°22′N, 44°18′E
Ra'isi 14°04′N, 44°05′E
Rajamah 14°58′N, 43°41′E
Rakwah, Wadi 17°12′N, 43°31′E
Ramādah 13°34′N, 43°51′E
Ra's, J. 14°03′N, 43°37′E
Rasyān, Wadi A major wadi of Taiz province arising north of Taiz and passing through Maqbanah and reaching the sea north of Yakhtul
Rawḍah, Ar 15°25′N, 44°13′E
Rawnah, Ar 13°45′N, 43°47′E
___, Wadi Ar 15°09′N, 44°20′E
Raydah 15°49′N, 44°02′E
Raymah, J. A large mountain area centred on 14°37′N, 43°45′E and including Ja'farīyah, Salafīyah and Bilad Aṭ Ta'am
Rāziḥ, J. 16°58′N, 43°15′E
Regma See Rajamah
Ribāṭ 14°12′N, 44°21′E
___, Suq Ar 14°40′N, 43°35′E
Riḥāb 14°13′N, 44°10′E
Rima', Wadi A major wadi flowing from Madīnat Al 'Abīd (14°38′N, 43°57′E) to Ḥusaynīyah (14°21′N, 43°19′E), dividing J. Raymah from 'Utmah
Riṣābah 14°42′N, 44°21′E
Roboa One of Forsskal's collecting localities, see Suq Ar Rubū', Waṣāb As Sāfal
Rubaysah 13°14′N, 44°09′E
Rubū', Suq Ar A place with a Wednesday market particularly
1. (Waṣāb As Sāfal) 14°17′N, 43°43′E
2. (Wasāb Al 'Ālī) 14°20′N, 43°53′E
3. ('Utmah) 14°30′N, 43°58′E
Rujaymah, Wadi 13°18′N, 44°15′E
Rujm (Rujum) 15°27′N, 43°38′E

Ṣaba', Wadi 15°26′N, 45°20′E
Ṣa'bah, Aṣ 15°23′N, 43°19′E
Sabalī, Wadi 14°28′N, 44°08′E
Sabbān, Wadi 14°17′N, 44°30′E
Ṣabir, J. 13°31′N, 44°03′E
Sabt, Suq As A place with a Saturday market particularly
1. (Ibb) 14°04′N, 44°10′E
2. (J. Bura') 14°53′N, 43°26′E
Ṣa'dah 16°56′N, 43°45′E
Saddah, As 14°07′N, 44°26′E
Sāfān, J. 15°04′N, 43°37′E
Sāḥlah 13°33′N, 44°02′E
Sahmān, Qa' 15°16′N, 44°03′E
Sa'īd, Banī 14°33′N, 43°35′E
Sa'īd Miṣr 14°33′N, 43°27′E
Salafīyah A district centred on 14°40′N, 43°51′E
Salam A district north of Taiz, centred on 13°47′N, 43°54′E and often considered part of Shar'ab
Ṣalīf 15°18′N, 42°40′E
Ṣalū, As A district centred on 13°22′N, 44°11′E
Sama' J. As 15°36′N, 44°18′E
Sāmi', J. 13°23′N, 44°06′E
Samīr, Suq 15°07′N, 43°55′E
Samsarah (lit. a caravansarai) 13°15′N, 44°00′E
Sana'a (*also* Sana') 15°21′N, 44°12′E
Sanabān 14°25′N, 44°29′E
Sanafān 14°16′N, 44°27′E
Sanḥān A district centred on 15°14′N, 44°15′E
Sāqayn 16°53′N, 43°30′E
Saqūl 13°14′N, 44°16′E
Sāra', Wadi 15°15′N, 43°31′E
Sawraq, J. 13°43′N, 44°20′E
Ṣayfarah, As 13°36′N, 44°01′E
Ṣayrah 13°13′N, 44°08′E
Sayyānī, As 13°49′N, 44°10′E
Sayyīd Sulayman One of Schweinfurth's collecting localities in the Tihama, exact location unknown but perhaps Sulaymanīyah, 14°49′N, 43°12′E
Selfije See Salafīyah
Sha'af 14°34′N, 43°36′E
Shadhbān, Wadi 15°05′N, 43°46′E
Shahārah 16°11′N, 43°42′E
Shāhil, J. Ash 15°51N, 43°24′E
Shaḥir 15°22′N, 43°21′E
Shalālah, Wadi 14°19′N, 44°27′E
Shamlān 15°24′N, 44°08′E
Shamrān, Bayt 15°04′N, 43°40′E
Shamsān (Maḥābishah) 15°56′N, 43°24′E
___, J. 15°28′N, 44°08′E

Shamsān, Ḥammām 16°32′N, 43°35′E
Shar'ab A district northwest of Taiz, centred on 13°45′N, 43°47′E
Sharafayn, J. The range of mountains centred on Maḥābishah 15°57′N, 43°26′E
Sharas, Wadi 15°44′N, 43°38′E
Sharāwah 17°21′N, 43°23′E
Shaybah, Banī 13°11′N, 44°02′E
Shibām 15°30′N, 43°55′E
—, J. 15°03′N, 43°43′E
Shijah, Naqīl One of Wissmann's collecting localities, its exact location not known
Shirnamah 15°36′N, 43°41′E
Shirq, J. Ash A large district centred on 14°45′N, 43°58′E
Shirqabī, Naqīl 14°27′N, 44°16′E
Shirqī, Suq 13°51′N, 43°57′E
Shūb 1. (Sana'a) 15°22′N, 44°12′E
2. (Maqātirah) 13°11′N, 44°08′E
3. (Maqbanah) 13°40′N, 43°09′E
Silf, Suq As 15°13′N, 43°56′E
Sihām, Wadi A major wadi separating Ḥaymah and the Ḥarāz from J. Ash Shirq and J. Raymah, flowing into the sea at 14°41′N, 42°58′E
Sina, Wadi One of Botta's collecting localities, said to be near Janad (13°40′N, 44°10′E)
Sirwāḥ District and town, 15°27′N, 45°01′E
Sūdah, As 15°57′N, 43°46′E
Sūdān, Wadi 13°46°N, 44°09′E
Suhayl, Dayr As 14°59′N, 43°07′E
Sukhnah 14°48′N, 43°26′E
—, Wadi 15°14′N, 43°50′E
Sumārah, J. 14°11′N, 44°11′E
Surdud, Wadi One of the major wadis of Yemen rising in Ḥaymat Ad Dakhilīyah and flowing past Khamīs Banī Sa'd, Kadān and Ḍaḥī to 15°10′N, 42°52′E

Ṭa'am, Bilad Aṭ 14°51′N, 43°32′E
Ṭa'if 14°37′N, 43°01′E
Taiz Provincial headquarters, 13°34′N, 44°01′E
—, J. 15°36′N, 43°46′E
Ta'kar, J. 13°52′N, 44°07′E
Ṭawīlah 15°29′N, 43°44′E
Thabāb 13°16′N, 44°26′E
Tha'bat, Wadi 13°34′N, 44°01′E
Thalūth, Suq Ath Any place with a Tuesday market especially that in the Wadi Lā'ah 15°32′N, 43°40′E
Thilā 15°34′N, 43°54′E
Tin'am 15°23′N, 44°27′E
Ṭuba', Aṭ 14°55′N, 43°26′E
Tuḥaytah, At 14°11′N, 43°13′E
Ṭūr, Aṭ 15°36′N 43°23′E
Turbah, At 13°13′N, 44°07′E

Uafad See Waṣāb, one of Forsskal's collecting localities
'Ubāl 14°59′N, 43°32′E
'Ubr Sa'd 14°53′N, 43°27′E
'Udayn An area of lush valleys centred on the suq at 13°58′N, 44°00′E
'Ujlī 16°47′N, 43°49′E
'Ukaybir 15°16′N, 43°20′E
'Umar, Banī A district centred on 13°10′N, 43°51′E
Um Laylah, J. 17°17′N, 43°27′E
'Urāj, Dayr 14°55′N, 43°24′E
'Urr, Al 15°14′N, 43°53′E
'Usil 15°03′N, 43°40′E
'Utān 15°19′N, 44°11′E
'Utmah A mountainous district centred on 14°30′N, 43°57′E

Walajah 15°21′N, 43°18′E
Wa'lān 15°04′N, 44°16′E
Waṣāb Two districts. Waṣab As Sāfal centred on 14°16′N, 43°35′E and Waṣab Al 'Ālī centred on 14°20′N, 43°52′E
Washah A district centred on 16°20′N, 43°24′E
Wāsiṭ 16°59′N, 43°48′E
Watadah 15°19′N, 44°48′E
Wuslah, Naqīl 15°10′N, 44°37′E

Yakhtul 13°27′N, 43°16′E
Yarīm 14°18′N, 44°22′E
Yarīsh, Bayt 15°13′N, 43°56′E
Yāzil 15°14′N, 44°00′E
Yislah, Naqīl 14°57′N, 44°16′E
Yufrus 13°27′N, 43°56′E
Yūr, Wadi The wadi separating J. Ḥufāsh and J. Milḥān c 15°18′N, 43°23′E
Yusafīn A district of Qabayṭah centred on 13°17′N, 44°23′E

Zabīd 14°13′N, 43°18′E
—, Wadi A very large wadi arising in Maghrib 'Ans, Ibb and Mudhaykhirah, flowing westwards and reaching the sea at Al Fāzah, 14°07′N, 43°06′E
Zafīn, J. 15°40′N, 43°47′E
Zahir, J. Ḥiṣn Az 15°07°N, 43°28′E
Ẓallamlam, J. 14°33′N, 43°04′E
Zaydīyah 15°20′N, 43°00′E
Zīl, Az 15°03′N, 44°21′E
Zuhrah 15°43′N, 42°59′E
Ẓula' 15°24′N, 44°06′E
Ẓulmah 14°06′N, 44°05′E

SELECT BIBLIOGRAPHY

The bibliography is designed to provide the reader with sources of further information on various aspects of the Yemen flora. With this purpose in mind it is annotated and divided into sections, and publications which do not add significantly to the information in the main text of this book are omitted. Further references, however, can be found in 'A Botanical Bibliography of The Arabian Peninsula', *Notes from the Royal Botanical Garden*, Edinburgh 40 (1): 43–61 (1982).

A. BASIC WORKS ON YEMEN PLANTS

BLATTER, E. (1919–36) *Flora Arabica*. Rec. Bot. Surv. India 8: 1–519 A checklist largely outdated by Schwartz (1939) but useful for Arabic names and biographical information.

DEFLERS, A. (1889) *Voyage au Yemen* (Somewhat unreliable taxonomically but with many species descriptions and a general description of the vegetation.)

FORSSKAL, P. (1775) *Flora Aegyptiaco-Arabica* (The basic work on the Yemen flora, containing numerous descriptions of new species; particularly valuable for Arabic names and medicinal uses of plants.)

SCHWARTZ, O. (1939) *Flora des tropischen Arabien*. Mitt. Inst. Bot. Hamburg. 10. 1–393 (The most recent checklist of Yemen plants used as a starting-off point in the present work.)

SCHWEINFURTH, G. A. (1894–99) *Sammlung arabisch aethiopischer Pflanzen*. Bull. Herb. Bois. 2 app. II 1–340 (An unequalled and very reliable source of information on Yemen plants.)

B. FLORAS OF NEIGHBOURING COUNTRIES

Numerous articles in botanical journals bear on the taxonomy of Yemen plants but they are too many to mention individually. Similarly many standard reference floras have been consulted but only one, *The Flora of Tropical East Africa*, is both up to date and directly relevant to the flora of Yemen. It is unfortunately still incomplete but the volumes dealing with grasses, *Cucurbitaceae*, *Aizoaceae* and *Convolvulaceae* are particularly valuable as most species found in Yemen also occur in the area of the *Flora of Tropical East Africa*.

Shorter general floras are also invaluable and the following may prove particularly useful in identifying plants hitherto unrecorded in Yemen:

AGNEW, A. D. Q. (1974) *Upland Kenya Wild Flowers* (Excellent, particularly on ferns.)

ANDREWS, F. W. (1950-56) *The Flowering Plants of the Anglo-Egyptian Sudan*. (Difficult to use but reliable.)

COLLENETTE, I. S. (1985) *An illustrated guide to the Flowers of Saudi Arabia*. (Excellent illustrations.)

FIORI, A. (1912) *Boschi e Plante Legnose dell Eritrea*. (Useful on woody plants.)

NAPPER, D. M. (1963–71) 'Cyperaceae of East Africa.' *Journal of East Africa Natural History* Nos. 106, 109, 110, 113 and 124

TACKHOLM, V. (1974) *Students' Flora of Egypt*, edn. 2. (Particularly useful for weeds and desert plants.)

C. PHYTOGEOGRAPHICAL WORKS

SCHWEINFURTH, G. A. (1911) 'Über die Florengemeinschaft von Südarabien und Nordabessinien.' *Verh. Ges. Erdk. Berlin* 9–10: 1–20

WICKENS. G. E. (1976) *The Flora of Jebel Marra*. (Contains a useful discussion of the main geographical divisions of the African flora and summarises recent theories on this topic)

ZOHARY, M. (1973) *Geobotanical Foundations of the Middle East*. 2 vols.

D. THE VEGETATION OF YEMEN

Passing references to the vegetation of Yemen are made in numerous books and reports but apart from Deflers (1889) the only important sources of information are the following:

AL-HUBAISHI, A. & MÜLLER-HOHENSTEIN, K. (1984). *An Introduction to the Vegetation of Yemen* (A general account with numerous colour illustrations and lists of Arabic names)

DEIL, U. (1984) 'Zur Vegetation der Gebirgstihama am Beispiel des Beckens von At Tur.' *Jemen Studien* 1: 225–236

—— (1986) 'Die Wadivegetation der nördlichen Tihama und Gebirgstihama der Arabischen Republik Jemen, in *Contributions to the Vegetation of Southwest Asia*, ed. H. Kürschner

—— & MÜLLER-HOHENSTEIN, K. (1984): 'Fragmenta Phytosociologica Arabiae-Felicis – Eine Euphorbia balsamifera Gesellschaft.' *Flora* 175: 407–426

—— & MÜLLER-HOHENSTEIN, K. (1985) 'Beitrage zur Vegetation des Jemen 1.' *Phytocoenologia* 13: 1–102

—— & MÜLLER-HOHENSTEIN, K. (1985) 'Die Gefässpflazen des Beckens von At Tur.' *Bot. Jarhb. Syst.* 106: 23–52. (A species list of the Aṭ Ṭūr region)

HEPPER, F. N. (1977) 'Outline of the vegetation of the Yemen Arab Republic.' *Publ. Cairo Univ. Herb.* 7 & 8: 307–322

HEPPER, F. N. & WOOD, J. R. I. (1979) 'Were there Forests in the Yemen?' *Proc. Seminar Arabian Studies* 9: 65–71

POPOV, G. A. & ZELLER, W. (1963) Ecological Survey Report on the 1962 survey in the Arabian Peninsula. *F.A.O. Report* UNSF/DL/ES/6

RATHJENS, C. & WISSMANN, H. von (1934) 'Sudarabien Reise vol 3: Landeskundliche Ergebnisse, in *Abh. Geb. Auslands. Hamburg* 40

SCHOCH, R. (1978) *Land Evaluation and Classification from Landsat-Imagery in the Yemen Arab Republic.* Swiss Technical Co-operation Service, Zurich. Very accurate in predicting where interesting plant communities occur

WESTINGA, E. & THALEN, D. C. P. (1980) *A survey and problem analysis of the rangelands of the Rada' district.* Rada Integrated rural Development Project

WISSMANN, H. von (1941) 'Arabien und seine Kolonialen Ausstrahlungen' in *Lebensraumfragen europäischer Volker*, ed. K. H. Dietzen et al., Vol. 2: 379–387

WOOD, J. R. I. (1982) 'Report on the vegetation of the Haraz Region for the Erosion Control and Afforestation Project in the Haraz Region.' Deutsche Gesellschaft für Technische Zusammenarbeit (unpubl.) (Includes a key to woody plants as well as a detailed discussion of valley forest)

WOOD, J. R. I. (1983), An Outline of the Vegetation of the Yemen Arab Republic in *USAID/Cornell University Soil Survey of North Yemen*. (Includes 18 maps and drawings)

WOOD, J. R. I. (1985) 'The vegetation of the Tihama' in the *Report of the 1982 Tihama Expedition*

E. THE VEGETATION OF NEIGHBOURING COUNTRIES

Accounts of the vegetation of neighbouring countries are of considerable interest in placing the vegetation of Yemen in a wider geographical context. The following give accounts of the vegetation in most nearby countries:

DEFLERS, A. (1895) 'Esquisse de Geographie Botanique. La Vegetation de l'Arabie Tropicale au-dela du Yemen.' *Revue d'Egypte* 1: 349–370, 400–430

FREITAG, H. (1971) 'Studies in the Natural Vegetation of Afghanistan' in *Plant Life of South-West Asia*, ed. P. H. Davis, D. C. Harper & I. C. Hedge. (Juniper woodland.)

GILLETT, J. B. (1941) 'Plant Formations of Western British Somaliland and Harar Province of Ethiopia.' *Kew Bulletin* 1941: 37–199

KÖNIG, P. (1986) 'Zonation of vegetation in the mountainous region of South-western Saudi Arabia' in *Contributions to the Vegetation of Southwest Arabia*, ed. H. Kürschner

KÖNIG, P. (1987) *Vegetation und Flora im südwestlichen Saudi Arabien*

MANDAVILLE, J. P. (1977) 'Plants in scientific results of the Oman Fauna and Flora Survey 1975.' *Journal of Oman Studies Special Report* 229–267

PICHI-SERMOLLI, R. B. G. (1975) 'Una Carta Geobotanica dell Africa Orientale.' *Webbia* 13: 15–132

POPOV, B. B. (? 1960) Vegetation of Arabia South of the Tropic of Cancer excluding Yemen. *F.A.O.* (unpubl.)

RADCLIFFE-SMITH, A. (1980) 'The Vegetation of Dhofar in Scientific Results of the Oman Flora and Fauna Survey 1977 (Dhofar).' *Journal of Oman Studies Special Report* 2: 59–86

VESEY-FITZGERALD, D. F. (1955) 'Vegetation of the Red Sea Coast south of Jeddah, Saudi Arabia.' *Journ. Ecol.* 43: 477–487. (Excellent.)

VESEY-FITZGERALD, D. F. (1957) 'Vegetation of the Red Sea Coast north of Jeddah, Saudi Arabia.' *Journ. Ecol.* 45: 547–562

WISSMANN, H. von (1972) 'Die Juniperus – Gebirgswalder in Arabien.' *Erdwiss. Forsch.* 4: 157–176

F. ARABIC NAMES

All the books listed in Section A (except Schwartz (1939)) and the papers by Westinga and Thalen (1980) and Wood (1981–3) in Section D, contain Arabic names of Yemen plants. Apart from these the main sources are the following, several of which also have interesting facts about the economic uses of Yemen plants:

AL-DIMYATI, M. M. (1965) *Mu'jam Asma' al Nabatat al Warida fi Taj al 'Arus lil Zabidi.* (Zabidi was a Yemeni and many of the names in his encyclopaedia may be of Yemeni origin.)

FLEURINTIN, J. & PELT, J. M. (1982) 'Repertory of Drugs and Medicinal Plants of Yemen.' *Journ. Ethonopharmocology* 6: 85–108

HAMIDULLAH, MUHAMED (ed.) (1973) *Kitab al-Nabat* part of the alphabetical section

KHATTAB, A & EL-HADIDI, NABIL (1971) 'Results of a Botanic Expedition to Arabia in 1944-45.' *Publ. Cairo Univ. Herb.* 4

LEWIN, B. (ed.) (1953) *The Book of Plants of Abu Hanifa el-Dinawari* (part of the alphabetical section)

MEYERHOF, MAX (1943–4) 'Sur un traite d'agriculture compose par un Sultan Yemenite du xive siècle. *Bull. Inst. Egypt* 25–26

SCHWEINFURTH, G. A. (1912) *Arabische Pflanzennamen* (The basic source for Arabic plant names used in Yemen)

SERJEANT, R. B. (1964) 'The Cultivation of Cereals in Medieval Yemen.' *Arabian Studies* 1: 25–74

G. NARRATIVE & TRAVEL ACCOUNTS

BOTTA, P. E. (1880) *Relation d'un voyage dans L'Yemen enterpris en 1837*

FORSSKAL, P. (1950) *Resa til Lycklige Arabien Dagbok 1761–63*. (Forsskal's diary.)

CHRISTENSEN, C. (1918) *Naturforskeren Pehr Forsskal*

HANSEN, T. (1964) *Arabia Felix: The Danish Expedition of 1761–67*

SCOTT, H. (1942) *In the High Yemen*. (Possibly the best book in English about Yemen)

INDEX TO FAMILIES AND GENERA IN THE TAXONOMIC PART

Plate 1. *Myrica salicifolia* (Myricaceae)

Plate 2A. *Commicarpus pedunculosus* (Nyctaginaceae)

Plate 2B. *Dianthus uniflorus* (Caryophyllaceae)

Plate 3. *Ochna inermis* (Ochnaceae)

Plate 4. (left) *Hibiscus deflersii*; (right) *Hibiscus purpureus* (Malvaceae)

Plate 5. *Adenia venenata* (Passifloraceae)

Plate 6. *Diplotaxis kohlaanensis* (Cruciferae)

Plate 7. *Reseda sphenocleoides* (Resedaceae)

Plate 8. *Primula verticillata* (Primulaceae)

Plate 9. *Kalanchoe yemenensis* (Crassulaceae)

Plate 10. *Crotalaria squamigera* (Leguminosae - Papilionoideae)

Plate 11. *Ormocarpum yemenense* (Leguminosae - Papilionoideae)

Plate 12. *Acacia hockii* (Leguminosae - Mimosoideae)

Plate 13. *Oncocalyx doberae* (Loranthaceae)

Plate 14. *Phyllanthus hodgelensis* (Euphorbiaceae)

Plate 15. *Jatropha variegata* (Euphorbiaceae)

Plate 16. *Caucanthus edulis* (Malpighiaceae)

Plate 17. *Turraea parvifolia* (Meliaceae)

Plate 18. *Acokanthera schimperi* (Apocynaceae)

Plate 19. *Ceropegia rupicola* (Asclepiadaceae)

Plate 20. *Ceropegia rupicola* (climbing form) (Asclepiadaceae)

Plate 21. Succulent Asclepiadaceae. A: *Echidnopsis scutellata*; B: *Caralluma cicatricosa*; C: *Huernia macrocarpa*; D: *Huernia macrocarpa* var. *arabica*; E: *Caralluma deflersiana*; F: *Caralluma shadhbana*.

Plate 22. Succulent Asclepiadaceae. A: *Caralluma subulata*; B: *C. edulis*; C: *C. wissmannii*; D: *C. sp. B*; E: *C. chrysostephana*.

Plate 23. *Plectranthus asirensis* (Labiatae)

Plate 24A. *Becium serpyllifolium* (Labiatae)

Plate 24B. *Craterostigma pumilum* (Scrophulariaceae)

Plate 25. *Lavandula citriodora* (Labiatae)

Plate 26. *Verbascum yemenense* (Scrophulariaceae)

Plate 27. *Campylanthus pungens* (Scrophulariaceae)

Plate 28. *Cistanche rosea* (Orobanchaceae)

Plate 29. *Barleria bispinosa* (Acanthaceae)

Plate 30A. *Galium kahelianum* (Rubiaceae)

Plate 30B. *Helichrysum arwae* (Compositae)

Plate 31. *Atractylis kentrophylloides* (Compositae)

Plate 32. *Centaurathamnus maximus* (Compositae)

Plate 33. *Kleinia pendula* (Compositae)

Plate 34. *Senecio harazianus* (Compositae)

Plate 35. *Vernonia bottae* (Compositae)

Plate 36. *Cyanotis nyctitropa* (Commelinaceae)

Plate 37. *Aloe pendens* (Liliaceae)

Plate 38. *Dipcadi filifolium* (Liliaceae)

Plate 39. *Kniphofia sumarae* (Liliaceae)

Plate 40. *Holothrix aphylla* (Orchidaceae)